Handbook of Research on Urban Tourism, Viral Society, and the Impact of the COVID-19 Pandemic

Pedro Andrade
University of Minho, Portugal

Moisés de Lemos Martins
University of Minho, Portugal

A volume in the Advances in Hospitality, Tourism, and the Services Industry (AHTSI) Book Series

IGI Global
PUBLISHER of TIMELY KNOWLEDGE

Published in the United States of America by
IGI Global
Business Science Reference (an imprint of IGI Global)
701 E. Chocolate Avenue
Hershey PA, USA 17033
Tel: 717-533-8845
Fax: 717-533-8661
E-mail: cust@igi-global.com
Web site: http://www.igi-global.com

Library of Congress Cataloging-in-Publication Data

Names: Andrade, Pedro de, editor. | Martins, Moisés de Lemos, DATE- editor.
Title: Handbook of research on urban tourism, viral society, and the impact of the COVID-19 pandemic / Pedro Andrade and Moises Martins, editors.
Description: Hershey, PA : Business Science Reference, [2022] | Includes bibliographical references and index. | Summary: "This book also analyzes how the tourism industry has handled global crises in the past and proposes business models for information and knowledge dissemination to appropriately handle disasters, presenting provides innovative research and perspectives on the revitalization of cultural tourism industries and services by addressing the creation of jobs in the areas of restoration, leisure, and culture"-- Provided by publisher.
Identifiers: LCCN 2022010639 | ISBN 9781668433690 (hardcover) | ISBN 9781668433713 (ebook)
Subjects: LCSH: Tourism--Portugal. | Hospitality industry--Portugal. | COVID-19 Pandemic, 2020---Economic aspects--Portugal. | COVID-19 Pandemic, 2020---Social aspects--Portugal.
Classification: LCC G155.P75 H36 2022 | DDC 338.4/791469--dc23/eng20220518
LC record available at https://lccn.loc.gov/2022010639

This book is published in the IGI Global book series Advances in Hospitality, Tourism, and the Services Industry (AHTSI) (ISSN: 2475-6547; eISSN: 2475-6555)

British Cataloguing in Publication Data
A Cataloguing in Publication record for this book is available from the British Library.

For electronic access to this publication, please contact: eresources@igi-global.com.

Advances in Hospitality, Tourism, and the Services Industry (AHTSI) Book Series

Maximiliano Korstanje
University of Palermo, Argentina

ISSN:2475-6547
EISSN:2475-6555

Mission

Globally, the hospitality, travel, tourism, and services industries generate a significant percentage of revenue and represent a large portion of the business world. Even in tough economic times, these industries thrive as individuals continue to spend on leisure and recreation activities as well as services.

The Advances in Hospitality, Tourism, and the Services Industry (AHTSI) book series offers diverse publications relating to the management, promotion, and profitability of the leisure, recreation, and services industries. Highlighting current research pertaining to various topics within the realm of hospitality, travel, tourism, and services management, the titles found within the AHTSI book series are pertinent to the research and professional needs of managers, business practitioners, researchers, and upper-level students studying in the field.

Coverage

- Travel Agency Management
- Tourism and the Environment
- Leisure & Business Travel
- Casino Management
- Health and Wellness Tourism
- Customer Service Issues
- Destination Marketing and Management
- International Tourism
- Hotel Management
- Service Management

IGI Global is currently accepting manuscripts for publication within this series. To submit a proposal for a volume in this series, please contact our Acquisition Editors at Acquisitions@igi-global.com or visit: http://www.igi-global.com/publish/.

The Advances in Hospitality, Tourism, and the Services Industry (AHTSI) Book Series (ISSN 2475-6547) is published by IGI Global, 701 E. Chocolate Avenue, Hershey, PA 17033-1240, USA, www.igi-global.com. This series is composed of titles available for purchase individually; each title is edited to be contextually exclusive from any other title within the series. For pricing and ordering information please visit http://www.igi-global.com/book-series/advances-hospitality-tourism-services-industry/121014. Postmaster: Send all address changes to above address.

Titles in this Series

For a list of additional titles in this series, please visit: www.igi-global.com/book-series

Employability and Skills Development in the Sports, Events, and Hospitality Industry
Vipin Nadda (University of Sunderland in London, UK) Wendy Sealy (University of Chichester, UK) and Emma Delaney (University of Surre, UK)
Business Science Reference • © 2022 • 260pp • H/C (ISBN: 9781799877813) • US $215.00

Handbook of Research on Sustainable Tourism and Hotel Operations in Global Hypercompetition
Hakan Sezerel (Anadolu University, Turkey) and Bryan Christiansen (Global Research Society, LLC USA)
Business Science Reference • © 2022 • 430pp • H/C (ISBN: 9781668446454) • US $315.00

Entrepreneurship Education in Tourism and Hospitality Management
Satish Chandra Bagri (Hemwati Nandan Bahuguna Garhwal University, India) R.K. Dhodi (Hemwati Nandan Bahuguna Garhwal University, India) and K.C. Junaid (Hemwati Nandan Bahuguna Garhwal University, India)
Business Science Reference • © 2022 • 330pp • H/C (ISBN: 9781799895107) • US $230.00

Global Perspectives on Strategic Storytelling in Destination Marketing
Ana Cláudia Campos (CinTurs, Universidade do Algarve, Portugal) and Sofia Almeida (Faculty of Tourism and Hospitality, Universidade Europeia, Portugal)
Business Science Reference • © 2022 • 300pp • H/C (ISBN: 9781668434369) • US $240.00

Promoting Social and Cultural Equity in the Tourism Sector
Priscila Cembranel (Universidade Sociedade Educacional de Santa Catarina, Brazil) Jakson Renner Rodrigues Soares (Universidad da Coruña, Spain) and André Riani Costa Perinotto (Parnaíba Delta Federal University, Brazil)
Business Science Reference • © 2022 • 302pp • H/C (ISBN: 9781668441947) • US $240.00

Challenges and Opportunities for Transportation Services in the Post-COVID-19 Era
Giuseppe Catenazzo (ICN Business School, France)
Business Science Reference • © 2022 • 268pp • H/C (ISBN: 9781799888406) • US $250.00

Handbook of Research on Cultural Tourism and Sustainability
Claudia Ribeiro de Almeida (University of the Algarve, Portugal) Joao Carlos Martins (University of the Algarve, Portugal) Alexandra R. Gonçalves (University of the Algarve, Portugal) Silvia Quinteiro (University of the Algarve, Portugal) and Maria Laura Gasparini (University of Bologna, Italy)
Business Science Reference • © 2022 • 482pp • H/C (ISBN: 9781799892175) • US $315.00

IGI Global
PUBLISHER of TIMELY KNOWLEDGE

701 East Chocolate Avenue, Hershey, PA 17033, USA
Tel: 717-533-8845 x100 • Fax: 717-533-8661
E-Mail: cust@igi-global.com • www.igi-global.com

List of Contributors

Table of Contents

Detailed Table of Contents

The contemporary pandemic conjuncture may lead to a 'viral society', an unprecedent social paradigm involving deep economic, political, and cultural transformations. To cope with this globally unstable situation, it is urgent to deconstruct and reconstruct knowledge (e.g., through a social and sociological Encyclopedia of Viral Tourism addressing a particular genealogy of research on tourism studies and to be disseminated via a Virtual Sociological Museum). The author invokes here recent debates and his own personal research. Epistemologically, the pertinence of a hybrid and open research and several theoretical topics and case studies on tourism are discussed, for example, tourism articulated with the COVID-19 pandemic and the encounter of this viral tourism conjuncture with the war in Ukraine, which defines a new breed of war tourism. Regarding methodologies, the author converses on Hybrid Discourse Analysis and Visual, Virtual, and Viral Methods, such as sociological comics, video papers, artistic sociology, and sociological augmented reality.

The approach the author takes on tourism is not related to economic analysis, nor even a political analysis stricto sensu. He configures tourism as a cultural activity. The perspective developed on tourism is here phenomenological and socio-anthropological. Hence, the author does not dwell on business (nec/otium – not/leisure; activity, work, commerce) but mostly on leisure (otium – time free from activity). First, he points out that tourism reflects our current obsession with one territory and traveling in one territory. The second element developed in this chapter is the experience of traveling, which is always a possibility of an encounter – an encounter with the other and an encounter with nature.

COVID-19 is not the first pandemic we suffered during the urban era, but it is the first one to force governments to respond with such large-scale severe restrictions. Alongside institutional measures are the proliferation of collective endeavors that respond to community demands emerging from the struggle to contain contagion. Individuals of different age groups and demographic profiles are crowdsourcing their own solutions for both local and national issues using online tools to map and respond to pandemic- and lockdown-related crises. In seeking to understand these technological appropriations during the COVID-19 crisis in the Brazilian context, where 87% of the population lives in cities, the authors employed an online ethnographic research method and conducted a thematic analysis on 41 crowdsourced mappings in the country. The results present five main groups of matters of concern conveyed by these maps as well as the role of digital media in addressing them.

Culture and creative industries are nowadays a most relevant dimension of cities' economies and a fundamental asset of the tourism sector. The rise of COVID-19 has strongly and negatively impacted these resources, triggering responses from international organizations and municipalities eager to address the emergent constraints and difficulties experienced by society. This chapter analyzes how UNESCO assisted state members in dealing with the situation through the dissemination of information, studies, and reports, and how cities reacted to lockdowns, reduction of revenues, and the need to support and encourage their citizens. The analysis of information provided by UNESCO Creative Cities and World Heritage Cities concerning activities implemented provide a broad image of cities' reactions to the pandemic and of creative field distinctiveness.

Prior to the pandemic, more than 42 million individuals visited Las Vegas annually, and 14% were from outside the U.S. The number of overseas visitors increased by more than a million people during the decade from 2009-2018. The greatest increase in international visitors came from Brazil, China, South Korea, Argentina, and India. This chapter sought to examine international visitors to Las Vegas with the overarching research question: Why has the international tourist market to Las Vegas increased among specific countries and how can this information be utilized for future marketing purposes in a post-COVID-19 world? The most significant findings highlighted the increases in GDP and GDP per capita of the visitor countries. The city of Las Vegas can better target growth patterns like these to plan for the return of international visitors, along with marketing plans to reinvigorate international tourists to the region following the pandemic.

The pandemic has impacted several aspects of urban life in cities, from the elementary and essential daily living in the public space of its resident community to the collapse of tourism, which, until 2020, was in ascending progress in the case of Braga and which has repercussions on the entire economy. In this context, several cities took the opportunity to experiment with 'pop-up' measures (i.e., opportunity-based, rapidly implemented, low-cost, temporary, and experimental) to transform public space by allocating it to pedestrians and urban living. This chapter explores two experiences tested in Braga, during the pandemic, which, by their latent capacity of catalysis in other areas, glimpse and enhance a transformation in the mobility of the city in a future scenario.

At a time when all public access spaces in the city were closed, including cafes, the Public Market of Braga remained an open space, albeit with strict entrance control rules. The authors thus advance with the hypothesis that the PM was fundamental to guarantee the routine mobility of Braga's citizens, especially the most disadvantaged and, with that, to ensure their sense of hope and confidence in the future. Likewise, the PM is a place of resistance also from the perspective of sellers, on the one hand, maintaining its economic survival, albeit with visible difficulties, and on the other hand, ensuring the maintenance of a vital communication space. To reflect on the specific ways in which this phenomenon has occurred since January 2021 is the main objective of this chapter.

This chapter analyses the tourism and urban dynamics transformations of the two leading cities in Portugal – Lisbon and Porto. In a (neo)liberal context and within a process of financialization of the economy, the authors address the political vision of each city, the local stakeholders and residents' evaluation, and public policies analysis. This chapter concludes that the implications of recent transformations are numerous. The process of residential and functional gentrification, resulting mainly from the touristification of the city, and the concomitant growth of the floating population, was slowed down by the COVID-19 pandemic. However, housing and services prices are still high, and the unavailability of housing is indeed a problem. Thus, integrated, strategic, and collaborative planning for sustainable urban development is assumed as a crucial issue for the next decade.

This chapter addresses the importance of the Roman Archaeological Ensemble of Merida as a tourist destination and the necessary revitalisation of the tourism sector through digital innovation. Despite the different policies to reactivate the sector implemented in the post-COVID-19 scenario, the authors consider that there is still much work. They propose how digital technologies play a crucial role in the conservation and dissemination of cultural heritage, and they are committed to mobile and augmented narratives as an ideal strategy for urban archaeological tourism. Throughout the chapter, they review key notions that articulate the essential elements of the tourist experience and comment on previous projects that could serve as a starting point.

The Venetian landscape is as much a product of its economic activities, past and present, as of its physical environment. Tourism is the main source of income for the city, and also a challenge: the enormous pressure exerted by visitors is creating distorted incentives for inhabitants to abandon the city to the point that depopulation and environmental harm are the most relevant issues in the city's political agenda. The chapter delves into the case of Venice to explore the themes of urban revitalization as a process of framing and social mobilization. In particular, the authors focus on the role of higher education institutions in triggering urban change and on the nature of actors engaged in the transformation of the social fabric of a city. Based on the experience developed within a European project aimed at revitalizing and rethinking cities, the chapter explores the potential of HEIs in transforming cities, the hurdles and critical factors that might inhibit their contribution, and the strategies that can facilitate their engagement in processes of social and economic transformations.

Section 4
COVID-19's Touristic Economy

The COVID-19 pandemic was a major phenomenon with strong health, political, economic, and social consequences. But the health crisis was also coupled with an equally powerful communication crisis. This double crisis has of course had an impact on the world of tourism. Since mobility has been partially authorized again in Europe, hoteliers are looking to win back their customers, especially on these same social media. What is the place of the pandemic in online communication? How can the subject of the crisis be raised without scaring customers? Are the modes of communication different depending on the category of the hotel? Based on 10 hotels of different categories taken at random in Paris, London, Berlin, and Rome, the study seeks to analyze the informational policies implemented online by individual hoteliers. The results show that the variables "country" or "category" are not significant; there are strong differences from one hotel to another, which highlight a greater or lesser capacity or interest in using these communication tools to promote the establishment.

Urban tourism has particularities that distinguish it from other forms of tourism. The urban tourist wants to enjoy the city, its monuments, its environment, its cuisine, its events. The tourist will always be an outsider to the city, someone who will enjoy an area that is not designed specifically for tourism. But the tourist does not have to be a stranger in the city. The pandemic affected urban tourism in a more "violent" way. However, the pandemic has not affected the pull factors of cities – the cuisine, the monuments, the events, the vibe are still there waiting to be enjoyed. They will be, and city tourism will return with full vigor. This chapter argues that the quest for sustainability and change supported by a disruptive set of new technologies and the consequent skill shifts will possibly affect urban tourism intensely. These new trends have the potential to give rise to different solutions than those that have been attempted in the past. We will see if we can better reconcile cities with their tourists to the benefit of all.

In this chapter, the authors present the current situation and the legislation produced as well as efforts done by medical doctors at the end of the 19th century in order to introduce hygiene as a natural as well as a cultural way of preserving physical health as well as mental health. The work done by doctors supported a general development of economic life as well as it saved the Portuguese society before laboratories took over our bodily management.

Section 5
Outbreak Politics, Policies, and Power Within Hospitality

Today all people become victims of harassment in one form or another. Harassment at work is a very hot topic nowadays as lot of women are objects during their work. The chapter is focused on presenting the essence of harassment and how it can take place in a kitchen. The first part discussed harassment and mobbing in the workplace. The methodology part presented a study conducted among women chefs and kitchen staff. The study was conducted through online questionnaire and included 39 respondents. Besides this, an interview with 13 women chefs has been performed. The study aims to understand if there is harassment in the kitchen and what forms it takes. It also aims to understand how women as chefs are perceived in their job positions and how they are harassed. The results show that women chefs have been victims of different forms of harassment in the kitchen.

This investigation aims to study the situation of human rights during the COVID-19 pandemic in Guinea-Bissau between January 2020 and January 2022. The research organized an inquiry to families and another to market sellers about the effects of the pandemic and the measures enacted by the government and presidency to contain it. A public hearing was also organized, with various entities from the high commission to trade unions, journalists, and the public order police, for information on how each institution saw its situation and action in this period. The conclusions of the analysis of all these qualitative and quantitative data allow us to affirm that the fragility of Guinea-Bissau has such a social weight that a disease that killed fewer people than malaria, diarrhea, or tuberculosis did not overcome the problems of human rights stemming from poverty, institutional fragility, and low incomes in general.

Museums were one of the cultural sectors most negatively affected by COVID-19. The restrictions introduced in the crisis period had multiple negative effects in the case of people's mobility and tourist travel. However, the temporary closure of museums also created opportunities. Museums have given greater relevance to the use of digital technology and virtual networks by increasing new forms of virtual communication with the public in the access to collections as well as the flexibility of activities. The proactivity and creative responses in moments of crisis, although of museums' initiative benefited from strategies promoted by cultural agents with contributions from the community itself. The crisis was also an opportunity to discuss and reflect on desirable changes for the museum of the future. The predictable socio-demographic changes in the public who frequent museums and the role that virtual communication will play were some of the concerns assumed by museums. The research focused on a restricted number of museums in the city of Lisbon, national museums.

Chapter 22

Since public art works are normally city collections, park installations, or landscape interventions, in order to achieve a rich, diverse, and representative sample, it is necessary to hit the road, to ramble in the park, and to explore the city. So, the main assumption is that public art value should also be judged by the consideration of its environment context, its social substratum, and the means used for its discovery. A literary feeling is required for its encounter. A journalistic curiosity is required for its inquiry. Travelling by car enables these aims, for it remains a reliable way of getting to know a region or a country. Besides that, on a period of pandemic or post-pandemic times, travelling by car is a defensive form of travelling for it avoids the agglomerations of airports and railway stations.

Chapter 23

The author describes her personal experience and insights on Portugal as a German artist who has lived and worked worldwide. She observes the changes from her first arrival during the European Football Championship in 2004, the experience of the WebSummit and various exhibitions in Portuguese Museums, up to the COVID-19 confinement and reopening in 2022. She tells the story of 18 years of development of the country, its tourism and economy, the spiritual path, and also the "zen exercises" the community has and had to offer: getting lost in Lisbon, finding faith in Fatima, sanctuaries in Santarem, and tabernacles in Tomar, as Portugal became her literal port of the grail (Port-u-gal). From children to education, poetry, prose, science, food, art, and religion, she paints a picture from the view of a foreigner mingling with the Portuguese and the ex-pat community. Isolated in the countryside during COVID-19, she connected online with Portuguese artists and created new alliances.

Foreword

This extraordinary collection of reports, taken as a whole, implicitly recognizes that tourism changes the urban social fabric in myriad ways, and COVID changes tourism. *Urban Tourism Viral Society* does not purport to offer an answer or a position on the ultimate meaning of this double shift. It does, however, provide the reader with observations and analyses needed to reflect upon and to assemble for oneself a foundation for understanding the entwined challenges of COVID and urban tourism, and to take a stand.

As Marx worked out the ways capitalism inserts itself into every detail of social life, he was compelled to confront its core contradiction and adopt a dialectical method to explain the simultaneous arrival of extreme wealth and extreme poverty. The case studies assembled here suggest that as tourism overwhelmingly inserts itself into the urban fabric, the dialectic breaks free of class relations and inhabits every transaction and every moment. Tourism brings wealth but it also disrupts local life, and spoils the very things tourists desire. Far from "containing the seeds of its own destruction," until 2020, with few notable exceptions, e.g., Venice, tourists and locals alike embraced the deteriorating status quo and seemed content to spiral down together.

COVID brought about the simultaneous need to respond to a deadly public health emergency and to survive the economic consequences of the immediate cessation of tourism. The chapters in this timely book respond to numerous aspects of this double challenge. Each contribution is poised above a reset of the ways urban tourism might be packaged after COVID. Does the temporary loss of all income from tourism suggest that local economies should diversify and become less dependent on tourism in the future? There is evidence here that some local people would be happier with fewer tourists who crowd them out of their public spaces and drive real estate and other prices up. Or, should the region double down on its tourist offerings and position itself to be among the first to recover from COVID? Several chapters offer strategies for doing just that including emphasis on sightseeing by automobile even while COVID restrictions are in place.

Several chapters helpfully bear down on the historic connections between the themes of the book that converge in dramatic fashion in 2020 but have always been lurking in the background. Carlos Sangreman, et al. (ch. 18) remind us that the necessary measures taken by the state to protect life and restrict tourism under COVID mirror those that have been used by corrupt political regimes to dampen civil unrest—closing borders, limiting gathering and association, etc. In a pivotal chapter, Manuela Hasse reminds us that disease prevention has shaped tourism well before COVID. She points out that before it was muted by modern medical practices, the city was widely regarded as a place of filth, disease, and ill-health. An early important form of tourism was the annual summer departure for the curative waters of rural spas and swimming in the sea, all to ward off dangerous infectious diseases.

An important discussion in *Urban Tourism Viral Society* concerns the future role of information technology, smart cities, virtual reality and augmented reality in urban tourism. Several of the chapters note that the COVID lockdown accelerated the development and acceptance of telecommuting, distance learning, and other ways of connecting humans to distant interlocutors, far away events like film festivals, and objects like museum displays. These chapters argue that tourism is about to be radically transformed by information technologies, especially immersive virtual and augmented reality. The proposed insertion of new technologies into the tourist experience range from improving the ways on-site information is delivered via smart phones and other mobile devices, to the construction of hyper-realistic three dimensional digital models of the world's attractions. Once the attractions have been reproduced in cyberspace, itineraries can be accessed without having to leave home via lightweight headsets connected to a computer. The programs would allow the at-home tourists to look around and wander through the attractions much as they would on site.

There are a number of fascinating speculations in *Urban Tourism Viral Society* about whether advanced technologies can deliver a tourist experience that is equivalent to "being there." There seems to be little question that removing travel from tourism would be good for the planet. There is no consensus about the good that might come from shifting the profit from the travel industry to the titans of technology.

One of the arguments in favor of the near total replacement of tourist travel by virtual experience is made by Andreas Montenegro (ch. 28). He suggests that virtual tourism is coming, but only after a shift away from hundreds of years of humans getting their information from words on paper, books and screens. He very honestly admits that this will be a major cultural transformation eventuating in everyone getting digital content planted everywhere in the world and accessed in real time through smart glasses.

An opposing argument is proposed by Fernando Contreras (ch. 20) who makes an immediate distinction between the "excessive virtualization of heritage objects" and their "rational contemplation." He suggests there may be limits to "digital projection into the social imaginary." A student of Walter Benjamin might suggest that the more perfect the reproduction, the more separated it is from its original. The dialectic does not stop at the silk cord.

If one of the desires of the tourist is to learn something about the culture of she is visiting, her most valuable resource is the response of the local people to her presence. The German born artist, Regina Frank (ch. 23) remarks that if you want to know a culture, taking your child with you is the best way. "In Portugal . . . in every restaurant, children are treated with respect and admiration. They are welcome everywhere and receive lots of love and attention." Of course, this is not true for every tourist mother and child. The way it was true for Regina and her daughter and not true for boorish tourists and their brats is exactly what gets left out of hyper-realistic immersive virtual tourism. For many, what gets left out is the essence of the tourist experience.

Dean MacCannell
University of California at Davis, USA

Dean MacCannell *is emeritus professor of Environmental Design and Geography at the University of California at Davis. His 1976 book The Tourist has never gone out of print and has been translated into 12 languages. He is also the author of Empty Meeting Grounds—The Tourist Papers, The Ethics of Sightseeing, and more than 100 articles on aspects of hypermodern society and culture.*

Preface

OVERVIEW: BRIEF STATE OF THE ART

What are the conditionings that COVID-19 pandemic prosecutes on the contemporary tourism condition? Are we witnessing a new social virulent contact and/or a novel social health contract? Are citizens becoming more controversial or more conversational? Are populations being more solitary or more solidary?

In order to try to answer, even partially, to these or to other urgent questions evoked by such unprecedent virus risk and global crisis, in the following paragraphs a *critical bibliography* is presented, including various aspects of the planetary encounter, clash and confrontation between two major processes of the 21st century: the tourism sector and the Corona virus.

Such state of the art includes works on the *social fabric* in general, but as well about economics, policies and politics, science, technology, culture, discourses and arts perspectives on this matter.

The *period of time* covers the pandemic spread since 2020 until 2022. The respective writings may be used as a panel for understanding the evolution of this phenomenon, step by step. Sources are organized by year, where some main themes, discussed within specific spaces and times, may be easily identified.

The works presented are not exhaustive. This is just a brief sample of the numerous reflections and case studies produced and disseminated through the last three years. Other clues may be found in each of this book author chapters, both in their 'References' and 'Additional Readings' sections.

Inside such works, the *social fabric* doesn't appear as uniform, but plural. In fact, it constitutes an arena where important entities and issues are daily tested, within specific social contexts such as touristic destinations. These localities are activated by diverse social agents, the tourist but also the visited citizens, through concrete practices, e.g., travelling and other leisure activities, but also other related actions across various social realms. Such deeds often present or represent core social structures, in this case mainly tourism and public health. Or other broader structures, e.g., the travel and all its types and modes of circulating and participation in the social everyday, with or without the COVID-19 pandemic presence and in(flu)ence.

ECONOMIES AND ECOLOGIES OF TOURISM WITHIN THE COVID-19 EPOCH

More particularly, in the *economic realm*, some of the most relevant studies are those closely connected with the *health organization and resources* of every country, that had a paramount impact on the struggles against the development of the pandemic worldwide, connected with *tourism* activities. Global statistics attest to the relevance of the COVID-19 pandemic impact on the travel, tourism and hospitality sector worldwide (Nhamo, 2020). In this text, COVID-19 is connected with the Sustainable Development Goals (SDGs), which deals with the 'global tourism value chain' including multiple tourism niches, and with international programs for disaster risk reduction. On his turn, Parag Verma (2021) reflects on the pre-and post-impact of the COVID-19 pandemic crisis, and dresses an inventory of research about computer-driven methodologies for the fight against the outbreak. In the context of economic development, Azizul Hassan (2021) points out some technological solutions to 'overtourism', which means the excesses, that the author calls 'decimations', perpetrated by the industry of tourism inside the visited sites, regions and countries, and considering the crisis situation provoked by Corona virus. Concerning foreign investments, Cristina Jönsson (2022) debates foreign direct investment (FDI) in tourism, within the conjuncture of the COVID-19 pandemic, and the respective economic consequences. On their turn, Lesego Stone and others authors carried out a study in 2022 on protected areas for tourism in Southern Africa and their impact on community livelihoods and sustainable development.

Some of the numerous works on *COVID-19 articulated with tourism*, since 2020, produced in several countries, are the following: Hebbard (2020); Helble & Fink (2020); more specifically, J. Hubert reflected about tourism enterprises in Lao and the effects of COVID-19 on them (2020); Insights West is a firm who interpreted resident's public perceptions facing Corona Virus (2020); Scheyvens et al. focused on the adaptivity of South Pacific populations to tourism in the era of COVID-19 pandemic (2020). Concerning the *health situation* caused by such COVID-19 spread, David Klooz (2020) criticizes the West and, in particular, the World Economic Forum, for promoting lockdowns, masks and 'social distancing' policies, which cause multiple negative consequences to public health. A study published by the International Monetary Fund, written by Serhan Cevik (2020), deeply articulates infectious diseases and tourism flows. Another work, sponsored by the Asian Development Bank (2021), analyses the global and Asian situation of the aviation, which suffered a notorious depletion since March 2020, as a central activity for tourism.

Other studies pay more attention to the *impact and recovery* from the virus outbreak. Mario Cimoli (2020) intends to suggest recovery measures for the tourism sector, in terms of sustainability and resilience in Latin America. Yanuar Nugroho (2020) try to understand the COVID-19 impact on micro, small and medium firms and tourism in Indonesia. Mahmu Demir (2021) writes about the consequences of COVID-19 on the tourism industry. Ogech Adeola (2021) proposes a study on COVID-19 recovery in Africa, inside a gender optics. Joey Erh, J. (2021) talks about COVID-19's economic effects on Tourism in Singapore. Niamh Espiner (2021) debates perceptions of the Corona virus impact on tourism and other recreational outdoor activities. Maximiliano Korstanje (2022) explore the traits of the Post-COVID-19 conjuncture. Another solution for tourism recovery is digital nomadism, according to Francisco Situmorang (2022), who aims to redesign tourism development and sustainability, to overcome the overtourism issue in Bali.

POLICIES, POLITICIANS, AND POLITICS DEBATING THE PANDEMIC AND TOURISM

Other lens pertains to the *policies and politics* in what regards tourism activities, under the aggression and possession by the Corona virus. Besinger Clemens (2020) notes that, to recover from the Corona virus pandemic in Egypt, it should be necessary more planning by the State, simultaneously with the re-opening of economy. On their turn, Kibrom Abay and Ibrahim Hosam (2020) notice that services that requiring face-to-face interaction, such as tourism activities, loose significantly with the pandemic, and information areas are 'winners'. Within the law realm, Sumathy Permal and others debate humanitarian, strategic, environmental, legal, and business postures regarding the impact of the pandemic on marine lives, its industries, and the marine environment in Malaysia. In India, the Union Finance Minister Nirmala Sitharaman announced measures to recover sectors affected by the second wave of COVID-19 Pandemic, such as a Credit Guarantee Scheme for Tourism Industry, a Free Tourist Visa and a Credit Guarantee Scheme for Public Health (Taxmann, 2021).

One bold strategy in environmental justice, within COVID-19 outbreak, is 'socializing tourism', according to Freya Higgins-Desbiolles (2021). This signifies to promote traditional ecological, cultural and knowledges rights and interests of local peoples, communities and living landscapes. A solution is to use digital data to complement classical sources, in order to inform government policies. In the context of global mobility, Maximiliano Korstanje and George Babu (2021) suggest that COVID-19 imposed a new type of terror and the 'end of hospitality' within the tourist industry. This situation was caused as well by restrictive governments policies, such as closure of borders and airspace, strict lockdowns and social distancing, and cancellations of domestic and international flights. The status of tourists changes from 'democratic ambassadors' to 'undesired guests'. A study from United Nations (2021) argues that Corona virus transformed the mountain tourism sector and provoked major losses for communities and businesses. To solve these issues, governments should invest in infrastructure, mountain tourism services; augment multi-level-governance, partnerships with active communities, assessments of the impact of tourism on mountains, management of waste and resources. Nian Peng (2021) takes a more political approach, in the present conjuncture or rivalry between the United States and China and the Corona virus pandemic, in what regards information for governments, companies and social organizations, in order to promote cooperation and investment in trade and public health. And what is occurring concerning the pandemic inside the already confined political ecology of islands, in the case of the Caribbean and the Pacific? Some consequences of the collapse of tourism are the revival of local agriculture and fishing, but also, in the politics and empowerment dimensions, the rebirth of self-reliance, citizen governance and even a 'new normal' (Campbell, 2021).

THE COURSE OF DISCOURSES ON SCIENCES, TECHNOLOGIES, CULTURES AND ARTS

Last but not the least, global and local institutional *discourses*, within *science, digital media, education, cultures* and *arts* projects, help to redefine the present *virulent conjuncture*, during and after the outbreak.

Communicative Tourism: Information and Knowledge on Travel and Viruses

Communicative tourism signifies a paradigm of tourist activities centered on the tourism production and dissemination of information and knowledge, targeting three of the main agents of transcultural contemporaneity: citizens, tourists and migrants displaced from peripheral societies into central societies.

Regarding knowledge in general, the case of India is paradigmatic (Balavantagol, 2020), in terms of huge loss of human capital for tourism communication, due to COVID-19 outbreak, inside the scientific and educational sectors, such as schools, universities, and related stakeholders, such as industries and services of tourism.

Within the *knowledge production* dimension, tourism and digital media are interrelated (Carvalho, 2020), in what concerns digital management and marketing applied to tourism and travel, mobile technologies applied to sustainable tourism, geotagging and tourist mobility, smart destinations, robotics in tourism, the tourist experience in a pandemic situation, generations and technology in tourism, big data and e-tourism 2.0. And the use of scientific tools connected with social instruments to delimitate global pandemics proves to be crucial (Chatterjee, 2021). In particular, the role of Historic narratives is paramount, as Christian and Jayne Rogerson outline, in order to counter 'present-mindedness' of mainstream urban tourism writings. This past may be devolved to the present historical moment for urban tourism, by debating issues such as gastronomic tourism, VFR travel, airportscapes, climate change, AirBnb and creative tourism. Moreover, 'Psychology of sustainability' (Sarfraz, 2022) related with sustainable development and global economy, is another scientific optics that may be proven useful for studying SARS coV-2 pandemic.

On its turn, *Social Sciences* gave a notable contribution for the hermeneutics of tourism within the COVID-19 era. For instance, the discussion of interdisciplinary prisms about the risks and effects of Corona virus pandemic (Omeraki, 2021). Or the 'digital nomad lifestyle' articulated with the traditional and modern job market in Egypt, analyzed via an online survey and netnography research (Abdelhady, 2021). Several socio-economic dimensions were studied considering the effects of Corona virus in Bangladesh, such as education, tourism, local and international trade, GDP, remittance, import & export of goods and employment (Hafsa, S., 2021). Consumer habitudes as well have been transformed. Considering that consuming less is the best way to save our planet, this author advocates for a world without shopping, as it happens partially with the COVID-19 surge, and giving more attention to ecology (MacKinnon, 2021). Multiple social dimensions conditioned by pandemic risks are analyzed by Indrajit Pal (2022), including tourism and the action of governments and development organizations, and their investments on communities. In all such knowledge production, digital methods are paramount for the study of SARS coV-2 outbreak, such as artificial intelligence (AI) and machine learning (ML), Internet of Things (IoT), and drone technologies, in articulation with one another (Panigrahi, 2021).

On the *knowledge dissemination* side, the communication of health messages relating to outbreaks, epidemics, and pandemics was analyzed by David Berube (2021), by mentioning both associated health theories and practices. Target audiences included both science literate and illiterate inside Western and developing economies. Detected obstacles to such communications were politics practices and norms, individual interpretations, and biases as mortality salience, news avoidance and quarantine fatigue.

Education

Management education and environment protection education were also metamorphosed by the COVID-19 pandemic. Inside the areas of tourism, aviation and hospitality industries, some new pedagogic themes are more flexible business models for development, in association with understanding resident's perceptions and emotional experiences, strengthening the human resource management, and improving the destination image for marketing (Yuhua, 2022).

Cultural Heritage

For instance, some States developed protocols fighting COVID-19 in the tourism sector related with cultural heritage (Ministry of Information, Tourism and Heritage, 2020). The management of world heritage articulated with sustainable tourism in the context of SARS-CoV 2 was interpreted by Kurt Luger and other authors (2020). The United Nations World Tourism Organization-UNWTO (2021) reflected on sociocultural consequences of Corona virus and suggested operative guides to be used within recovering strategies. Jarko Saarinen and Jayne Rogerson (2021) discuss the transforming relationships between demand, supply and location sector on the Global South, regarding tourism activities related to other sectors, such as heritage. Vladimir Geroimenko (2021) and hiss virtual team circumscribe the relations of Augmented Reality with Tourism, Museums and Heritage, e.g. their current status and future trends, informal learning and heritage preservation inside emerging post-COVID-19 museums and heritage sites. Some practical uses include augmented reality applications, mixed reality environments and immersive installations, cultural heritage education and tourism promotion, visitors with special needs, and On his side, Sajal Roy (2022) writes about the influence of COVID-19 on socio-cultural activities, marking the differences between the condition of underdeveloped and developed countries and the metamorphosis of the global order.

In order to comprehend such debates and share your informed opinion as readers, it is useful to consult, besides the above-mentioned referenced authors and works some of the testimonies announced and enunciated below, by the authors of this book.

A RESEARCH NETWORK ON URBAN TOURISM AND COVID-19 PANDEMIC

'Viral Society' is a concept that translates the fact that a novel societal paradigm seems to be emerging, based on multiple and profound economic, ecological, political and cultural transformations on the social fabric, after the spread of the Corona virus pandemic since 2020. For example, urban tourism has radically been modified. This book intends to comprehend the contours of the metamorphoses that occurred on tourism structures, activities and agents, via the spread of the COVID-19 and the subsequent social effects, through interrelated sociological, philosophical and epistemological lens.

For this aim, the strategy proposed in the present *Handbook for Research on Urban Tourism, Viral Society and the Impact of COVID-19 Pandemic,* constitutes one of the possible perspectives for implementing a sustainable hybrid and open research. A *hybrid research* signifies a posture, at a writing production stage, that pay attention to the multiple differences among social and scientific communities worldwide, and aims to contribute to a more profound dialog among that researchers' specificities and among the respective scientific productions. And an *open research* means a position, at a dissemination momentum, that intends to diffuse a hybrid scientific production to a global *research audience*, via an effective and affective open access. In a sentence, a sustainable hybrid and open research seeks to fusion and intermediate scientific creation and popularization.

Either the book's title and sections, or the authors and the texts convoked here, testify such objectives.

As for the title, it reveals a collective work undertaken by a solid network of knowledges and sensibilities, that rallies scientific, technological and artistic academic and specialist researchers, affiliated within some of the most prestigious universities and research centers worldwide.

Such reflexive and sensitive arena vehiculates a profound debate on a hybrid problematics around the origins and impact of COVID-19 pandemic on urban tourism, within contemporaneity.

This research book constitutes the first action organized by the *Observatory on Heritage, Digital Arts and Tourism*, a new collective team and international web, operating inside Communications and Society Research Centre, at University of Minho, Braga, Portugal.

TARGET USERS AND POTENTIAL USES

This book is mainly targeted to researchers, teachers, students and other social agents within scientific communities, in connection with the above-mentioned scientific purposes, applied to technological and social needs.

Therefore, the proposed texts intend to develop a scientific daily research *ethos* within the academy, but also disseminate content among external users, such as citizens, tourists, migrants and refugees, who hopefully may also become, as readers and critics, partly co-authors of knowledge produced collectively.

In fact, the present co-operative work aims to contribute, in some way, to revitalize tourism industries and services during and after the COVID-19 pandemic. This purpose can be achieved via the enhancement of knowledge transfer among universities, innovating industries (spin offs, startups), tourism agencies, museums, etc. And such dynamics may help the creation of jobs in the areas of hospitality, restoration,, leisure and culture, among others.

The book uses mainly critical knowledge and social narratives, meaning pertinent but also accessible conversations on how to circulate inside the virulent city contaminated by Corona virus.

PILLARS FOR A WORLDWIDE DEBATE AND IMPACT

The whole discussion is specified here, within the following ten main discussion pedestals, which may edify the correspondent nodes of this global knowledge web (Figure 1).

1. Pre / post touring within the viral society.
2. Journeys across virulent and violent urban territories.
3. Pandemic cases at touristic cities.
4. Covid's touristic economy.
5. Outbreak politics, policies and power within hospitality.
6. The 'virulization' of mobile heritage and culture.
7. Confined travels on arts worlds and museums.
8. Itineraries on education for deconfinement.
9. Contaminated cinema destinations.
10. New media for post-viral flows.

Inside such content sections and in the following pages, the texts forged by authors who collaborated in this collective work are introduced, through brief comments on their core contribution.

Each text is also presented via a *concept map* that clarify the respective text's main concepts, articulated with tourism processes and/or the COVID-19 pandemic, which are the two main ideas of this collective book. Note that such syntheses, inside the concept maps, doesn't illustrate all concepts and their distribution, discussed by each author inside her/his text. In order to identify these main concepts, consult the keywords provide by each author. Differently, the shown concept maps just demonstrates that tourism and pandemic may be considered two of the *mediating concepts* that connect, present and represent a plethora of intermediary social phenomena.

Figure 1. Book's chapters and authors research network

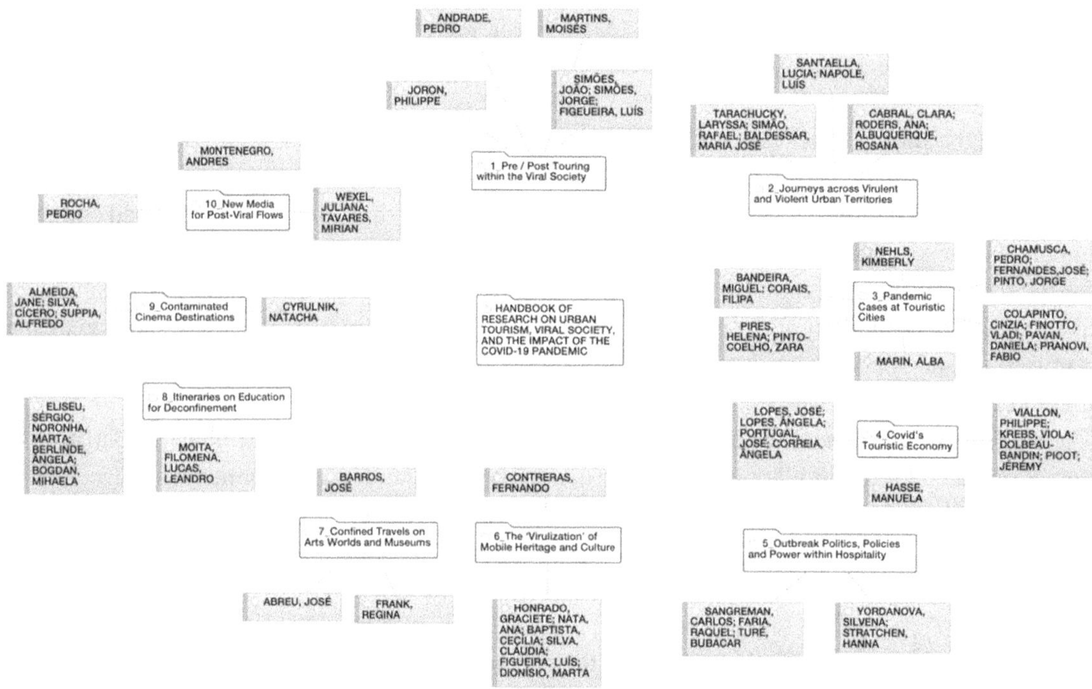

Consequently, within such concept maps, only 'tourism' and 'COVID-19 / pandemic' are considered as their central concepts, and therefore placed in the core of each map image. These maps denote sociological networks that have the form of a somehow 'social planetary system'. In other words, the concepts socio-semantically closest to the main term within a conceptual network, are those located in that central idea proximity, in the image. Note as well that the socio-conceptual space within these networks operate a divide between the social phenomena and respective ideas, that (re)present, at the network's left, anterior or antecedent phenomena in what regards the central process/concept, such as its conditions and causes. On the right side of the networks, are situated the social processes and the inherent sociologic terms that testify the posterior entities, consequences and effects that are related with the central and mediating concept, be it 'tourism' or 'COVID-19 / pandemic'.

These networks also practically work, for this book's reader, as a *visual table of contents*, concerning the book's substantive information. A table of contents is a *paratext*, as Gerard Genette (1997) puts it, meaning a text that introduces another text and makes its reading clearer and easier, such as an abstract. Inside the 'visual' text of a social concept network, the reader may easily relate the central sociological concept to its neighbors, and to the inherent social phenomena network.

Such social and sociological networks are also useful as an *author conceptual biography,* within the intertextuality activated via the continuous social and scientific flux process, where the reader lands to a certain author's text, or lifts from that text to another. This means that the reader may not just enjoy the fruition of the concepts revisited by each author, inside her/his texts, but she/he can as well know better the conceptual paths that each author subscribes, by following her/his interrelated and interactive engravings that are visible inside each socio-conceptual network.

Therefore, the ultimate aim here is to provide a diversified but simultaneously synthetic content, that may prove useful for future research and pedagogical projects. For instance, a possible *Virtual Sociological Museum,* that may gather contemporary issues, such as tourism in the age of the *Viral Society,* in order to be disseminated for a plethora of profiles and audiences, e.g., as previously noted, researchers, professors, students, citizens, tourists, migrants and refugees.

This augmented posture may lead to several short, medium and long term results, such as: a sound economic, social, political, ecological and cultural impact, through novel ways to promote the reconstruction of a sane urban fabric after the SARS CoV 2 pandemic; enhance local urban trade and industries; increase democratic and political conscience on pandemics; and promote more profound intercultural exchanges. Moreover, in order to reach worldwide contexts and target audiences, the book aims compliance with the following United Nations 2030 Agenda goals for a sustainable development, among others: Goal 3: Quality Health; Goal 4: Quality Education; Goal 9: Industry, Innovation and Infrastructures; Goal 11: Sustainable Communities and Cities.

In short, knowledge on Corona virus may be learned via daily urban experiences under and after pandemic. Practical recommendations suggested in this book, can be produced or applied by multiple stakeholders (state agents, universities, start-ups, corporations, ONGs, associations, communities), who would define new rules and ethics, in order to develop new competences and literacies, that will originate healthier urban lifestyles.

The next sections will clarify some dimensions of this book topics, as well as the purposes and methods utilized by the contributing authors.

PRE/POST TOURING WITHIN THE 'VIRAL SOCIETY'

The first debate will discuss some broad epistemological, theoretical and methodological issues, in what regards unprecedent social phenomena partly originated by COVID-19 pandemic.

Figure 2. Pedro Andrade,
Which Sociology of Urban Tourism, in the Day-After Viral Society?

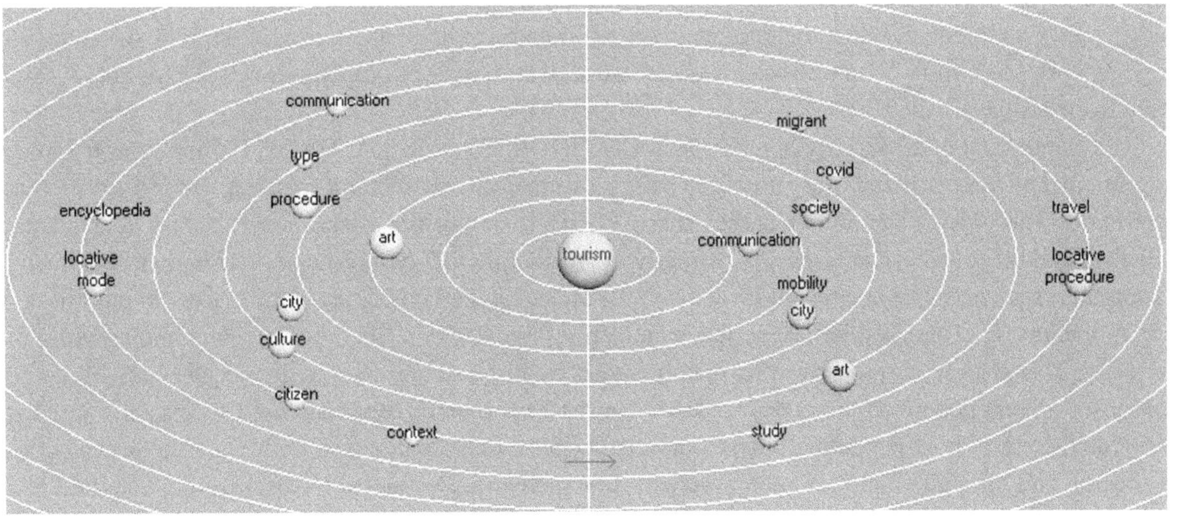

In the first chapter, Pedro Andrade, through a text named 'Which Sociology of Urban Tourism in the Day-After Viral Society?' tries to dress a genealogy of research on Tourism Studies, necessary for understanding the present 'Viral Society'. Epistemologically, a Hybrid/Open Research is discussed. And several theoretical and empirical tourism subjects are presented. As for methodology, the author converses about sociological Hybrid Methods, e.g. Hybrid Discourse Analysis; and he talks on Visual, Virtual and Viral Methods, e.g., Sociological Comics, video papers, Artistic Sociology and Sociological Augmented Reality. Finally, a proposed Virtual Sociological Museum includes a digital Encyclopedia of Viral Tourism, partly analyzed via an Alphabet of Interconceptual Relations.

Figure 3. Moisés Martins,
Tourism as Experience and Desire for Territory and Traveling

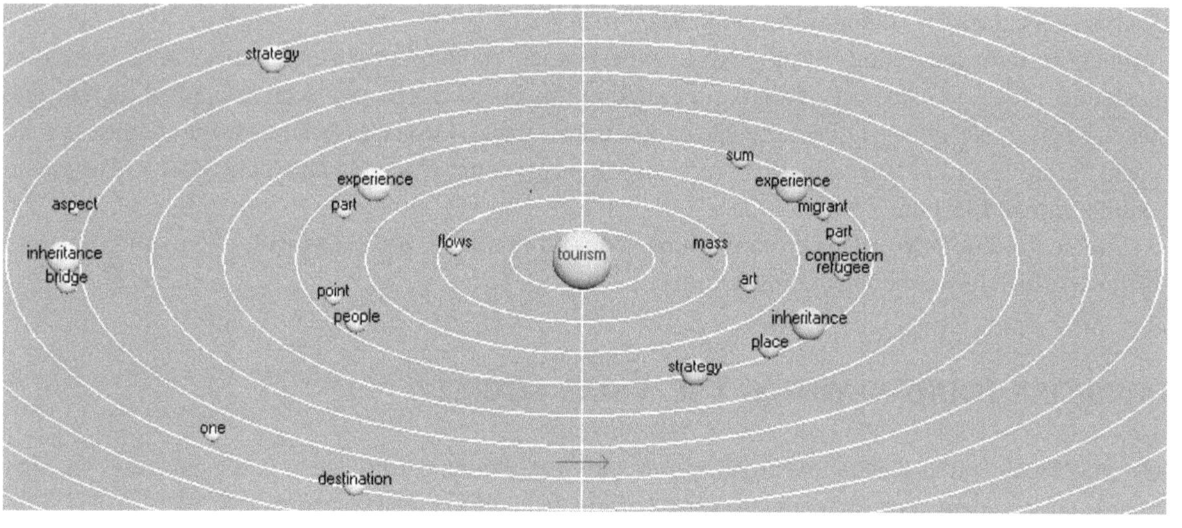

Within Chapter 2, with the title 'Tourism as Experience and Desire for Territory and Traveling', Moisés Martins takes an approach on tourism which is not related to economic analysis, nor even a political analysis *stricto sensu*. The author configures tourism as a cultural activity. The perspective here developed on tourism is phenomenological and socio-anthropological. Hence, Martins shall not dwell on business (*nec/otium* — not/leisure; activity, work, commerce) but mostly on leisure (*otium* — time free from activity).

In Chapter 3, 'Lethal Virality of Otherness: COVID-19, Tourism, and Travel', Philippe Joron founds a seminal debate on contemporary processes that are immeasurably metamorphosing our more profound certitudes and hopes. The author argues that COVID-19 is not the first pandemic, but, as it affects tourism and other social mobilities, is certainly the pandemic that compromises our human fulfilment. Therefore, such crisis, as the recent invasion of Ukraine, can be considered 'total crises or 'pancrises'.

At Chapter 4, 'Total Tourism Experience and Viral Environment: Epistemology and Management of Tourism Destination', a researchers team constituted by Jorge Simões, João Simões and Luís Figueira, proposes a lucid debate on the concept 'Total Tourist Experience', especially and specifically in what regards the era of post-pandemic. Such tourist experience may be detected both via a theoretical conversation and through empirical good practices. For such aim, the questions discussed are applied in a dedicated survey.

Figure 4. Philippe Joron,
Lethal Virality of Otherness: COVID-19, Tourism and Travel

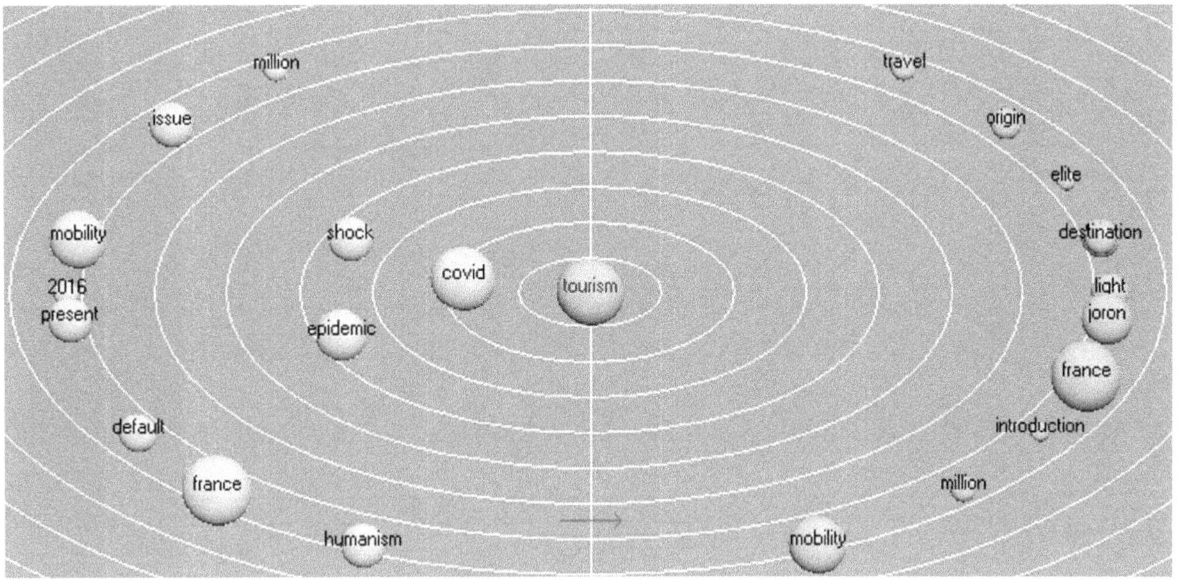

Figure 5. Jorge Simões, João Simões, Luís Figueira,
Total Tourism Experience and Viral Environment: Epistemology and Management of Tourism Destination

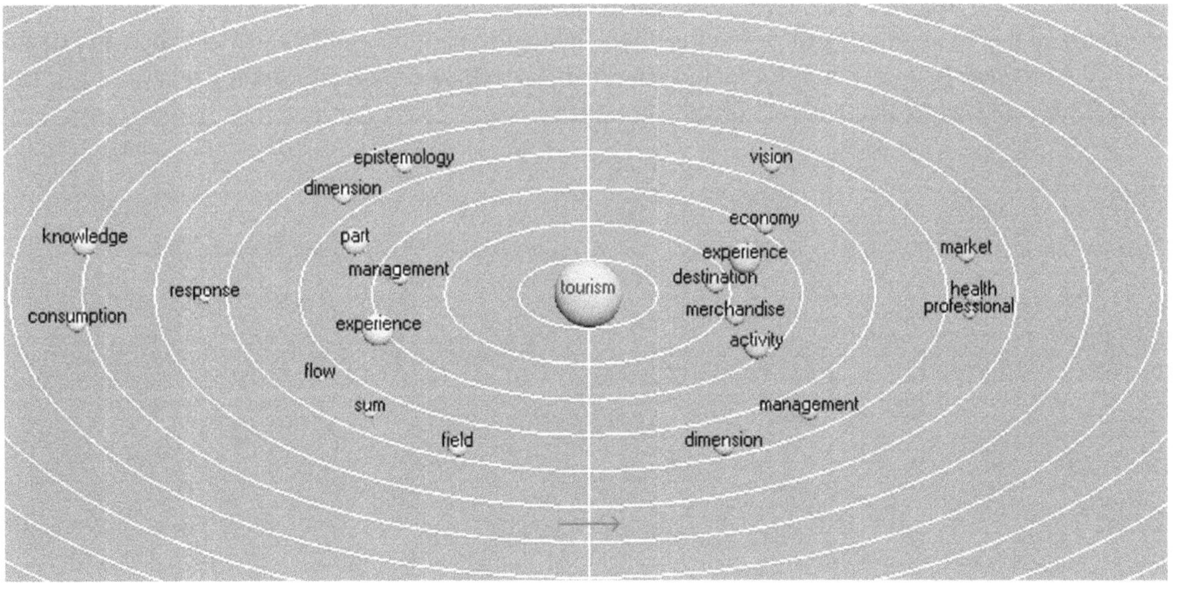

JOURNEYS ACROSS VIRULENT/VIOLENT URBAN TERRITORIES

The second grand theme is the territory and the role of emergent problematic urban scenes and scenarios, debated as social contexts of practices for multiple processes, including tourism, travels and health outbreaks.

Figure 6. Lucia Santaella, Luiz Felipe Napole,
The Ambivalences of Smart Cities

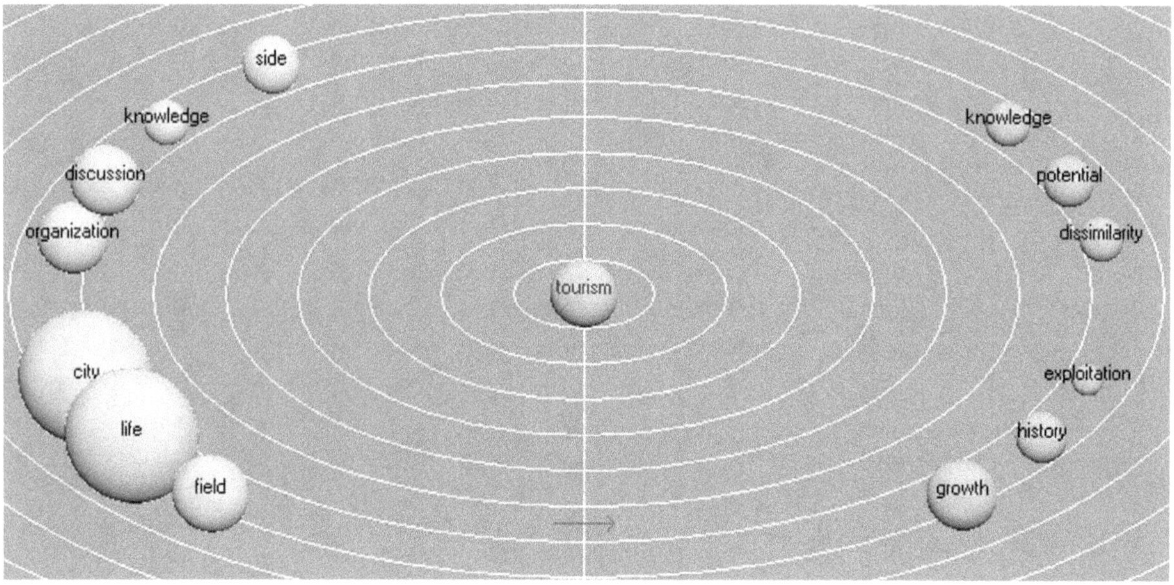

Via Chapter 5, named 'The Ambivalences of Smart Cities', Lucia Santaella and Luiz Felipe Napole focus their relevant discussion on smart cities, megalopolises that are contemporary with the COVID-19 pandemic. Smart cities ambivalences balance between their utility, such as a more efficient administration of the urban fabric, and the perils of intelligent technologies, e.g., more public surveillance on private lives. These smart high-tech entities include internet, surveillance cameras, and a plethora of sensors among other dispositifs that control our social everyday.

Figure 7. Laryssa Tarachucky, Rafael Simão, Maria Baldessar,
Digital Media and the COVID-19 Urban Crisis: What Can We Learn From the Crowdsourced Mappings in Brazil?

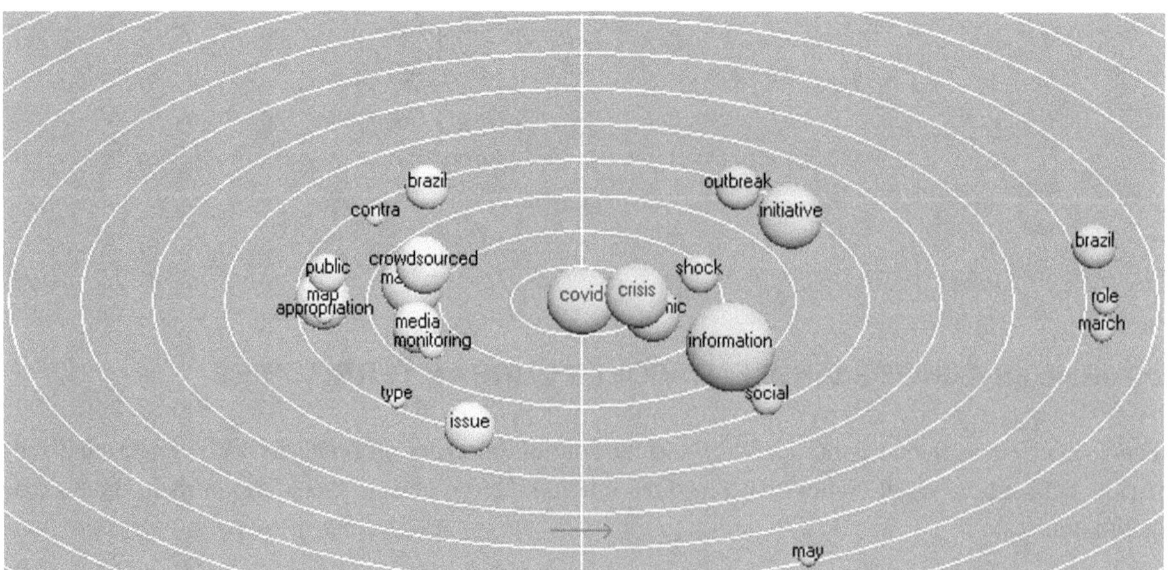

In Chapter 6, named 'Digital Media and the COVID-19 Urban Crisis: What Can We Learn From the Crowdsourced Mappings in Brazil?' Laryssa Tarachucky, Rafael Simão and Maria Baldessar rigorously trace the unprecedent measures taken by governments to combat the COVID-19 pandemic. However, original solutions for such crises emerged from the civil society. In fact, Brazilian communities provided crowdsourcing strategies to fight the lockdown, partly by using the internet. Methodologically, the authors present here an online's ethnographic research, including a content analysis on crowdsourced mappings around Brazil.

Figure 8. Clara Cabral, Rosana Albuquerque, Ana Pereira Roders,
Mitigating Pandemic through Creativity: UNESCO's Responses and Cities' Reactions

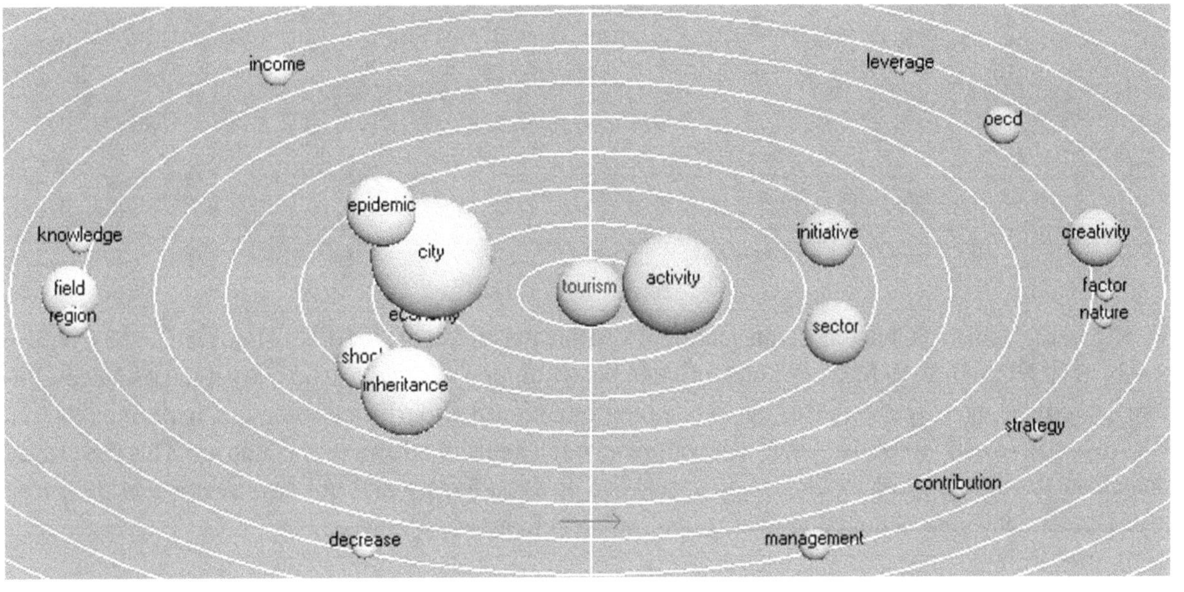

In Chapter 7, Clara Cabral, Rosana Albuquerquea and Ana Pereira Roders offer an interesting text with the title 'Mitigating Pandemic Through Creativity: UNESCO's Responses and Cities' Reactions'. In these pages, the author problematizes culture and creative industries as sound pillars of tourism and of the economic dimension of cities. Some solutions were originated from international organizations and municipalities. In particular, the role of Unesco on this matter, via reports and other studies diffusion among the State-members, is exposed. The response by cities is documented through Unesco Creative Cities and World Heritage Cities initiatives and via the respective information sources.

PANDEMIC CASES AT TOURISTIC CITIES

Section 3 is mainly about case studies including damages on touristic cities caused by the Corona virus and Sars CoV outbreak, and other related phenomena.

Figure 9. Kimberly Nehls,
International Visitors to Las Vegas: a 10-year Economic and Cultural Analysis of high-growth Markets

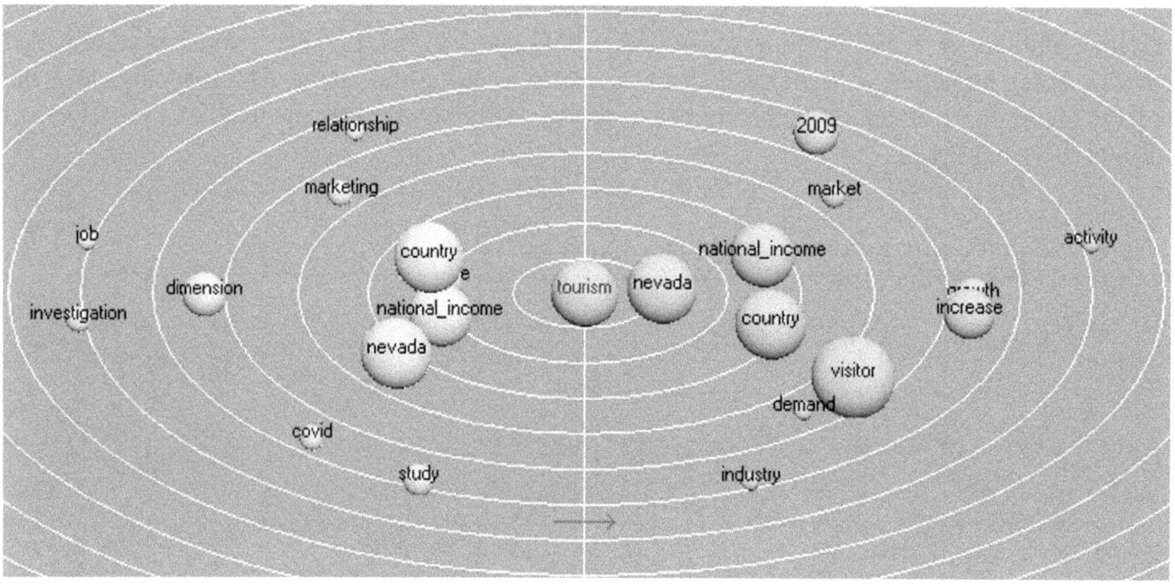

Kimberly Nehls, in Chapter 8, 'International Visitors to Las Vegas: A 10-year Economic and Cultural Analysis of High-Growth Markets', makes a notable testimony of the touristic market in Las Vegas, before and during the Covid era. This city constitutes an important touristic destination, including visitors who originate mostly from the American continent, but also from other world countries. This case study insists on the necessity of making regular analysis on such leisure and holidays sector and of tourist profiles, for achieving better economic results and development.

Figure 10. Miguel Bandeira, Filipa Corais,
Circumstance and Opportunity. Pop-Up Mobility in Pandemic Times: two Case-studies in Braga

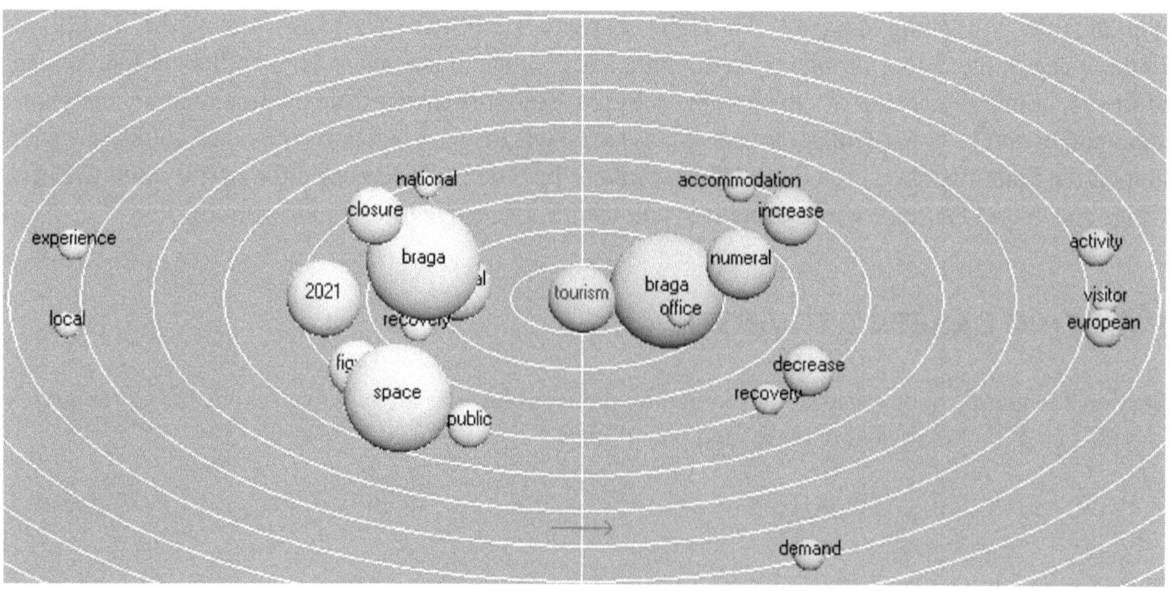

On their turn, Miguel Bandeira and Filipa Corais wrote a remarkable Chapter 9, called 'Circumstance and Opportunity. Pop-Up Mobility in Pandemic Times: Two Case Studies in Braga'. They underline the impact of the various virus variants across the territory and upon the spaces of the polis. The case of Braga, a touristic city in the Portuguese Northwest region, is symptomatic in what regards the fall of tourism venues. Some solutions emerged in the form of 'pop up' measures, including a growing centrality of pedestrians' mobility, and the corresponding enhancement of better living conditions.

Figure 11. Helena Pires, Zara Pinto-Coelho,
Everyday Mobility in an Urban Context of Uncertainty: Times of Resistance in Braga's Public Market

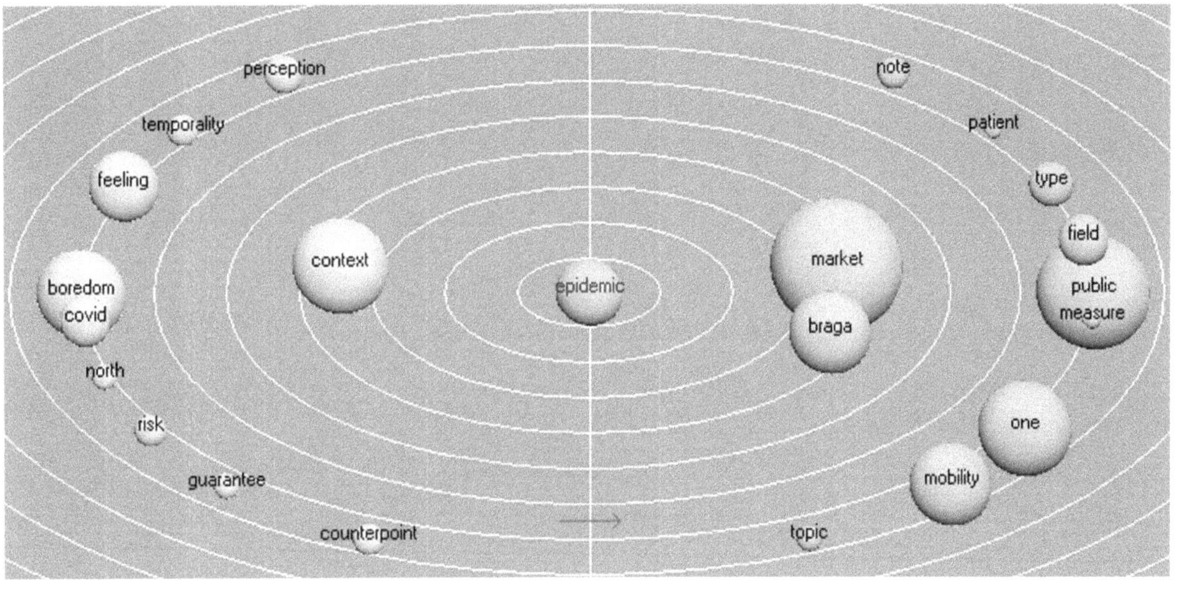

In Chapter 10, 'Everyday Mobility in an Urban Context of Uncertainty: Times of Resistance in Braga's Public Market', Helena Pires and Zara Pinto-Coelho debate the reconstruction of daily routines within urban mobilities, in order to face the insecurity caused by the Corona virus outbreak. Some of these routines took place at the Braga's Public Market, which was object of an opportune fieldwork study by the authors. Such urban resistance included not just the market consumers, but also the merchants, everyone mediated by the communicative space of the urban market.

Pedro Chamusca, José Fernandes and Jorge Pinto sign Chapter 11, 'COVID-19 Pandemic, (Over) Tourism and Urban Dynamics: Lessons From Lisbon and Porto'. These Portuguese major cities were the stage of an overwhelming tourism before the COVID-19 arising. Phenomena as 'touristification' provoked a significant increase of gentrification. The pandemic subsequent urban transformations needed urgent economic and administrative actions. These included sustainable urban strategic plans for development, recomposition of the travelling population, and rescue of these metropolis tourism activities and events.

Mérida is the urban arena where Alba Marin traces a sound reflection, in the Chapter 12, about 'The Necessary Digital Transformation for a Post-COVID-19 Scenario of Mérida's Roman City'. The author articulates this city's Roman archaeological site with the urgence of using digital technologies, for seriously preserving and analyzing such Historical and Cultural Heritage memory, in order to be used and enjoyed as a valid tourism destination. These technologies may enhance the production and dissemination of augmented and mobile narratives for urban archaeological tourists.

Figure 12. Pedro Chamusca, José Fernandes, Jorge Pinto
COVID-19 Pandemic, (Over) Tourism and Urban Dynamics: Lessons from Lisbon and Porto

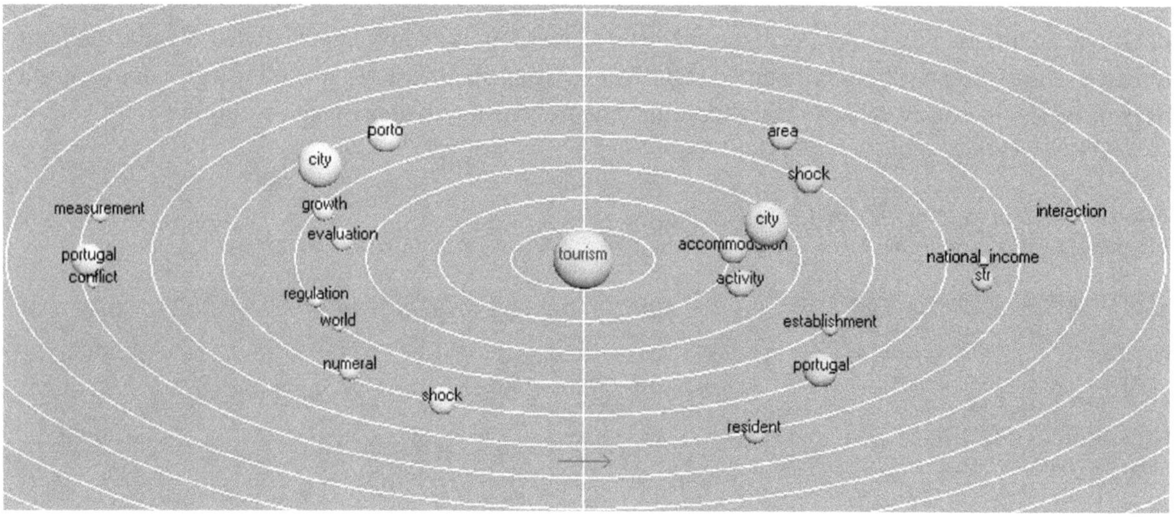

Figure 13. Alba Marin,
The Necessary Digital Transformation for a PostCOVID-19 Scenario of Mérida's Roman City

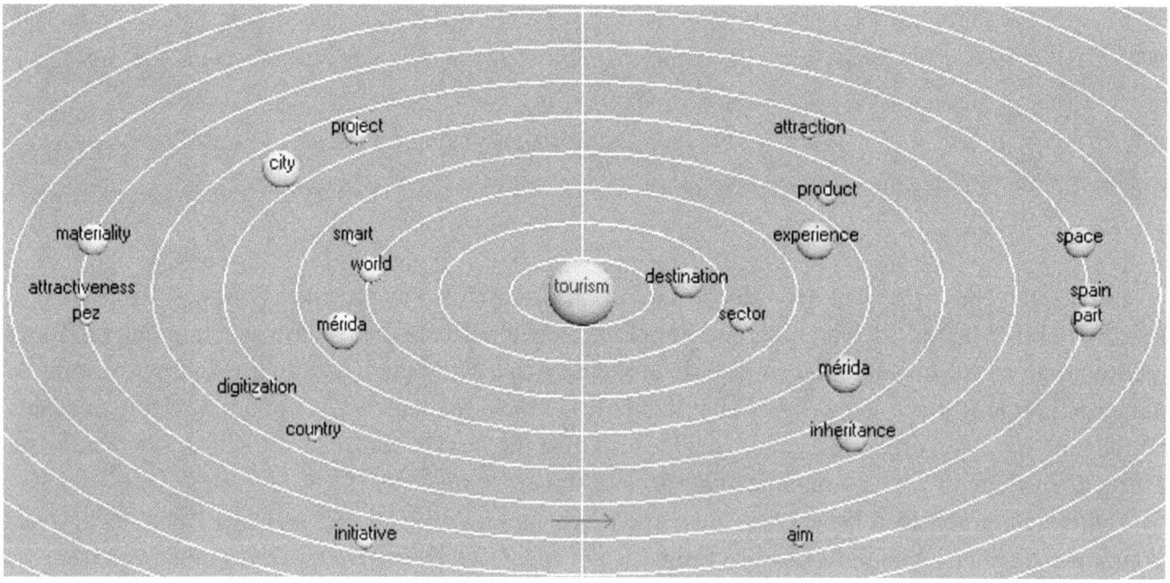

Chaper 13, 'Redefining and Reframing a City: Co-Designing a New City Model in Venice – Universities and Society to Envision a Resident-Friendly Future', was written by Cinzia Colapinto, Vladi Finotto, Daniela Pavan and Fabio Pranovi. This stimulant critical text insists on urban revitalization as a process for mobilizing the State and civil society for coping with the risks of overtourism. Venice is a city that has substantial incomes originated by tourism. However, recent phenomena besides overtourism, as COVID-19, aggravated the inhabitants abandon of the city, and the urban infrastructures themselves. Higher education institutions may play a decisive role on the necessary solutions to such an economic, social, political and cultural problem.

Figure 14. Cinzia Colapinto, Vladi Finotto, Daniela Pavan and Fabio Pranovi,
Redefining and Reframing a City, Co-Designing a new City Model in Venice: Universities and Society to envision a Resident-Friendly Future

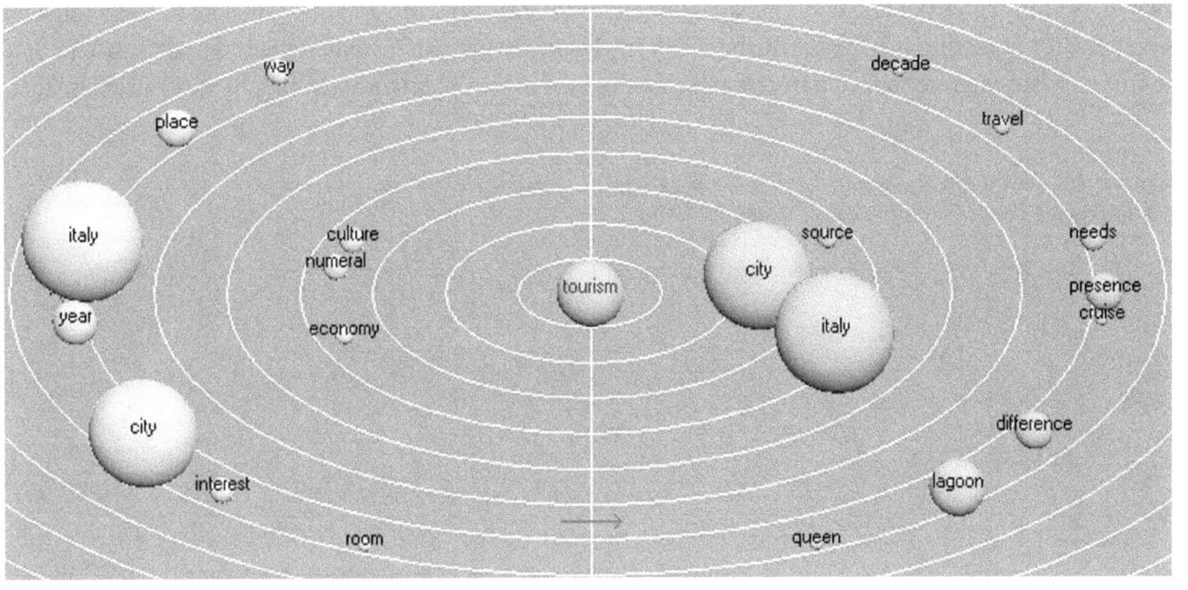

COVID'S TOURISTIC ECONOMY

The fourth book section works on the issues of the economic impact caused by the pandemic on tourism.

Figure 15. Philippe Viallon, Cécile Dolbeau-Bandin, Jérémy Picot, Viola Krebs,
Management of the Health Crisis on Social Medias by Hoteliers in Europe

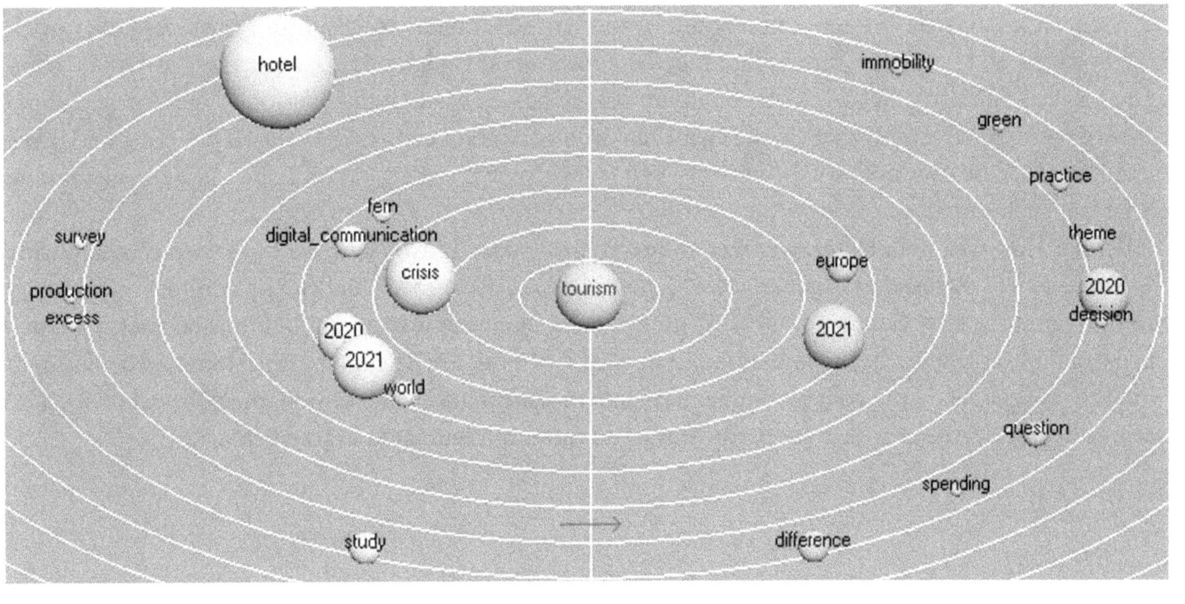

Philippe Viallon, Cécile Dolbeau-Bandin, Jérémy Picot and Viola Krebs, within Chapter 14, wrote an interesting text with the title 'Management of the Health Crisis on Social Media by Hoteliers in Europe'. According to these authors, the tourism sector suffered a double crisis, both economic and communicational. Within such dramatic context, social media was one of the most fruitful means, used by tourism hoteliers, for attracting lost clients during COVID-19 pandemic. Such study reveals that dissimilar communicative strategies conducted by these professionals within cyberspace resulted in diversified results.

Figure 16. José Lopes, Ângela Lopes, José Portugal, Antónia Correia,
Challenges for Urban Tourism in a Post-Pandemic World

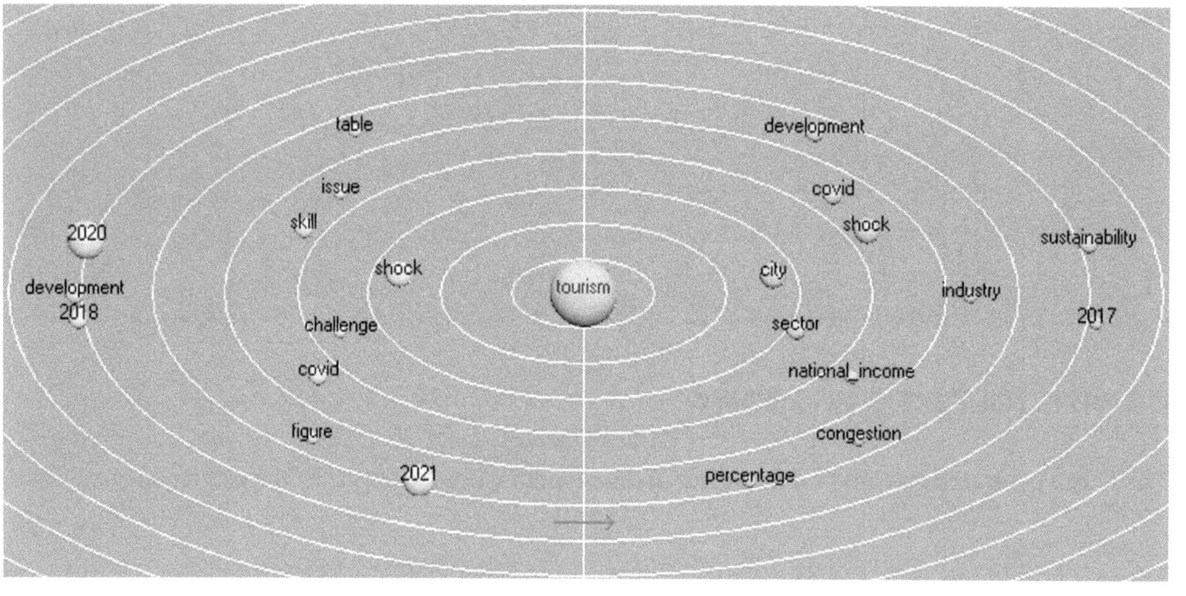

A decisive study on the economic realm of tourism was conducted by José Lopes, Ângela Lopes, José Portugal and Antónia Correia, that may be consulted at Chapter 15, 'Challenges for Urban Tourism in a Post-Pandemic World'. Urban tourism is a niche centered on the fruition of the city and about some of its major assets, such as monuments, environment and cuisine. COVID-19 has damaged some of these assets, but not entirely. Moreover, new technologies, among other factors, may contribute to transform such situation.

On her side, Manuela Hasse presents an invigorating social history of the body, before and during the Corona virus menace, in Chapter 16, with the title 'Body, Exercise, and Health: Culture and Society in Portugal at the 21st Century'. This text understands the present from the light of the past. For such purpose, it describes the health situation in Portugal since the XVIII century until today, articulating it with the present pandemic context. Some measures include nature valorization, open-air exercises, and various hygiene measures, that are partly similar to those implemented in past centuries.

Figure 17. Manuela Hasse,
Body, Exercise and Health: Culture and Society in Portugal at the XXIth Century

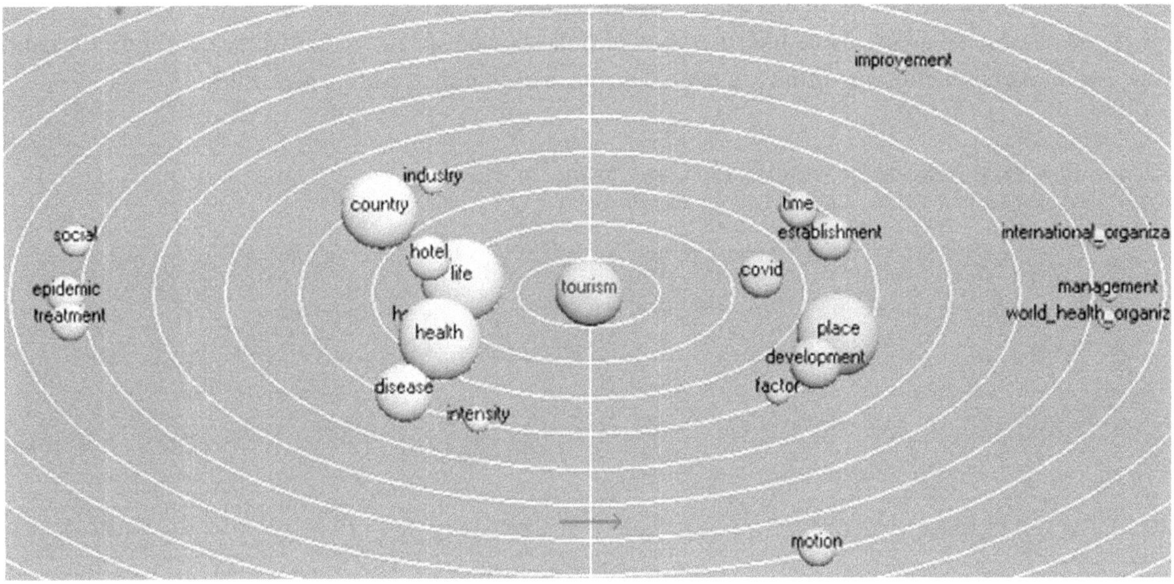

OUTBREAK POLITICS, POLICIES, AND POWER WITHIN HOSPITALITY

The fifth section problematizes some issues associated to citizen politics, state policies and power realms, related with the touristic sector under the influence of the Corona virus pandemic.

Figure 18. Silvena Yordanova, Hanna Stratchen,
Work Place Harassment of Restaurant Workers

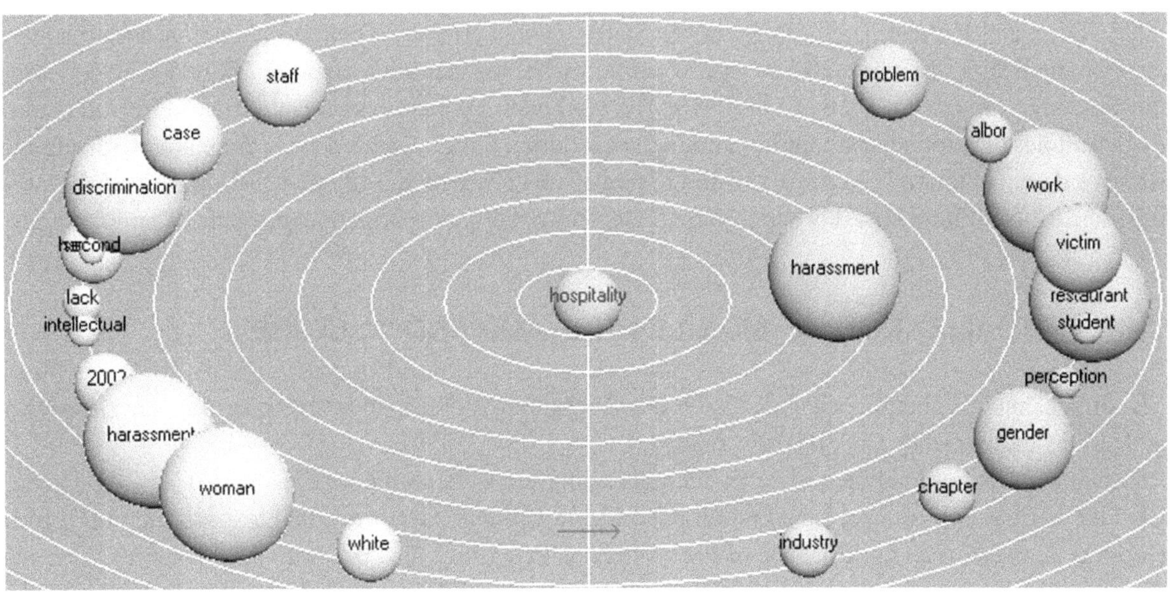

In Chapter 17, 'Workplace Harassment of Restaurant Workers', Silvena Yordanova and Hanna Stratchen enterprise a fair discussion about gender discrimination and bullying, targeting women working in a restaurant kitchen. Methods used in this study include interviews and a questionnaire, answered by women chefs. As results, the analysis and interpretation of data showed specific and diverse configurations and types of harassment, present within the everyday space of that inequal touristic culture.

Figure 19. Carlos Sangreman, Raquel Faria and Bubacar Turé,
COVID-19 and Human Rights in a fragile State Guinea-Bissau

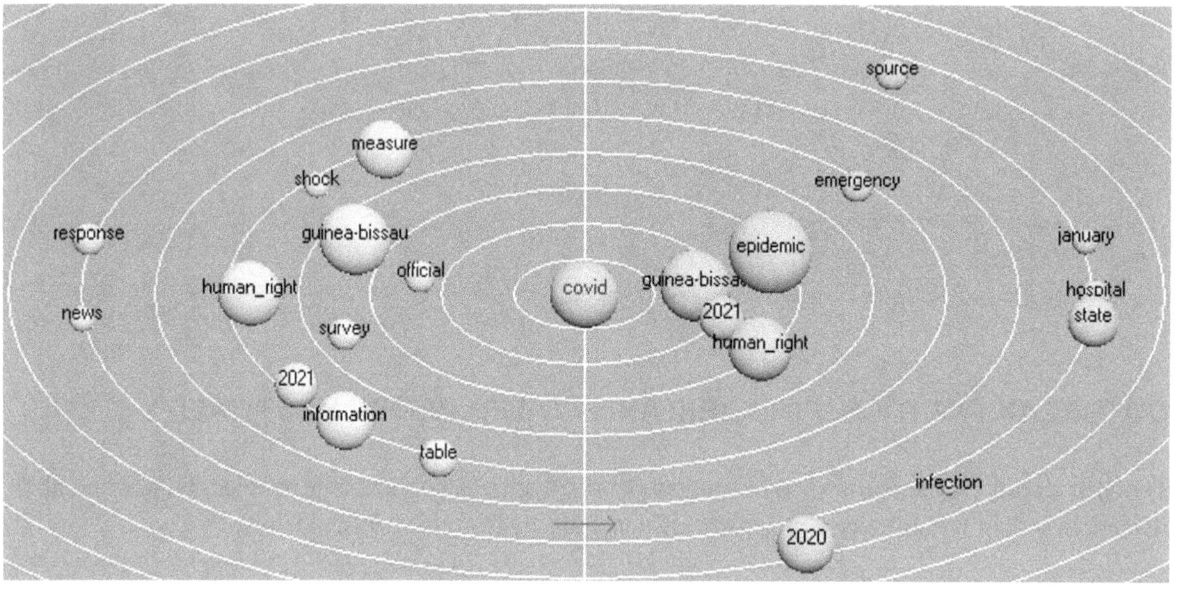

On their side, researchers Carlos Sangreman, Raquel Faria and Bubacar Turé wrote Chapter 18, 'COVID-19 and Human Rights in a Fragile State: Guinea Bissau'. Their investigation questions the actual conjuncture of Guinea-Bissau, specifically in what regards the situation of human rights, within the pandemic context caused by COVID-19. The methodology includes two inquiries addressed to two main samples of respondents: families and market sellers. The data gathered was compared with administrative documentation. The impact of the outbreak was perceived as another factor that corroborates the fragile infrastructures of this country, that reinforce poverty and a relative omission of human rights.

THE 'VIRTUALIZATION' OF MOBILE HERITAGE AND CULTURE

The dimension of cultural heritage is taken in account here, connected with the consequences, effects and affects motivated by COVID-19 spread.

Chapter 19 is outlined by a research team incuding Graciete Honrado, Ana Nata, Cecília Baptista, Cláudia Silva, Filipe Romão, Luís Figueira and Marta Dionísio. Their original text has the title 'From Heritage and Historical Significance of Places to New Directions for Post-Pandemic Tourism'. The authors suggest here the concept of "Total Tourism Experience-TTE", which may profoundly contribute to the debate on Tourist Studies. In fact, they circumscribe a 'value chain', in order to define the experience by the tourist, nurtured by the following means and methods: Travel, Accommodation, Restaurant, Entertainment and Leisure, Safety and Health, Access to Goods and Complementary Services. In addition, the recent situation provoked by COVID-19 pandemic is analyzed. Within this new conjuncture,

Figure 20. Graciete Honrado, Ana Nata, Cecília Baptista, Cláudia Silva, Filipe Romão, Luís Figueira, Marta Dionísio,
From Heritage and Historical Significance of Places to new directions for Tourism in Post-Pandemic

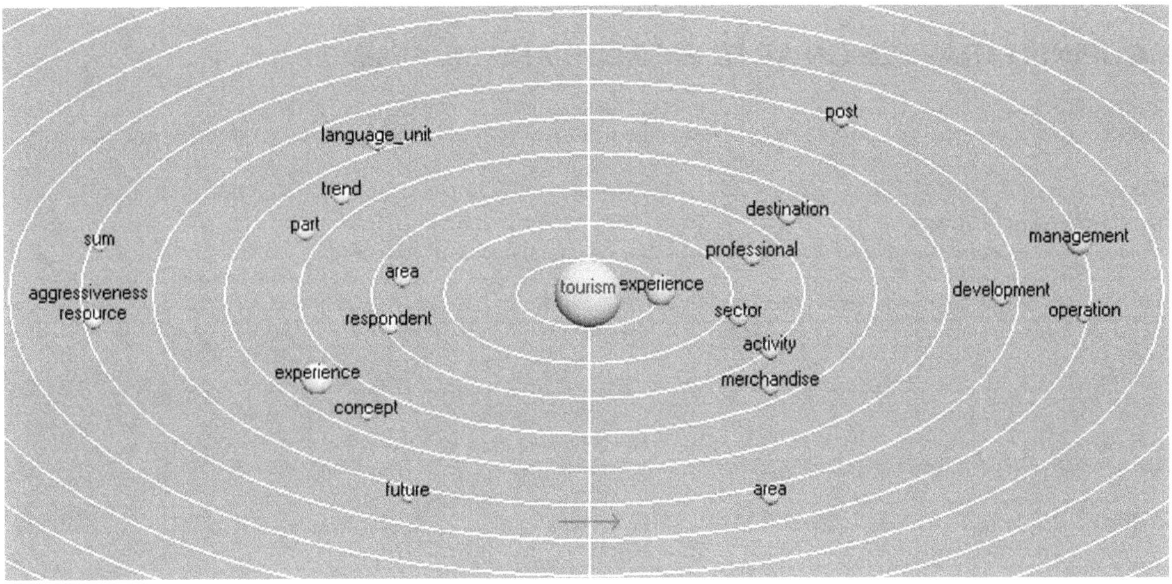

Figure 21. Fernando Contreras,
Thinking about Urban Heritage in the Times of Pandemic. Art Staging and Coordinatio

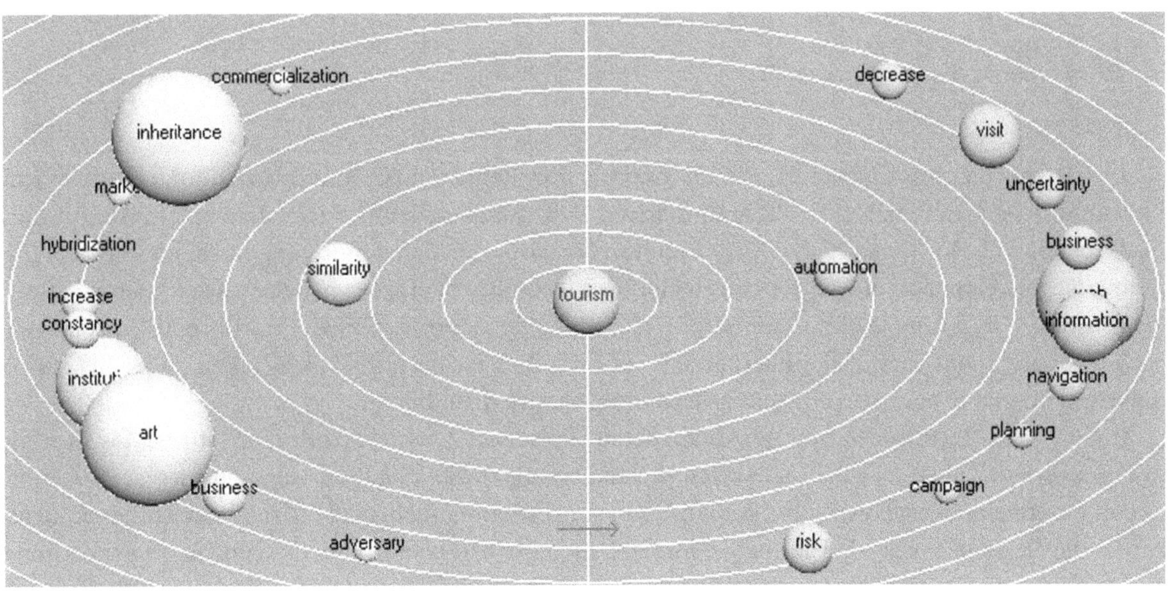

the precedent value chain underlying TTE is taken as an efficient 'opportunity response' to counter the virus outbreak.

On his turn, Fernando Contreras wrote a provocative essay in Chapter 20, 'Thinking About Urban Heritage in the Times of Pandemic: Art Staging and Coordination'. The author envelops his reflection within a Visual Culture knowledge and wisdom. He notes that strategies of displaying and consuming culture faced a significant transformation after the social and cultural confinement due to the emergence of COVID-19. This process was evident at museums, a huge locality serving cultural tourism. Such situation ended up in a virtualization of heritage, subsidiary to the capitalist paradigm of consumption.

CONFINED TRAVELS ON ARTS WORLDS AND MUSEUMS

This Section 7 pertains to a debate on touristic circulation across art worlds' contexts, media, and networks.

Figure 22. José Barros,
Impacts, Resilience and Creativity in Cultural Tourism and Leisure in a Time of Pandemic

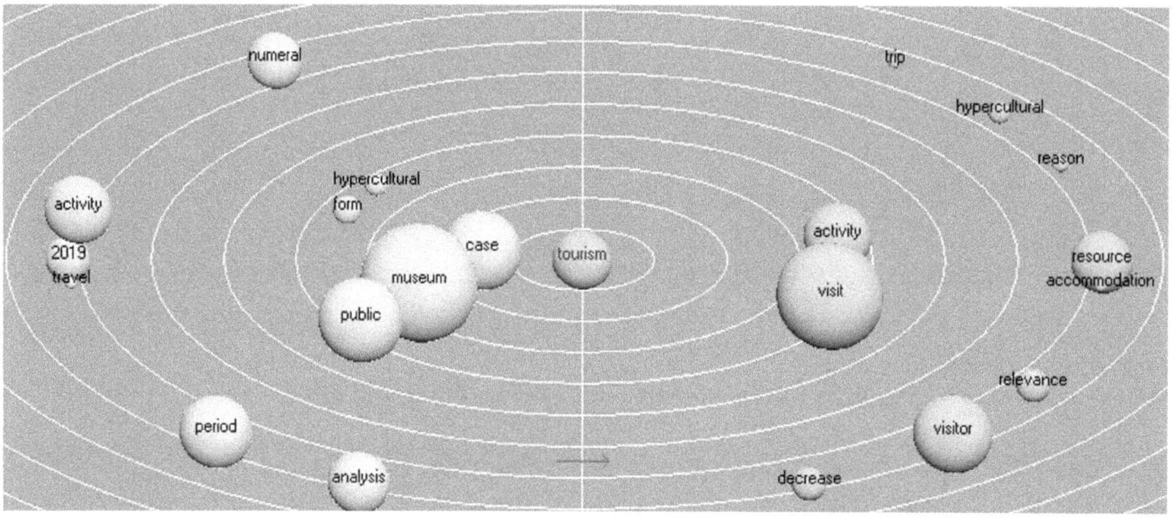

José Barros presents Chapter 21, named 'Impacts, Resilience, and Creativity in Cultural Tourism and Leisure in a Time of Pandemic'. The author drives a vibrant questioning on creative management using new technologies, for virtual visits by hypercultural tourists at national museums and at museums of the future. Negative impacts of Corona virus include less mobility of citizens and decrease of touristic travels. However, via a sample of museums at Lisbon, some major positive aspects of the outbreak emerge, such as unprecedent uses of the Internet and opportunities for debating issues about the efficacity of museums on giving access to collections to more diversified publics.

Chapter 22 was produced by José Abreu, about 'Portuguese Public Art Primary Itineraries: A Map of Discoveries'. This is an exhaustive scientific and artistic travel about touristic journeys. A collection of cultural itineraries and respective guidance and suggestions gathers many of the more authentic spots across the North of Portugal. The core of touristic interests and pleasure include public art expressions such as sculptures, monuments and other heritage sites or objects. The aim is to offer new cultural routes for cultural tourism, through at least three strategies: practising a defensive paradigm of travel by car against the perils of COVID-19; subscribing the logic of continuous discovery of the cities' memories, environment and social contexts; applying a literary and journalist approach to the places visited.

Figure 23. José Guilherme Abreu,
Portuguese Public Art Primary Itineraries: a Map of Discoveries

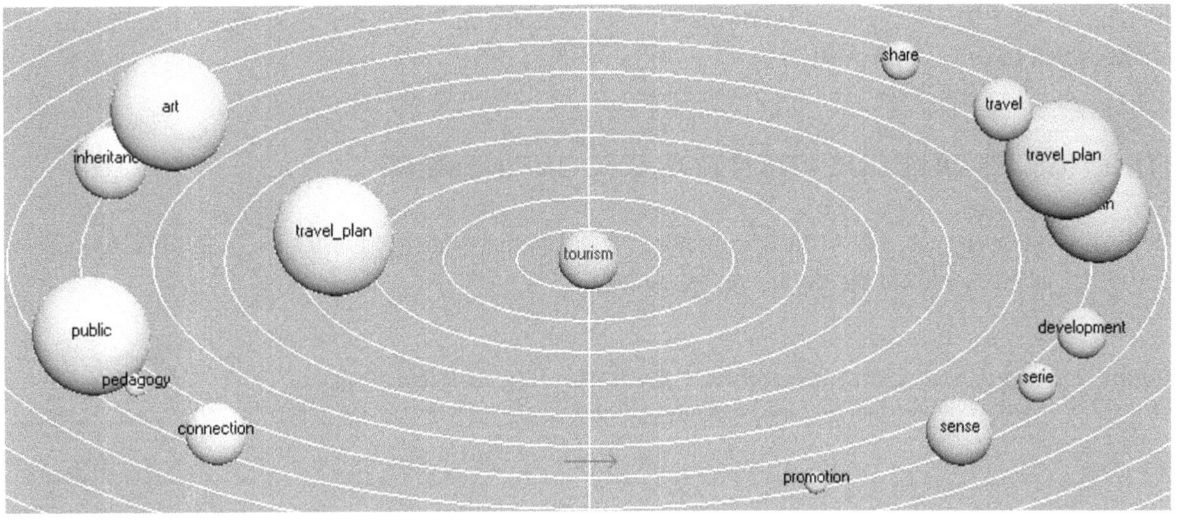

Figure 24. Regina Frank,
From Tourist Destination to Tabernacle

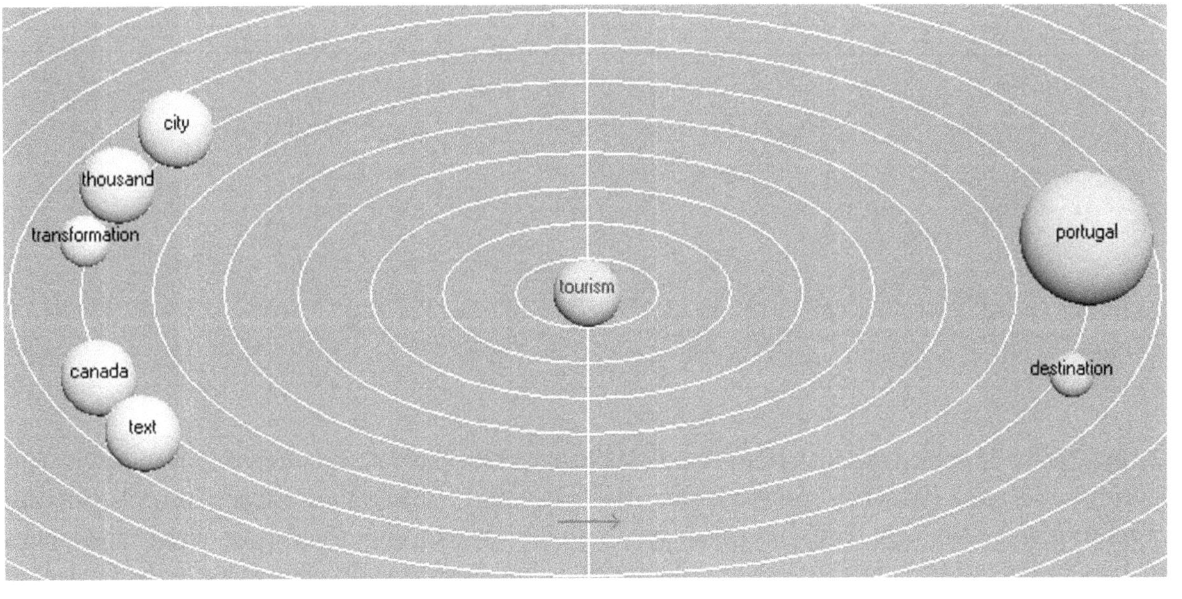

Chapter 23, 'From Tourist Destination to Tabernacle', is a genuine auto-biography of the artist understood as a traveller. Regina Frank narrates her multiples nomadic experiences within diverse spaces and times, as a German visual artist committed to the discovery of foreign but somehow familiar landscapes. Some of her art exhibitions testify such pulsion for the voyage, e.g., several exhibitions under the title 'The Artist is Present', from 1989 until 2010, and 'The Heart is Present' since 2011. She underlines the encounter and the inter-travel between German and Portuguese cultures, in particular inside the momentum of COVID-19 spread and at the heart (both in 'core' and 'love' senses) subjacent to the cultural changes that the virus provoked.

ITINERARIES ON EDUCATION FOR DECONFINEMENT

Education and its issues, within the conjuncture of confinement due to COVID-19, in addition to tourism activities, are presented at the book's eighth section.

Figure 25. Sérgio Eliseu, Marta Noronha, Ângela Berlinde, Mihaela Bogdan,
Virtual Tourism as a Pedagogical Practice: Adaptation of Curricular Content during and after a Pandemic

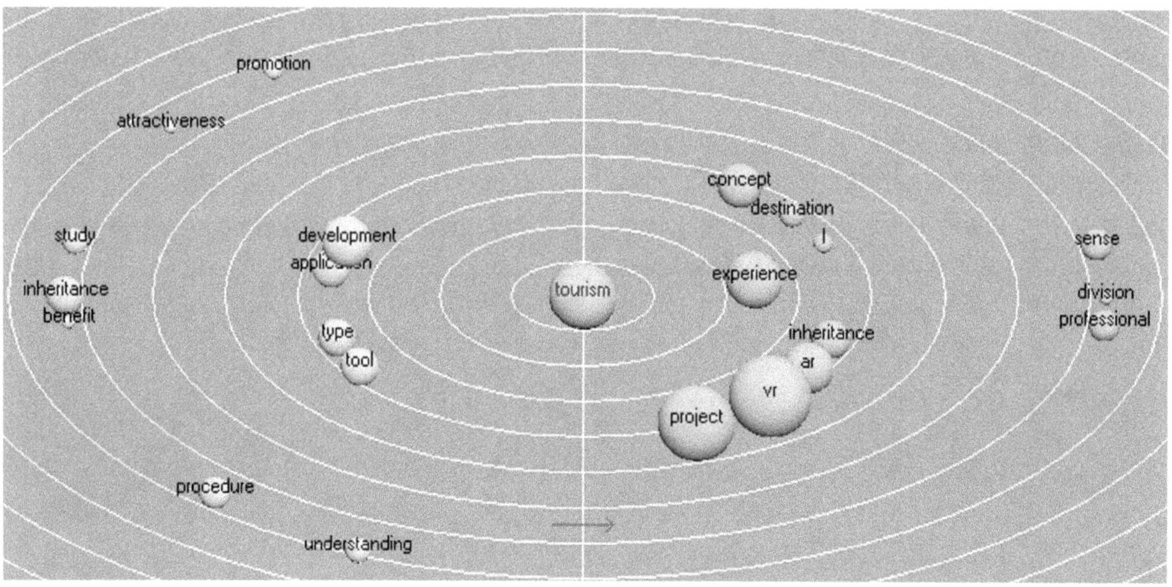

In Chapter 24, called 'Virtual Tourism as a Pedagogical Practice: Adaptation of Curricular Content During and After a Pandemic', Sérgio Eliseu, Marta Noronha, Ângela Berlinde and Mihaela Bogdan, promote a suggestive core thesis: post-pandemic solutions may be implemented via new technologies, such as virtual reality, augmented reality and mixed reality. One exemplificatory case is the Education sector, in order to prevent post-pandemic anxiety and potential future outbreaks within this pedagogic territory and actors. New mobile and personal technologies allow safer trips against possible contaminations. An experience on these matters demonstrates the upper hypotheses, via a VR device including training content consulted by Erasmus students.

Figure 26. Filomena Moita, Leandro Lucas,
The Cordel Pandemic: The Daily Life of Students from the Interior of Paraíba-Brazil told in Verse

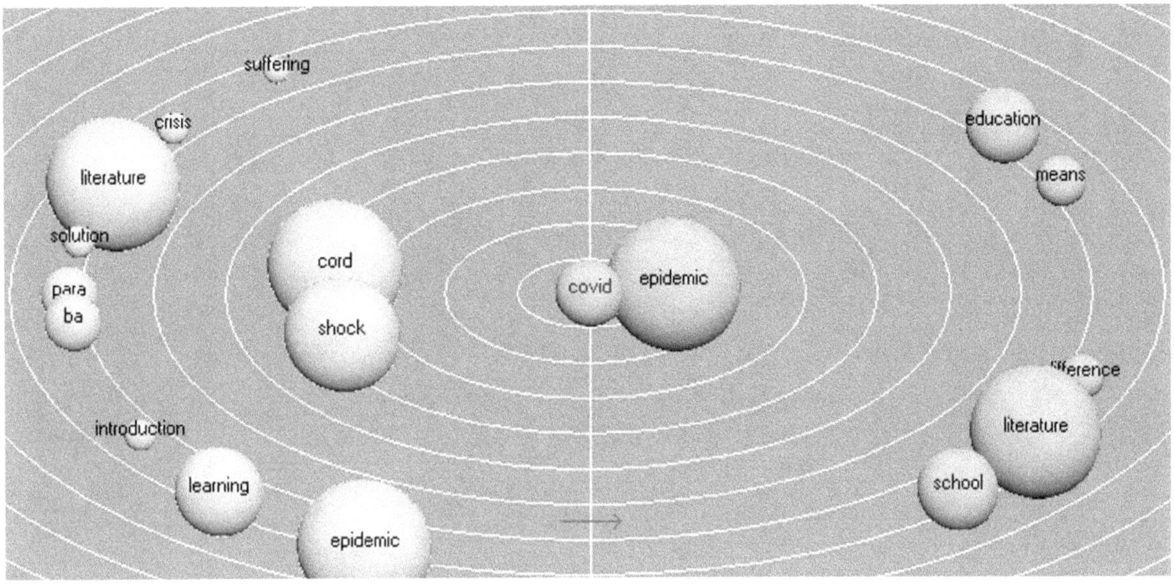

At Chapter 25, Filomena Moita and Leandro Lucas explore 'The Cordel Pandemic: The Daily Life of Students From the Interior of Paraíba-Brazil Told in Verse'. This is an exciting case study on the profound transformations perpetrated by Corona virus on pedagogical methods, both in teaching and learning. An experiment is conducted within an academic community, in order to delimitate the role of storytelling in pandemic times and spaces, through the correspondent content analysis. A chosen literary style for this aim is 'cordel', a genre characterized by the narration of daily life moments in verse, which were commented with the purpose to relate daily narrations with students everyday actions, within a context of virulent outbreak.

CONTAMINATED CINEMA DESTINATIONS

Research on Tourism articulated with Corona virus may be undertaken as well regarding other arts, such as cinema and the audiovisual sector (video, etc.).

In fact, in Chapter 26, Natacha Cyrunlik envisages the documentary both as a means, a media and a method, in order 'To Identify the Evolution of Tourism Practices in a Sensitive Way Through Documentary'. The author creatively filmed numerous sites in France where tourists circulate, and more recently some of the localities contaminated by COVID-19 pandemic. She opposes the potentialities of the documentary genre to the insufficiencies inherent to the postcard, a traditional resource for engraving the touristic memory and identity.

Equally within the cinematic arts, Jane Almeida, Cícero Silva, and Alfredo Suppia in Chapter 27 show a notable approach about 'Film Festivals in Contamination: Streaming Strategies in Brazil'. They conduct a reflection on the centrality of the festival event's paradigm, for the sustainability of the tourism sector. Inside these theoretical lens, some practical applications are suggested, to face and combat the

Corona virus threat. One of those is clarifying a typology of festivals, involving traditionalist, hybrid, mimetic, online and nostalgic strategies, articulated with different technological configurations. All this circumscribes novel ways to organize streaming platforms systems, for the dissemination of videographic media for polysemic audiences' fruition.

Figure 27. Natacha Cyrunlik,
To Identify the Evolution of Tourism Practices in a Sensitive Way through Documentary

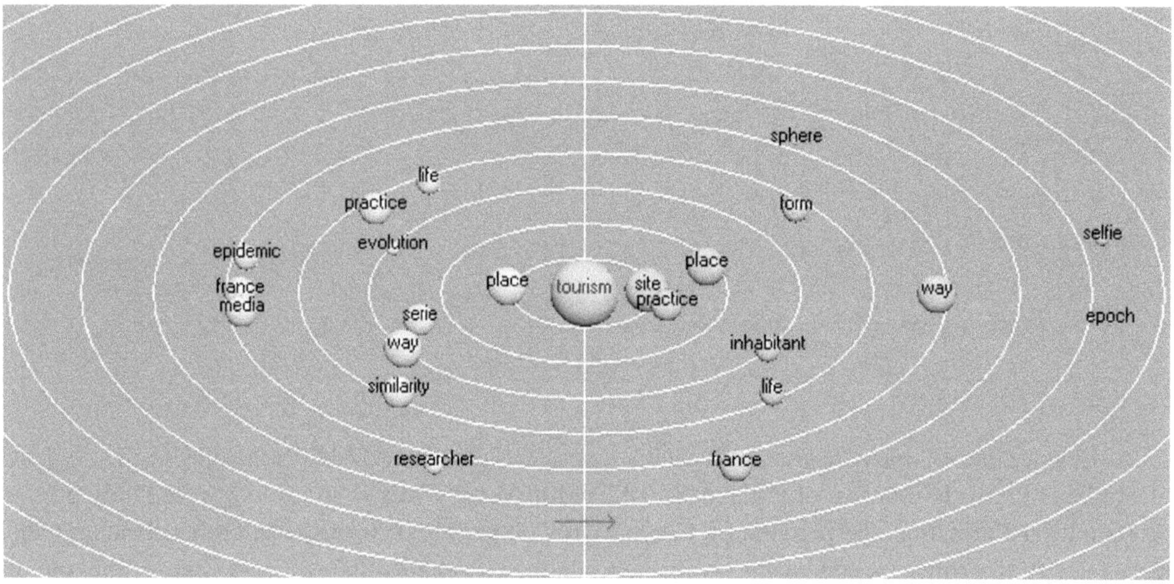

Figure 28. Jane Almeida, Cícero Silva, Alfredo Suppia,
Film Festivals in Contamination: Streaming Strategies in Brazil

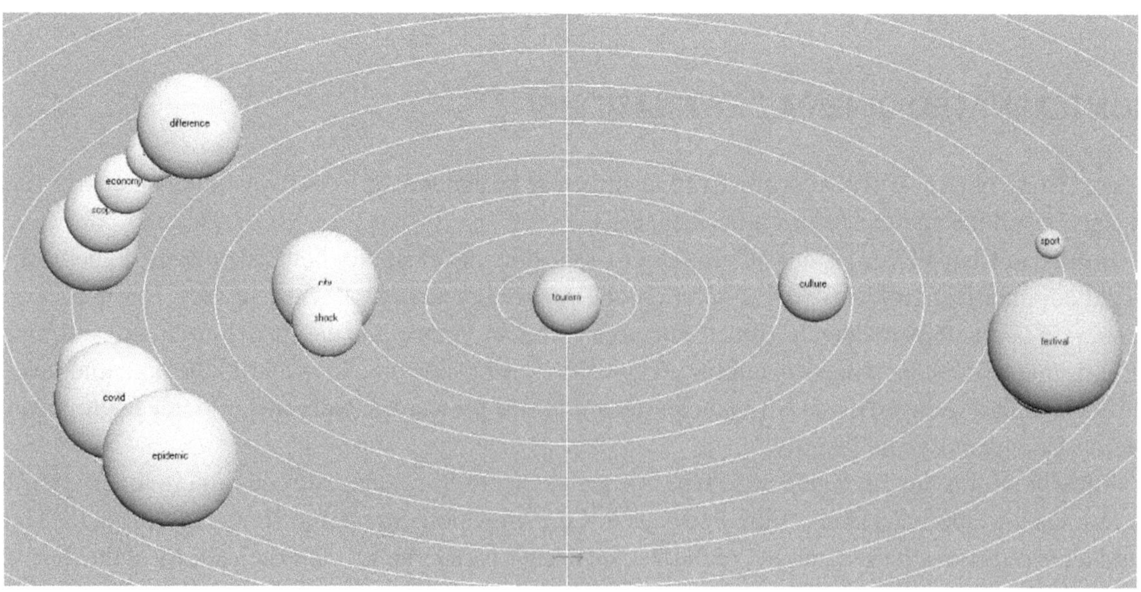

NEW MEDIA FOR POST-VIRAL FLOWS

Finally, the tenth book section takes care of the possible, probable and commutable futures, which may happen after the pandemic outbreak (if this will occur completely?), vehiculated by travels mediated by technologic devices and dispositifs.

Figure 29. Andres Montenegro,
Extended Reality, as the Immersive Environment Conjunction of Augmented Reality, Virtual Reality, and Mixed Reality: A Phenomenological Response facing Modern Culture

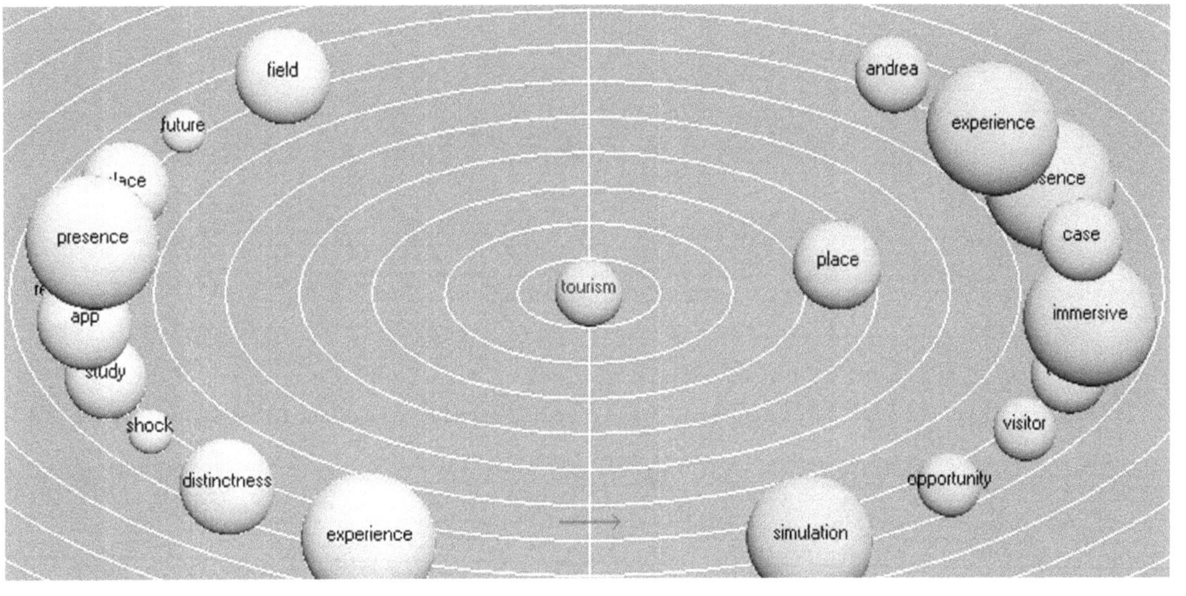

Within such troubled route, Andres Montenegro, in Chapter 28, predicts an eventual itinerary, through 'Extended Reality as the Immersive Environment Conjunction of Augmented Reality, Virtual Reality, and Mixed Reality: A Phenomenological Response Facing Modern Culture'. The author considers that 3D worlds metaverses, will constitute the standard interface of the future, in what regards the digital and virtual panorama. The core tool to this aim is Extended Realty, which will allow a more complete experience for citizens and tourists to interact using their whole body's senses, via phenomenological experiences aiming to understand contemporary social and cultural changes.

Similarly, Pedro Rocha, in Chapter 29, proposes some promising courses of action via 'Urban Tales Visible With Augmented Reality: A Street Exhibition in Campanhã (Porto)'. Using Augmented Reality as well, he suggests to edify 'augmented communities', who may communicate via information addressed by stories that convey more conscience of daily life to citizens and tourists. In this way, stories and History may work together as two central ingredients for a better urban everyday sharing. An example is presented, an exhibition where life stories are not just told, but also shown, inside a problematic urban environment, so that people may connect more deeply.

Figure 30. Pedro Rocha,
Urban Tales Visible with Augmented Reality: a Street Exhibition in Campanhã (Porto)

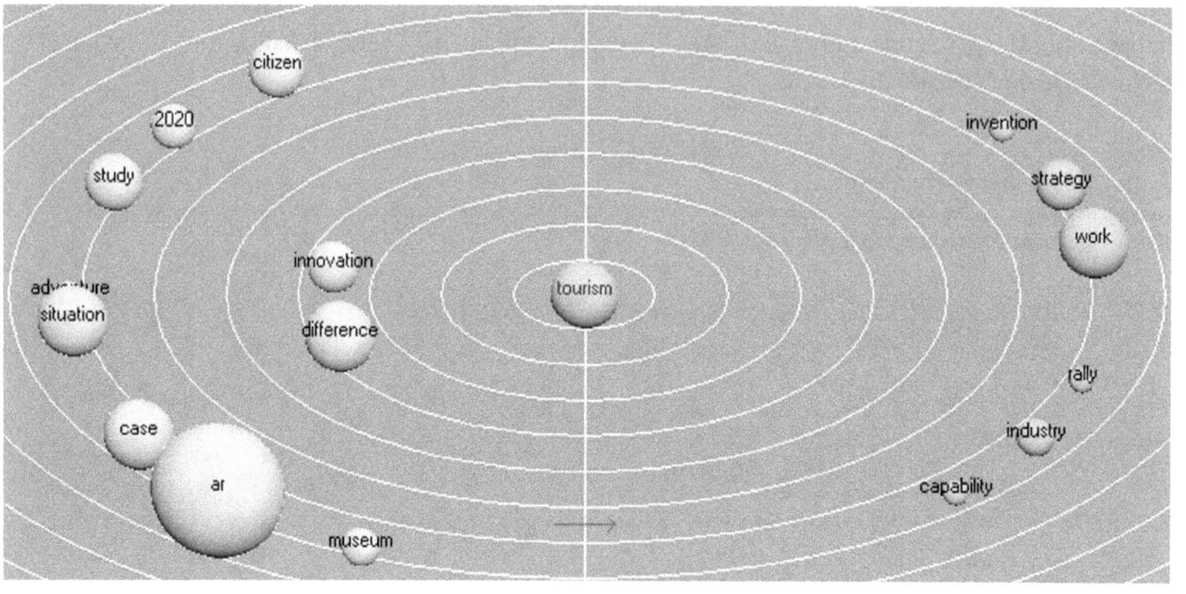

Figure 31. Juliana Wexel,
Street Art, Intersectional Feminism and Digital Media-Art: A Post-Digital and Post-Pandemic Cyberperformative Art

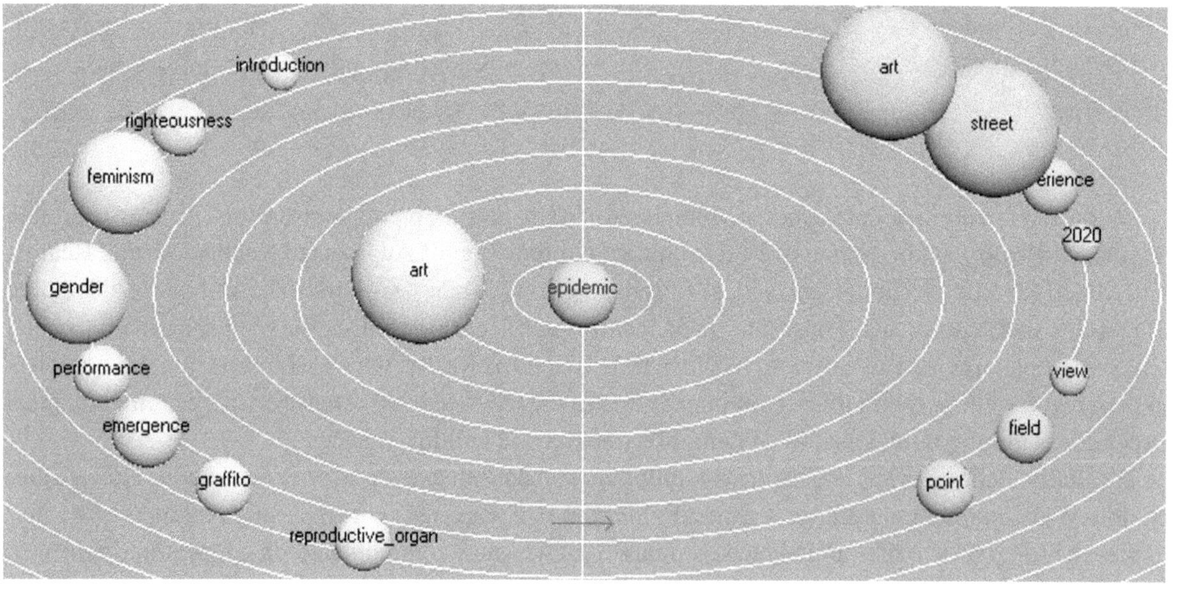

Last but not the least, Juliana Wexel, at Chapter 30, introduces an innovative project on 'Street Art, Intersectional Feminism, and Digital Media-Art: A Post-Digital and Post-Pandemic Cyberperformative Art'. This proposition retakes some pertinent concepts, such as 'gender equity' and 'cyberperformance', but also others less known and some original, e.g. 'Instagrammism', 'Selfiecity' or 'Vulva Art'. The present post-pandemic context and the inherent artivist posture, are partly built via street art, intersectional feminism and digital technologies. As examples of these ideas, several artist exhibitions show hybrid praxis including immersive environments made of survey information, auto ethnographic data and Augmented Reality, among other means, media and methods.

Pedro Andrade
University of Minho, Portugal

Moises de Lemos Martins
University of Minho, Portugal

REFERENCES

Abay, K., & Hosam, I. (2020). *Winners and losers from COVID-19: Evidence from Google search data for Egypt*. Intl Food Policy Res Inst.

Abdelhady, M. (2021). *Digital Nomad Lifestyle in Dahab, Egypt. Current development on the job market*. GRIN Verlag.

Adeola, O. (2021). *Gendered Perspectives on COVID-19 Recovery in Africa*. Springer Nature.

Balavantagol, K. (2020). *COVID-19 Pandemic: It's Inconspicous Multi-Dimensional Implications (Migrants, Industry, Agriculture, Tourism)*. Nitya Publications.

Berube, D. (2021). *Pandemic Communication and Resilience*. Springer Nature.

Campbell, Y., & Connell, J. (2021). *COVID in the Islands: A comparative perspective on the Caribbean and the Pacific*. Springer Nature.

Carvalho, J. V. D., Rocha, A., Liberato, P., & Peña, A. (2020). *Advances in Tourism, Technology and Systems*. Springer Nature.

Cevik, S. (2020). *Going Viral: A Gravity Model of Infectious Diseases and Tourism Flows*. International Monetary Fund.

Chatterjee, S. (2021). *COVID-19: Tackling Global Pandemics through Scientific and Social Tools*. Academic Press.

Cimoli, M. (2020). *Recovery Measures for the Tourism Sector in Latin America and the Caribbean Present an Opportunity to Promote Sustainability and Resilience*. United Nations, ECLAC.

Clemens, B. (2020). *COVID-19 and the Egyptian economy: Estimating the impacts of expected reductions in tourism, Suez Canal revenues, and remittances*. Intl Food Policy Res Inst.

Demir, M., Dalgic, A., & Ergen, F. (2021). *Impact and Implications of COVID-19 on the Tourism Industry*. Business Science Reference.

Erh, J. (2021). *COVID-19's Economic Impact on Tourism in Singapore*. ISEAS.

Espiner, N. et al. (2021). *Exploring the Impacts of the COVID-19 National Lockdown on Outdoor Recreationists' Activity and Perceptions of Tourism*. Lincoln University.

Genette, G. (1997). *Paratexts: Thresholds of Interpretation*. Cambridge University Press.

Geroimenko, V. (2021). *Augmented Reality in Tourism, Museums and Heritage*. Springer Nature.

Hafsa, S. (2021). *Impacts of COVID-19 Pandemic on Different Sectors & Overall Socio-Economic Conditions of Bangladesh: at a Glance: RMG, SME, Education & Tourism Sector*. Eliva Press.

Hassan, A., & Sharma, A. (2021). *Overtourism, Technology Solutions and Decimated Destinations*. Springer Nature.

Hassan, A., & Sharma, A. (2022). *Overtourism, Technology Solutions and Decimated Destinations*. Springer.

Hebbard, S. (2020). *Tourism, Travel & COVID-19*. KR Publishing.

Helble, M., & Fink, A. (2020). *Reviving Tourism Amid the COVID-19 Pandemic*. Asian Development Bank.

Higgins-Desbiolles, F. (2021). *Socialising Tourism*. Taylor & Francis.

Hubert, J. (2020). *The Impact of COVID-19 on Tourism Enterprises in the Lao People's Democratic Republic*. Asian Development Bank.

Insights West [Firm]. (2020). *BC Resident's Public Perceptions: COVID-19 Travel and Tourism Wave 5 Report.* Destination British Columbia.

Jönsson, H. C. (2022). *Tourism and Foreign Direct Investment*. Routledge.

Klooz, D. (2020). *The Reset.* David Klooz.

Korstanje, M., & George, B. (2021). *Mobility and Globalization in the Aftermath of COVID-19*. Springer Nature.

Korstanje, M. E., & George, B. P. (2022). *The Nature and Future of Tourism. A Post-COVID-19 Context.* Apple Academic Press Exclusively Company.

Luger, K., & Ripp, M. (2020). *World Heritage, Place Making and Sustainable Tourism*. StudienVerlag.

Luo, Y. (2022). *Tourism, Aviation and Hospitality Development During the COVID-19 Pandemic*. Springer Nature.

Ministry of Information, Tourism and Heritage. Zanzibar Comission for Tourism. (2020). *Standard Operating Protocols for COVID-19 in Tourism Industry*. M.I.T.H.

Ministry of Information, Tourism and Heritage. Zanzibar Commission for Tourism. (2020). *Standard Operating Protocols for COVID-19 in Tourism Industry*. M.I.T.H.

Nhamo, G., Dube, K., & Chikodzi, D. (2020). *Counting the Cost of COVID-19 on the Global Tourism Industry*. Springer Nature.

Nugroho, Y., & Negara, S. (2020). *COVID-19's Impact on Micro, Small, & Medium Enterprises and Tourism in Indonesia*. ISEAS-Yusof Ishak Institute.

Omeraki, S. (2021). *Handbook of Research on Interdisciplinary Perspectives on the Threats and Impacts of Pandemics*. IGI Global.

Pal, I., & Shaw, R. (2022). *Pandemic Risk, Response, and Resilience*. Elsevier.

Panigrahi, C. (2021). Computational Modeling and Data Analysis in COVID-19 Research. CRC Press.

Peng, N. (2021). *The Reshaping of China-Southeast Asia Relations in Light of the COVID-19 Pandemic*. Springer Nature.

Permal, S. & Mahmoud, H. (2021). *The Impact of COVID-19 Pandemic on Malaysia's Maritime Sectors and Way Forward.* Maritime Institute of Malaysia (MIMA).

Rogerson, C. M., & Rogerson, J. M. (2021). *Urban Tourism in the Global South*. Springer.

Roy, S., & Nandy, D. (2022). *Understanding Post-COVID-19 Social and Cultural Realities*. Springer Nature.

Saarinen, J., & Rogerson, J. M. (2021). *Tourism, Change and the Global South*. Routledge.

Sarfraz, M., Ozturk, I., & Shah, S. (2022). *Coronavirus Disease (COVID- 19)*. The Impact on Psychology of Sustainability, Sustainable Development, and Global Economy.

Scheyvens, R. (2020). *Development in a World of Disorder: Tourism, COVID-19 and the Adaptivity of South Pacific People*. Massey University.

Situmorang, F. (2022). *Tourism Recovery Based on Digital Nomadism Post-COVID-19 in Ubud Bali*. Syiah Kuala University Press.

Stone, L. (2022). *Protected Areas and Tourism in Southern Africa*. Routledge.

Taxmann. (2021). *Key Highlights of FM Nirmala Sitharaman's Press Conference to Boost COVID-19 affected Sectors.* Taxmann Publications Private Limited.

United Nations. (2021). *Mountain tourism – Towards a more sustainable path*. Food & Agriculture Org.

UNWTO. (2021). *UNWTO Inclusive Recovery Guide: Sociocultural impacts of COVID-19*. UNWTO.

Verma, P. (2021). *COVID-19*. CRC Press.

ADDITIONAL READING

Akhtar, R. (2021). *Coronavirus (COVID-19) Outbreaks, Environment and Human Behaviour*. Springer Nature.

Arie, S., & Dimitri, I. (2022). *Handbook of Tourism Impacts*. Edward Elgar Publishing.

Armand, F., Igor, R., & Arne, B. (2021). *Impacts and Implications for the Sports Industry in the Post-COVID-19 Era*. IGI Global.

Balasingam, A., & Ma, Y. (2022). *Asian Tourism Sustainability*. Springer Nature.

Balasubramanian, M. (2020). *Covid 19 - The New Age Pandemic*. Notion Press.

Balasundharam, V., & Koepke, R. (2021). *Diversion of Tourism Flows in the Asia & Pacific Region: Lessons for COVID-19 Recovery*. International Monetary Fund.

Bank, Asian Development (2021). *Impact of COVID-19 on CAREC Aviation and Tourism.* Asian Development Bank.

Bank, Asian Development (2021). *Sustainable Tourism after COVID-19.* Asian Development Bank.

Bank, Asian Development (2022). *Supporting Post-COVID-19 Economic Recovery in Southeast Asia.* Asian Development Bank.

Beirman, D. (2021). *Tourism Crises and Destination Recovery*. Sage (Atlanta, Ga.).

Borg, J. (2022). *A Research Agenda for Urban Tourism*. Edward Elgar Publishing.

Borges, A. et al. (2022). *Gastronomy, Hospitality, and the Future of the Restaurant Industry: Post-COVID-19 Perspectives*. IGI Global.

Bristow, R., & Jenkins, I. (2021). *Liminality in Tourism*. Routledge.

Caribbean, Economic Commission for Latin America and the (2020). *Report on the Economic Impact of Coronavirus Disease (COVID-19) on Latin America and the Caribbean.* United Nations.

Costa, P. et al. (2021). *Rebuilding and Restructuring the Tourism Industry: Infusion of Happiness and Quality of Life*. IGI Global.

Dash, T., Charman, D., & Kenneth, P. (2022). *Covid-19: The Economy and Society*. Allied Publishers.

Dogan, G. et al. (2021). *COVID-19 and the Hospitality and Tourism Industry*. Edward Elgar Publishing.

European Travel Commission. (2020). *European Tourism: Trends & Prospects Quarterly Report (Q3/2020)*. European Travel Commission.

Farmaki, A., & Pappas, N. (2021). *Emerging Transformations in Tourism and Hospitality*. Routledge.

Fotea, S.; Fotea &Văduva, S. (2022). *Navigating Through the Crisis – A special Issue on the Covid 19 Crises.* Springer Nature.

Giuseppe, C. (2022). *Challenges and Opportunities for Transportation Services in the Post-COVID-19 Era*. IGI Global.

Goretti, M. et al. (2021). *Tourism in the Post-Pandemic World.* International Monetary Fund.

Gowreesunkar, V. et al. (2021). *Tourism Destination Management in a Post-Pandemic Context*. Emerald Group Publishing.

Grum, D. (2021). *Psychological Characteristics of Tourist Behavior During a Pandemic and Expectations of Future Tourists*. IGI Global.

Hall, C., Lundmark, L., & Zhang, J. (2020). *Degrowth and Tourism*. Routledge.

Hashimoto, A., Harkonen, E., & Nkyi, E. (2020). *Human Rights Issues in Tourism*. Routledge.

Hudson, S. (2020). *COVID-19 and Travel*. Goodfellow Publishers Ltd.

Hudson, S. (2022). *International Case Studies on Tourism Destination Management and Covid-19*. Routledge.

Idema, O. (2021). *Covid 19 Disruptions and the Two Lover Birds*. Independently Published.

Kavoura, A., Havlovic, S., & Totskaya, N. (2021). *Strategic Innovative Marketing and Tourism in the COVID-19 Era*. Springer Nature.

Kholoud, K. (2021). *Public Health and Economic Resiliency in the Post-COVID-19 Era*. IGI Global.

Kulshreshtha, S. (2021). *Virus Outbreaks and Tourism Mobility*. Emerald Group Publishing.

Kusumah, A., et al. (2021). *Promoting Creative Tourism: Current Issues in Tourism Research*. Routledge.

Land, T. (2021). *Love in the Time of COVID-19*. Independently Published.

Lew, A., et al. (2021). *Global Tourism and COVID-19*. Routledge.

Lim, W., Kaur, S., & Cheong, H. (2021). *COVID-19, Business, and Economy in Malaysia*. Routledge.

Lucia, M., & Giudici, E. (2020). *Humanistic Tourism*. Routledge.

Mallya, D., & Silva, R. (2020). *Impact Of Covid – 19 Crisis nn The Global Economy And Other Sectors Worldwide*. Idea Publishing.

McComb, J. & Ward, L. (2020). *COVID-19 and Tourism in South Asia*. [s.l.].

Miller, M. (2020). *Economic Development at the Community Level*. Routledge.

Monaco, S. (2021). *Tourism, Safety and COVID-19*. Routledge.

Nunkoo, R., Wen, J., & Kozak, M. (2022). *Handbook on Tourism, Public Health and Wellbeing*. Edward Elgar Publishing.

Osabuohien, E. et al. (2022). *COVID-19 in the African Continent*. Emerald Group Publishing.

Perez, A. (2021). Blindness, Light, and the COVID-19 Pandemic. Frontiers Media SA.

Piera, B., Errichiello, L., & Micera, R. (2021). *Post Covid-19 Tourism: a Pathway Towards Sustainable Development in the Mediterranean Region*. Mc Graw Hill.

Rais, A. (Ed.). (2022). *Coronavirus (COVID-19): Outbreaks, Vaccination, Politics and Society: The Continuing Challenge*. Springer International Publishing.

Rezaei, N. (2021). *Coronavirus Disease - COVID-19*. Springer Nature.

Rotgerink, L. (2020). *An In-depth Study of the Tourism Situation in Dubrovnik, Taking Into Account the Effects of the COVID-19 Pandemic, and Its Future Recovery of the Tourism Flow by Utilizing Film-induced Tourism*. Breda University of Applied Sciences.

Santos, A. et al. (2020). *Behavioural Changes in Tourism in Times of COVID-19*. Publications Office of the European Union.

Seabra, C. et al. (2021). *Pandemics and Travel*. Emerald Group Publishing.

Séraphin. Hugues & Yallop, A. (2020). *Overtourism and Tourism Education*. Routledge.

Sharma, A., Hassan, A., & Mohanty, P. (2022). *COVID-19 and the Tourism Industry*. Routledge.

Sikali, A., Akedjol, S. & Kissougle, F. (2022). *Leisure activities amidst the Covid-19 health crisis*. Editions L'Harmattan.

Silva, C. (2022). *Local Government and the COVID-19 Pandemic*. Springer Nature.

Sloboda, B., & Yaya, S. (2022). *Economic Impact and Recovery Following a Global Health Crisis*. IGI Global.

Sukmana, O. et al. (2022). *Social and Political Issues on Sustainable Development in the Post Covid-19 Crisis*. Routledge.

Thams, A. et al. (2020). *An Initial Assessment of Economic Impacts and Operational Challenges for the Tourism & Hospitality Industry Due to COVID-19*. IUBH Internationale Hochschule.

Toubes, D., & Noelia, A. (2021). *Risk, Crisis, and Disaster Management in Small and Medium-Sized Tourism Enterprises*. IGI Global.

Trakic, A. (2021). *Covid-19 and Business Law*. Walter de Gruyter GmbH & Co KG.

Ujwary-Gil, A., & Godlewska-Dzioboń, B. (2022). *Challenges in Economic Policy, Business and Management in the COVID-19 Era*. INE PAN.

Verbin, I. (2020). *Corporate Responsibility in the Digital Age*. Routledge.

Verma, A., & Kumar, S. (2021). *Emerging Business Practices and Trends During COVID-19*. Book Rivers.

Webb, N. (2020). *The Healthcare Mandate: How to Leverage Disruptive Innovation to Heal America's Biggest Industry*. McGraw Hill Professional.

Wilks, J. (2022). *Tourist Health, Safety and Wellbeing in the New Normal*. Springer Nature.

Yeoman, I., McMahon-Beattie, U., & Sigala, M. (2021). *Science Fiction, Disruption and Tourism*. Channel View Publications.

Acknowledgment

The editors are grateful for the precious support received from Communication and Society Research Centre-CECS at the University of Minho, European Union, Portugal, in particular as the first public action undertaken by the Observatory of Heritage, Digital Arts and Tourism for Territory Policies, an emerging platform of this research unit.

Section 1
Pre-/Post-Touring Within the Viral Society

Chapter 1
Which Sociology of Urban Tourism in the Day After Viral Society?
For an Intercultural, Intermediary, and Inter-Methodological Hybrid and Open Research

Pedro Andrade
University of Minho, Portugal

ABSTRACT

The contemporary pandemic conjuncture may lead to a 'viral society', an unprecedent social paradigm involving deep economic, political, and cultural transformations. To cope with this globally unstable situation, it is urgent to deconstruct and reconstruct knowledge (e.g., through a social and sociological Encyclopedia of Viral Tourism addressing a particular genealogy of research on tourism studies and to be disseminated via a Virtual Sociological Museum). The author invokes here recent debates and his own personal research. Epistemologically, the pertinence of a hybrid and open research and several theoretical topics and case studies on tourism are discussed, for example, tourism articulated with the COVID-19 pandemic and the encounter of this viral tourism conjuncture with the war in Ukraine, which defines a new breed of war tourism. Regarding methodologies, the author converses on Hybrid Discourse Analysis and Visual, Virtual, and Viral Methods, such as sociological comics, video papers, artistic sociology, and sociological augmented reality.

INTRODUCTION: A VIRAL SOCIETY IS EMERGING?

Within Tourism Studies, one of the most influent and seminal debate is the discussion articulating, on one hand, the 'leisure class', a concept coined by economist Thorstein Veblen (1995 [1899]), with, on the other hand, the social condition of the tourist, taken as a protagonist of modernity, and even as a

DOI: 10.4018/978-1-6684-3369-0.ch001

paradigm of the modern man (MacCannell, 1976). This last author, subscribing an anthropologic and ethics posture, dresses a significant difference between travel and tourism social processes, and analyses several other phenomena related to these two concepts, e.g., 'staged authenticity', which means an imaginary but realistic destination world, artificially created for the tourist fruition. MacCannell's strategy is undertaken via a Semiotic of Attraction and an Ethnomethodology of sightseers. This reflection is enriched through multiple essays of unprecedent and/or meaningful phenomena, inside what the author names the process of 'cannibalism' of cultural and material goods into commodities (Idem, 1992). In particular, sightseeing is revisited, some years later, in a full developed anthropological, sociological and ethical critique (Idem, 2011).

Departing from these and other intuitions, it is possible to understand, more profoundly, our contemporaneity, in what concerns travels, tours, trips, journeys, voyages, flights, treks, wanders, passages, transits, peregrinations, itineraries, accesses, crossings, runs, and a plethora of other nomadic circulations inside the infinite routes across our local and global planet. The present text aims to propose one of these perceptions, interpretations, and critiques on travel and tourism, particularly in connection with another core phenomenon that is nowadays emerging, the pandemic or Corona virus.

Virus and pandemics are not phenomena that suddenly appeared in 2019-20, such as COVID-19, or in another isolated and unique historical moment. Also, these aren't processes that will vanish forever. On the contrary, these issues constitute and maintain health and social problems, that we must learn how to deal with periodically.

Within such instable context, the present essay departs from a question, addressed from previous studies, written by the author of tis text, in the very moment of the pandemic outbreak in 2020, which interrogate the possible arise of a novel paradigm of social fabric, which could be named 'Viral Society'.

The current pandemic crisis situation seems to indicate that Covid-19 is deconstructing the current network society, and reconstructing it in an emerging paradigm of society, the Viral Society. [Some viral society's characteristics are the following:] *Viral economies and technologies*, based on capitalism global fragilities and economic crises, caused not only by computer viruses, but also and increasingly, by biological viruses; *viral politics and politicians*, for example the possibility that certain States, institutions, organizations, associations, or other social agents, use various types of viruses as local or global weapons of threat, aggression or surveillance; *viral cultures and cults*, meaning modes of exercising Science and the arts, or other knowledge and leisure, through virulent strategies, that is, according to the idea that the more followers and / or friends (or, in certain cases, the more enemies) they articulate or branch with a given individual or collective subject, the more cultural and cult value that subject accumulates and / or distributes across the social fabric (Andrade, 2020c, p. 58).

OBJECTIVES: AN ENCYCLOPEDIA ON URBAN TOURISM AND VIRAL SOCIETY, USING HYBRID DISCOURSE ANALYSIS.

The central purpose of this chapter is a reflection on contemporaneity, first of all within the social laboratory of the present conjunctural and dramatic encounters and counter-clashes between tourism and Corona virus.

As more specific objectives, the text intends to propose some possible pillars for the edification of an *Encyclopedia on Urban Tourism under Viral Society,* by using sociological hybrid methods, such as *Hybrid Discourse Analysis* and *Visual/Virtual/Viral Methods*, explained below. Some illustrations of concepts that nurture this Encyclopedia, may be consulted at this chapter's Section with the title 'Key Terms and Definitions'. The Encyclopedia is one of the nodes of a broader research network, named *Virtual Sociological Museum.*

The construction of such Encyclopedia topics, includes two core articulated parts: (a) a research on tourism *conceptualization* and the respective *empirical work* in this area; and (b) a discussion of some *sociological methodology*, classical or digital, applied to Tourism Studies. These concepts and methods will be articulated with one another, and some of them may be hybridized into *mediating or hybrid concepts*. For instance, as for sociological terms, the hybrid concept 'viral tourism'. And, in what regards methodology, the *intermediary* or *hybrid methods* derived from the fusion of the visual, the virtual and the viral dimensions of social reality.

Such debate must be contextualized as well by broader Tourism Studies. Therefore, in the next pages, a debate is sought about a series of published and sketched research and sources, that may work as a background and present investigation strategy for the future. Not just individual, 'complete' or confined research, but also collective, 'incomplete' and open work in progress and syntheses, hopefully useful for a global and democratic debate.

Examples of concepts included in the Encyclopedia on Urban Tourism, Viral Society and Hybrid Discourse Analysis, are presented in the section 'Key Terms and Definition, in the final part of this chapter.

For those interested in more details, in general you may search and consult these or related themes and authors within the University of Minho Repository, and download the respective texts in PDF: http://repositorium.sdum.uminho.pt_. However, some particular references and their links to such collective archive are as well provided here, around the following discussion.

RESEARCH STRATEGY: HYBRIDIZATION AND OPENESS OF TOPICS AND OF METHODS

It is not possible to understand the COVID-19 pandemic without consider the situation before the correspondent coronavirus social outbreak. In particular, some reflections and case studies, produced and disseminated by the author, that exemplify the above-mentioned hybridization posture, are here discussed, often referred to other Tourism Studies writings.

The present chapter aims to debate a Sociology of Tourism founded on a *Hybrid* and *Open Research* (Andrade, 2003). At a writing moment, a *hybrid research* includes and fusion both theoretical reflections and empirical studies, traditional and digital methodologies, national and international teams, in short intercultural and transcultural research cultures. At a diffusion step, an *open research* aims to convey free access to content, by global and local *research audiences.*

Both modes of practicing research may be themselves articulated with each other. In other words, a Hybrid/Open Research is an investigative perspective and mode of research that is understood as a process of deconstruction of knowledge about society and the mixing or their dimensions, leading to their reconstruction, archiving and dissemination in the form of '*hybrid and open memories*'.

Therefore, research and the respective research heritage may be pursued through a hybrid and open style, that includes the hybrid production and open reception of investigation by both the researcher and the common citizen. This may involve multiple shared tasks: questioning the complex and hybrid social; organization and critique of polysemic sources and data; co-participation in the application and use of mixed and interdisciplinary methods; public discussions about work in progress and on research results, by culturally and geographically diverse researchers and research audiences.

For such aim, Hybrid/Open Research must articulate Social Sciences and Humanities to New Media, specially across digital social networks, both at Web 2.0 (the reading/writing Internet, e.g., Facebook) and at Web 3.0 (the so-called Semantic Web, for instance Freebase or Wikipedia).

In short, productive Hybrid and receptive Open Research and hybrid & open digital memories, may constitute some of the fundamental pillars of the emergent '*research society*'. This means a social paradigm where common citizens may become a sort of '*lay researchers*' and, in the process, reformulate contemporary expert's knowledge, domination and power, by hybridizing specialists and people knowledges (Andrade, 2008a).

Next, some examples of Tourism reflections and case studies on tourism and its contextual processes, will be presented.

CONTENTS: WHAT SOCIAL PROCESSES TO STUDY IN A HYBRID & OPEN SOCIOLOGY OF TOURISM?

Within the precedent epistemological optics, the author, within a sociological perspective focused on the cultural-touristic city, in the last decades developed research, published several journal articles, edited journal issues and co-organized several colloquia, on the problematics of critical cultural tourism, in particular at Lisbon and Braga (Andrade, 2020h).

Popular Tourism: Excursions and Dinner Groups (1981, 1984, 1986, 1991)

While student at a Doctorat de 3ème Cycle at École des Hautes Études en Sciences Sociales, Paris, under supervision of Marc Ferro, the author of the present text carried out fieldwork on *popular tourism* (1981, 1984), about *Excursions and Dinner Groups* (Grupos Excursionistas-Almoçaristas-Jantaristas), who organize collective excursions and dinners, departing from their headquarters situated within *taverns* (see Figure 1).

Such groups are leisure or social solidarity associations, that have a common box, supported mainly by quotas of members, in order to allow periodic collective meals and excursions. They also provide economic assistance and solidarity to those in need in the Group itself or to local inhabitants in their neighborhood, and are more frequent in cities, e.g., Lisbon, Porto, Setubal, etc.

This study included questions and hypotheses on *redistribution of goods* (from the tavern owner to clients) and *reciprocity* (peer help among customers and tavern habitués) in urban neighborhoods, corroborated by interviews, direct observation and content analysis of *Excursionist Art*.

Excursionist Art is a type of popular artistic practice, which mainly gathers paintings made by Excursionist Groups. Although the activity of these groups has led to the production of numerous other objects or processes – figurative collages, textual transcriptions, etc. - a reflection (Andrade, 1986) concentrates on circumscribing the areas of action of the objects of art, which are parallel and dialog with the reported activities around social solidarity and popular tourism.

Figure 1. Popular Tourism: Excursions and Dinner Groups at taverns
Source: Creative Commons

These art works are exhibited in the taverns that serve as the operational basis and headquarters for the Group. They are defined, at first glance, as symbolic emblems of these *sui generis* clubs. A sample of such art works is presented in both empirical and theoretical text above mentioned.

Related with the study of these groups, the construction of the social and theoretical object *'tavern'*, should not be framed inside sociological theories that are reducible or polarized in its presuppositions and principles, what the author calls 'dichotomous theories'. On the contrary, the social and sociological tavern was developed within 'mediating theories', that is, those that best perform the necessary negotiations to make a socio-scientific pact or contract among existing theories, and among them and the scientific community that produces / reproduces them. In this case, and at the conceptual level, it is central to articulate two fundamental mediating concepts, *everyday* and *habitus*, associated with each other and with other theoretical terms, intermediaries or not. The aim was the production of new categories, such as *'everyday habitus'* and *'habitus of travel'*, which are related within *'intercommunicating conceptual networks'* (Idem, 1985a, 1991).

Sociology of Travel (1993)

Inside the broader framework of a *Sociology of Travel* (research conducted at the University of Coimbra), the various significant *forms of travel* were categorized. In fact, it is necessary to observe that the phenomenon of travel is not exhausted in tourism, although this is its dominant form in modern societies, and especially within 'postmodern' or 'late modernity' societies. On the other hand, in exercising the criticism of certain types of tourism, and equally of an alternative, *post-tourism*, the author intended to extract the most recent phenomena that surpass them, e.g., those called *'critical tourist', 'counter-tourist'* and *'inter-journey'*. Fragments of other forms of travel, and increasingly omnipresent manifestations of genuinely daily rhythms, are visible in all of each trip modes (Andrade, 1993).

Science Museums Literacies and Travelling Publics (2000-2005)

Other theoretical and empirical studies were developed through projects on public profiles of visitors and tourists within Science museums (Idem, 2010a).

This research (cf. Figure 2) was carried out within the project *'Scientific and Technological Literacy and Public Opinion: the case of the science museums publics'*, funded by Foundation for Science and Technology, Lisbon (POCTI / SOC / 35279/2000). Here is gathered part of the questions and content related to the topics discussed within the works undertaken by a team of nine researchers from the Interdisciplinary Center of Science, Technology and Society at the University of Lisbon (CICTSUL), who authored a collective book. The project developed a theoretical reflection on communication and public understanding of Science, and undertook empirical studies about science exhibitions on dinosaurs' themes, at the National Museum of Natural History and the Vasco da Gama Tower, at Lisbon, from October 2000 until April 2005. The sample was constituted by 1846 respondents, including profiles of families and children, teachers and students and tourists, among others.

Some issues raised were as follows:

- What *representation of Science and Technology* is built by the scientist, in view of its public presentation? Representation of Science includes the production and registration of the scientific discovering process, in principle undertaken by the scientist or by a research team in a laboratory, university, research center or inside other scientific and technical institutions, companies, etc.
- What *presentation* (of the representation) regarding Science / Technology is shown at the public sphere, e.g., by a Science Museum? The public presentation / mediation of Science is a strategy for exposing a scientific object, idea, process, procedure or author to a non-specialist public. This is achieved by mediating institutions such as museums and science fairs, associations e.g., friends of a Museum, and generally through any oral, written, electronic or digital media.

Figure 2. Science and cultural heritages, literacies and museums, publics of public communication of art
Source: Pedro Andrade Sociological Collection

- What *re-appropri*ation (of the representation's presentation) of Science / Technologies is carried out by scientific audiences? The publics re-appropriation of science is understood as the critical reception and reformulation of scientific processes and contents. This includes a set of tactics operated by a non-specialist public/audience, or 'lay-scientists' in institutions, organizations and associations for the dissemination of science, in order to co-build democratic scientific knowledge and citizenship.

- What *scientific-technological literacy* emerges in each of the previous 3 steps?

Art Museums Publics and Tourists (2007-2011)

Other studies were undertaken on *audiences at art exhibitions*, e.g., at the Berardo Collection Museum. Such research (see Figure 2) is based on a project with the title *Public Communication of Art: the case of local and global art museums*, funded by Foundation for Science and Technology, Lisbon (Ref. PTDC/CCI/68595/2006). This project is revealed, in what regards its objectives, scientific questions and technologies, within its published book (Andrade, 2016a), at Part I, entitled 'What Theory and Methodology for Museum Studies?'. The considered perspective is an illustration of what is understood as Open Research, inspired in some way in the Umberto Eco's 'Opera Aperta' idea, and also in the attitude underlying the global availability of free and modifiable software such as 'open source', which, among other impacts, provoked a boom in planetary digital creativity.

The main problematics included questions about art worlds and universes / metaverses within the museum. E.g., debates on the urban space, museums and public art; museums, global crisis and cultural austerity; urban playfulness and culture; public communication of art and Museology; Sociology of museums: publics studies, planning and strategies; the audience of Berardo Collection Museum and of artist Joana Vasconcelos; segments of publics such as families, schools, tourists, visitors in lifelong learning; interviews of artists and gatekeepers' opinion on art worlds. Other series of questions pertain to artists or other gatekeepers of arts webs, and issues related with the new media: e.g., digital interactivity at museums; cultural heritage and digital technologies; new media, social networks and virtual museums; theories of digital cultures; culture, games and metaverses; digital social networks and Web 2.0.

Interaction and Interactivity as Virtual Wandering

In the precedent case studies, the strategies of *interaction at local communities* and *interactivity* around digital devices by *cultural audiences at exhibitions,* were analyzed, as well as public *viewpoints of augmented reality* consulted by tourists in different places in Portugal (Andrade, 2016b).

Such research discussed several theoretical and methodological aspects for the understanding of the mentioned problematics, and furnished details regarding the accomplishment of the investigation tasks carried out by the project team.

About the topics addressed, some central sociological concepts were redefined and starting points revisited, such as the following: what is a museum? What are the main objectives and activities of a museum? Who are and what do museum audiences want? What does 'museability' and 'musealization' mean? Which are the contours of 'public communication of knowledge' at the museum?

The Social and Sociological Touristic City (2013-2022)

In addition to the previous debates, within the perspective of the *cultural-touristic city*, which dialogues with the recent winning project of *Braga as a Unesco Media City*, in the last decades other research avenues were wandered and vagabondized, in particular within the cities of Lisbon and Braga (Andrade, 2018a).

The purpose of these studies is essentially critical but also pedagogical: that is, they aim to provide and discuss some sociological reflections on the contemporaneity of the urban fabric, e.g., Landry (2006) on 'the art of city making', or Giffinger (2007) about the '*smart cities*'. The target readers are researchers, teachers, students and other stakeholders within civil society, so that it may inspire informed conversations on the urban fabric, its structures, institutions and practices of its agents, and namely tourism.

Sometimes the city aspects that we judge to be insignificant or even infamous, reveal themselves as those that seem to be the most intimate and central to their inhabitants. Therefore, they are true indicators of the social changes and transformations that inevitably pervade our daily life, both in the case of the perception of the city by the citizen, tourists or migrants, and at the momentum of the interpretation of urban regularities and irregularities, by the social scientist or by the terrain philosopher. This is the case of leisure and touristic places and practices, that equally manifest themselves as *loci* of information, and essentially as contexts of practices and actions of urban knowledge, that establish and institute democratic and intercultural citizenship.

Urban Public Art for Tourism and Intercultural Communication

City public art may act as a tourism catalyst, if it is understood and practiced as conveying the following communication processes and actions within the urban spaces and scapes (see Figure 3):

Public artists (through art works), and sociologists (e.g. via case studies) can communicate, to Portuguese citizens and travelers to the city of Lisbon (tourists, migrants), a preview or *pre-vision* of such intercultural metropolis. This signifies to visualize it before tourists or migrants look at it and discover the plurality of daily life at Lisbon and of other cultural events (via sites, social networks, etc.). Public art works could as well disseminate a *post-vision*, i.e., a textual and visual visualization after tourist's and migrant's observation of the city *in situ* (Andrade, 2014).

Inside such perspective, an illustrative case study is about an intercultural event that gathers, dialogically, the above-mentioned social agents (citizens, tourists and migrants): the *Imminent Festival*, which may be presented as interculturality immanent to public art (Andrade, 2020b). Moreover, *public e-art* was discussed (Idem, 2020c), that is, public art that is produced and shared across cyberspace and cybertime.

And these artistic initiatives must be confronted with controversies, classic and current, about *mobile cultures* such as cultural tourism and, in particular, the tourist communication process. In the last previous mentioned essay, a brief glossary seeks to synthesize and define some sociological concepts that circumscribe central social phenomena, underlying the articulation and hybridization between urban public art and tourist communication.

Moreover, an interview of a public art artist, Luis Baldini, may clarify the precedent questions, through his biography, art works and concrete exhibitions. In fact, this artist provides, in first person, an interesting and informative view on the state of art of urban public art, particularly in the case of the graffiti culture and about its implications for city inhabitants and visitors (Andrade, 2020d).

Urban and Digital Social Networks for Tourism

Moreover, city and tourism phenomena should be partially analyzed through the processes of tourism practices and discourses within *cyber travels* (Andrade, 1997a); or via digital information targeted to tourists (Pirolli, 2018) and, in particular, *tourism social networks*.

Figure 3. Social / Sociological / interdimensional city and public art
Source: Pedro Andrade Sociological Collection

These tourism social networks are not just digital tourist networks, but also *pre-digital urban travel* and tourism networks, including social practices and unique interpretations of both: (a) the public space of *everyday urban life;* and (b) the daily lives inside cyberspace and *cybertime*, which involves a dialectical *creative destruction* of social phenomena, including touristic industries, modes of travel and profound changes in what regards tourist practices. To better understand such creative destruction conjuncture, and the respective role of network users as the tourist, see Andrade, 2018c, and the following considerations.

Contemporary economic and financial crisis is closely associated with social innovation. This process profoundly influences cyberspace's phenomena within our globalized communications paradigm. The correspondent debate on the articulation between crisis and innovation was reconceptualized by Marx, Nietzsche, Sombart, Schumpeter, David Harvey, etc. However, Manuel Castells elaborated an economic but also communicational explanation, which seems closer to the current crisis. Castells uses the notion of 'space of flows' created and shared by globalized capitalism, across information and communication networks at cyberspace.

When considering *tourism social networks*, it is necessary to reflect on '*touristic information and knowledge flows'* in the present crisis conjuncture. This is observable especially within digital social networks such as Facebook, where e.g., content privacy is being debated and engenders reluctance in what concerns public fidelity towards this or to other networks. To comprehend such process, note that not just consumption goods or information flows, but also *knowledge flows*, may became central means of production.

Within this dynamic, knowledge flows sometimes complement, or complete information streams directed to tourists. Other times though, knowledge competes with information, replacing or even destroying it, to produce something else, for example in what concerns *content curation*, useful for economic, social and cultural activities.

Such configurations of '*creative destruction'* processes are more visible within *Web 3.0* (i.e. the Semantic Web) than inside *Web 2.0* (or Social Web). Web 2.0 means a cyberspace strategy of production, distribution, exchange and consumption of information, where the user is not just reading and consuming information, but also writing/producing it, e.g. in a blog post or comment. The Semantic Web signifies an even newer paradigm of cyberspace/network, visible in sites where data is not just mainly shown and permuted (as in Web 2.0), but is also semantically clarified, via deeper explanations and interpretations. Such operations, among others, provoke a deeper transformation of information itself into knowledge.

Major companies are shifting to this innovative cyberspace based on knowledge flows, for example Google and Facebook, two central firms that contributed soundly to Web 2.0 public image, in the XXI century's first decade. Some years ago, Google introduced a new search engine, Knowledge Graph, which uses semantic heuristics that partially characterize Web 3.0. And Facebook, aiming to compete with Google in this lucrative search engines' business, proposed a 'graph search' engine that transforms Facebook's internal raw information inscribed by users, into knowledge about the relationships among diverse properties of the content shared by users. For instance, *tourism users* can ask where / when their friends are travelling, and have such quest translated into a visual semantic report.

Such processes may be interpreted sociologically as 'creative destruction', as noted supra. In fact, information flows associated with digital social networks are being dialectically annihilated in part, and often transformed into knowledge flows. This information flow crisis is manifested through the contradiction between the competing values of information on users and/or values of knowledge about users, including tourists. A political consequence of this process is the replacement of Web 2.0 discourse order by Web 3.0 mode of exercising power. Semantic Web new / innovative power paradigm is not based, anymore, just on the vigilance of information as it happens in Web 2.0, but also on control of knowledge concerning the content flow sent to users.

Through this strategy, institutions/organizations operating within Semantic Web may implement a sort of *Web 3.0 Panopticon*, where they may envision, more efficiently, knowledge about users, e.g., knowledge about users' private life, such as tourists. Thus, the Social Semantic Web 3.0 may control not just information on where, when and how users act or think, but also may observe the knowledge, in terms of why and what for, users receive or produce knowledge. Therefore, even if the social and participative dimension of Web 2.0 and Web 3.0 cultures is a relevant step for democracy and citizenship, several economic, political and cultural risks subsist, that are necessary to uncover.

One possible strategy to overcome this situation is to promote critical *desambiguation* of a given shared content. This is a process of differentiating and opining on several meanings of a given message, and can be done e.g. through research journalism, objective stories, and argumentative classifications, typologies, qualitative interpretations, quantitative explanations, fiction, arts, etc., by citizens in general, and by researchers in particular, but also by tourism agencies and other travel stakeholders (Andrade, 2018i).

A Mobile Tourism?

Within contemporaneity, other central starting questions emerge concerning Tourism Studies, such as the following: how to reflect and empirically analyze the practices of *mobile cultural tourism*, within the framework of *urban social mobilities*?

As one of the possible and probable conjectures, it is useful to demonstrate this regularity: the new urban mobilities, and in particular cultural tourism, have been profoundly transformed by *locative technologies*, such as mobile digital devices (mobile phones, etc.) (Sheller & Urry, 2004, 2006; Sheller & Isaacson, 2013).

The term 'mobility' holds a wide number of connotations and is... mobile. Not surprisingly, research directions on this subject proved to be incommensurable (Adey, 2009). And, in the diversified social and sociological contexts, it is urgent to revisit the discussion on the concept of *'tourism mobilities'* (Rickly, 2016).

In what regards cultural tourism, e.g., it is useful to discover the *mobile art flows* that this segment of tourists develops during its visits carried out at art museums. Such art flows, on one hand, are conditioned by the tourist socio-demographic characteristics (age, gender, etc.); on the other hand, art flows articulate the time course of the museum visit with its space courses (paths to an interesting painting, trajectories to museum's services such as the shop or the bar, etc.).

This discussion on *tourism and museum cultures* may be connected with urban mobilities and its related instruments, e.g. the mobile devices (cell phone, etc): (a) first of all, through contextualizing the genealogy of the historical and social phenomenon 'mobility'; (b) within a second stage, by tracing a discussion on the particular phenomenon of 'tourism mobilities'; (c) such debate helps to understand this central idea 'tourism mobilities', for instance defined as mobile art flows at the museum (Andrade, 2018d,e).

As another exemplificatory application of these reflections to the empirical field, the author carried out a case study on an intercultural space in Lisbon, at Rua Augusta, and about its potentialities for the sustainability of an urban tourism model. In particular, he performed the analysis and interpretation of a corpus of photographic/videographic sources and the respective discourses/representations, partly produced by and on mobile phones. These sources were obtained at that central street habitat, by direct observation of three types of urban agents working within tourist communication processes: inhabitants, tourists and migrants.

It was concluded that, currently, such societal actors in interaction, develop unprecedented strategies of *tourism communication* through the production of face-to-face behaviors, messages via mass media or by mobile phone, within the following various activities and traditional or recent tourist everyday practices: booking of journeys; predictions and previsions of the stay; daily experiences of alterity or mimicry with indigenous citizens or migrants in a given city; collecting, musealization and diffusion, among family or friends, in what regards memories (photographs, videos, etc.) registered at the country visited (2020g).

COVID-19 Pandemic and Urban Tourism (2020-2022)

The worldwide tourism sector suffered a huge and serious crisis in 2020, due to the spread of coronavirus. Within such context, what impact is COVID-19 having on society, in particular on urban cultural tourism? And how will it be possible to carry out the *de-confinement of cultural tourism events*? Among other initiatives (economic strategies, political decisions), cultural events may participate in the reinvention of the touristic field, and through this perspective contribute to the global struggle against the recent mischaracterization of multiple aspects of Portuguese and of other countries societies, caused by the pandemic. Figure 4 shows two interrelated processes on this behalf: a survey on preferred tourism destinations, expressed by four profiles of tourists, in the initial period of COVID-19 pandemic crisis, 17-18 July; and a socio-cultural process against coronavirus, pursued by a cultural and political resistance movement, named *SOS Art.pt*, where the author of this text was one of the co-founders, for solidarity with visual artists touched by the outbreak.

In fact, in the short term, an alternative is the intensive and extensive production, sharing and dissemination of social and cultural actions, events and works, as well as the correspondent sociological knowledge, in particular within cyberspace and cybertime, for example through websites, blogs and social networks. Some examples of such cultural resistance to coronavirus, in the form of cultural events and of solidarity activism targeting cultural professionals, are the *2020 Imminent Festival*, *Quarantine Museum* and *Pessoa Effect proj*ect (Andrade, 2020e).

Another goal lies in the construction of concrete knowledge instruments for clarification of public art applied to tourism activities. E.g. in the form of a digital *Encyclopedia of Public Art,* included in a *Virtual Sociological Museum*, targeted to audiences such as researchers, teachers, students, travel agencies, museums, galleries and other institutions, NGOs, cultural associations, as well as tourists, citizens and migrants. As a sketch to this purpose, a Glossary of public art includes some themes and questions enunciated or annunciated there, to be included in the Encyclopedia, such as the following: (a) as for current social processes: mobile cultures and arts, e-public art, urban popular tourism, museum's tourist mobilities, cybertime, City 3.0, Web 3.0 or social-semantics network, digital cultural tourism, Tourism 3.0, creative and innovative tourism, communicative tourism; (b) in what regards theories, concepts and new methodologies for the study of public art applied to the tourist communication: Sociology of travel, post-colonial representations; transculturalism, digital cultural heritage, Artistic Sociology, Hybridology, sociological exhibitions on Tourism 3.0/City 3.0 (Andrade, 2020 and the section 'Key Terms and Definitions', in the final part of this chapter).

Figure 4. Tourism destinations during Covid-19 pandemic crisis, and cultural resistance against Corona virus
Source: Pedro Andrade Sociological Collection

Viral Society and Viral Risk Tourism

Furthermore, and as stated above, it is credible that we are entering into a *viral society*, characterized, among other attributes, by an unprecedented economic global crisis, social confinement, social distance and augmentation of Internet and mobile phones' uses as alternatives for face-to-face interaction. For a more detailed understanding of this social process and the correspondent sociological concepts, consult again the present text's section 'Key Terms and Definitions'.

Figure 5. The debate on viral risk tourism
Source: Pedro Andrade Sociological Collection

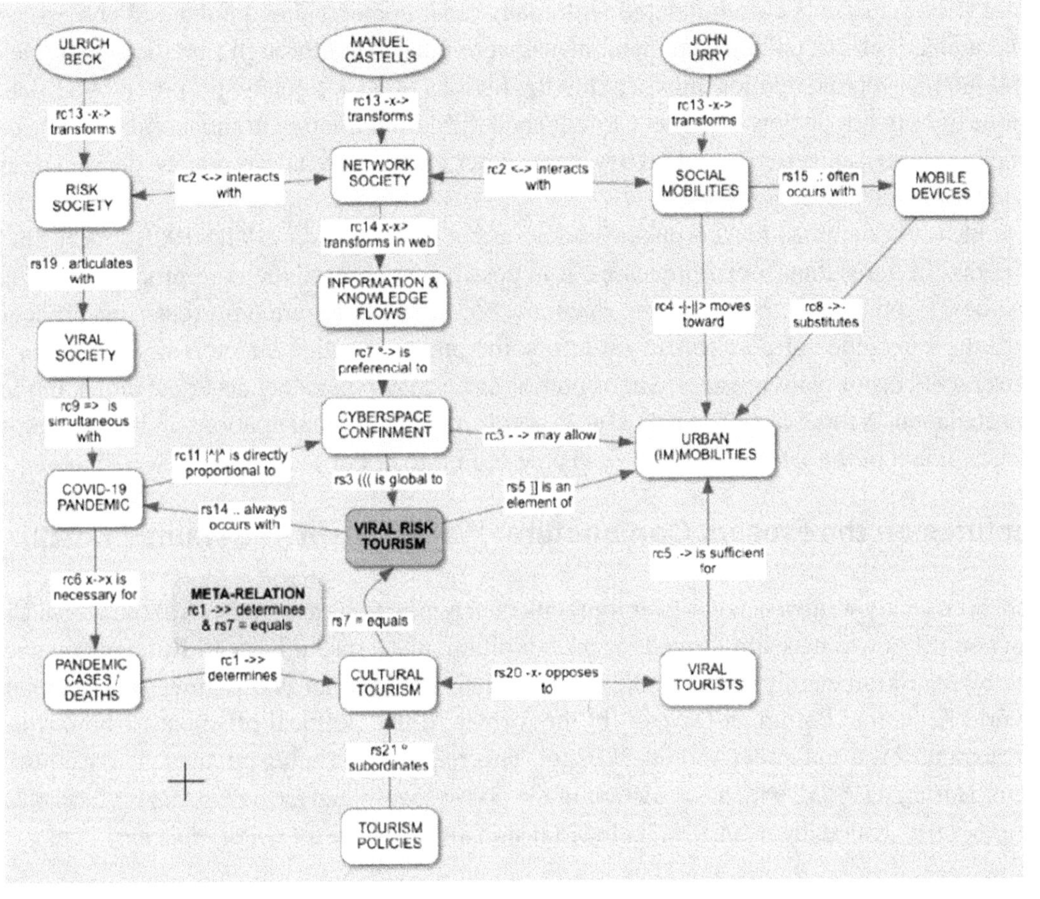

Within such viral context, the urban tourist, and especially the cultural tourist, are experimenting new ways of relating to the visited societies and cultures, in particular by substituting some practices of face-to-face tourism, into virtual actions in cyberspace, e.g. in order to search content related to his/her travel. In other words, a sort of *viral virtual tourism* is rising today (Idem, 2021c).

Some recent studies in the same direction of articulating tourism with the COVID-19 pandemic are the following: a reflection about processes of consumption globalization, connected with lifestyles, in the context of 'Viral Society' (Andrade, 2021b).

Moreover, applying several theories of economy (Marx's exchange value related with goods' use value, and Baudrillard 'symbolic value') an informative-textual value of commodities, represented by data inscribed in goods' labels (product origin and composition, etc.) was extracted within the context of viral society (Idem, 2022a).

In addition, the role of mobile digital games, in particular 'research games', was suggested for a more intense visiting at art museums by tourists, and in order to avoid COVID-19 effects (Andrade, 2022b). In other essay, the 'viral city' was interpreted in what regards the everyday life of citizens and tourists (Idem, 2022c).

Last but not the least, the process of *Viral Risk Tourism*, undertaken within viral society, was analyzed, through a debate on several authors who are relevant for the comprehension of this type of tourism. In fact, Viral Risk Tourism is closely related with many other contextual and collateral processes within the social arena. Some of these social phenomena, were interpreted through specific sociological concepts and interconceptual relationships. Taking the notions of Risk Society, proposed by Ulrich Beck, the critique of Network Society (Manuel Castells) and the delimitation of urban mobilities, tourist gaze, etc., which are terms suggested by John Urry, some other ideas and relations may be derived or created, such as those depicted in Figure 5.

These ideas were applied to a corpus of sources about tourism and COVID-19. By means of *sociological networks* translating social processes, it is possible to observe the core processes related with these themes, together with the respective relations. For instance, Figure 6 presents social phenomena related to the emergence of coronavirus: at left, some previous entities, factors or conditions and, at right, several effects or consequences. Such anterior or posterior elements never act alone, but in more or less articulation. A more complete list of these social and sociological relations and the correspondent frequencies, found in the selected sample, is visible in Figure 8. For more details, see Andrade, 2022d.

Conjectures on the Present Conjuncture: War Tourism in Ukraine? (2022)

Another text recently written debates 'war tourism', concerning two case studies: (a) the war in Ukraine, reflecting on (a1) a war museum formed via an exhibition, in a street at Kiev, of Russian tanks recently captured by the Ukraine army, and (a2) the named, 'House of Russian War Crimes', inaugurated at the 2022 World Economic Forum, at Davos.; (b) the former Aljube political prison at Lisbon, which was transformed into a war museum (Andrade, 2022e). This reflection is being pursued via the construction of a '*Dark Heritage Pedia*', which is a section of the above-mentioned *Encyclopedia on Urban Tourism and Viral Society*, which, by its turn, is included inside a *Virtual Sociological Museum*.

Figure 6. Sociological network about the social processes' network around 'COVID-19 pandemic
Source: Pedro Andrade Sociological Collection

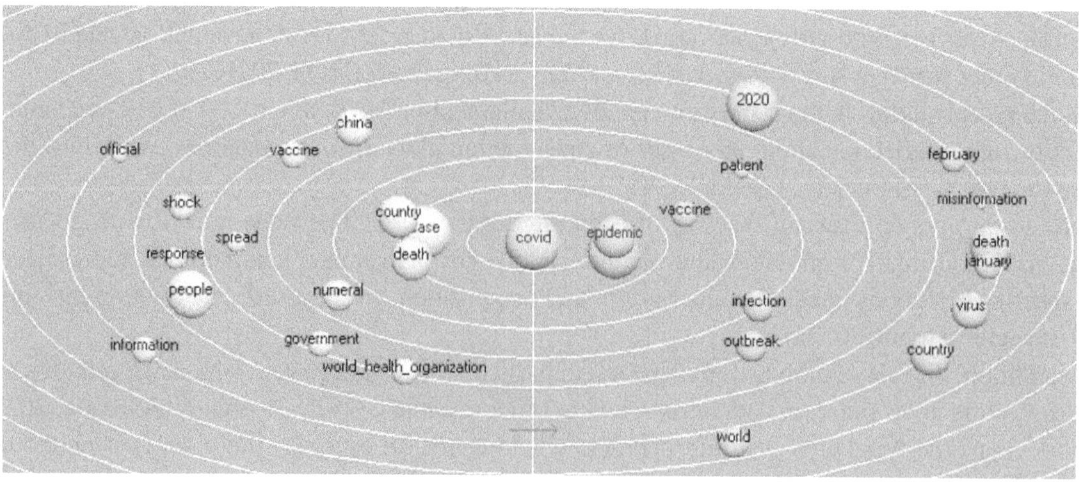

Dark Heritage Pedia is being constructed by using sociological digital methods and softwares, in order to interpret some of the contemporary economic, political and cultural conjuncture, e.g., in Ukraine, and, in particular, the recent impact by Covid-19 on Ukranian tourism.

Figure 7. Sociological network about the social processes' network around tourism related with Corona virus outbreak
Source: Pedro Andrade Sociological Collection

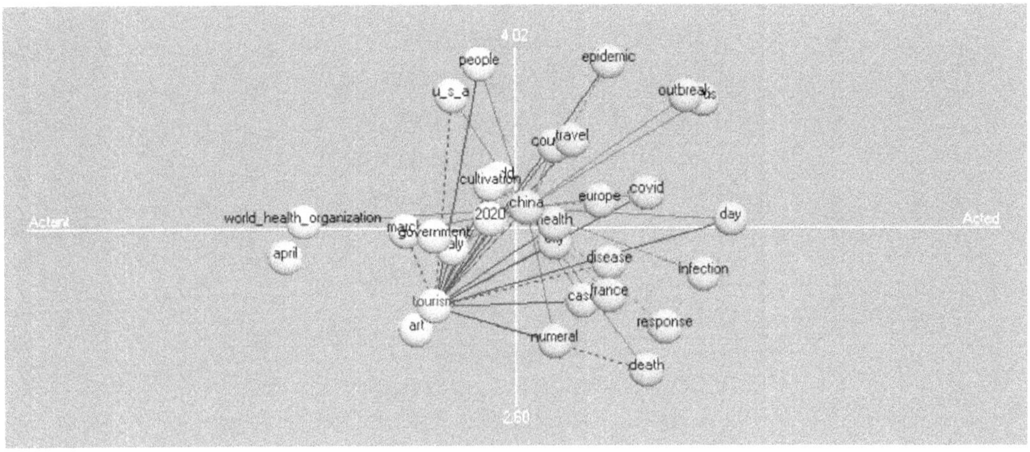

In what regards the Encyclopedia substantive themes, an applied Interdimensional Sociology distinguishes *social processes, structures, contexts, practices, agents and objects*, among other social facets of reality. As for method, it is possible to use *Hybrid Discourse Analysis* tools (see below the section 'Methodologies and Media'), and, specifically, the *Alphabet of Interconceptual Relations* (Andrade, 2007a). As an example of a previous use of this Alphabet, consult again Figure 8, which shows a list and frequencies of all the social and sociological relations among concepts, extracted from the corpus of the above-mentioned study on Viral Risk Tourism.

Such sociological instruments help to translate social processes, events and facts into the form of scientific *questions, hypothesis, conjectures and theses*. These may be tested within specific research, applied to sources resident in contemporary concrete situations, e.g. sites inside the internet or social networks.

In other words, the researcher here operates by using the Alphabet of Interconceptual Relations, through adequate sociologically *relational keywords,* in order to characterize some social phenomena, which include social entities and their relationships. Within the sources sampled, social processes told in common languages are here represented as *sociological propositions,* that articulate two or more concepts through *sociological relationships* extracted from the Alphabet of Interconceptual Relations. Some of these connections are illustrated below.

Firstly, Figure 9 presents selected *social and sociological questions* about Ukraine, and the respective results found in textual sources, by searching the whole Internet through Google on 9 june 2022 and using qualitative analysis softwares such as NVivo to organize the findings from the sources. Such questions are shown as *social and sociological interrogations*, in what regards the following types and modes of arguing:

Figure 8. List and frequencies of social and sociological relations among concepts within a corpus selected for a study on Viral Risk Tourism
Source: Pedro Andrade Sociological Collection

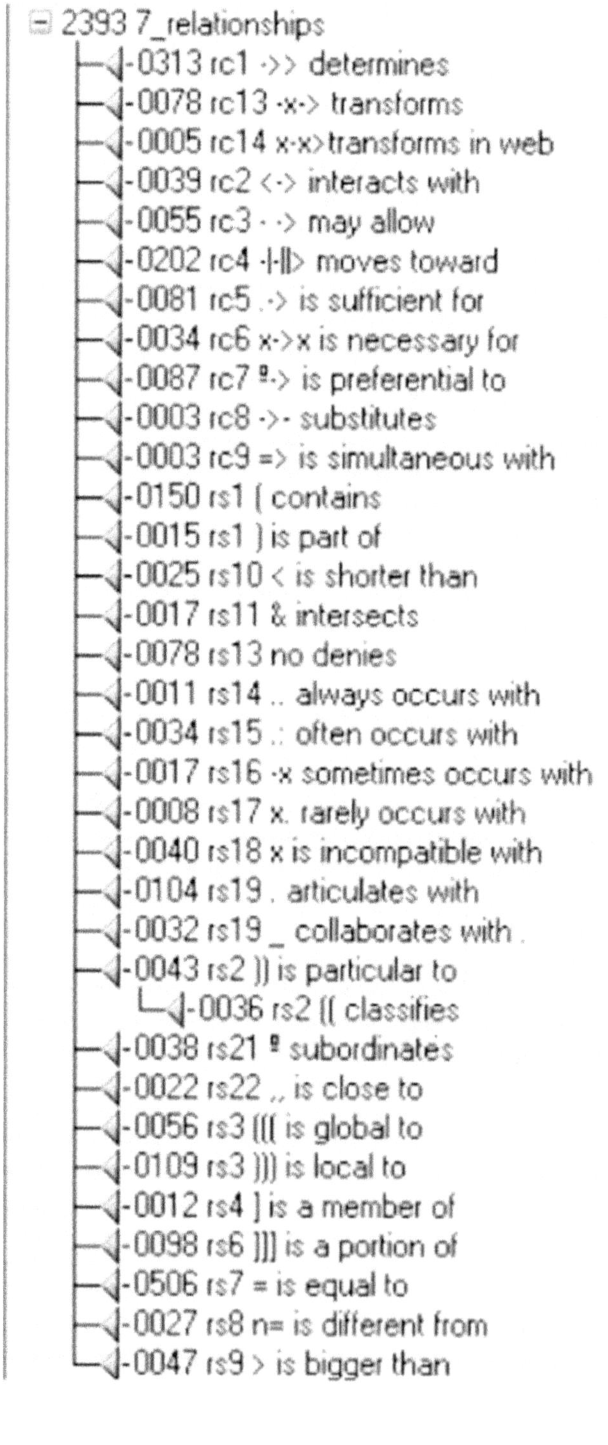

Figure 9. Sociological questions on Ukraine
Source: Pedro Andrade Sociological Collection

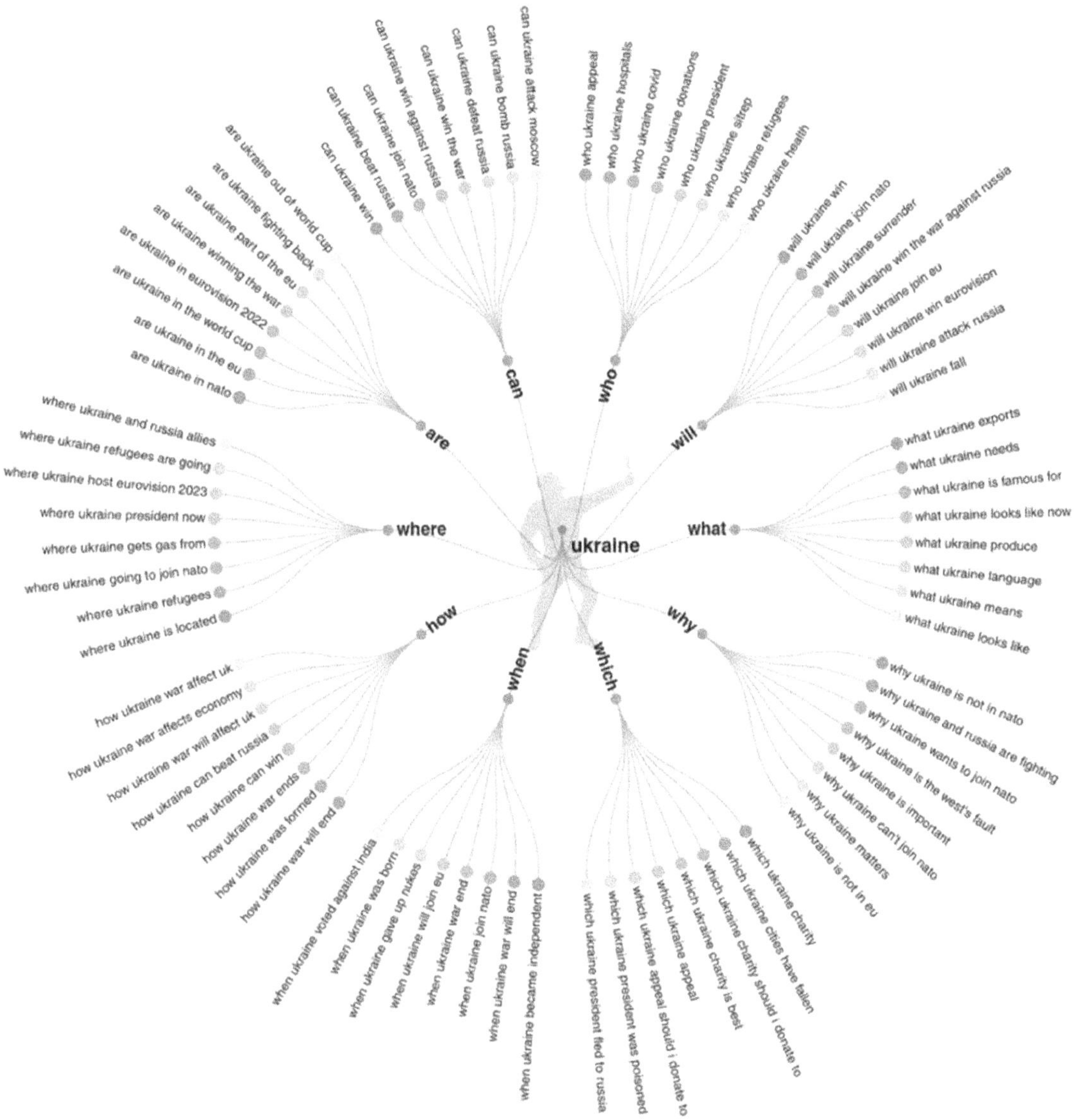

'What' demands are often (but not always) associated with *social processes*. For example: 'what Ukraine exports' (economy); 'what Ukraine needs (policies, politics)'; 'what Ukraine is famous for' (cultural heritage), etc.

'Where' questions refer to *social contexts* where action takes place. For instance: 'where Ukraine refugees are going', 'where Ukraine president [is] now', 'where Ukraine gets gas from, 'where Ukraine refugees [migrate]', etc.

'When' articulates with *historical or conjunctural times and periods*, such as: 'when Ukraine gave up nukes [nuclear weapons]', 'when Ukraine will join E.U.', 'when Ukraine war end', 'when Ukraine join NATO', 'when Ukraine became independent', etc.

'Who' interrogations reflect the *social agents* involved in those actions. E.g.: 'who [is] Ukraine president', 'who [are] Ukraine refugees', 'who [are] in Ukraine hospitals', 'who [in] Ukraine [had] covid', 'who [gives to] Ukraine donations', etc.

'How' is connected with *social practices, actions, ways, methods and procedures* to do something.

'Why' pertains to the *structural conditions and causes, or institutional and individual motifs, intentions*, etc. subjacent to some events. For example: 'why Ukraine is not at NATO', 'why Ukraine and Russia are fighting', 'why Ukraine wants to join NATO', 'why Ukraine is the West's fault', 'why Ukraine matters', etc.

By selecting each of these questions links, the researcher may find the respective sources in the Internet.

Secondly, it is useful to search for *social and sociological comparisons, that may work as hypotheses and conjectures*, translated by keywords that represent the respective relationships, such as the following (Figure 10):

'Like' is a relational keyword that represents a similarity among social entities. Examples are: 'Ukraine [is] like war in Asia (…) Ireland (…) Afghanistan (…) Cuban missile crisis (…) Syria (…) Vietnam'.

'And' means a relation that conveys an intersection or a (socio)logical addition, in what regards social entities. Exs: 'Ukraine and Russia map', 'Ukraine and Russia war', 'Ukraine and Russia news'.

'Or' signifies a (socio)logical relation that expresses alternatives to something: 'Ukraine or Russia who will win', 'Ukraine or the Ukraine BBC', 'Ukraine or Russia [who is] older'.

'Versus' is associated with the relation 'opposite to': 'Ukraine versus Russia military', 'Ukraine versus Russia news'.

Thirdly, other results take the form of more *assertive social and sociological propositions* (Figure 11). that may be utilized, in a given research, *as final theses* in investigation. E.g., via (re)search keywords as 'can', 'for', 'with', 'without', 'is', 'to'. In the referred Alphabet of Interconceptual Relations, it is possible to find the following correspondences:

'Is' signifies a textual keyword that represents, among other keywords, the structural relationship 'is equal to', and, other times, the relation 'is a type of'. Some illustrations: 'Ukraine is winning', 'Ukraine is part of E.U.', 'Ukraine is on fire'.

'With', 'without' are keywords that mean, in a textual corpus, the structural relationship 'co-ocurrence', the first one having a co-ocurrence of '1', and the second showing a co-ocurrence of ´0´. Instances: 'Ukraine [is] with NATO', 'Ukraine with open arms', 'Ukraine without a passport (…) Internet (…) Crimea and Donbass'.

'Near' is a common word that, in the Alphabet of Interconceptual Realtions', clarifies the structural relation 'sequence', 'affinity', among other meanings. Exs: 'Near country', 'near me'.

'To' is an indicator of the conjunctural relation 'scope', 'finality': 'Ukraine to English translation', 'Ukraine to win Eurovision'.

Moreover, in order to *articulate Ukraine with both tourism and COVID-19*, the researcher may put some other questions on a specific corpus. An example is given in Figure 12. Here, relations are scarcer, as the articulation of two or more terms, when using the search logical operator AND, gives less results in the sources.

However, several sociological questions may also be put in this case. For example:

'What': what is the Covid-19 situation in Ukraine? What are the covid restrictions in this country?
'Where': where to visit Ukraine?
'When': when can I travel to Ukraine again?

‘Who’: who can travel to Ukraine now?
‘How’: how tourists can get information? How to travel to Ukraine from UK?
‘Why’: why is Ukraine Covid safe?

For example, the question *‘What’: what is the Covid-19 situation in Ukraine?* leads to various pages on that issue, such as *Covid-19 pandemic in Ukraine: Current information about coronavirus and quarantine,* consultable at: https://covid19.gov.ua/en

Figure 10. Hypotheses through comparisons about Ukraine
Source: Pedro Andrade Sociological Collection

Figure 11. Propositional theses on Ukraine
Source: Pedro Andrade Sociological Collection

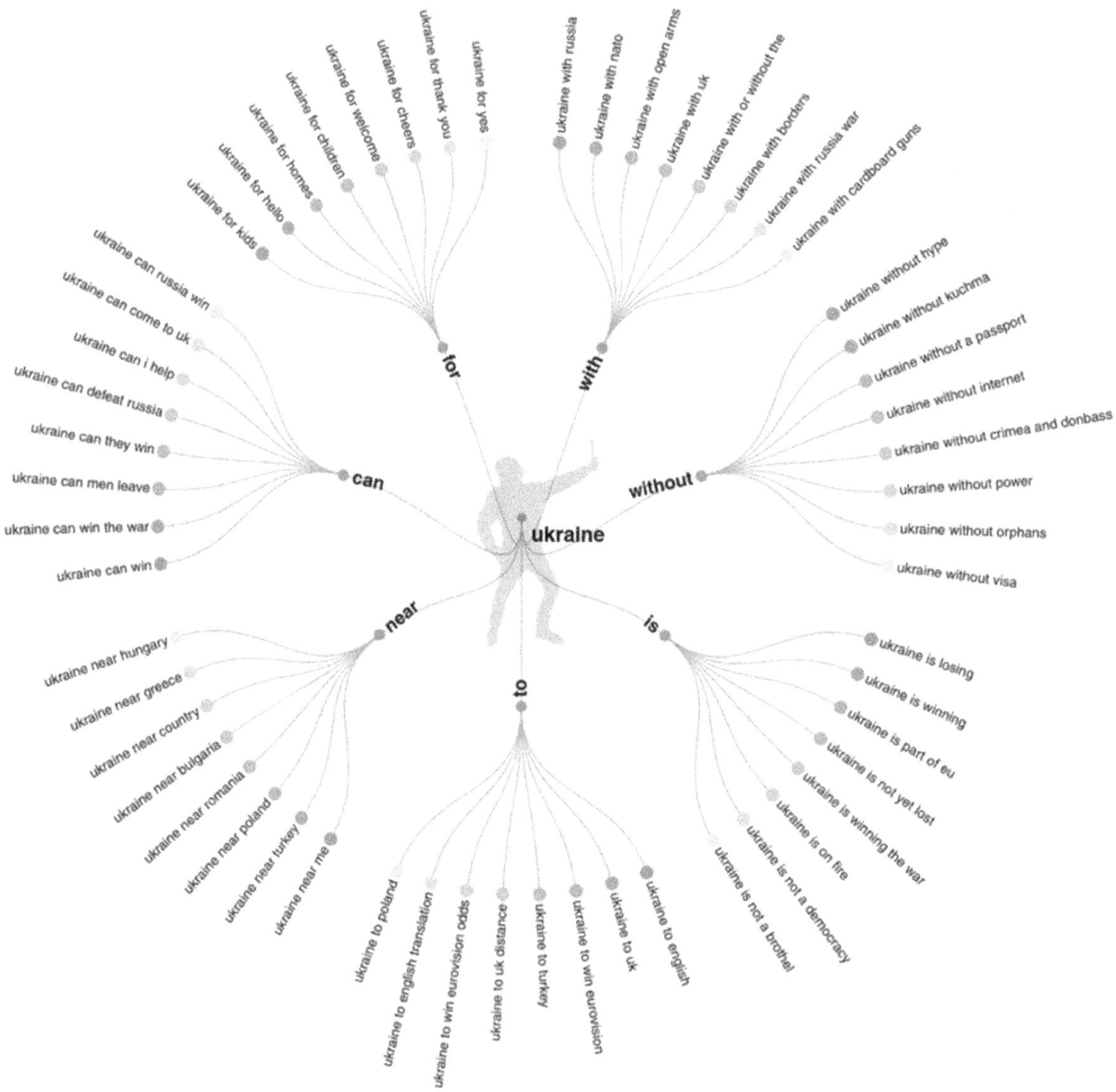

METHODOLOGIES & MEDIA: HYBRID DISCOURSE ANALYSYS (HAD) AND VISUAL / VIRTUAL / VIRAL METHODS

As stated in the Introduction, beside traditional sociological methodologies, new media are creating unprecedent conditions for the emergence of new Social Sciences methodologies. One of them is *hybrid research*, an investigation posture which is not just founded on text, but also on visual and virtual modes of knowledge, explained below with more detail.

Figure 12. Tourism and Covid in Ukraine: questions
Source: Pedro Andrade Sociological Collection

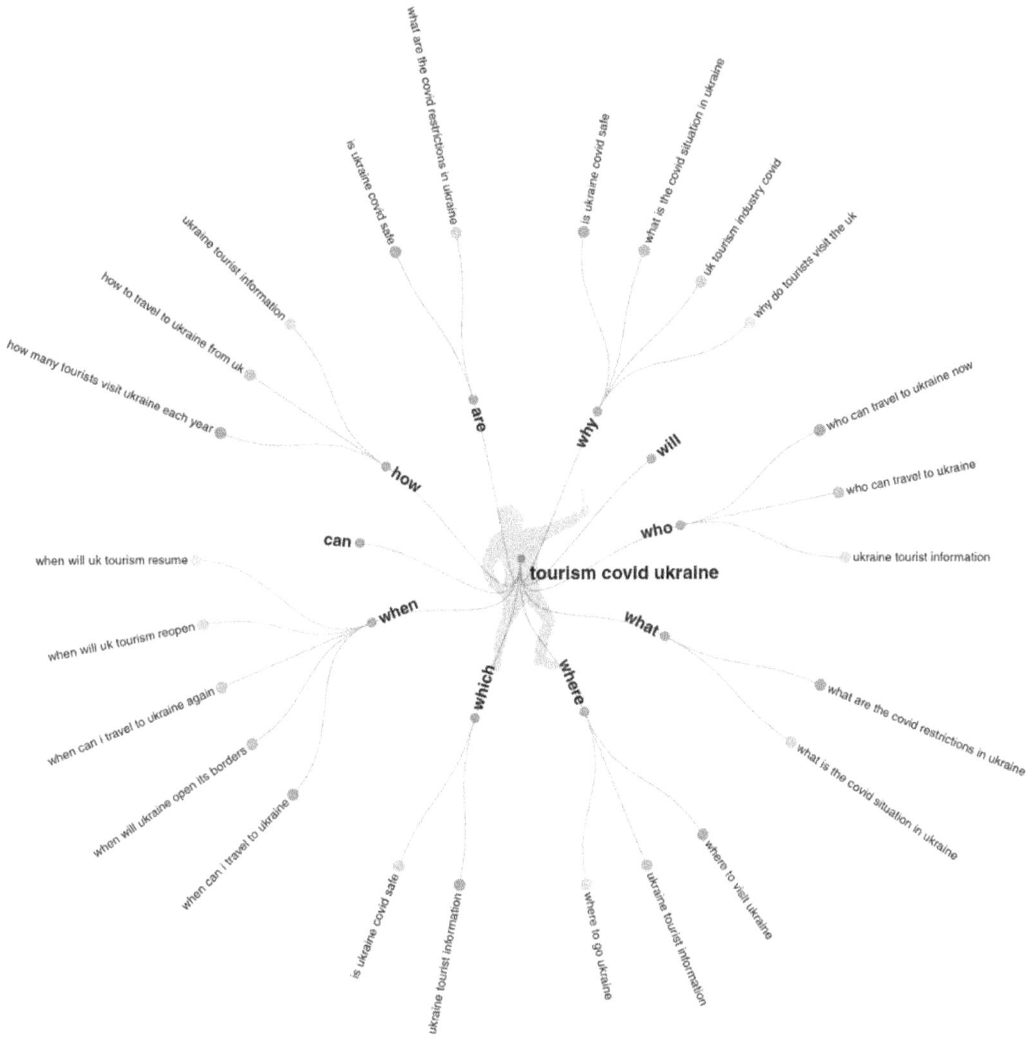

Epistemology of Tourism

With regard to *urban and tourist mobility* and *cultural e-Tourism*, recent research has suggested an *'epistemology of tourism'* (see Figure 13), aiming to circumscribe the debate and the clash among different *methodologies* and *research subjects* within Tourism Studies, such as the condition of the *citizen-researcher.* For a further definition and extended discussion on these issues and the following sociological concepts, see Andrade, 2018b, 2013a,b.

All epistemological, theoretical and methodological debates that aim at scientific credibility, cannot ignore the corresponding application to the social fabric. Inversely, action should always inform reflection. Therefore, tourism research may rationally demonstrate and sensorially exhibit the following: one of the sociological genres, *Artistic Sociology*, transports sociology and its scientific language, from the academia to creative extension activities, such as the exhibition of sociological knowledge within urban public space.

In the same way, artistic knowledge and language should contaminate sociological discussion through an innovative sensibility. This was possible through the insertion, within a sociological text not only of images from an art exhibition presented as 'Figures' (1,2 … n). In addition, the art exhibition itself can be understood as a social and sociological configuration, that is an organic part of the very body of the traditional sociological text.

Thus, a profound hybridization of knowledge is sought, which can enrich, but also subvert, both sociological debates and art exhibitions. Such purpose was accomplished in the above-mentioned 2018 essay via interconnected means: an epistemological approach between Artistic Sociology and Hybridology; the theoretical problematization of mobile cultures; the empirical field work in the context of urban communication and tourism communication in the context of *Tourism 3.0*. This emergent mode of tourism travel (Tourism 3.0) is defined based on the following traits: greater interest on intangible heritage by tourists; overcoming of the dichotomy between high culture and popular culture, a process for example witnessed by the opening of tourists to public art at the street; hybridization between cultural production and consumption; desire for authentic experiences across the tourist travel (Richards, 2000).

In other words, and more precisely, a previous practical demonstration of such strategy was undertaken by the author of the present text in 2017, at the exhibition 'New Art Fest' 17, as a field for the application of innovative sociological and artistic methodological approaches. A first concrete step was a *Sociological Exhibition on Tourism 3.0 / City 3.0* (see again Figure 9), that demonstrated and showed the urban and travel knowledge, circulating within the space of the art gallery. In a second phase, this knowledge tested through the exhibition audience, was reintroduced in a scientific journal article, inside 'Revista de Comunicação e Sociedade' journal (revisit Idem, 2018b). Such a double research movement hybridizes and confronts, in both 'originary' and 'original' forms, scientific and artistic knowledge and practices.

Figure 13. Epistemology of tourism: debate on Tourism 3.0 (left), 'Sociological Exhibition', an Artistic Sociology work by Pedro Andrade, including Virtual Sociological Gallery and Virtual Survey (right)
Source: Pedro Andrade Sociological Collection

Hybridology and Hybrid Methods: A Rapid Geneology of Hybrid Discourse Analysis

The *hybrid* is a core entity within the contemporary social fabric. Several schools of thought and authors have been debating this social process, in the last decades, namely Post-Colonial Studies (Appadurai, 2013; Bhabha, 1996, 2004; Canclini, 1995/2005; Said, 2008).

Other works concentrate on the potential of *hybrid methods*, which are scientific procedures that articulate and sometimes fusion the natures of methodologies themselves. For instance, the hybridization of: (a) on the one hand, classical traditional sociological methods, such as Discourse and Content Analysis; (b) with, on the other hand, digital methodologies e.g. Qualitative Discourse Analysis (QDA) used via NVivo software among others, 2-D or 3-D games, or virtual, augmented, mixed and hybrid reality. In particular, the author developed a specific method, *Hybrid Discourse Analysis*, whose definition and stages of development are clarified below.

Hybrid Discourse Analysis (HDA) (…) may be defined as a critical and hybrid way of undertaking discourse analysis through fusions between the different natures of things, while analyzing a given corpus of social data. HDA hybrid nature may be understood as: (a) an articulation between critical social/human sciences and arts/ literature [and/or other types of knowledge areas]; (b) a hybridization of methodological approaches, for instance, both quantitative and qualitative procedures; (c) a blending between traditional and experimental methods; (d) a dialogue between pre-digital and digital media; (e) a dialectics of, on one hand, 'classical' media relationships like hypermedia and, on the other hand, emerging digital combinations of media, such as *hybrimedia*. This last concept refers to the hybridization among media/literacies of different nature, for instance visual and touch through digital devices (…) (Andrade, 2016d, p. 159).

Since 1984, the author, through an epistemological reflection on the nature of concepts within Sciences, suggested the notion of *relation-concept.* This term signifies a special type of concept, that connect and hybridize other concepts, which was applied to content analysis of texts (Andrade, 1991, 2002). For instance, consider the following sentence: 'Covid-19 transformed contemporary tourism'. It is possible to understand 'Covid-19 as an 'anterior concept' (that is, a condition, a cause, etc.) in this sentence, related with a 'posterior concept' (meaning a conditioned phenomenon, a consequence, an effect, among other denotations). The relation between these concepts is a relation-concept, insofar as it hybridizes the precedent and the consequent concepts, into mediating concepts, which may be 'contaminated tourism', 'pandemic tourism', etc.

Moreover, the author of this text developed these ideas, integrated into a sociological methodology, the above referred *Interdimensional Sociology,* understood as a paradigm for the organization of theory related with sources about interconnected and *hybridized dimensions* of the social fabric. For such purpose, several instruments or steps were used within qualitative softwares: the syntactic segmentation of sources and of the semantic sentence; relation-concepts applied to the codification of the corpus; the construction of the mentioned *Interconceptual Relations Alphabet* and of the argumentative links alphabet (a first version was developed while student in 1984); the suggestion of the 'segments card' and the 'visual index'. All these tools constitute pillars for the design of 'bipolar and tripolar network maps' (Andrade, 2002).

Other direction is the application of hybrid ideas to Arts and Literature. A blog of a new kind was constructed in 2006, the *hybrilog*, which hybridize several blogs of different natures (e.g. a text blog, a video blog, a video poetry blog, a digital art blog, etc.). Such simultaneously social, cultural and *hybri-*

media initiative created a novel public space, the *hybridosphere*, that is, a cyberspace hybridized with a digital time, named *cybertime*. This last term means the set of rhythms and beats (the relatively long time of a web page publication, the almost instantaneous moment of a social network exchange, etc.), that an user experiments when circulating within the cyberspace arena inherent to the plural genres of blogs. All these cybertravels may forge art in blogs or blogs that behave like works of art, named *blogarts*, or even as a museum that has the form of a blog, the *bloseum* (Andrade, 2006a,b).

In what regards literature or other arts of the writing, several *modes of hybrid writing* may be typifyed:

In short, *trans-writing* transcends each one· of the media ways of writing that compose it. Therefore, hybrilog trans-writing is becoming a more and more deeper hybrid writing, that produces something different from the summing of the several media involved in its production-diffusion. To conclude, I would say that trans-writing transforms and surmount the writing processus itself, at least as we know it presently. This new mode of writing and talking in cyberspace and in cybertime, which constitutes an inedited cyberliteracy, engenders parallelly a singular sub-version of digital citizenship (Andrade, 2006b, p. 3).

The above-mentioned Alphabet of Interconceptual Relations, together with the Alphabet of Argumentative Relations (AIR and AAR, respectively), since their original versions back in 1984, were enhanced and augmented later, in an *Alphabet of Universal Relations (AUR)*. In effect, by 2007, after a debate on several systems of logic, the author stated this: 'the alphabet we intend to build is not formed by letters, but from logical relationships.

(...) For this purpose, it is inspired by different types of logic: a) analog logic (used by pre-rational systems or irrational thinking); b) formal logic (Aristotle, Descartes); c) dialectical logic (Heraclito, Hegel and Marx); d) Boolean logic; e) fuzzy logic (used in Artificial Intelligence), etc. Thirty seven different logical relations among concepts were circumscribed, based on social relationships, which constitute a solid pillar for the analysis and interpretation of (socio)logical sources. (Andrade, 2007a, p.145).

In addition to the manifestations of the hybrid already discuused in this text, the hybrid is present as well within *social literacies*. This concept denotes the incomensuable modes of listening/reading and talking/writing inside the social fabric. Such social processes represent other possible study targets for Social and Human Sciences, and in particular for Hybrid Discourse Analysis.

Literacy is understood as a set of reading and writing strategies underlying a given mode of knowledge (…) in addition to the reading and writing regimes of a national or mother tongue (…) Thus, it is possible to speak of scientific technological, artistic literacies, etc. At the confluence of these literacies, an unprecedent way of knowing emerges, *Hybridology*. This is a particular case of literacies, a kind of *meta-literacy* consisting of a reading and writing style that is simultaneously reflective and intuitive, which focuses on the hybrid entities and identities that proliferate in contemporaneity, especially within social networks. (…) Another hybridization takes place between, on the one hand, the ways of reading/writing by an specialist or professional in Social Sciences and, on the other hand, the options for reading/writing of the common citizen (Andrade, 2011a, p. 71-72).

All these reseraches reinforce the emeregenge of *Hybridology*, which is a genre of hermeneutics that substantively reflects on hybrid phenomena, and metodologically utilizes *hybrid methods* and *hybrimedia*.

Some practical sociological examples of these epistemological reflections are the following: a case study of audiences at the Berardo Collection Museum in 2010, through the *Multitouch Questionnaire,* a questionnaire that was activated inside a multitoch digital table. In this table, the *Game of Trichotomies* allowed the public to associate three related art works created by the artista Joana Vasconcelos (Andrade, 2011b). Other example of new sociological methodologies are the *GeoNeoLogic Novel* (Andrade, 2011c).

In all of them, the common citizen participates both in the consumption of knowledge and in its production. In short, nowadays the ordinary citizen rises up as a *lay scientist*, and makes possible to understand our current world as an immense *Society of Research*, where the boundary and power between specialist and non-expert blurs and dissolves inexorably (Andrade, 2008a ; Idem, 2011b, p.52).

Within the cultural heritage arena, other studies were conducted through the lens of a *Sociology of the Ruin*, within the context of social networks. Ruin, as all social phenomena, has a relational and hybrid nature. Therefore, it may be studied through Semantic-Logic Sociology, and inside of one of its tactics, *Social Hybridology*, that uses the *Hybrid Hermeneutics* style of interpretation, and in particular by *Hybrid Discourse Analysis*. For further clarificiation of these concepts, see Andrade, 2013a. Moreover, a synthetic debate on Hybridology, connected with the digital sphere and new media, can be found at Andrade, (2015c).

The anterior ideas were also tested in what regards the *digital public sphere* (2013b), inside Web 2.0 and Web 3.0, where *semantic-logic network maps* analyze social fields and their social relations. E.g., 'communication and media' (p. 207), 'digital social spaces and time's that constitute cyberspace and cybertime (p. 209); 'Politics and Society' (p. 211), 'communication and political' (p. 212); 'social network user and content' (p. 213), *thricotomies* (meaning 3 related concepts), in the case of Web 2.0 and Web 3.0 according to its 'social' meaning (p. 214).

Another interesting subject, within Post Colonial Studies, is the hybrid interpretation of post-colonial co-ordinary literature within 'transmediatic knowledge'. This is a profound reflection about the geneology of the philosohical concept 'hybrid', on diverse modes of social, cultural and literary *hybridizations*, and about the correspondent interpretation via Social Hybridology (Andrade, 2014).

Other works articulate, even deeply, hybridization and post-colonialism. Inspired by Edward Said e Homi Bhabha, some texts debate hibridism; hybrid intercultural societies; post-colonial literacies, common meaning networks, media hybridization, hybridosphere, processes interpreted via Social Hibridology (Andrade, 2014).

Yet another field for the application of Social Hybridology and Hybrid Discourse Analysis is cinema. While researcher at CECS, University of Minho, the author edified, in 2016, a Knowledge *Network on Transcultural Communication* including an Archive, a Knowledge Base and a Virtual Museum. The Knowledge Node 'Transcultural Cinema' included a variety of sources (texts, photos, videos, etc.) about Film Studies and movies directors, authors of critical cinema and resistance cinema. A Sociological perspective discussed postcolonial concepts and theories (e.g. clash of civilizations, conflict of meanings). In particular, the discursive resistance against colonialism is often based on mobilizing hybridizations. In fact, contemporary cultures are essentially *hybrid cultur*es (Andrade, 2016e).

In sum, hybridity may be used as a plural strategy within the social fabric. For example, an interesting study debates strategies for relating interactivity and multimodality, within digital and virtual media (Martins, 2012).

The Visual, the Virtual and the Viral

As for *visual knowledge,* it is becoming a central mode of understanding our digital societies (Paquette, 2010). In particular, a visual research strategy is *Visual sociology* (Harper, 2012). In fact, *virtual visual* processes are everywhere, and may be either instrument of social gaze sharing among social actors when looking/seeing one another, or an object of studies within a *Sociology of gaze*, inspired on Simmel (Andrade, 1997b).

Figure 14. The sociological comics technique on City 3.0 / Tourism 3.0: mass tourism, counter-tourist, local interactivity, mobile cultural tourism, intercultural networks of tourism
Source: Pedro Andrade Sociological Collection

Another method presented in several publications, possible to apply to the analysis and interpretation of cultural tourism, is *Sociological Comics* (Andrade, 2015a, b, 2016c, 2018b, g, 2020f and Figure 14). Such tool attempts, by means of a story in the form of comics, to express: (a) on one hand, the knowledge implemented by institutions located in the communicative, democratic and digital contemporary city, such as academia; (b) and, on the other hand, the knowledge produced by urban digital citizens. In fact, social problems may not just be narrated, but also analyzed by means of a sociological research that uses visual and textual arguments inside comic book panels. Thus, some purposes are: question social issues and to demonstrate sociological hypotheses; clarify theoretical concepts, such as City 3.0, Culture 3.0 and Tourism 3.0, using visual sources e.g., mobile phone's photos and videos.

Another emerging sociological method is *video paper* (Andrade, 2018h, 2020f). A video paper is defined as a scientific article produced and presented through the medium 'video', which can debate a question or opinion, present a case study or discuss scientific research.

In particular, sociological comics and video papers may be applied to cultural tourism mediated by mobile digital devices, for example mobile phones. Such processes are reconfigured in multiple modes and types, through tourist mobile practices seemingly neutral, but in truth controversial.

With regard to the expected results of these experiments, some are sociological archives and collections of hybrid sources on urban cultural tourism, which articulate the most traditional textual sources and messages with images, animations and sounds disseminated through digital videos, and via other digital and mobile technologies, such as virtual and augmented reality.

FUTURE RESEARCH AND RECOMMENDATIONS

Social and Intercultural Networks for Tourism Communication

In order to understand the multiple connotations of some concepts above debated, it is necessary to apply them more profoundly to social empirical fields. Sociological terms as *Tourism 3.0* and *Web 3.0* were defined in precedent sections. As for *Culture 3.0*, its genealogy may be elucidative, in particular the transformation of culture within the emerging of modernity. In other words, modern culture is a process derived from the 18th and 20th-century industrial economy (*Culture 1.0)*; next, the idea of culture as a cultural industry raised in the 20th century (*Culture 2.0*); finally, the notion of Culture 3.0 is closely associated with the new digital technologies, meaning that culture is an instrument for creating identity and values, stimulating social cohesion and encouraging creativity, by using digital tools (Sacco, 2011). In what regards 'City 3.0' or 'social semantic city', this notion signifies a localized and globalized territory and place, historically and structurally transformed into a geographical urban network, which uses the digital, social and also semantic networks characteristic of Web 3.0, for instance within the activities of Tourism 3.0 (for more details, see Andrade, 2018b, p. 253).

For fulfilling the precedent aims, it is important to understand the socio-cultural causes, constraints and effects that mobile urban tourism and its modes of communication exert on the city's everyday life and, indirectly, on the sustainability of the urban fabric, the surrounding territory and, in particular, on tourism. When tourists visit a different society and culture, in particular within the COVID-19 present conjuncture, what purpose do they aim and how do they communicate with the inhabitants of the country visited? Regarding such objectives, conscious or not, the *contacts* between visitor and visited can contribute to socio-economic, cultural and symbolic *contracts*. For example, a deeper and more sustained *intercultural dialogue* among communities and regional/local identities/differences of the actors and countries involved.

CONCLUSION

In sum, unlike some partial scientific postures, which may become deceptive illustrations of ethnocentrism and scientific fundamentalism, Hybrid and Open intercultural research intends to fusion complementary and opposite peoples, objects of study, methodologies and researcher's profiles.

Inside such strategy, the tourism sector and its activities, in the context of COVID-19 outbreak, should be more deeply studied as a social process that is manifestly central for the economies and societies of more and more countries. Among other reasons, this is due to the fact that such activity is transversal within local and global societies and cultures. In effect, tourism is related and sometimes hybridizes, more or less profoundly, multiple other phenomena still incommensurable. In addition to the ones referred above, within our instable times, an example of contemporary social, demographic and political processes relating to travel under the viral society, is the forced migration diaspora testified by the recent waves of refugees escaping from the Ukraine war.

REFERENCES

Adey, P. (2009). *Mobility*. Routledge. doi:10.4324/9780203875483

Adu-Ampong, E. (2012). *Managing Tourism at Heritage Sites*. Lap Lambert Academic Publishing.

Andrade, P. (1981-1984). *Les groupes 'Excursionistes'-'De Déjeuners' Portugais et leur Art* [Portuguese Excursionist-Lunch Groups and their Art]. Paris: École des Hautes Études en Sciences Sociales.

Andrade, P. (1985). *La taverne, son boire et leurs savoirs: ruses et muses mystico-profanes* [The tavern, its drinking and their knowledge: mystical-profane tricks and muses]. Paris: École des Hautes Études en Sciences Sociales.

Andrade, P. (1986). A Arte Excursionista [The Excursionist Art]. *Colóquio-Artes*, *68*, 5-11.

Andrade, P. (1991). A Taberna mediática, local reticular de negociações sociais e sociológicas [The Mediatic Tavern: reticular locality of social and sociological negotiations]. *Revista Crítica de Ciências Sociais*, *33*, 265-286.

Andrade, P. (1993). Sociologia da viagem: deslocações diárias e anti-quotidiano nómada [Sociology of Travel: daily trips and nomad anti-everyday]. *Revista Crítica de Ciências Sociais*, *37*, 51-77.

Andrade, P. (1997a). Navegações no cibertempo: viagens virtuais e vitualidades da ciberviagem [Navigations in cybertime: virtual journeys and cybertravel]. *Atalaia, 3*, 111-124.

Andrade, P. (1997b). Sociologia do Olhar: um novo olhar sobre a Sociologia? [Sociology of gaze: A new look at sociology?]. *Atalaia*, *3*, 197-201.

Andrade, P. (2002). Some synthetic ideas to work with qualitative analysis software. *Atalaia/Intermundos*, *10*(11), 153-165.

Andrade, P. (2006a). Net Art and Hybridosphere: the Hybrilog as Blogart [Arte da Net e Hibridosfera: o Hybrilog como Blogart]. In L. Dopico (Eds.), *Conferencia Internacional de Arte Electrónico e Dixital* (pp.137-143). Pontevedra: Artech.

Andrade, P. (2006b). Cybermuseus e cibermuseologia: o hybrilog como um Bloseum [Cybermuseums and cybermuseology: the hybrilog as a Bloseum]. In H. Gottesdiener & J. Vilatte (Eds.), *Culture and Communication* (pp. 553-557). Avignon: International Association of Empirical Aesthetics.

Andrade, P. (2006c). *A sub-version of writing and of citizenship: the blog's auto-writing and the hybrilog as trans-writing*. In VI Colóquio Internacional 'Discursos e práticas alquímicas', Guimarães.

Andrade, P. (2007a). O Alfabeto de Relações Universais (ARU) [The Alphabet of Universal relations (ARU)]. *Revista de Comunicação e Linguagens 'Mediação dos saberes', 38*, 143-55.

Andrade, P. (2007b). *Imagem e Web 2.0: algumas redes sócio-conceptuais mediadoras da Wikipédia* [Image and Web 2.0: some socio-conceptual networks mediating Wikipedia]. Paper presented at Conferência Internascional 'Imagem e Pensamento'. Centro Cultural de Belém, Lisboa.

Andrade, P. (2008a). A Sociedade da Investigação e do Jornalismo: boas práticas de cidadania participativa através da internet móvel e do social bookmarking [The Society of Research and journalism: good practices of participatory citizenship through the mobile internet and social bookmarking]. In *Anuário Internacional de Comunicação Lusófona* (pp. 307–312). Braga: SOPCOM.

Andrade, P. (2010a). Introdução. In P. Andrade (Ed.), *Museus, públicos e literacia científico-tecnológica: redes sociais de comunicação de significados no espaço interdimensional do museu* [Museums, publics and scientific-technological literacy: social networks for the communication of meanings within the interdimensional space of the museum] (pp.25-32). Lisboa: Ed Colibri. http://hdl.handle.net/1822/70456

Andrade, P. (2011a). A hibridação das literacias: métodos GeoNeoLógicos na Sociedade da Investigação [The hybridization of literacies: GeoNeoLogical Methods in the Research Society]. In S. Pereira (Ed.), *Atas do Congresso Nacional 'Literacia, Media e Cidadania'* (pp. 71-89). Braga: Centro de Estudos de Comunicação e Sociedade.

Andrade, P. (2011b). *Sociologia Semântico-Lógica da Web 2.0/3.0 na sociedade da investigação: significados e discursos quotidianos em blogs, wikis, mundos/museus virtuais e redes sociais semântico-lógicas* [Semantic-Logical Sociology of Web 2.0/3.0 in the Research Society: meanings and everyday discourses in blogs, wikis, virtual worlds/museums and semantic-logical social networks]. Lisboa: Edições Caleidoscópio.

Andrade, P. (2011c). *Novela GeoNeológica nº 1: um caso de Literatura Transmediática /1ª Novela da Web 3* [GeoNeoLogical Novel nº1: a case of Transmediatic Literature]. Lisboa: Edições Caleidoscópio.

Andrade, P. (2012). *Público e privado na época contemporânea.* Centro de Estudos Comparatistas, Faculdade de Letras da Universidade de Lisboa.

Andrade, P. (2013a). Sociologie Sémantico-Logique des Ruines: pour une hermeneutique hybride de la ruine à la Web 2.0 et à la Web 3.0. *Sociétés, 115*, 105-119.

Andrade, P. (2013b). Sociological ontology of the digital public sphere: the case of Web 2.0/3.0. *Comunicação e Sociedade, 23,* 202 – 216. http://hdl.handle.net/1822/29484

Andrade, P. (2014). Post-Colonial Co-Ordinary Literature And Web 2.0/3.0: thinking back within transmediatic knowledge. In M. Pope (Ed.), *New Literary Hybrids in the Age of Multimedia Expression: crossing borders, crossing genres.* John Benjamin Press. http://hdl.handle.net/1822/39995

Andrade, P. (2015a). *Austerity History through Sociological Comics: A Guide to Social Media and Networks against Austeritocracy for use by all Generations.* Create Space.

Andrade, P. (2015b, June). *Discourses and local public opinion in the urban and virtual public sphere: The case of struggles on austerity at Lisbon, 2013.* Paper presented at II Simpósio Internacional EDiSo: Discursos e sociedades em movimento, 18-20 junho 2015, Coimbra.

Andrade, P. (2015c). 'Arts Numériques et Musées Virtuels', 'Jeux Sociologiques', 'Hybridologie', 'Ontologies Visuelles'. In M. Veyrat (Ed.), *110 Notions sur les Arts Numériques.* Paris: Les Éditions de l'Immatériel.

Andrade, P. (2016a). Introdução [Introduction]. In P. Andrade (Ed.), *Comunicação pública da arte* [Public Communnication of Art] (pp. 21-32). Lisboa: Edições Caleidoscópio. http://hdl.handle.net/1822/70454

Andrade, P. (2016b). Que teoria e metodologia para os Museum Studies? [What theory and metodology for Museum Studies?]. In P. Andrade (Ed.), *Comunicação pública da arte* (pp. 33-48). Lisboa: Edições Caleidoscópio. http://hdl.handle.net/1822/70463

Andrade, P. (2016c, June). *Os Sociological Comics: ou as lutas sociais contadas por histórias sociológicas* [Sociological Comics: or social struggles told via sociological stories]. Paper presented at VI Congresso da Associação Portuguesa de Antropologia, 2-4 Junho, Coimbra.

Andrade, P. (2016d). GeoNeoLogical Novel and other Hybrimedia Experiments: or how to use Hybrid Methods such as Hybrid Discourse Analysis (HDA) within a Knowledge Base. *MATLIT, 4*(2), 157-181.

Andrade, P. (2016e). Transcultural cinema debated in a knowledge network: postcolonial hybrid meanings within resistance cinema. *Comunicação e Sociedade, 29*, 395 – 411.

Andrade, P. (2018a). A urbe social e sociológica: ou a cidade entendida como contexto e enquanto co-texto quotidianos [The social and sociological city: or the city understood as everyday context and co-text]. In *Culturas Urbanas e Contemporaneidade* (pp. 103-110). Centro de Estudos Lusíadas.

Andrade, P. (2018b). Epistemologia e Metodologia do Turismo Cultural Urbano: o caso da Sociologia Artística das culturas móveis e da comunicação turística em redes sociais urbanas [Epistemology and methodology of urban cultural tourism: the case of the artistic sociology of mobile cultures and tourism communication in urban social networks]. *Revista de Comunicação e Sociedade, 33,* 243-269. http://hdl.handle.net/1822/55275

Andrade, P. (2018c). Innovative role of users within digital economy: the case of information / knowledge flows at social and semantic networks (web 2.0/3.0). In L. Carvalho & P. Isaías (Eds.), *Handbook of Research on Entrepreneurship and Marketing for Global Reach in the Digital Economy* (pp. 269-290). IGI Global. http://hdl.handle.net/1822/70088

Andrade, P. (2018d). Mobile art flows: tourism mobilities at the museum. In E. Araújo, R. Ribeiro, P. Andrade & R. Costa (Eds.), *Viver em|a mobilidade: rumo a novas culturas de tempo, espaço e distância. Livro de atas* (pp. 193-198). Braga: CECS. http://hdl.handle.net/1822/55107

Andrade, P. (2018e). Turismo cultural móvel. Espaços, tempos e logos da arte [Mobile Cultural Tourism. Art spaces, times and logos]. *Revista Lusófona de Estudos Culturais, 5*(2), 349 – 363. http://hdl.handle.net/1822/57459

Andrade, P. (2018g). Digital Literature and Big Knowledge: the case of Sociological Comics. *MATLIT, 6*(2), 138-147.

Andrade, P. (2018h). *Ativismo Sociológico através do Video Paper: artes e imagens públicas publicitadas por imagens sociológicas* [Sociological Activism through Video Paper: public arts and images disseminated via sociological images]. Paper presented at Congress 'Arte Urbana no Plural', Leiria.

Andrade, P. (2018i). Cultural e-Tourism depicted by Digital Discourse: The Case of Portuguese Networks of e-Heritage and Discursive Innovation. In *Innovative Perspectives on Tourism Discourse* (pp. 1-17). IGI Global. http://hdl.handle.net/1822/70087

Andrade, P. (2020a). Mobile culture for tourism communication. In *Encyclopedia of Organizational Knowledge, Administration, and Technology* (pp. 1638-1650). IGI Global. http://hdl.handle.net/1822/70086

Andrade, P. (2020b). Arte pública urbana e comunicação turística [Urban public art and tourism communication]. *Revista Lusófona De Estudos Culturais, 7*(1), 39-59. doi:10.21814/rlec.2119

Andrade, P. (2020c). Cultural resistance to Covid-19: an Encyclopedia of Public Art, for artists and tourism publics? *Cadernos de Arte Pública, 2*(2), 57-74. http://hdl.handle.net/1822/70084

Andrade, P. (2020d). Um autor de arte pública urbana: Luis Baldini [An author of urban public art: Luis Baldini]. *Revista Lusófona de Estudos Culturais, 7*(1), 259-279. doi:10.21814/rlec.2677

Andrade, P. (2020e). Pessoa effect within viral society: viral sociological poetry using apps and e-books anti-virus. In *Proceedings of 2nd International Conference on Transdisciplinary Studies in Arts, Technology and Society, ARTeFACTo2020* (pp. 96-103). Faro: Universidade do Algarve, Centro de Investigação e Comunicação. http://hdl.handle.net/1822/70089

Andrade, P. (2020f). Cultural tourism and mobile digital devices: an analysis of tourists in Lisbon through sociological comics and video paper. In C. Sarmento & M. Quincé (Eds.), *Intercultural Routes across Multicultural Spaces: From Theory to Dialogue* (pp. 23-37). Porto: Centro de Estudos Interculturais. http://hdl.handle.net/1822/70091

Andrade, P. (2020g). Rua Augusta, an intercultural space: tourist communication via locative media for territory sustainability. In *XII Congresso da Geografia Portuguesa: Geografias de Transição Para a Sustentabilidade, Livro de Atas* (pp. 326-331). Guimarães: Universidade do Minho. http://hdl.handle.net/1822/70090

Andrade, P. (2020h). Site *Viral & Intercultural Tourism City*. Available from: https://www.sites.google.com/view/viral-tourism-city

Andrade, P. (2021c) Viral Urban Tourism within Viral Society: Sociological Touring Guides using Apps and e-Books. In *International Conference Connections: Exploring Heritage, Architecture, Cities, Art, Media. Amps Proceedings Series 20.1* (pp. 125-134). University of Kent/AMPS. http://hdl.handle.net/1822/74116

Andrade, P. (2022a). The Value of Products and Services in Tourism: the Case of Viral Society. In C. Sarmento (Ed.), *Concepts and Dialogues across Shifting Spaces in Intercultural Business* (pp. 83–106). Cambridge Scholar Publishing. Available at https://www.cambridgescholars.com/product/978-1-5275-7725-1

Andrade, P. (2022b). Cultural Mobile Tourism and Tourist Knowledge at Viral Society: Sociological Research Games Against Covid-19, Within A Big Knowledge Context. In *Amps Proceedings Series 25. Urban Assemblage: The City as Architecture, Media, AI and Big Data* (pp. 99-114). University of Hertfordshire. Available at: http://architecturemps.com/wp-content/uploads/2022/02/Amps-Proceedings-Series-25.pdf

Andrade, P. (2022c). Everyday Urban Life: Genealogy and Journal of Actors' Body in the Viral City. In *The City of the Senses: the Senses of the City* (pp. 35-65). Braga: Uminho Editora. http://hdl.handle.net/1822/76038 https://ebooks.uminho.pt/index.php/uminho/catalog/view/51/127/1255-1

Andrade, P. (2022d). Risk Tourism Within Viral Society: A Study Using Hybrid Discourse Analysis [Lusophone Journal of Cultural Studies]. *Revista Lusófona de Estudos Culturais, 9*(1), 127-151. Available at: https://hdl.handle.net/1822/78577

Andrade, P. (2022e). Dark Heritage Pedia: an Augmented Research Encyclopedia, for a Virtual Sociological Museum on the Insatiable/Unsafe/Insane Society. In *Dark Heritage Tourism in the Iberian Peninsula: Memories of Tragedy and Death*. Cambridge Scholars Publishing.

Appadurai, A. (2013). *The future as cultural fact: Essays on the global condition*. Verso.

Bhabha, H. (1996). *Negotiating Rapture: The Power of Art to Transform Lives*. Museum of Contemporary Art. Braga Creative City of Media Arts. Accessed March 15, 2020: https://en.unesco.org/creative-cities/braga

Bhabha, H. (2004). *The location of culture*. Routledge.

Canclini, N. (2005). *Hybrid cultures: strategies for entering and leaving modernity*. University of Minnesota Press.

Giffinger, R. (2007). *Smart Cities: Ranking of European medium-sized cities*. Centre of Regional Science.

Harper, D. (2012). *Visual sociology*. Routledge.

Harper, D. (2012). *Visual sociology*. Routledge. doi:10.4324/9780203872673

Landry, C. (2006). *The Art of City Making*. Earthscan Publications.

MacCannell, D. (1976). *The Tourist: A New Theory of the Leisure Class*. Schocken Books.

MacCannell, D. (1992). *Empty Meeting Grounds: The Tourist Papers*. Routledge.

MacCannell, D. (2011). *The Ethics of Sightseeing*. University of California Press.

Martins, M. L. (2012). Média digitais: Hibridez, interactividade, multimodalidade [Digital media: Hybridity, interactivity, multimodality]. *Revista de Comunicação e Linguagens, Lisboa*, *43/44*, 49–60. doi:10.3917oc.111.0017

Ministry of Health of Ukraine. (2022). *Covid-19 pandemic in Ukraine: Current information about coronavirus and quarantine*. https://covid19.gov.ua/en

Paquette, G. (2010). Visual knowledge modelling for semantic web technologies: models and ontologies. Hershey. doi:10.4018/978-1-61520-839-5

Pirolli, B. (2018). *Informing tourists in the digital age*. Routledge.

Richards, G., & Raymond, C. (2000). Creative Tourism. *ATLAS News*, *23*, 16–20.

Rickly, J., Hannam, K., & Mostafanezhad, M. (Eds.). (2016). *Tourism and Leisure Mobilities: Politics, Work, and Play*. Routledge. doi:10.4324/9781315686660

Sacco, P. (2011). *Culture 3.0: A new perspective for the UE 2014-2020 structural funds programming*. European Expert Network on Culture.

Said, E. (2008). *On late style: Music and literature against the grain*. Vintage.

Sheller, M., & Urry, J. (2004). *Tourism mobilities: Places to play, places in play*. Routledge. doi:10.4324/9780203340332

Sheller, M., & Urry, J. (2006). *Mobile Technologies in the City*. London: Routledge.

Sheller, N., & Isaacson, M. (2013). *Tourist mobility and advanced tracking technologies*. Routledge.

Veblen, T. (1994). *The Theory of the Leisure Class: An Economic Study of Institutions*. Penguin Books. (Original publication 1899)

ADDITIONAL READING

Adu-Ampong, E. (2012). *Managing Tourism at Heritage Sites*. Lap Lambert Academic Publishing.

Black, G. (2012). *Transforming Museums in the Twenty-first Century*. Routledge. doi:10.4324/9780203150061

Boniface, P., & Fowler, P. (1993). *Heritage and Tourism in The Global Village*. Routledge.

Buntinx, G. (2006). *Museum Frictions: Public Cultures/Global Transformations*. Duke University Press Books. doi:10.2307/j.ctv11cw1hd

Butler, R., & Suntikul, W. (Eds.) (2017). *Tourism and War*. Routledge.

Castells, M. (1996). *The Rise of the Network Society: The Information Age: Economy*. Wiley-Blackwell.

Chhabra, D. (2021). *Authenticity and Authentication of Heritage*. Routledge. doi:10.4324/9781003130253

Cohen, E. (1979). A phenomenology of tourist experiences. *Sociology*, *13*(2), 179–201. doi:10.1177/003803857901300203

Cohen, E. (1979). A phenomenology of tourist experiences. *Sociology*, *13*, 179–201.

Cohen, S. (2013). *Lifestyle mobilities: intersections of travel, leisure and migration*. Ashgate Publishing Company.

Cohen, S. (Ed.). (2014). *Understanding and governing sustainable tourism mobility: Psychological and behavioural approaches*. Routledge. doi:10.4324/9780203771501

Egger, R., & ... (Eds.). (2016). *Open tourism: open innovation, crowdsourcing and co-creation challenging the tourism industry*. Springer., doi:10.1007/978-3-642-54089-9

Elliot, A., & Urry, J. (2010). *Mobile Lives*. Routledge.

Faraj, R. (2014). *E-tourism Revolutionary Effect Studying E-tourism in Lebanon*. Saarbrucken: Lap

French, Y., & Runyard, S. (2011). *Marketing and Public Relations for Museums, Galleries, Cultural and Heritage Attractions*. Routledge.

Frihammar, M. (2019). *Heritage of Death*. Routledge.

Fullagar, S., Markwel, K., & Wilson, E. (Eds.). (2012). *Slow travel and tourism: experiences and mobilities*. Channel View Publications.

Gans, J. (2020). *Economics in the age of covid-19*. MIT Press First Reads.

Goodson, L. (Ed.). (2004). *Qualitative research in tourism: ontologies, epistemologies and methodologies*. Routledge.

Graburn, N. (1977). The Museum and the Visitor Experience. Roundtable Reports, 1-5. Retrieved from http://www.jstor.org/stable/40479310

Hall, C. (2011). *Fieldwork in tourism: methods, issues and reflections*. Routledge.

Hall, M., & Tucker, H. (2004). *Tourism and Postcolonialism: Contested Discourses, Identities and Representations* (M. Hall & H. Tucker, Eds.). Routledge.

Hanna, S. (Ed.). (2015). *Social memory and heritage tourism methodologies*. Routledge.

Hannam, K., Mostafanezhad, M., & Rickly-Boyd, J. (Eds.). (2016). *Event Mobilities: Politics, Place and Performance*. Taylor & Francis.

Hanusch, F., & Fursich, E. (Eds.). (2014). *Travel journalism: exploring production, impact and culture*. Palgrave Macmillan., doi:10.1057/9781137325983

Harrison, D., & Hitchcock, M. (Eds.). (2005). *The Politics of World Heritage: Negotiating Tourism and Conservation*. Channel View Publications.

Higgins-Desbiolles, F., Doering, A., et al. (2021). *Socialising Tourism: Rethinking Tourism for Social and Ecological Justice*. Routledge.

Hooper, G. (2019). *Dark Tourism: Practice and Interpretation*. Routledge.

Huntington, S. (2011). *The Clash of Civilizations and the Remaking of World Order*. Simon & Schuster.

Johnson, P., & Thomas, B. (1992). *Tourism, Museums and the Local Economy: The Economic Impact of the North of England Open Air Museum at Beamish*. Edward Elgar Pub.

Kaminski, J., Benson, A., & Arnold, A. (Eds.). (2014). *Contemporary issues in cultural heritage tourism*. Routledge.

Katsoni, V., & Stratlgea, A. (Eds.). (2016). *Tourism and culture in the age of innovation: second International Conference IACuDiT, Athens 2015*. Berlin: Springer.

Kerr. M. at al (Eds.) (2022). *Children, Young People and Dark Tourism*. Routledge.

Kirshenblatt-Gimblett, B. (1998). *Destination Culture: Tourism, Museums, and Heritage*. University of California Press.

Latour, B. (2007). *Reassembling the Social*. Oxford University Press.

Lavine, S. (1991). *Exhibiting Cultures: The Poetics and Politics of Museum Display*. Washington, DC

Lippard, L. (2000). *On the Beaten Track: Tourism, Art, and Place*. The New Press.

Lisbon Green Capital. Accessed July 20, 2020. https://lisboagreencapital2020.com

Lytras, M. (2011). *Digital culture and e-tourism: technologies, applications and management approaches*. IGI Global., doi:10.4018/978-1-61520-867-8

MacLeod, D. (2003). *Niche Tourism in Question: Interdisciplinary Perspectives on Problems and Possibilities*. University of Glasgow Press.

Maddrell, A., & Terry, A. (2015). *Sacred mobilities: journeys of belief and belonging*. Ashgate.

Matsuo, K. (Ed.). (2015). *Tourism Informatics: Towards Novel Knowledge Based Approaches*. Springer., doi:10.1007/978-3-662-47227-9

McDaniel, K. (2018). *Virtual Dark Tourism: Ghost Roads*. Palgrave Macmillan.

Molz, J. (2014). *Travel connections: Tourism, technology and togetherness in a mobile world*. Routledge.

Naik, A. (2020). *Heroes of a pandemic: Those who stood up against covid-19*. Indy Pub.

Nvelli, M., & ... (Eds.). (2022). *Handbook of Niche Tourism*. Edward Elgar Publishing.

Open tourism. Springer.

Papathanassis, A. (2011). *The Long Tail of Tourism: Holiday Niches and their Impact on Mainstream Tourism*. Gabler Verlag.

Peace, S. (2020). *Abstract Art Images #9, during Covid-19: Making Art while Cocooning*. [s.n].

Pessoa, F. (2015). *What the tourist should see.Livros Horizonte. Pirolli, B. (2018). Travel Journalism*. Routledge.

Pforr, C. (Ed.). (2021). *Consumer Tribes in Tourism*. Springer.

Pollet, M. (2020). Majority of French, Germans, Italians & Spanish scrap holiday plans amid coronavirus outbreak – poll. *Euronews*. Accessed July 24, 2020.https://www.euronews.com/2020/07/29/

Rakic, T., & Chambers, D. (2012). *An introduction to visual research methods in tourism*. Routledge.

Richards, G. (1996). *Cultural Tourism in Europe*. CABI.

Richards, G. (2011). Creativity and tourism: The state of the art. *Annals of Tourism Research*, *38*(4), 1225–1253.

Rickly, J., Hannam, K., & Mostafanezhad, M. (Eds.). (2016). *Tourism and Leisure Mobilities: Politics, Work, and Play*. Routledge.

Rio, R. (2020). *Turismo in quarantena. Gli effetti del coronavirus sul sistema italiano*. Tangram Edizioni Scientifiche.

Salazar, N. (2010). *Envisioning Eden: Mobilizing Imaginaries in Tourism and Beyond*. Berghahn Books.

Salazar, N., & Jayaram, K. (Eds.). (2016). *Keywords of mobility: critical engagements*. Berghahn Books.

Sassen, S. (1994). *Cities in a World Economy*. Sage (Atlanta, Ga.).

Seixas, X. (2021). *Sites of the Dictators: Memories of Authoritarian Europe, 1945-2020*. Routledge.

Sharda, N. (2010). *Tourism informatics: visual travel recommender systems, social communities, and user interface design*. IGI Global., doi:10.4018/978-1-60566-818-5

Sharpley, R., & Stone, P. (2009). *The Darker Side of Travel*. Channel View Publications.

Sheller, N., & Isaacson, M. (2013). *Tourist mobility and advanced tracking technologies*. Routledge.

Simmel, G., & Levine, D. (Eds.). (1972). *Georg Simmel on Individuality and Social Form*. University of Chicago Press.

Simpson, B. (2013). *From Heritage to Terrorism: Regulating Tourism in an Age of Uncertainty*. Routledge.

Slocum, S., & Kline, C. (Eds.). (2015). *Scientific tourism: Researchers as travellers*. Routledge.

Smithsonian Books.

Sweeney, S. (2000). *Internet marketing for your tourism business: Proven techniques for promoting tourist-based businesses over the Internet*. Maximum Press.

I. Tussyadiah, & A. Inversini (Eds.). (2015). Information and communication technologies in tourism 2015: *Proceedings of the international conference in Lugano*, Switzerland, February 3-6. Berlin: Springer.

Tzanelli, R. (2019). *Thanatourism and Cinematic Representations of Risk*. Routledge.

Urry, J. (2007). *Mobilities*. Polity Press.

Urry, J., & Larsen, J. (2011). *The Tourist Gaze 3.0*. Sage.

Verstraete, G. (2009). *Tracking Europe: mobility, diaspora, and the politics of location*. Duke University Press.

Verstraete, G. (2009). *Tracking Europe: Mobility, diaspora, and the politics of location*. Duke University Press.

Vu, T. (2021). *Factors Influencing Customer's Purchase Intention in Niche-Tourism E-Marketplace*. GRIN Verlag.

White, L., & Frew, E. (Eds.). (2016). *Dark Tourism and Place Identity*. Routledge.

World Tourism Organization-UNWTO. *International tourism and covid-19*. Accessed July 29, 2020. https://www.unwto.org/international-tourism-and-covid-1

World Travel Awards. Accessed January 6, 2020. https://www.worldtravelawards.com/nominees/2019/portugal

KEY TERMS AND DEFINITIONS

The concepts and their definitions presented below aim to contribute to a collective Encyclopedia on Urban Tourism, Viral Society and Hybrid Discourse Analysis, a section of the project Virtual Sociological Museum. All worldwide texts work via social and discursive webs that are woven by intertextual and rhizomatic networks of articulated meanings. An Encyclopedia is just a more synthetic rhizome. The following terms may be inter-related via the words written in Italics, located inside each definition, that orient the reader to the Encyclopedia concepts entries (i.e. the definition titles) noted in Bold, at the first words of each paragraph.

City 3.0 or Social-Semantic City: Globalized locality and configured in an urban geographic network that includes digital, social, and semantic networks, characteristics of *Web 3.0*, particularly in the activities of *Tourism 3.0*.

Communicative Tourism: This concept signifies a paradigm of tourist activities centered on the tourism production and dissemination of information and knowledge, targeting three of the main agents of transcultural contemporaneity: citizens, tourists and migrants displaced from peripheral societies into central societies.

Creative City: Model of urban space that brings together culture, creativity, and the transformation of the city. This stance highlights, among other processes and practices, intercultural cities, city psychology, creative bureaucracies, and the measurement of creativity in cities.

Creative Tourism: Tourism model related to the creativity of urban communities and small cities, in order to revitalize the economy, society and culture of these localities, through tourist activities combined mainly with small industry, small commerce and handicrafts.

Cultural Citizenship: Social fusion between *cultural policies* and *cultural politics*.

Cultural Heritage: It is a collection of archives and memories related to a people, a nation, a country or a community. The material cultural heritage mainly includes physical works, in areas such as architecture, arts, literature, etc. The intangible cultural heritage gathers non-physical works, for example, oral memories, cultural, intercultural and transcultural events, traditions, etc.

Cultural Policy: Strategy developed by the State and cultural institutions in order to promote cultural diplomacy, cultural governance by various social actors, artistic literacy, cultural inclusion, public art, the identities and differences of citizens.

Cultural Politics: This idea means the daily active, critical, and political participation by citizens in democratic cultural life.

Cultural Tourism: Type of *tourism* that is predominantly interested in cultural aspects of society (arts and their institutions or contexts), for example, a Museum of Public Art or other public cultural spaces, such as the street.

Culture 3.0: In addition to the understanding of culture as a product derived from the industrial economy in the 18th and 20th centuries (Culture 1.0), or the concept of culture as a cultural industry in the 20th century (Culture 2.0), the notion of Culture 3.0, associated with new digital technologies, means that culture is a means of creating identity and values, stimulating social cohesion and encouraging creativity.

Digital Cultural Tourism: Mode of *tourism* associated with cyberspace and cybertime, and the mobile culture conveyed by the tourist through the cell phone, for example, as an instrument for linking public arts to cyberculture.

Digital Public Art: Sub-genre of public art, created, operated, and disseminated in cyberspace/cybertime.

Hybrid Discourse Analysis (HAD): Is a critical and hybrid discourse analysis genre that undertakes fusions among the different natures of things while analyzing a given corpus of social data. For example, by hybridizing knowledge types, methodological approaches such as *hybrid methods*, pre-digital, digital media and *hybrimedia*, intercultural and interprofessional research teams, among others.

Hybrid Methods/Hibrimedia: Mixture, fusion or hybridization of diverse scientific, technological or artistic methods and media, for example, those that characterize the following modes of knowledge: social sciences (questionnaire, etc.), new technologies (interactive digital devices built in hypermedia) and the arts (object art, procedural art like installation and performance).

Hybridology: It consists of the scientific, technological, and artistic study of the hybrid entities that abound and somehow define our globalized contemporaneity, as in the case of the growing demographic hybridization in European societies, through decades of massive immigration. Other examples are *hybrid methods, hybrid discourse analysis* and *hybrimedia*.

Innovative Tourism: Type of *tourism* linked to social innovation, especially within the urban mobility characteristic of the *smart city*.

Mobile Culture: Way of exercising culture and social life in general, which is partly transformed into a digital life, linked to the rhythms and moving places of everyday urban life, in particular the urban mobility of the city 3.0, the smart city and the creative city, especially through portable devices, such as the laptop and tablet computer, or the iPod and mobile phone.

Mobile Locative Tourism: Configuration of *tourism* travel in which the information and knowledge about the trip (internet searches about the tourist destination, memories captured during the visit, etc.) are carried out mainly via mobile phone, in person at physical locations or virtually within sites online. The new mobilities characteristic of mobile locative and viral tourism redefine the current social mobilities. That is, they establish processes of remobility, which develop certain regularities already detected in the case of tourism communication via locative media and communicative tourism.

Public Art: Artistic manifestation produced, exhibited, perceived, judged, and practiced in the public sphere, e.g., in public urban sites (streets, squares, etc.) and across mass media, such as newspapers, radio and television; and within cyberspace/cybertime.

Public Art Publics: Audiences who include specific sociodemographic characteristics and develop particular communication careers, inside or outside their visits to public art sites and events. For example, when these audiences relate, on the one hand, works of public art and the space of the museum or other places of public art exhibition, such as the street, squares and other city locations, with, on the other hand, their own experience of the city, work, family and school. Some main segments of such audiences are these: families; students and teachers at an educational institution; an isolated visitor or groups that aim to carry out continuous training throughout their life; the tourist subscribing to cultural tourism, creative tourism, innovative tourism and communicative tourism. However, other profiles, still marginalized, must be included in artistic audiences and, in particular, in public art publics, such as pensioners, disabled people, immigrants and refugees.

Research Society: In contemporary times, ordinary citizens can openly search (research) and research (through open research), open information and knowledge, using global tools and devices, such as *Google* or cell phones, in various social scenes or arenas, including physical or virtual museum spaces, or the street. In doing so, ordinary citizens can construct concepts and definitions, for example, on *Wikipedia*, and thus, somehow compete with professional scientists and artists, on the production and of information and knowledge and plural, local or global knowledge.

Smart City: City paradigm that favors planning, monitoring and digital technologies to achieve greater predictability in urban restructuring, among other aspects of greater mobility, *innovative tourism* and security in public space. However, this ubiquitous view of the city and the citizen carries risks, from the intrusion into his private life, through disrespect for human rights, to the naturalization and uncritical acceptance of a generalized panoptism. It differs from *city 3.0* as this urban mode is more profoundly connected with *Web 3.0* and *Tourism 3.0*.

Social Inheritance: This idea means a dialectical and not static legacy, subjacent to social formations. A social inheritance includes not just cultural heritages, but also economic, ecologic and political heritages, among other social legacies, that may constitute sufficient pillars for necessary social transformations, within a given historical conjuncture.

Social Remobility: It is defined as a posture to combat Viral Society, aiming at overcoming it and the alternative recreation of social mobility processes amid contemporary social and communicative processes, such as communication among citizens, tourists and migrants. In a *viral society*, social de-mobility occurs. In other words, the mobile society, in which 'everything is on the move', as John Urry (1990, 2007) puts it, has partially transformed itself into a motionless society. Therefore, it is necessary to de-move it from its i-mobility, through social remobility, among other strategies. Some examples of these social re-mobilization processes are urban public art linked to mobile cultures, such as tourist culture and cultures inherent in digital social networks. Such cultures are founded and merged, today, in virtual-viral communities that circulate in cyberspace and cybertime, which are conflicting digital public spheres where, presently, pre-viral societies deconstruct and gradually reconstruct themselves as post-viral societies.

Social-Semantic Website: This type of digital site is closely associated with *Web 3.0*. Explicitly presents an explanatory paradigm or sections on its own semantic content (ideas, concepts, facts, events, etc.) and its logical relationships (connections between ideas within the site, or links between pages on the site or between these and places) external to him on the internet).

Tourism: Mode of travel dominant within industrial societies and modernity. Types of tourism include *cultural tourism, digital cultural tourism, innovative tourism, mobile locative tourism, Tourism 3.0, viral risk tourism.*

Tourism 3.0: It is defined based on the following traits: greater interest by tourists in intangible heritage; overcoming the dichotomy between high culture and popular culture, a process for example witnessed by the opening of tourists to public art on the street; hybridization between cultural production and consumption; desire for authentic experiences in the tourist trip; such a paradigm of modern tourism is revealed as one of the practical manifestations of *city 3.0*, which often allows the use of Culture 3.0 within the public cyberspace of *Web 3.0*.

Tourism Communication: Communicative paradigm around *tourism* communication founded on three distinct modes of communication, but also hybridized in contemporary times: the pre-modern mode of communication in co-presence (face-to-face conversations, etc.); the mode of mass communication characteristic of modern societies (press, radio, television); and the digital communication mode associated with postmodernity (cyberspace, cybertime).

Viral Society: Society paradigm defined by societal processes never seen before, such as viral economies and technologies; viral policies and politicians; viral cultures and cults, viral risk tourism. One of the processes that seek to overcome the viral society is *social remobility*.

Viral Risk Tourism: This process is defined as a new mode of traveling that inherits some of the characteristics of *viral society*, risk society and network society. For example, due to forced confinement, potential tourists increasingly choose to make virtual travels in cyberspace and cybertime rather than physical travel. Another attribute of viral risk tourism is the development of *mobile locative tourism*, which can become as or more important than presential tourism.

War Tourism: Such phenomenon registers a striking actuality. Conflictual and political phenomena observed by war tourists, includes world wars, civil wars, battles, and battlefields themselves. This is a process that is frequent for many centuries. It balances between a sort of pulsion of death deployed by voyeurs, and a more positive connotation, in the form of projects to remember traumatic periods of war, in order to remember and correct mistakes that shouldn't have happened and be repeated again, and to construct or reconstruct lost or depleted identities. Some contemporary illustrations are a public exhibition, in Kiev, May 2022, displaying Russian tanks captured by the Ukrainian army; and the named 'House of Russian War Crimes', inaugurated at the 2022 World Economic Forum, at Davos.

Web 2.0 (or Social Web or Reading/Writing Internet): Type of digital social network that allows an active posture on the part of the user: in addition to reading the information, he can write content such as articles (posts) or comments on a blog and share personal and professional information on digital social networks, such as Facebook, Twitter, or YouTube.

Web 3.0 (or Social-Semantic Web): Paradigm of digital social networks that is based, among other discursive devices, on social-semantic sites. For example, *Wikipedia*, Freebase, public art communication sites.

Wikipedia: A social semantic *Web 3.0* site, in the form of a global digital glossary, where common citizens may define ideas and concepts, even if evaluated by a *Wikipedia* commission, dedicated to controlling some anti-democratic practices and writings, such as racist postures, fake news, among others.

Chapter 2
Tourism as Experience and Desire for Territory and Traveling

Moisés de Lemos Martins
University of Minho, Portugal

ABSTRACT

The approach the author takes on tourism is not related to economic analysis, nor even a political analysis stricto sensu. He configures tourism as a cultural activity. The perspective developed on tourism is here phenomenological and socio-anthropological. Hence, the author does not dwell on business (nec/otium – not/leisure; activity, work, commerce) but mostly on leisure (otium – time free from activity). First, he points out that tourism reflects our current obsession with one territory and traveling in one territory. The second element developed in this chapter is the experience of traveling, which is always a possibility of an encounter – an encounter with the other and an encounter with nature.

A DESIRE FOR TERRITORY

First, I want to point out that tourism reflects our current obsession with one territory and traveling in one territory [1]. I say obsession because it is a movement that reflects strategies of returning to the territory, to a physical and spiritual territory, in the search for possibilities of encountering the other, when the current territories offer us, more and more, landscapes, atmospheres, and electronic environments. Our life is now a circumnavigation of digital and virtual territories (Maffesoli & Martins, 2011; Martins, 2012, 2017).

I shall begin by depicting tourism as a human strategy to return to the territory and travel in a territory (physical or spiritual), searching for a response to the acceleration of our time and its electronic mobilization. Such a response is made of hybrid strategies, sometimes ambiguous and always complex, as hybrid ambiguous and complex is the human in the technological crossroads it is summoned to make. Indeed, today one cannot dodge the debate about technique and the role the new technologies, including media, "have in redefining culture, that is, in delimiting the human" (Martins, 2010a, p. 84). Because

DOI: 10.4018/978-1-6684-3369-0.ch002

this is our time and we have no other time to live but this, that of the technological mobilization of the human, we need to be up to these circumstances (Martins, 2010b, 2011a, 2012).

Today, when we ask someone where they live, the most common answer is now given in time: 10m/15m away from their job by car; or half an hour away from the university by bus. A place measured in time is no longer essentially a territory, a space, a place where one lives. However, I think that the current tourism frenzy unveils a somewhat unhappy awareness, a certain uneasiness (Martins, 2015). A territory is now a place where one travels by GPS, with space being measured by coordinates. The experience we have of the territory is increasingly electronic. The territories' promotional films are electronic; web portals give us a glimpse of virtual landscapes, very detached from actual landscapes; we search for tours and travel destinations on tourism agencies' web portals; we buy tickets to tourism destinations on the internet.

There is no doubt that the electronic experience expands our ability to fictionalize, tell stories, and make narratives using resources from various media, including various technical formats and languages: photographs, picture postcards, selfies, videos, films, and other media (Andrade & Caeiro, 2020; Pinto-Coelho & Pinto, 2018; Martins, 2020).

Yet, there is something melancholic about this journey, which is largely digital and virtual, with electronic landscapes, atmospheres, and environments (Martins, 2011b, 2017). Traveling through websites, portals, socio-technical networks, digital repositories, and virtual museums is a somewhat melancholic experience because we enjoy images of them more than enjoying the things themselves; in selfies on Facebook or Instagram, we consume images that exhaust there.

Some call this electronic experience, simultaneously digital and virtual, "abandonment of territory", deterritorialization. For in the technological acceleration and mobilization, we no longer have a solid and secure foundation, a known and controlled territory, and a stable and guaranteed identity.

THE EXPERIENCE OF TRAVELING

The second element I will develop in this essay is the experience of traveling, which is always a possibility of an encounter — an encounter with the other and an encounter with nature [2]. Nevertheless, electronic traveling, encountering the other and nature, does not include the body of the other nor the roughness of nature. It includes mostly fictions of the other, simulacra, effervescences, and euphoria of the other. It includes the spectacle of the other — which, according to Guy Debord (1991), are nothing but "guardians of the sleep of reason" (p. 16, i. e., the reason of the other: of its difference, diversity, and authenticity) [3].

I would say that tourism is an art of connections, for we know from Plato's (1986) *Symposium* that the art of connections is erotica. Therefore, it is not untrue to say that tourism is erotica, that it has sex appeal, as Mario Perniola (2004) says of technical objects. Like their peoples and cultures, territories are experienced and lived by the tourist. As an experience, tourism today involves all aspects of the tourist's day-to-day existence in the place they travel: shopping, eating, drinking, and daytime and nighttime entertainment. As much as experiencing the world's monuments, the tourist also becomes part of the tourism product he or she experiences.

Culture is an ever-increasing source of experiences and possibilities in its broad, anthropological sense. Hence, we can say that all tourism is cultural. Tourism today is a "total social phenomenon" in the exact terms anthropologist Marcel Mauss (1966) characterized the phenomena that totalize existence.

In fact, today, tourism involves a tourist's whole life. We can imagine him or her going to the beach in the morning, a museum in the afternoon, and walking around the city's old town in the evening. It is quite likely that a tourist's first encounter with culture is a restaurant menu and the waiter and the food in that restaurant near a resort. The tourist's motivation is to learn, discover, experience, and consume the cultural attractions, tangible and intangible, at a tourism destination (Piñeiro-Naval & Sierra, 2018, pp. 10–16).

Moreover, these attractions, or products, refer to material, intellectual, spiritual, and emotional characteristics distinctive of a society. They include arts and architecture, historical and cultural heritage, culinary heritage, literature, music, creative industries, living cultures, value systems, beliefs, and traditions (Pinto-Coelho & Pinto, 2018). Here we have tourism as a total social phenomenon because it involves an individual's life and affects it in all its aspects.

TOURISM, THE MASSES, AND MACHINES

Tourism is a fruit of the civilization of "masses and machines", in the exact terms Ernst Jünger (1990, p. 108) characterized the 19th and 20th centuries. However, tourism being an event of modernity, it is fair to keep in mind the observation of the Mozambican sociologist, Elísio Macamo (2005), that Mozambique has ambiguously stepped into modernity since it did so through colonization.

As a mass phenomenon, tourism emerged in the mid-19th century, with industrial society and the distinction between working time and leisure time, and developed mainly after World War II, with the affirmation of the welfare state.

At first, it established itself around three "S": sand, sun, and sea. One might say that tourism is the experience the masses have of traveling. It is a mobilization through machines, from choosing destinations on the internet to booking accommodation and a flight online, with a travel agency, to remote, exotic, and archaic destinations, increasingly experienced as unique events: singular, different, and authentic. Finally, such experiences, which Pine and Gilmore (1998) call "memorable" (p. 97), will be immediately echoed and perpetuated as "eternal instants" through photographs, selfies, panoramic images, and videos. Quite significant and very appropriate, however, is the fact that Michel Maffesoli's book *L'Instant Éternel* (The eternal instant), published in 2000, has the subtitle "the return of the tragic in post-modern societies" because the tourism experience is, without a doubt, a somewhat melancholic experience. However, I will only develop this further in the next section.

THE DOOR AND THE BRIDGE

Here I will analyze tourism by drawing on two exemplary social forms that Georg Simmel (1988) identifies as the makers of sociality: the "door" and the "bridge" (pp. 161–168). The "door" allows the exit to meet the other. One must leave oneself, and one's home to encounter the other. The *polis*, the city of men, derives from this encounter. The collective good, meaning the idea of community as a totality, incorporating the margins (minorities, migrants, exiles, refugees, the frail, the impaired) springs from this movement that the "door" symbolizes as an altruistic, cooperative, and sympathetic movement (Martins, Marques, Ramos, 2020; Martins, 2019a).

The "bridge" is thus a resource of sociality, a mediation, where the path is blocked and dead-ended. It is a resource that establishes the flow between people and maintains the utopia of a community. Today, tourism assumes this condition: it allows circulation out of doors. As a bridge, it unblocks blocked paths and closed horizons. In the global culture, opened by information and communication technologies, flows predominate, including flows of people, "tourists, immigrants, refugees, exiles, visiting professionals, and other groups and individuals on the move" (Piñeiro-Naval & Serra, 2018, p. 10).

However, the way tourism makes itself a "door" and a "bridge," and also an "ethnic landscape" (Apadurai, 1996, p. 33), has, in my view, an illness. I do not see it opening to a sense of community because it does not have the *polis* as a horizon, it does not project a utopia, nor does it constitute a collective promise. Tourism does not exist for the sake of the community. Tourism materializes (consummates) in consumption. Hence its tragedy: it is an "eternal instant", in Maffesoli's (2000) expression, but it is not the configuration of human eternity, which only the promise of a community horizon could provide. The Argentine poet Jorge Luís Borges (1988) reminds us in *The Unending Gift*: "there is something immortal in the promise; in it, something lives forever" (p. 364). The *polis* demands citizenship, and tourism, by and large, is contrary to this demand.

Tourism is most often multitudes, masses, accelerated and mobilized for consumption by powerful transportation machines (by land, sea, and air: airplanes, cruises, trains, automobiles, among others), and powerful industries of hospitality, traveling, and vacationing.

I will, however, next refer to the possibility of community in tourism. It will be the object of the next and last point, where I articulate tourism activity with heritage.

TOURISM AND HERITAGE

In this last point, my purpose is to question the tourism strategy of returning to the territory through heritage, a strategy in which the tourism industry increasingly invests. The relationship between territory and heritage is evident, at least in three dimensions: heritage is by nature spatial; contemporary economic and cultural geography focuses on it; society takes it as an economic and cultural asset.

Let us now characterize heritage. Heritage encompasses a mesh of tangible and intangible assets, a mesh of material and tangible domain, natural landscapes and edified structures, sites, and monuments, and an intangible dimension, of oral traditions and social practices (Ashworth et al., 2007).

Using the past, or the practices of patrimonialization, derives from a selective social process of construction (or activation) and regulation, according to a scale of values and meanings, with material consequences (Carvalho & Fernandes, 2012). The current process of acceleration and technological mobilization of the human for the digital and the virtual, with the consequent loss of the secure foundation, the known territory, and the stable identity, to which I have already alluded, has the effect of a fierce reaction to preserve memories and create identities. I call this movement a new strategy of returning to the territory, a new strategy with new territorialization practices. In a fierce and desperate strategy, which gives vent to this obsessive desire to return to the territory, heritage comprises gardens, literature, industry, railroads, military and industrial architecture, gastronomy, and even comprises spaces of seclusion, suffering, tragedy, or deprivation (Coutinho et al., 2018).

CLOSING REMARKS

As I have underlined, territories are doors and bridges to the encounter with the other; they are possibilities of encounter. They are meeting points. Being meeting points, they represent possibilities for constructing the *polis* and possibilities for community construction. There is also the other in heritage, whether physical space heritage (tangible, material asset) or the mental space heritage (intangible, immaterial asset). That is why we can talk about encounters with heritage, which are, however, increasingly, encounters mediated by digital resources. Citizens use digital media to buy and sell heritage in online auctions and learn the history of the objects' places. It is a fact, for many people, and especially for young people, digital media are now the bridge and even the door to heritage experiences.

REFERENCES

Agamben, G. (2000). *Enfance et histoire* [Infancy and history]. Payot & Rivages. (Original work published 1978)

Andrade, P., & Caeiro, M. (Eds.). (2020). Public Art for Tourist Communication. *Lusophone Journal of Cultural Studies, 7*(1). Available at: https://rlec.pt/index.php/rlec/issue/view/128/fullissue7.1EN

Apadurai, A. (1996). *Modernity at large: Cultural dimensions of globalization.* University of Minnesota Press.

Ashworth, G., Graham, B., & Tunbridge, J. (2007). *Pluralising pasts: Heritage, identity and space in multicultural societies.* Pluto Press.

Baudrillard, J. (1981). *Simulacres et simulation* [Simulacra and simulation]. Galilée.

Benjamin, W. (1992). *Sobre arte, técnica, linguagem e política* [On art, technique, language, and politics]. Relógio d'Água.

Borges, J. L. (1988). The unending gift. In Elogio da sombra. Obras Completas (1952- 1972). II. Teorema. (Original work published 1966)

Carvalho, P., & Fernandes, J. L. (Eds.). (2012). *Património cultural e paisagístico. Políticas, intervenções e representações* [Cultural and landscape heritage. Policies, interventions and representations]. Imprensa da Universidade de Coimbra. doi:10.14195/978-989-26-0585-2

Celan, P. (1996). O Meridiano [The Meridian]. In *Arte Poética. O Meridiano e outros Textos* (pp. 42–64). Colibri. (Original work published 1971)

Coutinho, B., Baptista, M. M., Martins, M. L., & Rebelo, J. (2018). Portugal, land of tourism: Dissonances and touristic uses of the Estado Novo's heritage. *Lusophone Journal of Cultural Studies, 5*(2), 213–251. doi:10.21814/rlec.347

Debord, G. (1991). *A sociedade do espectáculo* [The society of the spectacle]. Mobilis in Mobile. (Original work published 1967)

Deleuze, G., & Guattari, F. (1972). *L'anti-oedipe* [Anti-oedipus]. Éditions de Minuit.

Deleuze, G., & Guattari, F. (1980). *Mille plateaux* [A thousand plateaus]. Editions de Minuit.

Deleuze, G., & Guattari, F. (1991). *Qu'est-ce que la philosophie?* [What is philosophy?]. Minuit.

Guimarães Rosa, J. (2011). *Grande sertão – Veredas* [The devil to pay in the backlands]. Editora Nova Fronteira.

Jünger, E. (1990). La mobilisation totale [Total Mobilization]. In *L'Etat Universel – suivi de La mobilisation totale*. Gallimard. (Original work published 1930)

Macamo, E. (Ed.). (2005). *Negotiating modernity: Africa's experience*. Zed Books Ltd.

Maffesoli, M. (2000). *L'instant éternel. Le retour du tragique dans les sociétés postmodernes* [The eternal instant. The return of the tragic in postmodern societies]. Denoël.

Maffesoli, M., & Martins, M. L. (2011). L'imaginaire des médias [The imaginary of the media]. *Sociétés, 111*(1), 5–9. doi:10.3917oc.111.0005

Martins, M. L. (2002). De animais da promessa a animais em sofrimento de finalidade [From animals of promise to animals in suffering for purpose]. *O Escritor, 18/20*, 351–354. Available at: http://hdl.handle.net/1822/1676

Martins, M. L. (2003). O quotidiano e os *media* [Everyday life and the media]. *Todas as Letras, 5*, 97–105. Available at: http://hdl.handle.net/1822/1064

Martins, M. L. (2007). Nota introdutória. A época e as suas ideias [Introductory note. The time and its ideas]. *Comunicação e Sociedade*, (12), 5–7. Available at: http://hdl.handle.net/1822/24115

Martins, M. L. (2009). Ce que peuvent les images. Trajet de l'un au multiple [What images can do. Journey from the one to the many]. *Les Cahiers Internationaux de l'Imaginaire, 1*, 158–162. Available at: http://hdl.handle.net/1822/24245

Martins, M. L. (2010a). Linguagem, verdade e conhecimento. As ciências da comunicação e o contemporâneo [Language, truth and knowledge. Communication Sciences and the contemporary]. In A. S. Silva, J. Cândido Martins, L. Magalhães, & M. Gonçalves (Eds.), *Comunicação, cognição e media* (vol. 1, pp. 76–86). Universidade Católica Portuguesa - Centro de Estudos Filosóficos e Humanísticos. Available at: http://hdl.handle.net/1822/24118

Martins, M. L. (2010b). A mobilização infinita numa sociedade de meios sem fins [The infinite mobilization in a society of means without ends]. In C. Álvares & M. Damásio (Eds.), *Teorias e práticas dos media. Situando o local no global*. Edições Lusófonas. Available at: http://hdl.handle.net/1822/24250

Martins, M. L. (2011a). *Crise no castelo da cultura. Das estrelas para os ecrãs* [Crisis in the castle of culture. From the stars to the screens]. Grácio Editor. Available at: http://hdl.handle.net/1822/29167

Martins, M. L. (2011b). Technologie et rêve d'humanité [Technology and the dream of humanity]. *Les Cahiers Européens de l'Imaginaire, 3*, 56–61. Available at: http://hdl.handle.net/1822/24245

Martins, M. L. (2012). Média digitais: Hibridez, interactividade, multimodalidade [Digital media: Hybridity, interactivity, multimodality]. *Revista de Comunicação e Linguagens, Lisboa, 43/44*, 49–60. doi:10.3917oc.111.0017

Martins, M. L. (2015). Cultural studies as the new humanities. *Lusophone Journal of Cultural Studies, 3*(1), 341–361. doi:10.21814/rlec.100

Martins, M. L. (2017). Prefácio: A circum-navegação de atmosferas e paisagens tecnológicas [Preface: The circumnavigation of technological atmospheres and landscapes]. In H. Pires, M. Curado, F. Ribeiro, & P. Andrade (Eds.), *Cibercultura: Circum-navegações em redes transculturais de conhecimento, arquivos e pensamento* (pp. 11–18). CECS. Available at http://hdl.handle.net/1822/49352

Martins, M. L. (2019a). A comunicação e a informação na cultura [Communication and information in culture]. *Maremagnum*, (23), 33–40. Available at: http://hdl.handle.net/1822/61843

Martins, M. L. (2019b). The "refugee crisis" in Europe –The "refugee crisis" in Europe – between totality and the infinite. *Comunicação e Sociedade*, (Special Issue), 37–52. doi:10.17231/comsoc.0(2019).3059

Martins, M. L. (2020). Tecnologia e literatura: As narrativas transmediáticas [Technology and literature: The transmedia storytelling]. *Letras de Hoje, 55*(1). doi:10.15448/1984-7726.2020.1.34786

Martins, M. L., Marques, M. A., & Ramos, R. (Eds.). (2020). Public discourses on migrants, refugees and minorities: the 21st century in debate. *Comunicação e Sociedade, 38*. https://revistacomsoc.pt/index.php/revistacomsoc/issue/view/154

Mauss, M. (1966). *The gift; forms and functions of exchange in archaic societies*. Cohen & West.

Perniola, M. (1993). *Do sentir* [From feeling]. Presença. (Original work published 1991)

Perniola, M. (2004). *O sex appeal do inorgânico* [Sex appeal of the inorganic]. Ariadne. (Original work published 1994).

Pine, B. J., & Gilmore, J. H. (1998). Welcome to the experience economy. *Harvard Business Review, 76*(4), 97–105. PMID:10181589

Piñeiro-Naval, V., & Serra, P. (2018). *Cultura, património e turismo na sociedade digital: Uma perspetiva ibérica* [Culture, heritage and tourism in the digital society: An Iberian perspective] (Vol. I). UBI – Labcom.

Pinto-Coelho, Z., & Pinto, R. (Eds.). (2018). Discourses and images of cultural tourism. *Lusophone Journal of Cultural Studies, 5*(2). https://rlec.pt/index.php/rlec/issue/view/99

Platão. (1986). *O banquete* [Symposium]. Europa-América.

Simmel, G. (1988). Pont et porte [Bridge and door]. In *La Tragédie de la Culture et autres essais* (pp. 161–168). Rivages.

Sloterdijk, P. (2000). La mobilisation infinite [The infinite mobilization]. Christian Bourgois Ed.

KEY TERMS AND DEFINITIONS

Cultural Heritage: Cultural heritage encompasses a mesh of tangible and intangible assets, a mesh of the material and tangible domain, natural landscapes and edified structures, sites, and monuments, and an intangible dimension, of oral traditions and social practices.

Cultural Tourism: A human strategy to return to the territory and travel in a territory (physical or spiritual), searching for a response to the acceleration of our time and its electronic mobilization. Cultural tourism is tourism as an art of connections and erotica. As much as experiencing the world's monuments, the tourist also becomes part of the tourism they experience.

Electronic Experience: The new technologies, including media, have a role in redefining culture, that is, in delimiting the human. The electronic experience expands our ability to fictionalize, tell stories, and make narratives using resources from various media, including different technical formats and languages: photographs, picture postcards, selfies, videos, films, and other media.

Melancholic Imaginary: Traveling through websites, portals, socio-technical networks, digital repositories, and virtual museums is a somewhat melancholy experience that expresses a tragic imaginary because we enjoy images of them more than enjoying the things themselves. In technological acceleration and mobilization, we no longer have a solid and secure foundation, a known and controlled territory, and a stable and guaranteed identity.

Territory: Territories are doors and bridges to the encounter with the other; they are possibilities of encounter; they are meeting points in the search for possibilities of encountering the other, the current territories, physical and spiritual territories, means, more and more, landscapes, atmospheres, and electronic environments. In fact, our life is now a circumnavigation of digital and virtual territories.

Tourism and Mass Culture: Tourism is the experience the masses have of traveling. It is today a mobilization through machines, from choosing destinations on the internet to booking accommodation and a flight online, with a travel agency, to remote, exotic, and archaic destinations, increasingly experienced as unique events: singular, different, and authentic.

Traveling Experience: Traveling is always a possibility of an encounter – an encounter with the other and an encounter with nature. In the global culture, opened by information and communication technologies, flows predominate, including flows of people, tourists, immigrants, refugees, exiles, visiting professionals, and other groups and individuals on the move. These travelers constantly update mobile cultures, developed through portable devices such as smartphones. Moreover, creative cities or *smart cities* are nothing but places and non-places of nomadic cultures.

ENDNOTES

1. The concepts of territorialization, deterritorialization, and reterritorialization were developed, primarily by Deleuze and Guattari (1972, 1980, 1991), first in *Anti-Oedipe* (Anti-Oedipus), and later in *Mille Plateaux* (A Thousand Plateaus) and *Qu'Est-Ce Que la Philosophie?* (What Is Philosophy?). My proposition here is that the current technological mobilization of the human, totally (Jünger, 1990) and infinitely (Sloterdijk (2000), for any given competition, any given market, and any given statistic, has led to a movement of supreme deterritorialization (Martins, 2007, 2010, 2019a; Martins, Marques, Ramos, 2020). The consequence is the loss of the three accents of time,

which Paul Celan (1971) argued as defining the human condition: the acute of the present time, the grave of historicity, and the circumflex, which is a sign of the expansion of time, of eternity.

2. We are all rehearsing the human, that is, rehearsing an idea of traveling. However, it is necessarily a troubled journey, full of dangers and obstacles to overcome. We are rehearsing traveling as wandering, enigma, labyrinth, and as roughness, viscosity, and uncertainty. We are all rehearsing traveling as doubt, although also as a memory of paths already traveled and experiences already lived. I borrow from João Guimarães Rosa, in *Grande Sertão - Veredas* (The Devil to Pay in the Backlands), the concept of "travessia" (crossing), which points to a particularly rough, uncertain, and difficult journey (Guimarães Rosa, 2011).

3 The notion of "crisis of the experience" originally had an equivalent in Benjamin (1992) in the concept of "crisis of the narrator", and today it seems unstoppable due to the technological acceleration of our time. Agamben (2000) speaks of the impossibility we find ourselves in today in appropriating our properly historical condition, making "our day-to-day existence unbearable" (p. 20). Whereas in characterizing the contemporary experience, Perniola (2004) introduces the concept of the "already felt" and questions the sex appeal of the inorganic, which is as fascinating as it is disturbing. As for Baudrillard (1981), we know his concept of the realization of the real as a simulacrum.

Chapter 3
Lethal Virality of Otherness: COVID-19, Tourism, and Travel

Philippe Joron
Université Paul-Valéry Montpellier 3, France

ABSTRACT

Sanitary issues and pandemics are in no way new in the history of our collective questioning concerning the organisation of the present and the arrangement of the future according to imperatives external to ourselves, the consequences of which we have been suffering ("black plague," "Spanish flu," "Ebola," "HIV AIDS," etc.). However, we are now taking for granted, with the appearance in 2019-2020 of COVID-19, the fact that we are no longer in a position to exist fully without external interference in the shared project of human fulfilment. The COVID-19 pandemic and its impacts on tourist mobilities can be analysed in the light of other mesological forms in a state of interdependence. Polemology uses the term "total wars," actual or potential. The recent invasion of Ukraine by Russia, with all its diplomatic, political, military, economic, migratory, humanitarian, and other consequences, is a tragic illustration of that. By terminological extension, it is now possible to speak of "total crises" or "pancrises."

INTRODUCTION

The study presented here in the form of an essay aims to question the essentiality of our socio-philosophical, political and economic inclinations to the generic elaboration of limits, separations, borders, walls, distinctions, waterproofing, etc., in regards to the sanitary treatment of Covid-19 and its implications on global tourist flows. This necessarily leads to the highlighting of a whole arsenal of pre-existing and collateral cultural dualities or dichotomies, which perpetuate our ambivalent relationship to otherness. The other can be an object of understanding, recognition, empathy, and even love, but he or she is also a target of rejection, misunderstanding, misrecognition, and even hatred, just as much as the subject who makes the other his or her identification medium.

In this case, if current news are in charge in the media, afterwards it is the humanities and social sciences that serve the cutlery and the contents of the meal, with a little analysis about the dishes served slipped under the napkin. The media are constantly scanning the news that feed them, always looking

DOI: 10.4018/978-1-6684-3369-0.ch003

for other nutrients that will suffer the same fate. Thus, the impacts of Covid-19 on tourism could prove to be relative if they were to be compared to those of the climatic, migratory, social and geostrategic emergencies that are shaking our contemporaneity. But these issues cannot be dissociated, because they structure us, because they shape our future.

On a methodological level, the use of “we” here is intended to be “wholehearted”, not exclusively academic but irrevocably social, without any compromise, close to a counter-hegemonic “we” ready for all kinds of arguments and reconciliations. This “we” is therefore equivalent that of the ordinary man, our substratum and our horizon, as a complement to the philosopher, i.e. the scientist who knows by convention and professional recognition. But the philosopher, through his literary elbowing and his mediatic rants, does not escape the worn-out scrapes of the bar counter that serves as a stage to him, just as the ordinary man cannot escape those of existential philosophy, which has a foothold in the spittoon of speech and invective.

This methodological positioning is an intellectual line of action that in no way demands the reader’s approval. It simply states the “I” open to research for continuity with what is not itself, or the purulent saturation of the knowing subject in the discovery of the object to be known, which feeds and echoes the interpellations of the first term. Thus, the “I” becomes “we”. A simple phenomenological posture that momentarily suspends the thesis of the natural attitude, since any position within or outside the brackets is worth an affirmation of the situation necessarily sought by the one who seeks to know (Husserl, 1950). Normally, according to what we have been used to, there can be no ego by evaporation or evanescence. It would have to be situated and dated (Monnerot, 1946). But the “defenestration” is emerging, open to all mediatic possibilities (Joron, 2011).

THE PARABLE OF THE SANDWICH AND OF TOTAL CRISIS

The issue of pandemic is in no way new in the history of our collective questionings concerning the organization of the present and the arrangement of the future. Nevertheless, the consequences of which we are suffering (‘Black Death’, ‘Spanish flu’, ‘Ebola’, ‘HIV AIDS’ etc.), we are now taking for granted, with the appearance in 2019-2020 of Covid-19 is to be linked to the fact that we are no longer in a position to exist fully in the shared project of our human fulfilment. In other words, we acknowledge the abyss before us, and acknowledge our damaged, perishable, fatally noxious situation in which we find ourselves.

George Simmel’s complex and powerful sociology, especially that of social forms and reciprocal interactions, teaches that the study of a phenomenon is always most effective and productive when it takes into consideration its relations to other formal wholes and their constituent totality (Joron, 2018). These relationships can be paradoxical, divergent, contradictory. This is also the cost of their conflicting harmony.

Thus, in a first overview of this health issue, the emergence of Covid-19 inevitably stems from a natural basis, i.e. the transmission of a viral disease from animals to humans: nature at work. A second overview suggests, through speculation, the result of unconfined laboratory manipulations: an interference of culture on nature, subject to all conspiracy theories. A third state of affairs reveals contamination on a human scale through the multiplicity of cultural exchanges and social mobilities: a globalist culture in action.

These forms of emergence and viral contamination, which are ultimately social, can be analyzed and understood in the light of other mesological forms in a state of interdependence. The language of polemology uses the expression ‘total war’, whether actual or potential. Russia’s invasion of Ukraine on the orders of President Vladimir Putin, with all the diplomatic, political, military, economic, migratory consequences that this entails at the global level, is a tragic example of the relevance of this concept of ‘total war’. By terminological extension, it is now possible to speak of “total crises”, of expansionism of crises in exponential additions of national and international layers that feed the disarray, anxiety and fear of the populations affected or potentially affected. A “pan-crisis” that wears out the desire to resolve each of them.

We are all familiar with Murphy’s Law, which based on the theory of the buttered sandwich, according to which the possibility of any system failing due to the weakness of at least one of its components inevitably leads to a probable if not certain failure of the whole. Failing to fall (first possibility), to turn over and land (second consecutive possibility) on the side of its ‘flap’ [1] or cover. There is also the possibility that a sandwich may remain in an initially stable state while suffering from a full overload that would have no choice but to spread out of it, through its overflows or vacuities of structure (i.e. through the fragmentary absences of the bread crumb or the famous holes in the racket which have nothing to do with the bread but which give a certain idea of the toast). It is a simple law, stated among others to give texture to the wholeness that allows society to hold together.

Laws, under the guise of physical or moral principles that are deemed immovable but nevertheless normative and therefore subject to competition and variation, exist only because they help us to think about otherness (the relationship to the other, to nature, to culture and to objects) and to organize ourselves within that scope. For ourselves and against others if they are not ourselves. For them, if they agree to enter our zones of influence. For our benefit, in a nature that is being abused.

Human cultures are like that, especially those that occupy our contemporaneity. Our planetary toast, spherical and not flat, our open-air playground of failure and conquest, serves as a banquet for our insatiable hunger for what we do not know. It inexorably pays the price, as do the guests and inevitably all those who never get the leftovers.

FROM PANDEMIC TO ENDEMIC VIOLENCE

And all of this is concomitant with the growing awareness in Western countries of endogenous violence, which is inherent to the human soil that constitutes them, but which is nevertheless reactivated, re-ignited or potentiated, depending on the situation, by cultural, political, religious and social sources outside their borders. In November 2005, the media treatment of a French news item concerning the death of two teenagers electrocuted in an EDF transformer in the town of Clichy-sous-Bois in the Paris region was taken at fault (Joron, 2006).

This tragic episode resulted in the contamination of violence in the *banlieue* areas of several French cities. The French then discovered what they had already sensed during the 2002 presidential elections, namely: the endogenous status of violence, beyond any kind of supposed importation. Everyone was finally aware that he was a weapon in patience, in a position of security with a finger on the trigger in a state of impatience. In other words, at the very least, there could be no permeability and capillarity of so-called exogenous violence without an autochthonous nutritive substratum. France was its own violence. Because it could not be otherwise in any state. The French exception was rightly undermined here.

In his *Reflections on Violence,* Georges Sorel indicated in 1908 that according to *bourgeois* philosophy, “violence would be a remnant of barbarism and would be destined to disappear under the influence of the progress of enlightenment". Faced with *bourgeois* cowardice which, according to Sorel, hardly dared to use force, i.e., the repressive means of the state directed against any kind of workers' agitation, Georges Sorel considered that the proletariat had a historical card to play by mobilizing all the arsenal of its own violence to achieve its ends of class struggle for the destruction of social injustices. The parsimonious force of the bourgeois state versus the unleashed violence of the oppressed proletariat.

In 1935, in his analysis of the rise to power of Nazism in Germany and its ideological dissemination in Europe, the Marxist philosopher Ernst Bloch put forward the concepts of "non-contemporaneity" and "dissimultaneity" to explain the absence of inscription of scientific Marxist discourse in the field of the utopia constitutive of revolutions. This lack gave a space to the energy deployed by the theoreticians of National Socialism to occupy the void thus left vacant in the construction of new myths, the frustration of the German petty *bourgeoisie*, pauperised and stunted by the economic and political consequences of the Treaty of Versailles signed on 28 June 1919 at the end of the First World War. These notions of "non-contemporaneity" (with a sociological horizon) and "dissimultaneous consciousness" (with a psychological scope), insist on the incursion of barbaric values and practices from the past, deemed obsolete and archaic, into the bosom of modernity, into the contemporaneity of the present of the time (Joron, 2012, 23-32).

These analyses, situated and dated, remain relevant nowadays. They still serve as points of support, as a line of force for understanding these 'variable constants' that invade our historical present. It orients our questionings about our common origin and destiny, our obsessions with the anthropological 'throw' that it is up to us to organize as best we can with the nature that serves as our foundation.

In 1984, Patrick Tacussel wrote his first important sociological work, *L'attraction sociale*, as a fighting and resisting exaction against the ideological rigidities of the human and social sciences of the time, following the Sixties' consciousness of 'everything is possible'. On the question of the spatial use of human communities, which afterwards occupies the present of our problems concerning social mobilities linked to the Covid-19 pandemic, tourist concerns, terrorist outpourings, climatic, economic and political migratory flows, but also the recent invasion of Ukraine by Putin's Russia. Patrick Tacussel invoked the changes in historical speed which, without being reduced 'to revolutionary events alone, as is too often the case, indicate states of spatial non-contemporaneity'. He indicated the sustained, irrevocable coexistence of paradoxical, even contradictory forces or powers between desires for mobility ("spaces for"), concerns for withdrawal ("spaces against"), and the imaginary circumvolutions that digital technologies have now taken over.

THE GREAT DISPLACEMENT OF FEARS

In this, the climatic issue has long been a part of the above-mentioned points, and tragically responds to the absence of solutions concerning our postures, our convincing ideologies and our unfolding imaginations. Questions about our conquering humanity in the face of its biotope, which pays the price. As a diffracting mirror of this state of affairs, we are still serving the soup of our desires, paid all on credit, with water, vegetables, meat, condiments and energy that will soon be lacking in the destiny of those who already regret knowing that they are trailing behind events, suffering the burden of our inconsistencies and cowardice out of short-term interest.

We all know the impact of our mobility and trade on climate. Our responses are still lagging behind the scope of our knowledge. It is to be feared that they will remain frozen for a long time following the international stalemate caused by Putin's imperialist aims towards Ukraine. Dependence of Europe, and especially Germany, through two gas pipelines linking the latter to Russia via the Baltic Sea (Nord Steam). Germany, which had already decided to limit its nuclear production in order to invest in renewable and decarbonized energies, and which now sees its supply capacities limited to coal and Russian gas. The United States, self-sufficient and an exporter of oil and shale gas, are of course in favor of cutting off the Nord Stream pipeline.

However, following the Paris climate agreement adopted on 12 December 2015, signed and ratified the following year by the European Union and 191 other states, it was agreed that by 2050 we would limit the increase in the average temperature of the planet to 2°C, by acting on the reduction of greenhouse gas emissions produced by industry, services, agriculture, domestic life and mobility, and by mobilizing financial means to achieve this objective. This agreement followed the declaration of principles of the United Nations Framework Convention on Climate Change (UNFCCC) adopted at the 1992 Earth Summit in Rio de Janeiro.

The IPCC (Intergovernmental Panel on Climate Change), established in November 1988 at the request of the G7, issued the second part of its latest report (AR6) on 28 February 2022. It also indicates that these climate changes, attributable to human action, are causing and will accentuate involuntary migrations and displacements of populations who are simply trying to survive by fleeing their precarious lives that have become deadly.

In France, in the context of the presidential elections of April 2022, the recurrent issue of immigration is once again brought to the fore by the extreme right. It is accompanied by social concerns about purchasing power and existential survival, already expressed by the *Gilets Jaunes* episode. It is imposed in the face of climate emergency that struggle to affirm its political centrality.

With the benefit of a recomposition of the ideological landscape instigated by the rout of the Socialist Party and the UMP in 2017, Emmanuel Macron has invested a previously unlikely political center, concentrating a large part of the fragmented left and right, emancipated from conventional political bipolarism. Here too, even more than in the recent past, the question of otherness, including our relationship to Nature and the Other, is making a difference. These questions are not specific to the French context. They are sharable and subject to influence. They are inevitably global.

Since its emergence at the beginning of 2020, the Covid-19 pandemic has considerably reduced our material exchanges and physical mobility. The more or less drastic measures taken by states in terms of movement and containment, at the international and various national levels, have lowered our alert threshold concerning the infringements of freedom of movement and the resulting acceptance of social control, notably via the problematic of vaccination.

They have also instilled in our consciences minute alarms, which had been dormant until then, about our effective dependence on elsewhere, on relocations (masks, vaccines, tests, etc.), on touching flesh to flesh, kiss to kiss, hand to hand, nose to nose, shoulder to shoulder, mouth to mouth. We were asphyxiating ourselves for lack of otherness, withdrawn philosophically and carnally into ourselves, on the lookout for any suspicious movement.

While periods of lockdown invited to self-reflection, self-care and sharing from a distance (reading, sport, social networking, DIY, Netflix), they also invited us to dismantle ourselves or, more precisely, the exact excesses of which we are made: domestic violence, alcoholism and other expedients, societal anorexia, consumer bulimia, paranoid conspiracies, etc. We were no longer able to exchange directly with the Other, and therefore with ourselves, but only with intermediate pretexts, with surcease mediation; other-object, thing-object, etc. We were fully our own defenestration, not only mediatic but also intimate. The confinement thus provoked the extradition of ourselves into inner, violent and precluded regions, by multiple effects and added to egotism in perdition. We have become what we are through existential reduction.

Cynically, there is nothing tragic about it, because everything is. We are simply trying to search for ourselves, to sniff ourselves out, to find ourselves with picks, invective and other weapons of destination. If our nature spurs us on by means of supposedly appropriate cultural positioning, Nature, the one we are not master of despite our defeated illusions, takes offence in return. And some of us, whether states, peoples, communities or individuals, pay the price of that in the short term. In the medium and long term, we are all impacted, permanently, until the dawn of a hope that is overdue.

In other words, we try to live with ourselves, or at least we try to envisage forms of sociation, gentle or violent as the case may be, capable of making our differences cohabit with a world that we suppose is still shareable. In this sense, our animality comes into comfort with our humanity by default, since our humanism is made in excess without recognition of our intrinsic flaws, since the Other still remains a target for our interest in confrontation, since the Other in its 'non-explainable difference' (Georges Bataille) projects into us its similarity to ourselves (Joron, 2015).

In order to get rid of this other "ourselves", i.e., our own principial nature, we make it an object of manipulation, domination, submission, denigration, love by denial. This other, which intimately holds us in spite of ourselves, this "ourselves" constitutive of the whole, inexorably traces its path. Perhaps we should accompany it, but we have chosen to make our own way without it, since we consider it as our own strangeness to be taken down. This is our policy of comfort, out of self-proclaimed blindness. Of course, our mattresses and/or benches would not have confirmation of their sleeping functions without a few blankets or extra heaters... We definitely care about ourselves, even if it leads to our downfall.

But in another sense, converging with what has just been indicated, we are trying to exist, to circulate, to move, to feed, to drink, to heat, to cool, to express our humanistic infirmities supported by the ideological exoskeleton of our sharing or competition. We try to do so, even if it means undoing ourselves permanently.

We are nearly 8 billion people in a precarious situation of globalization. And we still place ourselves, through blindness, in a hushed relationship with the reality of difference. More precisely, we refuse to see what could still be involved in a possibly responsible relationship with regard to the confrontation of our intrinsic differences.

We have so much to do in the humanitarian lands that propel us, that sometimes contain us, and inevitably condition us.

And in this stagnation of which we are at the origin, we still have a chance. In particular in the attempt not to be too far from ourselves, in the close perspective of this Other who makes us, of these others who put us into action. But these humanist shades, in the European community and the French nation, filled with ideologies of open benevolence in embassy or consulate affairs, giving way to political dikes that try to limit or cancel out our similarities, go against what makes us or seems to make us perfect.

In short, we lose ourselves in the abyss of our otherness because we keep coming back to what made us or will make us, without taking into account the hammering of our present concerning our leaks and losses in terms of necessarily amalgamated differences. We are, in fact, in a state of *disexistence*.

TOURISTIC SYNDROMES OF HUMAN REPRESSION

The Covid-19 pandemic, however fearsome it may have been at the dawn of 2022, has radically changed our relationship to distance and temporality. Our semblance of spatio-temporal fluidity, propelled by our technological conquests, has momentarily given way to the radicality of space that situates us and of time that dates us. The digital distancecomfortable in interpersonal relations, has however become a poor choice in the organization of human affairs. How to understand without touching? How to deal without weighing? How to negotiate without touching? How to love without touching? How can we deal with the virtual presence of the Other, which makes the reality of the encounter, which espouses the physicality of relationships, absent? The technique of sensations is at work.

According to the teachings of Auguste Comte, humanity is more importantly composed of the dead rather than the living. Factually, he is not wrong. Our fleshy materials and our historically recorded memories attest to this. In the overdraft of its positivist radicalism, Comte's sociology showed a surprising relativism at the end of its life. Reading it anew in the present tense shatters here and there our monolithic certainties as to its disciplinary legacy. Thus, the dead govern us, even if we surreptitiously eliminate them from our discordant memories. Since with time everything goes away, or almost. Since the almost does the rest, in a pile of heterological residues that ensure our foundations of existence.

There is an observation that is permeated with evidence but also with unknowns. Like a piece of so-called "noble" meat which, by the fact of its ingestion, persists in our overhanging postures, undermining our programmed collapse. Fat attracts us, because it perfects our societal preserves. The dead are our flesh, itself swollen from the moment it is brought to life. Pandering. But the living are also accountable for a certain responsibility for the bill of said humanity, in circumstances of which they call themselves workers. To keep our history books open, perhaps, and to allow some to choose the style and use the capitals that seem advantageous to them. The writing of the humble cannot, however, stop at the overhang of the blotter.

Individual and collective impostures, at all levels, can still be subjected to critical, contradictory analysis. Necessarily combative. And in this sense, sociology still has a future, provided of course that it remains faithful to its initial postulates of openness to other disciplines in charge of the same objects as its own. If it is said to be a fighting combat, it is in its constituent purity both an expression and an observation of recurrent conflicts between the one and the many. The essential thing here is a matter of shared or complementary perspectives, at the very least not ignoring the recognition permitted by the exchange of the parties involved.

In a study devoted to the possible junctions between tourism and terrorism, relentlessly situated and dated (Joron, 2019), we showed the disarray of political thought that relied on the thesis of the Great Replacement of European peoples and traditions defended by Renaud Camus (a former member of the Socialist Party in the 1970s and 1980s) (Camus, 2011), publicized by extreme right-wing identarian groups and users of a cacophonous and all-encompassing complosphere. Fear of the Other, more precisely of the Islamic East, which would threaten the Christian West through demographics. Éric Zemmour, candidate of the political movement *Reconquête*! in the April 2022 presidential elections, has made himself the filial champion of this fear while claiming for himself and his ideas the putative heritage of Gaullist political thought, thus relegating the de-demonization of the *Rassemblement National* wanted by Marine Le Pen (of the former *Front National* created by her father) to "soft" proposals on immigration.

This thesis of the Great Replacement is broadly based on that of the Great Displacement of fears, erecting barriers, fences, barbed wire, walls, corridors, watchtowers, weapons, death, suffering and tears between peoples who depend respectively on their political, economic and cultural ways of life, but who nevertheless seek to transfigure them into global exchanges, no doubt out of necessity.

And so, we can see that, in a "pancrisis" that serves as a cover for our existential toast, in the folds of these questions added in successive layers that envelop our lives, the issue of tourism became essential, but at the same time relative. It is essential because low-cost mobility is being called into question, because mass tourism that was made accessible by the democratization of access has become questionable, and probably will be for a long time. We are already limit our desired paradises to artificial, virtual and digital parodies, however sophisticated they may be. It is also relative, because its libertarian imprecations are now fading in the face of the ecological and humanitarian depletions that occupy us just as durably. But in return, there is a hint of hope for the societal commitment of our human platforms in favor of the unfortunates and the destitute who knock on our doors.

In 2016, the number of tourists coming to France remained stable, making France the world's leading tourist destination, ahead of the United States and Spain, with almost 83 million international tourists compared with 85 million in 2015, a record year [2]. Secondly, in terms of international mobility and in particular air traffic, it appears that the number of passengers transported by air numbered around 4 billion in 2017, confirming continued growth (+6.3% in 2016) despite fears linked to terrorism and the threats that continue to weigh on the security of aircraft and airports [3]. Since 2020, throughout the Covid-19-episode, data showed a collapse in air traffic. According to information provided by the International Air Transport Association (IATA) on 30 September 2021, demand for freight transport fell by 10.6% in 2020 compared to 2019 [4]. Airlines carried 1.8 billion passengers in 2020 compared to 4.5 billion in 2019. A 60% drop in traffic, mainly international but also domestic.

As a result, these traffic limitations in terms of freight and passengers have significantly increased the cost of air transport, while decreasing the impact of international and domestic mobility on global greenhouse gas emissions.

DES-EXISTENCE

According to its Latin etymology, existence simply means the fact of being. Des-existence, which is not its opposite, claims this state, while including the problems of survival and overexistence, which seem to cause a collapse of being.

But nothing is less certain.

Survival implies a lack of living that can be compensated for by immeasurable effort on the part of those who suffer it, without them ever being sure of its beneficial effects. It is the work of those who have little or nothing to lose in material terms, even if they are forced to take out loans and over indebtedness. Morale, and then morality in disarray, can accompany the same movement.

Overexistence is inevitably material and financial, but also psychological, scenic and virtual. It consists in asserting one's desire for existence in the eyes of all, against all odds. It was the prerogative of stars (Morin, 1972). It is now our conquest.

In fact, survival and overexistence are not necessarily opposed. They can intersect, distort and complement each other. Thus, it is possible to be in a state of material and mental survival while living in a state of overexistence projected onto others. Everything becomes possible in the midst of the globalization of exchanges, the individuation of shareable affects, and the digitalization of duplicitous alterities.

Out of minimal necessity, described as 'survival' for some and 'sub-life' for many others, we have to deal with what is imposed on us (history, culture, injustice, the social, the other, the different or the identical), what we want for ourselves, what we may want for our accepted or foreign acolytes on the one hand, and our aspirations for change or immobility on the other. So, we have to "make do" with our own "half-assed", "musty-smelling" situation (Bloch, 1978, 17), while at the same time inventing other supports or salvation that can provide some comfort.

Here as elsewhere, words have a meaning, and are also defensive weapons, but may also become offensive by destination.

Thus, on the contrary, out of maximum necessity, we invent our possible super-existence from day to day, because it is our project of conquest over a whole that we reduce to very little, because we suppose that it perfects us. Proof of this is our dithyrambic logorrhea on social media that provide us with a surplus of existence (Susca, 2011).

This has a most effective impact on the layout of our strategic coordination points, our defensive trenches and our offensive charges concerning the apprehension of the elasticity of the present that makes us, of our fallible contemporaneity, of what we can still do for an all-encompassing, mesological "ourselves". This, despite multiple circumstantial retreats, still dares to envisage lines of flight towards exogenous lands to be conquered, according to our personal interests that are more or less compatible with those of our declared groups of belonging.

More or less, because the 'us' component seems to be unravelling in these times of diffracted communication, beyond political, ideological and moral appearances that serve as screens for the reflections of our mirrors. Has it never existed, even in the immemorial past, which allowed us to suppose a kind of original continuity with beings, nature, the world and everything else? The "we" component is like a decoy around which some admittedly praiseworthy wills (institutional and/or intimate, collective and/or individual) are waving, while the "I" or "ego" entity continues to assert its partnerships of circumstance for a change of world that it continues, however, not to wish for. This world is already no longer ours. It no longer belongs to us. While the "we" seems to no longer exist, because it no longer expresses itself in any way other than schizophrenia, because it no longer has any recognition in the contract of representative democracy. Because it no longer has a framework. Because he is somewhere else. Because it has tasted the pleasurable weather of political and communicational defenestration. But then, will the "we" component be satisfied with participatory democracy?

NARROW MINDS AND BROAD FRUSTRATIONS

Under the pressure of digital technologies, we collectively have taken for granted an ultimate change in the relationship to the other and consequently to ourselves. We could then exist at a distance, in the "here and now" of a finally accessible spatiality and a now controllable temporality, under programmed orientation, independently of our physical materiality and our voluntary understanding, on the digital platforms of otherness.

We were only concerned not to be too far away from digital commensality and virtual ordination, or else we would no longer really exist. We lived several lives, in multiple versions of ourselves, on the screen of possible elsewhere in windows of access, while remaining more or less in conformity with the classicity of our exchanges in situ, in this netherworld of concrete experiences to which we ventured our basic necessities. We thought we were no longer pegged to our mass discontinuity, we dreamed of our original continuity in the affirmation of our singularities. We were our own stars, our own models, our own desires to be different in an irremediable erasure of contingencies.

We were still experiencing a kind of digital compensation with regard to the delimitation of our physical, individual boundaries. And so we thought we were one once and for all, in total symbiosis with these others and these ensembles that make us up in organic frictions.

We did, however, have strings attached to our physical and verbal exchanges, for all intents and purposes.

But strings proved elastic: confinement, curfew, social distancing, on the one hand, and then Putin's invasion of a democratic Ukraine that says in return, through the heroic acts of its citizens, that it is forcing resistance in the face of a Russian imperialism that summons up the past in a bid to try to exist.

Politics in the past and communication cannulated in propaganda. Sociology of *desexistence* and living together by forceps. In this heap of political, ideological, economic, digital and social fractures, we must search for a possible elsewhere, we test the fragrances of the exogenous.

It is also a question of leadership, of "suitcase carriers", of "arm-ballers" without baggage who allow themselves to be reassuringly submissive, malleable to the invitation to take part in the contract that is submitted to them. As if they were not aware of what they are. But they do know what they are doing. For themselves. Because we also measure the scope of their actions and they detect the intensity of our gaze on them. Because they are seeing us, on the wall of our clumsy tags. We understand them, because we are made of the same wood, the same breaks, the same knots, the same molds eating away at the fiber.

Under these conditions, how can we still choose a leader when many agree to strike out the candidacy of one or the other? Why would anyone want a leader at all? Handholding, whether it is original or opportunistic, sincere, calculated or manipulated, is now subject to the tremors of clamping. Because it is our burden. Because precision, ephemeral in its attainment, is to be sought in the shavings, in the rejects, in the scrapings, in what remains. So there is no fatality without struggle, just as there can be no struggle without hope.

These are the questions of our time, which we almost forget are constantly being reactivated as time goes by and as situations change: old-fashioned questions with a new look, which wear out their trouser bottoms on the projection of a new concrete floor for our living together, even though it is already sunken and in the grip of time; outdated findings.

While everything that remained to be founded was already ordered by the inventive deformities of our capacities to adapt to the reproduction of the identical. From this point of view, Pierre Bourdieu was quite right, supported by his followers who were on the lookout. Like many others before him, with him and against him on these questions. In his time, he focused the attention of the credulous on a simulacrum of good feelings slumped over the sufferings of those who tried to live again and again in a fog of contradiction, domination and symbolic violence rightly denounced (Bourdieu, 1993). Bourdieusian sociology was in the air at the time. In the same way as Sartre's existentialism (Sartre, 1996) on the barricades of the sheepish invective which, in the aftermath of its wadded-up nights lulled by Soviet illusions, harangued the workers' and students' mobilizations of 1968.

On these questions, the adaptable positions of mood, affect and resentment (don't we say that the absence of ideology marks our era) remain obviously divided in the extreme between collapsology and transhumanism.

End of human times and techno-genetic redemption, they say. War discharge, we see. Around this sociologically unfathomable in-between, at the heart of which each singularity of testimony invents a universe of prospection, itself permeated by radicalities in credit of scoring, certain signatures can bear the imprint of our determination to sign the blank check of participatory *laissez-faire* for simple solutions, however multiple and adaptable to each. These solutions become indefensible when they are confined to personal interests that are in search of collective legitimacy.

Aesthetics of populism? Beauty of singular deformities? Or the masterly clarity of a whole nourished by a few obscurities that make a difference? Don't we have a few examples from the past to definitively say what manipulations are at work in the present? Democracy? Already fallible in its principle, letting its detractors say and do what they want, even those who hold it in high esteem and defend it until they get their chocolate medals.

A basic and fully employed solution to oil the mechanisms of survival: to remain alone all together (Joron, 2020, 15-27), with the only alternative being a determination to be without succumbing too much to the pressure of confinement that keeps us at a distance from each other, while throwing each of us onto his own unbearability revealed by the lethal existence of those who become destructible sparring partners for him. This is the same distance that binds us together, just like the tears or wounds whose fulfilment we actively accompany (Bataille, 1970) [5].

In today's world, the Bourdieu's and Sartre's, whether ideological, political, egotistical or nobly disinterested in their commitment, would probably have been mere virtual blasters in a dormant inferno. They would perhaps not have existed any more than the digital articulators of the "Yellow Vest" revolt. No less so than those who castigate the intellectual, media, political and technocratic "*bienpensants*" towards whom their demarcation turns out to be infinitesimal, to a length of material and symbolic capital that is just as microsomal. They do what they denounce. By drawing the attention of others to their chosen targets. They are those they despise and they pretend to support those they drape in their leniency. They afflict the Other, the bearer of all evils, in order to symbolically unload their cultural and material heritage, which they want to preserve and make grow in *trompe-l'oeil*. Their reflection is their only horizon.

With one exception, however. Those who point out, denounce and scorn, imposing their rantings beak to break through the media, benefit from the interested ear of the same intellectual, media, political and technocratic "*bienpensants*" mentioned above. A carbon copy. All on the prowl in the same hunt for recognition of which they are their own prey. Counter-mirror. All the others, all the 'ill-thinkers', all the reflexive roundabouts and confinements are also trying the same maneuvers. And probably rightly so. They arrogate to themselves the right to do so (Joron, 2020, 67-81).

Adaptable freedoms. Convertible sincerities. Truths of the moment.

And then the free expression of the determination to be, quite simply.

But there are still a few things that should always be remembered, even if it means making of exhaustion an effective partner. We are our own strangeness, with its neighboring opportunities and ruin. And we generate both benevolence and hatred with regard to our similarities in a state of possibly hierarchical differentiation.

Possibly. It still remains the label or mark of our freedom in considering and/or recognizing the other.

Violence. Behind the curtain or under the ornaments of the representations that govern us, which give us the illusion of a concern for existence that is good against all odds, there remains the feeling of a danger of annihilation.

Without otherness, there is no singularity. In return for this existential boomerang, without a singularity that can be explored from every angle and in every pose, there is no overall texture.

Knitting and machining of social links. This amounts to stitching the edges of the holes that fill us, to reinforcing the contours of the void that binds us. With a confident turn of the hand, we orchestrate our postural tremors, both physical and psychic, in the face of what could undo us.

Thus, we depend on others to pay the cost of a priceless otherness with ourselves. The quest for recognition in the comices, even if it means subjecting ourselves to the tithe puncture that obliges us to periodically pay a debt to which affiliation would give the right.

We then try to be more or less presentable, year in, year out, in the blurring of our errors and weaknesses, in the affirmation of some incredible qualities, the most fanciful of them, kindly distorted by their recipients. The choice is plethoric. So many concentric, sometimes triangular circles of influence, entirely open to their gradual narrowing of interest, whose contractual aporias are subject to the inevitability of "successive sincerities" (Maffesoli, 1990, 57) that show the urgency of the personal positions of their affianced in the sidelining of those of others.

These circles, whether fraternal, humanist, conspiratorial, secessionist, associative, political, media, trade unionist, religious, intellectual, corporatist or even cooperative, sometimes finding their consistency in a mixture of circumstances, can only take root if they are activated the actions of those motivated by devotion or simple calculation. They only have influence on those who fear their influence. This inevitably feeds the flow of their sources and tributaries. They make this known, in a catimini that says, writes, puts itself on stage and tries to impose its conditions as if nothing had happened. According to a discretionary gathering that plays the card of hyperspectacularity (Machado da Silva, 2013). Paradoxes. But all this is not without compromises. Gift and counter-gift. Accepting to receive in order to better give in return. Beautiful ideas, when misused by those who sell them and those who subscribe to them, give ideals a hard time. Humanism by default and humanity by excess.

But also revolts. All fraternity is a community sieved through which the residual tears of its own displays of humanist ecumenism drip, leaving only the bitter salt of simple tasks disguised as essential, ethical and virtuous principles to be tasted. In the labyrinthine impasse mentioned above, friendship must be able to find a semblance of a path. Nothing is less certain. But the power of hope can still make a difference.

Confined, very narrow passage. With a double meaning. That of "making do" with resignation, but also indignation, confrontation or even the exploration of the benefits that can be derived from such situations. To no longer be at the mercy of those who think the world or society according to their sole conception of vertical debourbage applied to others, but also of those people with little means (Sansot, 1993) mired in chimeras of materialistic and identificatory possession, who serve as folding seats for the desires of the former in terms of stakes. The little hands make the big gestures, unwillingly. The small stories of some make the captive capitals of others. While the complaints and whining occupy the terrain of being in its human configuration, with the only hopeful course being a feeling of surplus value. Thus done.

As we sometimes say when invited to use ancient locutions, "better to hold than to see coming". A philosophy of common sense following its own Ariadne's thread, trained to the vicissitudes of a life that knows the solidity of the fibers from which its moorings are made.

And so it goes. In the Simmelian accompaniment of the conflict (Simmel, 1992), of the relationship with the other (Simmel, 1979, 53-79), is placed here in a situation of sociological problematization. The "for" only has real social efficiency in the designation, denunciation and reprobation of an "against" to be reduced or to be destroyed. The neutralization or annihilation of the latter is thus a condition for the affirmation of the former. Conflict expresses a shared, sometimes tragic recognition. Without conflict, without friction between contradictory or opposing forces, there is no existence in relief. No love either. Without confrontation, at least symbolic, I erase the difference that faces me.

These observations, however colorful in their grey zoning, apply to all the social classes or individual positions of opportunity (submissions, allegiances, rebellions, annihilations) that make us reflect upon, if not bend over backwards to, our unstable situations.

REFERENCES

Balzac, H. D. (1965). Eugénie Grandet. Éditions GP – Presses de la Cité. (Original publication 1834)

Bataille, G. (1970). *Œuvres Complètes* (Vol. 2). Gallimard.

Bloch, E. (1978). Héritage de ce temps (J. Lacoste, Trans.). Payot.

Bourdieu, P. (1993). *La misère du monde*. Seuil.

Camus, R. (2011). *Le Grand Remplacement*. Éditions David Reinharc.

Husserl, E. (1950). Idées directrices pour une phénoménologie (P. Ricoeur, Trans.). Gallimard. (Original publication 1913)

IATA. (2022). https://www.iata.org/contentassets/8ba31bb109324ae3973bb8b6560829be/2021-09-30-02-fr.pdf

Joron, P. (2006). La violence sacrificielle. *Les Cahiers de l'IRSA*, 245-261.

Joron, P. (2011). La sudation du quotidien ou les pores du réel médiatique. *Sociétés, 114*, 53-61.

Joron, P. (2012). La prime à l'émotionnel. In M. Maffesoli & B. Perrier (Eds.), *L'homme postmoderne* (pp. 23–32). Ed. François Bourin.

Joron, P. (2015). Comment la sociologie est-elle possible? *Sociétés, 127*, 13-24.

Joron, P. (2018). Georg Simmel et la sociologie du futile. Dans les anfractuosités du social et de l'intime. *Episteme*, *20*, 46–67.

Joron, P. (2019). Tourisme et terrorisme. Grand Tour et petites mains de la Terreur. *Sociétés*, *143*(143), 7–17. doi:10.3917oc.143.0007

Joron, P. (2020). Les réfléchissants. Gilets-Jaunes, ronds-points et nulle part ailleurs. In *Sociétés, 150* (pp. 67–81). De Boeck.

Joron, P. (2020). Sós todos juntos. Pele digital e fissuras digitais. In C. Freitas, J. Machado da Silva, & P. Joron (Eds.), *Laço social e tecnologia em tempos extremos Imaginário, Redes e Pandemia* (pp. 15–27). Sulina.

Machado da Silva, J. (2013). *A sociedade midíocre: passagem ao hiperspectacular.* Sulina.

Maffesoli, M. (1990). *Au creux des apparences. Pour une éthique de l'esthétique*. Plon.

Monnerot, J. (1946). *Les faits sociaux ne sont pas des choses*. Éditions de Minuit.

Morin, E. (1972). *Les stars*. Seuil.

Planetoscope. (2017a). https://www.planetoscope.com/tourisme/415-nombre-de-touristes-en-france.html

Planetoscope. (2017b). https://www.planetoscope.com/Avion/1837-passagers-transportes-par-avion-dans-lemonde.;html

Sansot, P. (1993). *Les gens de peu*. PUF.

Sartre, J. P. (1996). *L'existentialisme est un humanisme*. Gallimard. (Original work published 1946)

Simmel, G. (1992). *Le conflit*. Circé.

Simmel, G. (2019). Digression sur l'étranger. L'École de Chicago, naissance de l'écologie urbaine, 53-77. (Original publication 1908)

Sorel, G. (1990). Réflexions sur la violence. Seuil. (Original publication 1908)

Susca, V. (2011). *Joie tragique. Les formes élémentaires de la vie numériques*. CNRS Éditions.

Tacussel, P. (1984). *L'attraction sociale*. Méridiens Klincsieck.

KEY TERMS AND DEFINITIONS

Climate Change: Changes in ecosystems caused by both natural cycles and the cultural, ideological, and behavioural impacts of human agitation in a full supremacist expression towards nature.

COVID-19: An infectious and contagious disease caused by a coronavirus identified in 2019, leading to a global pandemic by early 2020. Measures taken to contain or limit the pandemic: urgent techno-pharmaceutical development of vaccines deemed effective; worldwide dissemination and use of these vaccines (with significant disparities); containment or restrictions on movement of populations at local, national, and international levels.

Des-Existence: A process of disengagement from the fact of existing or having a palpable and acceptable reality on the ontological and material level, provoked, and maintained by the multiple supports of digital super-existence.

Humanism by Default: Humanism is a set of philosophies or intellectual movements that find their full expression in the modern period of history, which places man and his values above all other political, economic, racial, and traditional considerations. Humanism aims at emancipation from animality, bestiality and savagery. The concept of "humanism by default" reflects this desire for the ideological and moral non-affiliation of humans with the status of animality.

Humanity by Excess: Humanity is defined as the set of human beings circumscribed according to supposedly superior specist characteristics (morality, intelligence, altruism, etc.) distinguishing them from other animal species. The concept of "humanity by excess" integrates the notions of animality, bestiality and savagery into the factual characteristics of humans, outside of any moral considerations.

Mesology: The science of *milieus*, the study of the relationship of living beings, and therefore of human beings, with their living spaces. The *milieu* is defined by what is distant from the periphery, from what is around, from the surroundings or from the extremities. It is therefore not synonymous with environment, but rather indicates what is specific to the mediation between a singular being and its natural and cultural soil.

Otherness: State or quality of what is other, of what is different or distinct, of what is in a situation of discontinuity. This state highlights mesological and phenomenological relations between the Ego and the Other, between the thinking/acting self and the rest to be thought/acted, necessarily implying reciprocal forms of alteration.

Total Crises: Breakdowns of equilibrium which were initially independent or dissociated in their management and which, under the impetus of the globalisation of material and virtual exchanges, have been added to each other, complementing each other and entering into a protocol of synergy which is difficult to control. The notion of "pancrisis" refers to this totality of crises, under reciprocal influence.

Tourism: The action of travelling, of travelling and visiting an "elsewhere" (cultural and/or natural) in response to the desire to experience difference and exoticism. Contemporary tourism has its origins in the Grand Tour undertaken by the aristocratic and bourgeois elites between various European holiday resorts or places of historical renown.

Violence: An act of aggression against the moral, psychological, cultural or physical integrity of the individual or collective Other. The practice and ethical reception of violence are considered to be normative, since they are variously appreciated (between condemnation and necessity) according to the cultural, historical, political or even legal modalities of human organisations.

ENDNOTES

1. Name given in certain provinces of France in the 19th century to the sweet or salty nutrients that were added to bread slices. Cf. Honoré de Balzac, *Eugénie Grandet*, Éditions GP – Presses de la Cité, Paris, 1965 [1834], p. 193.
2. [https://www.planetoscope.com/tourisme/415-nombre-de-touristes-en-france.html.] Consulted on 14/10/2017.
3. [https://www.planetoscope.com/Avion/1837-passagers-transportes-par-avion-dans-lemonde.;html.] Consulted on /10/2017.
4. [https://www.iata.org/contentassets/8ba31bb109324ae3973bb8b6560829be/2021-09-30-02-fr.pdf] Consulted on le 09/03/2022).
5. « Les êtres humains ne sont jamais unis entre eux que par des déchirures ou des blessures. »

Chapter 4
Total Tourism Experience and Viral Environment: Epistemology and Management of Tourism Destinations

João Tomaz Simões
https://orcid.org/0000-0003-3923-5555
Polytechnic Institute of Tomar, Portugal

Jorge Manuel Simões
https://orcid.org/0000-0002-7967-5216
Polytechnic Institute of Tomar, Portugal

Luís Mota Figueira
Polytechnic Institute of Tomar, Portugal

ABSTRACT

Tourism activities produce multiple stimuli in both the conception and enjoyment of travel. These travels, whose main purpose is to provide an experience to visitors, have not been properly studied, particularly in the post-pandemic era. There are a number of movements and schools of thought but not all aligned to one that is the total tourist experience. It is, therefore, that the authors propose in this chapter to raise questions that will lead to the production of ideas about this theme. These questions will be considered in the survey built throughout the investigation.

INTRODUCTION

Experience is for all People, a synthesis characteristic of their life, meaning both the actions and the resulting effects of the acts in which every single person is involved, by their own will or by other circumstances. Experience can be also a way for everyone to act individually or in a gregarious way, carrying out actions by using their senses applied in a certain context of place and time. If we consider

DOI: 10.4018/978-1-6684-3369-0.ch004

the term viral and relate it to the context of the health crisis and the context of tourism, we generate a scenario in which one and the other interact in a demanding way: that finding clear, fast and systemic communication becomes a case of scientific interest.

Travelling in a warm climate or eating in an icy, Nordic cultural dwelling always means using the senses and memory. Tourism planners rely on this human constancy. In this logic, each human being is the subject of his own reality, and of what he learns and apprehends by experiencing situations. The acquisition of knowledge through the practice of an activity which takes place naturally (in direct contact with Nature), at the mercy of the natural elements, or activated for scientific purposes (in direct contact with Culture), becoming a laboratory for experimentation, creates an explanatory hypothesis with experiential components. Experimentation and the awareness of knowing oneself as the bearer of knowledge composed of empirical culture, linked with theoretical culture, connects both in the experience of every human being. Productive activities aimed at raising the social and economic conditions of society put the human being to the test. The world that is experienced is the total world that we manage to consider, according to our possibilities, only conditioned by our limits. In that sense, the management of experimentation is fundamental to innovation in tourism, particularly in the thematic tourism segment.

In a hedonistic society, this is an unavoidable reality because human ambition is also total. Achieving a gratifying experience, as total as possible, is the consumer's goal in relation to the objects he acquires. The management of experiences in tourism takes a particular position and the mnemonic that we present, in the ETT (*Experiência Turística Total* - Total Tourism Experience) form, as we will explain, can configure a part of the response that has to be given for the tourism revival in times of health crisis. Thus, the present chapter (which configures the first part of a series of three future publications) intends to present an exploratory study that starts from an empirical approach, with the objective of establishing a new technical-scientific concept (ETT). It focuses on the evolution of the paradigm of consumption in tourism, accentuated in recent years, highlighting for this, the relationship of the main developers and managers of tourism, with the viral environment and the experience that is perceived by the tourist.

BACKGROUND

Tourism Experience as a Comprehensive Basis

Because it can generate both human and economic flows, based on fulfilling needs and desires of various kinds, the very condition of tourism requires material and human resources, tangible and intangible, so that the tourist can consume them. This consumption, which is very specific because it occurs at the same time as it is produced, is one of the unique characteristics of tourism. The sharing of acronyms and abbreviations is part of this system that links the production chain in the "Demand-Offer-Geographies-Operators" relationship. In the professional lexicon, speed of communication is a fundamental particularity as a central success factor in tourism: quality in the provision of services to the customer and in the exchange of goods between suppliers also depends on the communicational clarity of the system.

Experience in the field of tourism is, and always will be, highly sensitive. The time sought for relaxation and leisure means that those who adopt this behavior seek relaxation trough several activities or just pure idle, in other words, the most detachment possible from social and professional obligations sustained by the search for greater serenity and relaxation about daily life. We often draw, personally, webs of subterfuge and try to get away from possible problems arising from tiredness or mere evasion of the so-called "real life", seeking the tourism escapism that, as we know, sells dreams, (Hall, 2008). Seeking planned detachment is a way of settling accounts with tiredness and with the anxiety that dominates us when energies need to be renewed, (Netto & Gaetra, 2010; Oine & Gilmore, 2011).

In this sense, the freedom that we try to stick to the "touristic plan" thought at a personal level, in family or in a group of friends, works as a proactive space to accommodate the experiences that we wish to have or that, in the reality of real, sensitive life, we are meant to experience, (Amorim, et al., 2015). From the point of view of the structuring of tourism, which is necessarily based on a journey, on a displacement (theoretical, in preparation and in execution) we experience, respectively, feelings and physicality's.

It is debatable to assume the moment at which we can say that the tourist trip began (figure 1), but in terms of its organization and design, we know that there is consumption before its execution, and that this will even be part of the Tourism Satellite Account. In this logic, the excitement of future memory and its combination with the travel experience, as ballast of the life of every human being, is a motivating factor that determines that mental moment that "initiates" the trip. Thus, it is natural that each journey is determined by a before, a during and an after.

In Tourism and both upstream, in its preparation, and downstream, in its execution, we find multiple realities, agglutinated in the concept of experience, (Goeldner & Ritchie, 2012; Hall, 2008; Landry, 2010). This concept, which may be considered theoretically as scientific exploration based on the configuration of a hypothesis (to be experienced in practice), may also be situated, in objective practice, on a plan of knowledge about a specific and singular reality obtained through the senses and their activation process and memorial register.

Figure 1. Total Tourism Experience Environment (2022)
Source: Authors (2022)

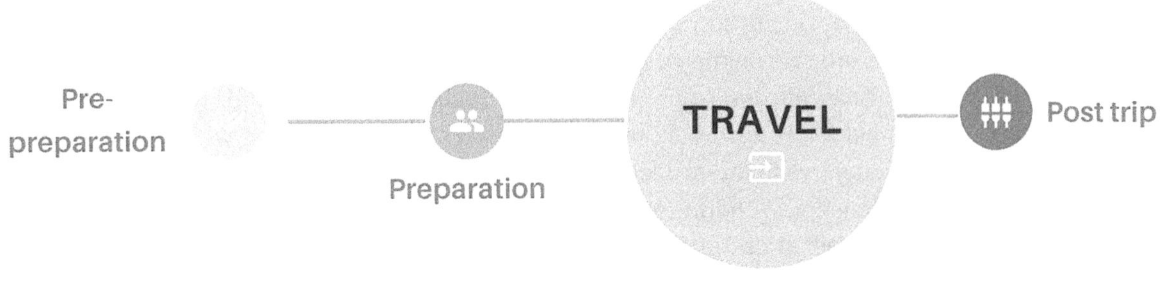

Tourism lives from the activation of resources that serve it and from consumption as well as from the memories it arouses. In this activation is a historical path which, in an epistemological analysis, is rooted in the oldest civilization, (Pearce, 2005). From cycle to cycle of civilization, from the Paleolithic and Neolithic periods to pre-classical societies and classical civilizations, especially Greece, with its Olympic activity, and Rome, with its imperial capacity to tear up the seas, but also marked by road building activity, the path of travel and its motivations was traced (Ramos and Costa, 2017).

If, perhaps, the Silk Road is one of the most profound witnesses of historical mobility and the pilgrimage routes of the Middle Ages show that seas, rivers and roads are natural and cultural elements to be appropriated by travel flows not necessarily motivated by warrior or commercial needs, it becomes evident that, in the Renaissance, the range of travel needs was amplified. The epistemological cut between the Middle Ages (and its theocentrism) and the Modern Age (and its anthropocentrism) explains much of the history of tourism, (Boyer, 2002; Reisinger, 2009; Ignarra, 2003).

Diplomacy, war, trade, art, manufacturing, luxury, exoticism and many other words can be used in this epistemological approach to travel as the nerve centre of tourism. Travel means, as a human constancy, mobility and, in it, the acquisition of more knowledge. As Bourdieu (2021:60) states:

(...) the habitus, as the word indicates, is an acquired knowledge and also a having, a capital (of a transcendent subject in the idealist tradition) the habitus, the hexis, indicates the embodied, almost postural disposition - but rather that of an agent in action (...).

Therefore, the knowledge obtained by practising an activity in a certain domain of each human being's life (ideal and lived), depends on individual experience, transmissible to a collective. Man is trained to gain some habits. The knowledge has for base, the creation of some events placed as problems that inserted in experienced and evaluated situations, allow to verify some hypothesis of solution. In these hypotheses, the individual and untransferable locution of "I liked it" and "I didn't like it" are scaled according to a line that goes from very bad to very well. This experience means that each human being creates his or her own living habits and consumption habits. On the other hand, curiosity, as a habit, enhances the desire of every human being to try things outside consolidated habits, Cadilhe (2006), Bourdieu (2021). In this restricted sense, if, for example, we start a journey, we incite an experimental situation which, of displacement from our habitual residence to what we call the destination, is evaluated, and validated both by us and by those who provide us with this professionally organised experience. The perfect match between our personal performance and the performance that we are obliged to perform in a group is the basis of professional travel agencies based on the need to travel with higher quality, (Beni, 1990; Crouch, et al., 2006; Firmino, 2007; Soares, 2015).

If, at the destination, we decide to stay overnight after having eaten, we will certainly have the opportunity to appreciate this hitherto unknown place and, if there is an opportunity to entertain ourselves by visiting museums or historical centres, use the offers of night-time activities for fun and relaxation in a situation of pleasant but lively leisure, in the various places of entertainment with safety and with public health guarantees attentive to the resident and the traveller and, also, if we urgently need to buy painkillers, buy a pair of trainers, withdraw money, rent a car, etc. are promptly satisfied, we find ourselves in a certainly mature tourist destination. The routine of the trip, however, is not the same as the routine of the traveller. The travel routine, however, does not erase travel anxiety, as proven by some studies, (McKercher & du Cros, 2002; Getz, 2007; Pérez, 2009; Joseph, 2010).

Epistemology of Travel as a Matrix Element

The theory of tourism knowledge is recent because only after the 2nd World War the activity gained greater expression, through the efficiency of the most evolved means of transport generating a growing tourist economy until very recently have been interrupted by the health crisis, (Leiper, 1990; Tribe, 2011; Netto & Nechar, 2014).

It is impossible to study tourism in its enunciative polysemy without placing, in our view, four field points. What we call the basic tourism system, according to the acronym POGO, consolidates these four elements: Demand (which opens markets), Supply (which responds to markets), Geography (which maps markets) and Operators (which supply products to markets).

In this cloud, originated by human displacement, motivated by travel, are aligned the economic, social and scientific agents of which this activity is aggregator: in this same cloud move all complementary activities to the goods and services made available and marketed. In this field also move technological innovations that digitalization and artificial intelligence are evidence for all agents, (Ramos & Costa, 2017). The consumption trends of the tourism market represent the constant challenges for the entire ranks of professionals dedicated to this economic activity. The need for new knowledge, even by what represent certain conjunctures such as the sanitary crisis that we still experience, encourages the search for answers for the present and to serve the prospective vision that is placed as a permanent challenge.

Tourism knowledge, i.e., the epistemology of tourism, is established, as in many other economic activities at two basic levels: the vernacular and tacit domain and the academic and explicit level, (Jafari, 2005; Lopes, 2010). Practice and theory intertwine, and the production of critical thinking animates much of what is innovation and entrepreneurship in this economic sector, (Urry, 1995). In a marketing dimension that is intrinsically inherent to it, the tourism phenomenon is part economic component and is, in part, social component of great influence on Visitors (Barreto, 2004; Getz, 2007).

The "drag culture" that sticks to travel and its travellers, modifies the world, changes the world, shapes the world, (McKercher & du Cros, 2002). The protagonists know this. The travellers-researchers, still know it, eventually, in even greater detail due to the need to advance the management models appropriate to each historical cycle, (Cadilhe, 2006). The idea and the reality that place Tourism as one of the most transversal domains in the life of human beings clearly explain this concept of "drag culture" because the analyses of this area of transport aimed at leisure and recreation are the result of the multiple and interpenetrating approaches provided by many disciplines of scientific and technical knowledge. The literature demonstrates the paradigm of transversality as able to assimilate the new epistemological contributions of currents of analyses to the practices of travellers and their tourist markets integrated in the global market, UN (2020).

All travellers face for the first time (and throughout their tourist travels) the anxiety of travelling, the expectations of what is going to happen, the mental planning of the places they are going to visit and how they foreshadow their lengthy stay in each of them. The traveller's routine has the tendency to become more anxious, due to the exhaustion of the "novelty". Hence, in a search for new and more fulfilling experiences, innovation and entrepreneurship mean open doors for the creative, (Peixoto, 2000; Florida, 2002; FCT, 2019; UN-United Nations, 2020).

Many authors wonder about this issue of "novelties" because the tourism market space lives from this uneasiness about consolidated and mature products and new products made available year by year, (Sundbo & Darmer, 2008; Tribe, 2011; Sol et al., 2017). From the initiation level to the routine level, Time is indeed the fundamental element to consider as a constructive element of responsible touristification, (Reisinger, 2009; Cravidão, 2011). Within this type of study Brito's (2002, pp. 1-2) argument on this fundamental basis illustrates tourism as:

(...) in fact, a growing sector - we all were and still are, in some situations, privileged actors in the tourist activity - either as a host community receiving flows of travellers or as visitors from more or less distant places, in search of tropical landscapes, exoticism and traditional cultures, for periods of time, certainly, variable. Also, depending on the situation, we have already felt the (in)direct benefits of the activity as well as its fewer positive impacts, sometimes even degrading. On the other hand, we have

also contributed to the positive increase of the tourism sector in some regions, but we have certainly contributed to a (in)voluntary way to environmental and/or cultural devastation.

In a grid marked between advantages and disadvantages, we understand that the tourist activity is never neutral. Links are positioned that, in the value chain of tourist operations, namely cultural tourism, gain more solidity when the governance of the tourist destination incorporates a kind of "performative-touristic arts", i.e., when it is oriented, in compromise planning with the local administrations, to what is the core of the tourist demand of these more alternative segments to mass tourism. The ERT- regional entities of tourism and the tourism region of the continent, as well as the regional secretariats of the Azores and Madeira, represent this effort.

The marketing plans of these promotional bodies under the authority of Turismo de Portugal, I.P. reflect these institutionally felt needs to offer so-called "unique" experiences. This reality, not being the objective axis of this text, has an institutional and public dimension that should, however, be studied in another circumstance. Alternative tourism is necessarily much more demanding in terms of the creativity of promoters and professionals in the tourist industry (Brito, 2002).

In alternative tourism each of its protagonists, including researchers, tourism professionals, residents, and travellers (tourists and/or hikers) have a role determined by the technical systemic inherent to each segment. The behaviour codes of the actors positioned on the Demand side as well as those positioned on the Supply side are concerted in the social procedures according to the rituals which in each Geography are imposed on all the Operators.

Coopetition (competition and collaboration) is required as a state and public behaviour solidarized with business and private behaviour. Tourist destinations are territories that require, to function in the supply of goods and tourism services, commitments cemented by governance models that know how to define, design and execute. As shown in the history of tourism epistemological knowledge contributes to improve each model of management of the tourist destination. The epistemology of tourism demonstrates this issue, (Netto & Nechar, 2014).

If one knows what preceded us, we understand what happens to us and, most likely, we will have successes in the predictions that this knowledge gives us. If we put ourselves in the place of the cultural tourism operator, we observe that what most marks this field is not the heritage and culture (contrary to what the stereotype immediately judges) but, precisely, the impact that each revelation in travel has on the traveller. The familiarity of all tourism actors with mnemonics where the coexistence with acronyms, abbreviations and other symbols sharpens communication, qualifies dialogues and fits in the communicational trends of new generations, Castells (2007).

Travelling in comfort, sleeping peacefully, eating surprisingly and, surprisingly, being entertained quietly or noisily, feeling safe and confident about the health of each place of transit or stay and, above all, finding what you are looking for in one emergency among the many that can be listed is, let's face it, every traveller's dream. The search for novelty in a comfortable environment is what most of the traveller's desire. The Market knows this, anticipates this demand, combines possibilities of offering related products, makes dreams come true and lives from the same dreams.

Therefore, the success of the commercial exploitation and tourist economy of each destination or each tourist corridor depends on the experiences, "sought-promised-lived-measured-validated-recommended". In this cycle of experiences that happen in each plan, consequently: demand-oriented; based on promotion; experienced with a sensitive scale; scrutinised according to the levels of satisfaction; evaluated under validation processes; disseminated in communication axes under the seal of recommendation, a theoretical knowledge and a perception about the pleasure or displeasure of each type of experience thus observed is generated.

In the planning field, all this information is extremely valuable, whether for improving and organising processes, or the products, services or attractions consumed, or for serving new proposals that accompany changing markets. The standardization and quality of these processes would take us to the field of certification, which although fundamental in the development of sustainable destinations, is not the object of this work. However, this contribution means that standardization contributes to greater effectiveness, efficiency and economy, productive are part of this vision of ours.

In turn, planners and executives need to organise profitable productive processes. In Tourism, profitability is always associated to the levels of satisfaction that customers make expressive when they refer to it through research or by their own initiative in statements of criticism and support or of repairs to brands and their abilities to provide experiences to their customers. In this management dimension (fig.2), naturally, the achievement of what is considered as total is a legitimate ambition of both parties: the Demand that wants to maximise the cost-benefit relation under the perspective of the "good business"; the Supply that intends to maximise the direct and indirect effects of its clients' stay.

Figure 2. Dimensions of the Total Tourism Experience (2022)
Source: Authors (2022)

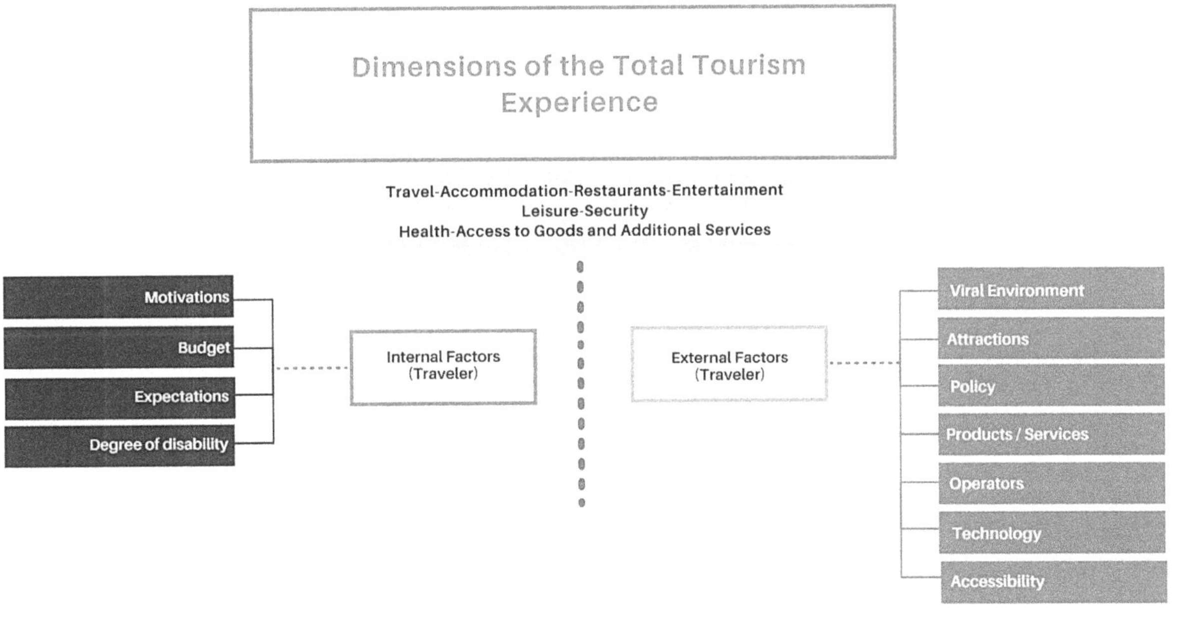

When considering that the Total Tourism Experience, being a mnemonic and acronym, concerning an operational reality of the tourism activity, is integrated, in the present world framework in the Viral environment, i.e. in a productive culture of tourist goods and services directly and indirectly dependent on the effects of the health crisis that is experienced globally, it is also assumed, logically, that the functional paradigm changes also mean new values and, therefore, new epistemologies. It is not by chance that Cadilhe (2006:12) says the following about his experience collected in the work "The moon can wait":

In this book appears only a small fraction of what I have seen, known and lived. A small fraction of what I have written and published. There are neither unpublished texts nor journeys yet to be made. It is, therefore, a personal anthology chosen from hundreds of my reports that appeared in the most varied titles of the Portuguese press.

Under this note that the Traveller attaches to his travel account it becomes clear that if epistemology can be considered as a theory of knowledge, in general, one can also intuit that any critical study on tourism will necessarily have to define itself from the elements of verist base, as is this example. Those who travel will know how to talk about travel, although travel fiction is a cultural reality that has inspired many creative people since Marco Polo. Material culture and immaterial culture are undeniably fundamental fields in the experience of the traveller in general and in tourist travel.

The values of epistemological basis (of the past of the activity), the values of the present (of contemporary practices) and the values attached to predictive visions (prospective and scenarios that may be deduced from critical analysis) are summoned to this discussion. In this way epistemology, as an auxiliary element of management, namely of tourist destinations, requires a vision on governance, both in critical-scientific and technical-practical terms. The standardization ETT assumed as a useful acronym to the field of tourism can integrate into this vision as we propose.

To accentuate and validate the principles, to try to manage the hypotheses, as well as to ascertain the results that are (always) expected from each type of analysis, namely the one we intend to ascertain in the question on the relevance of Tourism in a time of Health Crisis and in view of the need for tourist experience, forces us to consider the logical origin of tourism (travel?), the transdisciplinary value of leisure (entertainment and leisure?), the objective that imposes tourist behaviour on its actors (of accommodation and catering and other services?).

Gnoseology, if understood as theory on the one hand and as doctrine on the other, and in the case of what it structures what we consider as knowledge has, therefore, a preponderant role in the idea and interpretation of behaviours expected of people in different circumstances. Tourism planners have both concerns. The conditions imposed internally and externally on each human being and in the case, the human being who travels and becomes a tourist is the subject of studies since some profiles applicable to the tourist-consumer were defined. The tourism consumption trend communicated by the UNWTO (2020) at the end of each year in the form of a Report contributes to reinforce the tourism theory and, by addition, illustrate it with shareable examples.

MANAGEMENT OF TOURISM DESTINATIONS

We live, in fact, in a civilizational conjuncture that we can consider as a working hypothesis of observation according to the concepts of viral society. The brutal change that we have undergone all over the world and the way in which the concept of glocal is applied today in most analyses may also be associated with what some authors consider as a viral society. Living with the health crisis confirms this consideration. The assimilation of acronyms and acronyms is a reality of our daily lives and represents a new experience in society, receptive to new acronyms, some positive ones such as the ETT proposed here in the wake of the school of critical thought on the economy of experience that Pine and Gilmore (1999); 2011) and Oh *et al.* (2007) developed.

In this economy of experience, the question resulting from the spread of the virus that has become equally viral also arises. Managing anxieties and conflicts in destinations is becoming routine. Confine, wait, cancel, reopen, reconfine, etc. are active verbs that involve the tourist operation in all its modalities. As will be seen, they also affect the ETT (as a stenographic and current way of organizing the tourist product that, being composite, cements it), as well as the structuring of ethically positioned offers, according to the internationally accepted code (OMT, 1999).

How long will this situation of uncertainty, which only Nature commands, last, we do not know, nor will we know. The dispersion of people in the face of fears of contagion was also vitalized because the loose ideas that were being propagated by social networks and by both public and private "influencers", well-intentioned for their mission convictions, as well as the "profiteers" detached from any positional ethics created consequences. Public and published communication and the multimedia and dissemination supports of what was being written, said, reflected on, used, recycled, by an endless world of "followers" and "influencers" generated a polyphony of opinions, some credible, others non-credible, spreading a suggestive publicity stain whose contours are still difficult to define in a world moment under the restructuring of new trends in tourist consumption, (Ramos & Costa, 2017; Richards, 2018; Weston et al., 2019).

Thus, new thing implies new observation. The unexpected health crisis gave rise to unusual, unexpected, authentic responses, making it necessary to recover the creativity of tourism and, in tourism, (Richards, 2011; Beni, 2020 a). The propagation of measures to combat the crisis, the fear of viral spread, the massive dissemination of elements tending to be understandable, spread in a new and sometimes redundant communicational torrent, but well received by the audiences, was part of the therapy against viral fear, validated by some, mocked by others, insistently defended by many more, but, inexorably, a problem detected and integrated into the European and global agenda (Weston *et al.,* 2019; UN-United Nations, 2020).

In this logic, in a social constellation typified to each country, the value of each theory and/or each doctrine was scrutinized, but it is also what, temporarily, epistemology grants it. In this sense and with considerable depth, the theory that currently dominates the production of reflections on the "new tourism" grants all the scenarios that uncertainty provides in a constant reference to a limit: the viral line, its contextualization and the search for a new tourism theory has room for progression (Simmons, 2009).

The imposition of technical procedures comes from the confidence with which they are thought and planned on a proven scientific basis. This idea of relevance and the compulsive literature on conventional tourist destinations and their proven tourist dimension before the pandemic illustrates that many of them were managed within visions aligned with the continued growth of international arrivals (Ramos & Costa, 2017). We also know that Governance held a stable and increasingly relevant position. However, this paradigm has changed, Kotler *et al.,* (2017), requiring greater creativity in planning and management (Richards, 2018).

With the forced interruption caused by the pandemic, there were at least three fundamental moments and, this time, in a totally global way: the economies of the countries that send tourists, in comparison with the economies that preferentially receive tourists, both in contexts of international markets displaced from their routines and, therefore, with speeds of reaction and accommodation to new realities, very different from each other. The national tourist tradition was questioned and even in a situation of global economic crisis, it allowed embracing dreams (Lopes, 2010).

From the most recent literature positioned on the attentive observation of growths and decreases, but with much more attention to the problems of the intrinsic quality of tourist products, we witness the drift that, for the first time and after the boom of the Second World War, a movement apparent setback in theoretical reflection and engagement with a system of innovation and entrepreneurship that, at its base, comprises one of the elements of the ETT-Safety and Health. The viral environment is here to stay (Beni, 2020a).

The viral environment can be considered as the space in which humanity, living, appropriating, producing and consuming things, develops dominating a world that surrounds it and, from that world, builds what we call Culture. In a dimension of human experience in the face of a health crisis, the idea of each «viral thing» taking a place, affecting all the activities of this experiential dimension, experienced, in process, is common. Therefore, the idea of clusters takes on a more impressive dimension (Porter, 1998; Bovone, 2001; Cunha, 2020) and responds to current and conjunctural segmentations that, in their diversity, present themselves as a unit.

Thus, the cultural environment, generated with and in nature, carries the potential of being inserted as a great reference both to the natural heritage, provided by nature, and to the cultural heritage, Choay (2008), generated and maintained by the humanized vision of a world constantly under construction. In this construction, the acquired experience is part of this vision, because transdisciplinary has the potential to encompass the conservative system and the innovation system (Jafari, 2005). Tourism as an Industry of Peace, in fact, lives up to this reality that has been lost in the most remote times and is a world heritage site, giving rise to innovative cultural industries (Peixoto, 2000).

In this environment of tourism construction of the world, the ecological and biological basis coexisting in the natural elements and in the physical and chemical conditions, but also in the biological conditions and ecology determined by the geographies of the places, Ferrão (2014) determines if, inexorably, the ways in which the development of organisms and living beings is manifested, namely in their use for the valorisation to which they are subject, (Tilden, 2008; Coelho, 2017; Florida, 2019).

This manifestation is visible, both when these forms of growth are only confronted with natural conditions, as well as, when confronted with cultural conditions, they receive energies that allow them, for this very reason, to generate environments of comfort not only physical, but also metaphysical. Spiritual Tourism is, for the Authors, the climax of this material and immaterial construction.

The encounter of the natural world with the cultural world seen in a totality whose basis is its geography and humanity (human beings geographically identified and in encounters resulting from their travels) with the detail of geographic and territorial area (political power transforms geography into territory) and, therefore, social phenomena (which tourist activities most accentuate) involve individuals and are involved by them (Almeida & Abranja, 2009). Ethnic Tourism as a specific segment of cultural tourism is expressed in this encounter dimension (Pérez, 2009).

In this action-reaction game, the social areas, developed by the creatives associated with them, Florida (2002) in intersection with the cultural areas and, both, with the economic areas, give rise to adaptive scenarios that, between gains and losses, between advances and penalties, between goals achieved and others failed, between characteristics of the «contenders» and assimilations, influences, appropriations, submissions, conflicts, negotiations, strategies for pacification, diplomacy, life and death (Brito, 2002; Bourdieu, 2021), are stipulated as model.

In this logic there are model destinations. Thus, tourist destinations can also be seen as places that, in a game system, aggregate all the Demand segments that are nested in a kind of Social Tourism, because the social component of tourism is equivalent to its economic importance. This is a reality that tourist epistemology denotes, (Netto & Nechar, 2014). Ignoring the strength of the culture exchanged in tourism (visitor-visited) means ignoring their social potential at the expense of their economic potential: they are interdependent.

The environment, namely the work environment, is characterized by circumstances where the need for production involves the design of products and their effective production for consumption. The viability of tourist organizations depends on this basis (Amorim et al., 2015). The tourist product is simultaneously produced and consumed at the same time and in the same space and it is important to remember this starting line for the business (Almeida & Abranja, 2009; Cunha, 2020). This peculiarity makes the tourist product the object that awakens and sustains a unique experience. This experience shapes the tourist act that depends on the meeting between the Visitor and the Visited, Barreto (2004), that is, between the Tourist and the Resident, that is, between Demand and Supply.

From a perspective of generating narratives that can be associated with the experimentation of endogenous resources, which have become tourist attractions, Benjamim (2016), Gastronomic Tourism or Urban Tourism, Military Tourism or Scientific Tourism and many other segments accuse, in a good sense, the plasticity that promoters and consumers grant them, (Jafari, 2004; Beni, 2020 a).

In this logic of plasticity, an ideal tourist product is one that, by focusing on the tourist's needs, responds to their motivation, to their desire, to their motivation in full (Buhalis & Costa, 2006). However, as in other fields of human activity, perfection is a goal and never an object. Thus, the most important theoretical question is the one that it intends to answer, in execution, with the highest degree of performance that formally depends on the tourist potential of the future components of the tourist product (Cunha, 2008).

The tourist product designer tries to play the data on which he depends (supported, namely, on market studies on consumption trends in well-referenced geographies), (Cravidão, 2011; Ferrão, 2014). Each leader advances their idea of innovation, Kotler et al. (2017), and responds to socially felt needs (FCT, 2019). Therefore, in the idealization of the maximum satisfaction that any consumer seeks when purchasing a good or a service, the responsibility of service providers and operators who organize them, according to market studies making them tradable goods.

There is a set of issues inherent to this response to the consumer and, naturally, to his desire to obtain the maximum advantage in the cost-benefit ratio, (Hall, 2008).In a dimension of response by the organizations that structure the Offer of a tourist destination and, respective Governance, working in network in the current conditions both material and immaterial dependent on the health crisis, implies establishing bridges of basic nature, ICOMOS (2008) but without disrespecting the local cultures (spirit of place), and through strategies based on:

1. considering that the total tourist experience being localized, however, has its mandatory elements (Travel-Accommodation-Restaurant-Entertainment and Leisure-Safety and Health-Access to Complementary Goods and Services), which they are developed in order to satisfy the needs of tourist customers because, lacking one of them, this totality falls short of what was intended, weakening the experience.
2. integrating the dimension of the tourist's trip and stay gives rise to procedures, some routine and others, necessarily innovative because there are three periods of Tourism or, in other words, the ante-Covid 19 Time, the Covid 19 Time, the Post-Covid 19 time; In practice, this means that the organizers of tourism products take advantage of their competences as agents on the supply side, merging the experiences of a continuous flow of tourism (marked by the growing growth of international tourist arrivals), of an interrupted flow of tourism (marked by the market erosion in

the cycles of confinement in different countries), an intermittent flow of tourism (marked by the current problem of recovery and regression of tourist activity);

3. Granting that between 2020 and 2022 we live in a viral environment in the perspective of the Covid 19 disease and of written and multimedia viral communication focused on the discussion on the strengthening of social networks and their impact on the life of the territories, we can observe that the survey the triple type fits this quantification need.

In the subsequent phase of this work presented here, it is intended to develop the following points, namely:

4. establish a vision based on 6 questions, these questions corresponding to answers that, it is believed, can become indicative of simple samples in environments whose population, experiencing the same crisis, felt it almost homogeneously, as demonstrated by the literature.

Survey: Material and Methods

The survey, which is attached to this investigation, aims to allow an understanding of the penetration of ETT, from the perspective of its application as a new acronym and, at the same time, of professional decoding and in the surrounding markets. Thus, this survey will be applied to all Stakeholders who interact within a trip, aiming, through the results achieved, to carry out a study, investigation, in a longitudinal way.

Briefly, the survey will be presented in this investigation as a support instrument in this phase of ETT enunciation and, in the future, a new investigation will be presented that will demonstrate the results achieved through the survey presented here (longitudinal study, consolidating, over time, the proposal and its effective application). Finally, the researchers, through this investigation, and as a result that is stated as a result of the field work carried out in the meantime, will be available to carry out consultancy in this same area of knowledge, in order to demonstrate the relevance of the theme, for all potential and current stakeholders who have already collaborated with the authors in this preliminary phase.

As previously mentioned, this is an exploratory study that starts from an empirical approach, with the objective of establishing a new technical-scientific concept based on the evolution of the paradigm of consumption in tourism, accentuated in recent years, mainly due to the pandemic. One of the subsequent steps of this project is the collection of data related to the experience of Demand (tourists). For this, a questionnaire was developed that will be applied by a survey (presumably online), to perceive the way of acceptance of the proposal, surveys were carried out to a population directly related to Tourism to understand the penetration of ETT from the perspective of its application as a new acronym and, at the same time, a mnemonic for professional and market decoding.

The questions try to be clear, accuse the relevance of the objective, convey and raise a form of interaction between the authors and their deponents as an effective contribution for the mnemonic to become a common reference in professional terminology, but very relevant in the communication between the tourism actors not only in a national dimension, but, desirably, global. Are the following:

Question 1 – How do you rate, in terms of satisfaction, the trips you took in person during the health crisis period, from the last year to the present moment?

Question 2 – If you travelled digitally, what level of intensity do you put in the research you carried out?

Question 3 – What kind of Platforms did you favour when researching or booking your trip(s)?

Question 4 – How many platforms did you use to search or book your trip(s)?
Question 5 – Among the following elements in alphabetical order, please indicate which are the most important for you on a trip;
Question 6 – Do you consider that the use of acronyms in your work is important because it refers to an organization or a process?

Analysis of the Expected Results of the Survey

Based on the results obtained in 4, it will be possible to demonstrate the influence of the concept of «viral», and its evidence in the tourist discourses (of the promoters – of the operators – of the consumers) and, in a broad sense, of the resident populations (nor always passive) and groups of researchers (those in tourism, in general, and researchers in tourism sectors, in particular), UNWTO (2019), UN – United Nations (2020);

5. to conclude how, in fact, we went from a highly optimistic pre-pandemic tourist destination management model to the post-pandemic scenario, with all the epistemological problems and adaptation to new languages in the same management system that, in the tourism economy, went from a classic scarcity economy to a survival economy, Beni (2020 b);
6. Based on the European and Western political model of which the PRR – Recovery and Resilience Plan, Ministry of Planning (2021), is an example (despite not considering tourism as a strategic domain in the economy) there is, strategically, a response on the part of public administration, but also the shift towards a new paradigm in the tourism system which, paradoxically, in its restructuring, will have to validate this change, forced by the health crisis.

OVERFLOWING TOURIST LITERACY

The proposal intends to disseminate the ETT as something referentially useful providing greater efficiency, innovation, and entrepreneurship because the Authors believe that short, incisive and focused language is, in the scope of a digitalized society, an already present and growing reality. Automation and digitalization are imposing, in very short cycles, new acronyms, new mnemonics. In this logic, the search for alternative tourism, that is, for a new paradigm for the management of ongoing and future touristification processes, also makes sense, when seen in a global dimension. The symbolic power applied under policy measures aligned with the principles of the UN 2030 Agenda can support territorial legitimation actions that, from a perspective of accentuating the fight against the indifference generated by the type of non-responsible tourism, creates new conditions for a new economic paradigm, particularly in terms of the tourism economy.

Does the creation of mnemonics, called "stenography's", as we saw in Bourdieu (2021), increase tourist literacy? Could the creation of the ETT represent an operational acronym that contributes to this need? The Authors, for what they have exposed and adduced, think it is opportune to raise this central question, summarized in what the Total Tourist Experience means in a viral environment.

The answer, certainly needing more time and more empirical evidence, after this contribution, even to impose itself with greater scope, is to deepen the epistemology of tourism in a dimension of management of Tourist Destinations that, necessarily injured by the pandemic, undoubtedly need this new approach. Mandatory, no doubt.

Relating the Epistemology, Etymology and Terminology of a field of study and business activity such as the field of tourism, as the authors intend to demonstrate, may mean that the search for greater effectiveness, with economy and under the objective of efficiency in the tourist operation, in a broad sense, it lacks reflections that, innovating, simplify, if possible, the work processes and, equally, the leisure time. Thus, professionals and clients, when evoking the acronym, the acronym, the ETT mnemonic have a common reference that is advantageous for all.

CONCLUSION

In this brief essay we can conclude, through the methodology used, namely in two planes (literature review; traveller experience of each of the Authors), associated with a joint reflection. This was motivated by the constant search for greater knowledge on each day of professional activity in the fields of cultural tourism and heritage, marketing and geography of tourism and a shared vision of the importance of transdisciplinary management of tourism activity. In this logic we conclude that:

- What we set out to address in this chapter is organized here in an open text form and the founder of a path that effectively depends on the scenarios shaped by the current Covid-19 health crisis and its variants;
- Considering the teaching component of the Authors, from a perspective of the past, the experience accumulated by them in an environment of individual tourist consumption and their reflection (impossible to further describe due to the economy of text required in this article), is useful as an epistemic statement for generations of Students who are now entering the universe of polytechnic higher education.
- From a present perspective, the habitus, associated with the interventions of professionals in the sector, contains expected, routine variables, but considerably many more variables, disruptive, not expected at all, with the Authors who are spectators interested in this matter being poured into their lines of investigation.
- In a future perspective, it can be concluded that the Taylorian or Porterian planning models (among others) serving as theoretical and explicit ballast in relation to the goods and services markets, will contribute to new paradigms determined by the eventual model combination of "expansion-contraction" that the Authors understand can be considered in tourism.
    - As a model that tends to be viral as a solution, but also as an articulated response to the problems of confinement and social non-confinement that the economy in times of health crisis demands, the ETT, when perceived by tourism actors, can signify a new time and new procedures (communicative language associated with a possible consensus on the need to generate a «tourism shorthand»). This line of investigation becomes opportune in the context of the theoretical and practical reanimation that higher education, namely polytechnic, should be able to carry out through new proposals.
    - In a comprehensive scientifically useful and professionally facilitating dimension of tourist language, where acronyms such as UNWTO-UN-EU-ICOMOS-UNESCO- etc. refer us to readings and consequent actions, knowing the possible disruption that the proposal will

cause, the Authors understand its relevance: associated with clarity about the moment of resumption of tourist activity and feasibility of new proposals, such as the results presented, the objective set as a challenge has been achieved.

The development of this study, as mentioned before this preliminary phase, has already begun because, as was seen in the process of creating this text, the receptivity of both professional and consumer actors and the political and business, associative and research classes, showed openness to a second work phase to conclude and establish this proposal as a good practice, possibly in the form of a technical manual useful for the qualification of the economic sector of tourism.

This research received no specific grant from any funding agency in the public, commercial, or not-for-profit sectors, but received logistical and technical support from L-tour.ipt, belonging to the Polytechnic Institute of Thomar.

REFERENCES

Almeida, I. D., & Abranja, N. A. (2009). Turismo e Sustentabilidade. *Revista Cogitur*, 15-31. https://revistas.ulusofona.pt/index.php/jts/article/view/506/409

Barreto, M. (2004). Relações entre Visitantes e Visitados: um retrospecto dos estudos socio antropológicos. *Turismo em Análise, 15*(2), 133-149. https://www.revistas.usp.br/rta/article/viewFile/62663/65458

Beni, M. C. (1990). Sistema de Turismo - SISTUR Estudo do Turismo face à Moderna Teoria de Sistemas. *Revista Turismo em Análise*. https://www.revistas.usp.br/rta/article/viewFile/63854/66610

Beni, M. C. (2020a). Turismo e Covid-19: algumas reflexões. Rosa dos Ventos Turismo e Hospitalidade, 12(3), 1-23. doi:10.18226/21789061.v12i3a02

Beni, M.C. (2020b). *Saturação e Rejeição ao Turismo nas Destinações Turísticas*. São Paulo: Revista Brasileira de Pesquisa em Turismo. doi:10.7784/rbtur.v14i2.1847

Benjamin, W. (2016). *The Storyteller -Tales Out of Loneliness*. Ed.Verso.

Bourdieu, P. (2021). *O poder simbólico*. Lisboa: Edições 70.

Bovone, L. (2001). Os Novos Intermediários culturais – Considerações sobre a cultura pós-moderna. In Cidade, Cultura e Globalização. Oeiras: Celta Editora.

Boyer, M. (2002). El turismo en Europa, de la edad moderna ao siglo XX. *Historia Contemporánea, 25*, 13–34. https://eco.mdp.edu.ar/cendocu/repositorio/00924.pdf

Brito, B. R. (2002). *O Turista e o Viajante: Contributos para a conceptualização do Turismo alternativo e Responsável*. Lisboa: Associação Portuguesa de Sociologia. https://aps.pt/wp-content/uploads/2017/08/DPR462dea1a49422_1.pdf

Brito, B. R. (2002). *O Turista e o Viajante: Contributos para a conceptualização do Turismo alternativo e Responsável*. Lisboa: Associação Portuguesa de Sociologia. https://aps.pt/wp-content/uploads/2017/08/DPR462dea1a49422_1.pdf

Buhalis, D., & Costa, C. (2006). Tourism Business Frontiers - consumers, products and industry. In D. Buhalis & C. Costa (Eds.), *Tourism Business frontiers consumers, products and industry*. Elsevier B.V.

Cadilhe, G. (2006). *A Lua pode esperar*. Oficina do Livro- Sociedade Editorial Lda.

Castells, M. (2007). A Era da Informação: Economia, Sociedade – *O Poder da Identidade*. Lisboa: Fundação Calouste Gulbenkian.

Choay, F. (2008). *Alegoria do património*. Coimbra: Edições 70.

Cravidão, F. (2011). *Turismo, território e cultura: uma trilogia (sempre) em construção* (Vol. 29). Departamento de Geografia Humana da Universidade do Estado do Rio de Janeiro.

Crouch, D., Jackson, R., & Thompson, F. (2006). Introduction-The media and the tourist imagination. In M. Hall (Ed.), *Contemporary Geographies of Leisure, Tourism and Mobility* (pp. 1–13). Routledge.

Cunha, L. (2008). Avaliação do Potencial Turístico. *COGITUR Journal of Tourism Studies*. http://jts.ulusofona.pt/turismo01/pdf01/artigo_liciniocunha.pdf

Cunha, L. (2017). *Turismo e Desenvolvimento – Realidades e Perspetivas*. Lidel – Edições Técnicas, Lda.

Cunha, L. (2020). *Desenvolvimento do Turismo em Portugal: Os Primórdios. Fluxos & Riscos n.º1*. Faculdade de Ciências Sociais e Humanas Universidade Lusófona de Humanidades e Tecnologias. https://core.ac.uk/download/pdf/48576342.pdf

FCT – Fundação para a Ciência e Tecnologia. (2019). *Agenda Temática de Investigação e Inovação – Turismo, Lazer e Hospitalidade*. https://www.fct.pt/agendastematicas/docs/Agenda_Turismo_Final.pdf

Ferrão, J. (2014). *Políticas territoriais eficazes e participadas*. Lisboa: Instituto de Ciências Sociais, Universidade de Lisboa. https://www.animar-dl.pt/ponto-de-vista/704-politicas-territoriais-eficazes-e-participadas-joao-ferrao

Firmino, M. B. (2007). *Turismo – Organização e Gestão*. Escolar Editora.

Florida, R. (2002). *The Rise of the Creative Class. And how it's transforming work, leisure and everyday life*. Basic Books.

Florida, R. (2019). *The Innovations of the Creative Class Affect a Rural Area's Fortunes*. https://www.bloomberg.com/news/articles/2019-08-15/rural-innovation-links-to-creative-class-presence

Getz, D. (2007). *Event Studies: Theory, Research and Policy for Planned Events*. Butterworth-Heinemann.

Goeldner, C., & Ritchie, B. (2012). *Tourism. Principles, Practices, Philosophies*. John Wiley & Sons.

Hall, C. M. (2008). Tourism Planning Policies, Processes and Relationships. In *Policies, Processes and Relationships* (2nd ed.). Pearson Prentice Hall.

ICOMOS. (2008). *Declaração de Québec sobre a preservação do "Spiritu loci"*. https://www.international.icomos.org/quebec2008/quebec_declaration/pdf/GA16_Quebec_Declaration_Final_PT.pdf

Ignarra, L. R. (2003). *Fundamentos do Turismo* (2nd ed.). Thomson Learning.

Jafari, J. (2005). El turismo como disciplina científica. Política y Sociedad, 42(1), 39–56. doi:10.5209/POSO.24139

Joseph, J. (2010). *The Experience Effect: Engage Your Customers With a Consistent and Memorable Brand Experience*. Amacom.

Kotler, P., Kartajaya, H., & Setiawan, I. (2017). *Marketing 4.0: Mudança do tradicional para o digital.* Conjuntura Actual Editora – Edições Almedina.

Landry, C. (2010). *Creativity, Culture & the City: A question of interconnection.* http://www.forum-avignon.org/sites/default/files/editeur/ECCE_report.pdf

Leiper, N. (1990). Partial industrialization of tourism systems. *Annals of Tourism Research, 17*(4), 600–605. doi:10.1016/0160-7383(90)90030-U

Lopes, E. R. (2010). A constelação do Turismo na economia portuguesa-Portugal Desafios nos alvores do século XXI. Lisboa: Ed. O Sol é Essencial S.A.

McKercher, B., & du Cros, H. (2002). *Cultural Tourism – the partnership between Tourism and Cultural Heritage Management.* Routledge.

Ministério do Planeamento. (2021). *PRR – Recuperar Portugal, Construindo o Futuro.* https://recuperarportugal.gov.pt/wp-content/uploads/2021/10/PRR.pdf

Netto, A., & Nechar, M. C. (2014). Epistemologia do turismo: escolas teóricas e proposta crítica. *Revista Brasileira de Pesquisa em Turismo, 8*(1), 120-144. http://www.spell.org.br/documentos/download/30514

Netto, P. A., & Gaetra, C. (2010). *Turismo da Experiência.* SENAC.

Oh, H., Fiore, A. M., & Jeoung, M. (2007). Measuring Experience Economy Concepts: Tourism Applications. *Journal of Travel Research, 46*, 119-132. doi:10.1177/0047287507304039

OMT - Organização Mundial do Turismo. (1999). *O Código Mundial de Ética do Turismo.* Santiago do Chile: OMT. https://ethics.unwto.org/sites/all/files/docpdf/portugal.pdf

Pearce, P. L. (2005). The role of relationships in the tourist experience. In F. T. William (Ed.), *Global Tourism* (3rd ed., pp. 103–122). Butterworth-Heinemann. doi:10.1016/B978-0-7506-7789-9.50012-1

Peixoto, P. (2000). *O património mundial como fundamento de uma comunidade humana e como recurso das indústrias culturais urbanas, Coimbra.* https://www.ces.uc.pt/publicacoes/oficina/ficheiros/155.pdf

Pérez, X. P. (2009). *Turismo Cultural - Uma visão antropológica.* Santa Cruz de Tenerife: PASOS. https://repositorio.utad.pt/bitstream/10348/.../1/livro%20tc%20xerardo.pdf

Pine, B. J., & Gilmore, H. J. (1999). *The Experience Economy: Work is Theatre & Every Business a Stage.* Harvard Business School Press.

Pine, B. J., & Gilmore, J. H. (2011). The experience economy (updated edition). Harvard Business Review Press.

Porter, M. (1990). The Competitive Advantage of Nations. *Harvard Business Review.* https://hbr.org/1990/03/the-competitive-advantage-of-nations

Porter, M. E. (1998). Clusters and the new economics of competition. Watertown-Massachusetts. *Harvard Business Review, 76*, 77–90. PMID:10187248

Quivy, R., & Campenhoudt, L. (2008). *Manual de Investigação em Ciências Sociais*. Gradiva Lda.

Ramos, D., & Costa, C. (2017). Turismo: Tendências de evolução. *PRACS: Revista Eletrônica de Humanidades do Curso de Ciências Sociais da UNIFAP, 10*(1), 21–33. doi:10.18468/pracs.2017v10n1.p21-33

Reisinger, Y. (2009). *International tourism, cultures, and behaviour*. Butterworth-Heinemann.

Richards, G. (2011). *Creativity and Tourism: The State of the Art.* http://www.forum-avignon.org/sites/default/files/editeur/ECCE_report.pdf

Richards, G. (2018). Cultural tourism: A review of recent research and trends. *Journal of Hospitality and Tourism Management, 36*, 12–21. doi:10.1016/j.jhtm.2018.03.005

Rodrigues, J. C. B. (2012). *A Implantação do Automóvel em Portugal (1895-1910)*. Lisboa: Dissertação para obtenção do Grau de Doutor em História, Filosofia e Património da Ciência e da Tecnologia. Faculdade de Ciências e Tecnologia. Universidade Nova de Lisboa. https://run.unl.pt/bitstream/10362/9676/1/Rodrigues_2012.pdf

Simmons, A. (2009). *The Story Factor: Inspiration, Influence, and Persuasion through the Art of Storytelling*. Basic Books.

Sol, H., de Brito, M.P., Coelho, J.P., Pratt, C., & Lopes, E.R. (2017). Rooting a new event in its place: the case of Festa Templária, Tomar, Portugal. *International Journal of Event and Festival Management, 8*(3), 324-345. doi:10.1108/IJEFM-10-2016-0068

Sundbo, J., & Darmer, P. (2008). *Creating Experiences in The Experience Economy*. Edward Elgar. doi:10.4337/9781848444003

Tilden, F. (1957). *Interpreting Our Heritage*. University of North Carolina Press.

Tilden, F. (2008). *Interpreting Our Heritage* (3rd ed.). The University of North Carolina Press. https://is.muni.cz/el/1421/podzim2017/MUI_338/Interpreting_Our_Heritage__Chapel_Hill_Books_.pdf.

Tribe, J. (2011). *The Economics of Recreation, Leisure and Tourism*. Elsevier. doi:10.4324/9780080890517

UN – United Nations. (2020). *The Sustainable Development Agenda*. https://www.un.org/sustainabledevelopment/development-agenda/

UNWTO. (2019). *The Future of Work and Skills Development in Tourism – Policy Paper*. Madrid: UNWTO. doi:10.18111/9789284421213

UNWTO & IPSOS. (2019). *Global survey on the perception of residents towards city tourism: Impact and measures*. Madrid: UNWTO. doi:10.18111/9789284421107

UNWTO. (2020). *International Tourism Highlights – 2020 Edition*. https://www.e-unwto.org/doi/epdf/10.18111/9789284422456

Urry, J. (1995). *Consuming places*. Routledge.

Weston, R., Guia, J., Mihalič, T., Prats, L., Blasco, D., Ferrer-Roca, N., Lawler, M., & Jarratt, D. (2019). *Research for TRAN Committee – European tourism: recent developments and future challenges*. European Parliament, Policy Department for Structural and Cohesion Policies, Brussels. https://research4committees.blog/2019/10/29/european-tourism-recent-developments-and-future-challenges/

KEY TERMS AND DEFINITIONS

Experience Economy: Economic activities that support or are supported by experience-based products.

Gnoseology: The doctrine that focuses on knowledge, the theory.

Mnemonic: A technique or method for helping the memory (in this case, applied to ETT).

Total Tourism Experience: A set of experiences and practices that were experienced during the preparation, the trip, and the knowledge retained afterwards.

Tourism Experience: Knowledges and practices that were experienced during the tourism trip.

Tourism System: A set of processes or methods interlinked with tourism activities, to achieve specific goals in a systematic way.

Touristification: The process of transforming an asset into a tourism product.

Viral Environment: Living conditions under a viral conjuncture that demand urgent and temporary personal and professional new rules.

APPENDIX

Portuguese Questionnaire

ETT - Experiência Turística Total

Experiência Turística Total e ambiente Viral

* Obrigatório

* Este formulário irá registar o seu nome, por favor preencha seu nome.

1. Como qualifica em termos de satisfação, as viagens que realizou de forma presencial no período de crise sanitária, desde o último ano e até ao atual momento?

	Não viajei	Insatisfeito	Pouco Satisfeito	Satisfeito	Muito Satisfeito	Totalmente Satisfeito
Satisfação	○	○	○	○	○	○

2. Se viajou digitalmente que nível de intensidade coloca nas pesquisas que realizou?

	Nenhuma pesquisa	Pouca pesquisa	Alguma pesquisa	Muito pesquisa	Pesquisa intensiva
Pesquisa	○	○	○	○	○

3. Que tipo de Plataformas privilegiou na pesquisa ou reserva da(s) sua(s) viagens?

○ Online

○ Presenciais

2/8/2022

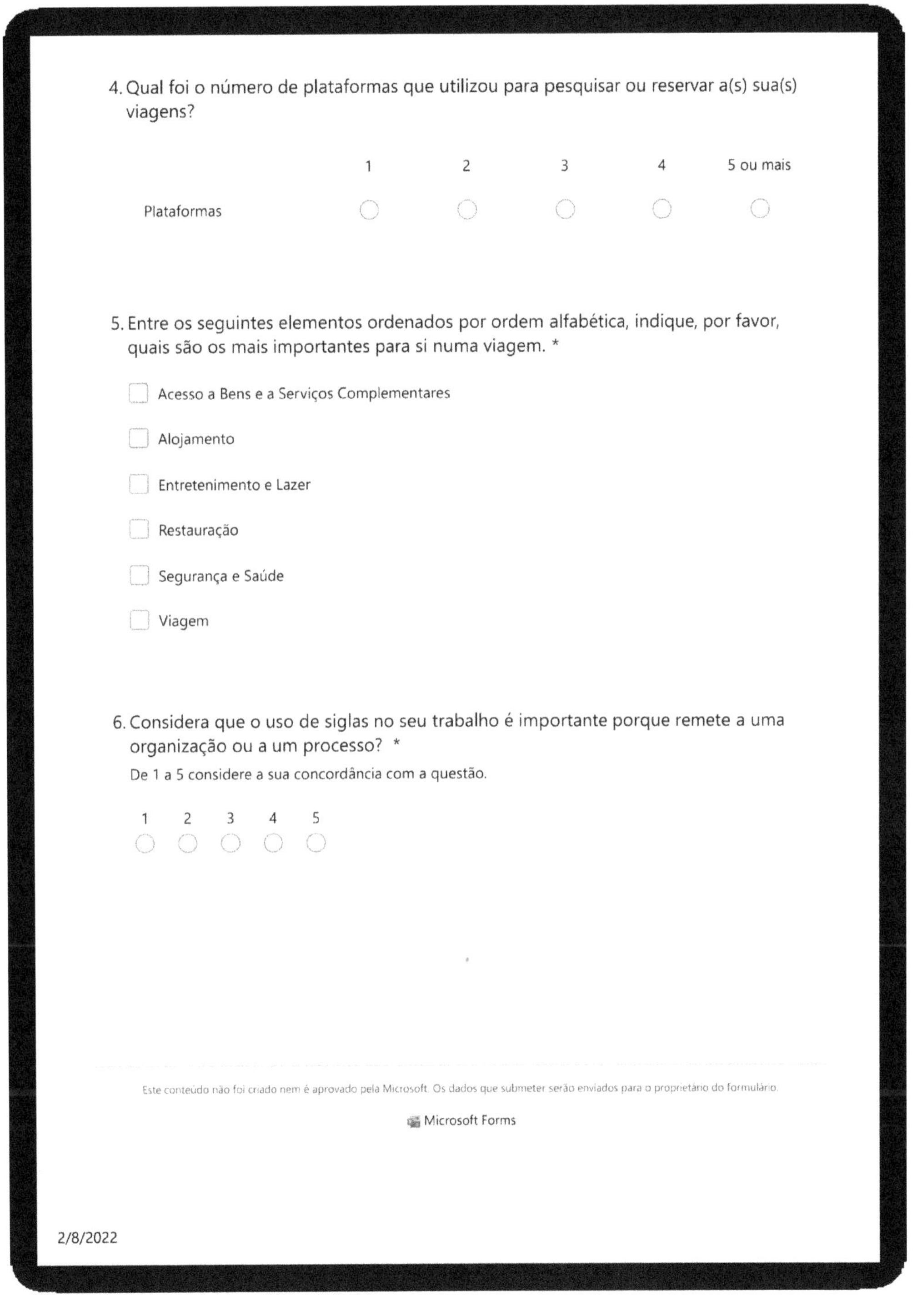

4. Qual foi o número de plataformas que utilizou para pesquisar ou reservar a(s) sua(s) viagens?

	1	2	3	4	5 ou mais
Plataformas	○	○	○	○	○

5. Entre os seguintes elementos ordenados por ordem alfabética, indique, por favor, quais são os mais importantes para si numa viagem. *

- [ ] Acesso a Bens e a Serviços Complementares
- [ ] Alojamento
- [ ] Entretenimento e Lazer
- [ ] Restauração
- [ ] Segurança e Saúde
- [ ] Viagem

6. Considera que o uso de siglas no seu trabalho é importante porque remete a uma organização ou a um processo? *

De 1 a 5 considere a sua concordância com a questão.

1	2	3	4	5
○	○	○	○	○

Este conteúdo não foi criado nem é aprovado pela Microsoft. Os dados que submeter serão enviados para o proprietário do formulário.

Microsoft Forms

2/8/2022

Source: Authors (2022)

Section 2

Journeys Across Virulent and Violent Urban Territories

Chapter 5
The Ambivalences of Smart Cities

Lucia Santaella
Pontifícia Universidade Católica de São Paulo, Brazil

Luiz Felipe Napole
Pontifícia Universidade Católica de São Paulo, Brazil

ABSTRACT

This chapter deals with the emerging urban spaces of the 21st century, at first named as sentient cities and then as smart cities, which result from the growth of megalopolises. This growth was accompanied by the expansion of disruptive technologies such as the internet, surveillance cameras, and sensors throughout the space interfering with domestic, public, corporate, and government affairs. The purpose of the chapter is to analyze the ambivalences that result from the inevitability of the use of intelligent technologies to guarantee the management and administration of the urban life, on the one hand, but the increase on the systems of control and surveillance, on the other side, to the point of converting the ambivalences into paradoxes.

INTRODUCTION

Specialists in topics related to culture have not failed to point out and discuss the effects of transformation that the digital revolution is causing in human life. One of the themes that stands out is human life in urban environments. In fact, in the last twenty years these environments have incessantly incorporated new technological devices aimed primarily at making the existence and mobility of citizens more reasonable. These transformations in urban life started to receive the denominations of sentient cities or also smart cities. Among the fields that have become more dependent on a coherent informational organization that facilitates urban mobility is the field of tourism. It is, therefore, a field that implies not only the installation of circuits of new technologies, but also implies a knowledge of the influence that intelligent organizations play today in urban life, especially for those who, in the position of visitors, seek in tourism the expansion of their knowledge about cultural diversity and its histories and tradition.

DOI: 10.4018/978-1-6684-3369-0.ch005

Considering that smart infrastructure and superstructures have become integral parts of contemporary cities, particularly those that stand out for their tourist potential, the objective of this article is to bring to the discussion the different facets that smart cities present, while highlighting the contradictions inherent to intelligence in the datafication stage of life in which we find ourselves today. The discussion is justified as a basis for tourism to be thought and carried out in the direction of sustainability, a goal to be reached unconditionally.

The topic of smart cities began to emerge in the 2000s to become a constant today. The reason for this lies not only in the disproportionate growth of megalopolises and in the continuous emergence of new metropolises, but also in the growing expansion of the internet in domestic, public, corporate and governmental spaces, and its corresponding incorporation into the necessary material infrastructure of cities. In fact, cities are getting smarter as algorithmic information processing capabilities are embedded, taking root, and spreading throughout urban infrastructure.

The field on which, in principle, the development of smart cities depends is the field of governance and government administration. Thus, e-governance and politics 2.0 refer to ubiquitous services of government, citizen engagement and participation, involving, for example, government semantic information services and ontologies. Today, most systems are computerized, and the population is increasingly connected to different media and social networks, which results in gigantic volumes of data that imply collection, storage of information and integration of many sources and technologies. For this, the technology park must be robust with a view to identifying and classifying essential information for decision-making.

If, on the one hand, urban intelligence has become inevitable for the management and administration of government services, on the other hand, all this constitutes a true apotheosis of the society of control and surveillance, as conceived by Gilles Deleuze (1992). This is precisely the theme that has occupied the thinking of critics of culture, because, although it is undeniable that the entry of artificial intelligence in the most diverse human activities (cities governance included) brings benefits, it precisely intensifies the processes of control and surveillance. Therefore, what follows aims at reflecting on the ambivalences of smart cities, in what is gained and in what is given blindly to obtain these gains.

WHY DID CITIES BECOME SENTIENT?

Cities are historic dwellings completely submitted to human projects. As Benevolo reminds us (1997, p. 9), cities did not always exist, "they began at a given moment of social evolution and may end, or be radically transformed at a given moment", as it seems to be the case of the current world, dynamized by human creativity, observed by the celestial electronic vision of artificial satellites, which have been orbiting the planet since 1957 (year of launch of the Soviet Sputnik program) and scaled by the telematic production of places, through the connection with digital communication networks since 1991 (year of the World Wide Web operation). Combined, intelligent technologies allow socio-communicative aspects to be overcome to achieve the "digitization of territorialities, ecosystems and their populations" as pointed out by Di Felice (2009, p. 12). As a result, they cause an alteration that goes beyond the socio-communicative sphere, because they cover living dimensions through sensors that capture movements, not only of our bodies, but also of microscopic particles in the air.

All these transformations gave rise to what came to be called "sentient city", a nomenclature that began to be used about a decade ago, causing disagreements and controversies. The adjective "sentient", which combines in one word "feeling" and "consciousness", refers to the capacity to feel and perceive subjectively, but does not include the faculty of self-awareness. It includes feeling, but not knowing. The first comes from the Latin *sentire* and the second comes from *sapere*. This would mean that a sentient city is able to feel everything that happens in it but is not able to know.

Since the 1980s, engineers and computer scientists have been conducting research aimed at embedding computational intelligence in environments. These researchers sought to go beyond the personal computer through ubiquitous computing that would move its processing to the periphery of our view. Small, inexpensive processors and wireless sensor networks would spread processing across the environment. As this information processing capability has been increasingly embedded throughout the urban environment, especially since, in 2008, Apple launched the 3G i-phone, urban computing and locative media, previously restricted to experimental artists and activists (Santaella 2010, p. 115-184) became a reality for any citizens endowed with mobile equipment with processing capacity.

In this context, why did the "sentient" attribute attached to the city generate discord? Most likely due to the remnants of a traditional humanism still faithful to the Cartesian dichotomies between the physical world and the world of consciousness. Claiming that the city is sentient destabilizes the ascetic separation of the qualities proper to the human essence and the rest of the universe stripped of them, including the universe of animals that, for Descartes, do not have sensations, but only present movements of matter. On the contrary, recent research has proven that animals are sentient beings. They have emotional capacity, that is, they feel pain, fear, stress, agony, expectation, pleasure, and joy, in addition to having memory and even homesickness. This applies to all vertebrate animals - mammals, birds, reptiles, amphibians, and fish. More than that, to the astonishment of the Cartesians, some recent studies have appeared around the ability to think of forests and plants (Kohl, 2013).

Therefore, there is no lack of reasons why the refutation of Cartesian dichotomies today extends to urban environments and, with the internet of things, also reaches what we used to call "things". As desktop computers became rarer, becoming miniaturized, to occupy and spread across the sidewalks, streets, and public and private spaces of cities, the information processing capacity became increasingly embedded and distributed throughout the entire urban space of everyday life. Faced with such conditions, there is no longer any reason to oppose the adjective "sentient" when applied to the city.

In fact, nowadays, urban infrastructure has become "able to feel and respond to the events and activities that transpire within the city. Imbued with the ability to remember, correlate and anticipate", the city acquires the power to reflexively monitor environments and our behavior in them, "becoming an active agent in the organization of everyday life in urban public space" (Shepard, 2011, p. 20; Santaella, 2013, p. 55-70; 2015). With this, the adjective "sentient" became euphemistic, and the city began to acquire the status of intelligence, hence the name "smart city" came into vogue. The adjective "smart" is doubly significant, firstly because smart is synonymous with intelligent, secondly because it works as an acronym for Self-Monitoring Analysis and Reporting Technology.

In this way, "a smart city uses digital technologies and networks seeking to be more sensitive to the environment, producing, consuming, and distributing a large amount of data in real time to help decision-making (of companies, governments and citizens)". With this, it seeks to "foster economic growth, innovation, expand forms of environmental sustainability, with a more efficient use of public resources, as well as citizen participation in the definition and discussions on urban problems" (Lemos, 2021, p. 73-74). To better understand the intelligence of cities, it is necessary to consider at least two aspects that need to be differentiated: the visible and the invisible ones.

VISIBLE ASPECTS OF SMART CITIES

First, it must be taken into account that sentient and intelligent cities have brought profound transformations to what we used to understand as urban spaces and public spaces. Although the material and physical spaces of the city have undergone few visible changes, although the buildings, parks, avenues, streets, houses, commercial enterprises are still there, there is a lot happening, new topologies emerging, beyond or below visibility. This is rule number one for understanding the meaning of a smart city. These new topologies have nothing to do with the traditional and still functioning underground sewage, water, lighting, etc. systems. It is true that, when we move through the streets, parks, malls, means of transport, we witness a choreography, which we learned to get used to, of people either with cell phones in their ears, or frantically strumming, or talking, apparently to themselves, as their devices flee to an immediate view. However, these choreographies are only understandable when we go beyond the most obvious layers of visibility.

Despite the few visible changes, there has always been a conflict between the traditional modern ideal of cities and the technological future. Previously conceived so that everything had a place to be, and a reason to be, it was intended to compose a way of life where interior (private) and exterior (public) were able to prevent any mixture. This design order meant regularity, repetition, predictability and was associated with a series of legal instruments based on scientific experiments, to model and regulate the urban form. The system of norms for the construction of urban spaces, for example, aimed at limiting the entry, exit, and circulation (of people and goods), which defined the status of certain individuals and social groups. There was a complex web of laws, regulating what is formal and what is informal. However, dominated by environmental and social problems on the one hand, and speculative and financial powers on the other, the city ended up resulting in a highly unbalanced set. For Secchi, the tradition of city politics declined by using heterogeneous materials in a different way, which resulted in different discursive formations, different representations, and different parts of the city. This gave rise to differences in the various parts of the planet" (Secchi, 2019, p. 46).

Therefore, the modern utopian city started to have no form, to escape from a designed image, from the moment the industry abandoned its concern with the urban fabric. This led the urban landscape to visibly concentrate a large volume of constructions, carried out in a short space of time, supplanting the idea of authenticity, previously used to evaluate cities: the place of the house, the place of work, the place of the institution. The space is now multipurpose, "multieverything". For Cacciari, in opposition to the Greek polis, the fundamental, "programmatic" characteristic of the Roman *civitas* was to grow; "there are no *civitas* that are not *augescens*, that do not *de-lire* (the lyre is the furrow, a sign that delimited the city, *delirium* means to leave the lyre, to go beyond the limits of the city)" (Cacciari, 2010, p. 16). It is the return of mixtures, where people live and work, all in one place, enhanced by technological improvements that dissolve barriers.

As for the notion of public space, it was customary to understand it in clear opposition to private space. However, according to Santaella (2007, p. 245-247; 2013, p. 71-92), since the explosion of the newspaper, magazines and, later, with radio and television, the clear boundary of separation between public and private began to be blurred, as journalism, and even more so, radio and television bring the outside world into the intimate precinct of private space.

On the opposite side of the public, what was considered as the private space? This would be linked to a sense of physicality, being in one's own space, linked, therefore, to the inviolability of personal space, including psychic space, synonymous with being anonymous and confidentiality. For this very reason, the prototype of the private space is the home. However, the separation between both spaces, cultivated in the 19th century, has already begun to be eroded by the newspaper and slipped when radio and television began to bring news and culture into homes. The slippage has intensified irrevocably with the internet, in which, if the user knows how to navigate with dexterity, infinite data, from multiple origins, for the most diverse purposes, move to the most intimate space of the home. With mobile technologies, the situation is reversed: the private begins to invade the public. Then all borders are blurred.

Therefore, Souza e Silva and Frith (2012, p. 186) claim that, with portable equipment, practices previously restricted to private spaces are now displaced to public spaces. This gives people the same sense of control and familiarity with public space that they have with private space. Therefore, instead of distancing people, current equipment should be seen as interfaces to the public space.

Before the era of mobility had imposed itself with the power it displays today, Flusser ([1991] 2015, p. 90-91) was already critically discussing the opposition between private and public space. For him, communication "consists in the fact that the flow of information commutes". Thus, information is no longer sought in public spaces, as it is delivered to the private space, "either by bundles or by network connectivity". At the same time, "there is no more private space", when our walls are pierced by visible and invisible cabling and the roof by antennas, which allow us to have the President of the United States in our kitchen "uninvited". In short, "there is no longer a private space in the Hegelian sense of the word, as the place where, by situating myself, I lose the world". Likewise, "the republic no longer exists, as the public space is completely taken over by cabling". If Flusser had lived a little longer and achieved mobile connectivity, he would undoubtedly be chorusing the studies of Souza e Silva and Frith today.

The movement between here and there are the forms of life that move freely in space. Computerized devices lead us through fluid space, a hybrid space, whose forms create their own language of world feeling, their own symbolism of space. The space is continuous, interconnected, and complex, presupposing the movement of the body that perceives it through scenarios, while these unfold in sequences and the eyes are directed to the cell phone screens. The broadest and most universal contents and forms of life are intimately linked to the most individual ones, since, for the big city, the fact that its interior life spreads out in waves over a wide territory is decisive.

What is briefly outlined above is what can be seen, the visible side of urban spaces. However, in smart cities, what matters is what is not seen, what makes it work in the invisibility of its informational infrastructures, but which, nevertheless, should not fail to be felt and thought about.

INVISIBLE ASPECTS IN SMART CITIES

Instruments of territorial control have always existed and aimed to create a sense of homogeneity when planning and executing the city on a global scale. McLuhan had already realized in 1969, that "the young student grows up in an electronically structured world. It is not a world of wheels, but of circuits, it is not a world of fragments, but of configurations and structures" (McLuhan, 2007, p.11). That's why the city's environment already presented itself as new through improved technologies that work as extensions of ourselves. If the machine was centralizing, superficial and fragmentary, the opposite of it constitutes the intended automation when the place becomes unimportant in the face of mobility.

Since the 2000s, we have been living in the age of mobility. Cell phones have become ubiquitous to the point of saturation. Today there are people who carry with them at the same time, a kindle, an i-pad, or any other type of tablet, a cell phone and even a notebook. E-mail has become entirely mobile with the advent of the mobile internet. Since the emergence of the i-pod, followed by Spotify, people carry with them, without weight and without measure, complete libraries of music. Gaming platforms have also gone mobile. With the Apple store, tiny phones can be loaded with new apps anytime, anywhere. In sum: providing answers about what the future of mobile technology will be, in addition to being reckless, is risky (Souza e Silva and Frith, 2012, p. 185-201).

What can be said, for now, is that studies on communication and mobile existence have spread across many disciplines and led to a shift towards a mobile paradigm in sociology. This paradigm takes devices into account, but places emphasis on the movements of people, objects, services, all connected by information. At the same time, criticisms have arisen in the humanities and in aesthetics about the ways in which public space is inseparable from surveillance, especially when the equipment is enhanced with geolocation capabilities (Schrock, 2012; Bruno, 2013).

The power of such technologies evidently does not come exclusively from themselves, but from the fact that they function as access routes in an urban ecosystem composed of human and non-human actors, or actants, according to Latour (2005), that is, those who, in a dynamic of occurrences, make the other do. Today informational networks surround us wherever we are. The environments are immersed in pulsating invisible clouds of information. We talk to people at a distance of hemispheres, and with the WhatsApp, without having to pay for it, we can also have access, with a few light touches of the fingers, to immeasurable information networks. In addition, location-sensitive, geolocated mobile technologies allow us to navigate online through urban spaces, transforming us into moving points of a two-dimensional map, while experiencing the three-dimensionality of the spaces we are contingently inhabiting. In short, we have come to live in a socially integrated and spatially contingent world. In Japan, for example, where all this reaches high levels of sophistication, the cell phone is used as a deterritorialized territorial machine (Shepard, 2011, p. 24).

For this reason, alongside studies on mobility, studies on the connective condition have been consolidated, as, for example, in Shaviro (2003), Spyer (2007), Jenkins et al. (2014). In Brazil, Di Felice's (2009) research on the new ways of living provided by connectivity has become notable. He says:

The dimension of atopic dwelling is elaborated by me in the paper "Post-Urban Landscapes", the end of the urban experience and the communicative forms of dwelling, in which I create the idea of the communicative forms of dwelling. It considers communication from an ecological perspective, in which the introduction of a certain type of communicative technology not only changes the form of communication between people, but also establishes a specific form of relationship with the environment. (ibid., 2016)

Housing in a broad sense is an extension of the bodily mechanisms of thermal control, while the city is an even broader extension of bodily organs, aiming to meet the needs of large groups. From the media, especially television, a few decades ago, McLuhan compared its social effects with the times of the village, a global village. Thus, he established the difference between the inhabitant of the city and the inhabitant of nature: "literate and civilized man tends to restrict space and separate functions, while tribal man freely projects the form of his body to encompass the Universe" (McLuhan, 2007, p.145). The concept of a global village seems interesting in describing the traditional arrangements of life in villages, in which what everyone is doing is constantly observed and discussed, incessantly.

Celebrating an organically united society, as in the village, in which authority relations are well established and everyone has a role to play for the social body to function, no longer seems to make sense in the current conjuncture in which a new informational layer results of uninterrupted data collection from millions of citizens. This is how smart cities reveal themselves as *civitas intellegens*, formed by inter (between) and *legere* (choose).

However, the devices for connection that, in fact, reach an ecological dimension and that give us free access to information and communication, in hypercomplex connective networks, present a double face: at the same time that we access, we are, in a way, accessed. Therefore, currently, in the urban space made up of networks of nodes, we are converted to mobile data packets. With each connection, information, unknown to us, is captured about our choices, preferences, tastes, location, frequency of places, personal connections, patterns of behavior. Finally, networks of increasingly fine information are captured about each person to the point that these networks know much more about each of us than the people we live with daily do. What results from this is an astonishing volume of data that, for some time now, has been processed by big data programs. That's why, to the age of mobility and connectivity, today is added – and ally with it – the latest era of big data.

BIG DATA AND ITS EFFECTS

What does big data mean? Translated literally means a gigantic volume of data. The field of big data is dominantly associated with business intelligence and analytics, which involves data manipulation and analysis, databases, machine learning, econometrics, data visualization and so on. It refers, therefore, to "techniques, technologies, systems, practices, methodologies and applications that analyze critical business data to help a company better understand its business and its market to make timely decisions" (Chen et al, 2012, p. 1166). Its high-impact applications in the business world include e-commerce and market intelligence.

However, there are other aspects more properly linked to e-government, health and safety systems, communication systems, etc. As can be inferred, although it is not customary to place market intelligence in the conditions of what is called smart cities, it is certainly involved there as well as e-government, security, and health. Let's see in a brief description, what each of these aspects seeks.

E-commerce, market intelligence and research on communication processes are aimed, for example, at monitoring and analyzing social media, crowd-sourcing systems, social and virtual games, whose analysis involves, among others, analysis of social networks, analysis of texts and the network, and even analysis of feelings and affections.

E-governance and politics 2.0 refer to ubiquitous services of government, citizen engagement and participation, involving, for example, government semantic information services and ontologies. The areas to be served by government services enhanced by big data are transport, public safety, health, including preventive health, and education. What matters ideally in all this is that governments find conditions to "unify systems, analyze data better, cross different indicators in different formats to define more assertively their objectives and strategies to improve service to citizens" (Elias, 1993).

Alongside the opportunities that big data offers for e-commerce, which are of interest to economics, administration and marketing, the health sector is facing a tsunami of content on health and health systems. In the field of security, data is fundamental in view of the dependence of the globalized society on digital training. As a result, the area of security informatics has been developed, whose main objective is "in the improvement of advanced computer technologies, systems, algorithms and databases for applications related to security through an integrated technological, organizational and police approach" (Chen, 2006, p. 7).

Big data analysis is basically based on data mining and data analysis algorithms. Embedded in e-commerce, mining algorithms were taken both for advances in neural networks, with a view to classification/prediction and clusters, and for genetic algorithms aimed at optimization and machine learning. Based on mathematical models and powerful artificial intelligence algorithms, statistical machine learning applies to data, text, and web analytics.

In addition, when it comes to big data, we commonly associate it with the five most relevant big techs in the Western global scenario (Alphabet, Amazon, Apple, Meta and Microsoft). Recently, the acquisition of Activision Blizzard (which develops, produces, and distributes video games), in January 2022 by Microsoft Games, involved a reported amount of US$ 68.7 billion, and appears to be about the games industry. However, we are talking about a database of billions of active users in an industry that strives to enable the mobile experience, the cell phones that have become smart as cities intend to be. Installing these applications means participating in a market that offers solutions to the public authorities responsible for managing cities.

To cite another example, selling cheaper in the case of Amazon means having more applications installed on these devices that have become companions in contemporary life. If the most latent element at first glance was not enough, which is the scale of the database, we should also highlight the result of a study published by the newspaper *El País* which points out that "it was already known that cell phones are at the limit of authorizations for use when collecting and sharing data. The novelty of the function of pre-installed applications lies in its extension, lack of transparency and privileged position within the cell phone" (Colomé, 2019). The problem, in the eyes of the researchers, is that Android, developed from a free code, created an "uncontrolled ecosystem", that is, there are more actors, besides Google, that are able to collect data from billions of people.

BIG DATA AND SMART CITIES

Previously made of concrete, glass, iron, avenues, streets, squares and parks, metropolises now house an infinity of devices and computer programs that, connected to the internet, constitute an immense nervous system supporting the daily life of billions of people. That intelligent machines and programs are ruling the world under our command is not just a technological revolution, "but a historic turning point in the ways we build and manage cities" (Townsend, 2013, p. xii). This is giving rise to a new type of city – powered by big data and cell phones, which have become the tiny platforms for reinventing the physical and managerial gigantism of contemporary cities.

According to Townsend (ibid., p. 4), along with big data, we witness the cacophony of billions of sensors tweeting from our pockets and bags, from the walls and sidewalks of cities, signaling minute by minute the location of vehicles, the room temperatures, possible seismic tremors and much more. The torrent of signs generated by the internet of things can exceed six petabytes (that is, one billion gigabytes) a year on our mobile networks alone. This will far surpass the informational volume of ten billion photos archived on Facebook that amount to more than one and a half petabytes. Programs for business, government, and citizenship services, through big data, are already and will be the immanent force to support urban life, in an indissoluble intersection of urbanization and ubiquitous computing.

Smart technology is not just a way of doing more with less, but it can be an opportunity to rethink and reinvent governance from a more open, transparent, democratic, and responsive model, in the sense of using social media as channels of communication with citizens, publishing government data on the web, unlocking public databases and much more.

In fact, public administration would be unfeasible today without the help of big data. Alongside governance, there are also many activists, entrepreneurs, and civic hackers who, through the gaps, seek new forms of utopia, when they create digital interfaces for people to see, touch and feel the city in completely new ways. Instead of proprietary monopolies, they build collaborative networks for different purposes in urban life. These bottom-up efforts may seem small-scale, but they have the potential to go viral across networks (Townsend, 2013, p. 9). However, such conditions are not able to dissolve the ambivalences that smart cities carry with them.

Simmel had already anticipated that through the accounting essence of money, in the relation of the elements of life, it was possible to arrive at a precision, a certainty in the determination of equalities and inequalities, a univocity in agreements and combinations comparable to the propitious diffusion of pocket watches, later wristwatches, and recently, the mobile phone itself.

Punctuality, accounting, exactitude, which coerce the complications and extensions of life in the big city, are not only in the most intimate nexus with its intellectualistic and economic-monetary character, but also need to color the contents of life and facilitate the exclusion of those irrational, instinctive and sovereign impulses and characteristics, which aim to determine the way of life from within, instead of receiving it from the outside as a universal form, schematically defined (Simmel, 2013, p. 316).

Simmel (ibid., p. 327) has observed that life has become easier as stimuli, interests, fillings of time and consciousness are offered to it from all sides. However, at the same time, when produced by organizations of things and powers, they suck life into a current in which it practically dispenses of all movement to swim. Thus, the metropolis, a phenomenon of modern social formation par excellence, has little internal cohesion and undefined limits, opening itself to infinite connections with outer space. What comes after is a programmatic discourse according to a new layer that overlaps the urban fabric, that is, the informational layer, distributed by devices that communicate with each other and compete for the attention of managers of public coffers. This is the case of smart cities and their developments from which technological improvement becomes a delirium for human beings as it goes beyond the limits of the body and, at the same time, helps us to connect with it.

Criticisms against smart cities do not stop there, nor are they limited to cities programmatically built for tourist exploitation and the growth of neoliberal capitalism, as is the case of Dubai and others of the same type. Criticism advances by pointing to questions still unanswered, such as:

Are cities that make massive use of technologies more flexible and adaptable? Does public participation, through applications and websites, establish two-way communication, or is it just a simulacrum of popular participation? With more data collected, what is actually done, how is it used in planning and concrete actions? Does the datafication of urban management generate effective actions, improving the population's living conditions? What are the instruments for controlling, auditing, and evaluating these projects? (Lemos, 2021, p. 75-76).

Amid these issues, everything related to the monitoring of big data by means of artificial intelligence algorithms has been placed under strict critical scrutiny. A discussion of the risks is provided by Acemoglu (2022) who begins by recognizing the transformative effects of artificial intelligence on the economy, society, politics as well as the impact of AI algorithms on social media, data markets, worker monitoring and automation of work. However, while technology platforms are used for the development of a variety of new products, services and production techniques, their dark potential must also be considered.

Thus, for the author, the demand of AI technologies for data and the ability of AI techniques to process vast amounts of data about users, consumers and citizens produces a number of troubling disadvantages, such as (a) breach of privacy by companies and platforms that collect and distribute excessive amounts of information about individuals for the processing and use of their data; (b) unfair competition: companies with more data can gain a strong advantage over their competitors, which allows them to extract more surplus from consumers. As a result, price competition in the market is relaxed, producing potentially deleterious effects; and (c) behavioral manipulation when machine learning techniques can allow companies to identify and exploit biases and vulnerabilities that consumers themselves do not recognize (Acemoglu, 2022, p. 2).

CONCLUSION

Among the harmful effects of artificial intelligence, the one that most affects the exercise of healthy citizenship is, without a doubt, the invasion of privacy, a factor that, together with issues related to sustainability, intensifies the ambivalences of smart cities to the point of transforming them into paradoxes, otherwise we will see.

While the inexhaustible collection of data related to individuals is fundamental to the administration and management of e-government for the treatment of security, health, locomotion, education, etc. of the growing populations of large metropolises, not to mention the billions of inhabitants of the globe, the aspect of surveillance and control has reached increasingly alarming levels since Deleuze (1992) wrote his text on the subject.

Whenever we enter the networks and go a little beyond the reactive responses to the system, an informational layer of another order of magnitude, unknown to us, is added to the networks. It records the profile history of our personal transactions: what we buy, our travels in urban space and its patterns, our personal profile (gender, age, zip code and related demographics such as area of residence, presumed social class, etc.). This involves data mining capable of establishing patterns that match known profiles of a likely type of consumer and that predict what we will buy at some point next and where. To do this, computers work with metadata that have been trained on what and where to look through neural network algorithms in which performance is measured in terms of the percentage of false positives or false negatives (Shepard, 2011, p. 31).

More than that, all real-time information is correlated with historical data on our mobility patterns, purchase history, personal interests and preferences which leads to accurate predictions about what our wants and needs may be or what actions we will take in the near future. As if that wasn't enough, let's imagine, for example, what happens when data from our profile and facebook timeline are added to this whole mix?

We cannot guarantee anything about preserving data privacy when all these inferential engines are put into action. When commercial interests and governmental and corporate agencies hold back these technological developments, along with them new forms of control, surveillance and consumption emerge. This is the price we pay, a high price, in exchange for expectations, not always fulfilled, of safety, health, quality public services, etc.

The recognition of the shadows that surround the exercise of citizenship also does not stop there. The world population is expected to increase by two billion people in the next 30 years, according to the United Nations projection. The total population of the planet should increase from the current 7.7 billion to 9.7 billion in 2050. The research, published by the Department of Economic and Social Affairs of the United Nations (UN), also states that the world population can reach its peak at the end of the century, approximately eleven billion people. However, we are talking about a world where about 2.9 billion people have never used the connection according to the International Telecommunication Union (ITU).

Elaborated in 2012 during the United Nations Conference on Sustainable Development Rio +20 and launched in September 2015 at the Sustainable Development Summit, the 2030 Agenda outlines a global plan with seventeen objectives and 169 targets for the 193 member countries to move towards sustainability, a balance between the economic, social, and environmental sectors. What was once ecological urbanism, sustainable urbanism, eco-urbanism, among others, today moves from groups of urban planners to groups of data scientists focused on capturing sensors spread across the globe. Cities become smart as the established goals are surpassed towards the goals listed by the ISO 37122:2019 standard. However, with the increase of entire populations moving to metropolises, migrations and immigrations, the question remains "why, for whom" resources and public policies are allocated to cities that will soon receive the increase of millions of people? (Santaella, org., 2016).

Let's look at the case of the city of São Paulo, which occupies the first place in the Brazilian ranking of smart cities at the moment. What are the city limits? Impossible to define, in most of the northern portion, due either to the land base, or to the disturbance processes between municipalities in the metropolitan region, as well as the western and eastern portions. So, we are dealing with public policies for which population, limited to which place? The three main rivers of São Paulo: Tamanduateí, Tietê and Pinheiros, together with their respective hydrographic basins, are unhealthy due to constant delays in sanitation projects. The city breaks records with hundreds of kilometers of traffic jams daily.

In addition, researchers Marcelo Batista Nery, Altay Alves Lino de Souza and Sergio Adorno published in 2019 a study on the urban-demographic patterns of the city of São Paulo, based on census sectors, and found that "there are at least eight different types of urban fragmentation on the micro/local scale", noting that "none of the groups from A to H is reflected in the existing political-administrative divisions, denoting a disconnection between public management and city structure". It is still possible to note that "the center-periphery dichotomy is inadequate for understanding the city and its good management, as it ignores or minimizes social and economic diversity and the uses of the urban fabric" (Nery, Souza, and Adorno, 2019, p. 29-30). Apparently, in the absence of other necessary studies, we created technological bubbles scattered throughout the urban fabric, islands of concentrated wealth and a lot of cameras and sensors that are mixed with urban furniture.

Faced with this imbroglio of difficulties, it is worth listening to Morozov (2019) in his proposal of alternative forms of public and community property in which open codes and free software stand out, in contrast to the current regime of data ownership of the technology giants, although, for the author, smarter cities is an IBM's advertising creation. In Brazil, for example, even though it is a developing country, it has traditionally moved the public debate positively, firstly, with the Civil Rights Framework for the Internet and, more recently, the General Law for the Protection of Personal Data. In addition, in the city of São Paulo, the Strategic Master Plan, prepared from decade to decade, is organized in such a way that the entire population participates, including through the internet.

A question that remains is: would companies that develop applications for cities (and these can be both giants and start-ups) renounce their copyrights for the benefit of the common good? Investments in billions of dollars and offering services on demand, subscription life (and we can highlight the reported case of Microsoft that will convert active users into billions of subscribers to its Game Pass) give us indications that the answer is no. The cinematographic representation of technological conglomerates in the humorous film "Don't Look Up" also makes us think negatively. As a possible alternative, perhaps the delegation of data protection and manipulation to universities and research centers would give companies the attribution of support to projects and their implementation in order to balance conflicts.

What can be concluded from all this is that smart cities are today taking the ambivalences, paradoxes and contradictions that have always been the hallmark of the digital universe to a paroxysm. The physical population growth of the metropolises seems to be inextricably accompanied by the growth in the complexity of the problems. This can only lead us to the idea that facing these problems should be an incessant task that depends on the multisectoral engagement of society and the activation of the innovative, critical, and creative capacities of the human being. After all, the shadows of the contemporary should not be able to hide all the flashes of light that must be activated in view of the ideals of sustainability that govern the spirit of a common life, a key principle of citizenship.

REFERENCES

Acemoglu, D. (2022). *Working Paper 29247. Bureau of Economic Research.* https://www.nber.org/papers/w29247NATIONAL.

Benevolo, L. (1997). *História da Cidade*. Perspectiva.

Bruno, F. (2013). *Máquinas de ver, modos de ser: vigilância, tecnologia e subjetividade*. Sulina.

Cacciari, M. (2010). *A cidade*. Gustavo Gili.

Chen, H. (2006). *Intelligence and security informatics for international security: information charing and data mining*. Springer. doi:10.1007/0-387-30332-4

Chen, H., Chiang, & Storey. (2012). Business intelligence and analytics: From big data to big impact. *Management Information Systems Quarterly*, *36*(4), 1165–1188. doi:10.2307/41703503

Colomé, J. P. (2019). *Como você é espionado por seu celular Android sem saber*. https://brasil.elpais.com/brasil/2019/03/17/tecnologia/1552777491_649804.html

Deleuze, G. (2002). Post-scriptum sobre as sociedades de controle. In Conversações (pp. 219-226). Editora 34.

Elias, F. (2013). *Big Data é oportunidade para o setor público*. http://www.guiadascidadesdigitais.com.br/site/pagina/big-data-oportunidade-para-o-setor-pblico

Felice, M. (2009). *Paisagens pós-urbanas. O fim da experiência urbana e as formas comunicativas do habitar*. Annablume.

Felice, M. (2016). *Entrevista concedida a Eduardo Weinhardt*. https://transobjeto.wordpress.com/2016/04/03/estrevista-massimo-di-felice/#more-1038

Flusser, V. (2015). *Comunicologia. Reflexões sobre o futuro*. Martins Fontes.

Harvey, D. (2009). *Condição Pós-Moderna*. Edições Loyola.

Jenkins, H., Green, J., & Ford, S. (2014). Cultura da conexão. Criando valor e significado por meio da mídia programável. *Aleph (Jerusalem)*.

Kohn, E. (2013). *How forests think. Toward an anthropology beyond the human*. California University Press.

Latour, B. (2005). *Reassembling the social. An introduction to actor-network-theory*. Oxford University Press.

Lemos, A. (2021). *A tecnologia é um vírus*. Sulinas.

Mcluhan, M. (2007). *Os meios de comunicação como extensões do homem*. Cultrix.

Morozov, E (2019). *A cidade inteligente. Tecnologias urbanas e democracia*. Ubu.

Nery, M. B., Souza, A. A. L., & Adorno, S. (2019). *Os padrões urbano-demográficos da capital paulista*. doi:10.1590/s0103-4014.2019.3397.002

Santaella, L. (2007). *Linguagens líquidas na era da mobilidade*. Paulus.

Santaella, L. (2013). A cidade e o corpo como interfaces. In *Comunicação ubíqua. Repercussões na cultura e na educação* (pp. 55–70). Paulus.

Santaella, L. (2014). *Dos não-lugares à cidade senciente*. https://periodicos.sbu.unicamp.br/ojs/index.php/rua/article/view/8638917

Santaella, L. (2016). *Cidades inteligentes. Por que, para quem?* Educ.

Schrock, A. R. (2012). *Mobile Interface Theory: Location-Aware Mobile Devices in Urban Environments*. https://ijoc.org/index.php/ijoc/article/viewFile/1727/783

Secchi, B. (2019). *A cidade do rico e a cidade dos pobres*. Âyiné.

Shaviro, S. (2003). *Connected. Or what it means to live in the network society*. University of Minnesota Press.

Shepard, M. (2011). *Sentient city. Ubiquitous computing, architecture, and the future of urban space*. The MIT Press.

Simmel, G. (2013). As grandes cidades e a vida do espírito. In A. Botelho (Ed.), *Essencial Sociologia* (pp. 311–329). Companhia das Letras.

Souza, E., Silva, A., & Frith, J. (2012). *Mobile interfaces in public spaces: Locational Privacy, Control, and Urban Sociability*. Routledge.

Spyer, J. (2007). *Conectado. O que a internet faz com você e o que você pode fazer com ela*. Zahar.

Townsend, A. M. (2013). *Smart cities. Big data, civic hackers, and the quest for a new utopia*. W.W. Norton & Company.

KEY TERMS AND DEFINITIONS

Big Data: Literally means a gigantic volume of data. Therefore, it is usually defined through five Vs: volume, velocity, variety, value, and veracity. Volume comes from the growing increase in data with which the Web is fed today. Velocity refers to the increased scalability of the computer that allows for faster processing of data. Variety accounts for the heterogeneity of data types. Value implies the concern with the reliability of the data and, finally, Veracity corresponds to the distinction that must be established between false and true data.

E-Governance: Consists of the use by governments of information and communication technologies in the provision of services and dissemination of information on the Web through or from public organizations. For this to be achieved, the information and digital infrastructure available to citizens and society in general must have a wide reach.

Internet of Things (IoT): Means the ability of different types of objects to connect to the Internet, from home appliances to cars. This is possible because objects have acquired, thanks to specific technologies, the ability to collect and transmit data from the cloud. IoT devices are already being used both in common situations of daily life and in the professional scope.

Privacy: When defining privacy, it is necessary to differentiate what is privacy in general from the concept that privacy has acquired in the digital universe. In this, privacy refers to informational privacy and is defined as a universal right of network users' data protection so that there is no damage to private life, honor or other damages involving legal issues.

Sentient Cities: A sentient being is defined by being able to feel and experience feelings, such as pain, anger, joy, etc. Although it is evidently not capable of experiencing feelings of this kind, a city can also be sentient because, thanks to the sensors and connectors with which it is provided, it becomes a great vector of communication between devices and humans.

Society of Control: Gilles Deleuze proposed that the society of control continues but is distinct from the disciplinary society studied by Michel Foucault. While this was governed by the disciplinary and confinement strategies of the institutions of modernity, the societies of control, in turn, are governed by the invisibility and nomadism that is typical of the functioning of informational networks.

Chapter 6
Digital Media and the COVID-19 Urban Crisis:
What Can We Learn From the Crowdsourced Mappings in Brazil?

Laryssa Tarachucky
University of São Paulo, Brazil

Rafael Soares Simão
https://orcid.org/0000-0001-5566-2456
Federal University of Santa Catarina, Brazil

Maria José Baldessar
Federal University of Santa Catarina, Brazil

ABSTRACT

COVID-19 is not the first pandemic we suffered during the urban era, but it is the first one to force governments to respond with such large-scale severe restrictions. Alongside institutional measures are the proliferation of collective endeavors that respond to community demands emerging from the struggle to contain contagion. Individuals of different age groups and demographic profiles are crowdsourcing their own solutions for both local and national issues using online tools to map and respond to pandemic- and lockdown-related crises. In seeking to understand these technological appropriations during the COVID-19 crisis in the Brazilian context, where 87% of the population lives in cities, the authors employed an online ethnographic research method and conducted a thematic analysis on 41 crowdsourced mappings in the country. The results present five main groups of matters of concern conveyed by these maps as well as the role of digital media in addressing them.

DOI: 10.4018/978-1-6684-3369-0.ch006

INTRODUCTION

In late 2019, an outbreak of pneumonia started in the city of Wuhan, China, rapidly spreading to other parts of the globe. Identified on January 7th of the following year, the new coronavirus (SARS-CoV-2) had consequences not only on public health systems, but also on several different aspects of urban life (Gao & Yu, 2020). Coping with the fact that the disease required cross-sectoral collaborations, the solutions were not only in health-related. The establishment of systems for remote work, the intensification of the digital infrastructure and the forming of alliances between local authorities and technology companies to help track the virus's spread and predict new outbreaks of the disease were just a few out of many and diverse responses to the pandemic. With them, some debates gained strength, such as concerns about the consequences of the surveillance power given, on an emergency basis, to companies and states over people's lives and its implications on privacy and the exercise of citizenship, given repeated violations by companies like Google and Facebook (Zuboff, 2018) and the technocratic approach city officials often take when adopting "smart city" technologies (Söderström, Paasche, & Klauser, 2014).

COVID-19 is not the first pandemic the world faces after the urban era, but it is the first one to cause governments to put in place such large-scale mobility restrictions (both within and between cities and countries) (Askitas, Tatsiramos, & Verheyden, 2020) spread in the context of consolidated internet access.

On the other end of these measures, given the delay - even unpreparedness - of public institutions to respond to community demands emerging from the struggle to contain contagion in cities around the world, is the proliferation of collective endeavors. Individuals of different age groups and demographic profiles have crowdsourced their own solutions for both local and national issues by using online tools to map and respond to pandemic and lockdown-related crises.

In this paper, we assume that the crowdsourced maps made by local networked publics can express matters of concern that are not met by official institutions. In seeking to understand these technological appropriations during the COVID-19 crisis, as suggested by Tarachucky (2021), the research presented here is guided by two questions: (i) how are digital media appropriated by citizens to create these crowdsourced maps? and (ii) what are the different types of crowdsourced maps emerged during the COVID-19 pandemic crisis in Brazil? To answer these questions, this research excludes mappings from official sources of information, such as state agencies or news outlets. In addition, to limit the scope of the research, the locus of analysis is limited to the Brazilian context.

We situate our work on the intersection between urban studies and media studies to make the following contributions: First, we provide an overview of the literature on the use of crowdsourced maps as tools to deal with urban crises. Second, we present an analysis of forty-one crowdsourced maps produced in Brazil between March and June 2020. Each of these maps highlight different community strategies to deal with multiple factors of an issue. Third, we classify these initiatives to provide a set of themes that express matters of concern that might be addressed in the context of similar pandemics. Finally, we reflect on urban studies and media studies in the context of public health crisis, seeing roles for urban planners, designers and policy makers.

The remaining of this paper is structured as follows. Section 2 provides an overview of the impact of COVID-19 on urban space and introduces the concept of crowdsourcing. In Section 3, the research methodology as well as the analysis lens are presented. Sections 4 analyzes the results obtained and Sections 5 and 6 offer an outlook for further research and summarize our key findings.

Background

The word "city", as discussed by Sennet (2018), has a double meaning: the physical space (ville) and the way in which people inhabit that space (cité), and both play important roles when dealing with sanitary crises. Throughout history, infectious diseases have influenced how cities are built and inhabited. The relationship between the built environment and the population's health condition is a recurring topic in the effort of eliminating various diseases (UN-Habitat, 2021). The bubonic plague in the 14th century and the Spanish flu in the beginning of the last century are examples that have mobilized urban planning and sanitation development in cities so that inhabiting them remained possible. And, as cities grow and become denser, sanitary conditions become an urgent matter that involves both directing the shape of the built environment, by creating regulations such as minimum street widths or sidewalk paving standards and trying to change how people act around the city, by establishing rules on how to behave within different environments (Sennet, 2018).

At the beginning of 2020, that debate was brought up again. On January 30th, 2020, the new coronavirus outbreak was classified by the World Health Organization (WHO) as a Public Health Emergency of International Concern. On March 11th, WHO declared COVID-19 a pandemic. By April 21st, 2020, 2.4 million people had been infected and 163,000 had lost their lives around the globe - numbers that, by the same date in 2021, had grown to 148 million and 3 million, respectively. Among these, there were 14 million cases and around 400,000 deaths in Brazil. The containment of the virus has demanded international efforts, not only for treating the infected (medical supplies, IPE, etc.), but also scientific cooperation on ways to improve treatment and reduce the COVID-19 contagion curve (Gao, & Yu, 2020).

On the institutional level, closing borders was among the first measures taken by governments - national or statewide - to block the virus from entering territories with no cases. In afflicted cities, contagion control measures frequently include social distancing policies, mask mandates, communication, and awareness-raising campaigns (e.g., stay at home), reduction or suspension of public transport systems, social distance markings in shared spaces or even temporary lockdowns. Aside from containing the number of cases, the spatial demands for treating and isolating those infected have led authorities to take a series of other measures. In Genova, Italy, authorities have converted a ferry into a hospital ship due to high demand. Hotel rooms have been used to house health workers and endangered individuals, while football stadiums have been converted into field hospitals in Brazil. Automated control and vigilance systems have also been put in place in different places: countries such as China, South Korea and Singapore have used robots, drones and big data to trace the outbreak, disinfect hospitals and deliver supplies. In Singapore, a robot would patrol public spaces, detect groups of people and ask them to keep their distance. Similar robots have been deployed in other countries to warn people about mask mandates and measure people's body temperature. Smartphone apps have been implemented to trace cases and warn people about cases in their surroundings as well as to encourage users to report lockdown rule-breakers (Cioffi, Lugi, & Cecannecchia, 2020).

On the individual side, life changed rapidly as lockdown measures were put in place and working from home became more frequent. At a first moment, changes in consumption behaviors became evident as individuals and communities were forced to find digital solutions to mitigate the impacts of movement restrictions (e.g., home office, delivery) and the negative effects (e.g., economic, social, and psychological) of social isolation. As time went by, a series of manifestations started taking place in virtual environments, such as artists' live performances on social media or quarantine shaming (e.g., calling out those who were attending parties during lockdown), and physical ones as well, such as balcony performances.

The COVID-19 pandemic has drastically altered the way people live and interact within cities. Such a crisis, in a context where over 50% of the world's population lives in cities (UN Department of Economic and Social Affairs, 2018), highlights the fact that public health is definitely an urban issue. The densification of the built environment may lead to a faster spread of infectious-contagious diseases (Sennet, 2018) and, since the new coronavirus is better transmitted in places where people are close together, contagion will likely occur more easily where there is higher population density.

In Brazil, a country where 87% of its population live in cities, this is aggravated especially in urban centers where lower income people travel long distances between their homes and work daily, sharing frequently overcrowded mass transit systems (Villaça, 2012). Low-income communities have become particularly vulnerable to the effects of the pandemic, as a consequence of the higher number of inhabitants per unit, lower access to sanitation and longer travel distances from public health units, which also contributes to expose those sharing mass transit to risk (Pereira *et al.*., 2020).

Despite institutional and individual efforts to ensure collective safety and wellbeing, COVID-19 revealed and intensified social challenges, from inequitable death rates between different social groups to the stretching of basic services (especially healthcare) to a breaking point (UN-Habitat, 2021). This time the responses to these challenges had the use of digital tools as a fundamental characteristic - something that did not exist in most of our historical parallels. This practice was particularly marked by the adoption of collaborative and distributed network approaches. Employing a series of different media, civil society members have organized spaces where anyone with access to the Internet can contribute, crowdsourcing information about matters of concern regarding the crisis caused by both the pandemic and its countermeasures and making them visible and actionable.

MAIN FOCUS OF THE CHAPTER AND METHODOLOGY

Crowdsourcing is a term originally used to describe the act of taking work once performed within an organization and outsourcing it to a network of people in the form of an open call for participants (Howe, 2006). It is often used to deal with humanitarian and urban crises (Tarachucky *et al.*., 2021). There are several cases where the international community has gathered online to help populations affected by disasters, especially in high-profile cases such as the Fukushima nuclear disaster (Kera, Rod, & Peterova, 2013), the search for remains of Malaysian Airlines flight MH370 (Xu *et al.*., 2017) or the collective effort to map and update geographic information after the Haiti earthquake in 2010 (Zook *et al.*., 2010).

Recent records show that crowdsourcing practices had already been employed in combating disease outbreaks such as Ebola (Abbott, & Barbosa, 2015; Brown *et al.*., 2015). During first months of the COVID-19 outbreak, on an institutional level, data and information collected through crowdsourcing was used to analyze the prevalence of the disease, its incidence among different groups and locations, and its evolution over time and space (UN Habitat, 2021). However, as the crisis worsened, communities started applying these techniques to map and respond to a multitude of community matters of concern unaddressed by public authorities (Kokshagina, 2021). These community crowdsourced mappings, in Brazil, are the focus of this chapter.

The methodological process began with the identification of collaborative digital mappings that emerged from community mobilizations related to the pandemic. From March to June 2020, we employed an online ethnographic research method (Tierney, 2013) to collect data from social media platforms, WhatsApp groups and webpages, which consisted of the following steps: (1) authors joined the Santa

Catarina state's Urbanismo contra o Corona (Urban Planners Against Coronavirus) WhatsApp Group and gathered shared information; (2) authors checked Google and news outlets to search for collaborative actions and COVID-19-related mappings between March and May 2020; (3) authors checked a series of online, open-to-all spreadsheets containing crowdsourced data, and each entry in those was manually checked for geographic information; (4) authors selected those that required users to enter at least the city each entry referred to; and (5) authors had each selected initiative, their characteristics and scope laid out in a matrix. [1]

This resulted in a set of forty-one crowdsourced mappings used as data for this research. Next, a thematic analysis (Braun & Clarke, 2006) was conducted on this corpus, according to six phases: (1) familiarization with the data, (2) generating initial codes, (3) searching for themes, (4) reviewing the themes, (5) defining and generating names for the themes, and (6) writing the report.

In order to determine how digital media are applied to COVID-19-related crowdsourced maps (research question 1), data was sorted according to the scope of information (local or national), the mode of data collection (active or passive), the types of media used for data collection, and the means by which results were disseminated. Then, the thematic analysis sought to determine the range of different mappings that Brazilians crowdsourced during the early stages of the COVID-19 pandemics (research question 2). We explored the interplay of objects and issues in collective endeavors that respond to pandemic and lockdown-related crises by looking at the crowdsourced mappings through the lens of matters of concern, a concept related to the notion of networked publics (Latour, 2008; Latour, & Weibel, 2005). It derives from the discussions aroused by the exhibition Making Things Public: Atmospheres of Democracy, which explored how objects can express the factors and consequences of an issue and make them accessible to a broader public.

FINDINGS AND DISCUSSION: CROWDSOURCED MAPS EMERGED DURING THE COVID-19 PANDEMIC IN BRAZIL

Mapping Characteristics

Unlike other examples in the literature, where volunteers from several countries participated in assisting a specific crisis, many of the mappings found in the present study came from the affected populations themselves. The number of initiatives helps visualize the dimension of the pandemic impact on the most vulnerable sectors of Brazilian society. Although the data are not standardized among the different initiatives, the fact that they are all catalogued at least by municipality - and in some cases by exact geographical location or address - makes it possible to use them in GIS software as well as to identify spatial relationships between matters and other variables. In this section we present general characteristics of the mappings identified, such as their scope, the different media used to capture data, and the ways in which information is disseminated.

Forty-one initiatives were surveyed, of which twenty-nine have a national scope, allowing the insertion of data related to any location in the country. Two out of the remaining twelve concentrate information on specific states and the other ten refer to metropolitan regions (Table 1), of which five serve the metropolitan region of São Paulo, the largest urban concentration in the country with 21.5 million inhabitants.

The data collection was voluntarily entered by users in the media made available by the organizers in most cases, except for the two apps intended to monitor the outbreak. In these two apps, location was entered automatically, based on the mobile device's GPS.

Table 1. Mappings reach

Initiative's reach	Number of initiatives
National	29
City level	10
State level	2

Source: The authors (2022)

In general, the initiatives used a single data collection channel, except one survey conducted by UOL, a Brazilian media portal, which accepted the submission of information via email, Twitter and Instagram. Data entry was mainly done by filling out predefined fields, using the Google Forms service or publicly available Google Sheets spreadsheets (Table 2).

Table 2. Digital media used for data collection

Media	Number of initiatives
Online form (Google Forms or Google Sheets)	23
Online form (website)	10
E-mail	4
Smartphone app	2
Facebook	1
Instagram	1
Twitter	1

Source: The authors (2022)

In some other cases, the website itself had embedded forms in which the user could fill in the data. For example, in the initiatives aimed at strengthening the local economy, users would register their establishment or that of others, providing information such as the establishment's name, address, contact numbers and the types of services or goods it sold, which would then serve as filters for visitors.

The initiatives surveyed also present different degrees of sophistication in the dissemination of information. Most of them (28) gather the information in lists or spreadsheets, publicly available, where users can either enter data or view previous entries. The remaining initiatives (13) use maps for data display and/or collection, making it easier to visualize the extent of the problems and the most affected regions.

Mapping Categories

The technological objects that express matters of concern are committed to the particularities of a situation, addressing a set of conditions in ways that align with the public's values and practices (Di Salvo *et al.*., 2014). The thematic analysis conducted has helped us identify five main groups of matters of concern which were addressed by these mappings: (1) monitoring the outbreak, (2) Monitoring COVID-19 social and economic impacts, (3) Aggregation of initiatives and opportunities, (4) Solidarity networks facilitation and (5) Strengthening the local economy. Each of the analyzed mappings express matters of concern for self- organizing Brazilian communities. They also convey lived experience and near-future desires. Each one of them covers a set of conditions emerging in these communities due to the pandemic context and communicates perspectives on how technology might address those conditions.

Monitoring the Outbreak

The first group reported here is made up of platforms built with the goal of monitoring the evolution of the pandemic in cities and containing the spread of the virus. The COVID-19 virus is primarily transmitted when a person breathes in respiratory droplets that are ejected when an infected person coughs or sneezes (WHO, 2020). These particles can cover a distance of 1.8m, putting any individual close to a carrier at risk. Measures to contain the outbreak include detecting those infected, isolating them and tracing their contacts (Chamola *et al.*., 2020), as well as individual care related to keeping distance between yourself and another person. On a community level, these measures included working remotely, canceling events, and even went as far as shutting down crowded establishments and mass transit systems.

While vaccines and specific drugs were not available to treat the illness, reducing the number of new infections was mandatory to ensure that large numbers of critically ill patients were not left without the possibility of medical care. The platforms identified in this group generally combine health data provided voluntarily by users with their mobile devices' location data. They contribute to the identification of points of contagion and can help authorities take specific measures for regions at risk of outbreak. They also help their users better understand how the virus is affecting their surroundings, so they may avoid hotspots, or even take more severe precautions.

Specifically, these platforms reveal the need to match the real state of the infection with data obtained by health authorities to have a more accurate picture of the outbreaks. The *Desviralize* platform is an example of this situation. It was designed to decentralize information and offer more accurate data about the behavior of the virus in spatial terms through crowdsourcing.

The *Brasil Sem Corona* and *Juntos Contra o COVID* initiatives used a similar logic of obtaining and providing data by building a collaborative platform that allowed users to answer a questionnaire about possible symptoms of the disease and provide information in case they had contact with someone who has contracted the virus.

In these cases, the platforms were concerned with bridging the gap between the actual state of contamination, based on anonymous information in real time, and the information that reached local authorities in order to allow for more accurate and rapid responses to the situation.

Monitoring COVID-19 Social and Economic Impacts

The impact of pandemic related restrictions was greater on certain populations and economic sectors. Eight of the identified mappings had the objective of locating and monitoring people and businesses most affected by these restrictions and used online forms and email to collect data.

These initiatives can be divided into three subgroups. The first one aims at collecting data about specific groups or economic sectors. Students from the University of São Paulo, for example, have mapped the damage to third sector activities and those dependent on them. The *Pelas Nossas Mães* project collected data about domestic workers' working conditions to write a manifesto demanding adequate policies to protect the category from abuse and unemployment, while the *Casa Mãe* Institute conducted a survey to detect mothers in vulnerable situations.

The second subgroup focuses on the assurance of fundamental rights during the pandemic. The *Observatório de Direitos Humanos* and the *Rede de Proteção e Resistência ao Genocídio* have created channels for reporting human rights violations and police violence in São Paulo periphery. *Pastoral Carcerária Nacional* has gathered data on jailed women's situation and the problems they faced amidst the sanitary crisis.

The status and quality of public services was the last subject monitored. The water supply was mapped by two initiatives, one limited to the city of São Paulo and another on a national level. Another mapping has looked upon the educational system, identifying places where on-site classes were paralyzed or partially and fully resumed.

Aggregation of Initiatives and Opportunities

The third group is made up of projects concerning the aggregation of initiatives and opportunities. As the new coronavirus began to spread throughout the Brazilian territory, different groups of people, with different needs, began to gather around common concerns and act on shared issues. The vast amount of information that people were exposed to in that time course and the sparse way in which that information was delivered to people, spread through several websites and social media profiles, made it hard for those who were in need to find help and opportunities.

The thirteen mappings identified in this group are actions that respond to the need to organize and centralize information as well as allow the targeted search for initiatives, bridging the gap between them and the communities they are intended for. Thus, they met the need that grassroots communities had for easy access to knowledge about activities provided by community or private entities, job opportunities and public notices for the funding of creative solutions.

The initiatives concerned with this issue contained items from the other mapped groups, such as solidarity campaigns or impact monitoring, but were broader and more varied, generally splitting the items in several categories. The largest of these initiatives, conducted by the *Ponte a Ponte* organization, had nearly three hundred contributions by May 2020 and resulted in a detailed report on the trends of the scenario, its main actors, the quality of communication and recommendations on how to better create new campaigns (Aoqui *et al.*., 2020).

Solidarity Network Facilitation

The need for social distancing, added to periods of lockdown and restrictions to mass transit capacity affected people's income generation and access to basic health services, especially among those who already faced financial difficulties before the pandemic, placing them in situations of social vulnerability. In some cases, the sharp and sudden drop in income made it difficult for people to have access to basic hygiene and food items. Seeking to alleviate these problems in the face of the government's delay in putting affirmative actions into practice, different organizations started to map information about communities and their needs, as well as information about groups that were distributing certain resources.

The concern with forming solidarity networks is the issue that attracted the greatest number of mappings, counting fourteen initiatives. They differ from those in the previous group (although many were identified from within those surveys) because they have an exclusive focus on solidarity actions, even though they cannot be characterized as direct actors. Any user with access to the spreadsheets or websites was able to contribute and, although the objectives are specific, ten of these mappings have a national scope.

These initiatives give visibility to community leaders, people in vulnerable situations and their demands, and research or volunteer and non-governmental organization initiatives, thus creating networks of cooperation and information sharing. The website *Movimentos contra a COVID-19*, for example, aims to form such networks and has received data from around one thousand and three hundred community demands and more than eight hundred actions from universities and research groups distributed all over Brazil.

Strengthening Local Economies

Strengthening the local economy, especially small businesses, was identified as a matter of concern in two initiatives, both websites with local coverage: *Fique em casa BH*, in Belo Horizonte, state of Minas Gerais, and *No bairro tem*, in Recife, state of Pernambuco, which display, on a map, small businesses close to the user, sorted according to user-selected filters. In these websites, it is possible to find restaurants, mini-markets and other products and services that are not offered by conventional location-based service apps, which dominate the delivery market and whose main focus is food or supermarket delivery, such as iFood, Rappi, Loggi, Uber Eats, etc.

The emergence of these alternatives reflects not only the impacts of social distancing and the movement restrictions, but also the discontent of establishments and delivery workers with the aforementioned apps. The high fees charged from establishments together with low wages paid to workers and the lack of workers' support by the apps before and during the pandemic has led to a series of strikes in several cities (Delgado, & Carvalho, 2020). By giving up the management of payments and nation-wide coverage, local platforms allowed for direct contact between establishments and consumers, who then negotiate rates and service provision conditions.

FUTURE RESEARCH DIRECTIONS

The characteristics of crowdsourced maps can vary significantly depending on the chosen methods and goals. Future research will need to reflect upon these variations more deeply if we intend to better understand the usefulness of crowdsourced maps within the messy space of public health crises. As we continue to explore these technological appropriations, collecting data about what kinds of challenges are faced and how they are addressed among people, as well as measuring tangible impacts of the crowdsourced maps, both for inhabitants and for managers, can provide useful insights on possible uses of crowdsourced mappings for participatory planning activities.

CONCLUSION

COVID-19 had a great impact on urban life, imposing diverse restrictions and forcing people to reorganize the most diverse aspects of their daily lives. Crises like this one involving lockdown measures, social distancing, and diverse public health issues, generate a situation of urges and emergencies that are detected at a hyperlocal level, a great number of issues that need to be solved quickly and, as a counterpart, an extraordinarily strong social creativity and sense of agency.

In particular, the COVID-19 crisis revealed the decisive role of digital technologies to respond to collective hyperlocal issues, offering the necessary tools for rapid, self-organized responses and serving as means to build long-term resilience. Even though COVID-19 is a global crisis that needs to be addressed by governments at all levels, citizens are often required to act upon a local scale in order to address different issues that are not met by the public institutions - either for lack of time, lack of human and material resources or also for negligence.

The generation of crowdsourced maps was frequently used by Brazilian publics to solve collective issues arising from the COVID-19 crisis. In this paper, we analyzed the matters of concern expressed in the mappings made by Brazilian publics during the early stages of the COVID-19 crisis. We found matters of concern as a useful lens to explore these collective endeavors. In assessing the matters of concern expressed in the crowdsourced mapping of issues present in small communities or wider urban spaces during the course of public health restrictions can help us to identify possible avenues of action during similar situations. This can be useful both in the case of new waves of the pandemics or in the case of a novel virus. In addition, it can help urban publics to better inform the planning of processes to give faster and more structured answers to localized issues.

The objects we analyzed, the crowdsourced maps, are artifacts made in a political sense - of positioning and/or facing the slowness in the response (or lack of it) to population's needs by Brazilian government - in the context previously reported and express five main challenges faced by Brazilian communities. Firstly, they express the need of an entire population for reliable data regarding the status of the virus in their communities and for guidance on how to proceed in relation to the most affected areas. Secondly, they reveal the need for an equal distribution of resources since the coronavirus has made social problems more evident and widened the gap between public authorities and those in situations of social vulnerability. They also suggest a call for city officials and urban planners to take factors such as population density, inadequate housing and distribution of basic supplies more seriously. Thirdly, they disclose the need for centralized reference sources to let people know where to offer help or seek support and to make sure initiatives reach their target audience. Finally, they raise awareness about the need for more horizontal economic models and offer the possibility of strengthening peer-to-peer and sharing networks.

There is a lot we still do not know about the SARS-CoV-2, especially the extent and scale of impacts that will emerge in cities and urban communities due to the pandemic. However, it seems that while previous public health crises have led to very strong changes in the *cité*, this pandemic offers new horizons of change fundamentally related to the *ville*. Tackling the pandemic has changed, at least temporarily, the way people interact with the world and with other individuals. Horizontal problem-solving structures have become necessary, challenging old practices. While people organize themselves to fight against the pandemic, questions are beginning to arise about which adjustments will be permanent and how the world will be like after it.

The COVID-19 crisis raised the question whether cities will ever be what they used to be after lockdown - and if they should be what they used to be. Much is speculated about the possibility of a stronger sense of society remaining as a collective whole, instead of a fragmented agglomeration of individuals - and that this could lead to a long-term increase in collective action towards balanced consumption, work, and socio-environmental practices. However, we argue that for digital media artifacts to be scalable and have effects that last beyond the immediate needs of a community during a public health crisis, its design must be explicitly concerned with building publics by means of articulating issues, shared values, and practices.

Treated as a matter of concern, the analyzed mappings can indicate new roles and courses of action for urban planners, designers and policy makers who are creating new approaches to the emerging urban issues related to health crises. On the one hand, when added to reliable forms of analysis, the data collected via crowdsourcing can be seen as common assets, useful to establish urban policies in order to provide a quick response to possible future pandemics, as well as to plan a sustainable recovery for the current one based on real community needs and concerns. Communities that organize to address social issues related to COVID-19, building monitoring platforms, mapping and collaboration networks, are generally well aware of local social demands. However, urban managers - responsible for decision-making at the city level - do not always have this direct view. For them, understanding the results regarding matters of concern can be a resource for developing actions and adjusting their governance strategies.

Moreover, the matters of concern expressed in the analyzed mappings are good sources of inspiration for designers who wish to engage with processes that deal with the near-future consequences of the above-mentioned situations, i.e., the need for monitoring and tracking diseases' outbreaks and impacts, aggregating initiatives and opportunities, facilitating solidarity networks, and strengthening local economies. They are useful not only to guide participatory design processes that respond to issues of survival, but also to guide design processes that think beyond (e.g., speculative, critical or reflective approaches on how post-pandemic cities should be).

ACKNOWLEDGMENT

This research was partially supported by *Coordenação de Aperfeiçoamento de Pessoal de Nível Superior* (CAPES) – Finance Code 001.

REFERENCES

Abbott, P., Barbosa, S. F. F. (2015). Usando tecnologia da informação e mobilização social para combater doenças. *Acta Paulista de Enfermagem*, *8*(1), iii-v.

Aoqui, C., Añón, R. (2020). *Os primeiros 60 dias de COVID-19 no Brasil em 60 fatos, reflexões e tendências em filantropia, investimento social e o campo de impacto social* [The first 60 days of COVID-19 in Brazil in 60 Facts, reflections and trends in Philanthropy, Social investment and the field of socia impact]. ponteAponte.

Askitas, N., Tatsiramos, B., & Verheyden, B. (2020). *Lockdown strategies, mobility patterns and COVID-19.* CESifo Working Paper (8338).

Braun, V., & Clarke, V. (2006). Using thematic analysis in psychology. *Qualitative Research in Psychology*, *3*(2), 77–101. doi:10.1191/1478088706qp063oa

Brown, B., Nasiruddin, M., Dao, A., & Halabi, M. (2015). Responsible use of pop culture and communication in the face of Ebola virus. *PLoS Neglected Tropical Diseases*, *9*(8), e0003890. doi:10.1371/journal.pntd.0003890 PMID:26248052

Chamola, V., Hassija, V., Gupta, V., & Guizani, M. (2020). A comprehensive review of the COVID-19 pandemic and the role of IoT, drones, AI, blockchain, and 5G in managing its impact. *IEEE Access: Practical Innovations, Open Solutions*, *8*, 90225–90265. doi:10.1109/ACCESS.2020.2992341

Cioffi, A., Lugi, C., & Cecannecchia, C. (2020). Apps for COVID-19 contact-tracing: Too many questions and few answers. *Ethics, Medicine, and Public Health*, *15*, 100575. doi:10.1016/j.jemep.2020.100575 PMID:32838002

Delgado, G. N., & Carvalho, B. V. (2020) Breque dos Aplicativos: Direito de resistência na era digital [Applications Breque: Right of resistance in the digital age]. *Le Monde Diplomatique Brasil.*

Disalvo, C., Lukens, J., Lodato, T., Jenkins, T., & Kim, T. (2014). Making public things: How HCI design can express matters of concern. In *Proceedings of Conference on Human Factors in Computing Systems* (pp. 2397-2406). Association for Computing Machinery. 10.1145/2556288.2557359

Gao, X., & Yu, J. (2020). Public governance mechanism in the prevention and control of the COVID-19: Information, decision-making and execution. *Journal of Chinese Governance*, *5*(2), 178–197. doi:10.1080/23812346.2020.1744922

Howe, J. (2006). The rise of crowdsourcing. *Wired Magazine, 14*(6).

Kera, D., Rod, J., & Peterova, R. (2013). Post-apocalyptic citizenship and humanitarian hardware. In R. Hindmarsh (Ed.), *Nuclear disaster at Fukushima Daiichi: Social, political and environmental issues* (pp. 97–115). Routledge.

Kokshagina, O. (2021). Open COVID-19: Organizing an extreme crowdsourcing campaign to tackle grand challenges. *R&D Management.*

Latour, B. (2008). *What is the style of matters of concern? Two lectures in empirical philosophy.* Van Gorcum Amsterdam.

Latour, B., & Weibel, P. (2005). *Making things public: Atmospheres of democracy*. MIT Press.

Pereira, R. H. M., Braga, C. K., Servo, L. M., Serra, B., Amaral, P., & Gouveia, N. (2020). *Urban mobility and access to the healthcare system by patients with suspected and severe cases of COVID-19 in the 20 largest cities of Brazil*. IPEA.

Sennet, R. (2018). Construir e habitar: Ética para uma cidade aberta [Build and Inhabit: Ethics for an open city]. *Record*.

Söderström, O., Paasche, T., & Klauser, F. (2014). Smart cities as corporate storytelling. *City*, *18*(3), 307–320. doi:10.1080/13604813.2014.906716

Tarachucky, L. (2021). *Framework conceitual para apoiar a formulação de estratégias de design de interação urbana envolvendo o uso de recursos de mídia locative digital* [Doctoral dissertation]. Federal University of Santa Catarina. Repositório UFSC. https://repositorio.ufsc.br/handle/123456789/226801

Tarachucky, L., Sabatini-Marques, J., Yigitcanlar, T., Baldessar, M. J., & Pancholi, S. (2021). Mapping hybrid cities through location-based technologies: A systematic review of literature. *Cities (London, England)*, *116*, 103296. doi:10.1016/j.cities.2021.103296

Tierney, T. F. (2013). *The public space of social media: connected cultures of the network society*. Routledge. doi:10.4324/9780203093863

UN-Habitat (2021). *Cities and pandemics: Towards a more just, green and healthy future*. Author.

United Nations Department of Economic and Social Affairs. (2018). *World urbanization prospects: The 2018 revision*. Author.

Villaça, F. (2012). *Reflexões sobre as cidades brasileiras* [Reflections on Brazilian cities]. Nobel.

World Health Organization (WHO). (2020). *Modes of transmission of virus causing COVID-19: Implications for IPC precaution recommendations: Scientific brief*. World Health Organization.

Xu, Z., Zhang, H., Hu, C., Liu, Y., Xuan, J., & Mei, L. (2017). Crowdsourcing-based timeline description of urban emergency events using social media. *International Journal of Ad Hoc and Ubiquitous Computing*, *25*(1-2), 41–51. doi:10.1504/IJAHUC.2017.083481

Zook, M., Graham, M., Shelton, T., & Gorman, S. (2010). Volunteered geographic information and crowdsourcing disaster relief: A case study of the Haitian earthquake. *World Medical & Health Policy*, *2*(2), 7–33. doi:10.2202/1948-4682.1069

Zuboff, S. (2018). Big other: Capitalismo de vigilância e perspectivas para uma civilização de informação. In F. Bruno, B. Cardoso, M. Kanashiro, L. Guilhon, & L. Melgaço (Eds.), *Tecnopolíticas da vigilância: Perspectivas da margem* (pp. 17–68). Boitempo.

ADDITIONAL READING

Brabham, D. C. (2013) Crowdsourcing: a model for leveraging online communities. In The Participatory Cultures Handbook (pp. 120-129). Routledge.

Brabham, D. C., Ribisl, K. M., Kirchner, T. R., & Bernhardt, J. M. (2014). Crowdsourcing applications for public health. *American Journal of Preventive Medicine*, *46*(2), 179–187. doi:10.1016/j.amepre.2013.10.016 PMID:24439353

Créquit, P., Mansouri, G., Benchoufi, M., Vivot, A., & Ravaud, P. (2018). Mapping of crowdsourcing in health: Systematic review. *Journal of Medical Internet Research*, *20*(5), 1–23. doi:10.2196/jmir.9330 PMID:29764795

Dąbrowska, J., Keränen, J., & Mention, A. L. (2021). The Emergence of Community-Driven Platforms in Response to COVID-19: GetUsPPE, a multi-sided platform that emerged during the COVID-19 pandemic, offers key insights on how to mobilize and leverage diverse actors to provide solutions in emergencies. *Research Technology Management*, *64*(5), 31–38. doi:10.1080/08956308.2021.1945347

Franzoni, C., Poetz, M., & Sauermann, H. (2021). Crowds, citizens, and science: A multi-dimensional framework and agenda for future research. *Industry and Innovation*, 1–34.

Ghezzi, A., Gabelloni, D., Martini, A., & Natalicchio, A. (2018). Crowdsourcing: A review and suggestions for future research. *International Journal of Management Reviews*, *20*(2), 343–363. doi:10.1111/ijmr.12135

Tarachucky, L., Sabatini-Marques, J., Yigitcanlar, T., Baldessar, M. J., & Pancholi, S. (2021). Mapping hybrid cities through location-based technologies: A systematic review of the literature. *Cities (London, England)*, *116*, 103296. doi:10.1016/j.cities.2021.103296

Vermicelli, S., Cricelli, L., & Grimaldi, M. (2021). How can crowdsourcing help tackle the COVID-19 pandemic? An explorative overview of innovative collaborative practices. *R & D Management*, *51*(2), 183–194. doi:10.1111/radm.12443

KEY TERMS AND DEFINITIONS

Crowdsourced Maps: Maps that are produced collaboratively, by the use of crowdsourced geographic information. In crisis contexts, crowdsourced maps are often used by online communities to gather, analyze, and disseminate critical information related to affected areas and/or populations.

Crowdsourcing: The practice of inviting, via open calls, individuals, or groups to contribute to the completion of online tasks by sharing data or knowledge, completing given pre-defined tasks, or helping fund initiatives. These tasks can be either done for free or for different types of compensation.

Digital Media: A means of transmitting or communicating data or information encoded in machine-readable format. Any form of media that uses electronic devices for distribution.

Matter of Concern: Latourian concept related to the notion of networked publics. Related to matters that express the factors and consequences of an issue, highlighting controversies, differentiation, and contrasts, and proposing a discussion of differences.

Networked Public: Group of actors who use networked technologies to explore issues, achieve goals, or solve shared problems. The term is an alternative to terms such as "audience" or "consumer", highlighting a more engaged stance.

Outbreak Prediction: Attempt to foresee a localized increase in the number of cases of a disease during a non-epidemic epidemiological situation.

Self-Organized Crisis Response: Represents the spontaneous emergence of a networked public acting proactively in disaster relief activities.

ENDNOTE

1 Data is available in Portuguese at: http://bit.ly/3n0idzH

Chapter 7
Mitigating the Pandemic Through Creativity:
UNESCO's Responses and Cities' Reactions

Clara Bertrand Cabral
Universidade Aberta, Portugal & Open University, Portugal

Ana Pereira Roders
Delft University of Technology, The Netherlands

Rosana Albuquerque
Universidade Aberta, Portugal & Open University, Portugal

ABSTRACT

Culture and creative industries are nowadays a most relevant dimension of cities' economies and a fundamental asset of the tourism sector. The rise of COVID-19 has strongly and negatively impacted these resources, triggering responses from international organizations and municipalities eager to address the emergent constraints and difficulties experienced by society. This chapter analyzes how UNESCO assisted state members in dealing with the situation through the dissemination of information, studies, and reports, and how cities reacted to lockdowns, reduction of revenues, and the need to support and encourage their citizens. The analysis of information provided by UNESCO Creative Cities and World Heritage Cities concerning activities implemented provide a broad image of cities' reactions to the pandemic and of creative field distinctiveness.

INTRODUCTION

Cities are hubs of innovation and inventiveness, places where nowadays most of the world's population live, thrive, and try to accomplish their hopes and dreams (Huybrechts, 2018; Lee & Rodríguez-Pose, 2014). Creativity is an essential part of people's and community's existence, helping to attain and strengthen

DOI: 10.4018/978-1-6684-3369-0.ch007

a common identity based in shared memories traditions and experiences, thus forging the 'spirit of the place' that enlivens historic urban landscapes (Colavitti & Usai, 2016; Rodwell, 2018; Rogers, 2017).

Due to their interactive networks of people, markets and activities, cities are considered important development fields for creativity (UNESCO & UNDP, 2013). In 2011, the *Recommendation on the Historic Urban Landscape* assembled in a single regulatory document the dispositions in previous charters, conventions, declarations, memorandums and recommendations on the subject of the conservation of historic areas (UNESCO, 2011, footnote 2), thus formalizing and improving the reflections and works initiated four decades before, prompted by reactions to decontextualized urban renewal (Davidoff, 1965; Jacobs, 1961) resultant from the modernist movement (Bandarin & van Oers, 2012; González Martínez, 2017; Jokilehto, 2014; Taylor, 2020; Veldpaus, Pereira Roders, & Colenbrander, 2013). This shift to a broader awareness of the city that recognized the need to manage both the historic buildings and its surroundings, as well as the importance of heritage, cultural diversity and creativity for society (Council of Europe, 2005; Fojut, 2009; Stenou, 2002; UNESCO, 1976), also valued the benefits of listening to the residents' demands, launching the seeds for the development of heritage participatory processes that would gradually became highly recommended in the following decades as enhancers of human rights, inclusiveness, social responsibility, sustainability and democracy (Rosetti et al., 2022).

The expanded attention to the contribution of people to the urban environment's sustainability and the safeguarding of cultural heritage entailed the growing awareness that creativity and creative industries are also essential to achieve not only the inclusive regeneration of cities, but also the well-being of its residents, a way to foster diversity, a source of income and economic development, and an incentive for tourism activities (Capello et al., 2020; Cerisola, 2019; Jelinčić, 2021; Lenzi & Perucca, 2020; UNESCO, 2016). The interlinkage of cultural heritage and creativity as a source of innovation (Cerisola, 2019; Cominelli & Greffe, 2012), the growing requirement and acceptance of bottom-up approaches to urban cultural heritage and creativity involving a multitude of stakeholders (Della Lucia & Trunfio, 2018), as well as demands for a more diverse and creative urban tourism sensitive to, and respectful of, the local context (Richards, 2020) are issues that should be taken into consideration and addressed by those involved in decisions regarding cities' management towards sustainable development.

Inspired in the 2005 UNESCO *Convention on the Protection and Promotion of the Diversity of Cultural Expressions* which emphasizes "the vital role of cultural interaction and creativity, which nurture and renew cultural expressions and enhance the role played by those involved in the development of culture for the progress of society at large" (UNESCO, 2005, preamble), the historic urban landscape approach proposed by the 2011 *Recommendation* encompasses creativity in the urban context by stating that "[u]rban areas are increasingly important as engines of growth and as centers of innovation and creativity" since "they provide opportunities for employment and education and respond to people's evolving needs and aspirations" (UNESCO, 2011, para. 1). Moreover, the *Recommendation* considers "cultural diversity and creativity as key assets for human, social and economic development" (UNESCO, 2011, para. 12), reminding that

> *[w]hen properly managed through the historic urban landscape approach, new functions, such as services and tourism, are important economic initiatives that can contribute to the well-being of the communities and to the conservation of historic urban areas and their cultural heritage while ensuring economic and social diversity and the residential function. (UNESCO, 2011, para. 18).*

These viewpoints are supported by studies and data from several sources (EY, 2015; UNESCO, 2015, 2016, 2018a, 2018b, 2020a, 2022; UNESCO & UNDP, 2013; UNHABITAT, 2020) pointing to the importance of cultural and creative industries (CCIs) in people's lives and cities' economies. Particularly in difficult times, CCIs play a very important role in dealing with cities' predicaments by aiding citizens to endure adversities, and by facilitating urban recovery (Bertoni et al., 2021; Grotenhuis & Fesel, 2020). Literature review focusing on the impact of the COVID-19 pandemic on creative industries identified the main themes that attracted authors attention: cultural entrepreneurs and economy, museums and libraries, information technology and software business, social media, music industry and festivals, and publishing and journals (Khlystova et al., 2022).

This chapter analyzes the United Nations Educational, Scientific and Cultural Organization's (UNESCO) initiatives to guide States Parties in the management of the current crisis and how urban UNESCO World Heritage sites associated with the Organization of World Heritage Cities (OWHC) reacted to the challenges they were confronted with.

UNVEILING CULTURAL AND CREATIVE INDUSTRIES

Industries pertaining to the cultural and creative sector should be perceived both as a specific sector and as an occupation (Campbell et al., 2019), since on the one hand they are sources of innovation, jobs and income that affect positively all other economic sectors, while on the other hand they also having significant social impacts in promoting health and well-being, fostering social inclusion, and enhancing local social capital (Albuquerque, 2013; OECD, 2021). Prior to analyzing how can CCIs positively address pandemic constraints it is important to understand their nature, aims and benefits.

Notwithstanding the many proposed definitions (Khlystova et al., 2022) and scopes (Belfiore, 2020; Campbell et al., 2019), according to the European Parliament, cultural and creative industries (Figure 1) can be defined as

those industries that are based on cultural values, cultural diversity, individual and/or collective creativity, skills and talent with the potential to generate innovation, wealth and jobs through the creation of social and economic value, in particular from intellectual property; they include the following sectors relying on cultural and creative inputs: architecture, archives and libraries, artistic crafts, audiovisual (including film, television, software and video games, and multimedia and recorded music), cultural heritage, design, creativity-driven high-end industries and fashion, festivals, live music, performing arts, books and publishing (newspapers and magazines), radio and visual arts, and advertising (European Parliament, 2016, para. 2).

Following the two 2008 and 2010 reports prepared by the United Nations Conference on Trade and Development (UNCTAD) and the United Nations Development Programme (UNDP) providing a framework for identifying and better understanding the growing significance of creative economy in international trade, the 2013 Special Edition of the UNESCO and UNDP *Creative Economy Report* while also addressing economic topics such as tourism, focused mainly on the non-monetary benefits of these multiple activities and their contribution for sustainability, providing a holistic view of human development, advocating the importance of cultural diversity, and highlighting the role of culture as a driver and an enabler of sustainable development (UNESCO & UNDP, 2013).

Additionally, the increasing relevance of cultural and creative industries' economic value cannot be overlooked. As emphasized by a UNESCO and The World Bank position paper on cities, global exports of creative goods grew from approximately US$200 billion to US$500 billion between 2002 and 2015, generating annual global revenues of $2,250 billion and exports of over $250 billion prior to the COVID-19 crisis, the equivalent to 3% of global GDP (UNESCO and The World Bank, 2021, p. 13), generating 29.5 million jobs worldwide corresponding to 1% of the world's active population (EY, 2015; UNESCO, 2018b). In this context, creative industries and the related field of cultural tourism become strategic assets for local economies, especially in urban areas (EY, 2015; OCDE, 2021), where more than half of the world population live (UNHABITAT, 2020) and where more than 80% of the global economy is generated (UNESCO and The World Bank, 2021).

Figure 1. Creative and cultural industries and subfields (examples)
Source: adapted from European Parliament, 2016, para. 2 and UNESCO & The World Bank, 2021, p. 31

Cultural and Creative Industries	Subfields	
Audiovisual and Interactive Media	▪ Film ▪ TV and Radio ▪ Streaming ▪ Podcasts ▪ Videogames ▪ Videomapping ▪ Virtual Visits	Cultural Tourism
Performing Arts	▪ Drama ▪ Live Music ▪ Dance ▪ Festivals ▪ Feasts and Fairs ▪ Theatre	
Intangible Cultural Heritage	▪ Festive Events ▪ Rituals and Social Practices ▪ Oral Traditions ▪ Knowledge, Skills and Traditional Crafts ▪ Culinary Traditions & Gastronomy	
Literature and Press	▪ Books ▪ Newspapers ▪ Magazines ▪ Libraries ▪ Book Fairs ▪ Literary Festivals	
Visual Arts and Crafts	▪ Fine Arts ▪ Photography ▪ Crafts	
Design and Creative Services	▪ Architecture ▪ Fashion Design ▪ Graphic Design ▪ Interior Design ▪ Advertising	
Cultural and Natural Heritage	▪ Museums ▪ Historic Places ▪ Cultural Sites ▪ Archaeological Sites ▪ Natural Sites ▪ Cultural Landscapes ▪ Industrial sites	

Culture and creative industries are also a valuable tool to address 'overtourism', understood as a state of affairs in which the impact of tourism exceeds the carrying capacity (physical, ecological, social, economic, psychological, and/or political) of a place (Mihalic, 2020; Peeters et al., 2018; UNWTO, 2018), a situation extremely difficult to control and manage that most of the times has very negative impacts for the environment and the quality of life of local communities (Adie et al., 2020; Calle-Vaquero et al., 2021; Dodds & Butler, 2019). Due to their attractiveness urban heritage, and especially World Heritage monuments and historic centers, are particularly affected by the excess of visitors (Adie et al., 2020; De Luca et al., 2020) and the promotion of urban CCIs (such as creative cities' programs) may be a way to divert visitors to different cultural attractions – hence pointing to the relevance and benefits of adopting a holistic stance to city planning and management as foreseen in the historic urban landscape approach (UNESCO, 2011) in order to achieve sustainable and responsible tourism, enhance cities' livability and improve people's well-being (Smith & Ripp, 2020).

The above considerations highlight the fundamental role that cultural and creative industries play in our societies, even though this is feebly recognized (Sacco et al., 2018), suggesting that cities, cultural and creative industries, and tourism activities are three contemporary factors deeply interconnected and, in several aspects, interdependent. They also hint to the fact that the cultural and creative industries sector was severely affected by the COVID-19 pandemic.

ADDRESSING THE PANDEMIC CONSTRAINTS

Global Answers

As earlier mentioned, the pandemic affected negatively simultaneously cultural and creative industries, cities, and tourism. Travel restrictions, lockdowns, mandatory confinements, and a general sense of uncertainty kept visitors away from cultural facilities, indoor heritage sites, theaters, and other spaces where cultural and creative industries usually take place, affecting all those involved, from professionals to employers and employees of the entire creative value chain – creation, production, distribution, and access. In the OECD regions alone, tourism and cultural and creative sectors were the most affected, with jobs at risk of disappearing ranging from 0.8% to 5.5%, possibly with lasting effects. The coronavirus crisis has, in fact, brought to light the several fragilities of the CCI sector, especially concerning inadequate financing and economic support to take advantage of technological innovations, precarious employment situations (many artists work informally, part-time, or with precarious contracts), and inexistent or inadequate social security benefits (ILO, 2020; OECD, 2020).

A most perceptible result of lockdowns was the increase of the already existing (Landry, 2016) use of Information and Communications Technology (ICT) to reach the public through online contents which, in many cases, became free of charge. Digitalization of literary works, on-line transmissions of events, webinars, virtual visits, augmented reality, among others turned heritage and creative resources available to an increasing larger audience, who welcomed this new way of accessing to cultural contents (ILO, 2020; OECD, 2020). Furthermore, online participation fulfilled people's needs of connecting to others during confinement replacing face-to-face contacts and physical travel with internet mediated interactions, virtual visitation to heritage sites, sharing of memories, engaging in social media, and co-creating a whole new set of endeavors that helped getting over hard times and sometimes loneliness (Ginzarly, 2021).

Education has also benefited from internet-mediation, and online schooling became a reality for many students of all ages, which might have been an opportunity to establish closer links between the education and cultural and creative industries sectors, to raise awareness to cultural issues and improve young generation's capacities to appreciate art (OECD, 2020).

As for tourism, while in 2019 the sector generated 7% of global trade, employing one in every ten people globally, in 2021 arrivals were still 72% down comparing to pre-pandemic levels, increasing poverty and inequality and threatening heritage conservation (United Nations, 2020; UNWTO, 2022). Since urban areas are clusters of creative industries and cultural (and natural) heritage, which are highly attractive to visitors, cities with a revenue based primarily on tourism activities were particularly affected, and the consequences may be long-lasting (Goodger & Ferran, 2020; UNWTO, 2021). Nevertheless, rebuilding tourism may also be an opportunity for transformation, taking advantage of visitors' decrease to encourage the reformulation of tourism businesses through innovative means such as the enhanced use of new technologies, the establishment of partnerships among stakeholders, and the implementation of measures to foster communities' resilience and sustainability (UNSDG, 2020).

To address the pandemic impacts on society, UNESCO has been developing since the beginning of the COVID-19 an intense activity to assist Member States in addressing emerging constraints and to help surpass the profound impact that the pandemic is having on the culture sector. UNESCO information materials, studies, surveys, and reports provide, on the one hand, an overview of the rapidly evolving situation, help raising awareness about the impact of the crisis on economic revenues of professionals and cities, disseminate examples of how countries around the world adapted to the situation, and propose measure to deal with the pandemic's impacts. On the other hand, cities' efforts to adapt to new constraints and routines that hindered the usual vibrancy of urban places deliver wide-ranging examples of practices and experiences that may be useful for the future.

UNESCO's RESPONSES

As the only United Nations agency with an explicit mandate in the field of culture (UNESCO, 2013, p. 11) and, to the best of our knowledge, also the only worldwide intergovernmental organization to address cultural issues, UNESCO envisages achieving though culture-related initiatives the objectives stated in its Constitutive Act of contributing "to peace and security by promoting collaboration among the nations through education, science and culture in order to further universal respect for justice, for the rule of law and for the human rights and fundamental freedoms which are affirmed for the peoples of the world, without distinction of race, sex, language or religion, by the Charter of the United Nations" (UNESCO, 1946).

This "collaboration among nations", which are now UNESCO's 193 Members States and 11 Associate Members, includes jointly addressing the worldwide most pressing social matters. COVID-19 undoubtedly falls into this category, and UNESCO has been developing since 2020 an intense activity to inform about the impacts of the pandemic on the different Organization's sectors, conventions, programs and activities, recommending actions, developing initiatives and, in general, guiding States Members and organizations linked to UNESCO networks in the management of this atypical and unusual situation. A dedicated website informing about the different UNESCO actions (UNESCO, 2020b) and an online platform disseminating initiatives and experiences from UNESCO networks (UNESCO, 2020c) were developed and are freely accessible.

Regarding issues related to cities, the UNESCO Cities Platform gathers in a single online page the Organization's networks and programs concerning urban issues (UNESCO, 2020a), including the following eight networks and programs: UNESCO Creative Cities Network; UNESCO Global Network of Learning Cities; Megacities Alliance for Water and Climate; Disaster Risk Reduction and Resilience; International Coalition of Inclusive and Sustainable Cities; World Heritage Cities Program; Media and Information Literacy Cities; UNESCO-Netexplo Observatory Cooperation on Smart Cities. Some played a central role in the global response to the pandemic as regards cultural and creative industries.

The following projects and resources give a brief insight on UNESCO responses to the COVID-19 pandemic in the fields of culture and creativity, including in urban areas.

Culture and COVID-19: Impact and Response Tracker

Brochure on Culture Issues

UNESCO published weekly, between April and July 2000, a snapshot brochure entitled "Culture & COVID-19: Impact and Response Tracker", which provided an overview of the rapidly evolving situation. This brochure made it possible to inform about the immediate impact of the crisis on cultural and natural sites, monuments and museums, access to culture, livelihood of works in the culture and creative sector (CCS), travels and tourism, illicit trafficking of cultural property and looted artifacts, disseminating examples of how countries around the world adapted. In September 2020 it was replaced by the more encompassing brochure *Culture and Public Policy Tracker*, that also monitors the role of culture in public policies regarding the UN Sustainable Development Agenda.

https://en.unesco.org/news/culture-COVID-19-impact-and-response-tracker_
https://en.unesco.org/culture-development/tracker

Resiliart

Project Aiming at Stakeholders' Participation

As stated above, the pandemic affected with particular intensity the professionals linked to the creative industries, and the confinement and restrictions imposed by governments profoundly disrupted the activities related to the entire creative value chain – creation, production, distribution, and access – and aggravated the already fragile economic situation of artists. To support them and ensure everyone's access to culture, UNESCO launched in April 2000 the Project ResiliArt, a global movement that gathered cultural professionals from around the world to discuss the current state of the creative industries through virtual discussions. Cultural industries professionals were also encouraged to join the movement and replicate ResiliArt's discussion sessions in their respective regions, resulting in a set of videos where artists from all over the world express their views and feelings.

https://en.unesco.org/news/resiliart-artists-and-creativity-beyond-crisis

Resiliart and the UN International Year of Creative Economy for Sustainable Development

The United Nations declared 2021 the 'International Year of Creative Economy for Sustainable Development', to be implemented by the United Nations Conference on Trade and Development (UNCTAD), in consultation with UNESCO and other relevant UN entities. Resiliart contributed to the Year by organizing a debate on "How we can Rebuild the Creative and Cultural Industries in the Aftermath of the COVID-19 Pandemic" to examine how to revive the creative economy and make the creative sector more resilient in the face of future crises, as well as how and to design cultural policies that are better aligned with the 2030 Agenda.

https://en.unesco.org/news/international-year-creative-economy-sustainable-development

Resiliart x Mondiacult

As part of the pandemic recovery and in preparation of the 2022 Mondicult conference, UNESCO is expanding the ResiliArt movement through the ResiliArt x Mondiacult initiative, aiming at collecting information on the needs, gaps and opportunities on the cultural and creative sector by encouraging the organization of debates among stakeholders, followed by the response to a survey concerning its outcomes. UNESCO will afterwards analyze the answers to accordingly develop regionally categorized overviews and trends, thus contributing to the high-level discussions at the Mondicult conference in September 2022. https://www.unesco.org/en/mondiacult2022

2005 Convention COVID-19 Response

Online Platform to Share Good Practices

The COVID-19 pandemic brought to light the pre-existing volatility of creative industries (UNESCO, 2019a; UNESCO, 2020d), compelling States Parties to take urgent measures to protect and promote the diversity of cultural expressions and its practitioners. In this context, governments reacted by approving and implementing new policies and measures with the assistance of non-governmental organizations and the private sector, as advocated in the UNESCO 2005 Convention on the Protection and Promotion of the Diversity of Cultural Expressions (UNESCO, 2005) and the Recommendation on the Statute of the Artist (UNESCO, 1980). These practices were progressively introduced in the interactive policy monitoring platform of the 2005 Convention, where nowadays a description of a significant number policies and measures reported by States Members are shared for reference and for inspiration classified into three main categories: direct support for artists and cultural professionals; support for sectors of the cultural and creative industries; strengthening the competitiveness of the cultural and creative industries (UNESCO, 2020e).

https://en.unesco.org/creativity/COVID-19

UNESCO Creative Cities Network responses

Online Meeting

Among other objectives, the UNESCO Creative Cities Network (UNESCO, 2020f) aim at fully integrating culture and creativity into local development strategies and plans by developing hubs of creativity and innovation, as well as strengthening the creation, production, distribution and dissemination of cultural activities, goods, and services. The Network held in June 2020 the online meeting "Urban Solutions: Learning from the cities' responses to COVID-19", which was attended by mayors, local policy makers, professionals, practitioners, and representatives of NGOs and international institutions, who discussed the actions to be taken, including the immediate response to the crisis, the recovery after the pandemic and the construction of a more resilient future for cities. A special session explored the impact of the pandemic on the tourism sector and how sustainable tourism should be sought in the future to support the resilience and sustainability of cities in the post-COVID-19 period.

https://en.unesco.org/creative-cities/content/cities-response-COVID-19
https://en.unesco.org/urban-solutions-Learning-from-cities-responses-to-COVID19

World Heritage City Webinar

Webinar

As a result of the pandemic, 90% of World Heritage Sites endured total or partial closures at the pandemic height, with significant losses of visitors, revenues, and staff, whose negative impacts will predictable have long-lasting effects (UNESCO, 2021a; UNWTO, 2021). The webinar 'Re-thinking Urban Heritage for Resilience and Recovery' organized by the World Heritage Cities Programme (UNESCO, undated) explored how heritage can be integrated in and help support recovery, resilience, and sustainable development in World Heritage urban sites.

https://whc.unesco.org/en/events/1526/

World Museums Report 2021

Publication and Online Debate

The closure of the vast majority of the world's museums had considerable economic consequences, since many museums are dependent on the income generated by visitors. The results of a study developed by UNESCO seeking to understand how museums adapted to the constraints imposed by the pandemic and to explore ways to support them based on the contributions of states and museum professionals intitled "Museums around the world in the face of COVID-19" (UNESCO, 2021b) identified more than 800 museum actions in response to the COVID-19 crisis around the world. Many actions focused on the promotion of museums, the digitalization of contents and the dissemination of virtual visits. The online debate 'Reflections on the Future of Museums' envisaged the preparation of the High-Level Forum for Museums, whose second edition will take place in September 2022.

https://www.youtube.com/watch?v=Ht9QdjqlnEU

Living Heritage survey

Platform on Living Heritage Experiences and the COVID-19 Pandemic

Intangible cultural heritage elements and practitioners were particularly affected by the suspension of festivals and festivities and the closing of fairs and markets where handcrafters sell their products (UN-SDG, 2020). The accounts from more than 70 countries showed the devastating impact of the COVID-19 pandemic on living heritage worldwide, as well as the role it can play in maintaining social connectivity and solidarity during physical distancing and confinement. UNESCO launched an online survey in April 2000 to include a wide range of stakeholders, with particular emphasis on the communities at the heart of the 2003 Convention for the Safeguarding of Intangible Cultural Heritage. The survey invited respondents to share their experiences in their own words, resulting in a collection of testimonies in English, French and Spanish, which are available in their original language through an online platform. A set of operational principles and modalities to safeguard intangible cultural heritage in emergency situations has also been envisioned (UNESCO, 2020g) and is being developed by UNESCO Field Offices.

https://ich.unesco.org/en/platform-on-living-heritage-experiences-and-COVID-19-01123

These actions entailed reactions from the cities belonging to UNESCO's networks, namely the UNESCO Creative Cities Network, and associated partners, such as the Organization of World Heritage Cities (OWHC, 2022a), which have been shared trough publications and online platforms.

URBAN REACTIONS

The information concerning urban reaction to the COVID-19 pandemic here presented was retrieved from the UNESCO publication 'UNESCO Creative Cities' response to COVID-19' (UNESCO, 2020h) and the webpage 'COVID-19 pandemic Initiatives' compiled by the Organization of World Heritage Cities (OWHC, 2022b), which together provide information concerning 100 cities in 43 countries (Figure 2) from five regions[1] (Figure 3).

Figure 2. Number of cities per country reporting actions in response to COVID-19

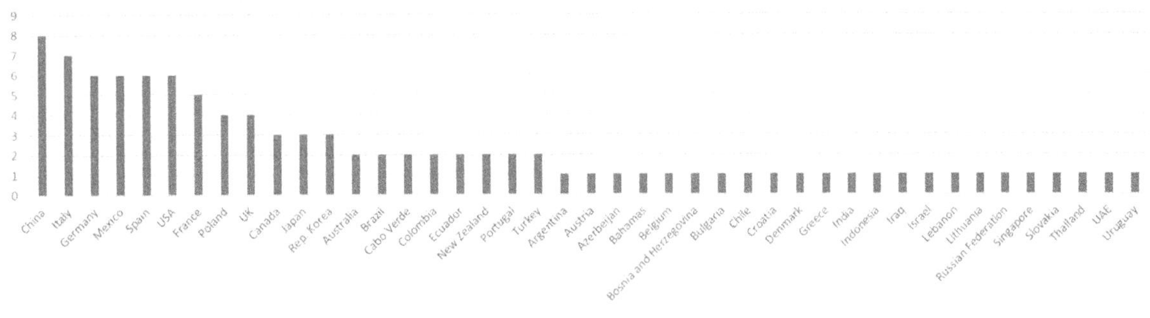

The UNESCO Creative Cities Network (UCCN) was implemented in 2004 to promote cooperation among cities that had identified creativity as a strategic factor for sustainable urban development in the fields of Crafts and Folk Art, Design, Film, Gastronomy, Literature, Media Arts and Music. This UNESCO program assembles cities working together to place creativity and cultural industries at the heart of their development plans at the local level, while actively cooperating at the international level (UNESCO, 2020i). The information regarding the activities developed in response to the COVID-19 pandemic was transmitted to UNESCO in response to a call by the UCCN Secretariat in March 2020.

The Organization of World Heritage Cities was created in 1991 to facilitate the exchange of information and expertise among cities inscribed on the World Heritage List, as well as to foster cooperation on matters concerning urban heritage (OWHC, 2022a). The initiatives addressing the pandemic were voluntarily shared by OWHC cities in an online platform requesting information on the recovery of economy, tourism and cultural and educational activities (historic sites, museums, events, etc.), as well as indication of any other measures designed specifically as a reaction to the COVID-19 outbreak (OWHC, 2022b).

Figure 3. Cities in the study by program, country and region
Africa: 2 | Asia and the Pacific: 25 | Arab States: 1 | Europe: 57 | Latin America and the Caribbean: 16
Crafts and Folk Arts: 10 | Design: 15 | Film: 8 | Gastronomy: 7 | Literature: 13 | Media Arts: 12 | Music: 13 | World Heritage: 22 | (World Heritage and Creative: 2)

UCCN Creative Field / WH Site	City	Country	Region
Crafts and Folk Arts	Ballarat	Australia	APA
Crafts and Folk Arts	Barcelos	Portugal	EUR
Crafts and Folk Arts	Biela	Italy	EUR
Crafts and Folk Arts	Carrara	Italy	EUR
Crafts and Folk Arts	Gabrovo	Bulgaria	EUR
Crafts and Folk Arts	Hangzhou	China	APA
Crafts and Folk Arts	João Pessoa	Brazil	LAC
Crafts and Folk Arts	Kanazawa	Japan	APA
Crafts and Folk Arts	Nassau	Bahamas	LAC
Crafts and Folk Arts	Sharjah	UAE	ARB
Design	Detroit	USA	EUR
Design	Kortrijk	Belgium	EUR
Design	Bangkok	Thailand	APA
Design	Beijing	China	APA
Design	Bilbao	Spain	EUR
Design	Buenos Aires	Argentina	LAC
Design	Curitiba	Brazil	LAC
Design	Kaunas	Lithuania	EUR
Design	Mexico City	Mexico	LAC
Design	Montreal	Canada	EUR
Design	Saint-Étienne	France	EUR
Design	Singapore	Singapore	APA
Design	Torino	Italy	EUR
Design	Wuhan	China	APA
Design / WH City	Querétaro	Mexico	LAC
Film	Bradford	UK	EUR
Film	Bristol	UK	EUR
Film	Łódż	Poland	EUR
Film	Rome	Italy	EUR
Film	Sarajevo	Bosnia and Herzegovina	EUR
Film	Terrassa	Spain	EUR
Film	Valladolid	Spain	EUR
Film	Yamagata	Japan	APA

UCCN Creative Field / WH Site	City	Country	Region
Gastronomy	Bergamo	Italy	EUR
Gastronomy	Chengdu	China	APA
Gastronomy	Gaziantep	Turkey	EUR
Gastronomy	Hyderabad	India	APA
Gastronomy	Macao	China	APA
Gastronomy	Portoviejo	Ecuador	LAC
Gastronomy	San Antonio	USA	EUR
Literature	Beirut	Lebanon	APA
Literature	Bogota	Colombia	LAC
Literature	Bucheon	Rep. Korea	APA
Literature	Dunedin	New Zealand	APA
Literature	Heidelberg	Germany	EUR
Literature	Iowa	USA	EUR
Literature	Melbourne	Australia	APA
Literature	Montevideo	Uruguay	LAC
Literature	Nanjing	China	APA
Literature	Norwich	UK	EUR
Literature	Seattle	USA	EUR
Literature	Slemani	Iraq	APA
Literature	Wonju	Rep. Korea	APA
Media Arts	Austin	USA	EUR
Media Arts	Braga	Portugal	EUR
Media Arts	Guadalajara	Mexico	LAC
Media Arts	Košice	Slovakia	EUR
Media Arts	Linz	Austria	EUR
Media Arts	Tel Aviv-Yafo	Israel	EUR
Media Arts	Toronto	Canada	EUR
Media Arts	York	UK	EUR
Media Arts	Enghien-les-Bains	France	EUR
Media Arts	Lyon	France	EUR
Media Arts	Qwangju	Rep. Korea	APA
Media Arts	Viborg	Denmark	EUR
Music	Ambon	Indonesia	APA
Music	Auckland	New Zealand	APA

UCCN Creative Field / WH Site	City	Country	Region
Music	Bologna	Italy	EUR
Music	Hamamatsu	Japan	APA
Music	Hannover	Germany	EUR
Music	Katowice	Poland	EUR
Music	Lliria	Spain	EUR
Music	Mannheim	Germany	EUR
Music	Medellin	Colombia	LAC
Music	Metz	France	EUR
Music	Morelia	Mexico	LAC
Music	Pesaro	Italy	EUR
Music	Praia	Cabo Verde	AFR
Music / WH City	Kazan	Russian Federation	EUR
WH City	Aranjuez	Spain	EUR
WH City	Bamberg	Germany	EUR
WH City	Bordeux	France	EUR
WH City	Cidade Velha	Cabo Verde	AFR
WH City	Cuenca	Ecuador	LAC
WH City	Dubrovnik	Croatia	EUR
WH City	Icherisheher	Azerbeijan	APA
WH City	Istambul	Turkey	EUR
WH City	Guanajuato	Mexico	LAC
WH City	Krakow	Poland	EUR
WH City	Oaxaca	Mexico	LAC
WH City	Pengzhou	China	APA
WH City	Philadelphia	USA	EUR
WH City	Quebec	Canada	EUR
WH City	Regensburg	Germany	EUR
WH City	Rhodes	Greece	EUR
WH City	Suzhou	China	APA
WH City	Tarragona	Spain	EUR
WH City	Valparaiso	Chile	LAC
WH City	Warsaw	Poland	EUR
WH City	Wismar	Germany	EUR

The distribution of cities among the five regions is extremely uneven, ranging from 57 European cities to only one city from Arab States and two African ones both in the same country, hindering the possibility of accomplishing inter-regions' comparisons. The number of cities per field is less imbalanced, especially among creative cites, ranging between seven and fourteen, while World Heritage cities are more strongly represented with 21 cities. Finally, two cities, are both Creative and World Heritage cities.

The content analysis of the more than two hundred activities provided by these two sources reveals the ways in which cities addressed the constraints posed by the pandemic. The information concerning the activities developed by each one of the cities included in the study was retrieved through an inductive content analysis (Carlomagno & da Rocha, 2016; Krippendorff, 2004) of the texts provided by cities, which uncovered the different resourceful initiatives developed in response to COVID-19. Nine categories emerged from the data analyzed, five pertaining to virtual initiatives, namely online activities connected to cultural facilities, online version of events, online education activities, online activities for children and young people, and generic online activities; one concerning awareness raising and planning actions; two regarding support, distinguishing between directly financial and non-financial; and finally, activities related medical and health issues (Figure 4). The information was afterwards quantified, and the results analyzed.

Figure 4. Number of activities provided by cities per categories

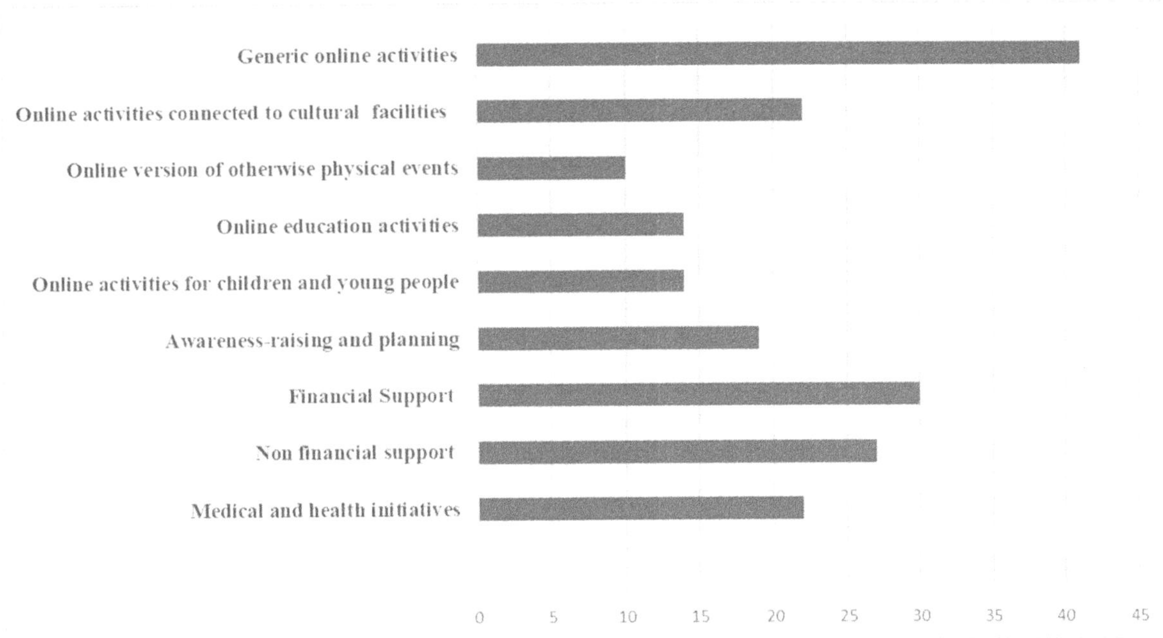

Among virtual initiatives either in websites or social media, which together correspond to slightly more than half of the reactions to COVID-19 retrieved (101/51%), the *generic online activities* are the most common (41/21%), encompassing activities such as the production and sharing of films, videos and photos (Ambon, Bradford, Heidelberg, Katowice, Mannheim, Łódź, Lyon, Oaxaca, Philadelphia, Praia, Qwangju, Rome, Terrassa, Valladolid, Viborg), online talks and reading and ebooks (Bogota, Braga, Enghien-les-Bains, Iowa, Kazan, Oaxaca, Seattle, Slemani), broadcasting of music performances and compilation of playlists (Bologna, Hamamatsu, Kazan, Llíria, Lyon, Morelia), sharing of art and various creative activities (Bucheon, Guanajuato, Lyon, Medellín, San Antonio), assorted cultural contents (Buenos Aires, Mexico City, Pesaro, Querétaro), dissemination of online databases connected to projects and newsletters (Detroit, Kortrijk, Melbourne), creation of virtual spaces for dialogue (Enghien-les-Bains, Guadalajara, Torino), interactive maps and virtual excursions (Kazan, Oaxaca, Philadelphia).

Some activities were developed jointly by two cities of the same field, namely the creation of an online database with design projects by the design creative cities of Detroit and Kortrijk, and the sharing of videos by artists explaining why culture is important in pandemic times by the music creative cities of Mannheim and Katowice.

Even though only two cities were identified in this category as presenting virtual exhibitions (Kanazawa, Suzhou), some others have developed similar initiatives in the context of museums and other cultural facilities. The same happens with pandemic related materials created by artists (Ballarat) and the streaming of specific cultural activities (Kaunas), since more specific similar actions fall under other categories.

The results concerning generic online activities point to a clear preference for the production and dissemination of visual contents such as films/videos and photos, which were shared by 15 cities, even though only 8 film creative cities communicated activities. In this sense, creative cities of music shared videos of music performances (Ambon, Praia) and of musicians telling their experiences (Mannheim and Katowice, in a common project), Media Arts cities provided videos of different cultural expressions (Lyon, Qwangju) or COVID-19 related (Viborg), a Literature city produced videos by artists (Heidelberg) and World Heritage cities disseminated documentaries on heritage (Philadelphia) or provided videos among other cultural contents (Oaxaca). Moreover, another two film activities reported were included in other categories since one was directed to youth and the other one was related to education.

Online *activities connected to cultural facilities* are the second most developed (22/11%), with offers of virtual visits, access to services, digitalized collections, publications, videos, and other contents and activities such as workshops by museums and art centers (Aranjuez, Beirut, Bergamo, Biela, Bologna, Bordeaux, Enghien-les-Bains, Gabrovo, Gaziantep, Icherisheher, Krakow, Nassau, Qwangju), provision of online performances and streaming services by music centers and theatres, (Aranjuez, Bologna, Hannover, Icherisheher, Lyon), free access to online books, music, collections and materials by libraries (Lyon, Norwich, Saint-Étienne, Wonju), access to online to cultural contents in general (Aranjuez, Kazan, Metz), virtual tours to cities with specific themes, sharing of photos and videos (Bradford, Rhodes), and exhibitions (Aranjuez) provided by different institutions.

Unsurprisingly, most of the museums and art centers provide online access to an assorted set of contents which include not only virtual tours and visualization of the collections, but also activities such as online games and workshops. One of the museums provides access to online intangible cultural heritage resources developed with the participation of communities (Krakow), while another disseminates videos on the impact of COVID-19 illustrated with artworks (Nassau), quite in the spirit of new museology, thus highlighting the social mission of museums and similar institutions (Mayrand, 2014). Moreover, the results obtained attest museums' broad capacity to encompass CCIs, since creative cities of literature, gastronomy, crafts and folk arts, music, media arts and World Heritage indicate museums' activities as means to address pandemic related constrains.

Going online was the solution found by some cities concerning the *organization of activities and events that would normally happen physically* (10/5%), from the inventorying cultural heritage online (Bordeaux) and the sharing of videos and interviews of local artists to replace visits to artists' studios (Carrara) to the organization of online encounters (Guanajuato), festivals (Mexico City), concerts (Valladolid), theatre performances (York), competitions (Hamamatsu), commemoration of festive (Llíria), and international (Valladolid) days, or just the transmission of past events (Nassau).

Several cities provided online *education initiatives* (14/7%) in different formats such as online courses (Hangzhou, Kazan, Norwich, Singapore, Yamagata), some through MOOC[2] (Enghien-les-Bains), free online education videos and tutorials (Macao, Morelia, Sharjah), workshops (Bucheon, Guanajuato), supply of education materials (Philadelphia, Terrassa) including free online software (Terrassa), and unspecified online education (Toronto).

Even though most initiatives are directed to the general public, some of these activities were intended to specific audiences, such as tutorials on video making for artists (Morelia) and training for women craft artists (Sharjah).

To overcome the closing of schools, *children and young people* were the target of dedicated online activities (14/7%) through making available online education contents for schools and implementing distance learning systems (Bamberg, Gaziantep, Kazan), as well as the organization of reading sessions and literature related events (Beirut, Bucheon, Gaziantep), writing (Dunedin, Norwich), drawing (Philadelphia, Valladolid) and gaming activities (Biela), as well as other online programs for this specific public (Beirut). Some cities also disseminated online the contents produced by young people (Guanajuato, Łódź, Sarajevo, Terrassa).

As indicated before, culture and creative industries were severely affected by the economic impacts of the COVID-19 pandemic, a situation that led many cities to delineate an apply measures to address this matter. A significant number of cities offered *direct financial support* to artists, practitioners and cultural associations (30/15%), including the attribution of grants, advancement of funds and payments, (Barcelos, Hannover, Heidelberg, Kanazawa, Katowice, Krakow, Melbourne, Metz, Montreal, Regensburg, Toronto), employment of artists to develop various initiatives (Austin, Morelia) or organization of online campaigns to support artists via fundraising actions, incentives to purchase art products or encouragement to disregard refunds of tickets of cancelled shows (Auckland, Košice, Linz, Mannheim, Medellín).

General pandemic measures such as the exemption or postponement of taxes, flexibility in application of penalties, suspension or reduction of rents and free subscriptions (Bordeux, Kazan, Macao, Quebec, Querétaro, Rhodes, Pesaro, Tarragona, Warsaw), as well as other support measures directed to vulnerable groups (Dubrovnik, Singapore) and small businesses (Philadelphia, Singapore) may have also benefited people connected to cultural and creative industries.

Non-financial support (27/14%) to address CCIs needs encompass a variety of measures including free publicity (Norwich, Oaxaca, Portoviejo), the creation of information desks for artists and cultural operators (Pesaro, Sharjah), distribution of necessity bags to artists (Austin), provision of spaces and materials for artists (Bangkok), online promotions (Icherisheher, San Antonio), special permits (Beirut), competition prizes (Istambul), extension of deadlines (Macao), cooperation among culture, creative, and non-culture sectors (Krakow).

Likewise financial support, non-financial measures directed to the population at large such as free parking (Bordeux), dissemination of official information (Saint-Étienne, Valparaíso), support messages (Bordeux), free meals (Bamberg, Hyderabad, Medellín, Montevideo, Philadelphia), help by volunteers (Kazan), provision of temporary work for the unemployed due to the pandemic (Oaxaca), enhancement of online business (Regensburg), also aided CCI employers and employees.

Some cities reported the adoption of support programs and measures without specifying its nature (Cidade Velha, Linz, Saint-Étienne, Warsaw), some directed to vulnerable groups (Tarragona). The distinction between financial and non-financial support is not straightforward, and together the two categories as responsible for nearly one third (57/29%) of all the activities reported, evidencing the pragmatic measures taken by cities to counteract COVID-19 economic impacts.

The activities regarding *awareness-raising and planning actions* (19/10%) are also wide-ranging. Awareness- raising initiatives include campaigns and materials about the pandemic with messages of hope and recommendations for people to stay at home (Nanjing, Saint-Étienne, Portoviejo, Valladolid, Viborg, Wuhan), encouraging the development of projects concerning tools and devices supporting remote work and study, or facilitating access to local goods and services (Torino), online newsletters about creative issues (Lyon), creation of awards (João Pessoa) and even a project delivering cultural experiences and activities from inhabitants' courtyards (Kaunas). Good practices were also addressed through the implementation of a project to track innovative programs and projects (Detroit and Kortrijk) and the encouragement for artists to set up more sustainable and resilient processes through the use of new materials, services, and experiences (Singapore).

Planning for recovery after the pandemic was also a concern, and cities developed several actions such as the organization of conferences to discuss supporting measures to different fields, including creativity (Metz, Tel Aviv-Yafo), the establishment of an action plan for economic reactivation after the pandemic (Oaxaca, Quebec), the creation of a catalogue of measures to support local industries and businesses (Hannover), the launching of social media campaign to encourage tourism in the future (Rhodes). One city also aimed at launching a post pandemic special funding program (Macao).

Many cities reported *medical and health initiatives* (22/11%) that will not be discussed here, even though the installation of art contents in hospitals and other medical facilities (Wuhan) and the dissemination of videos on sterilization methodology to artisans (United Arab Emirates) deserve to be mentioned.

The analysis of the distribution of the activities per category from the perspective of creative cities' fields and world heritage sheds light over the specificities concerning each cluster (Figure 5).

Figure 5. Distribution of activities per category from cities' fields and WH perspective

Number of cities	Categories - UCCN Creative Field or WH City	Generic online activities	Online activities connected to cultural facilities	Online version of events	Online education activities	Online activities for children and young people	Financial Support	Non financial support	Awareness-raising and planning	Medical and health initiatives	TOTAL
10	Crafts and Folk Arts	2	3	2	2	1	2	1	1	1	15
15	Design	6	1	1	1	0	3	2	6	5	25
8	Film	6	1	1	2	4	0	0	1	0	15
7	Gastronomy	1	2	0	1	1	1	4	2	2	14
13	Literature	7	3	0	2	4	2	3	1	0	22
12	Media Arts	5	3	1	2	0	4	2	3	2	22
13	Music	10	4	2	2	1	9	3	2	1	34
22	WH City	6	6	3	3	4	11	13	3	13	62
100	TOTAL	43	23	10	15	15	32	28	19	24	209

The activities developed by *Crafts and Folk Arts* Cities are evenly distributed though all categories, with special incidence in online activities connected to cultural facilities, namely museums, which also provide online learning and gaming activities for children and young people, and an art foundation, that fostered the production of a video on the impact of COVID-19 illustrated with artworks. Among the online version of events, the sharing of videos and interviews of local artists replacing visits to artists' studios and the delivery of materials to artisans are valuable and useful initiatives to support and promote both practitioners and their works. The medical and health initiative of disseminating videos on sterilization methodology to practitioners is certainly a good practice to prevent the pandemic dissemination.

Creative Cities of *Design* are well represented in the study, and reported mainly generic, awareness raising and planning, and medical and health activities. Nevertheless, most activities reported are quite unspecific and design unrelated, like the creation of an online platform with diverse cultural contents, the streaming of initiatives in inhabitants' courtyards, the creation of music playlist from local musicians or the organization of online festivals. Among the few design related initiatives are the joint projects from the cities of Detroit and Kortrijk to create an online database with design projects from UNESCO Creative Cities' Network and to track innovative programs and projects from design creative cities, as well as several awareness raising activities such as, for instance, the creation of posters taking advantage from designers' expertise. The provision of spaces and materials for artists seems an excellent support measure that probably helped overcoming the difficulties experienced.

Generic online activities are also the most reported by *Film* Creative Cities, and all activities reported, with no exception, are related to image and multimedia, including those directed to children and young people. As previously indicated, other creative fields mentioned many activities based in films as, for example, the videos encouraging people to stay at home or distance learning video tutorials, implying that film related creativity is transversal to other fields. The absence of support initiatives, financial or non-financial, suggest that this creative field was not strongly affected by the pandemic.

The same cannot be said about *Gastronomy* Creative Cities, that indicated several activities unrelated to this creative field. The most relevant reported are the creation of an online platform with information on gastronomy and several support measures, which include the deferment for payment of loans related to projects and rents, the extension of deadlines to finish projects, a campaign to promote restaurants that provide take-away, delivery and drive-through food services, the identification and promotion of places with good hygiene controls to deliver take-away and the distribution of free healthy meals to locked down residents and medical staff.

Creative Cities of *Literature* reported mainly generic online activities followed by programs directed to children and young people. Most activities described are field related and aimed both at writers and the general public. The examination of the activities' description points to a strong focus on the dissemination of all sorts of literature, from poetry to comics, through all means available including eBooks, social media, online newsletters, and the distribution of books from local authors in food baskets, as well as the involvement of readers of all ages, especially the younger, in reading sessions, discussions, courses and competitions. Non-financial support measures include a special permit for bookshops to open during the confinement and the recommendation of independent bookstores for online purchases.

Media Arts Cities reported extremely unspecific initiatives covering general activities related to other fields, suggesting that media arts can conveniently support and disseminate culture and creative industries. The generic online activities described include online talks, workshops and webinars, the creation of virtual space for dialogue dedicated to creative economy, the provision of online resources such as music, films, performing arts, dance, theatre, archeology, animation videos with information about health protective measures, among others, and financial support measures are directed to artists in general. The most media arts related activities seem to be medical and health related, namely the lightning of buildings to thank medical staff accomplished by Lyon, and the participation of tech community in a collaborative platform to find solutions to the needs of vulnerable population, including health professionals, implemented by Braga.

Most activities reported by Creative Cities of *Music* are related to this field, especially generic online activities, online versions of events and activities related to music centers. Conversely, only one online education activity and one financial support measure from the nine described is specifically directed to musicians. However, several music related activities were reported by cities from other fields, highlighting its transversal character.

World Heritage Cities were requested to inform on measures directed to the recovery of the economy, tourism, cultural and educational activities, and other actions implemented specifically in reaction to the COVID-19 outbreak, which explains the high number of financial and non-financial support measures and medical health initiatives. Providing support to the population by addressing the most pressing economic and health needs seem to be the major concern of world heritage cities, which may be explained by the fact that municipalities are frequently the managers of urban world heritage sites. Even though cultural activities were reported in several categories, world heritage issues are not directly mentioned, implying that the protection, conservation, and management of cultural and natural heritage was relegated to a secondary position during the COVID-19 pandemic.

PROSPECTS FOR THE FUTURE

Many solutions are being proposed to improve the conditions of the cultural and creative industries sector and prevent future crisis, among which the resource to digitalization and virtual environments to establish the connection between creativity providers and the public (ILO, 2020; OECD, 2020).

As regards the increased use of the internet, the encouraging results achieved point to its continuous use in the future, to support and complement physical face-to-face interactions (ILO, 2020; OECD, 2020; UNESCO and The World Bank, 2021). However, notwithstanding the positive experiences, the use of virtual mediums as a privileged way to reach people also uncovered the digital divide among countries, regions, social classes and generations, a situation that should be urgently addressed (OECD, 2020). Furthermore, the post-pandemic permanence of internet mediated CCIs may lead to a reduction of employment in sectors related to the physical exercise of cultural expressions (ILO, 2020).

This study does not address some relevant issues pertaining to the CCIs sector such as social classes and mobility (Brook et al., 2018), power relations among groups involved (Belfiore, 2020) or the relevance of the different capitals (intellectual, social and cultural) it encompasses (Townley et al., 2009), which would certainly enrich the analysis by unveiling other dimensions and difficulties. However, a shared challenging problem to tackle may be the general impoverishment of people at large after the pandemic due to the loss of jobs and economic constraints, leading to, on the one hand, to a decrease in cultural and creative industries consumers such as heritage visitors and performances' audiences and, on the other hand, a correspondent reduction in the economic resources and funding available to support such activities, whether the protection and conservation of heritage assets or the production of creative items (OECD, 2020).The attribution of a Universal Basic Income for cultural workers might be a solution to deal with the most pressing needs (Doustaly & Roy, 2022).The recovery of cities will certainly be slow and arduous, largely depending on their capacity to adapt and willingness to innovate (Pereira Roders, 2019), but the pandemic experience and the pressures upon urban areas may provide the opportunity to improve their livability and the well-being of its residents by giving cultural and creative industries a prominent place in historic urban landscapes as regards planning, conservation, design, and implementation of development policies (OECD, 2020; UNESCO, 2011). This can only be achieved through a more efficient and inclusive application of participatory processes in which stakeholders are involved in decision-making (UNESCO, 2022), which in turn will often require the adoption of better regulatory instruments (Rosetti et al., 2022), resulting in the enhancement of participative and deliberative democratic models (Bertrand Cabral et al. 2022). Citizen participation during the pandemic in European cities, the most represented in this study, seem to be short-time practices led by public authorities consisting mainly in the provision of information, sharing of knowledge, and public campaigns (Falanga, 2020), reducing participatory practices to passive forms that, although useful and legitimate, do not comply with recent investigation on the subject identifying consultation, intervention, partnerships and decision-making as more democratic, inclusive and empowering ways of participating (Bertrand Cabral et al., 2022; Rosetti et al. 2022).

An urban environment favorable to CCIs requires further enablers to translate culture and creativity into spatial, economic, and social benefits, namely basic infrastructures, services, and amenities; individuals and groups duly skilled and inclined to innovation; supportive networks and proper financing; and a unique combination of features that cultivate and attract creative talent and interested audiences (UNESCO and The World Bank, 2021). Overcoming the pandemic may be an opportunity for tourism activities to become more sustainable by transforming the sector's relation with nature, climate, and the economy, as well as by establishing effective partnerships with all stakeholders, from authorities to operators and communities, to avoid falling into unmanageable situations such as overtourism, with all its negative impacts, thus contributing to achieve some of the 2030 Agenda Sustainable Goals (UNSDG, 2020).

The UNWTO inclusive recovery guide for cultural tourism (UNWTO, 2021) recommends several measures such as an open and broad discussion about the meaning of conscious and responsible cultural tourism; exchange good practices and lessons learned to enhance responses and recovery; creation of participatory governance structures where artists and cultural professionals, local communities, the private sector and destination representatives can discuss and reach consensus on issues concerning cultural destinations; enhancement of the role of culture in destinations; encouragement of travelers to re-discover their local culture and creative industries; the promotion of urban tourism through culture by leveraging creativity and cultural diversity to boost entrepreneurship and innovation in urban destinations; support digital access and digital literacy, especially of women, youth, indigenous peoples and marginalized communities; ensuring that communities and practitioners are main beneficiaries and have a leading role in tourism strategies and management planning; develop community-centered tourism initiatives that actively engage practitioners of local and traditional knowledge and exploring how living heritage in tourism can improve the livelihoods of communities and practitioners, while safeguarding the social functions and cultural meanings of that heritage; educating visitors to be respectful and mindful of both the natural and cultural values properties and protected areas they visit, among many others.

RECOVERING THROUGH CCIs

"Building back better" is the United Nation's moto for recovering that must be systematically applied by all countries and regions, to all sectors of society and areas of activity, seizing the opportunity posed by the COVID-19 pandemic to direct our efforts to setting up a more sustainable world and achieving the 2030 Agenda Sustainable Development Goals and Targets. Maintaining 'business as usual' when we are facing such difficult challenges as climate change, destruction of cultural heritage and natural ecosystems, extreme poverty, worldwide desperate migrations, and a disquieting propagation of autocratic governance systems (Alizada et al., 2021) is a reckless attitude that might impair the future of humankind.

As an innovative and transformative-prone sector, cultural and creative industries are a soft power capable of both activating change and surveilling its effects. As long as committed policies are adopted, enabling environments are created and partnerships among public and private stakeholders are established, CCIs can renovate our societies for the better, valuably contributing to raise awareness about, and strengthen the capacities to, understand current challenges and encourage responses by individuals and societies as a whole (UNESCO and The World Bank, 2021). Likewise, CCIs may also act as supervisors of policies and projects, assessing if they are targeting sustainable development and blowing whistles when they are not. However, this can only be achieved with everyone's civic engagement in the construction of a better happier world, more inclusive, respectful, and sustainable.

The sustainable management of tourism activities and the diversification of topics of interest is mandatory to avoid the massification of pre-pandemic years. In this sense, creative tourism is another area to be developed by adding higher value to activities and improving visitor's enjoyment through fruitful interactions between visitors and host communities, and the co-creation of tourism experiences (OECD, 2020).

It is unclear some online activities described by cities which ones already existed before the pandemic and which were developed in response to the needs of users. The absence of internationally comparable statistics to reveal the full economic and social impacts of cultural and creative industries means that the sectors remain largely undervalued in the policy debate and that their many benefits for society are frequently underrated (OECD, 2021; Ost & Saleh, 2021). To overcome this absence UNESCO developed in 2009 a first set of culture for development indicators and is currently testing a group of 22 thematic indicators for culture in the 2030 Agenda through pilot-projects in several countries (UNESCO, 2019b).

It is still too soon to foresee what consequences the COVID-19 pandemic will have on humanity, either economic, environmental, social, cultural, geopolitical balance or even in long-term physical and psychological health, but we believe that it will impose hard and yet unknown strains over society, and that our way of living will never be the same. Some predict that a challenging crisis may arrive in the coming years, arguing that, although serious, it will be nothing more than the sixth wave of a series of innovation cycles that began more than two centuries ago with the industrial revolution, and that a renovated more sustainable world will emerge (Ost & Salleh, 2021), but further investigation on the impacts of the pandemic is necessary to fully understand how cities were affected and how can culture and creative industries contribute to recovering.

CONCLUSION

This study aiming at understanding the role of culture and creative industries in cities' reaction to the COVID-19 pandemic through the analysis of UNESCO's initiatives and the activities developed by UNESCO Creative Cities and urban World Heritage sites highlights the importance of CCIs in supporting cities in different domains, from informing and raising citizen's awareness to the adoption of prevention measures, from supporting education to enabling access to culture.

The analysis of cities' information highlights the transversal character of some cultural and creative industries like film and music since they are used as valuable resources by cities from several creative domains. Other creative fields such as crafts and folk arts and literature are quite self-focused, contrasting others such as gastronomy and design, that reported mainly generic activities poorly related to their fields, while Media Arts seems to be the most generic creative field, acting as support to the other six fields. Finally, World Heritage cities' concerns during the pandemic seem to be quite far from heritage issues.

The sustainable recovery from these difficult times largely depends on cities' capacities to adapt, innovate, and embrace democratic principles such as civic participation. Culture and creative industries may assist in this endeavor, for instance, by contributing with expertise, mobilizing individuals and groups to action and supervising policies and projects. Overcoming the uncertainty posed by a virus that will probably share our lives for many years requires hope in the future and trust in our unlimited adaptive capacity to address the new challenges ahead, knowing that cultural and creative industries will always be here to help, support and inspire us.

ACKNOWLEDGMENT

This research received no specific grant from any funding agency in the public, commercial, or not-for-profit sectors.

REFERENCES

Adie, B. A., Falk, M., & Savioli, M. (2020). Overtourism as a perceived threat to cultural heritage in Europe. *Current Issues in Tourism, 23*(14), 1737–1741. doi:10.1080/13683500.2019.1687661

Albuquerque, R. (2013). *Associativismo, capital social e mobilidade: contributos para o estudo da participação associativa de descendentes de imigrantes africanos lusófonos em Portugal* [Associativism, social capital and mobility: contributions to the study of the associative participation of descendants of Portuguese-speaking African immigrants in Portugal] (Vol. 41). Observatório da Imigração, ACIDI, IP.

Alizada, N., Cole, R., Gastaldi, L., Grahn, S., Hellmeier, S., Kolvani, P., Lachapelle, J., Lührmann, A., Maerz, S. F., Pillai, S., & Lindberg, S. I. (2021). *Autocratization Turns Viral. Democracy Report 2021*. Varieties of Democracy Institute (V-Dem). https://www.v-dem.net/static/website/files/dr/dr_2021.pdf

Bandarin, F., & Van Oers, R. (2012). *The historic urban landscape: managing heritage in an urban century*. John Wiley et Sons. doi:10.1002/9781119968115

Belfiore, E. (2020). Whose cultural value? Representation, power and creative industries. *International Journal of Cultural Policy, 26*(3), 383–397. doi:10.1080/10286632.2018.1495713

Bertoni, A., Dubini, P., & Monti, A. (2021). Bringing Back in the Spatial Dimension in the Assessment of Cultural and Creative Industries and Its Relationship with a City's Sustainability: The Case of Milan. *Sustainability, 13*(19), 10878. doi:10.3390u131910878

Bertrand Cabral, C., Pereira Roders, A., & Albuquerque, R. (2022). Formas de participação nos normativos patrimoniais. Ensaio de contextualização nos modelos da democracia participativa e deliberativa [Forms of participation in property regulations. Contextualization essay no participatory and deliberative democracy models]. Political Observer – Revista Portuguesa de Ciência Política, 16, 137-166. DOI: PCP2021.16 doi:10.33167/2184-2078.R

Brook, O., O'Brien, D., & Taylor, M. (2018). *There was no golden age: social mobility into cultural and creative occupations*. doi:10.31235/osf.io/7njy3

Calle-Vaquero, M. D. L., Garcia-Hernandez, M., & Mendoza de Miguel, S.Applications in Historic Centres. (2021). Urban planning regulations for tourism in the context of overtourism. Applications in historic centres. *Sustainability*, *13*(1), 70. doi:10.3390u13010070

Campbell, P., O'Brien, D., & Taylor, M. (2019). Cultural engagement and the economic performance of the cultural and creative industries: An occupational critique. *Sociology*, *53*(2), 347–367. doi:10.1177/0038038518772737

Capello, R., Cerisola, S., & Perucca, G. (2020). Cultural heritage, creativity, and local development: A scientific research program. In *Regeneration of the built environment from a circular economy perspective* (pp. 11–19). Springer. doi:10.1007/978-3-030-33256-3_2

Carlomagno, M. C., & da Rocha, L. C. (2016). Como criar e classificar categorias para fazer análise de conteúdo: uma questão metodológica [How to create and classify categories to do content analysis: A Methodological Question]. *Revista Eletrônica de Ciência Política*, *7*(1).

Cerisola, S. (2019). A new perspective on the cultural heritage–development nexus: The role of creativity. *Journal of Cultural Economics*, *43*(1), 21–56. doi:10.100710824-018-9328-2

Colavitti, A. M., & Usai, A. (2016). How cultural and creative industries are redefining policies for the historic urban landscape. The experience of Cagliari City as ECoC 2019's finalist and Italian Capitals of Culture. *Cultural Management Education in Risk Societies-Towards a Paradigm and Policy Shift?!*, 66.

Cominelli, F., & Greffe, X. (2012). Intangible cultural heritage: Safeguarding for creativity. *City. Cultura e Scuola*, *3*(4), 245–250. doi:10.1016/j.ccs.2012.10.003

Council of Europe. (2005). *Council of Europe Framework Convention on the Value of Cultural Heritage for Society*. https://www.coe.int/en/web/conventions/full-list2?module=treaty-detail&treatynum=199

Davidoff, P. (1965). Advocacy and Pluralism in Planning. *Journal of the American Institute of Planners*, *31*(4), 331–338. doi:10.1080/01944366508978187

De Luca, G., Shirvani Dastgerdi, A., Francini, C., & Liberatore, G. (2020). Sustainable cultural heritage planning and management of overtourism in art cities: Lessons from atlas world heritage. *Sustainability*, *12*(9), 3929. doi:10.3390u12093929

Della Lucia, M., & Trunfio, M. (2018). The role of the private actor in cultural regeneration: Hybridizing cultural heritage with creativity in the city. *Cities (London, England)*, *82*, 35–44. doi:10.1016/j.cities.2018.05.003

Dodds, R., & Butler, R. (Eds.). (2019). *Overtourism: Issues, realities and solutions* (Vol. 1). Walter de Gruyter GmbH & Co KG.

Doustaly, C., & Roy, V. (2022). A Comparative Analysis of the Economic Sustainability of Cultural Work in the UK since the COVID-19 Pandemic and Examination of Universal Basic Income as a Solution for Cultural Workers. *Journal of Risk and Financial Management*, *15*(5), 196. doi:10.3390/ jrfm15050196

European Parliament. (2016). *Report on a coherent EU policy for cultural and creative industries (2016/2072(INI))*. https://www.europarl.europa.eu/doceo/document/A-8-2016-0357_EN.pdf

Eurostat. (2019). *Culture statistics — 2019 edition*. doi:10.2785/118217

EY. (2015). *Cultural Times. The First Global Map of Cultural and Creative Industries*. Study on Behalf of the International Confederation of Societies of Authors and Composers (CISAC). https://unesdoc.unesco.org/ark:/48223/pf0000235710

Falanga, R. (2020). *Citizen participation during the COVID-19 pandemic: Insights from local practices in European cities*. Friedrich-Ebert-Stiftung. https://repositorio.ul.pt/handle/10451/45726?locale=en

Fojut, N. (2009). *The philosophical, political and pragmatic roots of the convention. In Heritage and beyond*. Council of Europe.

Ginzarly, M., & Srour, F. J. (2021). Cultural heritage through the lens of COVID-19. *Poetics*, *101622*. Advance online publication. doi:10.1016/j.poetic.2021.101622 PMID:34725533

González Martínez, P. (2017). Built heritage conservation and contemporary urban development: The contribution of architectural practice to the challenges of modernisation. *Built Heritage*, *1*(1), 14–25. doi:10.1186/BF03545666

Goodger, D., & Ferran, K. (2020). *City tourism outlook and ranking: Coronavirus impacts and recovery*. Tourism Economics.

Grotenhuis, F. J. B., & Fesel, M. (2020). *Perspectives on cultural and creative industries*. https://www.grotenhuisadviseert.nl/wordpress/wp-content/uploads/Frits-Grotenhuis-Perspectives-on-Cultural-and-Creative-Industries.pdf

Huybrechts, E. (2018). The historic urban landscape and the Metropolis. *Built Heritage*, *2*(4), 20–30. doi:10.1186/BF03545681

ILO. (2020). *COVID-19 and the media and culture sector*. ILO Sectoral Brief. https://www.ilo.org/sector/Resources/publications/WCMS_750548/lang--en/index.htm

Jacobs, J. (1961). *The Death and Life of Great American Cities*. Random House.

Jeannotte, M. S. (2021). When the gigs are gone: Valuing arts, culture and media in the COVID-19 pandemic. *Social Sciences & Humanities Open*, *3*(1), 100097. doi:10.1016/j.ssaho.2020.100097

Jelinčić, D. A. (2021). Indicators for Cultural and Creative Industries' Impact Assessment on Cultural Heritage and Tourism. *Sustainability*, *13*(14), 7732. doi:10.3390u13147732

Jokilehto, J. (2014). *Evolution of the normative framework. In Reconnecting the City: The Historic Urban Landscape Approach and the Future of Urban Heritage*. Wiley.

Khlystova, O., Kalyuzhnova, Y., & Belitski, M. (2022). The impact of the COVID-19 pandemic on the creative industries: A literature review and future research agenda. *Journal of Business Research*, *139*, 1192–1210. doi:10.1016/j.jbusres.2021.09.062 PMID:34629569

Krippendorff, K. (2004). *Content analysis: An introduction to its methodology*. Sage publications.

Landry, C. (2016). Culture and the digital city: its impact and influence. *Culture urban future: Global report on culture for sustainable urban development*, 157-164.

Lee, N., & Rodríguez-Pose, A. (2014). Creativity, cities, and innovation. *Environment & Planning A*, *46*(5), 1139–1159. doi:10.1068/a46202

Lenzi, C., & Perucca, G. (2020). Urbanization and subjective well-being. In *Regeneration of the Built Environment from a Circular Economy Perspective* (pp. 21–28). Springer. doi:10.1007/978-3-030-33256-3_3

Mayrand, P. (2014). The New Museology Proclaimed. *Museum International*, *66*(1-4), 115–118. doi:10.1111/muse.12048

Mihalic, T. (2020). Conceptualising overtourism: A sustainability approach. *Annals of Tourism Research*, *84*, 103025. doi:10.1016/j.annals.2020.103025

OECD. (2020). *Culture shock: COVID-19 and the cultural and creative sectors*. OECD. https://www.oecd.org/coronavirus/policy-responses/culture-shock-COVID-19-and-the-cultural-and-creative-sectors-08da9e0e/

Ost, C., & Saleh, R. (2021). Cultural and creative sectors at a crossroad: From a mainstream process towards an active engagement. *Built Heritage*, *5*(1), 1–16. doi:10.118643238-021-00032-y

OWHC. (2022a). *Introduction and mission*. Organization of World Heritage Cities. https://www.ovpm.org/all-about-owhc/introduction-and-mission/

OWHC. (2022b). *COVID-19 Pandemic Initiatives*. Organization of World Heritage Cities. https://www.ovpm.org/COVID-19/initiatives/

Peeters, P., Gössling, S., Klijs, J., Milano, C., Novelli, M., Dijkmans, C., Eijgelaar, E., Hartman, S., Heslinga, J., Isaac, R., Mitas, O., Moretti, S., Nawijn, J., Papp, B., & Postma, A. (2018). Research for TRAN Committee - Overtourism: impact and possible policy responses. European Parliament, Policy Department for Structural and Cohesion Policies, Brussels.

Pereira Roders, A. (2019). The Historic Urban Landscape approach in action: Eight years later. In *Reshaping Urban Conservation* (pp. 21–54). Springer. doi:10.1007/978-981-10-8887-2_2

Richards, G. (2020). Designing creative places: The role of creative tourism. *Annals of Tourism Research*, *85*, 102922. doi:10.1016/j.annals.2020.102922

Rodwell, D. (2018). The historic urban landscape and the geography of urban heritage. *The Historic Environment: Policy & Practice*, *9*(3-4), 180–206. doi:10.1080/17567505.2018.1517140

Rogers, A. P. (2017). Historic urban landscape approach and living heritage. *Conserving Living Urban Heritage: Theoretical Considerations of Continuity and Change*.

Rosetti, I., Bertrand Cabral, C., Pereira Roders, A., Jacobs, M., & Albuquerque, R. (2022). Heritage and Sustainability: Regulating Participation. *Sustainability*, *14*(3), 1674. doi:10.3390u14031674

Sacco, P. L., Ferilli, G., & Tavano Blessi, G. (2018). From Culture 1.0 to Culture 3.0: Three Socio-Technical Regimes of Social and Economic Value Creation through Culture, and Their Impact on European Cohesion Policies. *Sustainability*, *10*(11), 3923. doi:10.3390u10113923

Smith, B., & Ripp, M. (2020). *Guidelines for Sustainable Cultural Tourism*. Heritage Europe, OWHC. https://www.ovpm.org/wp-content/uploads/2020/07/tourismguidelines2020.pdf

Stenou, K. (2002). *UNESCO Universal Declaration on Cultural Diversity: a vision, a conceptual platform, a pool of ideas for implementation, a new paradigm*. UNESDOC Document code: CLT/2002/WS/11.

Taylor, K. (2020). The ideology of the urban cultural landscape construct. In *The Routledge Handbook on Historic Urban Landscapes in the Asia-Pacific* (pp. 48–66). Routledge.

Townley, B., Beech, N., & McKinlay, A. (2009). Managing in the creative industries: Managing the motley crew. *Human Relations*, *62*(7), 939–962. doi:10.1177/0018726709335542

UNESCO. (1946). *UNESCO Constitution*. http://portal.unesco.org/en/ev.php-URL_ID=15244&URL_DO=DO_TOPIC&URL_SECTION=201.html

UNESCO. (1976). *Recommendation concerning the Safeguarding and Contemporary Role of Historic Areas*. UNESCO. http://portal.unesco.org/en/ev.php-URL_ID=13133&URL_DO=DO_TOPIC&URL_SECTION=201.html

UNESCO. (1980). *Recommendation concerning the Status of the Artist*. https://en.unesco.org/creativity/governance/status-artist

UNESCO. (2005). *Convention on the Protection and Promotion of the Diversity of Cultural Expressions*. https://en.unesco.org/creativity/convention

UNESCO. (2010). *The Executive Board of UNESCO: 2010 edition*. UNESCO. UNEDOC Document code: SCX-2010/BROCH/CONSEIL EXECUTIF.

UNESCO. (2011). Recommendation on the Historic Urban Landscape. *Records of the General Conference, 36th session, Paris, 25 October - 10 November 2011, v. 1: Resolutions. Resolution 41*, 50-55. UNESCO. https://unesdoc.unesco.org/ark:/48223/pf0000215084

UNESCO. (2013). *Culture and UNDAF: A UNESCO Handbook*. UNESDOC Document code: CLT.2013/WS/9.

UNESCO. (2015). *Global Report 2015 - Re|Shaping Cultural Policies*. https://unesdoc.unesco.org/ark:/48223/pf0000242866

UNESCO. (2016). *Culture: urban future; global report on culture for sustainable urban development*. UNEDOC Document code: CLT-2016/WS/18.

UNESCO. (2018a). *Global Report 2018 - Re|Shaping Cultural Policies - Advancing creativity for development*. https://unesdoc.unesco.org/ark:/48223/pf0000260592

UNESCO. (2018b). *Investing in Creativity*. UNESCO. https://en.unesco.org/creativity/publications/investing-creativity

UNESCO. (2019a). *Culture & working conditions for artists*. https://en.unesco.org/creativity/publications/culture-working-conditions-artists

UNESCO. (2019b). *Culture | 2030 Indicators*. https://unesdoc.unesco.org/ark:/48223/pf0000371562

UNESCO. (2020a). *UNESCO for Sustainable Cities*. https://en.unesco.org/unesco-for-sustainable-cities

UNESCO. (2020b). *COVID-19* Response. https://en.unesco.org/creativity/COVID-19

UNESCO. (2020c). *Explore initiatives & stories from UNESCO networks*. https://en.unesco.org/covid19/initiatives

UNESCO. (2020d). *Freedom & Creativity: Defending art, defending diversity*. https://unesdoc.unesco.org/ark:/48223/pf0000373357

UNESCO. (2020e). *Culture in Crisis: Policy guide for a resilient creative sector*. https://unesdoc.unesco.org/ark:/48223/pf0000374631

UNESCO. (2020f). *Creative Cities Network*. https://en.unesco.org/creative-cities/

UNESCO. (2020g). *Resolution of the General Assembly: 8.GA 9*. https://ich.unesco.org/en/Resolutions/8.GA/9

UNESCO. (2020h). *UNESCO Creative Cities' Response to COVID-19*. https://unesdoc.unesco.org/ark:/48223/pf0000374264?posInSet=7&queryId=1f26fc80-270a-4c54-bb11-17e83ea8888b

UNESCO. (2020i). *Mission Statement*. Creative Cities Network. https://en.unesco.org/creative-cities/sites/default/files/Mission_Statement_UNESCO_Creative_Cities_Network_1.pdf

UNESCO. (2021a). *INF.5A.2 Impact of the COVID-19 pandemic on World Heritage and responses by the Secretariat*. Document WHC/21/44.COM/INF.5A.2. https://whc.unesco.org/document/187761

UNESCO. (2021b). *UNESCO report: museums around the world in the face of COVID-19*. https://whc.unesco.org/en/cities

UNESCO. (2022). *Re|shaping policies for creativity- addressing culture as a global public good*. https://unesdoc.unesco.org/ark:/48223/pf0000380474.locale=en

UNESCO & The World Bank. (2021). *Cities, Culture, Creativity: Leveraging Culture & Creativity for Sustainable Urban Development & Inclusive Growth*. https://www.worldbank.org/en/topic/urbandevelopment/publication/cities-culture-creativity

UNESCO & UNDP. (2013). *Creative Economy Report 2013. Special Issue*. http://www.unesco.org/culture/pdf/creative-economy-report-2013.pdf

UNHABITAT. (2020). *World Cities Report 2020*. https://unhabitat.org/sites/default/files/2020/10/wcr_2020_report.pdf

UNSDG. (2020). *Policy Brief: COVID-19 and Transforming Tourism*. https://unsdg.un.org/resources/policy-brief-COVID-19-and-transforming-tourism

UNWTO. (2018). *'Overtourism'? Understanding and Managing Urban Tourism Growth beyond Perceptions, Executive Summary*. World Tourism Organization Centre of Expertise Leisure, Tourism & Hospitality; NHTV Breda University of Applied Sciences; and NHL Stenden University of Applied Sciences, UNWTO. . doi:10.18111/9789284420070

UNWTO. (2021). *UNWTO Inclusive Recovery Guide. Sociocultural Impacts of COVID-19. Issue 2: Cultural Tourism*. UNWTO. https://www.e-unwto.org/doi/book/10.18111/9789284422579

UNWTO. (2022). *Tourism grows 4% in 2021 but remains far below pre-pandemic levels.* UNWTO Barometer. https://www.unwto.org/taxonomy/term/347

Veldpaus, L., Pereira Roders, A. R., & Colenbrander, B. J. F. (2013). Urban Heritage: Putting the Past into the Future. *The Historic Environment*, *4*(1), 3–18. doi:10.1179/1756750513Z.00000000022

ADDITIONAL READING

Gustafsson, C., & Lazzaro, E. (2021). The Innovative Response of Cultural and Creative Industries to Major European Societal Challenges: Toward a Knowledge and Competence Base. *Sustainability*, *13*(23), 13267. doi:10.3390u132313267

Healey, P. (2004). Creativity and urban governance. *disP-The Planning Review, 40*(158), 11-20. doi:10.1080/02513625.2004.10556888

Prezioso, M., & D'Orazio, A. (2020). *Tourism and Cultural Heritage for regional development Synergetic relations between Cultural Heritage (CH) and Tourism as driver for territorial development.* Topic Paper. https://www.espon.eu/sites/default/files/attachments/TNO%20Topic%20Paper%20on%20tourism.pdf

Remoaldo, P., Serra, J., Marujo, N., Alves, J., Gonçalves, A., Cabeça, S., & Duxbury, N. (2020). Profiling the participants in creative tourism activities: Case studies from small and medium sized cities and rural areas from Continental Portugal. *Tourism Management Perspectives*, *36*, 100746. doi:10.1016/j.tmp.2020.100746 PMID:32953432

UNESCO. (2018). *Investing in Creativity.* https://en.unesco.org/creativity/publications/investing-creativity

UNESCO. (2020). *Urban Solutions: Learning from Cities' responses to COVID-19.* https://en.unesco.org/urban-solutions-Learning-from-cities-responses-to-COVID19

URBACT. (2021). *How to make urban tourism more sustainable in post-Covid Europe.* https://urbact.eu/how-make-urban-tourism-more-sustainable-post-covid-europe

KEY TERMS AND DEFINITIONS

Creativity: Ability to create, to express himself/herself, or to respond, inventively and adaptively, to new approaches and challenges.

Cultural and Creative Industries (CCIs): Industries whose main purpose is the production, reproduction, promotion, distribution, or commercialization of goods, services, and activities with cultural, artistic, or heritage value.

Cultural and Creative Sector (CCS): Sector encompassing activities are based on cultural values, or artistic individual or collective creative expressions.

Historic Urban Landscape: The urban area resulting of a historic layering of cultural and natural values and attributes, incorporating the historic centers, the broader urban context, and its geographical setting.

Participation: The act of taking part in something through information, consultation, intervention, partnership, and/or involvement in decision-making.

Sustainability: Sustainability is a paradigm for thinking about the future in which environment, social, and economic factors are balanced in the pursuit of an improved quality of life.

Sustainable Tourism: Tourism that takes full account of its current and future economic, social, and environmental impacts, addressing the needs of visitors, the industry, the environment, and host communities.

ENDNOTES

1 In this study the countries are grouped following UNESCO distribution of Member States (UNESCO, 2010): Africa (AFR – Cabo Verde), Asia and the Pacific (APA - Australia, Azerbaijan, China, India, Indonesia, Iraq, Japan, Lebanon, New Zealand, Republic of Korea, Singapore, Thailand), Arab States (ARB – United Arab Emirates), Europe and North America (EUR- Belgium, Bosnia and Herzegovina, Bulgaria, Canada, Croatia, Denmark, France, Germany, Greece, Israel, Italy, Lithuania, Poland, Portugal, Russian Federation, Slovakia Austria, Spain, Turkey, United Kingdom, United States of America), Latin America and the Caribbean (LAC - Argentina, Bahamas, Brazil, Chile, Colombia, Ecuador, Mexico, Uruguay).

2 MOOC stand for Massive Open Online Courses, which are free online courses available for anyone to enroll.

Section 3

Pandemic Cases at Touristic Cities

Chapter 8
International Visitors to Las Vegas:
A 10-Year Economic and Cultural Analysis of High-Growth Markets

Kimberly Nehls
https://orcid.org/0000-0001-5792-3525
University of Nevada, Las Vegas, USA

ABSTRACT

Prior to the pandemic, more than 42 million individuals visited Las Vegas annually, and 14% were from outside the U.S. The number of overseas visitors increased by more than a million people during the decade from 2009-2018. The greatest increase in international visitors came from Brazil, China, South Korea, Argentina, and India. This chapter sought to examine international visitors to Las Vegas with the overarching research question: Why has the international tourist market to Las Vegas increased among specific countries and how can this information be utilized for future marketing purposes in a post-COVID-19 world? The most significant findings highlighted the increases in GDP and GDP per capita of the visitor countries. The city of Las Vegas can better target growth patterns like these to plan for the return of international visitors, along with marketing plans to reinvigorate international tourists to the region following the pandemic.

INTRODUCTION

The tourism sector is the most important economic driver in Las Vegas, Nevada, U.S.A. More than one-fourth of Southern Nevada's gross domestic product (GDP) and nearly one-third of jobs in the area are directly attributable to the tourism industry according to the Las Vegas Convention and Visitors Bureau [LVCVA] (2019). These numbers eclipse other cities in the country for the largest share of GDP and jobs from the tourism and hospitality sector and rival any high-tourist area in the world. Nationally, tourism accounts for 4.1 percent of GDP in the U.S. whereas it is greater than 20 percent in Las Vegas.

DOI: 10.4018/978-1-6684-3369-0.ch008

Among other top urban destinations in the United States, such as Los Angeles, New York, and Miami, the tourism industry accounts for no more than 5 percent of GDP. Las Vegas's economic dependence on tourism cannot be overestimated.

Prior to the COVID-19 pandemic, the 2019 calendar year counted more than 42 million individual visitors to Las Vegas, and 14 percent were from outside the U.S. Given the proximity to the U.S., a large percentage of international visitors came from North American neighbors, Canada and Mexico, but an increasing number of international visitors were from overseas. In fact, the number of visitors from Canada and Mexico was nearly stagnant over the past decade prior to the pandemic, while overseas visitors increased by more than 67 percent (>1.24 million visitors) during the same time. The greatest increase in international visitors in the past ten years have been from Brazil (+395%), China (+238%), South Korea (+214%), Argentina (+162%), and India (+124%). The growing international segment of Las Vegas's tourist population is an important one to note, especially because international visitors tend to stay longer and spend more than domestic visitors in the market (LVCVA, 2019; Yoo, Singh, & Ghaharian, 2017). Therefore, the purpose of this manuscript is to examine the increase of international visitors to Las Vegas and better understand the motivations for international tourism behavior in Las Vegas as an opportunity for future marketing from both an economic and cultural perspective in a post-covid world. The overarching research questions guiding this study are: Why did the international tourist market to Las Vegas increase among specific countries? And how can this information be utilized for future marketing purposes in a post-covid urban tourism market? These questions will be analyzed by focusing on the change in visitors over the 2009-2018 decade by determining the tourist country-of-origin's GDP, GDP per capita, innovation, Geert Hofstede dimensions, and world happiness index. By better understanding these visitor country inputs, the city can better target growth patterns to market for future growth and economic benefit, especially in a post COVID-19 world.

Theoretical Framework

Resource Dependence Theory [RDT] guided the work of this study. The definitive reference on this theory is *The External Control of Organizations* by Pfeffer and Salancik (1978). More than 40 years after publication, the tenets remain true: the procurement of external resources that are financial, physical, and informational in nature are a necessary principle for entities to survive and thrive. In a review of RDT research by Hillman, Withers, and Collins (2009), the authors reported how resources "depend on contingencies in the external environment" (p. 1404). In terms of this study, resource dependency theory is extended to the city of Las Vegas as an entity dependent upon external resources for survival and the devastating impact of the pandemic, especially in the 2020 calendar year. In order to return to pre-pandemic numbers, especially among international visitors, the city needs to have a comprehensive understanding of where urban tourism marketing efforts should be concentrated. It is unquestionable that the hospitality industry is the lifeblood of the city. According to the LVCVA (2019):

When measuring economic dependence on tourism by its share of Gross Domestic Product (GDP), it is again clear that Southern Nevada is more tourism-dependent than other major destinations... In absolute terms, the $22.5 billion of direct, tourism-related economic activity in Las Vegas ... was larger than the total reported for four of the seven cities that saw more visitors during the year. Orlando, the most visited U.S. destination in 2017, welcomed 72 million visitors compared to 42.2 in Las Vegas, but *that visitation generated $15.3 billion in economic activity, or $7.2 billion less than visitors to Southern Nevada contributed to the economy (p. 1).*

International visitors are particularly essential to the Las Vegas market, as international visitors spend twice as much on their visits, based upon total expenses (LVCVA, 2019; Velotta, 2012, Lim & Won, 2020). The relevance of resource dependency theory is based upon "how important the resource is to the final result," according to Andersson and Getz (2008, p. 146). Visitor data indicate that international travelers typically stay longer and spend more on gaming, shopping, and sightseeing while in town. International visitors are also more likely to gamble and have larger gambling budgets for their Las Vegas trips as well (Yoo et. al, 2017). Therefore, the international market – while only 14 percent of the overall number of visitors to Las Vegas – is a key demographic and an essential component of the local economy. According to Velotta (2012), "$1 of every $4 spent by tourists in Las Vegas was by a foreign visitor, and $1 of every $3 spent in nongaming purchases was by an international traveler." Las Vegas is dependent upon this resource contingent on the external environment, thus need to be resolute in the return of international visitors once it is safe to do so.

LITERATURE REVIEW AND HYPOTHESES

Tourism has grown to become one of the top five global industries (Hsu, Woodside, & Marshall, 2013; UNWTO, 2020). According to the United Nations World Tourism Organization (2020), international tourist arrivals worldwide grew four percent in 2019 to realize 1.5 billion visitors. Remarkably, international tourism has grown faster than world GDP growth since the 1950's (Lanza, Temple & Urga, 2003), the exception being this year (2020) during the global pandemic (UNWTO, 2020).

Copious amounts of research already focus on forecasting tourism demand for its effects on the city or region being studied (e.g., Chu, 2014; Witt, Newbould, & Watkins, 1992; Yoo et. al., 2017), as well as the resulting economic growth of regions due to inbound tourism (e.g., Balaguer & Cantavella-Jorda, 2002; Kumar et. al, 2020; Lanza et. al., 2003; Terzić, 2018). Studies of this nature highlight the impact of tourism on a city, region, or country, typically with a focus on job, trade, and tax growth. According to a meta-analysis of tourism and GDP, "the main thrust of research on the economic impact of tourism has been to measure direct and secondary impacts on local and regional economies… Models consider a causal relationship between tourism and economic growth" (Castro-Nuño, Molina-Toucedo, & Pablo-Romero, 2013). In the same meta-analysis, the authors concluded that a positive elasticity exists between the receiving region's GDP and tourism. Relatedly, Terzić's (2018) results confirmed strong positive connections between tourism and GDP as well as tourism and job creation.

Terzić's (2018) research, along with Chu's (2014) identified public policies and decisions that have encouraged tourism, including assessments of infrastructure, accommodations, entertainment, and other services to attract visitors while also supporting local individual entrepreneurs and private businesses to support new tourist demand. Other than public policies, foreign currency exchange is also a driver of international travel; when the U.S. dollar is considered weak, like it was following the recent economic recession, the number of international visitors increased as they saw Las Vegas and other U.S. cities as a good value. Cheng (2012), Crouch (1994), and Balaguer and Cantavella-Jorda (2002) wrote three of the many studies that found significant effects for the nominal exchange rate on tourism demand. Crouch (1994) found that unfavorable exchange rates resulted in less travel abroad as well as a decline in spending and length of stay, whereas favorable exchange rates increased tourist spending and new, first-time visitors to destinations. For all three papers, tourism in countries with a weaker currency experienced more visitor activity from countries with a stronger currency. Again, these results primarily look at the tourist-receiving country rather than the tourist-sending country.

This study is unique in that it examines the countries of origin of tourists and the countries' unique economic growth and cultural dimensions as potential drivers or impacts of travel. Rather than study the economic impact of tourism on Las Vegas – which has been well documented (e.g., LVCVA, 2019 and Lim & Won, 2020) – this research is the first of its kind to study the countries with the highest number of international visitors to Las Vegas. Specifically, we examine international tourist countries' economies and cultures with the intention to better understand and explain the patterns of foreign visitor growth in Las Vegas for potential impact on future marketing opportunities.

Bearing in mind that a weak dollar contributes to visitor growth because international spending goes further and greater disposable income increases travel, the first two hypotheses predict that the highest-growth countries for numbers of international tourists over the past decade will show economic growth over the past decade as well. In other words, if a country achieves economic growth as measured by their gross domestic product (e.g., GDP growth) and greater disposable income (e.g., GDP per capita growth), then a relative increase in visitors to Las Vegas may also occur.

Hypothesis 1: An increase in a country's GDP from 2009-2018 is related to an increase in visitors from that country to Las Vegas during the same time frame.

Hypothesis 2: An increase in a country's GDP *per capita* from 2009-2018 is related to an increase in visitors from that country to Las Vegas during the same time frame.

Additionally, international growth is partly due to new, nonstop long-haul flights into the area marketed between visitor bureaus in respective cities. For example, when Hainan Airlines created a temporary nonstop flight to Las Vegas from Beijing, nearly 20,000 passengers arrived directly from the Chinese capital in 2017, partly due to marketing awareness campaigns in mainland China (LVCVA, 2018). Furthermore, in-airport marketing adapted to the new Chinese visitors through Las Vegas's McCarran Airport's "China Ready" program in order to culturally welcome new guests. Examples include friendly tourism ambassadors to welcome passengers and help Chinese visitors to navigate the city, welcoming signage in Mandarin, and a WeChat and Alipay presence in the casinos and beyond (Raz, 2016). WeChat is a widely used social networking and mobile wallet app in China which was introduced in several strip hotels for ease of doing business. Las Vegas tourism officials had been pursuing a direct route from Beijing for many years, as Chinese travelers previously were one of the fastest-growing and largest-spending groups of international visitors to Las Vegas, despite the previous lack of nonstop flights. The Hainan Airlines flight and the >1000% increase in visitors direct from mainland China is just one example of how helpful marketing impacted the international visitor increase. If this example is an indication of a successful tourism partnership, then it appears that marketing activities and the welcoming nature of the tourist-receiving country positively impacts the number of tourists from the tourist-sending country. Therefore, wellbeing and happiness of tourist-sending country may have a relationship with an increase in tourism.

Hypothesis 3: Higher happiness ranking is related to an increase in visitors from that country to Las Vegas from 2009-2018.

Another example of how international visitors interact in the Las Vegas region is through the Consumer Electronic Show or CES. CES is annually marketed as one of the largest technologies and innovations conventions in the world with more than 180,000 attendees at the Las Vegas Convention Center and surrounding hotels. CES is the world's gathering place for new technology, gadgets, and inventions, and every major tech company on the planet participates in CES. Remarkably, of the 180,000 attendees, nearly 65,000 are international visitors to Las Vegas and represent 155 different countries (CES, 2019). The international rate of 35 percent of CES attendees eclipses the overall international visitor rate of 14 percent. Since one-third of the conference-goers are from out of the country, CES is very intentional about reaching its out-of-USA guests. A CES website was dedicated to international attendees https://www.ces.tech/international.aspx with information about navigating the city and the major conference events. Chinese visitors held the top spot for largest number of CES visitors outside of the US (15,383 in 2018), followed by South Korea (7197), Japan (6434), France (5474), Canada (4638), Taiwan (2809), Germany (2785), UK (2711), and Mexico rounding out the top ten at 1927 (CES, 2019). With numbers like these at CES focused on innovation, and the overall international tourist market exploding in Las Vegas, this is an area that deserved study. Thus, countries which ranked highly on innovation scales would be related to higher visitor volume to Las Vegas.

Hypothesis 4: Higher innovation ranking is related to an increase in visitors from that country to Las Vegas from 2009-2018.

A final consideration of visitor volume is the cultural makeup of the tourist-sending countries. Geert Hofstede's typologies of culture is the standard bearer to assess culture in a given country: "Hofstede's national cultural framework may be the most influential conceptual foundation in cross-cultural studies" (Hsu, Woodside, & Marshall, 2013, p. 680). Hofstede defines culture as "the collective mental programming of the human mind which distinguishes one group of people from another," and in this sense, a study of visitor volume to Las Vegas and cultural dimensions is warranted (Hofstede Insights, 2020). Hofstede identified six national cultural dimensions: (1) Power Distance, the degree of inequality or hierarchy between superiors and subordinates in society; (2) Individualism, the degree of independence versus collectivism in a society; (3) Masculinity, the dominance of values such as competition, assertiveness, and material awards versus consensus-building values; (4) Uncertainty Avoidance, the degree of how comfortable people are in ambiguous or unstructured situations; (5) Long-Term Orientation, the degree to which people in a society prepare for the future through means such as savings and education; and (6) Indulgence, the degree of free gratification of basic and natural human desires versus restraint and strict normative behaviors. Each of these dimensions is graded from 0-100, with societies at the low end of cultural scales closer to 0 and societies at the high end of each scale rate closer to 100. These typologies of culture vary from country to country with six dimensions at 100 on each scale for 600 options per nation. For example, the U.S. exhibits Power Distance: 40 / Individualism: 91 / Masculinity: 62 / Uncertainty Avoidance: 46 / Long-Term Orientation: 26 / Indulgence: 68. This means that the U.S. ranks highest on individualism and lowest on long-term orientation or future planning. The six dimensions may provide insight into an increase in international tourism to Las Vegas and how to better market to these cultural differences.

Hypothesis 5: There is a relationship between the six Hofstede dimensions and an increase in international tourists to Las Vegas from 2009-2018.

These five hypotheses in sync help to better understand the tourist-sending countries to Las Vegas to recognize and adapt to country preferences. In a post-covid environment, a better understanding of these areas will aid marketers where to direct their tourism efforts.

Table 1. International visitor growth to Las Vegas from 2009-2018 among 27 countries

Origin market	Visitors 2009	Visitors 2018	09-18%visitor change
Canada	1,461,400	1,597,280	9.30%
Mexico	1,082,940	1,146,980	5.91%
U.K.	629,866	736,660	16.96%
Australia	180,371	370,150	105.22%
China	70,112	236,970	237.99%
Germany	142,004	219,610	54.65%
Japan	157,326	212,630	35.15%
South Korea	59,414	186,610	214.08%
Brazil	36,557	180,850	394.71%
France	63,661	96,300	51.27%
Ireland	50,788	74,600	46.89%
Netherlands	40,326	59,380	47.25%
India	21,649	48,440	123.75%
Switzerland	33,603	45,330	34.90%
Taiwan	20,256	40,110	98.02%
Spain	29,568	38,280	29.46%
Italy	33,266	37,500	12.73%
Philippines	29,196	36,670	25.60%
Sweden	16,724	34,840	108.32%
Argentina	13,048	34,160	161.80%
Belgium	19,599	24,180	23.37%
Denmark	13,450	22,640	68.33%
Hong Kong	15,030	17,530	16.63%
Norway	8,838	14,980	69.50%
Israel	12,115	14,380	18.70%
Russia	7,095	13,660	92.53%
Austria	9,213	13,570	47.29%

METHOD

Sample and Variables

The Las Vegas Convention and Visitors Authority (LVCVA) provided data on international visitor trends to the city. LVCVA data included 27 countries that showed the greatest increase in visitors during the decade from 2009-2018. The dependent variable in all analyses was the percent change in international visitors during the decade from 2009-2018 on a country-by-country basis. The percent change over a ten-year period was calculated in aggregate to separate patterns from random fluctuations in tourist or economic activity on a year-to-year basis. The percent change was calculated for each of the countries and ranged from Brazil, at the high end, with nearly a 400 percent visitor volume increase over the ten-year period, to Mexico, with only a 5.91 percent increase. The average number of visitors increased 60 percent among these 27 countries.

A simple box plot of these data confirmed that one significant outlier existed, Brazil, with a 395 percent increase in visitor volume over the 10-year period. Brazil's increase in visitor volume of was seven times larger than the average increase in international visitors; thus, Brazil was removed from the analyses because of its unusual properties. Additionally, the overall percent visitor change was positively skewed (Skewness = 2.256, Std Er = .448), so a log transformation was necessary to realign these data to get a 1.000 significance-level for the Shapiro-Wilk test. Although this dependent variable was consistent throughout each analysis, the hypotheses were analyzed using a variety of independent (explanatory) variables and methods. Table 2 summarizes the variables and methods for each of the hypotheses in this study. Simple linear regression was the primary methodology for the first three hypotheses as it was best to assess the two interaction terms to determine how one may affect the other. Correlation shows the relationship between two variables and was utilized for the fourth hypothesis, while multiple regression was used for the fifth and final hypothesis since there were six predictor variables in that analysis.

DATA ANALYSIS AND RESULTS

Hypothesis 1 and 2 aimed to test the relationship between the percent change in visitor volume for the 27 countries and the countries' gross domestic product nationally and per capita. Similar to visitor volume, the percent change of GDP was calculated over a ten-year period so as to separate patterns from random fluctuations in economic activity on a year-to-year basis. World Bank[1] supplied the data on GDP per country, with the exception of Taiwan which was not disaggregated from China in World Bank data. In that instance, Economy Watch[2] provided the economic information on Taiwan. The average GDP increase among all 27 countries from 2009-2018 was a remarkable 39 percent, averaging nearly 4 percent GDP growth per year for ten years. It should be noted that not all of the international markets showed an increase in GDP over the 10-year period: Japan, Spain, and Italy all had slight 5% decreases in GDP from 2009-2018. Conversely, India (+103% GDP increase) and China (+167% GDP increase) exhibited the greatest GDP increases over the same time span, so certain countries skewed data more favorably toward an overall increase in GDP.

Table 2. Variables and Methods Organized by Hypotheses

Hypothesis	Dependent Variable	Independent Variable(s)	IV data source	Method of Analysis
H1: An increase in a country's GDP from 2009-2018 is related to an increase in visitors from that country to Las Vegas during the same time frame.	Percent change in international visitors to Las Vegas per country between 2009-2018.	Percent change in country GDP between 2009-2018.	World Bank	Linear regression
H2: An increase in a country's GDP per capita from 2009-2018 is related to an increase in visitors from that country to Las Vegas during the same time frame.	Percent change in international visitors to Las Vegas per country between 2009-2018.	Percent change in country GDP per capita between 2009-2018	World Bank	Linear regression
H3: Higher happiness ranking is related to an increase in visitors from a country to Las Vegas from 2009-2018.	Percent change in international visitors to Las Vegas per country between 2009-2018.	Happiness Index ranking	UN Sustainable Development Solutions created Happiness Index based upon six variables: logged GDP per capita, social support, life expectancy, freedom to make life choices, generosity, and corruption.	Confirmatory factor analysis for model fit, and regression for analysis
H4: Higher innovation ranking is related to an increase in visitors from a country to Las Vegas from 2009-2018.	Percent change in international visitors to Las Vegas per country between 2009-2018.	GII score	World Intellectual Property Organization (WIPO)'s 2019 Global Innovation Index	Correlation
H5: There is a relationship between the six Hofstede dimensions and an increase in international tourists to Las Vegas from 2009-2018.	Percent change in international visitors to Las Vegas per country between 2009-2018.	The six GH dimensions: Indulgence, Uncertainty Avoidance, Masculinity, Individualism, Long Term Orientation, and Power Distance	Hofstede's six national cultural dimensions	Multiple Regression

In regard to GDP per capita, the data were again garnered from World Bank, and descriptive statistics were similar to the overall nation GDP. The average per capita GDP among all 27 countries in this analysis increased 24 percent from 2009-2018, suggesting that the countries with an increase in visitors to Las Vegas also had an increase in GDP per capita. However, ten countries, mostly in Western Europe, showed single digit *declines* in GDP per capita from 2009-2018: Germany, Japan, Spain, France, Netherlands, Norway, Denmark, Austria, Italy, and Belgium. These minor declines were offset by very large increases in GDP per capita in other countries, most notably in Asia and South America. The largest GDP per capita increases among the 27 countries were exhibited by India, South Korea, and Argentina, each with double-digit GDP per capita growth, and China demonstrated triple-digit GDP per capita growth amongst its billion-plus population (+130%) over the timeframe studied.

Regression analyses were run on both country GDP percent change and country GDP per capita percent change as the continuous independent variables and percent change in visitors from 2009-2018 as the continuous dependent variable to determine significance levels. As indicated in Table 3, the ANOVA table for the GDP regression is F (2.23) = 11.113, p< .01. When the percent change in visitors was predicted, it was found that a percent change in overall country GDP (Beta = -0.604, p < .01) and GDP per capita (Beta = 0.679, p < .01) were both significant predictors. The overall model fit for change in GDP was R^2 = 0.365 (Table 4) and the R^2 = 0.461 (Table 5) for change in GDP per capita. Therefore, both Hypothesis 1 and Hypothesis 2 were supported. As GDP and GDP per capita increased, the visitor volume to Las Vegas also increased from the 27 countries studied. Specifically, for every one percent increase in GDP among these 27 countries, it induced a .91 percent increase in tourists to Las Vegas, and for every one percent increase in GDP per capita among these 27 countries, it induced a 1.162 percent increase in tourists to Las Vegas, as indicated in Tables 4 and 5.

Table 3. ANOVA for Percent Change GDP per capita and GDP among 27 countries from 2009-2018.

ANOVA						
Model		Sum of Squares	df	Mean Square	F	Sig.
1	Regression	46144.838	2	23072.419	11.113	.000[b]
	Residual	47751.866	23	2076.168		
	Total	93896.704	25			
a. Dependent Variable: 09-18%visitor change						
b. Predictors: (Constant), 09-18%GDPpercapita change, 09-18%GDP change						

Table 4. Linear Regression Coefficients for Percent Change GDP and Percent Change in Visitors from 2009-2018

Regression Coefficients[a]								
Model		Unstandardized Coefficients		Standardized Coefficients	t	Sig.	95.0% Confidence Interval for B	
		B	Std. Error	Beta			Lower Bound	Upper Bound
1	(Constant)	31.037	13.857		2.240	.035	2.437	59.636
	09-18%GDP change	.913	.246	.604	3.715	.001	.406	1.421
a. Dependent Variable: 09-18%visitor change								

Table 5. Linear Regression Coefficients for Percent Change GDP per Capita and Percent Change in Visitors from 2009-2018.

Regression Coefficients[a]								
Model		Unstandardized Coefficients		Standardized Coefficients	t	Sig.	95.0% Confidence Interval for B	
		B	Std. Error	Beta			Lower Bound	Upper Bound
1	(Constant)	39.368	10.936		3.600	.001	16.798	61.938
	09-18%GDPpercap change	1.162	.256	.679	4.535	.000	.633	1.691
a. Dependent Variable: 09-18%visitor change								

Figure 1.

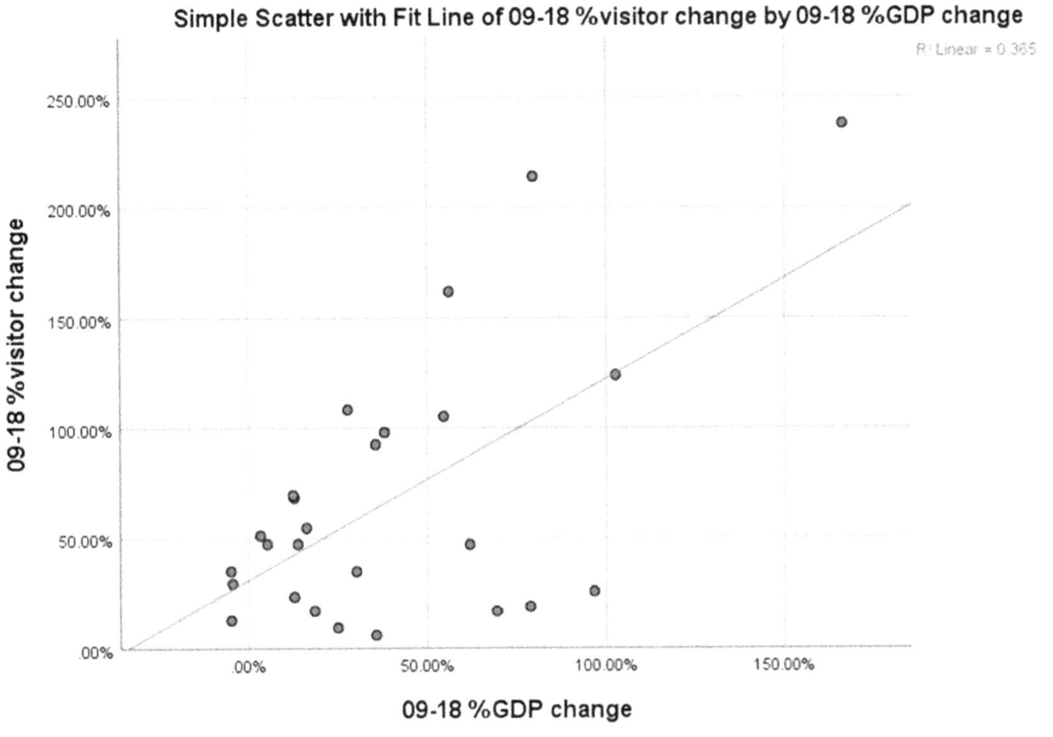

Figure 2.

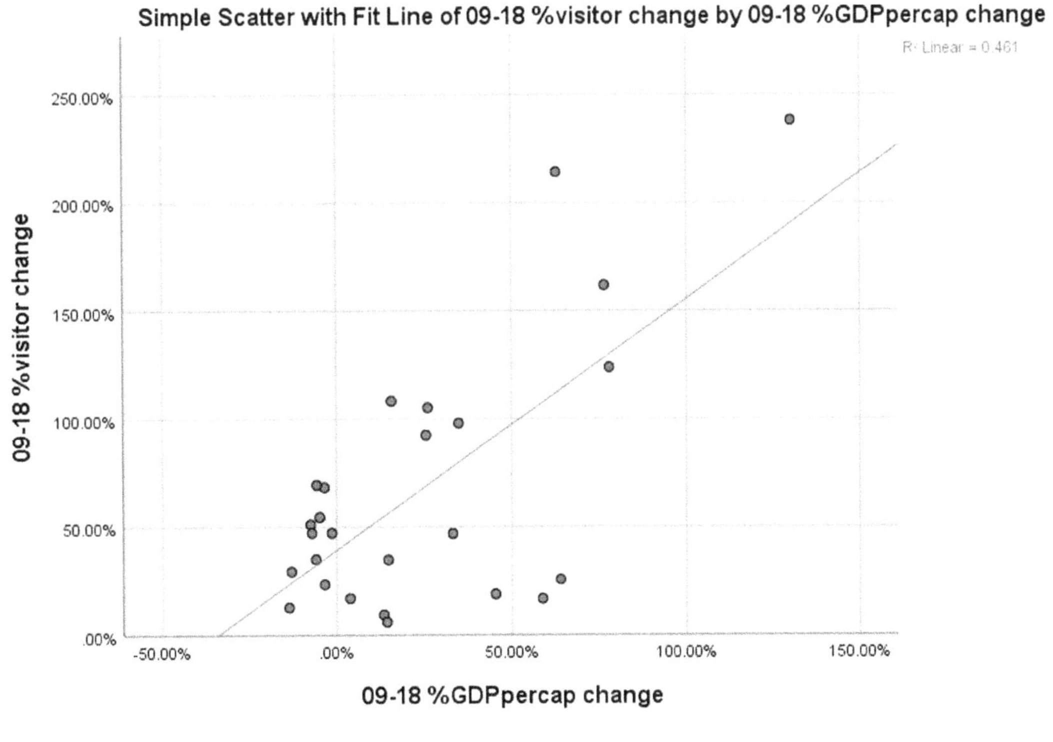

Hypothesis 3 recognized that happiness may have a relationship with an increase in visitors to Las Vegas. Happiness data are compiled annually by UN Sustainable Development Solutions Network into the World Happiness Report. One hundred fifty-three countries received a ranking on the Happiness Index based upon six variables: logged GDP per capita, social support, healthy life expectancy, freedom to make life choices, generosity, and perceptions of corruption. In the 2019 World Happiness Index, Finland ranked at the top of the list for two years, with a recent score of 7.769. Descriptive statistics for the 27 countries with high visitor volume to Las Vegas in this study determined a happiness index mean of 6.55. Of the 27, India received the lowest rating at 4.015 and Denmark had the highest rating at 7.6 (std. dev. 0.884). Each of the individual happiness variables was examined, and the value for corruption was reverse computed to load positively like the other variables. All variables in the happiness index were then transformed as a standardized value (z score). Z score is simply the raw score minus the population mean, divided by the population standard deviation, to be placed on a normal distribution curve, and these scores allowed for a Cronbach's alpha of .768 suggesting an acceptable level of internal consistency among these variables.

A single factor analysis of all 153 countries represented in the index was included with the key Z score variables; generosity was seen as the one variable not loading correctly or cross-loaded on multiple factors and was removed from the remaining analyses; as such Cronbach's alpha increased to 0.835, indicating a good fit for the variables. Modification indices recommended error term connections (Figure 3). Chi-square = 2.839, Degrees of freedom = 3, Probability level = .242, and all factor loadings are positive and significant at the .000 level. CFI = .998 and RMSEA = .053, both indicating a good fit.

Figure 3. One factor analysis of Happiness Index variables

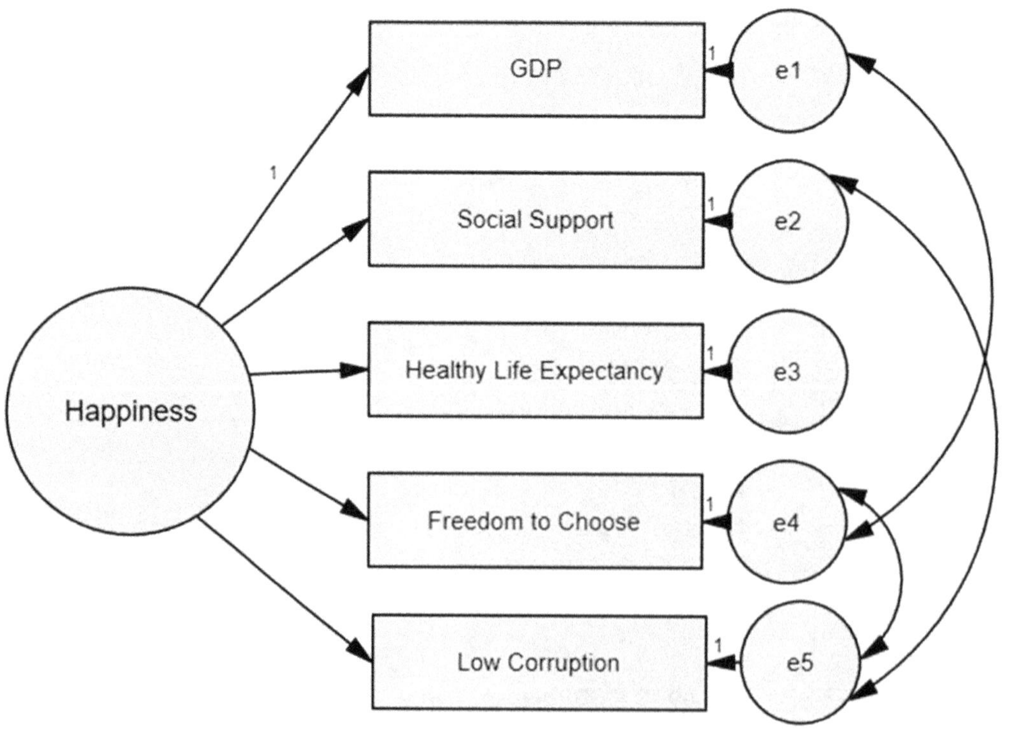

Following the 153-country analysis, the same model was used just for the 27 countries in the Las Vegas visitor analysis, and similar results were shown. Chi-square = 2.504, Degrees of freedom = 3, Probability level = .286, again with all factor loadings positive and significant, but CFI = .995 and RMSEA = .098, indicating not quite as strong of a fit for just 27 countries with RMSEA higher than 0.08.

Therefore, Confirmatory Factor Analysis with two factors of physical and psychological wellbeing were used to evaluate the factor structure of happiness among all 153 countries (Figure 4). The two-factor model had a good model fit with X^2 (2) = 1.664, probability = .435 and CFI =1 and RMSEA = .000. Although not significant, the two-factor model had a better fit than the one factor model: Delta X^2 = 2.839-1.861= 0.978, n.s. Physical wellbeing and psychological wellbeing were significantly correlated R = .897, p<.001 indicating a close relationship between the variables. All variables loaded positively and significantly, indicating that all are important factors when determining the likelihood of increased tourism.

Figure 4. Two factor analysis of Happiness Variables

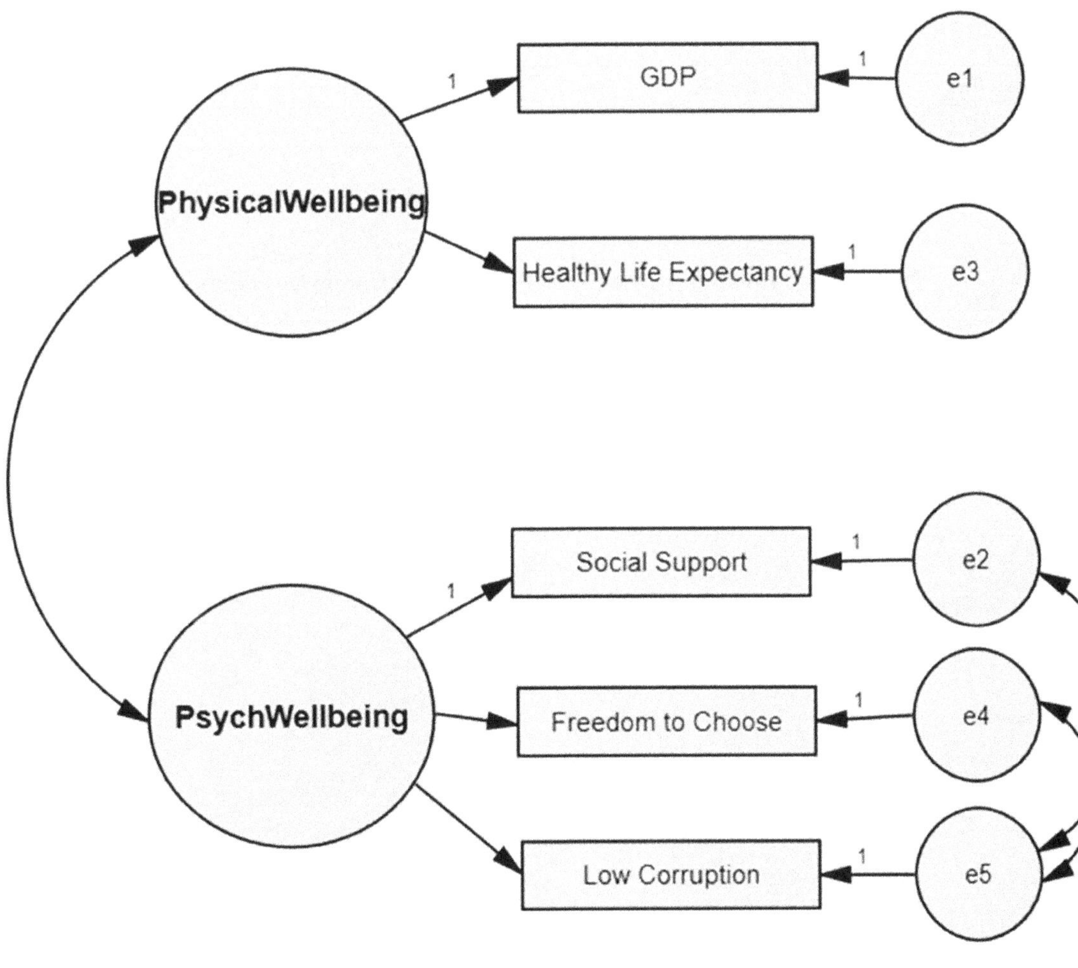

However, while these individual factors loaded positively when studied independently, Table 6 below shows the significance of the happiness index on the percentage change in visitors from 2009-2018, albeit not in the way anticipated. A higher happiness ranking is related to a lower percent change in visitor volume. Therefore, Hypothesis 3 was not supported. When the percent change in visitors was predicted, it was found that a percent change in Happiness Index rating (Beta = -0.393, $p < .05$) was a significant predictor, however opposite the way hypothesis predicted. The overall model fit for Happiness Index was R^2 = 0.154, which is weak. As blatantly indicated in the scatterplot with fit line, a country's greater ranking on the happiness scale did not indicate a greater increase in the country's visitors to Las Vegas.

Table 6. Linear Regression Coefficients for Percent Change in Visitors to Las Vegas from 2009-2018 (Dependent variable) and Happiness Index ranking (Independent variable).

Regression Coefficients[a]								
Model		Unstandardized Coefficients		Standardized Coefficients	t	Sig.	95.0% Confidence Interval for B	
		B	Std. Error	Beta			Lower Bound	Upper Bound
1	(Constant)	245.837	85.998		2.859	.009	68.347	423.328
	Happiness Index	-27.219	13.014	-.393	-2.092	.047	-54.077	-.360
a. Dependent Variable: 09-18%visitor change								

Figure 5.

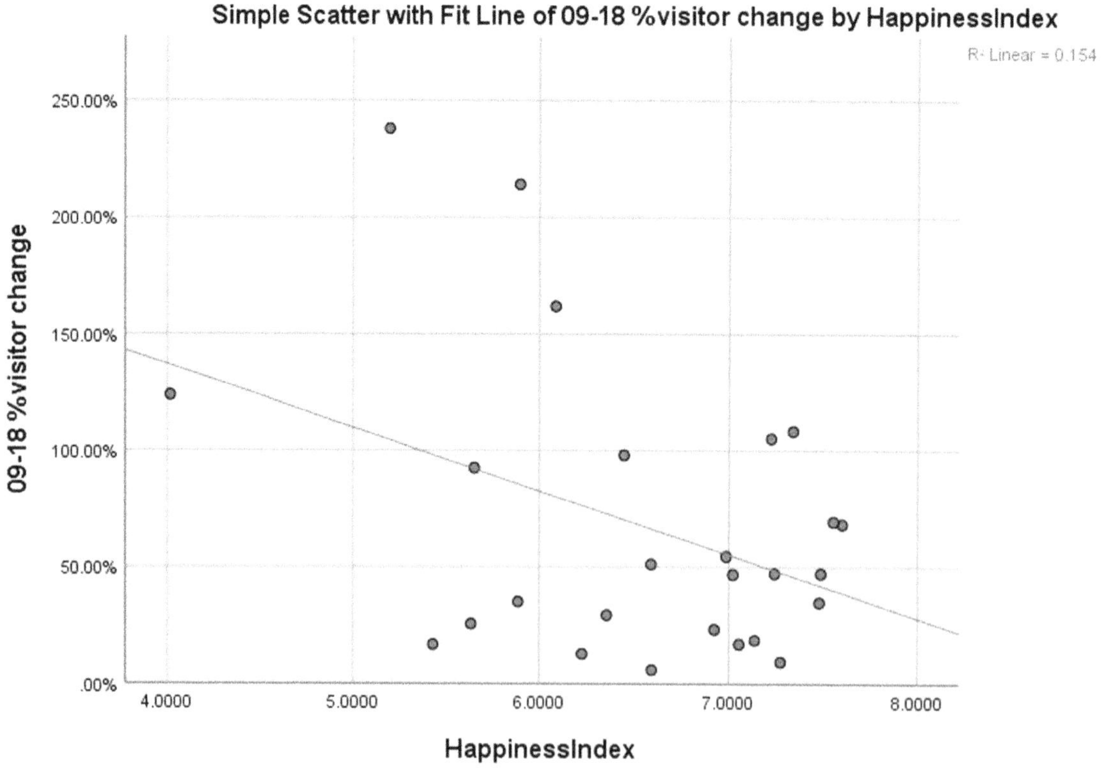

Hypothesis 4 was concerned with innovation in the tourist-sending country, suggesting that an increase in innovation ranking may be associated with an increase in visitor volume over a ten-year period. Innovation was defined and categorized by the World Intellectual Property Organization (WIPO) in their annual Global Innovation Index[3]. The Global Innovation Index (GII) ranks 130 countries on more than 80 indicators including patents, political environment supporting research and development, infrastructure, and business sophistication to coalesce into the GII. Countries are ranked from 0 to 100, with the highest scores indicating the most innovative countries. Taiwan was not included in the GII as it was included in China's ranking. The average score among the remaining 26 countries in this study was 50.89, std. dev. 9.87. Switzerland ranked the highest with a score of 67, and Argentina ranked the lowest in the group with a score of 32. As indicated by Table 7 below, GII was not significantly correlated to a percent change in visitors to Las Vegas from 2008-2019.

Table 7. Correlation between GII and Percent Change in Visitors to Las Vegas

Correlations			
		GII 2019	09-18%visitor change
GII 2019	Pearson Correlation	1	-.318
	Sig. (2-tailed)		.113
	N	26	26
09-18%visitor change	Pearson Correlation	-.318	1
	Sig. (2-tailed)	.113	
	N	26	27

Given the non-significance of innovation on visitor change, Hypothesis 4 is not supported. The final hypothesis questioned if there was a relationship between the six Hofstede dimensions and an increase in tourism to Las Vegas from 2009-2018. All six Hofstede dimensions were gathered for each of the 27 countries in this analysis, apart from one unknown variable for Indulgence in the country of Israel. The remaining 161 variables were analyzed through multiple regression with the GH dimensions as the six independent variables and the percent change in international visitors as the dependent variable (Table 8).

Table 8. Summary of the Multiple Regression with Geert Hofstede dimensions as predictor variables

Model Summary[b]					
Model	R	R Square	Adjusted R Square	Std. Error of the Estimate	Durbin-Watson
1	.466[a]	.217	-.030	89.19413%	1.492
a. Predictors: (Constant), GH Dimension Indulgence, GH Dimension Uncertainty Avoidance, GH Dimension Masculinity, GH Dimension Individualism, GH Dimension Long Term Orientation, GH Dimension Power Distance					
b. Dependent Variable: 09-18%visitor change					

As demonstrated in Table 9, only one GH dimension was statistically significant in the analysis: Individualism, which was negatively correlated, indicating that more collectivist societies are likely to have had an increase in visitors to Las Vegas from 2009-2018. None of the other Hofstede dimensions loaded significantly, suggesting that cultural differences between countries are not as relevant to increasing tourism demand to Las Vegas as other factors like economic growth.

Table 9. Correlation between Percent Change in Visitors to Las Vegas and the 6 GH Dimensions.

Correlations								
		09-18% visitor change	GH Dimension Power Distance	GH Dimension Individualism	GH Dimension Masculinity	GH Dimension Uncertainty Avoidance	GH Dimension Long Term Orientation	GH Dimension Indulgence
09-18%visitor change	Pearson Correlation	1	.207	-.422*	-.186	-.086	.285	-.317
	Sig. (2-tailed)		.311	.032	.362	.675	.158	.123
	N	26	26	26	26	26	26	25

*. Correlation is significant at the 0.05 level (2-tailed).

DISCUSSION

This research project sought to examine international visitors to Las Vegas with the following research questions: Why did the international tourist market to Las Vegas increase among specific countries? And how can this information be utilized for future marketing purposes in a post-covid urban tourism market? Support was found for Hypotheses 1 and 2, and not found for Hypotheses 3 and 4, and only for one dimension of Hypothesis 5. This suggests that GDP and GDP per capita are the strongest indicators of a percent change in visitor volume to Las Vegas from 2009-2018. Happiness, innovation, and cultural dimensions were not as relevant to visitor volume as GDP growth. Of all the analyses run, data indicated that only collectivist societies are likely to have had an increase in visitors to Las Vegas from 2009-2018. None of the other Hofstede dimensions loaded significantly, suggesting that cultural differences between countries are not as relevant as economic ones when it comes to tourism to Las Vegas. Notably, collectivist societies like India and China also had major economic growth. Therefore, for marketing purposes, it is germane for Las Vegas to focus its efforts on countries showing economic growth instead of any specific cultural factors such as innovation or happiness. The Las Vegas Convention and Visitors Bureau would be wise to watch GDP trends throughout the world and advertise in the countries with increased GDP and GDP per capita. Citizens of these countries could have more disposable income, and thus, more ability to travel internationally especially to Las Vegas.

The overall economic result of the coronavirus on the world economy and international travel in particular is yet unknown; borders are slowly opening and international travel is having a slight uptick, but many travelers are still cautious about going overseas. Ironically, in January 2020 before the worldwide spread of the virus, the United Nations World Tourism Organization (UNWTO) predicted an increase in international tourism for the year: "Based on current trends, economic prospects and the UNWTO Confidence Index, UNWTO forecasts a growth of 3% to 4% in international tourist arrivals worldwide in 2020" (UNWTO, 2020). This statement seems discouraging at this point, as many countries closed borders and quarantined to deter the virus spread immediately after that statement was made. The full economic impact of the virus is yet unknown, but it undoubtedly has affected international travel worldwide. For a city like Las Vegas that is incredibly resource dependent upon tourism – international and domestic – the results were devastating. Notably, Lim and Won (2020) studied the Las Vegas tourism economy during the 2007-2009 recession, and they indicated,

Even during economic downturns, some destinations experienced stable or increasing international tourism demands. For instance, international visitation to Las Vegas from 2007 to 2009, increased by 12.7%, while the overall international travel demand decreased by 1.9% for the same period. Las Vegas could survive the significant drop in domestic tourism demand thanks to the increasing international visitor volumes during the global economic recession of 2008. Las Vegas was one of the few destinations that benefitted from the increasing long-haul international visitors during the recession (p. 2).

Unlike the recession of the early 21st century, international visitors during the pandemic were non-existent. All hotel and casino resorts on the Las Vegas strip closed temporarily more than two months, including major properties such as Wynn and MGM Grand (Komenda, 2020). Local airport volume was minuscule during the same time. The loss of revenue and jobs, even short term, was difficult to comprehend, and international visitor volume was nill during 2020. Yet the city must continue to plan for the future, and this study could help gauge future travel plans of international visitors. Again, marketing managers on the strip would be wise to focus on countries with the greatest increases in GDP and GDP per capita as a future source of tourism – especially focusing on long-term economic growth rather than just one or two years; these data will help them plan for post-covid tourism. Marketing can also focus on making international tourists feel welcome once they do arrive since word-of-mouth marketing among international tourists is considered vitally important to encourage repeat visitors to the city.

In addition to managerial and marketing implications, this study makes two contributions to the literature on urban tourism for international visitors. First, it extends the application of resource dependency theory to the city of Las Vegas's international tourism market specific to the growing number of international visitors. The $22 billion USD in tourism revenue is highly dependent on international visitors, and better understanding this market from a cultural and economic perspective is essential for current and future marketing purposes. Secondly, this study was unusual in that it focused on the countries of origin of international tourists rather than the well-documented perspective of the benefits to the receiving country. This topic of international visitor characteristics has received little systematic scholarly attention, and more is warranted. The purpose was to better understand the consistencies among the 27 countries with the greatest amount of visitor growth to Las Vegas over the past decade. In aggregate, overseas visitors increased by more than 67 percent from 2009-2018, and analyzing countries' percent change over this timeframe also allowed for a longer-term view rather than a year-over-year change which could see unusual fluctuations. Empirical research that considered ten years of international travelers was essential for understanding what economic and cultural aspects of the countries most influenced the international travel to a destination like Las Vegas.

LIMITATIONS AND CONCLUSION

Although a limitation for this study was that it only focused on one city's international visitor population, Las Vegas typically receives more than five million international visitors each year, and it is greatly dependent upon this population of visitors more than any other city in the nation. Future research could extend this study to explore international visitors in other areas, and the hope is that this study will stimulate future research on international visitor patterns to improve our knowledge and practice about how we can better market to tourists worldwide in a post-covid era. Few studies examine both economic and cultural issues together; perhaps this study can also be a catalyst for using the same dependent variable such as visitor growth along with various predictor variables. The goal of this work is to comprehensively understand international visitors so that scholars and practitioners alike can better market and prepare for these guests once it is safe for international travel again. It will also be essential to explore motivations of international visitors in a post-COVID world. As tourism slowly starts to rebound to Las Vegas and other major destinations, it will be worthwhile to study which countries are sending visitors and where commonalities lie among those countries to better equip marketers with a plan for attracting such tourists. In a pre-COVID world, economic growth, as measured by GDP and GDP per capita, were the most influential aspect of international tourism to Las Vegas. It is unknown if this will remain the case, especially as the United States has the highest rates of coronavirus, and wealthy nations may pursue travel elsewhere. With the right planning and marketing in Las Vegas, as well as continued emphasis to embrace the international visitor market, hopefully these measures will bring the city back to its intended neon glory.

REFERENCES

Andersson, T. D., & Getz, D. (2008). Stakeholder management strategies of festivals. *Journal of Convention & Event Tourism*, *9*(3), 199–220. doi:10.1080/15470140802323801

Balaguer, J., & Cantavella-Jorda, M. (2002). Tourism as a long-run economic growth factor: The Spanish case. *Applied Economics*, *34*(7), 877–884. doi:10.1080/00036840110058923

Castro-Nuño, M., Molina-Toucedo, J. A., & Pablo-Romero, M. P. (2013). Tourism and GDP: A meta-analysis of panel data studies. *Journal of Travel Research*, *52*(6), 745–758. doi:10.1177/0047287513478500

Cheng, K. M. (2012). Tourism demand in Hong Kong: Income, prices, and visa restrictions. *Current Issues in Tourism*, *15*(3), 167–181. doi:10.1080/13683500.2011.569011

Chu, F. L. (2014). Using a logistic growth regression model to forecast the demand for tourism in Las Vegas. *Tourism Management Perspectives*, *12*, 62–67. doi:10.1016/j.tmp.2014.08.003

Consumer Electronic Show [CES]. (2019). *CES Fact Sheet.* Retrieved from https://cdn.ces.tech/ces/media/pdfs/ces-2019-fact-sheet.pdf

Crouch, G. I. (1994). Price elasticities in international tourism. *Hospitality Research Journal*, *17*(3), 27–39. doi:10.1177/109634809401700304

Hillman, A. J., Withers, M. C., & Collins, B. J. (2009). Resource dependence theory: A review. *Journal of Management*, *35*(6), 1404–1427. doi:10.1177/0149206309343469

Hofstede Insights. (2020). *Description and Country Comparison*. Retrieved from https://www.hofstede-insights.com/product/compare-countries/

Hsu, S. Y., Woodside, A. G., & Marshall, R. (2013). Critical tests of multiple theories of cultures' consequences: Comparing the usefulness of models by Hofstede, Inglehart and Baker, Schwartz, Steenkamp, as well as GDP and distance for explaining overseas tourism behavior. *Journal of Travel Research*, *52*(6), 679–704. doi:10.1177/0047287512475218

Komenda, E. (2020). COVID-19: Las Vegas Strip resorts are now closing, includes MGM, Wynn properties. *Reno Gazette Journal*. Retrieved from https://www.rgj.com/story/news/2020/03/15/las-vegas-strip-resorts-now-closing-first-up-wynn-and-encore/5056092002/

Kumar, N., Kumar, R. R., Patel, A., Hussain Shahzad, S. J., & Stauvermann, P. J. (2020). Modelling inbound international tourism demand in small Pacific Island countries. *Applied Economics*, *52*(10), 1031–1047. doi:10.1080/00036846.2019.1646887

Lanza, A., Temple, P., & Urga, G. (2003). The implications of tourism specialization in the long run: An econometric analysis for 13 OECD economies. *Tourism Management*, *24*(3), 315–321. doi:10.1016/S0261-5177(02)00065-1

Las Vegas Convention and Visitors Bureau [LVCVA]. (2018). *Arriving visitors on direct international flights 2013-2017*. Retrieved from https://assets.simpleviewcms.com/simpleview/image/upload/v1/clients/lasvegas/2017_Top_Air_Markets_Intl_b930157b-ced0-4acd-9f31-ee90cb89e43a.pdf

Las Vegas Convention and Visitors Bureau [LVCVA]. (2019). The relative dependence on tourism of major U.S. economies. *Economic Impact Series Brief.* Retrieved from https://assets.simpleviewcms.com/simpleview/image/upload/v1/clients/lasvegas/EIS_Tourism_Dependence_Jan_2019_FINAL_e332cdc1-7e25-4179-b4e9-63e1890adefe.pdf

Lim, J., & Won, D. (2020). preprint). How Las Vegas' tourism could survive an economic crisis? *Cities (London, England)*, *100*, 102643. Advance online publication. doi:10.1016/j.cities.2020.102643

Pfeffer, J., & Salancik, G. (1978). *The external control of organizations: a resource dependence perspective*. Harper & Row Publishing.

Raz, N. (2016). Inaugural Hainan Airlines flight from Beijing to Las Vegas lands at McCarran. *Las Vegas Review Journal*. Retrieved from https://www.reviewjournal.com/business/tourism/inaugural-hainan-airlines-flight-from-beijing-to-las-vegas-lands-at-mccarran/

Terzić, L. (2018). How tourism destination competitiveness contributes to GDP growth –The case of selected European economies. *Acta Oeconomica Pragensia*, *2018*(3), 24–34. doi:10.18267/j.aop.606

United Nations World Tourism Organization [UNWTO]. (2020). *World tourism barometer, No. 18*. Retrieved from https://unwto.org/world-tourism-barometer-n18-january-2020

Velotta, R. (2012) International tourists in Las Vegas outspend domestic travelers, analyst says. *Vegas Inc.* Retrieved from https://vegasinc.lasvegassun.com/business/real-estate/2012/jul/10/international-tourists-las-vegas-outspend-domestic/

Witt, S. F., Newbould, G. D., & Watkins, A. J. (1992). Forecasting domestic tourism demand: Application to Las Vegas arrivals data. *Journal of Travel Research*, *31*(1), 36–41. doi:10.1177/004728759203100108

Yoo, M., Singh, A. K., & Ghaharian, K. (2017). Foreign exchange rate effect on international gaming demand: An examination from an upscale Las Vegas casino. *UNLV Gaming Research & Review Journal*, *21*(1), 1–23.

KEY TERMS AND DEFINITIONS

Geert Hofstede Dimensions (GH Dimensions): The GH dimensions are a series of six measurements of national culture: Power Distance, Individualism, Uncertainty Avoidance, Masculinity, Long Term Orientation, and Indulgence. Most countries of the world are rated on these six characteristics to provide a comprehensive typology of culture for each nation. Learn more at geerthofstede.com.

Gross Domestic Product (GDP): This is a comprehensive measure of economic activity for a country, state, city, or region. As one example, as of the writing of this chapter, the GDP of France was over $2.5 trillion USD, and the economy of the Paris metropolitan area generates about one-third of the French GDP.

Las Vegas Convention and Visitors Authority (LVCVA): According to their website, the LVCVA is a professional organization designed to "promote Las Vegas as the world's most desirable destination for leisure and business travel." They create marketing for the region, as well as events, advocacy, and research to be the travel and tourism industry leaders of Southern Nevada.

Resource Dependence Theory (RDT): Resource Dependence Theory was identified in 1978 by Pfeffer and Salancik to indicate how firms, organizations, and other entities rely on the procurement of external resources to survive and thrive. Dependent resources may be

financial: physical, and/or informational in nature, and firm behavior may adapt for the acquisition of these resources.

ENDNOTES

1 https://datacatalog.worldbank.org/dataset/gdp-ranking

2 https://www.economywatch.com/economic-statistics/economic-indicators/GDP_Growth_Constant_Prices_National_Currency/

3 https://www.globalinnovationindex.org/

Chapter 9
Circumstance and Opportunity: "Pop-Up" Mobility in Pandemic Times - Two Case Studies in Braga

Miguel Sopas de Melo Bandeira
University of Minho, Portugal

Filipa Assis Corais
https://orcid.org/0000-0003-0777-7771
University of Minho, Portugal

ABSTRACT

The pandemic has impacted several aspects of urban life in cities, from the elementary and essential daily living in the public space of its resident community to the collapse of tourism, which, until 2020, was in ascending progress in the case of Braga and which has repercussions on the entire economy. In this context, several cities took the opportunity to experiment with 'pop-up' measures (i.e., opportunity-based, rapidly implemented, low-cost, temporary, and experimental) to transform public space by allocating it to pedestrians and urban living. This chapter explores two experiences tested in Braga, during the pandemic, which, by their latent capacity of catalysis in other areas, glimpse and enhance a transformation in the mobility of the city in a future scenario.

INTRODUCTION

The times we live in, especially, for the flow of the central territories we inhabit, will certainly be identified by the constraints of the pandemic, initially called "covid-19", and which has been manifesting variants throughout these two years. As a pandemic, it has a planetary expression, which can be temporally marked between its focus of emergence, announced in December 2019, in China; considered by the World Health Organization (WHO) as an International Public Health Emergency, on 20 January 2020, and classified as a pandemic on 11 March (in Portugal, when the first confirmed cases occurred on 2 March 2020); until today, when it is consensually admitted to already be under control.

DOI: 10.4018/978-1-6684-3369-0.ch009

This reflection is based on the hypothesis that this context is conducive to the paradigm shift for the promotion of sustainable urban mobility long claimed and determined in the Paris Agreement. In fact, several cities have tried to respond to the constraints generated by the pandemic in terms of imposing social distances, limiting the capacity of public transport and enclosed spaces, among others. In this sense, all over the world there has been the closing of traffic lanes to ensure more space for pedestrians and other soft modes. In fact, the pandemic can be considered an opportunity for a paradigm shift and, therefore, this chapter aims to analyze whether this circumstance ensures the consummation of a redistribution of public space in cities in a more equitable way by the various modes of transport and whether it ensures the promotion of sustainable urban mobility.

BACKGROUND

Covid-19

The scope of COVID-19's impact, which can be said, without any reservation, to be multi-sectorial, gives rise to the most diverse reflections on the worldview of all the assertions with which it may be approached. Therefore, after the moment of generalized shock and the emerging radicalism with which it quickly imposed itself, the great question that arises today will always start from establishing whether everything will go back to the way it was, or, on the contrary, whether nothing will be the same as before. In fact, this dilemma is nothing new whenever a deep rupture occurs in the daily habits of a society, as is happening now. If prudence leads us to admit that there will be as much of one as the other, there will undeniably always be the circumstance for the acceleration of the processes underway. However, it does not necessarily tend to be for the benefit of only those perspectives considered to be more progressive or more proactive.

It is true that circumstances generate opportunities, dramatize the tactics over the strategy, valuing the most intangible areas of the process, such as the prognostic vision, the capacity for initiative and, naturally, the determination and conviction that is placed in the actions undertaken.

Hence, the concrete opportunity, in terms of urban mobility and of the collective design to proceed with the gradual transition to a sustained urban development, promoting decarbonization, the improvement of the environment and the pursuit of the quality of life of the populations, in this case, of the city of Braga.

As from the emergency measures of March 2020, issued by the Portuguese government in the sense of general confinement, we immediately witnessed a brutal cut in road circulation in our cities, a drastic reduction in the use of public transport and the closure of hotel units and tourist travel, having, in turn, resumed neighborhood and proximity relations, and started to witness an increase in pedestrian travel and the use of light and clean means of locomotion, being able to enjoy the liberation of the road network. Thus, the favorable conditions were created for the anticipation of some of the measures envisaged and, in the same sense, the possibility of implementing actions capable of favoring the inversion of the mobility pyramid, without implying conflicting tensions with the rooted advantages enjoyed by car circulation.

Sustainable Urban Mobility

In the 20th century, the use of car transport was boosted, formatting our cities according to its maximum efficiency, allocating to this mode of transport the largest percentage of public space, compared to other uses, so that today it is consensual to understand that some of the main constraints to urban living and mobility, in general, emanate from this. Given that the transport sector contributes more than 30% to the total amount of greenhouse gases, this should be the subject of deep reflection by all of us, in our daily lives, and should stimulate the transformation of habits and behaviors of mobility.

In 2016 the Paris Agreement was signed by 55 countries which committed to the objectives of almost total decarbonization of society by the end of the century. As an intermediate goal, the 2030 objective was set to reduce greenhouse gas emissions in 1990 by 40%.

Currently, we are witnessing very different local realities, some of which are very close to each other, and time lags in terms of modal distribution, both on a European and global scale. In Portugal, despite the recent investments in mobility and the application of some European trends and impositions to achieve the inversion of the mobility pyramid (replacing the car by the pedestrian at the apex), specific knowledge and scientific research in this area are still scarce (C. S. de Oliveira, 2018), although there is a predominance of the car (corresponding to a weight of about 60% of the total modal split). However, if we assess the distribution of public space by each of the modes of use this discrepancy is even more pronounced.

Portuguese cities are thus going through a pioneering transition moment for the inevitable change in the mobility paradigm. Given the current state of affairs, however, fundamental changes will be necessary for Portugal to achieve the goal it has committed to for carbon neutrality by 2050.

The paradigm shift towards the promotion of Sustainable Urban Mobility (SUM), inverting the aforementioned mobility pyramid, will be addressed here through the transformations that need to be introduced in the Public Space, to catalyze the population's behavioral change that, in the meantime, will be necessary to legitimize it. With this purpose, it is also intended to demonstrate the potential of Architecture for the paradigm shift of the SUM conceptually subordinated to the design/project (Saelens et al., 2003; Zuniga-Teran et al., 2016) of the System of Public or Collective Spaces.

With this purpose, Banister & Hickman (2007), based on the previous work of Marshall (2001), establish the main contrasts between the perspectives of the past, conveyed by the technical-conceptual hegemony of traffic engineering, and the new approaches in defense of Sustainable Urban Mobility. That is, in a certain sense, the introduction of the design of the territory and the urban form integrated with the functionalist and automating criteria of human flows. Simultaneously, in the approach for the promotion of Sustainable Urban Mobility, the following are affirmed the social dimension coupled with the spatial one, given that the focus is unavoidably placed on the basis of people and, therefore, on the pedestrian mode, which comes to occupy the top of the hierarchy; the scale of action is the local scale, although imbued with relationships with the global scale (Morin, 2005, 2014); and, finally, the streets are given the role of street and urban living space (Gehl, 2010, 2011; Gehl & Svarre, 2013), in which the integration and coexistence of the various modes of travel is privileged. To this end, according to Banister & Hickman (2007), it is advocated the reduction of traffic speeds and demand management, benefiting the reliability of travel times (Corais, 2021, p. 20).

MAIN FOCUS OF THE CHAPTER

A Testimony of Reflection-Action

The case study presented in this brief reflection focuses on the city of Braga, a city located in the Northwest of Portugal, and refers to the impacts generated by the recent pandemic period. It will relate what happened and how it was responded, facing the same challenges, in other Portuguese and European cities.

The counterpoint between the measures planned for mobility in such disparate periods as pre- and post-Covid-19, will serve as a framework model to consider opportunities for redistributing public space and promoting sustainable urban mobility and walkability in cities, which are provided in unexpected moments of contingency and/or disaster. Covid-19 is not the first pandemic to cause us to ponder the role of planning and the importance of public space in promoting public health ((Honey-Rosés et al., 2020), so it is up to technicians and policy makers to prepare the ground for the realization of promoting sustainable urban mobility. In this context, Eltarability & Elghezanwy (2020) conducted a literature review that relates the impact of epidemics on urban design. Examples include the Black Death, in the 14th century, which led to the opening of public spaces throughout Europe; cholera, in the 19th century, caused by the mixing of drinking water and wastewater, leading to the need for waste management, as well as to the definition of greater ventilation and luminosity in streets; and also the Spanish Influenza, in the first quarter of the 20th century, which, like the current pandemic, to prevent contagion, valued the pedestrian mode over public transport.

The impact of the COVID-19 pandemic on the physical space of cities is a reality that tested the influence of conventional mobility patterns and the lifestyles of its residents, determining their resilience (Bereitschaft & Scheller, 2020) to the unknown situations created in the meantime. The ensuing crisis, both socio-economic and in public health, implied transformations in urban design (Honey-Rosés et al., 2020, p. 12) and a spontaneous functional redistribution of the same, inducing change, both in mentalities/perceptions and in ingrained mobility behaviours. Walking or cycling can thus assume here the multiple functionality of mobility promoter, decarbonization driver, social increment factor, rehabilitator of physical and psychological activity, and guarantee of compliance with the social distancing imposed by the pandemic, in addition to the intrinsic potential to regenerate public space (Eltarabily & Elghezanwy, 2020). In summary, pandemic can be a real catalyst of transition processes.

Green public spaces (e.g. parks, riverbanks, etc.), well connected, executed and scaled to ensure the social distances imposed by pandemic, are increasingly sought after by the population, especially visible at the time of the lifting of confinement restrictions. These routes, initially designed for recreational purposes, are now, along with other public spaces, a new potential for the use of sustainable active mobility for home-work-home or home-school-home journeys during unlockdown (Bereitschaft & Scheller, 2020). The growing demand by the population for this type of space during the lifting of the restrictions resulting from the pandemic shows the importance that society attaches to these spaces to ensure their physical and psychological well-being. This observation leads to reflection on the resizing, redistribution and redesigning of green spaces on a city scale.

On the other hand, all over the world, resilience plans and temporary interventions in public space have been implemented (as was the case of Milan and Vienna), whose timeliness prepared the ground for perennial transformations in mobility. Milan was the first city to announce permanent measures for the widening of pavements and the construction of about 35 km of bicycle paths, as well as the definitive closure of streets to car traffic, in order to mitigate the impact of the pandemic and, at the same time, ensure compliance with the mandatory minimum distance between people in public space. In fact, the pandemic brought about, or reinforced, new patterns of use of public space, the imposed social distancing, and the restrictions on the capacity limitation of the interior spaces of establishments with public attendance. It has transformed the undersized pavements into authentic lobbies for access to them, defining large queues near their doors that impose the resizing and redesign of these public spaces (Figure 1).

Figure 1. Formation of queues of people in the external access of establishments due to the restrictions imposed on their capacity by COVID-19 (Braga)
Source: [own source] (2020.04.24)

In this way, we intend to analyse the impact that COVID-19 had, on the one hand, on the transformation of the public space implemented by technicians and political decision-makers in cities and, on the other hand, how these measures contributed to changing the perception and mobility behaviours of the populations. The relevance of this reflection is also reflected in its applicability to future mobility planning and design at city scale. It is important to draw as many lessons as possible from the context in order to reprogram the unavoidable sustainability measures in a more integrated and integrative way.

After reviewing the literature on the interaction between public space, mobility and social well-being, and the respective articulation with the propagation of the SARS-CoV-2 virus, we will focus on two case studies implemented in the city of Braga, which configure the circumstance and the opportunity of our condition, and we will proceed to their respective analysis by comparison, we would say, at the level of a benchmarking, with other measures recently implemented in other cities.

The evaluation of the transformations operated in Braga public space was checked through a quantitative survey, based on the counting of people and vehicles, before, during and after the confinement periods, as well as the comparison with the previous situation of the availability of public space for the population's enjoyment, during those periods. Regarding the evaluation of the impact of the physical measures implemented in the public space on urban life, interviews were conducted and reactions were collected from the various stakeholders.

Impacts and Restrictions Arising from the Pandemic on Mobility and Public Space

Portugal

In March 2020, Portugal announced the first restrictive measures arising from the pandemic COVID-19. Since it is an airborne disease, which is easily transmitted through the respiratory tract, social distancing was one of the first and main measures imposed. The table 1 below and in figure 2 summarises the most relevant dates of the pandemic in Portugal (more detailed table in appendix 1):

Figure 2. Graph showing the evolution of new cases confirmed with infection
Source: https://coronavirus-portugal-esriportugal.hub.arcgis.com/apps/evolu%C3%A7%C3%A3o-do-covid-19-em-portugal/explore (consulted at 2022.02.12)

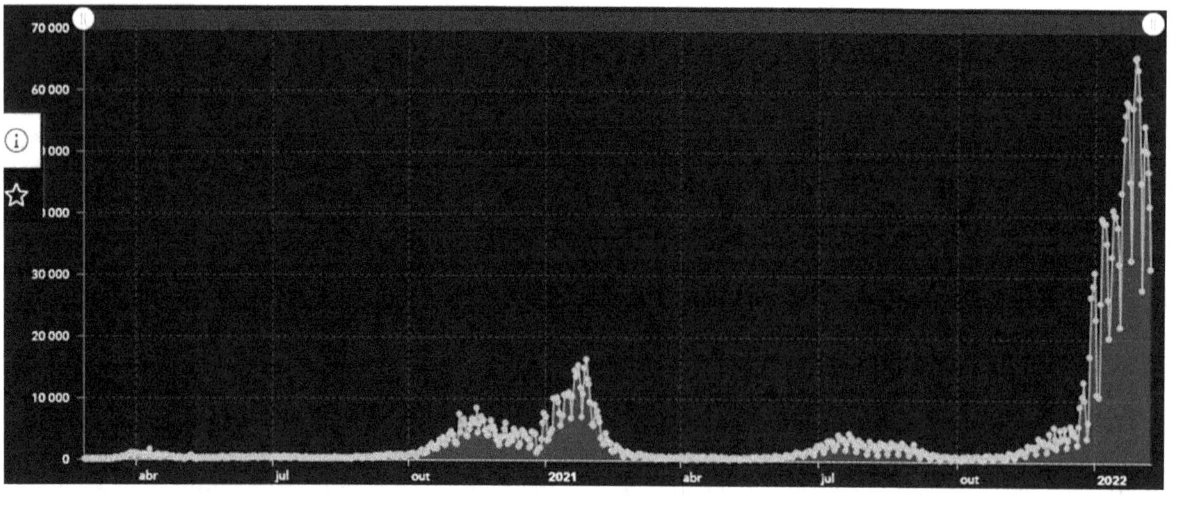

Table 1. Synopsis of the Evolution of the Pandemic in Portugal and the main measures.

Date	Number of positive cases- Portugal	Description	Main Measures
2020.03.02	2020.03.02.-2; 2020.03.16–331 (+86); 1st death - Portugal	1st COVID-19 case in Portugal	Teleworking; online classes in schools; cancellation of flights; suspension of events; reduce crowding in restaurants.
2020.03.18	2020.03.18 – 642 (+ 194); 2nd death; 2020.04.10 -15.472 (+1516)	Beginning of the State of Emergency and 1st Lockdown	State of Emergency. Mandatory lockdown and restrictions on circulation on public roads.
2020.05.03	2020.05.03 – 25.282 (+92) 2021.05.13 – 28.132 (+219)	End of state of emergency and start of calamity situation. (1st phase of the unlockdown plan)	Teleworking continues, but cultural services start to open. Some high street shops open (up to 200 m2), but with restricted opening hours and capacity; public transport at 2/3 of the capacity and masks are compulsory; outdoor sports allowed. No gatherings of more than 10 people.
2020.05.16	2020.05.16 – 28.810 (+227) 2020.05.29 – 31.946 (+350)	2nd phase of the unlockdown plan	Opening of restaurants and cafés with restricted opening hours and limited to 50% capacity, and the start of classes for 11th and 12th grade students. Nursery schools can start to open.
2020.06.01	2020.06.10 – 35.600 (+294)	3rd phase of the unlockdown plan	Opening of shopping centres and pre-schools; childcare centres, cinemas and theatres. 01.07 Teleworking is no longer compulsory and gymnasiums open with rules. 04.06 the Economic and Social Stabilization Programme (PEES) is approved. Until 30.06 borders with Spain closed.
2020.07.01	2020.07.15 – 47.426 (+375) 2020.07.26 – 50.000 (+209)	End of calamity situation and beginning of alert situation	Borders with Spain reopen. 21.07.07 financial package of aid to EU Member States to mitigate the effects of the pandemic and the Government publicly presents the "Strategic Vision for the Recovery Plan 2020/2030.
2020.09.15	2020.09.18–(+770) 2020.10.01 – (+854) 2020.10.08– (+1.278)	Portugal is now in a state of emergency	Back to school and face-to-face work, with the requirement to wear a mask. Gatherings limited to 10 people and shops can only open between 10am and 11pm. In cafés and restaurants near schools, groups may not exceed 4 people. Sports grounds remain closed to the public.
2020.10.15	2020.10.30 –141.279 (+4.656)	Portugal is now in a state of calamity	No gatherings of more than 5 people in the public space and family events may not have more than 50 people. No academic festivities.
2020.11.09	179.324 (+4.096)	State of Emergency	Obligatory night-time collection (23h-05h) during the week and from 13h on weekends, in 121 councils (70% of the population of Portugal).
2021.01.15	+10.663	2nd lockdown	The Government has determined a set of extraordinary measures aimed at limiting the spread of the pandemic and protecting public health, including: cultural and leisure activities, sporting activities and spas; the operation of restaurants for take-away service only; a ban on celebrations and other events.
2021.03.15	+256	1st unlockdown phase	Opening of day care centres, pre-school, primary and secondary schools, religious celebrations and trade to the wicket.
2021.04.05	159 New cases of infection	2nd phase unlockdown	Openings for 2nd and 3rd cycle schools and child day care centres; shops up to 200 m2 with doors onto the street, monuments, art galleries, etc.; terraces with a maximum limit of 4 people per table.
2021.04.19	220 New cases of infection	3rd phase unlockdown	Start of classes for secondary and higher education; opening of all shops and shopping centres; restaurants, cafes and pastry shops (with restricted opening hours and capacity); opening of cinemas, theatres and concert halls.
2021.05.03	387 New cases of infection	4th unlockdown phase	Restaurants, cafes and bakeries with no time limits, increased capacity; reopening of gyms and holding of large outdoor events.

continues on following page

Table 1. Continued

Date	Number of positive cases- Portugal	Description	Main Measures
2021.12.26	+ 3.732 Casos; 2021.12.30 - +30.829 casos	Further restrictive measures	Compulsory telework; suspension of classes; compulsory negative test for air travel, access to restaurants, cinemas and concert halls; discotheques closed.
2022.01.10	20.212 New cases of infection	Unlockdown	Start of face-to-face classes; compulsory teleworking regime until 14 January; reopening of bars and nightclubs on the 14th.
2022.01.14	40.090 New cases; 2022.01.27 - 65.706 New cases (new peak)		Teleworking regime goes from mandatory to recommended; reopening of bars and nightclubs; The limits regarding the occupation of commercial establishments are maintained - maximum indicative occupation of one person per 5m2 of area);
2022.02.09	34023 New cases; 2022.02.28 – +3.423		PCR or antigenic tests are no longer mandatory for international flights. Relaxation of isolation rules in positive cases (7 days if without symptoms; 10 days if with symptoms). No. of people fully vaccinated 8,996,933

Own preparation based on: https://www.dn.pt/sociedade/doze-meses-doze-momentos-na-evolucao-da-pandemia-em-portugal-13406881.html ;https://www.consilium.europa.eu/pt/policies/coronavirus/timeline/; https://graphics.reuters.com/world-coronavirus-tracker-and-maps/pt/countries-and-territories/portugal/ (consulted 2022.02.20).

Through this diachronic sequence of events, we can highlight, in this study, all kinds of restrictions with an impact on the use of public space and/or on mobility, namely, the mandatory minimum physical distances, the limitations to the gathering of people, the capacity limits in public transport and, as a consequence of teleworking and distance learning, the reduction of the number of people circulating in the cities. On the other hand, it is important to note the evolution of the pandemic, with its sharp growth gradients, its peaks, the periods of softening; from the social psychological perspective, the way people's behaviour, the media, the political decision-makers reacted, from the initial stupefaction, reacting more vigorously to the facts, to the periods of recession, causing social upsets that were the basis for new outbreaks.

According to the INE - *Instituto Nacional de Estatística*, tourist accommodation in Portugal fell by 63%, with the equivalent figures at international level reaching almost 75%, according to the World Health Organization (WHO) [1] taking the tourism slump and its recovery after the 2008-2009 economic crisis as an example, the "Reactivate Tourism Action Plan" [2] believes in the rapid recovery of tourism demand, after the pandemic was brought under control, and urges Portugal to create the best possible conditions to be at the forefront of this recovery of tourism activity.

Braga

In March 2020 Braga was one of the first municipalities in Portugal to take restrictive measures to contain the contagious outbreak of the pandemic. We remember, with emphasis, for the actions that had incidence in the regime of opening to the public of the commerce (table 2).

Table 2. Summary of the number of positive cases confirmed in Braga.

<table>
<tr><th>Date</th><th>State in Portugal</th><th>Measures in Braga</th><th>No. of positive cases in Braga</th></tr>
<tr><td>2020.03.24</td><td rowspan="10">State of Emergency and 1st Lockdown</td><td rowspan="10">2020.03.09 - Suspension of municipal sports activities; 2020.03.10 - Postponement of the 53rd edition of Agro - International Agricultural Fair; 2020.03.10 - closure to the public of municipal spaces, including the Tourism Office. 2020.03.15 - closure of non-essential services; 2020.03.16 - municipality implements "horário Zero" - restricts the opening hours of all establishments of commerce and services except essential services; 2020. 2020.03.17 - Braga Urban Transports (TUB) announced that will not charge monthly passes during the month of April, or parking fees to residents and merchants. 2020.03.19 - Municipality provides an email address for clarification of doubts. 2020.03.30 - Cancellation of all municipal activities until June. 2020.04.01 - Municipality supports local businesses. 2020.04.08 - Campsite provides accommodation for health professionals. 2020.04.30 - Continued closure of municipal public spaces until May 18.</td><td>21</td></tr>
<tr><td>2020.03.26</td><td>98</td></tr>
<tr><td>2020.03.27</td><td>131</td></tr>
<tr><td>2020.03.29</td><td>208</td></tr>
<tr><td>2020.03.31</td><td>220</td></tr>
<tr><td>2020.04.01</td><td>245</td></tr>
<tr><td>2020.04.08</td><td>423</td></tr>
<tr><td>2020.04.15</td><td>692</td></tr>
<tr><td>2020.04.22</td><td>903</td></tr>
<tr><td>2020.04.29</td><td>1012</td></tr>
<tr><td>2020.05.03</td><td>End of state of emergency and start of calamity situation with phased application of Lockdown measures (1st phase)</td><td rowspan="6">2020.05.11 - Braga screen against COVID-19 the employees of nursery schools and kindergartens; 2020.05.13 - lockdown plan of the municipality that will come into force from 18 May. 2020.10.28 - it becomes mandatory to wear masks in public spaces. 2020.10.31 - civic duty of recollection. The months of July and August recorded very positive figures for the number of tourist visitors compared to expectations and to previous months. In fact, after the Lockdown and with the opening of the borders and air corridors, many emigrants and French people returned, as well as Spanish tourists. There was also an upturn in demand from organised groups from France and Poland. After September, the new phase of contention limited the circulation between municipalities, reducing the number of tourists again.</td><td>1105</td></tr>
<tr><td>2020.05.16</td><td>(2nd phase of the unlockdown plan)</td><td>1153</td></tr>
<tr><td>2020.06.01</td><td>(3rd phase of the unlockdown plan)</td><td>1225</td></tr>
<tr><td>2020.07.01</td><td>End of calamity situation and beginning of alert situation</td><td>1256</td></tr>
<tr><td>2020.09.15</td><td>Portugal is now in a state of contingency</td><td rowspan="4">without access to information</td></tr>
<tr><td>2020.10.15</td><td>Portugal is now in a state of calamity</td></tr>
<tr><td>2020.11.09</td><td>State of Emergency</td><td>2020.11.25 - The Municipality of Braga and Agere (Municipal Water and Sanitation Company) implement an additional set of measures aimed at mitigating the impacts on their customers and support the economic tissue. 2020.12.03 - Announced the start of vaccination for January 2021; accommodation for health professionals (in Pousada da Juventude).</td></tr>
<tr><td>2021.01.15</td><td>2º Lockdown</td><td>Commerce and services closed; general duty to stay at home; restaurants and cafes take-away only. 2021.01.21 - suspension of face-to-face teaching activity.; 2021.01.25 - municipality supports catering and commerce by waiving municipal fees and licences.</td></tr>
</table>

continues on following page

Table 2. Continued

Date	State in Portugal	Measures in Braga	No. of positive cases in Braga
2021.03.15	1st stage of unlockdown	Implementation of national measures.	without access to information
2021.04.05	2nd stage of unlockdown	The recovery of the tourism sector, in Braga, began to be felt from April onwards, becoming more pronounced in July and August. The last quarter of the year recorded a new drop, due to the restrictions imposed by the worsening of the pandemic indices worldwide. Portuguese tourists had a greater expression in the market, going from 2nd place (in 2019, behind Spain) to first.	
2021.04.19	3rd stage of unlockdown		
2021.05.03	4th stage of unlockdown		
2021.12.26	Further restrictive measures		
2022.01.10	Unlockdown	Implementation of national measures.	
2022.02.09	Unlockdown	Implementation of national measures.	

Source: Own preparation based on, *(CMB, 2020);* https://www.jornaldenegocios.pt/economia/coronavirus/detalhe/o-mapa-e-todas-as-estatisticas-da-evolucao-do-coronavirus-em-portugal *(consulted at 2022.02.12);* http://covid19.cm-braga.pt/ *(consulted at 2022.02.12)*

In 2019, the classification of Braga as the second best European destination, collecting over 59 thousand votes from the initiative, "European Best Destination", as well as the election of *Bom Jesus do Monte* as a Unesco World Heritage Site, contributed to Tourism, in Braga, reaching maximum levels (385 145 attendances at the Tourism Office and an increase in the number of overnight stays of about 9.77%, compared to the previous year, which corresponded to the highest increase at a national level and about 10.7 million overnight stays) (CMB, 2019). However, this growth was interrupted as a consequence of the pandemic (see figure 3 and 4). In fact, in 2020, in Braga, despite the fact that in the months of January and February (months with no record of Covid-19 cases in Portugal) there was a growth in the number of visitors of about 6% and 8%, compared to the previous year, the development of the pandemic in the following months in Portugal and the consequent Lockdown, closure of services, air corridors and borders, led to a drop of about 77% in the number of tourists who were attended in the Braga Tourism Office and a decrease of about 58% in the number of overnight stays, compared to 2019 (CMB, 2020).

In 2020, 46 new local accommodations were registered in the National Tourism Register, in Braga, and 40 registrations were cancelled (CMB, 2020).

In 2021 (figure 5), Tourism in Braga, despite lockdown periods and restrictive measures implemented throughout the year, grew about 28.1% in relation to 2020 (CMB, 2021).

Proving the tendency for the upturn in Tourism, 79 new local accommodation units were registered in the National Tourism Register, representing a 71.7% growth compared to the year 2020 (CMB, 2021).

One of the main contributions to relaunch Braga's Tourism was its classification as European Best Destination 2021.

Figure 3. Evolution of the number of nights spent in Braga (2019-2020).
Source: (CMB, 2020)

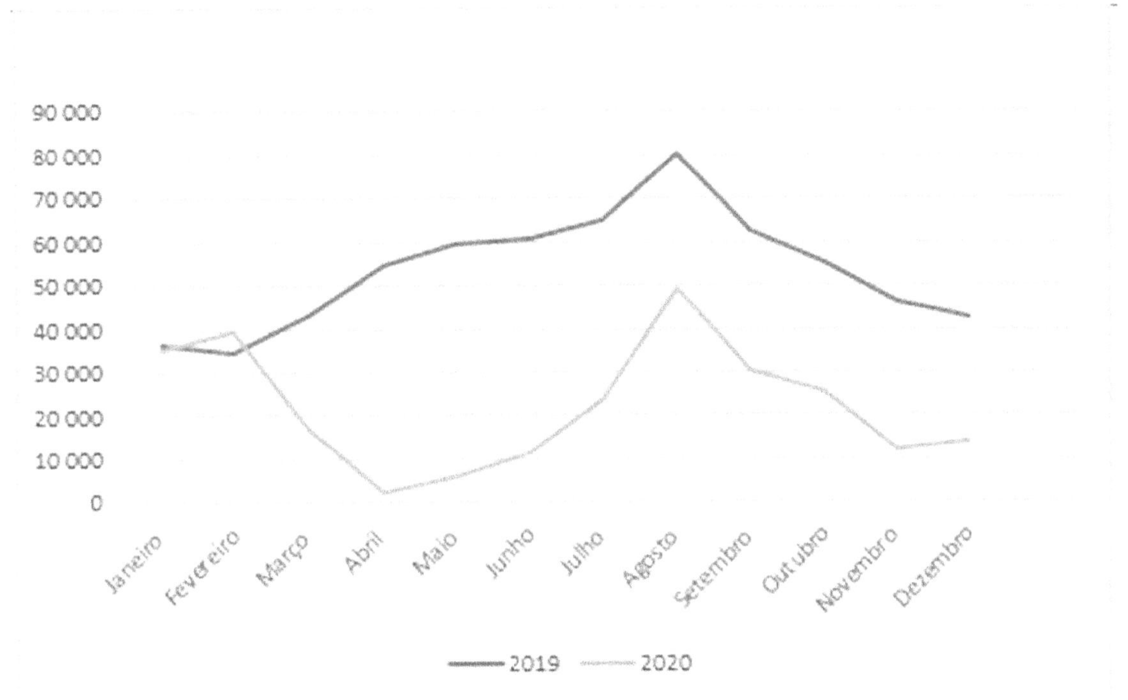

Figure 4. Visitors in the last 10 years in Braga.
Source: (CMB, 2020)

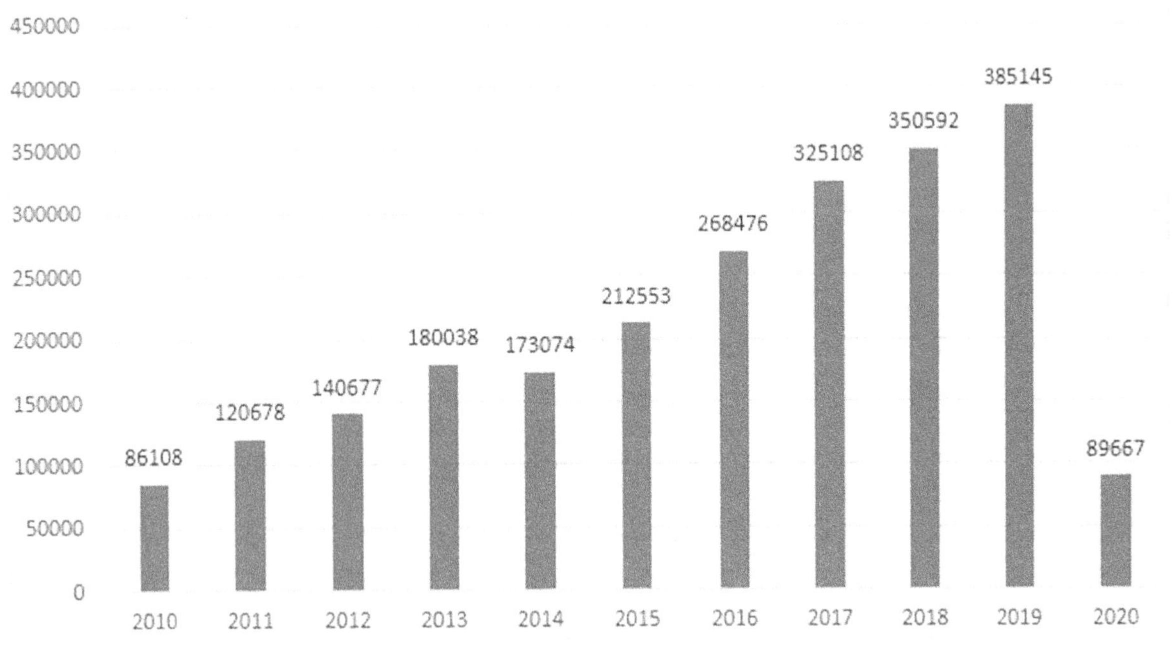

Figure 5. Evolution of the number of overnight stays in Braga (2020-2021).
Source: (CMB, 2021)

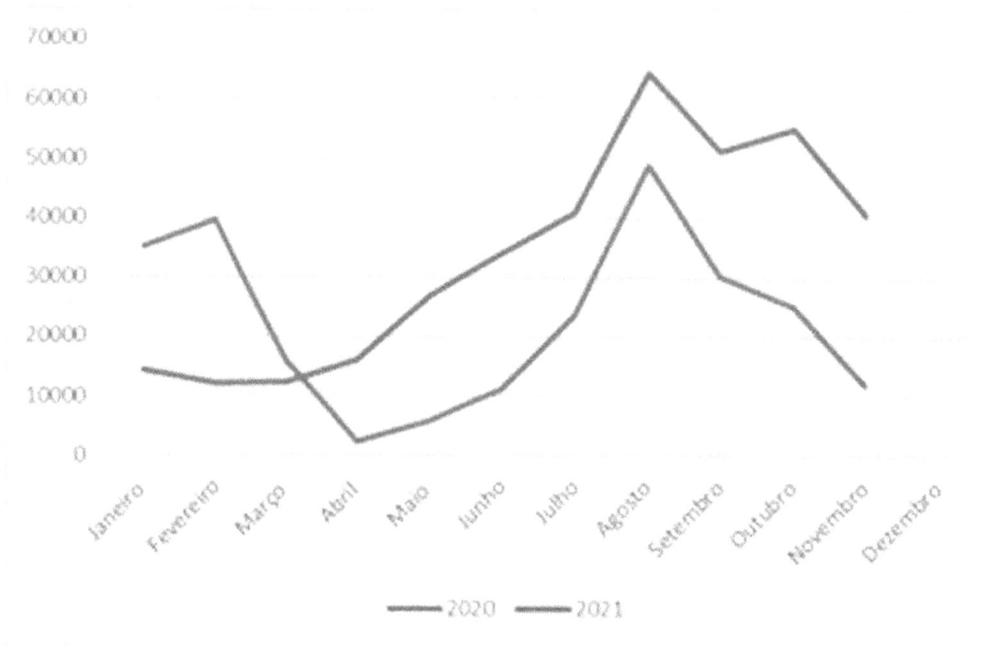

Road Traffic

During the Lockdown periods, verified during the COVID-19 propagation peaks, the reduction of traffic in the cities was notorious (figures 6 and 7). As a way of gauging the impacts on mobility and circulation in general arising from the new reality, the municipality of Braga has a network of road traffic counters in the main accesses and structural urban crossing roads capable of constituting a reliable indicator as to the change in flows.

Braga has six active traffic counters since 2016 (figure 8), located on the main supra-municipal and municipal traffic crossing roads that are listed below:

- Counter 70: *Fojo* Avenue- direction Braga -> Póvoa de Lanhoso.
- Counter 71: *Imaculada Conceição* Avenue- direction Rotunda Santos da Cunha -> Liberdade Avenue.
- Counter 72: Frei Bartolomeu dos Mártires Avenue - direction Continente -> Rotunda das Piscinas.
- Counter 73: João XXI Avenue - direction 31 de janeiro Avenue -> Liberdade Avenue. (installation and start of counts in 2017).
- Counter 74: Fojo Avenue - direction Póvoa de Lanhoso -> Braga.
- Counter 75: Frei Bartolomeu dos Mártires Avenue- direction Rotunda das Piscinas -> Continente.

These counters allow the construction of a database (with data referring to each minute) with, namely, the counting of vehicles and speeds, with the average speeds reached, the percentage of infractions and the 85th percentile.

Figure 6. Photo of a traffic distributor road in Braga (Av. Frei Bartolomeu dos Mártires) during the first Lockdown in 2020.
Own source.

Performing a comparative analysis of traffic counts on these counters, in the pre-pandemic periods and during the Lockdown periods and post Lockdown, one can conclude the following:

- In the 1st Lockdown period the count of the number of vehicles at all counters reaches the lowest values since the start of the counts (in September 2016);
- Regarding the average speeds, there is a slight increase in the two Lockdowns at all counters;
- In the phases after each Lockdown there is a gradual rise in all 6 counters.

Table 3. Average daily traffic in 2020 and 2021.

Fixed counters	ADT em 2020	ADT 2021
70	4480	5100
71	3863	4497
72	8181	8636
73	4162	4832
74	4208	4887
75	7765	8119

Source: CMB (2021).

Figure 7. Photo of a traffic distributor road in Braga (Av. Frei Bartolomeu dos Mártires) after Lockdown in 2022.02.22.
Own source.

The counts performed in 2020 and 2021 (Bandeira et al., 2021a), at the 6 fixed counters/radars installed at fixed points in Braga, show that the Average Daily Vehicle Traffic increased slightly (on average it increased by about 550 vehicles per day, at each count point), according to table 3.

Figure 8. Location of fixed traffic counters, located in the city of Braga.
Source: [own source]

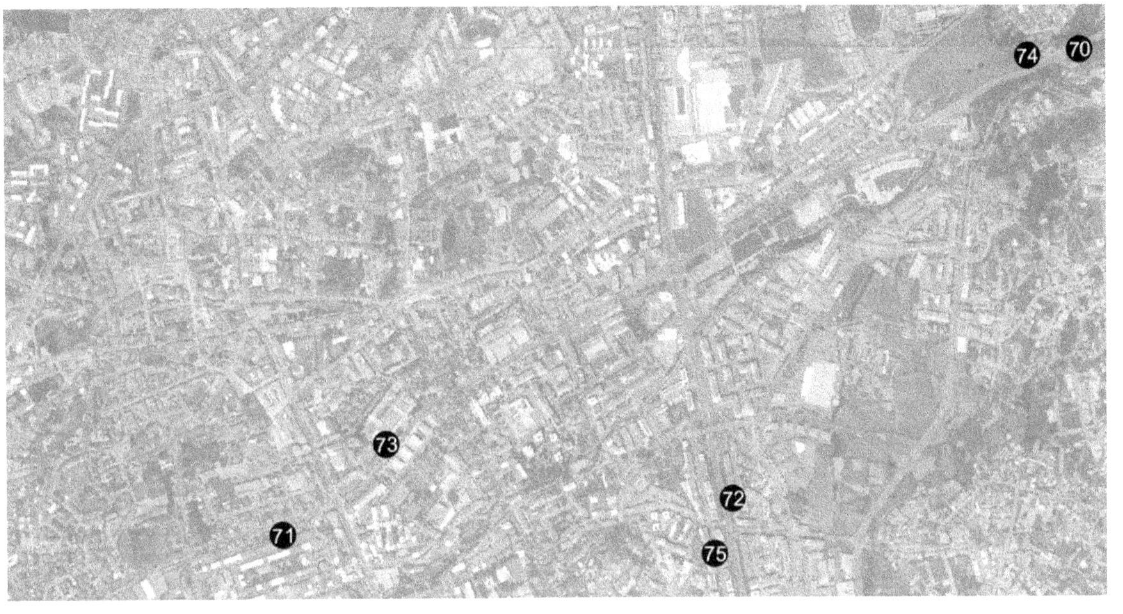

Public Transport

The pandemic caused a fearful effect in the use of public transport, in general, which, associated with the increase in the number of people teleworking, led to a substantial decrease in the users of this mode of transport, and Braga was no exception. Despite the efforts and measures implemented by the Municipal Company, Braga Urban Transport (TUB), to increase safety, public health and user confidence, there was a sharp decline in the number of passengers carried. This was all the more remarkable given that the number of passengers had been increasing steadily in previous years.

During the first period of Lockdown, TUB continued to guarantee the public transport service, defining, however, a reduction and rationalization of the service. Thus, from 23 March 2020, on working days, they offered the population a service equivalent to Saturdays, ensuring 49.7% of km's, in order to continue to guarantee the transport of essential workers (Hospitals, health centres and companies that continued to work). Line 95 (which connects the *Minho Center* and *Nova Arcada* shopping centres) and 96 (which connects the interface located in the *E.Leclerc* shopping centre to the residential and service area located in the parish of *Nogueiró*, on *Luís António Correia* Street), as they cross the city in the North-South and East-West directions, respectively, were exceptions, guaranteeing a frequency of every 40 minutes. The main measures implemented and widely publicised by the transport operator, during that period, consisted of the implementation of a contingency plan which, in summary, envisaged the following:

- Disinfectants are available inside the vehicles;
- Isolation of front seats to ensure social distance between passengers and drivers;
- Passengers entering and leaving the vehicles exclusively through the back door;

- During the lockdown, no validation of transport tickets was carried out, nor were tickets sold, having defined that travel was free of charge throughout the entire TUB network, from 14 March until the end of April, 2020;
- Reinforcement of the daily cleaning and disinfection of the vehicles with recourse to more human resources;
- Steam hygienization at a temperature of 120º, of the vehicles in circulation.
- Free parking spaces for residents and shopkeepers in April 2020 (note that, in Braga, it is the TUB that explores the concessionary parking in public spaces);
- Free parking for residents in their respective zones of residence between March and April 2020.

According to the accounts report of Transportes Urbanos de Braga (TUB), in 2020 there was a decrease in the number of users in the order of 43.92%, compared to 2019. In fact, in 2019, there were approximately 12,413,299 passengers carried by TUB, and in 2020 (table 4), this number decreased to 6,961,913 passengers (TUB, 2021).

Table 4. Number of passengers carried by TUB in 2019 and 2020 and respective variation.

2019	2020	Variação
12 413 299	6 961 913	-43,92%

Mês	2019	2020
janeiro	1 089 010	1 088 192
fevereiro	1 031 969	1 078 630
março	1 104 462	504 340
abril	977 885	2 673
maio	1 195 879	185
junho	913 257	459 007
julho	951 574	567 593
agosto	734 166	501 692
setembro	1 043 915	677 810
outubro	1 280 356	799 141
novembro	1 132 722	707 304
dezembro	958 104	575 346
Total	12 413 299	6 961 913

Source: (TUB, 2021, p. 90)

Public Space Allocated to Pedestrians

In 2020, with the beginning of the pandemic, COVID-19, and in 2021, with the pandemic state continuing, the municipality of Braga advanced with a programme of support to restaurants and local traditional commerce, called *Porta Aberta* (Open Door). Thus, this programme, having as its motto an identifying symbol of the city of Braga ("Braga with an Open Door") mobilised the exceptional opening of the public space, from the extended functioning of catering and drinking establishments, providing for the total exemption of fees for owners. In this way, it was possible to reconcile the demands of social distancing and reduction of capacity inside establishments, with the economic sustainability of local commerce. On the other hand, this measure guaranteed increased safety and public health, in the use of cafés and restaurants, and stimulated the local economy.

Thus, with support representing a loss of income estimated at around 500 thousand euros (cm-braga.pt, consulted on 2022.01.30), the municipality of Braga exempted the occupation of public space, namely, the installation of terraces, platforms and windshields, showcases and displays, chests and ice cream machines, flower boxes, among others.

These measures have encouraged businessmen to comply with the new rules stipulated at national level, reducing the need for inspections. On the other hand, they have speeded up the urgency of their implementation. This situation did not arise in a similar way in some Portuguese cities, where, for example, due to lack of initiative or economic desperation on the part of owners, the reduction in the number of tables in establishments was not guaranteed, nor was the distance between tables increased.

Other Examples

The abrupt reduction of the presence of cars in cities during periods of confinement, especially in the first wave of the COVID-19 pandemic, led administrations to consider the alternative role of soft modes in cities and the opportunity to redistribute public space, which was mostly allocated to the car. On the other hand, the demands for limited capacity inside shops and services led to the proliferation and extension of esplanades, the formation of queues of people at the doors of the most popular establishments, making public space complementary to these commercial facilities, and causing greater pressure on the need to consider the resizing of public space. In fact, spaces that were only designed for the circulation of pedestrians began to take on new functions, such as esplanades, urban living spaces or simply as a "reception hall", complementary to shops, restaurants, cafés, bakeries and other services.

Thus, many cities have transformed traffic routes into cycle or pedestrian paths, through temporary measures (Corais et al., 2022), called pop-up, which gradually became definitive measures, thus giving impetus to the paradigm shift of mobility and urban experience of public space.

Safety being a basic premise for the increase of active modes, in February 2020, 130 nations supported the Stockholm Declaration, which imposed 30 km/h as the maximum speed in more susceptible areas such as urban centres.

ELTIS, Europe's main observatory on urban mobility [3], has been monitoring the impact of the pandemic on mobility. This observatory has shown that before COVID-19, car journeys in Europe represented 47% of all urban journeys, public transport 21%, with pedestrians and cyclists 25% and 7% respectively. These values differ significantly in comparison with Portugal, with the values of table 5, for Lisbon, Porto and Braga.

Table 5. Modal split for the metropolitan areas of Lisbon, Oporto and the city of Braga.

	car	Public transport	cycling	Pedestrian
Lisbon	45%	35%	0,21%	19%
Porto	51,4%	24,6%	0,1%	21,6%
Braga	66%	12%	0.1%	18%

Source: [own elaboration, based on consultation of *INE - Instituto Nacional de Estatística, 2018* and INE, 2011.]

Thus, the impact of intervention measures in the public space aimed at changing the modal split becomes more important in Portugal, and it will be pertinent to monitor changes in mobility behaviour in the near future. Notably because the pandemic situation, with its imponderability, has not ceased to be a laboratory to be studied and compared with the return to normality, in order to better understand people's behaviour, also with regard to mobility.

According to ELTIS, before COVID-19, around 5.4% of Europeans worked from home, and this percentage rose to 57% with COVID-19.

As a consequence of the pandemic, society has had the opportunity to reflect on the importance of the private domestic space, where it has been confined for a long time, and has also, as a result, given public space, safety, health and air quality new attributes.

ELTIS reports that during the first lockdown, nitrogen dioxide (NO2) concentrations fell by 40 to 50% in Europe compared with 2019, and by around 10 to 20% for July and August 2020 compared with the same months in 2019.

The contrast felt, a little all over the world, in terms of the presence of the car and in terms of air pollution and noise, was particularly valued. Cities such as Milan, Paris and Vienna adopted exemplary measures in record time, which ensured lasting and irreversible transformations in the configuration and distribution of public space. The following table (table 6) systematises some pop-up measures, implemented in several cities during the pandemic period, aimed at redistributing public space in favour of sustainable urban mobility and urban living:

Authors Kraus & Kolch (2021) collected and compiled data from 106 European cities between March and July 2020 on the sizing of cycle paths and impact on their use and user growth during the pandemic, and concluded that the number of cyclists increased in cities where a cycle network was built, between 11 and 48% more than in cities that did not.

THE PANDEMIC AS A CATALYST FOR A PARADIGM SHIFT IN MOBILITY

Braga Case Studies

Framing the City of Braga

In approaching the case study of Braga, its role at a supra-municipal territorial scale cannot be neglected. In fact, Braga, as district capital, assumes a strong polarization (MPT & CMB, 2019; Quadrilátero, 2013) over its surroundings, attracting city users for the most diverse functions of the city (e.g. those consummated by the university, schools, central hospital, court, trade and services, among others) (Corais, 2021, p. 27).

Table 6. Pop-Up measures implemented during the pandemic.

CITY	MEASURES / PLANS
Lisbon	Implementation of a pop-up cycling network with the potential to become a permanent measure (e.g. Rua Castilho, Marquês da Fronteira, Berlim Avenue); incentives for the purchase of bicycles (fund of 3 million); covered parking for bicycles in Lisbon Mobility and Parking Company (EMEL) car parks and at public transport interchanges; measures to ensure greater distance between people in public space; increase in the area of esplanades by suppressing parking.
Milan	Since the Stockholm Declaration, it has imposed 30 km/h as the maximum speed throughout the city. In 2020, the Strade Aperte plan provided for 35 km of cycle paths in the city of Milan; in 2022, the Cambio Biciplan project, proposes to connect Milan to the outskirts through 24 lines, for a total of 750 km of cycle paths, by 2035.
Paris	Through Plan Vélo, they plan to invest more than 250 million euros and build more than 180 km of cycle lane and create more than 130,000 bicycle parking spaces; the 52 km of pop-up cycle lanes (created with pictograms and bollards) will become permanent.
London	It created 30km of temporary cycle lanes and recorded a 54% increase in cycling between May 2019 and May 2020. At the same time, it announced plans to turn the city into one of the largest car-free zones in the world. Of particular note is the increase of around 5,000 m2 of pedestrian space.
Liverpool	It has recorded a 161% increase in cycling trips between May 2019 and May 2020.
Scotland	Created Spaces for People funding package for temporary walking and cycling routes.
Brussels	It has built 40 km of new cycle paths and created incentives for the purchase of bicycles.
Budapest	Budapest is considering transforming the pop-up cycle paths into permanent facilities, as the Grand Boulevard is used by some 3000 cyclists daily. Similarly, it is planned to transform the ring road connecting Margaret Bridge and Petőfi Bridge to include definitive cycle paths.
Thessaloniki	It has temporarily widened the pavements to allow the installation of terraces for restaurants.
Philadelphia	They closed to auto traffic 7.5 KM of Martin Luther King Jr Drive, reallocating space for the soft modes.
Oakland	10% of all city roads (119 km) were closed to car traffic.

Source: [own elaboration] based on: https://www.dn.pt/dinheiro/pandemia-acelera-rede-de-ciclovias-na-europa-12365728.html; https://www.labrua.org/blog/cidadenapandemia/; https://ecf.com/dashboard; https://www.eltis.org/topics/covid-19; https://www.am-lisboa.pt/documentos/1593462996P3iJW6mg2Wt76ZU1.pdf

At the supra-municipal level, the role of the city network formed in 2008, with the *Quadrilatero* [4], which is composed of Barcelos, Braga, Guimarães and Famalicão, stands out (Quadrilátero, 2013). These 4 cities "form a territorial continuum with about 600 thousand inhabitants, of which 210 thousand reside in the central urban perimeters." (Quadrilátero, 2013, p. 5). This territory also presents very strong relationships with the surrounding territories (NUT III of Cávado and Ave) aggregating, at least, 14 municipalities that, according to the Quadrilátero (2013) corresponded to almost 1 million inhabitants in 2007 (Corais, 2021, p. 38).

Braga, being the District capital, presents a great capacity for fixing public services, when compared to the surrounding municipalities. Therefore, it is there that the strategic activities are concentrated, at the agglomeration scale (Corais, 2021, p. 40).

Thus, the city of Braga, inserted in a network of complementary cities, assumes a central position in the agglomeration which, in turn, is characterised by a significant demographic and economic dynamism. Due to its capacity of population and economic attraction, religious prominence, historical and administrative factors and the University itself, it can be considered that its area of influence goes beyond the municipality itself (Corais, 2003).

The polarising character of the county is confirmed by the daily registration of 147,248 movements that have the county as destination (of which 66.7% by active individuals and 33.3% by students) and that correspond to 84% of the total movements to and from Braga (175,213 daily commuting movements) (MPT & CMB, 2019).

Braga presents an area of 183 km2, in which lived, in 2017, about 181,382 inhabitants of which about 66% of the population lives in the centre, which, associated with the concentration of equipment and services reinforces the centripetal character of the city (CMB, 2015; Quadrilátero, 2013). This situation is even more relevant if combined with the one considered by the attraction of publics from other even more distant territories (through, for example, tourism, heritage, culture and international networks of cities, among others), reason that highlights the importance of having to reinterpret the city centre (1st and 2nd crown) according to the change of paradigm. That is, according to the promotion of the SSM, considering the transversality of scales present (Corais, 2021, p. 42).

Reinforcing the centrality of Braga, it is in the central parishes that the highest proportions of commuter journeys are recorded with destination in the parish of residence itself and it is in these that a pedestrian modal share above 60% is recorded (MPT & CMB, 2019). The analysis of the average travel time of the displacements in the municipality of Braga shows that about 58.6% are less than 15 minutes, 89.1% less than 30 minutes, and that the long duration trips have a residual weight (2.1%), which, associated with the fact that most of the movements present distances less than 4km, allowed the Sustainable Urban Mobility Plan - PMUS (2021) to estimate for this territory a strong capacity of captivation of the population for soft modes. Likewise, in this document, it was found that it is in the central areas that there is a greater preponderance of people's movements, associated with the concentration of trip generating poles, and the respective proximity, in general, to places of residence (Corais, 2021, p. 43).

On the other hand, the orographic platform on which the city of Braga was installed, since its origins and the subsequent expansion areas, is characterized by being relatively flat, which is a characteristic conducive to the use of soft modes of transport in the city. The increase of pedestrianization, as Sales Oliveira (2018) refers, incites a culture of living in the public space. This author highlights the cultural differences existing at this level between, for example, Portugal and Spain, highlighting that, in the latter, the urban experience is combined and stimulated by pedestrian commuting movements home-work and home-school.

Braga, and the urban dynamics inherent to it, assumes very specific characteristics in its territorial context. If, on the one hand, it has the particularity of being a centripetal territory, in a disperse and diffuse context, on the other hand, the housing dynamics and policies shaped an unbridled urban growth in the 80s and 90s, of the 20th century. Relegating the public space and the urban continuities in function of this design (Corais, 2003). These two specificities had consequences, also antagonistic, on the city's mobility conditions. The challenge now is to define new urban continuities for all regular or temporary users, residents or city-users (especially for pedestrians and vulnerable people, or those with reduced mobility), stimulating proximity and neighbourhood relations. Adding to this functional design, the structuring of the city, the qualification of the system of collective spaces and the promotion of quality of life. The city of Braga has several "scars" (materialised by the presence of road variants within the city, which constitute barriers to free movement and pedestrian crossing), which hinder the connectivity of pedestrian and cycling routes, due to urban barriers defined in the times when individual transport was the priority in traffic engineering policies (MPT & CMB, 2019). These barriers are flanked by interdependent and inseparable urban spaces, whose experiences require interventions that are not compatible with the annual budgets made available to municipalities. The challenge is, therefore, to define new urban continuities for All (especially for pedestrians and people with reduced mobility). In fact, according to data from INE 2011, around 30% of the population have at least one mobility difficulty. By responding to the mobility needs of these groups, the city is thus guaranteeing the conditions for the circulation of All. And to this functional design for the structuring of the city allies the qualification of the public space and the promotion of the quality of life (Corais, 2021, p. 44).

The population and functional concentration of Braga, as well as, its scale and mobility patterns verified in the *Quadrilátero* studies (2013) and in the SUMP of Braga (MPT & CMB, 2019) stimulate the use of soft travel modes (pedestrian and Cycling) due to the preponderance of short distances that need to be covered. However, the configuration of the public space/public domain conditions the transformation and the required paradigm shift of Sustainable Urban Mobility (Corais, 2021, p. 45).

The territory has the tools to act in favour of Sustainable Urban Mobility and redistribute the public space more equitably among the various modes and types of users, thus giving the opportunity for those who are receptive to join more sustainable modes of travel to do so safely and comfortably. From the knowledge of the history of the city and its urban morphology (Lamas, 1993; V. Oliveira, 2014, 2018) these mutations, of the Mobility patterns with interference at the level of the urban design of the public space, can and should be based on a respect and continuity for the identity characteristics of the city, enhancing them. In fact, the first urban matrix was based on sustainable means of transport and the whole city functioned on this assumption. The current stage consists in evaluating the evolution of the city and the times and ways of being, giving back to the city and the pedestrian its endogenous role (Corais, 2021, pp. 46, 47).

With the pandemic situation, we immediately became aware of the opportunity that the circumstances offered for the implementation of pop-up measures (Corais et al., 2022). After the most rigorous days of confinement it was possible to perceive, if not an increase, of pedestrian circulation and also of clickable circulation, although in the latter with less expression, an inviting reality to the realization of new initiatives and the acceleration of some sectorial measures planned or already under construction. At first sight, it seemed immediate to increase the level of implementation of ongoing projects: in terms of accessibilities, especially the "I can already pass here!" Project (Bandeira et al., 2021b), implying traffic calming measures near the main schools, with raised zebra crossings, podotactile surfaces, signalling, etc.; and to increase the extension of the cycling network, in particular, the systems of valorization of coexisting soft modes, even projecting the widened extension of the Fojo lane, in expeditious conditions of implementation. More than the good will, the stimulus and, also, the resource constraints - the pop-up measures imply large expenses, more than it may seem at first sight - the procedural obstacles arising from public procurement procedures, the bureaucratic and centralist slowness to which the municipalities are subject to, associated to a deficient culture regarding sustainable urban mobility, limited the potential opened by the temporal window of opportunity that the pandemic offered.

Case Study 1: Closure of Gonçalo Pereira Street to traffic

The *Gonçalo Pereira* Street, between *D. Afonso Henriques* Street and *S. Paulo* Square, is contiguous to the pedestrian area defined by the historical centre of the city of Braga, has about 55m of extension and about 8m of width. About 11 families live in this stretch of the street and there are three catering and drinking establishments, a bar, a law office, a beauty salon, a travel agency, a clothing shop, a furniture shop and a church (S. Tiago's Church). According to the traffic data made available by the Braga City Council, with recourse to counts carried out in the period between 17.02.2022 and 21.02.2022, an average of 1,566.2 vehicles/day pass through this section of the street (corresponding to an average of 19 cars by 15 minutes, with a maximum of 55 cars having been registered in a 15-minute period at the end of the day, between 5 p.m. and 5.15 p.m.) and the street makes available approximately 17 public parking spaces (figure 9).

Figure 9. Photo of Gonçalo Pereira Street open to traffic.
Source: [own source] (2022.02.24)

Braga has joined the European Mobility Week since 2015 and, as part of the programming of this event, since 2016 it has proposed to close this section of *Gonçalo Pereira* Street to traffic (figures 10 and 11). Over the years it has been intensifying and improving the use of this street during Mobility Week. Thus, in 2017 and 2018, the closure of the street was associated, in addition to the location of terraces of adjacent establishments, with a series of evening concerts and an entertainment programme.

In 2020 and in 2021, during the pandemic, through the programme implemented in Braga, "Porta Aberta" (Open Door), this section of the street was closed to traffic for a longer period of time than before (about 6 months each year), transforming notably a street that daily supports the circulation of crossing traffic and the parking of many cars, into a street for pedestrians and for the promotion of urban living (figures 12, 13 and 14).

For an adequate perception of the impact on the community and local agents of this measure of closing the street to car traffic, interviews were conducted with the President of the Parish Council, residents (about 50% of the existing houses) and businessmen of *Gonçalo Pereira* street.

Figure 10. Closure of Gonçalo Pereira Street to traffic during European Mobility Week 2017.
Source: authors (2017.09.22)

Figure 11. Closure of Gonçalo Pereira Street to traffic during Mobility Week 2017.
Source: authors (2017.09.22)

Figure 12. Photo of Gonçalo Pereira Street in August 2021
Source: authors

In a synthetic way, it is possible to report the following:

In the interview with the President of the parish council, he mentioned that the mode of transport he uses most often is the car. When questioned about the type of participations he received during the closing of the street, he referred that he received more complaints, due to the constraints in the circulation and parking. He also mentioned that he considers that the street can accommodate the two solutions throughout the day, i.e., during the day it can allow the circulation of vehicles and at the end of the day and at night, it can be closed to traffic and allow urban living and complementarity with restaurants and bars. He alerted, however, for the concern with the conditioning, provoked by this type of activities, to the respect for the limits of noise and for the quality of life of the inhabitants. He also highlighted the need to complement road closures with other street activities and dynamics.

Figure 13. Photo of Gonçalo Pereira Street in August 2021.
Source: authors

Figure 14. Photo of Gonçalo Pereira Street in August 2021
Source: authors

As for the interviews with the restaurant and beverage entrepreneurs of this stretch of street, they were unanimous about the advantages of closing the street to traffic. They consider that the traffic constraints are low, given the existing alternatives in the surroundings, and that the growth of the adjacent pedestrian area, for this street section, is an inevitability despite the interruption verified by *D. Afonso Henriques* street. The possibility of widening the narrow pavements is another advantage in the connection of the *S. Paulo* square and *S. Tiago* square to the Cathedral. In the general opinion of the businessmen interviewed, the main obstacle to the definitive closure of this street to traffic are political issues and the opposition of the Parish Council to this happening. They mentioned that opposition to this type of measure at the beginning of implementation is normal, but that they believe that over time the advantages would prevail over the disadvantages.

They said that the closure of the street to traffic had increased business by attracting more people passing by on foot, who were potential customers. They consider that the size of the esplanade was adequate, despite it being the first space in the establishment to be occupied.

The most common means of transport used is by car, scooter and pedestrianised transport. When they drive to their establishment they park in the streets surrounding the establishment or in covered car parks in the area.

In all periods of closure of the street to traffic (Mobility weeks of 2016, 2017, 2018, 2020 and 2021, in the last two years for the extended period from April to September), the entrepreneurs of this stretch of street, collaborated with the municipality of Braga in the decoration and dynamization of the space. In the interviews they mentioned that, besides the definition of esplanade areas for restaurants and cafés, they had the concern to create resting areas (with garden benches, for instance) in order to create exterior conviviality and public access spaces. These measures together with the live music concerts cooperated to make the street more attractive and to keep people longer in this street.

According to the data provided (in early September 2021) by one of the entrepreneurs of a restaurant and drinking establishment in this street, there was a sharp increase in the number of customers on the terrace during the "Open Door" programme (Figure 15), a testimony that claims the success of the measure, both from the financial point of view, for the entrepreneurs of this street, and from the environmental and decarbonization point of view, through the promotion of the pedestrian mode and the urban experience provided in a space that is usually used as a mere crossing space for car traffic.

Figure 15. Number of customers of one of the restaurants and drinking establishments in Gonçalo Pereira Street, during the "Braga de Porta Aberta" period (April to August 2021).
Own elaboration based on the data provided by the entrepreneur.

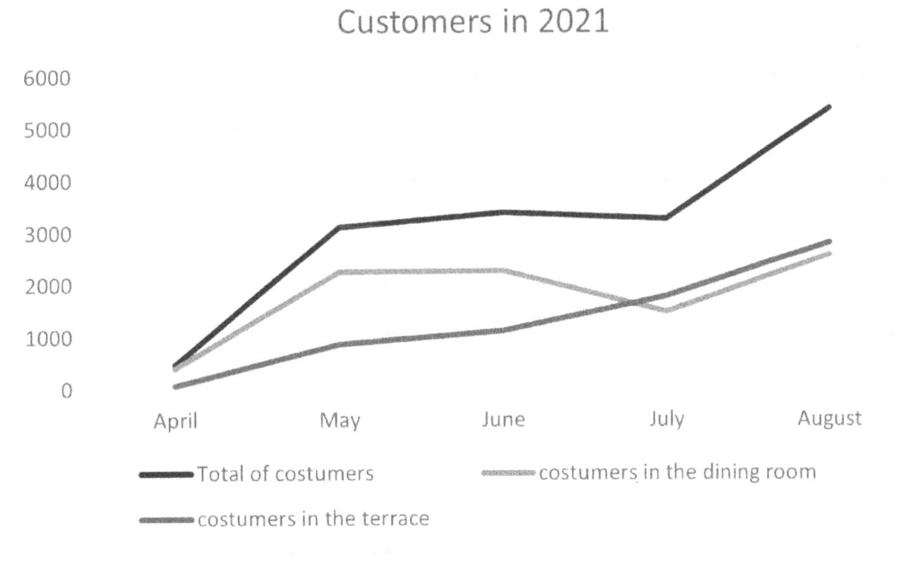

The survey revealed that all respondents use the car as a mode of transport and that when they go to their residences they park on the street, or in the surrounding streets or in private or collective garages (only one of the residences has its own garage). The impossibility of accessing the municipal leasing system on the public highway, due to the fact that parking on this street is not leased, was the main objection of the residents. This is the only argument that could lead them to agree with a possible permanent closure of the street, since the municipal regulations provide for the possibility of granting concessions in the surrounding streets. They find that, since the street has no concessionary parking (parking meters), this situation (unlike what happens in other nearby streets) encourages long-term parking, and they sometimes find that the same car is parked for days at a time.

Only a minority of the surveyed residents prefer the street without car traffic, as traffic causes noise and as improper parking limits pedestrian mobility. On the other hand, they consider that if the street is pedestrianised, in accordance with municipal regulations, they can purchase a permit to park on adjacent streets.

Those residents who prefer the street with traffic refer to the problems related to the noise at night caused by the concentration of people in the esplanades and in the bar existing in the street, as well as, the constraints derived from the existence of esplanades and of many people in the street, who leave rubbish on their doorsteps and who make the daily tasks of accessing residences and transporting shopping difficult. They warned that the majority of the population of this street is already old and has reduced mobility. As for the suggestions they made in case the street is definitely closed to traffic, they mentioned the need to provide the possibility of parking in the proximity for the residents (parking tickets).

Regarding the perception of the main obstacles for not having yet closed this section of the street to traffic permanently, opinions are divided between the scarcity of parking in the centre and political reasons.

Case Study 2: Pop-Up Measures to Promote Coexisting Soft Modes

Braga Municipality considers it determinant to reduce road accidents and, for such, the car circulation speed has to be compatible with soft modes. Thus, in the city centre, in residential areas, as well as in the proximity of facilities, such as schools, the speed of circulation should be less than 30 km/h.

According to the 2011 Census, the cycling mode in Braga is residual and, although the number of users is estimated to increase, in the 2021 Census, such increase shall not be as expressive as desired. In line with the strategy advocated by the municipality of Braga, of inverting the mobility pyramid, giving priority to soft modes (e.g. walking and cycling) and to public transport modes, some projects to promote soft modes integrated in a specific urban network have been publicly presented.

Until August 2021, Braga's cycling network was composed of the following axes:

- *D. João II* Avenue (in rehabilitation phase from the University of Minho Campus (UM) to the junction with EN 309) - about 3 km;
- *Nova de Santa Cruz* Street, from the University of Minho Campus to the *Padre Júlio Fragata* Avenue (segregated in the East-West direction and coexisting with car traffic in the opposite direction) - about 0.6 km;
- *D. Pedro V* street (in coexistence with public transports in the East-West direction and in coexistence with car traffic in the opposite direction) - about 0,8 km;
- East Eco route (from Lusíadas Avenue to Ponte Pedrinha street with sections in coexistence with pedestrians) - about 3.9 km (1 km in a bidirectional cycling path segregated from the pedestrian route - between Lusíadas Avenue and frei Bartolomeu dos Mártires Avenue and the remaining about 2.9 km in coexistence with pedestrians.

The Pop-up measures implemented in August 2021, to promote coexisting soft modes, are part of the coherent cycling network strategy for the city, preparing or announcing the permanent and integrated network, through provisional measures (figure 16).

Figure 16. Plan of the cycling network (in red the pop-up interventions).
Source: CMB.

In Portugal, according to the Portuguese Highway Code (article 90.3)

"Bicycle drivers should move along the right-hand side of the road, keeping a sufficient distance from the kerbs or pavements to avoid accidents".

Thus, in the specific case of bicycles, the natural circulation space for them is, in fact, in the same channel as cars. This, in fact, was a pop-up measure that pandemic conditions invited us to rehearse and move forward more boldly.

Figure 17. Implementation of Pop-up Measures to Promote Soft Modes in Coexistence
Source: [own source]

In this alignment, and due to the pandemic context, which reinforced the global impositions for decarbonization, through the need to respect physical distances and to respond to the reduction in demand for public transport in commuting, due to the fear of contagion, the promotion of soft modes was imposed as a way to join the useful to the opportune. Thus, in August 2021, the Municipality of Braga implemented pop-up measures on arteries connecting the main schools located in the city of Braga with the urban centre, consisting in the demarcation of the bicycle and scooter route (using red lines, marking the bicycle sign accompanied by a double arrow, and the reproduction, on horizontal signs, of the vertical sign for Maximum Speed 30km/h), on routes where this mode should circulate in coexistence with the remaining vehicles. This was a measure to promote road safety and to make motorists aware of the need to respect the most vulnerable vehicles, which aims to embody the characteristic of sharing intrinsic to the roads (between cars and bicycles), in accordance with the Highway Code. These markings create a very prominent visual effect that raises awareness among motorists and increases their attention to soft modes. In other words, they only reinforce the provisions of the Highway Code, alerting motorists to the need to coexist with bicycles (figure 17).

In this way, the initial strategy of promoting an awareness of sharing on the lanes, in which the soft modes should always be respected with special care, ensured, in turn, the establishment of more appropriate conditions for the circulation of these vehicles in the city, enhancing their increase and reducing road accidents.

This project is part of a global strategy defined for the implementation of the Council's cycling network and intends to be the pronouncement of the same, in the definition of a safer, more direct and coherent network, building a more diverse and democratic city in the forms of mobility.

FUTURE RESEARCH DIRECTIONS

The evaluation of the impact on the promotion of sustainable urban mobility of the pop-up measures carried out in various cities during the pandemic will need to be monitored and rigorously carried out over time, after the activities have stabilized and the pandemic fears have evaporated. When the world returns to true normality will be, therefore, the right time to verify whether the changes provided in the public space during the pandemic had long-term effects and definitive changes in culture and mobility behaviors of the population. In this sense, it will be relevant, after this interregnum of time, to perform this analysis and evaluation. In other words, it should be verified what direction the cities gave to the pop-up initiatives implemented during the pandemic and whether these were matured into definitive and society-changing measures, or whether they fizzled out and the opportunities generated by this circumstance were lost.

CONCLUSION

When the development of ongoing projects is confronted with imponderables of great impact, as is the case of the implementation of urban mobility measures in the face of the pandemic constraints we are experiencing, the necessary readjustments are imposed and, naturally, new opportunities to reflect on the transition processes open up. Such was the case with the implementation of the so-called "pop-up" measures, and the motivation behind the present reflection that we now share with our readers.

In principle, a reference to the undeniable willingness to reconcile the technical and planning perspectives to the screening of the policy decision, simultaneously, accompanied by the analytical scrutiny inherent to the research process itself. But, not everything is linear as it might at first sight appear. All the more so because the same solutions do not have the same applicability in different places. We are already aware of the feasibility of many of the ideas that are being defended as unavoidable for the consolidation of welfare conditions and quality of life in our cities. The transition process is difficult, it has to make its way, but it is possible.

The impacts felt in terms of tourism also have repercussions on mobility in general, but also on sustainable mobility. Certainly, the cities which are better qualified in terms of accessibility, but also bearers of a correct and careful urban road design, in particular of the public space, harmonious, diversified and, naturally, humanly scaled in terms of proximity experiences, will have better tourist attraction conditions and will gather the conditions to recover more quickly from the crisis we are going through.

Braga is one of the Portuguese medium-sized cities which has the best physical and dynamic conditions to increase the desirable transition process. On the one hand, its topography is suitable for walking and light means of locomotion; on the other hand, there is the need to contain the expansion of short-distance car circulation and, at the same time, it is urgent to discipline parking.

Having explained the evolution of the conditioning measures concomitantly with the pandemic trends, the country and the city adapted reality to practical demands. Various ideas and intervention projects soon emerged, in the "pop-up" context, either as a result of the projects underway or by the example that was arriving of practices followed in other cities. However, beyond the voluntarism and commitment of the technicians and political decision-makers, the opportunity/effectiveness of the execution times collided with several practical obstacles, which prevented it from going further. Unavoidable issues of regulatory nature, from the procedural requirements of public procurement, to the lack of human and financial resources, the structural articulation between the organic technical components; to the public clarification times of the interventions, conditioned in time the agility of useful implementation of this type of interventions. Note that these assumptions are common to most of the Country.

Even so, it was possible to rehearse some "pop-up" type interventions that served as an example of what was being done in Braga in an identical way to other Portuguese and European cities.

In fact, any conclusion on the real impact of the implementation of these measures on the modal split of the city of Braga is still premature due to the short time elapsed. However, it is already possible to gauge some conclusions arising from the punctual change in the use of each of the intervened spaces. Thus, it was verified, in relation to the first case study, that during the road closure period (about 6 months) about 1,566.2 vehicles/day stopped passing through that street, increasing the use of the public space by pedestrians. This transformation, in turn, ensured the decarbonization and the increase of the urban coexistence index of this street.

With regard to the second case study, this worked as a measure to raise awareness among motorists to the responsible coexistence with soft modes (e.g. bicycles and scooters), to reduce speeds and increase the use of scooters on the carriageway, as well as the proportional reduction of the use of this mode of transport on the pavements adjacent to these streets.

However, more than the practical result of the interventions, which had their circumstantial application, what stood out was the demonstration of the possibility and the reach of this type of action. Something always remains that, being a test of innovation, increases the capacity of critical mass to proceed with the necessary future transformations.

ACKNOWLEDGMENT

This research was funded by Fundação para a Ciência and a Tecnologia, IP (FCT)/ Massachusetts Institute of Technology (MIT-Portugal), grant number SFRH/BD/151416/2021, in the context of the doctoral thesis of Filipa Corais, with the title "THE CITY "WALKING" TO 2050. BRAGA AS A LABORATORY FOR A RESILIENT & SUSTAINABLE SYSTEM".

The authors would also like to thank the reviewers and the guidance team (Marta Labastida and Cecília Silva) of the aforementioned PhD thesis (Corais, 2020) of the document for their valuable contributions.

REFERENCES

Bandeira, M., Corais, F., & Pereira, F. (2021a). An urgent urban design challenge: time, sustainable mobility, and the humanization of collective space. In *Sustainable Urban Development. Topics, trends and solutions* (pp. 1–15). IOP Publishing Ltd. doi:10.1088/978-0-7503-3971-1ch3

Bandeira, M., Corais, F., & Pereira, F. (2021b). Coexistence of urban regeneration policies and sustainable mobility: reflections on, and practical cases of, urban acupunture. In *Sustainable Urban Development. Topics, trends and solutions* (pp. 1–11). IOP Publishing Ltd. doi:10.1088/978-0-7503-3971-1ch5

Banister, D., & Hickman, R. (2007). How to design a more sustainable and fairer built environment: transport and communications. *IEE Proceedings - Intelligent Transport Systems, 153*(4), 276–291. 10.1049/ip-its

Bereitschaft, B., & Scheller, D. (2020). How Might the COVID-19 Pandemic Affect 21st Century Urban Design, Planning, and Development? *Urban Science*, *4*(4), 56. doi:10.3390/urbansci4040056

CMB. (2015). *Plano Diretor Municipal de Braga*. Câmara Municipal de Braga.

CMB. (2019). *Síntese de resultados estatísticos do Turismo em Braga - 2019*. CMB - Divisão de Economia e Turismo.

CMB. (2020). *Síntese de Resultados Estatísticos do Turismo em Braga 2020*. CMB-Divisão de Economia e Turismo.

CMB. (2021). *Síntese de Resultados Estatísticos do Turismo em Braga - 2021*. CMB - Divisão de Economia e Turismo.

Corais, F. (2003). *Dinâmicas Territoriais na Aglomeração Urbana de Braga. Tese de Mestrado*. Faculdade de Arquitetura da Universidade do Porto.

Corais, F. (2021). *A Cidade a "Caminhar" para 2050. Braga como Laboratório para um Sistema Urbano Resiliente e Sustentável (Projeto Tese)*. Universidade do Minho.

Corais, F., Bandeira, M., Silva, C., & Bragança, L. (2022). Between the Unstoppable and the Feasible : The Lucid Pragmatism of Transition Processes for Sustainable Urban Mobility : A Literature Review. *Future Transportation*, 86–114. doi:10.3390/futuretransp2010006

Eltarabily, S., & Elghezanwy, D. (2020). Post-Pandemic Cities - The Impact of COVID-19 on Cities and Urban Design. *Architecture Research*, *10*(3), 75–84. doi:10.5923/j.arch.20201003.02

Gehl, J. (2010). *Cidade para pessoas*. Perspectiva.

Gehl, J. (2011). Life Between Buildings: Using Public Space. In *Landscape Journal*. IslandPress., doi:10.3368/lj.8.1.54

Gehl, J., & Svarre, B. (2013). Jan Gehl & Birgitte Svarre. In How to Study Public Life. doi:10.5822/978-1-61091-525-0

Honey-Rosés, J., Anguelovski, I., Chireh, V. K., Daher, C., Konijnendijk van den Bosch, C., Litt, J. S., Mawani, V., McCall, M. K., Orellana, A., Oscilowicz, E., Sánchez, U., Senbel, M., Tan, X., Villagomez, E., Zapata, O., & Nieuwenhuijsen, M. J. (2020). The impact of COVID-19 on public space: An early review of the emerging questions – design, perceptions and inequities. *Cities & Health*, *00*(00), 1–17. doi:10.1080/23748834.2020.1780074

INE - Instituto Nacional de Estatística. (2018). *Mobilidade e funcionalidade do território nas Áreas Metropolitanas do Porto e de Lisboa : 2017*. https://www.ine.pt/xportal/xmain?xpid=INE&xpgid=ine_publicacoes&PUBLICACOESpub_boui=349495406&PUBLICACOESmodo=2

Kraus, S., & Koch, N. (2021). Provisional COVID-19 infrastructure induces large, rapid increases in cycling. *Proceedings of the National Academy of Sciences of the United States of America*, *118*(15), 1–6. doi:10.1073/pnas.2024399118 PMID:33782111

Lamas, J. M. G. (1993). Os elementos morfológicos do espaço urbano - CAP2. In Morfologia Urbana e Desenho da Cidade (pp. 79–110). Academic Press.

Morin, E. (2005). *Introduction à la pensée complexe* (Vol. 534). Edition Points.

Morin, E. (2014). Pensée complexe et pensée globale. In Conférence "Tout se Transforme… (pp. 26–31). Fondation Calouste Gulbenkian. Délégation en France. doi:10.3917h.342.0026

MPT. (2019). *Plano de Mobilidade Urbana Sustentável de Braga*. Author.

Oliveira, C. S. de. (2018). Mobilidades em português : paradigma, cultura e potencialidades. *Centro de Estudos Comunicação e Sociedade*, 10–21.

Oliveira, V. (2014). Diferentes abordagens no estudo da forma urbana. *Revista de Morfologia Urbana*, 3–4. http://pnum.fe.up.pt/files/7614/0188/9302/RMU 2.1 Editorial.pdf

Oliveira, V. (2018). Article. *Revista de Morfologia Urbana*, *2018*(6), 1–75.

Quadrilátero. (2013). *Programa estratégico de cooperação*. https://www.cm-braga.pt/docs/CMB/Documentos/PECQuadrilatero.pdf

Saelens, B. E., Sallis, J. F., Black, J. B., & Chen, D. (2003). Neighborhood-Based Differences in Physical Activity: An Environment Scale Evaluation. *American Journal of Public Health*, *93*(9), 1552–1558. doi:10.2105/AJPH.93.9.1552 PMID:12948979

TUB. (2021). *Relatório de Contas 2020-TUB*. Transportes Urbanos de Braga. https://tub.pt/templates/frontoffice/enterprise/pdf/rl2020.pdf

Zuniga-Teran, A. A., Orr, B. J., Gimblett, R. H., Chalfoun, N. V., Going, S. B., Guertin, D. P., & Marsh, S. E. (2016). Designing healthy communities: A walkability analysis of LEED-ND. *Frontiers of Architectural Research*, *5*(4), 433–452. doi:10.1016/j.foar.2016.09.004

KEY TERMS AND DEFINITIONS

City-Users: Represent the group of people who despite not being residents of a certain city are users of that city for other purposes.

Coexistence of Transport Modes: Refers to routes in which segregation of the various modes of transport is not expected, but rather the coexistence of the various modes of transport in the same physical space, giving precedence to the most vulnerable modes (soft modes-pedestrian and cycling).

Paradigm Shift Toward Sustainable Urban Mobility: Changing from a modal split that favors the automobile to one that gives relevance to soft modes (walking and cycling).

Podotactile Surfaces: Support surfaces for the referral of blind and low-vision people.

Redistribution of Public Space: Presupposes the reorganization of public space in an equitable manner among the various modes of transportation, favoring the soft modes that are currently disadvantaged.

Social Distance: Refers to the distances imposed on society during the pandemic period (e.g., 2m distance between people in queues; distance between tables in restaurants and cafes).

Urban Living (or Livability): Refers to the urban valence conferred by the daily use of public space, intuitively promoted by the community, which guarantees the humanization of the public space.

ENDNOTES

1 (http://www.turismodeportugal.pt/SiteCollectionDocuments/estrategia/plano-acao-reativar-o-turismo-construir-o-futuro.pdf).

2 (http://www.turismodeportugal.pt/SiteCollectionDocuments/estrategia/plano-acao-reativar-o-turismo-construir-o-futuro.pdf)

3 (https://www.eltis.org/resources/case-studies/impact-urban-mobility-result-pandemic)

4 Quadrilátero - Special Purpose Municipality Association composed by the cities of Braga, Barcelos, Famalicão and Guimarães.

APPENDIX

Table 7. Detailed Synopsis of the Evolution of the Pandemic in Portugal and the main measures.

Date	Number of positive cases- Portugal	Description	Main Measures
2020.03.02	2020.03.02.-2; 2020.03.14 – 169 (+57) 2020.03.16–331 (+86); 1st death - Portugal	1st COVID-19 case in Portugal	Public employees on telecommuting or prophylactic isolation without loss of salary. On 05.03 TAP announces the cancellation of over a thousand flights in March and April. On 10.03, all flights to Italy are suspended and it is recommended to suspend events in open spaces with more than 5000 people or enclosed spaces with more than 1000 people. Visits to hospitals, nursing homes and hospices are suspended. In the Algarve, 60% of hotel reservations are cancelled. On 12.03 All schools of all grades suspend classroom activities. Closure of discos, museums and monuments and reduction of maximum capacity of restaurants. Masses are suspended. A 15.03 Ban on sports or leisure activities on the beaches. Ban on the consumption of alcoholic beverages on public roads and on holding events with more than 100 people. On 16.03 air traffic between Portugal and Spain is suspended, as well as rail links. On 18.03 air links to and from outside the EU are suspended.
2020.03.18	2020.03.18 - 642 (2nd death); 194 New cases of infection 2020.03.22 - 1,600 (320 new cases of infection) 2020.03.31 - 7,443 (1035 New cases of infection) 2020.04.10 -15,472 (1516 New cases of infection)	Beginning of the State of Emergency and 1st Lockdown	State of Emergency. Compulsory confinement and restrictions on movement on public roads. 02.04 - Ban on travel outside the county of residence during the Easter period and closure of all airports.09.04 - announcement that until the 9th year the 3rd period in distance learning, with assessment but without assessment tests or exams, maintaining support for families with children under 12. On 20.04 start of classes for elementary school students through RTPmemória (#EstudoEmCasa). Between 01 and 03 May, travel between municipalities is prohibited.
2020.05.03	2020.05.03 - 25.282 (92 new cases of infection) 2021.05.13 - 28,132 (219 new cases of infection)	End of state of emergency and start of calamity situation. (1st phase of the unlockdown plan)	Teleworking continues, but cultural services begin to open. Some street stores open up to 200 m2, but with restricted hours and capacity; public transportation at 2/3 capacity and masks are required; outdoor sports allowed. Gatherings of more than 10 people are forbidden. On 04.05 masks will become mandatory in the RA. The celebrations of 13.05 in Fátima take place behind closed doors. On beaches, it is compulsory to keep 1.5 meters between groups and 3 meters between umbrellas. Use of masks in schools and TP compulsory for over 10 years. Prime Minister urges the return to the streets, stores, restaurants and cafes.
2020.05.16	2020.05.16 - 28.810 (227 new cases of infection) 2020.05.22 - 30,000 (288 New cases of infection) 2020.05.29 - 31,946 (350 Newly infected cases)	Some restriction measures are withdrawn (2nd phase of the unlockdown plan)	Prime Minister urges a return to the streets, stores, restaurants and cafes. Restaurants and cafes open with restricted hours and 50% full and face-to-face classes begin in the 11th and 12th grades. Day care centers may begin to open. 20.05. The pandemic had an impact of 680.2 M€.
2020.06.01	2020.06.10 - 35,600 (294 new cases of infection)	Some restriction measures are withdrawn (3rd phase of the unlockdown plan)	Religious ceremonies are allowed; opening of shopping centers and pre-school centers; childcare centers and movie theaters and performance halls. 01.07 Teleworking is no longer mandatory and gymnasiums open with rules. 04.06 the Economic and Social Stabilization Program (PEES) is approved. Until 30.06 borders with Spain are closed. 26.06 - Pandemic cost 1,820.5 M€.

continues on following page

Table 7. Continued

Date	Number of positive cases- Portugal	Description	Main Measures
2020.07.01	2020.07.15 - 47,426 (375 new cases of infection) 2020.07.26 - 50,000 (209 Newly infected cases)	End of calamity situation and beginning of alert situation	Borders with Spain reopen. 03.07 Portugal is excluded from the United Reunion air corridors that do not require British nationals to be quarantined. 21.07 the European Council reaches agreement on a financial package to help EU Member States mitigate the effects of the pandemic and the Government publicly presents the "Strategic Vision for the Recovery Plan 2020/2030". On 30.07 bars and discos start operating under the same rules as cafes and pastry shops. Resumption of pavilion sports, but without public. 31.07 INE reports that the Portuguese GDP fell by 16.5% in the 2nd quarter compared to the same period last year.03.08 First day without Covid-19 fatalities since the beginning of the pandemic.22.08 the United Kingdom reopens to Portugal the air corridor that exempts travelers from mandatory quarantine on arrival. 01.09 is launched the App "StayAwayCovid". 09.09 Portugal records the highest number of daily contagions since April 20, with 646 new cases.
2020.09.15	2020.09.18 - (770 New Infections) 2020.10.01 - (854 New cases of infection) 2020.10.08 - (1,278 New cases of infection)	Portugal is now in a state of emergency	Back to school and face-to-face work, with the requirement to wear a mask. Gatherings are limited to 10 people, and businesses may only open at 10am and must close between 8pm and 11pm. In cafes and restaurants near schools, groups may not exceed four people. Sports venues are still not attended. AML and AMP - time lags and team rotation between telework and face-to-face work. United Kingdom puts Portugal back on the list of countries required to quarantine on arrival.10.10 reaches the maximum number of daily infections, 1646.
2020.10.15	2020.10.16 - (2,608 New cases of infection) 2020.10.22 - (3,270 New cases of infection) 2020.10.24 - (3669 New cases of infection) 2020.10.28 - (3960 New cases of infection) 2020.10.29 - (4224 New Infections) 2020.10.30 -141,279 (4,656 new cases of infection)	Portugal is now in a state of calamity	No gatherings of more than 5 people on the public road and family events cannot have more than 50 people. Forbidden academic festivities. 16.10 - New record daily cases, 2,608 and on 22.10 - 3,270 and on 24.10, 3669 and on 28.10, 3960 and on 29.10, 4224 and 30.10, 4656. 22.10 Ban on movement between municipalities between 30.10 and 03.11. On 25.09 the DG of the Budget reveals that the pandemic has cost until the end of August 2,521.7 M € . The 28.10 - use of mask is now mandatory in public spaces.
2020.11.04	7,497 New cases of infection	New restrictive measures	Partial lockdown in counties with more than 240 cases per 100,000 inhabitants in the last 14 days (121 counties), with mandatory lockdowns, new hours in establishments, mandatory teleworking, and fairs and markets banned, if the counties so decide. Throughout the continent restaurants with limitation of 6 people per table.
2020.11.06	2020.11.06 - (5,550 New Infections) 2020.11.07 - (6,640 Newly Infected Cases)	State of Emergency	The state of emergency allows restrictions on the freedom of movement and use by public authorities of resources, means and health facilities in the private sector.
2020.11.09	179,324 cases of infection(4,096 new cases of infection)	New restrictive measures	Obligatory night-time collection (23h-05h) during the week and from 13h on weekends, in 121 councils (70% of the population of Portugal).
2020.11.21	6,472 New cases of infection	New restrictive measures	Prohibition of traffic on the public highway and closing of commercial establishments after 1 pm on weekends.
2021.01.15	10,663 New cases of infection	2nd lockdown	
2021.03.15	256 New cases of infection	1st unlockdown phase	Opening of day care centres, pre-school, primary and secondary schools, religious celebrations and trade to the wicket.

continues on following page

Table 7. Continued

Date	Number of positive cases- Portugal	Description	Main Measures
2021.04.05	159 New cases of infection	2nd phase unlockdown	Openings for 2nd and 3rd cycle schools and child day care centres; shops up to 200 m2 with doors onto the street, monuments, art galleries, etc.; terraces with a maximum limit of 4 people per table.
2021.04.19	220 New cases of infection	3rd phase unlockdown	Start of classes for secondary and higher education; opening of all shops and shopping centres; restaurants, cafes and pastry shops (with restricted opening hours and capacity); opening of cinemas, theatres and concert halls.
2021.05.03	387 New cases of infection	4th unlockdown phase	Restaurants, cafes and bakeries with no time limits, increased capacity; reopening of gyms and holding of large outdoor events.
2021.12.26	3,732 new cases of infection; 2021.12.30 - 30,829 new cases of infection	Further restrictive measures	Compulsory telework; suspension of classes; compulsory negative test for air travel, access to restaurants, cinemas and concert halls; discotheques closed.
2022.01.10	20.212 New cases of infection		Start of face-to-face classes; compulsory teleworking regime until 14 January; reopening of bars and nightclubs on the 14th.
2022.01.14	40.090 New cases; 2022.01.27 - 65.706 New cases (new peak)	Unlockdown	Telework regime is changed from mandatory to recommended; reopening of bars and nightclubs; the limits regarding the occupation of commercial establishments are maintained - maximum indicative occupation of one person for every five square meters of area); In schools: End of isolation of classes after detection of a positive case; Testing of teachers and non-teaching staff in the first two weeks after returning to school. The digital certificate is now mandatory for access to: Restaurants; Tourist establishments and local accommodation; Cultural shows; Events with marked seats; Gymnasiums. The presentation of a negative COVID-19 test result is now mandatory for access to Visits to nursing homes; Visits to patients hospitalized in health care facilities; Large events and events without assigned seats or in improvised venues Sports venues (unless decided by the DGS). Remember that, due to the new DGS norms, there are also changes in isolation: Isolation is now applied only to positive cases and their cohabitants; People with a booster dose are exempt from isolation; Isolation is for 7 days.
2022.02.09	2022.02.28 - 3423 New cases		PCR or antigenic tests are no longer mandatory for international flights. Relaxation of isolation rules in positive cases (7 days if without symptoms; 10 days if with symptoms). No. of people fully vaccinated 8,996,933

Prepared by the author based on: https://www.dn.pt/sociedade/doze-meses-doze-momentos-na-evolucao-da-pandemia-em-portugal-13406881.html ;https://www.consilium.europa.eu/pt/policies/coronavirus/timeline/; https://graphics.reuters.com/world-coronavirus-tracker-and-maps/pt/countries-and-territories/portugal/ (consulted at 2022.02.20).

Chapter 10
Everyday Mobility in an Urban Context of Uncertainty:
Times of Resistance in Braga's Public Market

Helena Pires
Communication and Society Research Centre, University of Minho, Portugal

Zara Pinto-Coelho
Communication and Society Research Centre, University of Minho, Portugal

ABSTRACT

At a time when all public access spaces in the city were closed, including cafes, the Public Market of Braga remained an open space, albeit with strict entrance control rules. The authors thus advance with the hypothesis that the PM was fundamental to guarantee the routine mobility of Braga's citizens, especially the most disadvantaged and, with that, to ensure their sense of hope and confidence in the future. Likewise, the PM is a place of resistance also from the perspective of sellers, on the one hand, maintaining its economic survival, albeit with visible difficulties, and on the other hand, ensuring the maintenance of a vital communication space. To reflect on the specific ways in which this phenomenon has occurred since January 2021 is the main objective of this chapter.

INTRODUCTION

Participation in the public sphere, in what is common, according to Arendt (1958/2001, pp. 64-65), is a condition, a guarantee of the reality of the world and of our own reality. Furthermore, in everyday contexts, it is in the public sphere that subjects find, through their social interaction (which implies, in one way or another, the experience of mobility), ways of dealing with their awareness of finitude, of the world, of loneliness, in a word, the "fear of death": "that it's not the usual fear, let's say, for dying, but a kind of tiny, subterranean and permanent terror that takes over life…as if death, as the opposite of life, as absolute lethargy, definitive rigidity, paralysis and abyss, came to occupy the terrain of our

DOI: 10.4018/978-1-6684-3369-0.ch010

daily times. It is against the tendency to be captured by such a feeling of fear that we must fight — precisely, keeping ourselves active and concerned about others and the social life of which we are a part" (Gil, 2020, p. 97).The vital routine itineraries, such as the practice of 'going shopping', the sites where such practices occur are spaces of social communication, more or less intense, and of economic and symbolic exchange of experiences and emotions. It is also, and especially in the pandemic context, in these spaces (and not just on the street) that the city is constituted as a "city in motion" (Nel.lo, 2015), as an instance of the exercise of "citizen responsibility" in times of deep social crisis. "In crisis, space matters", says Nel.lo (2015), and collective heritage and goods can also include social practices or, in the author's words, the exercise of "citizenship in motion". Stavrides (2021) also discusses the potential of new ways of looking at common spaces, spaces that are inscribed in a city understood as a "common world in formation" and not as a reality or consummated spatial order. In this sense, it can be considered that crisis' contexts impose the urgency of reinventing new and common spaces. In the preface to the work by Stavrides (2021), *Common space. The city as a collective work*, Joana Braga clarifies that for the author "cities… are systems of spatial relations in continuous mutation, invested and mobilized also by the daily practices of its inhabitants" (Braga, 2021, p. 7). Do we now have reasons to believe in the definitive "decay of the public man", as Sennett (1974/2017) would argue? Or can we consider the role of transformations, for example at the level of urban design, in redefining new ways of building a "happy city", as stated by Montgomery (2015)?

This research focus on the veiled practices of infrapolitics, from the perspective of Scott (1992/2013). Alongside the new configurations of collective action in the public space, namely the action of social movements and even activists, the authors are interested on various forms of subtle resistance in everyday contexts of urban life, approaching public markets as institutions that are essential to the production of a sense of community. As Scott argues, "the discreet struggle that subordinate groups exert every day is, like infrared rays, beyond the visible part of the spectrum" (Scott, 1992/2013, p. 253). In what way are public market sites of hidden resistance, symbolic and material? To what extent do mobility itineraries in the context of everyday life, and in marketplaces, combine the reproduction of the established order with wandering and, as such, the marginal exercise of freedom(s)? Summing up, the aim of this chapter is to illustrate the unique way in which individuals resist in the contexts of everyday urban life, taking public markets as a vital public space, where such an exercise is carried out in a particular way, in articulation with the production of meaning of community.

Within this framework, we advance with the hypothesis, *based* on an ongoing ethnographic-inspired research, that the Public Market of Braga (north of Portugal) was fundamental to guarantee, during the COVID-19 pandemic, the routine mobility of Braga's citizens, especially the most disadvantaged, such as the elderly and lonely people, and, with that, to ensure their sense of hope and confidence in the future. Likewise, the Public Market was a site of resistance also from the perspective of sellers. On the one hand, maintaining its economic survival, albeit with visible difficulties. On the other hand, ensuring the maintenance of a vital communication space, shared between sellers, buyers/visitors and workers in the Public Market. The main objective of this chapter is to provide a description and analysis of the specific ways in which this phenomenon has occurred, during the second COVID confinement period in Portugal, with a view to contributing to a broader discussion on the role of public spaces in the resistance to "untamed time" (Gil, 2020), in particular phases of deep social crisis.

BACKGROUND

Everyday Mobility in an Urban Context - Public Spaces (Places), Boredom and Resistance

Urban Public Spaces (Places)

The authors of this chapter are interested in the anthropologization of urban space, its transformation into a lived place, an experienced place. Going back to Simmel (1903/1997), the space of the city appears associated with the idea of freedom, in turn evocative of an "attitude of reserve and indifference", adopted by individuals within the "crowds" and characterized by wandering, the refusal to fixation at a given point. This attitude is embodied in Simmel (1903/1997; 1908/1990) by the figure of the foreigner. Urban culture is thus defined as a singular culture, determined by mobility, by the specific way in which the foreigner is articulated with proximity and distance (namely, the "crowd"). In Baudelaire, we find the *flâneur*, a figure that was later taken up by Benjamin (2019). Following Simmel and Baudelaire, Benjamin explored, especially in the work *Paris, capital of the 19th century*, the specificity of the city and urban culture, associated with ephemerality, the transitory and moving character of social practices (these ideas were already foreshadowed in Baudelaire). Silvano (2001/2007, p. 21) summarizes this central idea, based on the authors in question, admitting "mobility as a factor in the organization of the city's culture and space".

The transformation of space into a lived place also presupposes the assumption that space is a social production, that is, constituted by social practices, as mentioned by Lefebvre (1974/1986). In this research public spaces are seen as sites of social interaction, spaces lived in a social context. Today, urban space is defined as a "relational space", naming "the spatial experience characteristic of 'reflexive modernity', as the pre-given nature of social space and the taken-for-granted contours of subjectivity are increasingly withdrawn in favor of the ambivalence of mobile spatial configurations and ephemeral individual choices" (McQuire, 2008, p. 22). In this chapter, we consider relational space as a condition of social and urban space, in its implications in terms of temporal acceleration of experience and spatial contraction. Paul Virilio, in *La vitesse de libération*, and not forgetting the contribution of Castells (1996), among others, clarified, in a particularly clairvoyant way, the effects of mobility technologies in the transformation of the daily experience of urban space. The consequent deterritorialization, the disconnection of the sensitive body from the surrounding environment, through the erasure of duration, can, in this sense, be perceived as phenomena that have altered, in contemporary times, the relational and social dimensions in urban life environments.

The urban space is, as a social and relational reality, fundamental for the actualization of public life. Several authors, including Sennett (1974/2017), defending the importance of public life, namely in the well-known work *The fall of public man*, have discussed the risks of the "death of public space". Sennett argues that narcissism, as a principle of the Protestant ethic of modernity, imposes, in public spaces, "the tyrannies of intimacy". Thus, paradoxically, community life is compromised: "the closer people come, the less sociable, the more painful, the more fratricidal their relations" (Sennett, 1974/2017, p. 417). Edgar Allan Poe, in *The man of the crowd*, describes the city, precisely, as an environment dominated by solitude. Sennett (1974/2017, p. 419) contends: "the city is the instrument of impersonal life...". As a counterpoint to this perspective, Tönnies (1887/1988) approaches the idea of "community" as an antidote to indifference. Is it possible to build a sense of community in urban public spaces, even if these are characterized by the experience of mobility and the transitory?

Sennett (2018) and Montgomery (2013/2015) have discussed how urban design, architecture and urbanism in general, including urban mobility issues, have affected access to urban space as a crucial instance for citizens' participation in public life. In particular, the phenomena of gentrification (economic, social...) and touristification have been the subject of extensive investigation and discussion (Harris, Williams & Griffin, 2012; Theobald, 2012; Urry, 1995). It is also relevant and urgent to think about the feeling of insecurity, uncertainty and risk that plague daily urban life (Kovadloff, 1998/2004). The routines that permeate everyday life, the itineraries that are repeated, like going to work and shopping, going to school and picking up our children, or meeting at the coffee-shop as a ritual, the daily walks with our dogs, all these repetitive, ordinary and predictable practices, the sameness, in a word, allow us to experience the possibility of trust in tomorrow and in the idea of the future. Says Kovadloff (1998/2004, p. 18): "Whoever has a future has a routine. Only a regular life can affirm the experience of a future". Hence, everyday mobility, in an urban context, is vital to overcome the extraordinary, the unpredictable, the threats of different kinds that also lurk in common life. The sudden break with the habit and the accidental rupture of routines are perceived as disturbing experiences of the emotional and social orders, both at the individual and social levels.

Everyday Boredom and Happiness

The problem of everyday life in an urban context leads us to summon the experience of boredom, born with modernity, as expressed in the writings of Charles Baudelaire, namely in *The painter of modern life* or *The flowers of evil*, defined as a paradoxical experience. On the one hand, a kind of everyday experience of non-life, the experience of stopped time, of non-event, ultimately, the experience of death in life. On the other hand, the fact that, being the experience of our own finitude, and therefore gaining the expression of everyday anguish, it can only be overcome by death itself. Svendson (1999/2006, p. 49) clarifies this idea: "In boredom, time is not simply something we earn, but a prison. Boredom is related to death, but it is a paradoxical kinship because, while profound boredom is experienced as a kind of death, death presents itself as the only possible total break with boredom. Boredom is a matter of finitude and nothingness. It is a living death, an unlife. The inhumanity of boredom allows us to see our own humanity in perspective". It is important to stress the daily experience of the lived body, which Lefebvre (2004) talks about, meaning that the experience of urban life, and specifically the experience of boredom, is both symbolic and experiential. Based on the Lefebvre's notion of "rhythmanalysis", which is "always concerned with plural rhythms" — considering, on the one hand, the 'lineal' rhythmicity (the regulating rhythms of modern social organization), and, on the other, the 'cyclical' rhythms (the rhythms of nature and biology) —, Highmore adds that "we need to return to the critical potential of rhythmanalysis, its concern with the bodily shocks, stresses and strains that might well point in the direction of a specifically urban and industrial rhythmanalysis" (Highmore, 2005, p. 147). Mobilities that structure the public space in everyday life are produced intermittently, sometimes corresponding to displacements, pedestrians or by different types of transport, sometimes to suspension compasses that retain individuals in queues, the metro or bus stops, or voluntarily seated on a terrace. They also include virtual navigations, often permanently, which accompany all types of occupations of the body and transform cities into deterritorialized places, into true cybercities or "globalized screens" (Lemos, 2004; Lipovetsky & Serroy, 2007; McQuire, 2008).

Today, when we speak of boredom in contexts of everyday urban life, we are speaking, therefore, of an intermittent experience, of boredom and, at the same time, of an escape from boredom. Resisting boredom becomes, from modernity onwards, a vital purpose. This means running away from the everyday experience of non-life, while running away from the idea and experience of a kind of pre-announced death. Hence the ambiguous nature of the very notion of "novelty", in Baudelaire's terms, in its association with the idea of modernity. The permanent desire for *Novelty* (as a new 'sacred' value in modernity) is what mobilizes the modern individual, while simultaneously the *New* transforms itself, in the immediacy that consumes it in the present, into an inevitable and always renewed death sentence. The idea of the *New* always refers to a future committed to its finitude. In the *New*, we already see its own end.

Commoning as Resistance

Therefore, how to resist to the social organization of everyday time that inscribes individuals in the discipline of the capitalist productive system, depersonalizing them and machinizing their routines? How to escape the death of boredom and, at the same time, how to find in the *blasé* attitude a way to reinvent some margin of individualization, and, as such, of freedom and critical sense? Svendsen (1999/2006, p. 27) argues that "boredom always implies a critical element, insofar as it is an expression of deep dissatisfaction, both in the face of a given situation and in relation to existence as a whole". From this point of view, Fernando Pessoa at his book *Livro do Desassossego* can be understood as a true philosopher of modernity and everyday urban life. Furthermore, says Svendsen (1999/2006, p. 26), "boredom is a 'privilege' of modern man". It is also in large metropolises that the vital need arises to resist the absolutism of the paradigm of rationality, the principles of control and surveillance, as discussed by Foucault (1975/1984). In this perspective, interrogating the urban space is to question its relationship with power. Individuals are thus subject to the social forces that produce a certain spatial organization and certain logics of emplacement. As Braga (2016/2021, p. 7) claims, "cities, structured systems of stable places and regular flows, contribute decisively to the shaping of the current forms of social and political organization, at the same time to the articulation of relations of power corresponding to these forms of governmental rationality are spatially translated, conditioning urban life". On the other hand, considering that the city can be understood as an open system, how can we today believe and build the idea of a happy city (Montgomery, 2013/2015)? In fact, as an utopian project, the city has always been imagined as a place of happiness. Currently, the idea that a happy city provides happy lives seems to spread: "...the happy city message is taking root around the world" (Montgomery, 2013/2015, p. 6).

It is important to link the issue of happiness in the context of urban life with the question of the city determined as a "common space" and "collective work" (Stavrides, 2016/2021). To live in an environment of happiness, from this point of view, means to live in connection with the community. The action that deserves special investment is the action of making common, commoning, according to Stavrides (2016/2021). Cities are, therefore, common places in formation, in turn produced by different structures and instances of communalization.

It should also be underlined the role of markets in this process of building the city as a collective work. According to Mumford (1982/2004), markets are one of the original components of the city: "even in the crudest economy, some way of distributing surpluses and exchanging special products of limited demand should be found, either by simple exchange, through gifts, or at banquets. The old urban consumers depended not simply on the peasant's products, but on the activities of the fisherman, the poultry farmer, the potter, the weaver, the blacksmith and, after all, this uninterrupted concentration of a single activity was one of the differentiating marks of the new urban economy…" (Mumford, 1982/2004, p. 86). Its ancestral location is indistinguishable from the location of the temple, which immediately indicates the importance of markets in their social and symbolic dimension.

Urban Public Markets: Spaces of Conflict, Communality and Mingling

Emil van Eck, Rianne van Melik & Joris Schapendonk (2020), reflecting on the effects of the COVID-19 crisis on Dutch marketplaces, show that the institutional ambiguity of different and sometimes contradictory dynamics of cancellations and regulations observed in one of their main fieldwork sites, Valkenswaard' Market, strongly resisted by the Rotterdam traders, transformed its status from a crucial site of recurring socio-spatial interactions and easy sociability into a public space of "rational and immediate economic interactions" utility (p. 382), that is, of immediate buying on site, without the typical "rubbing along" (Watson, 2009) and sonorities (Vedana, 2018) of marketplaces. Concomitantly, this situation encouraged alternative practices or new forms of social interaction (e.g., through the market's Facebook-page) and trading (e.g., switch to home delivering, a mobility strategy change) on the ground. According to the cited authors, this contradictory dynamic shows that, during the "temporary death" of public space of the corona crisis, the communal dimension of marketplaces remained alive highlighting mutual interdependencies of people in these kind of site as "one of the most critical public spaces in cities and villages" (p. 382).

The financial, economic, and social impact in the long run of the corona crisis on marketplaces around the world is still unknown. But we do know as Harvey and others (Harvey, 1973; Weber, 2002/1922) have demonstrated, markets and cities are two sides of the same coin. Despite their historical interrelation, the dual task of local authorities — both facilitating the market as well as policing the marketplace — as Kaplan (1984) shows — endow the relationship of the city authorities with their markets with contradictory meanings: they need and at the same time hate each other.

The COVID crisis added an additional challenge to public marketplaces, which are increasingly becoming, across various geographies, sites of a variety of forms of "urban contestation" between actors with different interests (Gonzalez, 2020). Caught between contradictory discourses of decline or decay and revival, public marketplaces, that often have a very central and strategic location in cities, have been pushed towards a 'gentrification frontier' (González & Waley, 2013). In this view, the transformations of marketplaces in recent decades in European cities such as in the case of Barcelona, Madrid (Hernández Cordero & Andreeva Eneva, 2016), but also in the global south, have been linked to new trends for food consumption, such as the search for the authentic and the gourmet, in interaction with wider processes of turistification and patrimonization and also processes of dispossession, displacement, privatization and social exclusion (Rivlin & Gonzalez, 2018). This strand of literature approaches the gentrification of public marketplaces, which is seen as a diverse and uneven process, as one of the dimensions of the neo-liberal urbanism that manages the city under the imperative of profit.

Within this debate, there is scholarship that highlights the various forms of urban conflict around redevelopment projects of public marketplaces which were contested by different actors attaching diverging values to them. Observable around the world, this kind of urban conflict gained a multitude of diverse forms: in Mexico, the dispute for the markets of La Merced took place between market traders and the local government, against an attempt to incorporate them into the touristic cultural circuits of authentic/original/autochthon food (Delgadillo, 2016); in Brazil (Juazeiro do Norte, North East Brazil), the fight of the vendors was against the privatization of the marketplaces, blamed for the precarization of work conditions and infrastructural obsolescence (Pereira, 2016); in the UK, market users and traders of traditional retail markets leaded campaigns in the name of the inclusionary potential for local communities of this kind of public space (Gonzalez & Dawson, 2015). As documented by Gonzalez (2020), what is at stake in these kind of struggles around public marketplaces often goes beyond the material interests of market traders, shifting the debate towards their public nature in cities and the collective right of the dispossessed and marginalized to the city (Harvey, 2012).

The discussion in the academic field around the exclusionary nature of the recent upscaling of traditional public marketplaces is counterbalanced by a literature line that underlines the well-known inclusionary potential of these kind of urban spaces as sites of continuation of "traditional" selling and shopping practices that are more dependent on interpersonal communication and physical interaction with places, goods, and people. Under this view, marketplaces, approached as one of critical sites of public space in the cities, are explored as spaces of social interaction and connection, social inclusion and for "the mingling of different cultures and the building of a sense of local community" (Watson & Suddert, 2006 p. 3). This academic work problematizes and complexifies longstanding common bipolar oppositions, such as the public/private, formal/informal, economic/festive, innovative/traditional divide and questions the myth that equalises marketplaces with trade or mercantile exchange.

There is a growing recognition, across a range of disciplines, that all economic transactions are, to some extent at least, embedded within a wider social framework and involve interpersonal processes, culture, meaning and sentiments, and everyday practices, as Zelizer, among others, has shown (Zelizer, 2005, 1994). As Avner Offer (1997, p. 451) puts it, "Exchange is not only an economic transaction, it is also a good in itself, a 'process benefit', usually in the form of a *personal* relationship driven by the grant and pursuit of *regard",* that is, responses as "acknowledgement, attention, acceptance, respect, reputation, status, power, intimacy, love, friendship, kinship, sociability", an attitude of approbation, for which reciprocity is expected. Such attitude needs to be communicated. Offer (1997, p. 454) recalls, llanguage and non-verbal signs are a "vehicle of regard", and "conversation is a gift economy ".

Based on an experience of more than 10 years of fieldwork in different public marketplaces of Europe and Latin America, the anthropologist Juan Robles (2008, pp. 20-21) gives examples of reciprocity mechanisms at play in this type of public spaces, based on his research about public markets in Madrid, Barcelona, and Valencia. Among them, the sell on credit, anchored on mutual trust due to long lasting and even intergenerational relations between traders and clients; the gift or lowering the prices to costumers living in a vulnerable situation, a practice that also works as a mechanism of economic redistribution; the exchange of information, favors and goods between neighbors, practices that stand over mutual trust and are returned in exchange by the clients with fidelity and loyalty, as attested by the presence of the lifelong costumer. Reciprocity is significant in relationships among vendors and between vendors and buyers. Robles shows that the legitimacy thus acquired by market traders gives them a major role in giving a sense of place (Massey, 1995) to these public spaces. Resonating with Robles position is the view of De La Pradelle (2006, p. 42), according to which "the market is the city as we dream it to be ... human scale ... where disorder is not bothersome ..., tight physical proximity gives rise to warm relations, [and] standing elbow to elbow with strangers induces a feeling of amicable solidarity".

The fieldwork carried out by Rivilin and González (2018) in Leeds, UK, documents a practice of an ethic of care towards other traders and customers carried out by the traders of Kirkgate. Traders ask about the well-being of customers that have not frequented the market for a while and watch each other's stalls while they take a rest. In their ethnography on eight different marketplaces in economically deprived urban neighbourhoods also in the UK, Watson and Studdert (2006) give more evidence attesting the importance of marketplaces as focal points of social inclusion, particularly for low-income groups and older people, by providing opportunities to escape isolation in the home or elsewhere, while also providing an economically inclusive space. As the authors show, marketplaces perform different roles for different social groups, and their importance and use among the population can vary according to age, gender and other factors. Social mixing is another function underline by this work, as some markets work as critical sites of social encounter (not necessarily positive or negative) and interaction across different socioeconomic, demographic, and ethnic/racial groups, by providing the opportunity of "rubbing along", a concept that Watson (2006, 2009) explains as a limited and informal way of recognizing that helps to improve the fear of others and anxiety facing difference.

This substantial body of work approaches public marketplaces as bounded by communal relationships, that is, as being public spaces that can serve as home territories or "familiarized locales" and feed a mixture of "intimate-secondary, quasi-primary relationships" containing — as Lofland (pp. 10-13) would put it — bubbles of the private and parochial (that is, of the world of neighbors, workplace, or acquaintance networks) realms or social territories. In this sense, marketplaces are hybrids, in-between-spaces. Where else "in the city can people talk to complete strangers and neighbours, discuss their health and political concerns, and exchange a recipe?", asks Rachel Black (2012, p. 8) in her book *Porta Palazzo: The Anthropology of an Italian Market*. In the tension between the desire to experience spaces where one can act and feel like a "local", and where one can concurrently experience the challenge of fleeting relationships with relative strangers in urban settings, is precisely where one can locate third places (Oldenburg & Brisset, 1982).

METHODS

The Research Field

Markets and fairs have a very long history in the city of Braga and have been key focal points of its centre. Since the roman times, Braga (with a population of 193 349 inhabitants in 2021) is known as a "market town".

The history of its covered public market goes back to the end of the nineteen century and includes several moments of equipment renovation and change of location in the urban centre. The public market has worked as the main hub of exchange between the city and its outlying rural areas since its installation in 1876.The ongoing renovation plan by the city council, who owns and runs the place, is justified by a discourse of decline and revival. The revival is branded as a combination of "tradition" (e.g., the selling of local products) and "innovation" (quality and diversity of goods, services, and installations), and calls to itself the role of a main cultural and touristic agent in the domain of gastronomy and culinary traditions, within an overall plan of urban regeneration.

Known as the "Praça – Mercado Municipal de Braga", the market is located in a plaza surrounded by several shops, near to the historical centre, in one of the biggest Parish of the urban area of the city, an urban area with a strong residential component. The architectural design of the renovated building follows the typology of a "square", with a central space combining traditional greengrocers, bakers, florists, handmade craft, shops on the periphery (e.g., butchery, charcuterie, "healthy" food), opening to the inside square, a fish and a live chicken/rabits sites located near the secondary entrance of the building, and a set of restaurant stalls siding the main entrance, with terraces opening to the outside of the building. The market hosts around 120 businesses with various tenancy agreements, from casual day to long-term leases, and the restaurant area is privately run and managed. On Tuesday, Wednesday and Saturdays, the market functions as an outlet for local grocery producers (usually a family-owned business) and nearby residents who also grow food. The space reserved for this type of selling is the floor of the corridors that circumvent the area between the peripheral shops and the central space of the market.

Researching Urban Public Markets

To understand how the Coronavirus crisis impacted on the ways people went about their daily rounds of living, specifically on their emplaced practices and experiences of doing and making urban public marketplaces, this chapter focuses on a concrete public marketplace, the Public Market of Braga in a particular moment of its long-life story — its opening as a renovated market. It provides the results of a snapshot study carried out in the context of an ongoing long term ethnographic-inspired research project (2021-2024). The project focuses on the role of the Public Market of Braga as a key site of social communication, that is, of communality.

Attending to the contexts of exchanges and situating themselves and their work "in place", theoretically as well as physically, the authors presume that something singular occurs in the public, face-to-face encounters of market traders and customers, of marketgoers, observers, and participants, in the concrete site under scrutiny that can be researchable by ethnographic and interpretive forms of analysis. Also, the authors share the conviction that whatever overarching dynamics of power, ideology and meaning inform public markets, "their bundle of practices and material arrangements" and "their network capillaries" (Everts, Jackson, Juraschek, 2021, p. 4), these are made manifest in some observable manner as people go about their everyday lives.

The authors take as points of departure the experiences, practices, and embodiments in and of the public market of Braga and situate themselves and their work in this marketplace. The following section draw on our fieldwork in the Public Market of Braga, Portugal, from the beginning of January 2021 till the end of March of the same year.

The fieldwork, that includes weekly regular visits to the site, engages with the idea of mobile methods, using walking as a means and a method — a way of getting to know the research site and gathering qualitative data through the process. The authors share the conviction that the rhythm and practice of walking leads to a different understanding of the research site from a static one. Walking through "the square" and the corridors of the market stalls and around them, the authors engage with the physical setting, in encounters and interactions with things and others — smell, feel and purchase the goods, walk along and chat around, observe repeatedly those others milling about buying, looking, selling, dickering, joking. Researching the field also includes writing field notes, conversations with key informants (vendors and representatives of the city local authority on the market), digital photographing and sound recording and archival research on the history of the market. These "polymorphous engagements" in the field stand on an idea of mobility in fieldwork, where the journeys one makes walking "also make their places and, as a corollary, the ethnographic field itself" (Lee & Ingold, p. 67). The information collected using these resources gave the ground to the description and analysis that is presented in the next sections.

ANALYSIS

A Singular Public Place

During January/February 2021, when Portugal suddenly became the country in the world with the worst infection and mortality rate by COVID[1], the public market of Braga opened its doors, after a lengthy process of revitalization, becoming the only open public facility in the city center: "...perhaps the market is one of the few and rare points of attraction in the center, since the commerce is closed, with the windows displaying the 'frozen' products since the beginning of the confinement, last January" (Fieldnote, HP, 11.02.21).

While cafes, restaurants and shops were closed in the vicinity, the Market resisted in its function as a public space. In this context, we can consider that two vital functions were fulfilled there. On the one hand, the daily supply of fresh food (vegetables, fruit, eggs, meat, fish...), essential for the survival of both sellers and consumers. On the other hand, the sociability, which is especially important in the context of a pandemic, given that social interactions had been drastically reduced, due to the widespread confinement.

Figure 1 The Public Market in Braga

In the researchers' fieldnotes during this period, the contrast between the ambiance, lived outside and inside the Market, stands out. The differences between the highest concentration of movement and the voids were clear. As the authors approached the market neighborhood, on their weekly walks, they could see comings and goings, bags and shopping carts in circulation, back and forth, along the sidewalks or near the bus stops. As they moved away, the landscape was dominated by the shopwindows suspended in time since the sales period, the closed cafes and the terraces composed of stacked chairs, the crazy and the frantic homeless people walking through the empty streets, the few impromptu meetings in the street corners, especially with elderly people, chattering, seeing each other wearing a mask and with no place to be. As the following excerpt expresses, "...on the way to the Market I didn't see anyone. The only stores that were open were one bakery/pastry shop with one or two customers and a butcher, with no one" (Fieldnote, ZPC, 01.20.21). In another of the fieldnotes, it can be read: "I went towards the Market following the usual path, starting at the *Rua de São Marcos* (São Marcos Street). I usually take the shortcut through the Castle, but this time the City Council's jet cleaning car is there, forcing me to divert to the Arcade. Few people pass each other (with everything closed, I find myself trying to guess their destiny) ...The coffee-shop *A Brasileira* has a ghostly appearance, with the tables and chairs on the terrace piled up and wet from the rain. Dwelling in the *Rua dos Capelistas* (Capelistas Street), I see some shopping carts coming from there to here, especially middle-aged women... (They got up early to go to the Public Market, it crosses my mind)" (Fieldnote, HP, 01.28.21).

In terms of access to the Market, admissions and queues were controlled by the City Council employees. At each entrance, there were two and sometimes three or four of them — "" There I see an employee identified with a badge... and who asks me to wait while two customers leave through the exit area. I notice that four men are there commanding entrances and exits, two in the entrance area and two in the exit area" (Fieldnote, HP, 02.6.21)" —, who communicated by telephone between the two posts, whenever some adjustment was justified in what concerns authorizing (or not) entries. There was no top-down control of the distances between customers while they were in the queue, especially on Saturday mornings. Customers kept their routine of going to the Market with family members, friends, or neighbors, which accentuated some crowding in some of the nodes in the queue. The use of a mask was mandatory inside the building, while alcohol-gel containers with pedal and a standup thermometer encouraged hand disinfection and temperature control, respectively. At the entrance, blue plastic fences, clearly visible, demarcated the circuit for the movement of people. In a conversation with one of the employees of the City Council at the entrance, the following was recorded: "I asked him about his business. He said that, due to confinement, there is less public. He thinks the market is less busy due to the situation. There can only be 120 people inside and on Saturdays it is difficult to control the flow. Queues are created" (Fieldnote, ZPC, 5.02.21).

Inside of the market, contrary to what was observed in other European markets, circulation was free, not subject to control in terms of social distance — both sellers and consumers, all wearing a mask, had complete freedom to approach each other, or in terms of pre-defined itineraries. No mechanisms to encourage distancing were introduced, nor by the authorities, nor by the traders. The stalls disposition in the central space and the squares marked on the floor in the local producers' area did not change. Agglomeration and proximity were not avoided. Visitors were still allowed to hang out inside of the market. But everyone was cautious to not come too close to each other.

Figure 2. The main entrance of the Public Market

Social Communication Strategies

Alongside the commercial exchanges, the Public Market was during the second confinement in Portugal a fundamental *medium*. In the context of buying and selling, or of interaction between employees who work in the Public Market, there is a range of diverse communication practices, at an interpersonal and social level, including through social media, the local press, and other channels, internal and external to the Public Market.

For customers and sellers, in addition to being a site for commercial interaction, the Public Market is also a space for entertainment. In one of the researchers' fieldnotes, one can read: "I keep going around the seller's area. I approach one of them, with cabbages and eggs on the floor. I apologize by saying that 'I'm just watching'. She watches me closely and looks eager to talk. I find out that she is over eighty years old and has been selling in the market since forever. The perception that some customers went there just for getting out is also highlighted in another passage: "At this stop, I still learn that in her opinion, many walk there just for a walk and that there are elderly people who go there every day, so they don't have to stay at home" (Fieldnote, HP, 11.03.21).

Figure 3. Inside the Public Market

Regarding to face-to-face communication, it is important to note that entrances queues were immediately used for interaction: "Well-groomed ladies, as if they had been prepared to go to the Market, in pairs. For some being there is being at home and, at the same time, being on the street, a place where familiar faces meet and conversation is exchanged" (Fieldnote, ZPC, 02.20.21). The Market is located, as mentioned before, on a threshold, halfway between the home and the street, the private and the public, summoning the notion of "third place" mentioned by Oldenburg & Brisset (1982) or the notion of *inter-esse*, expression used by Arendt (1958/2001), in reference to the social sphere where the private condition and the political vocation intersect each other.

Exchange interactions were also a pretext for casual conversations on the most varied subjects of day-to-day, from the pandemic situation to the new conditions of sale and purchase in the Market, the experiences and impressions related to the products, and conversations on personal matters and confidences. About sharing recipes, it can be read: "Look, just these days I made a side dish at night, just with a drizzle of olive oil, in the oven, so it doesn't catch. I made some soup for myself and my husband, but there my boy has a lot of appetite and I had to do something else, to accompany some chops" (Fieldnote, HP, 6.02.21). Or this other stretch, expressive of the importance of interpersonal communication: "There is more movement there than in the central area of the Market and animated conversations… old complicities" (Fieldnote, HP, 02.25.21).

These communicative interactions illustrate the variations in familiarity that customers have with each other and with sellers, as well as traders with each other. At the same time, it is also interesting to note that visitors weave circumstantial communicative interactions with each other, sometimes extended to the interaction with traders: "Someone asks, ' and cauliflower, do you have it?'. Mrs. Carminda reacts as if she doesn't know. 'I'll show you,' she hurries to say, as she takes a small, lacy-looking cabbage from a bag on the ground. 'How pretty, I say, it looks like a flower'. 'It's from Mr. José', she points to a stall at the top, in the entrance area. 'There, there are different products and little pumpkins that are really good. Cale-kale (curly kale) is very good for cancer prevention', she explains to me, 'go and check it out on the net, I'll put it in the soup'. Two other clients listen to the conversation…" (Fieldnote, HP, 6.02.21). And there is also the following passage, which took place at the fishmonger's: "the man-customer asked me if I used to go shopping there. I replied that lately yes and that I liked it very much. He then commented that he got lost there and that he spent a lot more money because he bought more, but that he liked to come there, even because of 'the conversations with the saleswomen'" (Fieldnote, HP, 02.25.21).

Figure 4. Outlet zone

While they were in front of the same stall choosing the products, the customers made comments about them, exchanging recipes, or extending the conversation established with the trader to those who were around them. These are occasional communicative practices, of short duration, but which, in the case of regular visits to the Market, build relationships of trust and proximity. The researchers themselves were not excluded from this same experience, having built, throughout the fieldwork process and regular visits to the Market, closer relations although uneven, with different sellers. The exchange of knowledge and the sharing of experience on how to cultivate fresh vegetables was, among others, one of the topics of conversations between and with shoppers: "We exchanged some comments between the two of them, I and that customer. She advises me to pass the turnip greens with olive oil and garlic, after they are cooked, and I find out, after this explanation, that she comes to get products to plant in her backyard…" (Fieldnote, HP, 02.18.21).

At the level of non-verbal language, the simple co-presence can be understood as one of the modes of communication, of putting in common, in the sense of implicit sharing of space and experience. Especially the elderly and the lonely, whose condition of isolation has been accentuated by the pandemic, visited the Market not only to shop, but also to walk around. Simply seeing the environment, walking aimlessly around the Market, in the style of a *flâneur*, observing others, the profusion and color of the products, always with variations according to seasonality, the different ways of dressing, walking, interacting, is in itself a daily hobby, according to what the researchers witnessed, sometimes associating the practice of shopping with leisure: "... the customers, some shopping in pairs, as if it were a friend's entertainment..." (Fieldnote, HP, 28.01.21).

Figure 5. The Public Market as an aesthetic co-experience

In addition to being a place, the Market thus becomes a landscape. A non-fixed, multisensitive, immersive and, in a way, cinematic landscape (Friedberg, 2009), since the site is a space that does not invite prolonged fixed stays (there are no chairs or any type of structure to sit on), but, on the contrary, invites to circulate. In contrast to the urban ambiances outside the Market, emptied of life, and the feeling of boredom, of eternal repetition of the same that characterized, at that time, the pandemic context in general, the Market offered itself as a site of regularities, but also of variations, arousing — this hypothesis can be advanced — both the collective feeling of trust arising from repetitions (Kovadloff, 1998/2004; Gil, 2020), as well as the excitement inherent in the appreciation of small novelties that animate the sameness of everyday life.

Likewise, the appreciation of the appearance of the products, the comments on the changes in the space resulting from the revitalization process, constituted a kind of relational and shared aesthetic experience (Bourriaud, 1998/2009) that simultaneously fulfilled its communicative function. It prevailed the general recognition, among traders and shoppers, that the new Market was now more beautiful, when compared to the old one. At the same time, complaints about functional problems were also plentiful: the rain and wind that entered the sides of the roof, the failure to fix the sales sites in the local producers' area (which were — and still are — obliged to book their selling point every week and the obligation to comply with schedules, morning and afternoon, from seven in the morning to five in the afternoon, from Monday to Friday (on Saturdays the Market closes at 1:00 p.m. in the afternoon).

It is important to remark that the pandemic was a topic of the everyday conversations also in the Market — "In one of the stores I overheard part of a conversation, between owners and customers, about the death toll and how frightening the present scenario was" (Fieldnote, ZPC, 02.29.21). But the general perception at that stage of the pandemic was that it would affect, above all, the elderly and vulnerable or patients with different types of pathologies, namely living in institutions such as nursing homes, and at the time there weren't many cases of COVID from close people to report. The disease was perceived as a real threat, but that happened to others. The feeling of fear (Gil, 2020), even so, spread and the researchers did not fail to take special care: "Physical contacts with money multiply and the concern about the risk of the pandemic with it. Also the need to be careful with hand disinfection" (Fieldnote, HP, 01.20.21); "...the saleswoman insists that I should try it and I find myself putting a grape in my mouth, which I do without thinking, already regretting it, but with no return, after reminding myself that I hadn't even disinfected my hands" (Fieldnote, HP, 01.28.21). The little movement in the market is, to some extent, explained by the pandemic: "people are afraid, and they should be even more afraid" (Fieldnote, HP, 02.11.21). In addition, the celebration of religious festivities was interrupted, and this is expressed in the following fieldnote: "'And we won't have Easter, as we did not have Christmas', they tell me. 'I have three children in Switzerland, but this year they did not come" (Fieldnote, HP, 18.02.21).

Temporalities in Counterpoint

Resisting the pandemic implies resisting a new spatio-temporal experience. On the one hand, it was important to resist the experience of confinement, which forced restrictions on urban mobility, with restaurants, services and non-essential commerce closed, in addition to the imposition of mandatory teleworking (where applicable). Public spaces became phantasmatic scenarios, suspended in time, and for many the management of daily time became a new problem. How not to succumb to sameness, to the experience of repeating the same, day after day, to the tedium of a routine with no foreseeable end on the horizon?

In this context, the Market offered itself as one of the few public spaces where it was possible to find, simultaneously, the experience of face-to-face communication, of being together and of co-presence, including the feeling of belonging to a community. In the Market, regular shoppers are familiar faces, they have known the sellers for a long time and often by their name, finding there a place of reciprocity (Picón, 2010). Exchange, as already mentioned, is a constant back-and-forth movement of words, confidences, advice, outbursts, products, money, recipes...

Different temporalities were experienced in the Public Market. Firstly, the *clock time*, the geometric time, divided into equal parts, the regular time that disciplines routines and determines occupations. This time was punctuated by the days of going or not going to the Market, selling, or buying, by the different atmospheres that characterized each day of the week, mornings, or afternoons, with more or less people, more stalls open or closed, by the presence or not of the peasants, due to the greater or lesser diversity of products and seasonality. Thus, linear time both produced a feeling of monotony and opened the possibility of small variations within the framework of the same repetitive structure.

Secondly, this same linear time was in turn contradicted by the *still time*, the empty time of everyday life, whether this corresponded to moments when salespeople did not have customers and seek to reinvent strategies for occupying time (reading the newspaper, having fun with social networks, rearranging the products at the stall over and over again.... The salespeople's conversations among themselves were also one of the hobbies on these occasions. On a particularly busy day, the following excerpts can be read: "I noticed the saleswoman... She seemed very comfortable talking with the butcher" (Fieldnote, HP, 01.20.21), and further, "Some saleswomen talk to each other, and are sometimes sitting, sometimes standing" (Fieldnote, ZPC, 01.20. 21).

Thirdly, the time of the present-future was crossed by the *mnemonic time*, the time of individual and collective memories. The perception of the new market evoked the memory of the old market and the confrontation between before and after was a constant. Furthermore, commercial interactions gave rise to the permanent updating of memories, about flavors, smells, ways of cooking and the arts of planting, emphasizing everyday practices and the practices of the arts of making, in Certeau's terms (1998), places, memories of past generations, parents and grandparents...

Fourthly, it is important to emphasize the *mythical time* or the time out of time. It means the time of narratives that give expression to beliefs, to a knowledge that does not fit within the limits of modern scientific rationality and that does not exclude contemporary mythologies from everyday life, as Barthes argues (1957/2020):

"The figure of Saint Anthony is displayed at the top of the shelf above the meat hanger on the left (I only noticed the aforementioned image to which the butcher calls my attention later). It used to be made of wood and had green glass eyes. Mr. João describes the figure with a broad gestural expression, drawing two large eyes with his hands. I get the impression that such a figure was remembered from childhood, I imagine, and that as such a child's perspective is described, as a dazzling and frightening figure. I ask where the original figure is, but the question does not get a convincing answer. He hastens to say that it is well kept, it is a family heirloom" (Fieldnote, HP, 02.11.21). On another field note, one can read "The butcher answers me, saying that Saint Anthony is the Saint of animals – 'He is known for talking to them' – and that he has to face inwards, towards what he blesses" (Fieldnote, HP, 11.03.21).

Finally, it is important to consider the *virtual space-time*, the migrations operated, in the Market, to the digital. On the one hand, the City Council immediately sought to encourage online buying/selling, even creating a remote shopping service project for this purpose (not yet available[2]) and even providing a small van, decorated with the new visual identity of the market, for home deliveries. On the other hand, sellers started to market their products using mobile phone and Facebook pages, in response to orders. The fact that many of the businesses are family and transgenerational ran, made the task easier with the younger ones helping with the digital procedures. The practice of home deliveries has also become a new reality. In conversation with one of the sellers, the researchers noted: "Now, with the pandemic, the fish is delivered at home" (Field note, ZPC, 01.20.21) and also "I noticed that this lady, while weighting my things, was on her mobile phone and setting down the things that someone was asking her to deliver. An order" (Fieldnote, ZPC, 5.02.21).

Figure 6. The Saint Anthony

Figure 7. The Saint Anthony

CONCLUSION

Marketplaces act as important economic and vital public spaces throughout the world. The chapter has addressed the current transformations of urban public markets, which are linked to complex processes of social change, in a particular moment of their already precarious life as urban public spaces: the corona crisis. Many marketplaces either closed or remained open under control of strict corona-preventive regulations.

In this chapter, the authors gave a snapshot of a long-term research project based on the Public Market of Braga. It addressed the impact of the COVID crisis on this site, during the second confinement in Portugal, which was also the moment where the renovated building opened its doors to the public. By observing and participating on the actual social practices of going to the market and being on the market on the ground, we conclude that urban public markets fulfill a fundamental function from the point of view of social interaction and communication. These are public spaces of special relevance to the most vulnerable, such as the elderly and the lonely. In the specific context of the pandemic, the Public Market of Braga became a unique open public space, as it remained the only public space for social interaction, in the center of the city, during the second confinement in Portugal. The authors observed on the ground the way customers, sellers and mere visitors transformed this public space into a place of daily living, of compliance with routines, fundamental to the feeling of individual and collective security, but also into a place of resistance to fear and unpredictability, as well as to a feeling of boredom that the pandemic's restrictive measures, generalized to the entire population, accentuated.

The face-to-face interpersonal communication, the co-presence, the shared aesthetic experience, as well as the daily experience of different temporalities (the linear time, the still time, the mnemonic time, the mythical time and the virtual space-time), not forgetting the use of Facebook and mobile phones, constituted a fundamental set of more or less veiled strategies of resistance in times of the COVID pandemic. Also, it should be added that the daily mobility, concerning the Public Market of Braga, in the physical and virtual space, worked as a counterpoint to the generalized feeling of emptiness and stillness of the lived space-time.

Has the fundamental role played by the Public Market of Braga as a critical public space changed after the specific moment of time covered by the snapshot study just presented? Does the effusive reception of its recent upscaling turned out to be an ephemeral one? Is its inclusionary potential under threat due to the exclusionary nature of its redevelopment project? These are some of the questions that are guiding our present fieldwork.

FUTURE RESEARCH DIRECTIONS

The authors first research findings highlight several areas for future research, from ethnographic and urban culture investigation to social communication studies.

REFERENCES

Arendt, H. (2001). A condição humana. Relógio d'Água. (Original publication 1958)

Barthes, R. (2020). *Mitologias*. Edições 70. (Original publication 1957)

Benjamin, W. (2019). *As passagens de Paris*. Assírio & Alvim.

Black, R. E. (2012). *Porta Palazzo: The anthropology of an Italian market*. University of Pennsylvania Press. doi:10.9783/9780812205794

Bourriaud, N. (2009). Estética relacional. Martins Fontes. (Original publication 1998)

Braga, J. (2021). A cidade como terreno comum: limiares urbanos e a invenção de novos horizontes do possível. In S. Stavrides (Ed.), Espaço comum. A cidade como obra colectiva (pp. 7–36). Orfeu Negro. (Original publication 2016)

Castells, M. (1996). *The rise of the network society*. Blackwell Publishers.

Certeau, M. (1998). A invenção do quotidiano. 1. Arte de fazer. Petrópolis.

Delgadillo, V. (2018). Markets of La Merced: new frontiers of gentrification in the historic centre of Mexico City. In S. Sarah González (Ed.), *Contested markets, contested cities: Gentrification and urban justice in retail spaces* (pp. 19–35). Routledge.

Eck, E., Melik, R., & Schapendonk, J. (2020). Window on the Netherlands marketplaces as public spaces in times of the covid-19 coronavirus outbreak: First reflections. *Tijdschrift voor Economische en Sociale Geografie*, *111*(3), 373–386. doi:10.1111/tesg.12431 PMID:32836485

Everts, J., Jackson, P., & Juraschek, K. A. (2021). The socio-material practices of the transformation of urban food markets. *Area Royal Geographical Society*, *53*(2), 389–397. doi:10.1111/area.12707

Foucault, M. (1975/1984). *Surveiller et punir*. Gallimard.

Friedberg, A. (2009). *The virtual window. From Alberti to Microsoft*. The MIT Press.

Gil, J. (2020). *O tempo indomado*. Relógio d'Água.

González, S. (2020). Contested marketplaces: Retail spaces at the global urban margins. *Progress in Human Geography*, *44*(5), 877–897. doi:10.1177/0309132519859444

Gonzalez, S., & Dawson, G. (2015). *Traditional markets under threat: why it's happening and what traders and customers can do.* https://eprints.whiterose.ac.uk/102291/

Gonzalez, S., & Waley, P. (2013). Traditional retail markets: the new gentrification frontier? Antipode: A Radical. *The Journal of Geography*, *45*(4), 965–983.

Harris, R., Williams, P., & Griffin, T. (2012). *Sustainable tourism*. Routledge. doi:10.4324/9780080498362

Harvey, D. (1973). *Social justice and the city*. Edward Arnold.

Heidegger, M. (1986/2005). *Ser e Tempo. Parte I*. Editora Vozes.

Hernández Cordero, A., & Andreeva Eneva, S. (2016). Mercados, museos o malls? La gentrificación de los mercados municipales en Barcelona y Madrid. EntreDiversidades. *Revista de Ciencias Sociales y Humanidades*, *6*, 143–173.

Highmore, B. (2005). *Cityscapes. Cultural readings in the material and symbolic city*. Palgrave Macmillan.

(2018). Introduction: Studying markets as spaces of contestation. InGonzález, S. (Ed.), *S. Gonzalez, Contested markets, contested cities. Gentrification and urban justice in retail spaces* (pp. 1–18). Routledge.

Kaplan, S. L. (1984). *Provisioning Paris: merchants and millers in the grain and flour trade during the Eighteenth Century*. Cornell University Press. doi:10.7591/9781501731426

Kovadloff, S. (2004). Sentido y riesgo de la vida cotidiana. Emecé Editores. (Original publication 1998)

Lee, J., & Ingold, T. (2006). Fieldwork on foot: perceiving, routing, socializing. In Locating the field-space, place and context in Anthropology (pp. 67-86). Berg.

Lefebvre, H. (1986). La production de l'espace. *Anthropos. (Original publication 1974)*

Lefebvre, H. (2004). Rhythmanalysis: Space, time and everyday life. *Continuum*.

Lemos, A. (Ed.). (2004). *Cibercidade. As cidades na cibercultura*. Ed. E-papers.

Lipovetsky, G., & Serroy, J. (2007). *L'écran global*. Seuil.

Lofland, L. (1998). *The public realm. Exploring the city's quintessential social territory*. Aldine de Gruyter.

Massey, D. (1995). The conceptualization of place. In M. Doreen & P. Jess (Eds.), *A place in the world? Place, culture and Globalization* (pp. 45–85). Oxford University Press.

McQuire, S. (2008). *The media city. Media, architecture and urban space*. Sage Publications. doi:10.4135/9781446269572

Montgomery, C. (2015). Happy city. Transforming our lives through urban design. Penguin Books. (Original publication 2013)

Mumford, L. (2004). A cidade na história. Suas origens, transformações e perspectivas. Martins Fontes. (Original publication 1982)

Nel.lo, O. (2018). *A cidade em movimento. Crise social e responsabilidade cidadã*. Livraria Tigre de Papel.

Offer, A. (1997). Between the gift and the market:the economy of regard. *The Economic History Review*, *L*, 3, 450–476.

Oldenburg, R., & Brissett, D. (1982). The third place. *Qualitative Sociology*, *5*(Winter), 265–284. doi:10.1007/BF00986754

Pereira, C. S. (2017). Mercados públicos municipales: Espacios de resistencia al neoliberalismo urbano. *Ciudades*, *114*, 39–46.

Rivlin, P., & González, S. (2018). Public markets: Spaces for sociability under threat? The case of Leeds' Kirkgate Market. In S. González (Ed.), *Contested markets contested cities. gentrification and urban justice in retail spaces* (pp. 133–149). Routledge.

Robles, J. (2008). Comercio urbano em espacios metropolitanos. Mercados, vidas y barrios. *Distribución y Consumo*, (Sep-Oct), 19–31.

Scott, J. C. (2013). A dominação e a arte da resistência. Discursos ocultos. Letra Livre. (Original publication 1992)

Sennett, R. (2017). The fall of public man. W.W. Norton & Company. (Original publication 1974)

Sennett, R. (2018). *Building and dwelling. The ethics for the city*. Penguin Books.

Silvano, F. (2007). Antropologia do espaço. Uma introdução. Celta Editora. (Original publication 2001)

Simmel, G. (1990). Digressions sur l'étranger. In Y. Grafmeyer & I. Joseph (Eds.), *L'école de Chicago* (pp. 53–59). Aubier. (Original work published 1908)

Simmel, G. (1997). A metrópole e a vida do espírito. In C. Fortuna (Ed.), *Cidade, cultura e globalização. Ensaios de sociologia* (pp. 31–43). Celta. (Original work published 1903)

Svendson, L. (2006). Filosofía del tedio. Tusquets Editores. (Original publication 1999)

Theobald, W. F. (2012). *Global tourism*. Routledge. doi:10.4324/9780080478043

Tönnies, F. (1988). *Community and society*. Routledge. (Original work published 1887)

Urry, J. (1995). *Consuming places*. Routledge.

Vedana, V. (2018). Escutar no som: Gravação e edição de etnografias sonoras a partir de um paradigma ecológico. *Ilha*, *20*(1), 117–144. doi:10.5007/2175-8034.2018v20n1p117

Virilio, P. (2000). *A velocidade de libertação*. Relógio D'Água.

Watson, S. (2009). The magic of the marketplace: Sociality in a neglected public space. *Urban Studies (Edinburgh, Scotland)*, *46*(8), 1577–1591. doi:10.1177/0042098009105506

Watson, S., & Studdert, D. (2006). *Markets as sites for social interaction. Spaces of diversity*. Policy Press.

Weber, M. (2022). *Economia e sociedade. Fundamentos da sociologia compreensiva*. Edições 70. (Original publication 1920)

Zelizer, V. (1994). *The social meaning of money*. Basick Books.

Zelizer, V. (2005). *The purchase of intimacy*. Princeton University Press.

ADDITIONAL READING

Adami, E. (2018). Shaping the social through the aesthetic of public places: The renovation of Leeds Kirkgate Market. In E. S. Tonnessen & F. Forsgren (Eds.), *Multimodality and aesthetics* (pp. 89–111). Routledge.

Blackledge, A., & Creese, A. (2019). *Voices of a city market: an ethnography*. Routledge.

Cook, D. (Ed.). (2008). *Encounters with value in marketplaces on five continents*. Palgrave MacMillan.

Counihan, C. (2015). Ethnography of farmers markets: studying culture, place and food democracy. In C. L. Swift & R. Wilk (Eds.), *Teaching food and culture* (pp. 113–128). Left Coast Press.

Robles, J. (2010). Pequeños comerciantes: Mediadores urbanos. *Revista Chilena de Antropología Visual*, *15*, 164–190.

KEY TERMS AND DEFINITIONS

Everyday Resistance: The more or less veiled strategies that individuals or groups develop in their daily lives, in order to minimize the oppressive effects resulting from the social and political dominant structures to which they are subject.

Insecurity: A feeling of lack of confidence in the surrounding urban environment, of uncertainty regarding the present-future generated by the perception of risk that threatens the regularity of the daily life.

Marketplaces: *Situated* phenomena; assigned to precise sites—in space and time—and social frameworks.

Mobility: The set of itineraries, displacements. and drifts that individuals experience in the context of everyday urban life, both in physical and virtual space.

Place: A meaningful site. Space is transformed into "place" when humans invest meaning on it and then become attached to it. Place is constructed and reconstructed over time by different groups of people. It is thus a socially produced space.

Routine: The monotonous fulfillment of the predictable, the regularity of the practices that make up everyday active life.

Social Communication: The process of putting in common, in a specific social setting.

Urban Public Markets: Long-standing urban institutions owned and managed by public authorities, located in a geographically distinct city location, which encompass specific social relationships and frameworks through which economic transactions take place and generators of social and cultural activity. Urban settlements which are a part of the public urban space.

ENDNOTES

1 Retrieved from: https://www.rtp.pt/noticias/mundo/portugal-e-o-pior-pais-do-mundo-nas-taxas-de-infecao-e-de-mortalidade_v1291729

2 Retrieved from: https://praca.cm-braga.pt/o-mercado-na-tua-casa/

Chapter 11
COVID-19 Pandemic, (Over) Tourism, and Urban Dynamics:
Lessons From Lisbon and Porto

Pedro Chamusca
https://orcid.org/0000-0001-6921-8057
Communication and Society Research Centre, University of Minho, Portugal

José A. Rio Fernandes
FLUP, Geography and Spatial Planning Research Centre, Portugal

Jorge Pinto
https://orcid.org/0000-0002-0195-3853
Higher Institute of Business and Tourism, Portugal

ABSTRACT

This chapter analyses the tourism and urban dynamics transformations of the two leading cities in Portugal – Lisbon and Porto. In a (neo)liberal context and within a process of financialization of the economy, the authors address the political vision of each city, the local stakeholders and residents' evaluation, and public policies analysis. This chapter concludes that the implications of recent transformations are numerous. The process of residential and functional gentrification, resulting mainly from the touristification of the city, and the concomitant growth of the floating population, was slowed down by the COVID-19 pandemic. However, housing and services prices are still high, and the unavailability of housing is indeed a problem. Thus, integrated, strategic, and collaborative planning for sustainable urban development is assumed as a crucial issue for the next decade.

DOI: 10.4018/978-1-6684-3369-0.ch011

INTRODUCTION

The tourism sector is a fundamental social, cultural and economic activity with a relevant capacity for generating wealth and employment creation. Moreover, it has been increasing its relevance in several places in the world before 2020. Between 2010 and 2019, the average annual growth rate in Portugal was 7.2% in overnight stays, which translated into an increase from 37 million nights in 2010 to just over 70 million in 2019. Tourist revenues also saw an average annual change rate of 10.3% over the last ten years, increasing from 7.6 billion to 18.4 billion (INE, 2021a).

Being the last year before Covid, 2019 registered several records for tourism in Portugal. That is the case of tourist accommodation units (hotels, local accommodation, rural tourism), which welcomed almost 27 million guests, 60% of them coming from abroad. The Tourism Satellite Account, released by the National Institute of Statistics (INE, 2021b), indicates that in that year i) tourist demand was equivalent to 15.3% of the national gross domestic product (7.6% more than in 2018); ii) Gross Added Value (GAV) generated by tourism represented 8.4% of the total GAV of the national economy; iii) employment in the activities directly related with tourism, measured in full-time equivalent, represented 9.4% of the national total, increasing 7.4%, and largely surpassing the employment growth in the national economy (3.1%).

However, as it is well known, the COVID-19 pandemic would prove to be highly damaging (Gretzel et al., 2021; Abbas et al., 2021; García-Gómez et al., 2021; Gössling et al., 2021; Rather, 2021; Soliku et al., 2021). According to the World Tourism Organization (World Tourism Organization, 2022), world tourism in 2020 suffered a decrease in international arrivals of 74%, and global destinations received one billion fewer international arrivals, compared to the previous year, due to travel restrictions and an unprecedented fall in demand.

In line with the world situation, Portugal recorded a sharp drop in demand, for 25.9 million overnight stays (-63.0%), whose total is equivalent to the numbers registered in 1994. There was also a sharp decrease in international demand, with 12.3 million overnight stays for foreigners (-74.9%) due to restrictions imposed on cross-border travel for most of 2020. The national market recorded 13.6 million nights (-35.4%). The decrease in revenues (-57.6%) meant a loss of 10 billion euros for the national economy in 2020 compared to 2019 (INE, 2021a). According to the Tourism Satellite Account (INE, 2021b), the Gross Added Value (GAV) generated by tourism represented 4.6% of the national GAV. In comparison, tourism activity was equivalent to 8.0% of the Gross Domestic Product (GDP) (INE, 2021b). The activities that were the more significant contributors to tourism GDP, such as accommodation services, cafes and restaurants, transport (especially air transport) and rental services, were amongst those that suffered the most severe impacts of the COVID-19 pandemic. In numbers, the loss in tourist GDP generated by these activities was between 48% and 55%.

Spatially, the "touristic cities" were the most affected, as people avoided the most crowded places. There was a sharp difference in the economy and atmosphere in several cities, especially in the biggest and those that used to depend more on tourism. In Portugal, Lisbon and Porto had a more significant impact from the breakdown caused by the pandemic. In 2019 it was reported that Lisbon had more Airbnb than Barcelona, Paris, Rome or Amsterdam, being the European capital with more local accommodation per inhabitant (according to Moody's data, which counted more than 30 homes registered on Airbnb per thousand inhabitants, cited by Chamusca, 2021). According to the Institute of Tourism Planning and Development, which calculated the tourist pressure on Portuguese cities and compared them to other European cities, Lisbon (9) and Porto (8) had more tourists per resident than London and Barcelona (INE, 2021b), two of the European cities with more annual tourists.

The impact of COVID-19 showed, once again, the fragility of tourism and leisure for territorial-based sustainable development. Several enterprises faced immediate problems, and some of the most attractive and desired places looked strangely void. Suddenly, cities faced new dilemmas when controlling and managing the pandemic was central. How to maintain the vitality of urban centers was/is the question. How would it be possible to support the local economy so dependent on a disappeared touristic activity? How to combat the depopulation (of both inhabitants and tourists) of some areas in the city?

In this article, the authors look at the tourism and urban dynamics transformations of the two leading cities in Portugal – Lisbon and Porto. Research concentrates on recent times and particularly on the effects of the COVID-19 pandemic, focusing on the role of public policies.

The authors consider that urban management plays a central role in a (neo)liberal context and within a process of financialization and globalization of the economy. Therefore, the analysis of public policies over the past four years is critical to understanding the (asymmetrical) effects of COVID-19, especially in urban areas that are very dependent on tourism. The authors discuss the political vision for each city, how the objectives are communicated and how they are implemented and considered by local stakeholders and residents.

Our research seeks to shed light on four research questions:

1) What was the role of tourism in the economic, social and cultural transformation of Lisbon and Porto before the pandemic? Furthermore, how is it perceived? What is the role of public policies?
2) What is the impact of the COVID-19 pandemic on these cities? What is the relation between tourism and public policies on how it impacts?
3) What has been done to manage the pandemic and minimize its effects on economic activity? What were the practical effects on local stakeholders and inhabitants? How was it perceived?
4) How are urban management strategies communicated, understood, and evaluated?

This paper is structured as follows. After this introduction, in section 2, the authors discuss the main theoretical concepts supporting this research. In section 3, the results of this research are presented. In section 4, the authors discuss solutions and recommendations for the issues addressed, namely urban tourism management in a (post) Covid period. In section 5, the researchers present the major conclusions.

BACKGROUND

The relevance of urban areas has been much discussed and analysed, as they are considered the main drivers of regional development and concentrate much public and private investment. With 74% of people living in urban areas in Europe (United Nations, 2019), cities are experiencing constant urbanisation and transformation. Therefore, one of the critical challenges that urban planners and researchers face is how to create sustainable cities that are planned around people's needs while reducing the impact of the use of resources and pressure on the environment (Frantzeskaki & Rok, 2018; Köhler et al., 2019).

Although in the last few decades, the concept and objective of sustainability have influenced many urban policies, namely those oriented by an ecological perspective, new challenges keep emerging. Within these new urban dynamics, notably in the realm of the environment and ecology preoccupations, a strong interest in looking at cities as sites of transition (Bulkeley et al., 2010) has developed. Consequently, theoretical apports linked to sustainability support this research, including the degrowth perspective (Schneider et al., 2010; D'Alisa et al., 2015; Kallis & March, 2015; Cosme et al., 2017) and the limits that economic systems have to grow on a planet with finite resources (Steffen et al., 2015). The sustainability transitions approach considers changing societal behaviour, especially in what is connected with tourism and consumption, and their urban and geographical sensitivity. Nowadays, that is a vital research concern (Köhler et al., 2019). The discussions on the post-modern city have been going on for a long time now (Harvey, 1989), highlighting the role of tourism-related activities to revamp de-industrialising urban economies related to neoliberalism and unlimited growth.

Moreover, various urban dimensions are in a constant transformation, being influenced by decisions taken at different levels and, also, by unexpected phenomena. The COVID-19 crisis is one of these, calling for multiple and urgent global urban challenges that must be associated with the Sustainable Development Goals (Vardoulakis et al., 2020; Gössling et al., 2021; Rather, 2021; Soliku et al., 2021). Many countries banned travel, closed borders, or introduced quarantine periods that profoundly changed daily lives and routines. International, regional and local travel restrictions had a direct and enormous impact on national economies, significantly affecting the tourism systems. This impact was especially relevant in urban economies that were dominated by tourism.

Tourism was severely missed in many places, mainly where the pre-COVID-19 urban political and scientific debate very much focused on touristification, gentrification, and real estate internationalisation in several cities.

Touristification is when a place becomes an object of tourist consumption, adapting its main characteristics and economic tissue to visitors, often neglecting the social, economic, and cultural problems that affect the local inhabitants (Ashworth & Page, 2011; Cochrane, 2020). The intensity of changes associated with touristification includes increased housing prices, increased carbon emissions, and the degradation of local sites, among others. As tourism becomes cheaper than ever in the past (especially for travelling and accommodation), and the technological revolution offers more accessible ways to travel with apps and sites, many authors identify the touristification of cities as one of the significant urban problems (Shaw, 2018; Aall & Koens, 2019; Fields & Rogers, 2019; Irigoy, 2019; Montezuma & McGarrigle, 2019; Zhang et al., 2019).

As the number of tourists in a given urban area continues to grow, many authors defend that there are too many visitors in several cities, assuming that they face overtourism problems (Cocola-Gant, 2016; Koens et al., 2018; Aall & Koens, 2019). The World Tourism Organisation (2022) defines overtourism as the impact of tourism on a destination that excessively influences citizens' perceived quality of life and quality of visitor experiences negatively. It is often presented as opposed to sustainable tourism. It includes pressures placed on local resources and facilities due to growth in numbers, changes in culture and loss of authenticity, being also associated with feelings of irritation and annoyance due to the presence of tourists (Goodwin, 2021). It is not easy to measure "how much is too much". However, overtourism is not confined to big cities, peak season or great tourist destinations. Instead, it is measured by the perceptions of different stakeholders (Singh, 2018).

As tourism pressures on urban spaces became more and more perceived, protests and anti-tourism feelings have developed in several cities, leading to some tourismphobia movements (Milano, 2018; Postma & Schmuecker, 2017; Pérez-García & García Abad, 2018; Simas et al., 2021). According to Milano (2018, p. 1), tourismphobia is "a feeling of rejection towards tourism that manifests in the form of assaults to restaurants, businesses and yachts; attacks on tourist buses, bikes damaged in tourist spots, and other acts of vandalism".

Both overtourism and tourismphobia have highlighted the discussion regarding the necessary balance between tourists' rights to travel and residents' rights (Perkumienė & Pranskūnienė, 2019). In Portugal, this debate has focused on the need to avoid transforming historic centres into places that expel residents and transform neighbourhoods into "tourist only" places, with much work on overtourism and residential and functional gentrification (Mendes, 2016; Tulumello, 2016; Fernandes et al., 2018a; Carvalho et al., 2019; Chamusca, 2021).

One of the leading factors in the rise of touristification and gentrification in Portugal – as in several cities across Europe – is the short-term rental market, namely Airbnb (Fernandes et al., 2018b; Chamusca et al., 2019; Fernandes et al., 2019; Cocola-Gant et al., 2021). The short-term rental (STR) market has grown exponentially in Lisbon and Porto, leading to solid growth of the floating population (Carvalho et al., 2019) expanding to other cities and regions (Fernandes et al., 2019). According to the National Registration of Tourism in Portugal, the legal registration of new STR went from 1555 registrations/year in 2010 to 13089 in 2019. In the same period, the number of new travel and tourist agencies per year went from 0 to 288 and the new tourist entertainment agents from 241 to 940.

While the real-estate prices were increasing, gentrification as a process of urban change and displacement driven by tourism became more important. Residential gentrification often goes hand-in-hand with functional change, i.e., the modification in the structure of commercial activity in a neighbourhood (Helbrecht, 2017). New users signal market demand for new sorts of urban services vis-à-vis previous ones, eventually leading to functional polarisation and market-driven disappearance of traditional shops and restaurants (Burnett, 2016), the latter being replaced by post-modern and hybrid solutions, or, in the case of tourism-led variants, "just another" guesthouse, coffee-shop or souvenir boutique.

Therefore, the management of the process of change is essential. Planning, too, the authors sustain. Thus, it is fundamental to analyse these dynamics and their consequences (overtourism, tourismphobia and gentrification), including the role of the COVID-19. Within the relevance of the management process, some international documents should be highlighted. Firstly, the 2011 UNESCO Recommendation on the Historic Urban Landscape (UNESCO, 2011). This tool, presented as a "soft-law" to be implemented by Member States on a voluntary basis, establishes a set of guidelines to integrate conservation and management of cultural heritage in cities and settlements with policies and practices for sustainable urban development. Secondly the UNWTO Recommendations on Urban Tourism. The UNTWO Secretary-General, Zurab Pololikashvili, assumes that the recent crisis "is an opportunity to rethink the tourism sector and its contribution to the people and planet; an opportunity to build back better towards a more sustainable, inclusive and resilient tourism sector that ensure the benefits of tourism are enjoyed widely and fairly". Thus, the UNTWO assumes that the way forward is based on the restart of tourism, a more resilient and sustainable one, fostered through: i) the promotion of investments that boost economic growth, digital skills, job creation and sustainability in tourism to public and private capital sources; ii) harnessing innovation and digital advances to improve inclusiveness, local community empowerment and efficient resource management; and iii) education of all the stakeholders involved (UNTWO, 2021). Finally, the Agenda for Sustainable Development. Tourism has the potential to contribute, directly or indirectly, to all objectives of the Sustainable Development Goals. It was included as a goal in Objectives 8, 12 and 14: inclusive and sustainable economic growth, sustainable consumption and production, and sustainable use of oceans and marine resources, respectively. In the 2030 horizon, the focus is on promoting responsible, sustainable, and universally accessible tourism. As one of the largest and fastest growing economic sectors in the world, tourism is well positioned to promote economic growth and development at all levels and generate income through job creation.

TOURISM, COVID-19 AND URBAN DYNAMICS IN LISBON AND PORTO

Methods

In developing this research, different methodological approaches were used to collect, analyze, and treat quantitative and qualitative information. This article examines scientific literature and policy documents related to sustainable urban development, urban policy and urban tourism. Lisbon and Porto were selected for a case study, considering that they are the two most important urban areas in Portugal, concentrating 44.5% of the Portuguese inhabitants in their metropolitan areas. Moreover, the more relevant touristic indicators registered in 2019, just before the COVID-19 pandemic, were concentrated both in Lisbon and Porto municipalities: 31.7% of tourism revenues, 30.3% of guests, and 15.9% of touristic accommodation units.

For the two cities, a detailed analysis was developed, and a variety of methods was used, namely:

- § Statistical analysis of indicators directly associated with the evolution of tourism (guests, overnight stays, profitability, among others), made available by the National Institute of Statistics and Tourism of Portugal.
- § Analysis and cartographic treatment of the evolution of STR accommodation, mainly what is associated with Airbnb internet platform available at Airdna and Inside Airbnb.
- § Evaluation of mobility patterns and their evolution, considering the road congestion data provided by the Tomtom Traffic Index.
- § Analysis of the demographic and social transformation, considering housing, rehabilitation and real-estate prices and other data available in the National Institute of Statistics site.
- § Analysis of policy documents related to urban management, particularly the regulation of tourist activity and pandemic management.
- § 2 open Interviews with the municipal public administration entities (written interviews, conducted by email on January 2022).
- § 161 questionnaires applied to local stakeholders on January and February 2022:
  - i) face-to-face with retailers and personal services covering different geographical areas and economic activities (42 in Lisbon, 34 in Porto). The 76 validated surveys are the result of 99 contacts, in which 20 refused to participate in the project and 3 did not complete the response in its entirety. The inquiry script is available attached. The selection of establishments to be surveyed considers the territory defined as "center" in official documents of the two municipalities and a functional survey carried out in these areas, which allowed to evaluate the number of existing establishments and the type of economic activities. In Lisbon, the representativeness by type of activity is as follows: personal items (28.6%), personal services (28.6%), accommodation, cafes and restaurants (28.6%) and other activities (14.2%). In Porto: personal items (29.4%), personal services (29.4%), accommodation, cafes and restaurants (29.4%) and other activities (11.8%);
  - ii) via email with social and cultural associations (7 in Lisbon, 4 in Porto);
  - iii) face-to-face with randomly selected inhabitants in the main streets of both cities (41 in Lisbon, 33 in Porto). These inhabitants have indicated to live permanently in the central areas and do now own or explore any economic activity within the city center.

These questionnaires were all identical, considering the following questions: i) How vital was tourism to the new urban dynamics of the city until the pandemics? ii) How do we evaluate the impact of tourism in different political, economic and social areas? iii) How do we evaluate the public authorities' strategies to manage overtourism? iv) How do we evaluate the public authorities' strategies to manage the COVID-19 pandemic? v) How do we evaluate the impact of the COVID-19 pandemic in different political, economic and social areas?

From Tourism to Overtourism

Over the last few years, the leading Portuguese cities have experienced a significant period of transformation, with new economic, social and urban dynamics, which some authors classify as a new life of the old urban centres (Fernandes & Sposito, 2013). In this transformation and economic revitalization process, tourism is identified as central. In 2019, the last year before the COVID-19 pandemic, tourism played an essential role in local and national economies. In Lisbon, tourism-related enterprises (accommodation, restaurants and similar) represented 11.8% of the total enterprises and 3.8% of the financial volume of business. In Porto, 10.2% of the enterprises were related to tourism activities, totalling 6.5% of the financial volume of business.

The increase is quite expressive and continuous between 2010 and 2019 (tables 1 and 2). Among the causes of this growth, it is essential to mention, first, the importance of the international airports of Lisbon and Porto (Francisco Sá Carneiro). More and more airlines, connections, and low-cost flights serve these well-located facilities near the city centres. Thus, it allowed expanding the flow of passengers, arriving directly and without stopovers, from increasingly diverse destinations. The available data (INE, 2021a) show a growth of 150% in the number of passengers disembarked in Porto and 121.7% in Lisbon in this period, with the volume of arrivals exceeding 6.5 million in Porto and 15 million in Lisbon in 2019.

At the same time, the accommodation has also increased. In Lisbon, the number of tourist accommodation establishments grew 269.4%, and the accommodation capacity increased 84.8%, representing a capacity of 128.2 guests per thousand inhabitants. In Porto, the numbers are equally expressive. The number of tourist accommodation establishments grew by 292.6%, and the accommodation capacity more than doubled (113.6%), making the city capable of accommodating 107.8 guests per 1,000 inhabitants. This data is even more significant if we consider that the universe of local accommodation registered on platforms such as Airbnb is not included here. In 2018, several parishes of Lisbon (Madalena, São Nicolau, Santa Justa, Encarnação, Santo Estêvão, São Paulo, Sacramento, Sé, Santiago, Mártires, Castelo, São Cristóvão e São Lourenço, Santa Catarina, São José, São Miguel and Coração de Jesus) and Porto (Santo Ildefonso and Sé) could receive more tourists than residents only in Airbnb STR units(Fernandes et al., 2018b; Fernandes et al., 2019).

Although the average stay is close to two nights per guest, there was a significant increase in the number of guests (110.5% in Lisbon and 147.9% in Porto) and the number of nights (125.9% in Lisbon and 171.3% in Porto). In addition, there is also a general increase in guests of international origin, which corresponded to more than 79% in 2019 in the two cities. In short, there is a considerable growth (of 184.7% in Lisbon and 246.9% in Porto over the last decade) of the tourism income, surpassing the billion euros in Lisbon in 2019.

Despite being economically very interesting – at least in the STR market – this uncontrolled urban tourism growth raised the discussion on overtourism. Both cities suffered negative impacts, and conflicts between inhabitants and tourists became more common.

Table 1. Touristic indicators change in Lisbon (2010-2019). Source: Own elaboration, using INE (2021a) data

	2010	2011	2012	2013	2014	2015	2016	2017	2018	2019	Change (%) 2010-2019
Accommodation capacity (no.)	35258	35618	35890	40235	43505	47627	51627	55598	59428	65156	84,8
Accommodation capacity per 1000 inhabitants (no.)	75,1	65,1	68,5	74,7	84,8	94,4	102,3	110	117,3	128,2	70,7
Overnight stays (no.)	6190588	6419256	6789166	7454655	9008523	9999851	11066130	12553476	13184470	13985262	125,9
Tourist accommodation establishments (no.)	193	193	197	255	301	356	396	497	576	713	269,4
Average stay (days)	2,2	2,2	2,3	2,3	2,4	2,4	2,4	2,4	2,4	2,3	4,5
Guests (no.)	2840550	2856574	2949579	3176138	3824011	4166183	4567931	5218386	5510934	5980014	110,5
Proportion of foreign guests (no.)	67,5	69,6	71,4	73,5	74,4	74,7	75,7	77,9	79,2	79,9	18,4
Room income from tourist accommodation establishments (thousands of €)	270886	294391	284541	310994	389704	466970	542564	700389	791081	851171	214,2
Total income from tourist accommodation establishments (thousands of €)	366163	389437	377133	415638	504761	597729	692260	872891	965632	1042581	184,7
Rooms (no.)	16917	17027	17419	18617	20673	22159	23741	25438	26955	29306	73,2
Net bed occupancy rate (%)	48,7	50,3	52	52,3	57,8	58,7	59,8	62,8	62	59,7	22,6
Passengers disembarked at the local airport (no.)	7000921	7389662	7999861	7999861	9092464	10066374	11261085	13273328	14446684	15523044	121,7

Table 2. Touristic indicators change in Porto (2010-2019). Source: Own elaboration, using INE (2021a) data

	2010	2011	2012	2013	2014	2015	2016	2017	2018	2019	Change (%) 2010-2019
Accommodation capacity (no.)	10896	10933	11912	12905	14605	15041	16025	18296	19901	23270	113,6
Accommodation capacity per 1000 inhabitants (no.)	53,2	46,2	52,4	54,6	66,3	70,1	74,8	85,4	92,6	107,8	102,6
Overnight stays (no.)	1690918	1783781	1815157	2102481	2548591	2879833	3300011	3782858	4091975	4587237	171,3
Tourist accommodation establishments (no.)	95	96	99	128	170	164	179	239	281	373	292,6
Average stay (days)	1,9	1,9	1,9	1,9	2	2	2	2	2	2	5,3
Guests (no.)	905758	945427	952185	1090066	1294066	1459060	1638128	1876720	1996461	2245291	147,9
Proportion of foreign guests (no.)	56,6	61,1	63,4	66,5	67,6	69,2	71,8	74,4	76,9	79,1	39,8
Room income from tourist accommodation establishments (thousands of €)	61038	62011	63096	69302	85349	108830	139124	179518	209055	243871	299,5
Total income from tourist accommodation establishments (thousands of €)	84208	83851	85169	93594	111706	139524	174864	218536	252643	292135	246,9
Rooms (no.)	5198	5283	5668	5781	6707	6947	7360	8157	8806	10272	97,6
Net bed occupancy rate (%)	42,5	44,3	42,5	44,9	48,8	53,7	56,7	58,3	57,2	55,1	29,6
Passengers disembarked at the local airport (no.)	2609663	2963976	2987538	3150224	3418940	4003902	4655436	5359258	5927374	6522895	150,0

Related Urban Dynamics

The increase in tourism has made these cities, especially its urban centers, more attractive to private investment and urban rehabilitation (often supported by EU funding). Thus, real estate speculation becomes an issue, with local inhabitants and traditional retail being forced to leave, as rents become unaffordable and legislation (following the 2008 financial crisis) makes eviction much easier. Gradually, social and functional tourism-led gentrification becomes common in both cities.

The consolidation of a globalized and tourist-oriented city highlights the debate on gentrification, a widely disseminated concept as a process of urban change marked by the installation of new residents (or users), usually from creative or economically higher classes, in some areas of the cities (Glass, 1964; Zuk et al., 2018). Although it is not a new process in Lisbon or Porto, gentrification has acquired a strong rhythm and particular characteristics, associated with a solid real estate pressure, intense replacing of residents for temporary users, and traditional economic activities for services and retail that depend on tourist activity.

The effects of this touristification (with overtourism) of cities is expressed mainly in two dimensions: the diversification of users, with a relevant floating population, and the increase of real estate prices, associated with social and functional gentrification.

The first dimension can be considered connected with a process of diversification and complexification of the concept of city inhabitant. Urban analyses based on statistical data of the resident population are increasingly incapable of understanding the new urban dynamics. That has led to discussing who lives and the dual view of city residents and city users (Hutchinson, 2010), resident and non-resident inhabitants, or permanent and floating population (Carvalho et al., 2019).

In cities where tourism is relevant and where universities and internationalized companies too, an increase in the floating use of the city is evident, despite the stagnation or decrease of the resident population. That is the case of Porto and Lisbon: between 2011 and 2021, in a period of economic revitalization, Porto lost 2.4% and Lisbon 1.2% of its resident population. The heavy loss of inhabitants is associated with a history of negative migratory and natural balances that characterize almost all the first twenty years of the 21st century. The year of 2016 marks for Lisbon and Porto a positive migration balance, allowing them to finally reverse a long depopulation process initiated in the last century and register an increase in the resident population (INE, 2021a).

Furthermore, it should be noted that the growth of the floating population has outweighed the loss of residents. Thus, it is essential to note that this two-decade period is characterized by increasing and diversifying the floating population (tourists, visitors and students). It happens at a time of solid compression and temporal acceleration of space, with greater mobility and individualization of the forms of appropriation and experience of the city (Fernandes & Chamusca, 2014). The presence of university students from various countries and the traditional installation of students from different places in Portugal has been very significant. The dissemination of international exchange programmes (notably Erasmus) with the homogenisation of the various cycles of higher education in Europe (associated with the Bologna reform) has led to a significant increase in student flows. Congresses and events also played an important role in this process, while the favorable climate and the historical sites attracted the most exigent "place consumers".

The second is associated with intense pressure on land prices, especially in the central area of the cities. According to INE (2021a), sales prices increased 40.6% in the two municipalities between 2017 and 2019. In the same period, rental prices grew 24.3% in Lisbon and 27.5% in Porto. Thus, market dynamics began to create all the conditions for the improvement of buildings, without the need for public support (direct) that was needed immediately after the Lehman Brothers and national crises.

Thus, the city center is transforming with growing touristification and social and functional gentrification. Its consequences are multiple: the physical transformation of the city, with rehabilitation and renovation; an economic activity shift of the area, visible, for example, in the multiplication of terraces and the reinforcement of dedicated transport (e.g. hop-in-hop-off buses and various types of a tuk-tuk), parallel to the increase of accommodation; several traditional establishments are unable to keep up with rising prices and are forced to close, although traditional trade support programmes were approved in 2017. In their place, new shops opened, mainly restaurant and retail units associated with international capital and oriented with tourism.

The Impact of the COVID-19 Pandemic

The pandemic had a significant impact on the economy of the cities of Lisbon and Porto. The decline in tourism was considerable. Between 2019 and 2020, Lisbon recorded a 75% drop in the number of guests, 47% in the number of new hotel establishments and 80% in income associated with tourism. In Porto, the scenario was similar: 72% in the number of guests, 37% in hotel establishments and 78% in the income generated by these activities.

The pandemic containment measures defined nationally and internationally have significantly impacted the urban environment of cities. The debate on overtourism practically disappeared. At the same time, there are significant changes in mobility and movement patterns as teleworking and digital commerce increase. Road congestion becomes less frequent (figures 1 and 2). In the Portuguese capital, in 2021, people lost less 69 hours in rush hours throughout the year compared to 2019. The decrease is smaller but equally significant in Porto, with 46 hours. In 2021 the figures increased, but they are still far from 2019.

Figure 1. Lisbon congestion during rush hour throughout 2021 in comparison to 2020 and 2019
Source: TomTom Traffic Index (https://www.tomtom.com/en_gb/traffic-index/lisbon-traffic/)

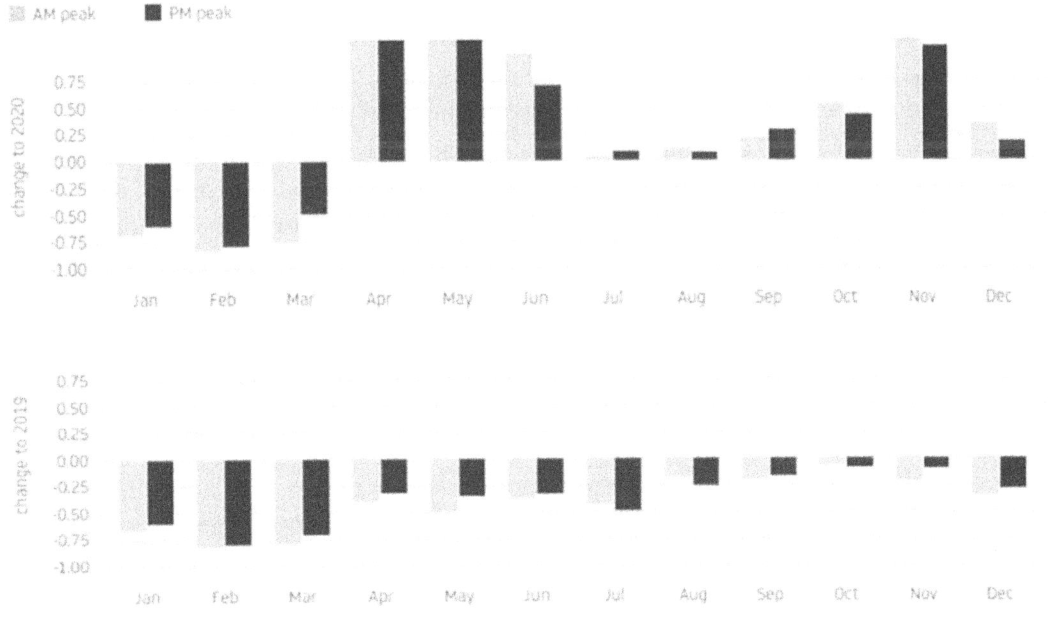

Figure 2. Porto congestion during rush hour throughout 2021 in comparison to 2020 and 2019
Source: TomTom Traffic Index (https://www.tomtom.com/en_gb/traffic-index/porto-traffic/)

Public Policies and City Management

Before the pandemic, the debate on the need to regulate tourism accommodation, particularly local short-term rental (STR), gradually intensified. Problems of overtourism and touristification were associated with the rising of the real estate prices and consequent evictions of residents and the closure of traditional shops.

The growing demand for real estate has determined a sharp rise in rents and the growth of precarious housing situations, forcing the Portuguese government to legislate. Through Law No. 62/2018 of August 22, the national government has assigned new powers to municipal bodies, namely the capacity to regulate the installation of new local accommodation establishments in places where their presence could be considered excessive: i) to preserve the social character of neighbourhoods and places; ii) to defend the stock of permanent housing; iii) to prevent gentrification processes; iv) to promote affordable rentals.

At an early stage, the two cities Mayors' speech was one of denial. In Lisbon, Fernando Medina, a social democrat (PS), said he did not know what "too much tourism" was. In Porto, liberal Rui Moreira (leader of an independent group) chose to catalogue gentrification as a "boring idea of the reactionary left". The growing protests and the evident signs that tourism had severe impacts on many people's lives forced the two municipalities to recognize later the need to act.

A municipal study in Lisbon concluded that there were areas of the city where one in four houses were already occupied by short-term rental accommodation. Based on that evidence, a political agreement was celebrated between the two left-wing parties in power at the local government (Socialist Party [PS - Partido Socialista] and Left Block [BE - Bloco de Esquerda]), and a Municipal Local Accommodation

Regulation was approved in 2019. Based on this regulation, two types of containment areas were created.: areas of absolute containment that correspond to homogeneous tourist areas with a ratio between STR establishments and the number of housings of 20% or more; relative containment areas that correspond to tourist areas where this ratio is between 10% and 20%. Within this criterion, a full suspension of new STR units was imposed in the central area of the city (Bairro Alto, Madragoa, Castelo, Alfama, Mouraria, Colina de Santana, Baixa, Avenida da Liberdade, Avenida da República and Avenida Almirante Reis), and restrictive measures applied to Graça, Colónias and Arroios.

In Porto, another study (also commanded by the City Council) would determine the need to limit the opening of new STR units. At the end of 2019, a Municipal Local Accommodation Regulation was approved. This regulation determines two different areas: the sustainable tourist areas, where the STR pressure (ratio between STR and housing) is under 25%, where no restrictive actions would be taken; the conditioned tourist areas where the STR pressure is over 25%, where several restrictive actions would be adopted, depending on if they conditioned containment areas (STR pressure over 50%), preventive containment areas (STR pression between 37,5% and 50%) or compromise containment areas (STR pression between 25% and 37,5%). Most of these areas are located within Porto's historical centre of downtown.

Despite this regulatory effort, the two cities presented lighter solutions than those approved by several European cities to mitigate the problems associated with overtourism and the massification of STR. For example, Amsterdam stopped its promotion as a tourist destination. Madrid decided to set a limit of 10,000 apartments in STR. Barcelona limited the number of permits in some areas of the city and set up a supervisory office. In Berlin, the initial ban on local accommodation was replaced by a limitation of 90 days a year per apartment per owner.

The COVID-19 pandemic would introduce some changes. In Porto, just two months after the pandemics first case was signalled in Portugal (on March 2020), the Municipal Local Accommodation Regulation was suspended for six months. In an interview to a national newspaper (Publico, 23rd March, 2020) the councillor of economy, tourism and commerce, Ricardo Valente, assumed that this suspension did not intend to deregulate the activity of STR but to translate, in practice, the understanding that public policy has to adapt to the circumstances. It added that this measure involves removing market constraints and not allowing individual entrepreneurs operating the Local Accommodation business (which in Porto account for 48% of the market) to see their taxation increased by 43%, as stated by the 2020 State Budget. Currently, the municipality is working on a new regulation proposal with a new definition of areas of tourist sustainability.

In Lisbon, in December 2021, left-wing parties approved an expansion of the suspension of new STR units to the entire municipality until the amendment to the municipal regulation of this activity came into force. The pandemic did not determine the suspension of the regulation adopted in 2019. However, with the change of power - in the elections of September 2021, the governance passes to the PSD (centre-right) - a new process for revising the municipal regulation started.

For the post-pandemic period, both municipalities consider it essential to regulate STR and tourism-related activities without damaging their positive effects on local economies. However, in Lisbon, the recently approved temporary suspension of new STR units is considered very harmful by the new Mayor. The Lisbon authorities argue that it is very damaging to the local economy and many families living from these incomes. The vision for 2030 is based on growth in tourism, with more excellent supervision in cases of disrespect by local communities, articulating this growth with a robust housing policy for the affordable rental market. Thus, sustainable growth of tourism activity should be defined and public investment strengthened. The crucial areas for public policies and investment refer to supporting the local economy and residents, ensuring good access to the housing market and various goods and services of education, health, and culture.

In Porto, the vision for 2030 considers tourism a crucial factor for the city's future, highlighting its role in the recent recovery and creation of a new dynamic. It recognizes the need to identify areas of sustainable tourism development, highlighting the strong commitment of the municipality to increase the market for affordable housing. To this end, several projects are being developed under the Recovery and Resilience Plan approved by the Portuguese government and EU, while several initiatives are being developed to strengthen animation, economic boosting, and sustainable mobility.

Stakeholders' Perception and Evaluation

Tourism had a significant impact on the economic and social revitalization of the cities of Lisbon and Porto over the last decade, accounting for much of the new dynamics observed (figure 3). That is the dominant opinion of the various local stakeholders, regardless of the group (economic agents, association movement or citizens). There is greater scepticism in Porto, especially on the part of the inhabitants, where almost 20% believe that tourism had a reduced or no importance at all in the transformation of the city, considering that this was mainly due to the effort of local residents and activities.

Figure 3. Stakeholder's evaluation of tourism importance for the new urban dynamics
Source: Own elaboration

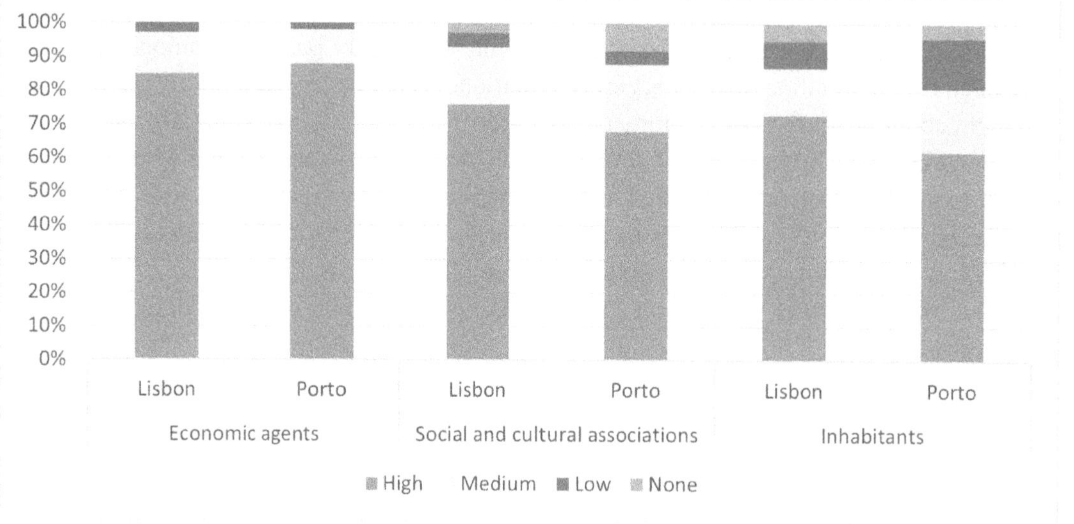

In assessing the effects caused by tourism in different areas, there are slight differences between the two cities (figure 4 and 5). Although the local agents of Lisbon and Porto identify common negative externalities (in the housing market and the identity and authenticity of the cities) and in the advantages (a boost to the local economy, more strength in security), the assessment of the governance context is quite different. In Porto, there is a positive evaluation of the relationship between tourism and city management, with a clear recognition that the city council has managed to capitalize on the city's attractiveness and thereby economically boost the city and solve some of the problems of abandonment and degradation that existed. The most critical are economic actors, especially the smaller ones, who criticize weak support for ongoing economic speculation. In Lisbon, there is a less favourable evaluation. Local stakeholders differ in the assessment, with the majority claiming that the government has not adapted to the growth of tourism and failed to manage much of the problems or leverage economic returns.

Figure 4. Stakeholder's evaluation on the tourism impact (Lisbon)
Source: Own elaboration

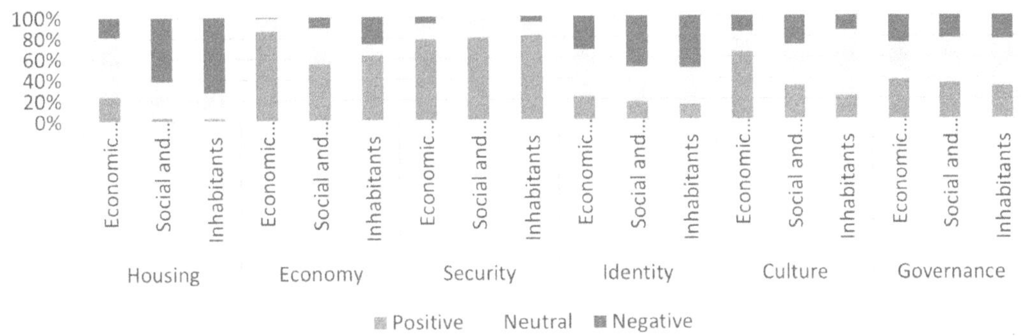

Figure 5. Stakeholder's evaluation on the tourism impact (Porto)
Source: Own elaboration

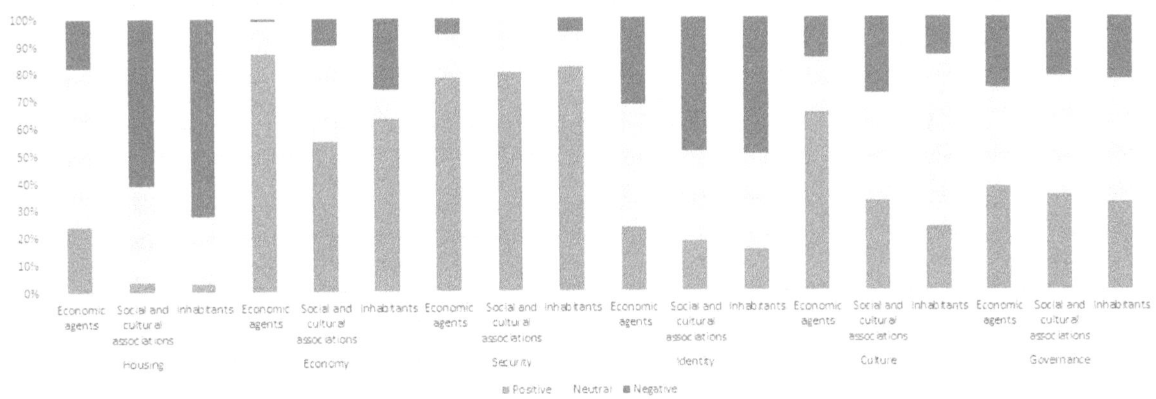

When asked to evaluate the strategies adopted to contain the growth of tourism (figure 6), the result is similar. The evaluation is more favourable in Porto, gaining greater prominence in classifying the strategy in Lisbon as inappropriate. For pandemic management (figure 7), the evaluation is generally favourable. All groups of local agents consider that the context was complicated and that the two municipalities sought to reconcile health issues with those of a social and economic nature. The most critical are the residents of Lisbon, who consider that the city council was absent. In both cities, the communication strategy was praised, stressing a very close contact with city council staff, especially with retailers and their associations.

Figure 6. Stakeholder's evaluation on the overtourism management
Source: Own elaboration

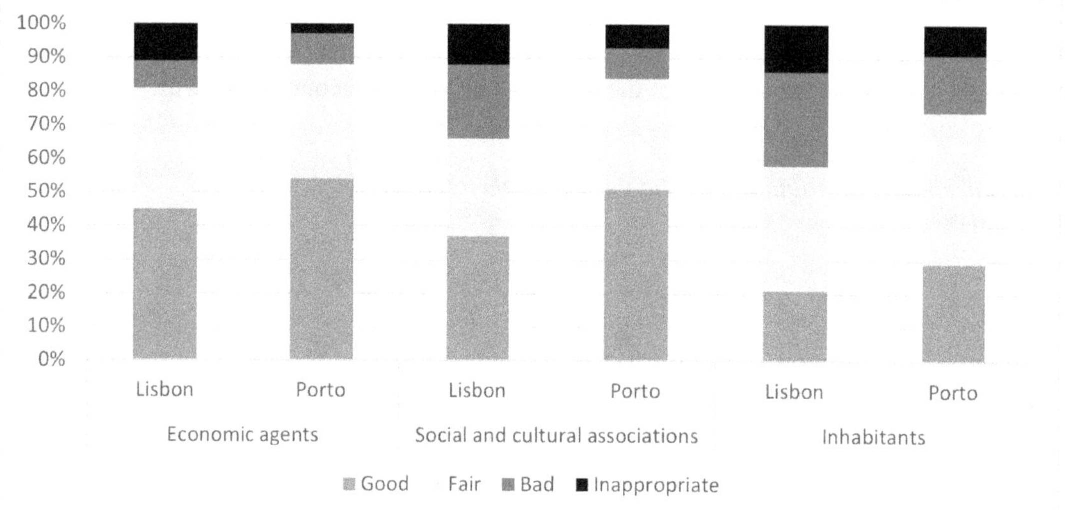

Figure 7. Stakeholder's evaluation on the COVID-19 management
Source: Own elaboration

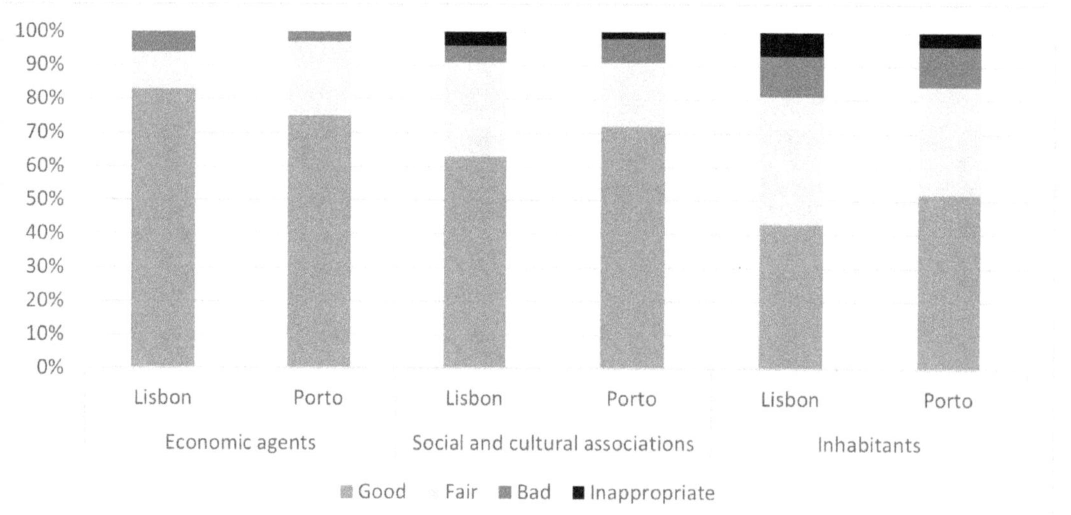

Finally, in assessing the effects of the pandemic (figures 8 and 9), the evaluation coincides with the identification that COVID-19 was highly detrimental to the local economy and the housing market. There is also the widespread feeling that this health crisis has made it possible to introduce some limitations to tourism and reinforced some mechanisms and strategies for social support. The growing attention on the housing (namely long-term renting) problems is especially noted by local inhabitants.

Figure 8. Stakeholder's evaluation on the COVID-19 impact (Lisbon)
Source: Own elaboration

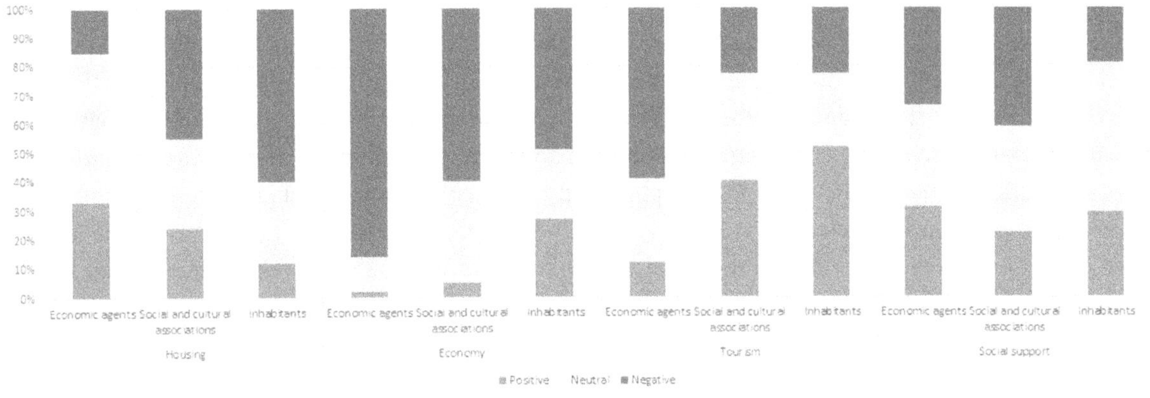

Figure 9. Stakeholder's evaluation on the COVID-19 impact (Porto)
Source: Own elaboration

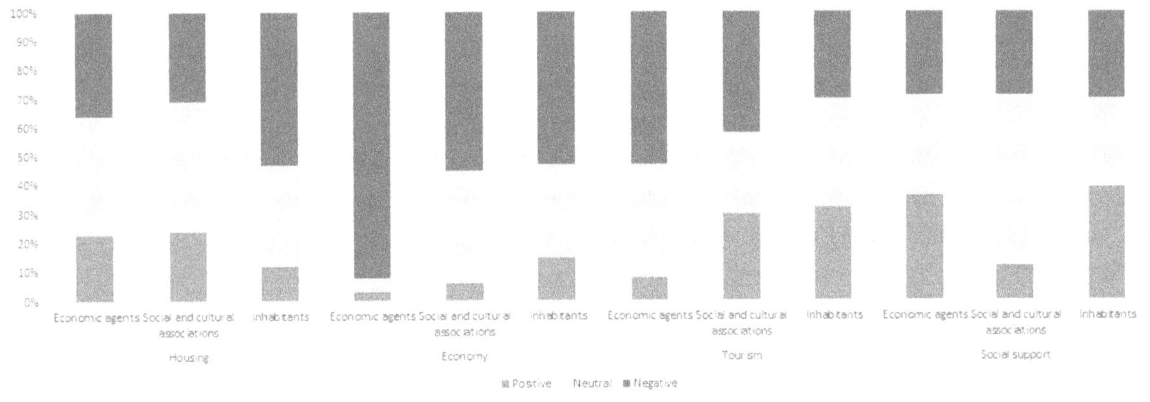

PROPOSALS AND RECOMMENDATIONS

There is no perfect formula for managing tourism growth in a specific territory. In addition, the difficulties in understanding when the tourist activity is too much, and its adverse effects outweigh the positive externalities, hampers the entire planning and management process. Furthermore, measuring tourism depends on people's position in society (a resident, an investor, a tourist) and the view you have for the future of a specific city. The presented data clearly shows this divergence, with more criticism in Lisbon, despite sharing the same dynamics and problems.

Even so, this study allows us to identify some of the impacts associated with the unregulated growth of tourism. Furthermore, it is also possible to identify that the public manager's vision does not always coincide with the interests and expectations of the different local agents, namely entrepreneurs, associative movements and civil society. The development of a collaborative and strategic planning process looks decisive in urban management for territorial development and the quality of life for all city users. The various dimensions of urban life – environmental, economic, social and cultural – are interwoven, and success depends on integrated approaches. Thus, physical urban renewal or tourism management must be combined with actions that promote education, economic development, social inclusion and environmental protection.

In Portugal, several of the Sustainable Development Goals of making cities and human settlements inclusive, safe, resilient, and sustainable is essential. In particular, urban planning for sustainability should focus on promoting sustainable urban transport systems, enhancing inclusive and sustainable urbanisation and capacity for participatory and integrated planning and management, protecting and safeguarding the world's cultural and natural heritage and supporting positive economic, social and environmental links between urban, peri-urban and rural areas.

The authors believe that collaborative and strategic planning should address five essential topics: infrastructures and decentralisation; local involvement and capacitation; spatial planning and regulation; smart tourism; and tourist-resident interaction.

In the first topic, improving services to a growing floating population is vital, assuring that destinations can accommodate the growing number of visitors without compromising their safety, cleanliness and identity. Therefore, urban management should be able to favour the connection and cohesion of the different parts of the urban area/region, making more extensive parts of the urban area attractive to visitors by reinforcing the attractiveness of new points and articulating them.

The second dimension is at the core of collaborative planning. Public-private collaboration should be improved while creating strategic partnerships and networks that consider all the urban dimensions instead of focusing on economic revenues only. Therefore, social networks and community resilience should be strengthened by decisive participation in the planning process: listening to communities' expectations and needs; fostering cooperation among different stakeholders, with effective involvement in the decision-making processes; promoting a sense of belonging by the residents and a good relationship with visitors; and empowering the local stakeholders' skills.

Planning should be associated with regulation to identify concrete measures to control tourism growth. Quantitative and qualitative measurement of tourism is needed by means of indicators and impact evaluation. That can be translated into short-term accommodation regulations, tourist taxes, visitor limitations, landscape protection, or determining legal minimums for the regular renting market or visits to specific places.

Strategies addressing smart tourism refer to planning and managing approaches that develop systems for collecting information on destinations' social capacity. In short, it consists of creating measurement and evaluation mechanisms associated with political and scientific studies that promote a strong identification of tourism and its impacts. This dynamic database will help develop a more digitalised and responsible economy while informing planners of the immediate impacts of the measures and programs they implement. That is much needed in both Lisbon and Porto.

Finally, the tourist-resident interaction is vital to assure that sustainable urban development is promoted with no conflicts or tourismphobia related problems. Therefore, cities should address the conflicts inherent to tourism negative impacts adequately. Thus, one possibility is that local public entities should develop initiatives and collaborative spaces for dialogue and interaction between all the city-users, educating visitors about their impact on destinations and sharing cultural and social practices.

Through these initiatives, the authors believe that it would be possible to develop a more integrated and participatory planning, considering all the dimensions and impacts (positive and negative) of tourism in urban areas, and so manage a "city with tourists", avoiding the creation of a "city of tourists". This approach had to consider the specificities of each urban area, and monitorization and evaluation of urban dynamics and tourism externalities. To this end, all the process has to be based on local agents' empowerment and effective participation in the decision-making process, at all planning stages. City management must consider collective expectations, including the discussion of the vision for the future of the city, ensuring that the city can welcome all its users sustainably and without associated conflicts.

FUTURE RESEARCH DIRECTIONS

The number of studies and regulations addressing the impacts of tourism in specific destinations, especially those related to overtourism, has significantly increased over the last decade. Within this study, we discuss some of the actions undertaken to tackle the problem in Lisbon and Porto, focusing on its social-cultural impacts, namely gentrification, loss of residents' quality of life and the increase in social and spatial inequalities. The authors also consider the impact of the COVID-19 pandemic with an approach that is quantitative and qualitative, identifying possible measures to manage the significant dependence on tourism by Lisbon and Porto, which we believe to be innovative in the Portuguese context.

By doing this with this study, the authors want to help increase awareness of the different dimensions of impacts caused by tourism pressure and the measures that can be undertaken to manage tourism and contribute to sustainable urban development. Furthermore, we believe that the stakeholders' perception and evaluation of tourism and related public policies should be vital to all local or regional regulations, as measures should be adjusted to people's needs and expectations. Therefore, the qualitative and quantitative analysis should be further developed and associated with scenario, collaborative and strategic planning.

CONCLUSION

After going through a process of decline at the end of the twentieth century, the centres of various European cities of medium and large dimensions have recently seen new social and economic dynamics associated with the growth of the floating population, with conflicts between tourists and residents and gentrification. Tourism development brought various economic benefits and improved the physical environment of cities. However, it also raised risks regarding the diversity of the cities and their territorial identity, with several implications for urban development policy.

Portugal cities share many of these problems. The examples of Lisbon and Porto show that economic activity has changed significantly in the last decade. In addition to the critical increase in tourist activity, new economies have emerged, whether from technology-based companies with national and international capital and a wide range of cultural and leisure, such as art galleries, bars, and night-life establishments. As a whole, these transformations have brought a whole new economic dynamic to the city, especially to the city-centre, as well as a set of new users that have radically altered the physical and social environment.

The implications of these transformations are numerous. The process of residential and functional gentrification, resulting mainly from the touristification of the city, and the concomitant growth of the floating population, was slowed down by the COVID-19 pandemic. However, housing and services prices are still high, and the unavailability of housing is indeed a problem. The evaluation of local actors underlines the importance of tourism in the revitalization of cities and points to some of its problems, particularly in areas such as housing and loss of identity, with risks of standardization and trivialization of urban values. In this context, even with the slowdown in tourist activity, economic and residential replacement processes are maintained. The public authorities have not yet found any solution. The prices remain incompatible with the income of the generality of the local population, attractive only to international investment funds.

In this context, the challenge is integrating tourism in the city's global planning, as limiting the installation of STR is undoubtedly not enough. Planning should be inclusive, collaborative and strategic, considering the need to manage infrastructures, the local involvement and capacitation, the spatial planning regulations while promoting smart tourism, tourist-resident interactions and sustainable development.

ACKNOWLEDGMENT

This work was financed by national funds through FCT – Foundation for Science and Technology, I.P., on the scope of the project UIBD/00736/2020. It seeks to analyse communication, territories, organizations and social dynamics, with a special focus on the role o public entities strategies to promote well-being, territorial development and cohesion.

REFERENCES

Aall, C., & Koens, K. (2019). The Discourse on Sustainable Urban Tourism: The Need for Discussing More Than Overtourism. *Sustainability*, *11*(15), 4228. doi:10.3390u11154228

Abbas, J., Mubeen, R., Iorember, P., Raza, S., & Mamirkulova, G. (2021). Exploring the impact of COVID-19 on tourism: Transformational potential and implications for a sustainable recovery of the travel and leisure industry. *Current Research in Behavioral Sciences*, *2*. doi:10.1016/j.crbeha.2021.100033

Ashworth, G., & Page, S. J. (2011). Urban tourism research: Recent progress and current paradoxes. *Tourism Management*, *32*(1), 1–15. doi:10.1016/j.tourman.2010.02.002

Bulkeley, H., Broto, V. C., Hodson, M., & Marvin, S. (Eds.). (2010). *Cities and low carbon transitions* (Vol. 35). Routledge. doi:10.4324/9780203839249

Burnett, K. (2014). Commodifying poverty: Gentrification and consumption in Vancouver's Downtown Eastside. *Urban Geography*, *35*(2), 157–176. doi:10.1080/02723638.2013.867669

Carvalho, L., Chamusca, P., Fernandes, J., & Pinto, J. (2019). Gentrification in Porto: floating city users and internationally-driven urban change. *Urban Geography*. https://www.tandfonline.com/doi/abs/10.1080/02723638.2019.1585139?journalCode=rurb20 doi:10.1080/02723638.2019.1585139

Chamusca, P. (2021). Urban planning and policy in Portugal: An overview on the role of EU funds and guidelines. *Urban Research & Practice*, 1–22. Advance online publication. doi:10.1080/17535069.2021.1980607

Chamusca, P., Fernandes, J., Carvalho, L., & Mendes, T. (2019). The Role of Arbnb creating a "new"-old City Center; facts, problems and controversies in Porto. *BAGE*, *83*(83). Advance online publication. doi:10.21138/bage.2820

Cochrane, A. (2020). Urban policy. In A. Kobayashi (Ed.), *International Encyclopaedia of Human Geography* (pp. 93–96). Elsevier. doi:10.1016/B978-0-08-102295-5.10229-X

Cocola-Gant, A. (2016). Airbnb, buy-to-let investment and tourism-driven displacement: A case study in Lisbon. *Environment and Planning A: Economy and Space*. doi:10.1177/0308518X19869012

Cocola-Gant, A., Jover, J., Carvalho, L., & Chamusca, P. (2021) Corporate hosts: The rise of professional management in the short-term rental industry, *Tourism Management Perspectives, 40*. https://www.sciencedirect.com/science/article/pii/S2211973621000921 doi:10.1016/j.tmp.2021.100879

Cosme, I., Santos, R., & O'Neill, D. W. (2017). Assessing the degrowth discourse: A review and analysis of academic degrowth policy proposals. *Journal of Cleaner Production*, *149*, 321–334. doi:10.1016/j.jclepro.2017.02.016

D'Alisa, G., Demaria, F., & Kallis, G. (2015). *Degrowth: A Vocabulary for a New Era* (G. D'Alisa, F. Demaria, & G. Kallis, Eds.). Routledge.

Fernandes, J., Carvalho, L., Chamusca, P., Mendes, T., & Gago, A. (2018b). *O Porto e a Airbnb*. Book Cover.

Fernandes, J., Carvalho, L., Chamusca, P., Mendes, T., & Gago, A. (2019). *Lisboa e a Airbnb*. Book Cover.

Fernandes, J., Carvalho, L., Chamusca, P., & Pinto, J. (2018a). *Gentrification in Porto: problems and opportunities in the past and in the future of an internationally open city. Revista de Geografia e Ordenamento do Território (GOT), 15(Dezembro)*. doi:10.17127/got/2018.15.008

Fernandes, J. R., & Chamusca, P. (2014). Urban policies, planning and retail resilience. *Cities (London, England)*, *36*, 170–177. doi:10.1016/j.cities.2012.11.006

Fields, D., & Rogers, D. (2019). Towards a critical housing studies research agenda on platform real estate. *Housing. Theory and Society*, 1–23. doi:10.1080/14036096.2019.1670724

Frantzeskaki, N., & Rok, A. (2018). Co-producing urban sustainability transitions knowledge with community, policy and science. *Environmental Innovation and Societal Transitions, 29*(June), 47–51. doi:10.1016/j.eist.2018.08.001

García-Gómez, C., Demir, E., Díez-Esteban, J., & Bilan, Y. (2021). The impact of COVID-19 outbreak on hotels' value compared to previous diseases: the role of ALFO strategy. *Heliyon*, *7*(8). doi:10.1016/j.heliyon.2021.e07836

Glass, R. (1964). *London: aspects of change*. MacGibbon & Kee.

Goodwin, H. (2021). City destinations, overtourism and governance. *International Journal of Tourism in Cities*.

Gössling, S., Scott, D., & Hall, C. M. (2021). Pandemics, tourism and global change: A rapid assessment of COVID-19. *Journal of Sustainable Tourism*, *29*(1), 1–20. doi:10.1080/09669582.2020.1758708

Gretzel, U., Fuchs, M., Baggio, R., Hoepken, W., Law, R., Neidhardt, J., Pesonen, J., Zanker, M., & Xiang, Z. (2020). e-Tourism beyond COVID-19: A call for transformative research. *Information Technology & Tourism*, *22*(2), 187–203. doi:10.100740558-020-00181-3

Harvey, D. (1989). *The Condition of Postmodernity*. Blackwell.

Helbrecht, I. (2017). Gentrification and displacement. In Gentrification and Resistance: Researching Displacement Processes and Adaption Strategies. Springer Gachmedien Wiesbaden GmbH. doi:10.1007/978-3-658-20388-7

Hutchinson, R. (Ed.). (2010). Encyclopaedia of Urban Studies. SAGE Publications, Inc. doi:10.4135/9781412971973

INE. (2021a). *Territorial Statistics. Several indicators and years*. Available at https://www.ine.pt

INE. (2021b). *Tourism Satellite Account*. Available at https://www.ine.pt

Irigoy, I. (2019). Rent gap reloaded: Airbnb and the shift from residential to touristic housing in the Palma Old Quarter in Mallorca, Spain. *Urban Studies (Edinburgh, Scotland)*, *56*(13), 2709–2726. doi:10.1177/0042098018803261

Kallis, G., & March, H. (2015). Imaginaries of Hope: The Utopianism of Degrowth. *Annals of the Association of American Geographers*, *105*(2), 360–368. doi:10.1080/00045608.2014.973803

Koens, K., Postma, A., & Papp, B. (2018). Is Overtourism Overused? Understanding the Impact of Tourism in a City Context. *Sustainability*, *10*(12), 4384. doi:10.3390u10124384

Köhler, J., Geels, F. W., Kern, F., Markard, J., Onsongo, E., Wieczorek, A., & Wells, P. (2019). An agenda for sustainability transitions research: State of the art and future directions. *Environmental Innovation and Societal Transitions*, *31*(January), 1–32. doi:10.1016/j.eist.2019.01.004

Mendes, L. (2017). Gentrificação turística em Lisboa: neoliberalismo, financeirização e urbanismo austeritário em tempos de pós-crise capitalista [Tourist gentrification in Lisbon: neoliberalism, financialization and austeritarian urbanism in post-capitalist crisis times]. *Cadernos Metrópole*, *19*(39), 479–512. doi:10.1590/2236-9996.2017-3906

Milano, C. (2018). Overtourism, malestar social y turismofobia. Un debate controvertido. Pasos: Revista de Turismo y Patrimonio Cultural, 16(3), 551-564.

Montezuma, J., & McGarrigle, J. (2019). What motivates international homebuyers? Investor to lifestyle 'migrants' in a tourist city. *Tourism Geographies*, *21*(2), 214–234. doi:10.1080/14616688.2018.1470196

Pérez-García, A., & García Abad, L. (2018). Turismofobia: Presencia, impacto y percepción del concepto a través de los medios de comunicación impresos [Turismofobia: Presense, impact and perception of concept through the printed media]. *adComunica – Scientific Journal of Strategies, Trends and Innovation in Communication, 16*, 201–219.

Perkumienė, D., & Pranskūnienė, R. (2019). Overtourism: Between the right to travel and residents' rights. *Sustainability*, *11*(7), 2138. doi:10.3390u11072138

Postma, A., & Schmuecker, D. (2017). Understanding and overcoming negative impacts of tourism in city destina-tions: Conceptual model and strategic framework. *Journal of Tourism Futures*, *3*(2), 144–156. doi:10.1108/JTF-04-2017-0022

Rather, R. (2021). Demystifying the effects of perceived risk and fear on customer engagement, co-creation and revisit intention during COVID-19: A protection motivation theory approach. *Journal of Destination Marketing & Management, 20*. doi:10.1016/j.jdmm.2021.100564

Schneider, F., Kallis, G., & Martinez-Alier, J. (2010). Crisis or opportunity? Economic degrowth for social equity and ecological sustainability. Introduction to this special issue. *Journal of Cleaner Production, 18*(6), 511–518. doi:10.1016/j.jclepro.2010.01.014

Shaw, J. (2018, October). Platform real estate: Theory and practice of new urban real estate markets. *Urban Geography, 17*. Advance online publication. doi:10.1080/02723638.2018.1524653

Simas, T., Le Cocq de Oliveira, S., & Cano-Hila, A. (2021). Tourismophobia or touristification? An analysis of the impacts of tourism in Poblenou, Barcelona. *Ambiente Construído, 21*(3), 117–131. doi:10.15901678-86212021000300541

Singh, T. (2018). Is over-tourism the downside of mass tourism? *Tourism Recreation Research, 43*(4), 415–416. doi:10.1080/02508281.2018.1513890

Soliku, O., Kyiire, B., Mahama, A., & Kubio, C. (2021). Tourism amid COVID-19 pandemic: impacts and implications for building resilience in the eco-tourism sector in Ghana's Savannah region. *Heliyon, 7*(9). doi:10.1016/j.heliyon.2021.e07892

Steffen, W., Richardson, K., Rockström, J., Cornell, S. E., Fetzer, I., Bennett, E. M., Biggs, R., Carpenter, S. R., de Vries, W., de Wit, C. A., Folke, C., Gerten, D., Heinke, J., Mace, G. M., Persson, L. M., Ramanathan, V., Reyers, B., & Sörlin, S. (2015). Planetary boundaries: Guiding human development on a changing planet. *Science, 347*(6223), 1259855. Advance online publication. doi:10.1126cience.1259855 PMID:25592418

Tulumello, S. (2016). Reconsidering neoliberal urban planning in times of crisis: Urban regeneration policy in a "dense" space in Lisbon. *Urban Geography, 37*(1), 117–140. doi:10.1080/02723638.2015.1056605

UNESCO. (2011). *Recommendation on the Historic Urban Landscape* (36 C/ Resolution 41). UNESCO World Heritage Centre, UNESCO. https://whc.unesco.org/en/hul

United Nations. (2019). World Urbanization Prospects. The 2018 Revision (ST/ESA/SER.A/420). Department of Economic and Social Affairs. Population Division.

UNTWO. (2021). *The Economic Contribution of Tourism and the Impact of COVID-19*. UNTWO. doi:10.18111/9789284423200

Vardoulakis, S., Salmond, J., Krafft, T., & Morawska, L. (2020). Urban environmental health interventions towards the Sustainable Development Goals. *The Science of the Total Environment, 748*, 141530. doi:10.1016/j.scitotenv.2020.141530 PMID:32827895

World Tourism Organization. (2022) *'Overtourism'? Understanding and Managing Urban Tourism Growth beyond Perceptions*. WTO. https://www.e-unwto.org/doi/pdf/10.18111/9789284420070

Zhang, T., Bufquin, D., & Lu, C. (2019). A qualitative investigation of microentrepreneurship in the sharing economy. *International Journal of Hospitality Management, 79*, 148–157. doi:10.1016/j.ijhm.2019.01.010

Zuk, M., Bierbaum, A. H., Chapple, K., Gorska, K., & Loukaitou-Sideris, A. (2018). Gentrification, Displacement, and the Role of Public Investment. *Journal of Planning Literature*, *33*(1), 31–44. doi:10.1177/0885412217716439

KEY TERMS AND DEFINITIONS

Cities: Area of population concentration, with markedly urban morphological, functional, and social characteristics. They are the main areas of population and economic settlement, concentrating conditions of administrative, economic, accessibility, and symbolic centrality.

Governance: Territorial management model in which the planning, management and action functions of democratically elected governments are complemented by networks and partnerships with the private sector and civil society, promoting their empowerment and effective participation in decision-making processes.

Inhabitants: A group of people who permanently reside in a city or territory.

Local Stakeholders: A group of actors, individual or collective, who, due to their connection to the city (as residents, entrepreneurs, workers, members of associations, or other), are part of the process of territorial transformation and can participate in decision-making processes.

Planning: A set of actions that, starting from the analysis of reality and the definition of some objectives to be achieved, supposes the use of a plan elaborated with scientific, technical or political criteria, and its application through a opportunely organized procedure.

Public Policies: A public policy is an action promoted by public entities, with a sense and intention that involves a complex set of decisions and operators, and a planned, non-casual decision-making process that takes place in time and aims to achieve goals through certain averages. It presupposes conflicts, compromises, coalitions, and convenient transactions.

Sustainability: Balanced model of management and development of territories and societies. It presupposes a rational use of existing resources, ensuring the balance of the environment and the development capacity of future generations.

Tourism: A specific type of leisure characterized by moving from one's usual place of residence and involving a minimum duration, unlike other recreational activities of a domestic nature, usually daily.

Chapter 12
The Necessary Digital Transformation for a Post-COVID-19 Scenario of Mérida's Roman City

Alba Marín
https://orcid.org/0000-0003-0285-7086
Universidad de Seville, Spain & Universidad de Extremadura, Spain

ABSTRACT

This chapter addresses the importance of the Roman Archaeological Ensemble of Merida as a tourist destination and the necessary revitalisation of the tourism sector through digital innovation. Despite the different policies to reactivate the sector implemented in the post-COVID-19 scenario, the authors consider that there is still much work. They propose how digital technologies play a crucial role in the conservation and dissemination of cultural heritage, and they are committed to mobile and augmented narratives as an ideal strategy for urban archaeological tourism. Throughout the chapter, they review key notions that articulate the essential elements of the tourist experience and comment on previous projects that could serve as a starting point.

INTRODUCTION

In the postCovid-2019 scenario, the city of Mérida has just been recognised Smart Tourism Destination, which means a package of support to turn the city into a smart destination. The Group of World Heritage Cities (UNESCO) in Spain developed a project to revive cultural tourism with the Braintrust Tourism Observatory. Mérida is one of the 15 cities that make up the Group of World Heritage Cities and offers one of the most relevant Roman sites in Europe. As a Roman archaeological site, it has been considered a World Heritage City since 1993 and has the most important Roman art collection in Spain in its National Museum of Roman Art. The urban heritage complex includes the Theatre and Amphi-

DOI: 10.4018/978-1-6684-3369-0.ch012

theatre, the Amphitheatre House, the Temple of Diana, the Portico del Foto, the Arch of Trajano, the Alcazaba, the Roman Circus, the Aqueduct of Miracles, the Archaeological Complex of the Morerías, the Baths of San Lázaro and the Proserpina Reservoir, the largest reservoir built by the Romans in the whole of the Mediterranean area.

The tourism reactivation project aims to provide the cities that make up the group with strategies and tools to face the post-covid 19 future. One of these lines is the increase of digitalisation for tourist consumption. The project results suggest that the digitalisation of the tourism value chain is essential for the development of Mérida as a Smart Tourism Destination. For this reason, the city has been trying to become a smart destination for some years. At the present time, with funding packages of millions of euros and following lines and ideas for activities such as the creation of a virtual tourism office, the development of augmented reality works in heritage spaces, the technological surveillance systems, the photogrammetry in monuments, the installation of digital screens, the creation of virtual brochures and interactive maps, among other ideas.

In the following pages, the authors review the city of Mérida, the public policies of the tourism sector at a national and international level, the post-pandemic context and the creative options for the digitisation of heritage and urban tourism. To do so, the reader will get to know the city geographically and contextually, since we start from the assumption that, although the history of Mérida is of international relevance, it is not particularly well known. Following, it delves into the phenomenon of Smart Tourist Destinations and what it means for the tourism sector in Spain. Next, the text focuses on another of the elements we articulate: the digitisation of cultural heritage and the need to attract and entertain the tourist and museum visitor, passing through exciting projects that enrich our narrative. Finally, the ideas converge, once again, in the field of the Roman City of Mérida.

BACKGROUND

A Roman Archaeological Complex

In Extremadura, there are four main tourist areas with different visitor flows: Cáceres and its surroundings, Badajoz, Mérida and its surroundings, Plasencia and the north of Extremadura, Zafra and the south of Extremadura. Mérida is the capital of a region with a surface area of 41.635 km^2, the second largest in Spain. This large area means that the geographical configuration has different visitors depending on the tourist areas.

The growth of tourism was wildly uneven in Spain, with some areas, such as Extremadura, whose integration into the tourist market was slow and timid, in contrast to what was happening in some coastal regions where the travel industry developed more prematurely (Campesino et al., 2010). For this reason, while the coastal areas have, since the second half of the 20th century, attracted a large part of domestic and international tourist flows, and became the model for mass tourism, the inland territories, except for the state capital, had very modest flows of travellers. (Rengifo-Gallego & Sánchez-Martín, 2021)

The city of Mérida is in the north of the province of Badajoz and also has a considerable extension (865 km2) with a population of 59,335 habitants (INE, 2020). Its main tourist attraction is its history, founded by Octavian Augustus in 25 BC. It was the capital of the ancient Lusitania, an enclave of great importance throughout history, from the Visigothic period to the Catholic Monarchs. This roman archaeological site is internationally renowned and the city also has the most extensive collection of Roman art in Spain in the *Museo Nacional de Arte Romano.*

Mérida has essentially urban tourism focused on the interest of its historical heritage. Urban tourism appeared in the 1980s as a complex, dynamic phenomenon with great economic potential. However, at the crossroads of two concepts, the authors must understand the nature of tourism in Mérida: urban tourism and archaeological heritage. The World Tourism Organization defines the tourism product as:

The combination of tangible and intangible elements, such as natural, cultural and anthropic resources, as well as tourist attractions, facilities, services and activities around a specific element of interest, which represents the essence of a destination's marketing plan and generates a holistic tourism experience, with emotional elements, for potential customers. A tourism product is marketed through distribution channels, which also set the price, and has a life cycle.

Urban tourism activity takes place in an urban space, where a series of cultural, architectural and social experiences and products are concentrated, which together make up the tourist offer (UNWTO - The World Tourism Organization). Tourism in this city is not only urban tourism but archaeological urban tourism; of which cities such as Rome and Athens are international benchmarks. In other words, archaeological sites go from being of scientific interest to being of tourist interest, to being part of the public and museum sphere (Morère, 2007). For Moreno-Melgarejo and Sariego-lópez (2017), archaeological tourism can be considered a set of tourist trips that aim to visit archaeological resources. And these archaeological resources, in order to be touristic, have to be prepared as such, including explanatory information for the tourist about the points of interest: guided tours, tourist applications explaining the archaeological heritage and didactic information.

POSTCOVID TOURISM SCENARIO, CULTURAL-HISTORICAL HERITAGE AND DIGITISATION

Tourism has a great weight in the Spanish economy, so much so its contribution to GDP in recent decades is around 12%. In 2019 Spain closed a cycle of growth that placed it at number 2 in the ranking of countries by international tourism revenues according to the World Tourism Barometer (2020) of the World Tourism Organization. This trend was drastically cut short at the onset of the pandemic, as in the rest of the world economy. As a result, the sector suffered falls of -28% in the first quarter and -95% in the second quarter of 2020, according to the data collected in Romero-Dexeus, C., & Paton, J. (2020). The percentage contribution of tourism to the GDP of the Spanish economy fell to 4.3%, a drop of more than 8 points, making tourism one of the sectors most affected by the crisis. International comparative studies, including the IMF's External Sector Report (August 2021) and the European Economic Forecast (November 2021), pointed out that the different dependence on tourism in different countries was one of the factors explaining the asymmetrical impacts of Covid on different territories. The Spanish economy has been particularly affected, since in 2020 it experienced the most significant fall among the Eurozone countries, precisely because of the singularly relevant weight of tourism in our economy. For this reason, any approach to a solid and sustainable recovery of the Spanish economy depends on recovering the potential of tourism.

The latest report published *Innovación turística y especialización inteligente en España. Palancas imprescindibles para la recuperación* (Fomeros-Dexeus, & Paton, 2020) provides a detailed review of the competitive situation of the tourism sector at an international, national and regional level, as well as a structural analysis of the sector and innovation in Spain. The report also shows the identification of the technologies used by regions and the areas of technological convergence. Finally, the report points to tourism innovation and digitalisation as one of the main active elements for the next steps.

In this scenario, public policies to boost tourism, which was already being developed before the pandemic, take on particular relevance. Following the pre-pandemic trends of transforming and improving the sector, these policies take on a special dimension in the current context, focusing now on the recovery and revitalisation of this precious sector. The country had a highly competitive position as a global tourist destination and planned to promote tourism. Different strategic plans:

- *Plan Turismo 2007* (Ministerio de Industria, Turismo y Comercio). On the need to permanently reinvent tourism products.
- *Plan Nacional e Integral de Turismo* (Secretaría de Estado de Turismo, 2012). Promotion of tourism products to diversify the offer.
- *Plan Turístico de Extremadura 2017-2020* (Junta de Extremadura 2017). Stresses the need to promote the creation of new tourism products.

The Plan de Recuperación, Transformación y Resiliencia, (PRTR) is the Spanish strategy designed to channel the funds earmarked by Europe to repair the damage caused by the Covid-19 crisis. The Government channels the investment following a series of urgent measures to modernise the Public Administration and implement this recovery plan through calls for tender, subsidies, and aids from the different ministries and other public agents. Within this Recovery Plan, we find the *Componente 14: Plan de modernización y competitividad del sector turístico*. The published call is part of line 2 of investment: "Programme of digitalisation and intelligence for destinations and the tourism sector", whose objective is to implement actions to promote the digitalisation of tourism destinations and companies. The call for projects for the scientific community aims to acquire knowledge that will help respond to the challenges of tourist destinations and implement a centralised platform for smart destinations. One of the main final objectives of the plan is precisely the creation of an integrated and smart destination platform, which provides interoperability to the supply of public and private services in the sector. A Tourism Intelligence System and the Spanish Network of Smart Destinations promotion complemented all of this. In short, a line of investment of millions of euros to provide solutions, renew and recover the sector.

Smart Tourism Destination (DTI in Spanish) Phenomenon

According to López de Ávila & García (2013), Smart Tourism Destinations (STD) are defined as innovative consolidated spaces with completing technological infrastructure. The focus on innovation, technologies and sustainability underpin the foundations of these destinations, complemented by the notions of participation and governance. Segittur (a Spanish state company dedicated to the management of innovation and tourism technologies) proposed its definition that was approved by Subcommittee 5 of DTI of the AENOR Technical Committee for Standardisation AEN/CTN 178 on Smart Cities:

A smart destination is an innovative tourist space, accessible to all, consolidated on a cutting-edge technological infrastructure that guarantees the sustainable development of the territory, facilitates the interaction and integration of the visitor with the environment and increases the quality of their experience in the destination and the quality of life of the residents. (López de Ávila & García, 2013)

Technology is one of the fundamental axes that articulates the structure of the STD; technology appears in all the links of the tourism value chain, becoming a key factor of competitiveness and smart governance. On the other hand, innovation is another of the fundamental axes of the STD structure and of the regeneration of the tourism sector, which is inescapably linked to the digital environment. The recently published Call of 30 December, approving the regulatory bases for the programme of aid to research and knowledge dissemination organisations for innovation and development projects to respond to the challenges of tourist destinations, and approving the call for applications for the 2021 financial year, within the framework of the *Plan de Recuperación, Transformación y Resiliencia*, shows the growing national interest in the revitalisation of the sector and the funding earmarked for this purpose within the budgets. In line with the objectives of Component 14 of the PRTR, the digitisation of the tourism sector is key to the attractiveness of tourism in the context of an increasingly hyper-connected society. One of the challenges this component address is the Digitisation and Tourism Intelligence Programme, which builds on the importance of digitisation for attracting tourists. In addition, the Recovery, Transformation and Resilience Plan, Component 14, (2021) indicates in this respect that "it is one of the areas where the margin for improvement is widest, as Spain ranks 27th in the "ICT readiness" section according to the World Economic Forum's report on tourism competitiveness".

The number of destinations in the Network has grown by 3 in recent years and constitutes an international reference space for the promotion of digitisation, innovation, accessibility or better governance of destinations, being a platform for the dissemination of good practices and the generation of knowledge. (Plan de Recuperación, Transformación y Resiliencia, Componente 14, 2021, p.6)

Line 2 of this component aims to "implement actions to promote the digitisation of tourist destinations and tourism businesses", with the Digital Transformation of Tourism Destinations being the main objective of sub-measure 1.

Mérida's Bid to become a Smart Tourist Destination

Before the pandemic, Mérida was ranked 2nd in Extremadura as a whole regarding the number of travellers and overnight stays. (Extremadura Tourism Report, 2021). In the last decade, the commitment to the development of tourism and the trends of the last few years in public policies and the challenges achieved by the city can be seen:

- Mérida as *Ibero-American capital of gastronomic culture* in 2016
- Mérida Holy Week as *Fiesta de Interés turístico Internacional*
- Organisation of the sector in Extremadura: Tourism Cluster in Extremadura (2009)
- Specialised training: Escuela Superior de Hostelería de Mérida, National Reference Centre for Agrotourism since 2010.

- Consolidation of the network of guest houses and centralised management by the public company Genidexa.
- Network of Protected Areas of Extremadura (Mérida).
- Increase in the number of *Bienes declarados de Interés Turístico Cultural* in Mérida.

Since 2018, the city has been developing the Smart Tourism project, an initiative linked to the National and Integral Tourism Plan 2012-2015, whose objective was to generate mechanisms to facilitate the incorporation of innovations in destinations. Furthermore, as part of the *Grupo Ciudad Patrimonio de la Humanidad en España*, the reactivation of the cultural tourism project was created together with the Braintrust Tourism Observatory. Its results point to the need to digitalise the tourism value chain for the Development of Mérida as a DTI. The commitment of the la *Delegación de Turismo de Mérida* for the revitalisation of the sector has led them to recently (2022) achieve the recognition of DTI, a first step to start implementing all those measures included in the Evaluation and Diagnosis Report carried out within the framework of the DTI methodology, as well as in the Action Plan. All of this to finally consolidate Mérida as a DTI.

The Post-Pandemic Context as an Opportunity for Digital Creation

The pandemic triggered the use of digital social networks, audio-visual entertainment platforms, live content and instant messaging, and the unexpected adaptation to teleworking and the boom in tools dedicated to it: messaging, videoconferencing, content management, task management, etc.). So remote working relationships became crucial during the pandemic that legislation regulating telework was needed in Spain. Royal Decree-Law 463/2020 of 14 March regulated and defined telework as work carried out at the worker's home, using computer systems, for a minimum period of 30% of the working day (Ruiz-Torres, 2021). Much of life has moved behind the screen: online education, online shopping, telemedicine, etc. This forced increase in remote activity made society aware that virtual services existed and worked, helping to demystify virtual environments. This normalisation, which has led to the identification of certain advantages such as overcoming spatial-temporal limitations, leaves us with an ideal scenario for adopting initiatives based on virtualisation, digital culture, and remote communication. One example is how the pandemic has accelerated Mark Zuckerberg's announcement of Meta. The post-pandemic context is an opportunity to implement technologies such as virtual or augmented reality, 3D scanning, videogrammetry and the development, for example, of virtual museums that may now be better received than before the pandemic.

Precisely the growing focus on the humanities and digital culture means that the organisations involved are paying more and more attention to technological innovations and how they can change the tourism sector through innovative models. Tourist destinations in general and museums, in particular, are now necessarily linked to digital creation. So much so that they have been including audio-visual departments and applying the new formats appearing throughout recent history. For example, the audio-visual documentary genre covers spaces such as museums or art galleries (art institutions), urban spaces, or academic spaces. Around the twentieth century, audio-visual formats entered the art and museum institution, where they grew in parallel to media and technological evolution, giving rise to multidimensional practices such as installation, visual performance, polyvision, multiple projection and transmedia narratives.

The Danish project *Flight for Life* (2017) exhibited at The Royal Arsenal Museum, which is part of The Danish National Museum, is an example of using transmedia and interactive storytelling to exhibit ethnographic work, bringing together different digital media and audio-visual practices. The exhibition brought together photographs, objects, traditional video and immersive video to engage the visitor in an ethnographic transmedia story.

Several international initiatives and projects are joining the redefining the museum experience for the digital age. Among the results of EU-funded research projects is a strand on how digital technologies play a crucial role in preserving and disseminating Europe's cultural heritage. The website [1] features the results of 16 projects that reveal digital innovations that help to achieve this goal: techniques to overcome accessibility problems in cultural spaces, digital resources to encourage citizens to reinterpret history, digital methods for newspaper research, new technologies for underwater cultural heritage or the creation of 3D models to explore built cultural heritage in person and remotely are just some of the topics highlighted. For example, the European GIFT project uses mobile apps to enhance the museum visitor experience by providing hybrid virtual museum experiences. It is not about introducing visitors to virtual worlds but combining both spaces and enriching them. The project also created the GIFT Box, a set of free and open-source methods and tools that museums can enrich visitor experiences.

Moreover, the transit of the works between physical space, digital space and even virtual space (in the Virtual Reality documentary) is not an expansion determined by the characteristics of the digital medium but linked to the interdisciplinary nature of the creations and the ambitions of the creators. Works are increasingly approaching interdisciplinary projects with objectives that go beyond media representation. The disciplinary relations and crossovers between technologies and expanded, interactive narratives and the museum institution are reflected in avant-garde works that experiment with historical, artistic and cultural heritage in new formats to bring the public closer and capture their attention.

In the past, museums were supposed to perform their function of preserving and disseminating heritage, but now they must also ensure that their content becomes experiences that seduce, educate, surprise and entertain visitors. For example, the documentary experience The Scream, 2018, by Sandra Paugam and Charles Ayats uses the famous painting by Edvard Munch to create an immersive experience in which the viewer is immersed into the world around this artwork and interacts with its elements around him. The canvas suddenly releases the Munch's demons, and chaos invades every millimetre of the personal (virtual) space. Similarly, the immersive documentary series Hors-Cadre immerses the viewer in some of most famous Swiss paintings (Arnold Böcklin, Féliz Vallotton and Ferninand Hodler). *The Opening of the Diet 1863* (2018) by The National Museum of Finland or the immersive experience in the UK of London's Tate Moderns in their Modigliani retrospective (2017/2018); these are just a few examples of the use of interactive audio-visual media for the creation of entertaining and surprising heritage experiences.

The introduction of ICT in the tourism sector is evident beyond audio-visual cultural experiences. For example, we see it in applications for managing reservations and flights, searching for services and opinions of other users, or consulting digital tourist guides. In recent years, virtual and augmented reality technologies have burst onto the scene, creating hybrid spaces to complement information and increase the attractiveness of the tourist experience.

In recent years, virtual and augmented reality technologies have burst onto the scene, creating hybrid spaces to complement information and increase the attractiveness of the tourist experience. For example, augmented Reality Agents (AuRA) are digital conversational agents that function as guides for visitors in museums. These agents are exciting elements that increase interaction with tourists, either virtual or hybrid environments. They have been implemented in museums such as the Museum of Science (Boston), the Heinz Nixdorf Museums Forum and the Smithsonian's National Museum of American History (virtual) (Martínez-Juárez, Puga-Nathal, & Martínez-Miranda, 2019).

With Augmented Reality, the user can explore or observe 3D elements in the physical world, creating an interactive hybrid space according to the functionalities of each application. This technology, integrated in products or services, could be a valuable tool to improve the value of the customer experience, adding virtual content to the real world. authors propose the following modalities:

- Personalised audio-visual information to visitors in museums, indoors and outdoors.
- Providing information about monuments and historical buildings in a city. Application for AR glasses. Enhancing or complementing (augmenting) the information of the real scene with virtual information (text, audio, audio-visual or 3D).
- Explore information and multimedia content about the city using the mobile camera.

Similarly, the virtual museum concept, which is still in research, is in the same vein. Martínez-Juárez, Puga-Nathal, & Martínez-Miranda (2019) suggest in their study to increase the cultural options and products offered to tourists as a way of bringing cultural heritage closer to visitors, facilitating knowledge of the city's archaeological offer and generating new cultural experiences enriched by technology. According to Polycarpou (2018), technologies aim to support, not replace traditional museum practices but even enhance them. The benefits of digitisation also include the digital visualisation of heritage elements that could not otherwise be disseminated in such a comprehensive way. Increased access through digital visualisation of artefacts that are complex to display, which for reasons of size or physical inaccessibility would be impossible to transfer to the viewer.

The use of digital approaches in managing cultural heritage may overcome these issues, and furthermore enhance the fruition of artefacts through new ways of communicating and interacting with visitors leading to virtual museums. (Guidi, Trocchianesi, Pils, Morlando, & Seassaro, 2010).

These museums do not necessarily refer to complete virtual museums but can be defined as additional ways of disseminating content and providing knowledge about cultural heritage, emerging from the current hybridisation process between museums and technologies. As simplified by Guerardini, Santachiara and Leali (2019); Digital technologies open up the possibility of large-scale virtual access with two main effects: empowerment of the role of visitors from passive to active subject and empowerment of the territory, promoting its knowledge and safeguarding processes.

The Active Visitor's Position and the Generating Engagement

Since the development of interactive audio-visual productions, the spectator develops an increasingly active position to the extent that interaction with the work is necessary for its enjoyment. This can be seen, for example, in interactive narratives in which the user's action is essential for the continuation of the story. From interactive audio-visual narrative to immersive works or participatory creations, we can identify the growing trend of viewer inclusion (Marin, 2021), resulting in audio-visual experiences that go far beyond viewing. User or viewer inclusion refers, in this context, to the surrender of part of the power of a representation to the user (audience, visitor), who is to some extent part of the creation process. It should be borne in mind that in interactive works, creation does not only take place while the work is in progress and developed, but a creative process takes place during the reception. The final work culminates with the spectator's contribution, with decisions and actions at the moment of reception and not only with the reading interpretation. For example, in works in which the spectator decides the order of visualisation of elements, the visual discourse is fragmented until the moment in which they decide the order, and that is when the visual discourse will end. And that discourse will be an individual experience since it depends on the personal decisions of each spectator. Interactivity, participation, personalisation or immersion are strategies used to increase the spectator's inclusion and the level of engagement with the audio-visual creation (Marin, 2021). Thus, depending on the technology, the strategies used to include the spectator will give rise to different forms of content exposure and narrative. The content, the possibilities of technology and formats, and the user's role will ultimately determine the type and level of viewer inclusion.

Reception can become a multisensory and individual experience, and the viewer determines the visual discourse and the final temporality of reception of the work. Audio-visual productions (on the subject of museums, look for specific terminology) are constantly changing and evolving at the media level. You can see it from the entry of audio-visual narratives into museums as video installations to their leap into urban space, GPS technology, projections and alternative realities (Virtual Reality, Augmented Reality and Mixed Reality). The digital image and reality merge and overlap, audio-visual representation is experienced and inhabited in immersive and mixed formats.

Virtual reality or 360-degree video make the user move into a digital space (immersive image). With augmented reality, it happens differently: it is the image that passes to the physical plane through an AR display device, either in glasses or through a screen superimposed on the physical plane. The result is that the viewer and the image are illusorily in the same dimension. An illusion that increases exponentially in cases where the screen disappears, as the viewer loses the notion of mediation (Chabert, 2012). This hyper-sensorialisation of reception seeks to offer the most vivid and realistic experiences possible to viewers.

Augmented reality (AR), as with virtual reality in other contexts, is increasing engagement in the museum visitor experience (Shin, 2019). In the recent research *Augmented reality in the tourism industry: A multi-stakeholder analysis of museums*, Serraballe et al. (2019) evaluated the application of AR in museums through their visitors and concluded that the use of this technology brings significant benefits to all stakeholders. Like other combinations of digital realities, Augmented Reality achieves value creation by making the experience more attractive to the tourist and more personal. A value that is created in the museum itself and reaches the social, technological and political environment thanks to the possibility of sharing content. This technology thus appears as an added value not only because it makes the cultural experience offered by the institution more attractive but also because the technological experience itself is an attraction for the tourist, something that adds to the main attraction of the institution. On the other hand, it makes it possible to attract a more extensive profile of tourists, since the offer is broadened through the technological experience, reaching those tourists who are not attracted by the traditional museum experience alone or even attracting tourists who are only interested in the experience of AR works.

Serraballe et al. (2019) also point to the shared experience as another critical factor in the promotion and diversification of the offer. Thus, interactive digital formats, especially those based on Extended Reality technologies, can not only be in themselves a tourist attraction or an added value to the museum experience but also achieve greater attention and interest from tourists, given that they promote a dynamic, interactive and participatory relationship between the cultural artefact and the visitor.

Along the same lines, the research-creation by Gherardini, Santachiara and Leali (2019) was based on the integration of two phases: the 3D digital reconstruction of real artefacts and their integration into an Augmented Reality application, to extend the physical experience with digital elements and information. This was done using Roman objects: a Roman mosaic floor in Modena (Italy) and two Roman funerary sculptures. It is one of many examples of digital reconstructions carried out and implemented in museums. Another example is the interactive augmented reality film created at the Natural History Museum of London, a participatory and entertaining experience that won the British Universities Film and Video Council's Learning on Screen Premier Award.

Empowerment of the Territory and Promotion of Knowledge

Heritage located in a specific environment enhances the identity and uniqueness of the place, revaluing aspects of territorial identity. "In a globalised world, the focus on preserving, valuing and explaining what is close to us has an extraordinary significance for citizens, who also have the opportunity to participate as members of the community" (Prats-Joaniquet, 2009, p. 85).

As we know, AR allows the visualisation of data of different nature embedded in the physical world, thus complementing the visualisation of the real space. AR systems can complement geolocation and portability systems, offering users augmented mobile cultural experiences embedded in the urban space. For example, the Museum of London designed the Street Museum app to navigate the city. Visitors walk through the streets, the app recognises points in the location and overlays historical photographs over the view of the site. With these mobile narratives, the museum experience moves beyond the museum and offers the visitor an urban cultural experience centred on historical heritage and with a narrative that integrates all points of interest. The narratives in movement play with the displacement of the spectator in the physical space and the superimposition of spaces and elements (physical and digital). The city can be used as a narrative space and the city tour as the story's driving force. The cultural use of public space favours reflection on the appropriation, exploitation and control of collective space as a place of representation, thus configuring new territorialisation, mobility and cultural uses of urban space (Lemos, 2008).

In the city of Salamanca, in 2015, researchers Joo-Nagata, García-Bermejo and Martínez-Abad developed a Virtual Territorial Heritage project, with the implementation of Augmented Reality and Mobile Pedestrian Navigation. The objective was to construct a mobile system within a process of educational training in the framework of territorial information on the historical urban heritage of the city. This work, which went beyond the construction of the mobile application, concludes on the importance of evaluating the experience the need to establish the devices' strengths, weaknesses, and limitations. It also points to the personalisation of heritage content to improve learning processes and local cultural identity.

Brief Previous Experiences

There are some previous initiatives along these lines, but they do not manage to configure an integrated tourist experience for the visitor:

- App Imageen Mérida [2]: It is an app for visiting the Roman theatre and amphitheatre of Mérida in virtual reality. It offers a digital version of both as they were 2,000 years ago, as well as their context. Imageen Mérida consists of 20 minutes of 360-degree content about the monuments and the city's archaeological process initiated in 1910.

- Qasr Mérida (Digital Drimz) [3]: In 2013, the private Andalusian company Digital Drimz took the initiative to develop an augmented reality application for tourist sites. The Smartphone application allows tourists to visit a city and its tourist monuments by expanding the information at each point of interest. Qasr was presented to the Consortium that manages the Monumental City of Mérida in demo format, but it already offered visitors the possibility of expanding information on each monument.
- EPIGRAPHIA 3D projet [4]: A selection of Roman inscriptions from the National Archaeological Museum (Madrid) and the National Museum of Roman Art (Mérida) are available to the public through 3D models on the project's website. All the content can also be found in a free app. The result is part of several scientific innovation projects that combine basic research in the Humanities with graphic engineering to bring Roman inscriptions closer to younger students and the general public: Deciphering Roman inscriptions in 3D. Virtual Epigraphic Science and Roman Inscriptions of Augusta Emerita in 3D: from the Museum to mobile devices, funded by the Spanish Foundation for Science and Technology - Ministry of Economy and Competitiveness and the University of Las Palmas de Gran Canaria. The website is a database with two galleries of 3D elements, one for each Museum worked on. The user can access each model, explore it in 3D and consult the primary data of each one. There is also "resources" with information on different aspects of epigraphy and the Roman world. However, the user has no integrated or narrative experience to explore, as the operation and user experience is somewhere between a database and a catalogue.

SOLUTIONS AND RECOMMENDATIONS

Because of the above, we consider that the archaeological tourist complex of Mérida is ideal terrain for implementing strategies to improve the tourist-cultural experience with digital tools, especially augmented reality creations. Mérida's initiative to develop as a Smart Tourist Destination included the photogrammetric survey of the monuments. The mayor of the city, Rodríguez Osuna, declared in 2020 that the intention "is to have a base with a high-quality 3D survey of the monuments". The intention was to promote virtual visits in augmented reality, with the creation of an application with an offline operation to "launch a journey to another reality with the inclusion of augmented reality at the service of tourism. This tool shows archaeological sites and historical monuments as they were originally built" (Europa Press, 2020).

- The creation of an augmented reality application that offers tourists a tour of the archaeological site in Mérida is particularly relevant for several reasons:
- The richness of the archaeological site, an example of what was once the capital of a Roman province, with 29 archaeological monuments and three museums: the *Museo Nacional de Arte Romano* in the first place and the Visigothic and Prehistoric Collections.
- The location of the archaeological elements distributed throughout the city facilitates the creation of mobile and augmented experiences.
- It strengthens the concept of the *Consorcio Ciudad Monumental* of Mérida "Making City". According to this, monuments, archaeological remains, buildings and infrastructures are not in the city but are the city. Some of the actions carried out in this spirit are: to avoid isolating the

archaeological remains with screens that prevent their contemplation from the street, to facilitate their accessibility and visibility so that they form part of the urban reality, to bring the remains closer to the citizens and to offer public information so that passers-by can increase their knowledge of them.

As we can see, it seems that the political spirit exists, but, for the moment, it has not materialised.

CONCLUSION

One mistake that museums often make when incorporating digital audio-visual technologies is the lack of a narrative that facilitates understanding the content and the visitor's enjoyment. On the other hand, there must be coherence between the exhibition of the content, the use of technologies and the touristic richness for the result to be genuinely beneficial and for all the elements to be valued. Sometimes, augmented reality, immersive experiences or transmedia narratives are used only as a tourist attraction. For example, the transmedia exhibition in Salamanca did not offer a transmedia experience, nor did it contemplate a truly transmedia narrative. The technologies used, such as Qr codes, substituted information that could appear on posters. If we want the institutions to bet on innovation and the introduction of other technologies or narratives, such as transmedia or XR, their application must be functional and profitable; they must provide an added value to the traditional type of exhibition, reaching where they do not reach.

We found that, in Mérida, despite the previous attempts, there is no digital communication strategy in this sense that is done in a planned way and that integrates all or a large part of the archaeological site. Therefore, we urge the agents involved to consider the successful projects of other cities with similar tourism models and integrate the whole into a digital, augmented and mobile archaeological tourism experience. An initiative of this type would enable the Consortium's commitment to integrated heritage management to be put to good use, considering the city as a single site, using a single methodological unit for archaeological recording and documentation, thus facilitating research, communication and dissemination of the results.

ACKNOWLEDGMENT

This research was funded by the European Union - NextGenerationEU.

REFERENCES

Chabert, G. (2012). Les espaces de l'écran. *MEI, Médiation et information, ecran et médias, 34*, 203-213. http://www.mei-info.com/wp-content/uploads/2015/03/MEI_34_14.pdf

European Economic Forecast. (2021). *Economic Forecast: From recovery to expansion, amid headwinds*. European Commission.

Gherardini, F., Santachiara, M., & Leali, F. (2019). Enhancing heritage fruition through 3d virtual models and augmented reality: An application to romanArtefacts. *Virtual Archaeology Review, 10*(21), 67–79. doi:10.4995/var.2019.11918

Guidi, G., Trocchianesi, R., Pils, G., Morlando, G., & Seassaro, A. (2010). A Virtual Museum for Design: New forms of interactive fruition. *16th International Conference on Virtual Systems and Multimedia*, 242-249. 10.1109/VSMM.2010.5665977

Joo-Nagata, J., García-Bermejo, G., & Martínez-Abad, F. (2015). Patrimonio Virtual del Territorio: Diseño e implementación de Reursos Educativos en Realidad Aumentada y Navegación Peatonal Móvil. *VAEP-RITA, 3*(1), 46–51.

Lemos, A. (2008). Mobile Communication and New Sense of Places: A Critique of Spatialization in Cyberculture. *Galáxia (São Paulo), 16*, 91–108.

López de Ávila, A. & García-Sánchez, S. (2013). Destinos turísicos inteligentes. *Harvard Deusto Business Review, 224,* 58-57.

Marín, A. (2021). The immersive documentary: Understanding the phenomenon in nonfiction stories through a typology proposal. *El Profesional de la Información, 30*(4). https://doi.org/10.3145/epi.2021.jul.11

Martínez Juárez, I. H., Puga Nathal, M. E., & Martínez-Miranda, J. (2019). Desarrollo de un guía virtual conversacional para museos en un entorno inmersivo de realidad virtual y aumentada. *Research in Computing Science, 148*(8), 387–395.

Moreno-Melgarejo, A., & Sariego-López, I. (2017). Relaciones entre Turismo y Arqueología: El Turismo Arqueológico, una tipología turística propia. *PASOS. Revista de Turismo y Patrimonio Cultural, 15*(1), 163–180.

Morère, N. (2007). Análisis del turismo arqueológico en España: Un estado de la cuestión. *Estudios Turísticos, 171*, 115–139.

Plan de modernización y competitivida de sector turístico. (2021). *Gobierno de España.* https://www.lamoncloa.gob.es/temas/fondos-recuperacion/Documents/16062021-Componente14.pdf

Plan de Recuperación. Transformación y Resiliencia. (2021). *Gobierno de España.* https://planderecuperacion.gob.es/

Plan Nacional e Integral de uturismo 2012-2015. (n.d.). *Secretaría de Esado de Turismo, Instituto de Turismo de España.* https://turismo.gob.es/es-es/servicios/Documents/Plan-Nacional-Integral-Turismo-2012-2015.pdf

Plan Turístico de Extremadura 2017-2020. (n.d.). *Dirección General de Turismo, Consejería de Economía e Infraestrcutura. Junta de Extremadura.* http://www.juntaex.es/filescms/con02/uploaded_files/planes/Plan_Turistico_Extremadura.pdf

Polycarpou, C. (2018). *The ViMM Definition of a Virtual Museum.* https://www.vi-mm.eu/2018/01/10/thevimm-definition-of-a-virtual-museum/

Prats-Joaniquet, C. (2009). Museos y territorio: Del recurso cultural al producto turístico. *HER&MUS, 2*, 80–89.

Real Decreto-Ley 463/2020, de 14 de marzo, por el que se declara el estado de alarma para la gestión de la situación de crisis sanitaria ocasionada por la COVID-19. *BOE,* 67/2020. Ministerio de la Presidencia, Relaciones con las cortes y Memoria Democrática. BOE-A-2020-3692

Rengifo-Gallego, J. I., & Sánchez-Martín, J.-M. (2021). Conceptos básicos del turismo. Aproximación a la evolución reciente de las políticas turísticas en Extremadura. In Recursos turísticos, territorio y sociedad en Extremadura: catalogación, nuevos usos y perspectivas. Universidad de Extremadura, Servicio de Publicaciones.

Report, E. S. (2020). *Global Imbalances and the COVID-19 Crisis*. International Monetary Fund.

Romero-Dexeus, C., & Paton, J. (2020). *Innovación turística y especialización inteligente en España. Palancas imprescindibles para la recuperación*. Segittur.

Ruiz-Torres, P. (2021). Covid-19: Teletrabajo en tiempos de pandemia. *Historia y comunicación social, 26*, 11-18. doi:10.5209/hics.74237

Sánchez-Oro, M. (2021). Aproximación sociológica al turismo de Extremadura. In Recursos turísticos, territorio y sociedad en Extremadura: catalogación, nuevos usos y perspectivas. Universidad de Extremadura, Servicio de Publicaciones.

Sánchez-Oro, M., García-García, Y., & Ramos-Vecino, N. (2021). *Informe sobre turismo Urbano*. Observatorio de Turismo, Extremadura & Universidad de Extremadura.

Serraballe, F. (2019). Augmented reality in the tourism industry: A multi-stakeholder analysis of museums. *Tourism Management Perspectives, 32*(October). https://doi.org/10.1016/j.tmp.2019.07.002

Shin, D. (2019). How does immersion work in augmented reality games? A user-centric view of immersion and engagement. *Information Communication and Society, 22*(9), 1212–1229. https://doi.org/10.1080/1369118X.2017.1411519

Turismo 2020, Plan del Turismo Español Horizonte 2020. (2017). *Ministerio de Industria, turismo y Comercio*. https://www.tourspain.es/es-es/Conozcanos/Documents/HistoricoPoliticaTuristica/PlanTurismoEspanolHorizonte2020.pdf

World Tourism Barometer. (2020). *UNWTO*. doi:10.18111/wtobarometereng

ADDITIONAL READING

Black, G. (2005). *The Engaging Museum: Developing Museums for Visitor Involvement*. Routledge.

Caracciolo, M. (2014). Experientiality. In P. Hühn, J. Pier, W. Schmid, & J. Schönert (Eds.), The living handbook of narratology (pp. 1–13). Hamburg University.

Lévy, P. (1995). *Qu'est-ce-que le virtuel?* Découverte.

Martínez-Sánchez, M., Martínez-García, J., González-Fernández, R., & Flores-García, A. (2021). El objeto histórico: Del museo a internet a través de la fotogrametría. *Boletín del Museo Arqueológico Nacional*, *40*, 379–394.

Park, W., Heo, H., Park, S., & Kim, J. (2019). A Study on the Presence of Immersive User Interface in Collaborative Virtual Environments Application. *Symmetry*, *11*(4), 1–19. doi:10.3390ym11040476

Rose, M. (2018). The immersive turn: hype and hope in the emergence of virtual reality as a nonfiction platform. *Studies in Documentary Film, 12*(2), 132-149.

Spinney, J. (2015). Close Encounters? Mobile Methods, (post) Phenomenology and Affect. *Cultural Geographies*, *22*(2), 231–246. doi:10.1177/1474474014558988

Tsichritzis, D., & Gibbs, S. (1991). Virtual museums and virtual realities. *Proceedings of the International Conference on Hypermedia and Interactivity in Museums.*

TucciG.CiniD.NobileA. (2011). Effective 3D digitization of archaeological artefacts for interactive virtual museum. *International Archives of the Photogrammetry, Remote Sensing and Spatial Information Sciences, XXXVIII-5W16*, 413–420. doi:10.5194/isprsarchives-XXXVIII-5-W16-413-2011

KEY TERMS AND DEFINITIONS

360-Degree Video: Is a video format that captures the scene at a 360° angle, the subsequent reception of which allows the image to be viewed in an immersive way.

Augmented Reality (AR): Is a digital system that makes it possible to visualise digitally created elements on a real physical scenario, generally through a digital device such as augmented reality glasses or camera viewers on mobile phones or tablets.

Engagement: In this context, it refers to the commitment between spectator and work, visitor, and museum, not only measured by the interactivity between both elements but also to an affective, cognitive, and behavioural state.

Mixed Reality (MR): Generally understood as the merging of the real and virtual worlds to produce new environments and visualisations where physical and digital objects coexist and interact in real time, derives from the combination of AR and VR systems.

Mobile Narratives: Stories in movement and anchored to physical spaces through which the spectator moves, personalised in many cases and generally using geolocation or augmented reality technologies.

Transmedia Narratives: A narrative process in which the elements of a story are dispersed across multiple distribution channels to create a unified and coordinated entertainment experience.

Videogrammetry: A videographic technique derived from photogrammetry for the three-dimensional reconstruction of objects by capturing them from various points of view.

Virtual Epigraphic: Digital reconstruction and study of ancient engraved inscriptions or writings.

Virtual Reality (VR): A digital system that creates an artificial world where the user can be, navigate and manipulate objects.

ENDNOTES

1 https://cordis.europa.eu/article/id/413473-how-digital-technologies-can-play-a-vital-role-for-the-preservation-of-cultural-heritage

2 https://www.imageen.net/

3 http://standard3dprinter.com/

4 http://www.epigraphia3d.es/

Chapter 13
Redefining and Reframing a City:
Co-Designing a New City Model in Venice - Universities and Society to Envision a Resident-Friendly Future

Cinzia Colapinto
https://orcid.org/0000-0003-1211-8033
Ca' Foscari University of Venice, Italy

Vladi Finotto
Ca' Foscari University of Venice, Italy

Daniela Pavan
Ca' Foscari University of Venice, Italy

Fabio Pranovi
Ca' Foscari University of Venice, Italy

ABSTRACT

The Venetian landscape is as much a product of its economic activities, past and present, as of its physical environment. Tourism is the main source of income for the city, and also a challenge: the enormous pressure exerted by visitors is creating distorted incentives for inhabitants to abandon the city to the point that depopulation and environmental harm are the most relevant issues in the city's political agenda. The chapter delves into the case of Venice to explore the themes of urban revitalization as a process of framing and social mobilization. In particular, the authors focus on the role of higher education institutions in triggering urban change and on the nature of actors engaged in the transformation of the social fabric of a city. Based on the experience developed within a European project aimed at revitalizing and rethinking cities, the chapter explores the potential of HEIs in transforming cities, the hurdles and critical factors that might inhibit their contribution, and the strategies that can facilitate their engagement in processes of social and economic transformations.

DOI: 10.4018/978-1-6684-3369-0.ch013

INTRODUCTION

Venice landscape is as much a product of its economic activities, past and present, as of its peculiar physical lagoon environment. The Venice lagoon is a good example of an environment wherein human communities and the surrounding ecosystem evolved interdependently: the shaped venetians behaviour and traditions, but it also was deeply modified by human intervention. The lagoon maintains the presence of a certain level of ecological processes that still deliver a good level of environmental sustainability that benefits the local community's wellbeing and that guarantees the city's touristic attractiveness.

Indeed, tourism is the main source of income for Venetians, to the point that many commentators talk about the "touristic monoculture" of the city and point to the problems inherent in such a dominance, namely the depopulation of the city centre and its unsustainability from a social and environmental point of view. The recent Covid-19 pandemic was an eye-opener, apparently, or just a pause in a process that will continue to grow unchanged as soon as international mobility is restored. Prior to lockdowns, 2019 was equally tough for the city: in November 2019, Venice was hit by the worst flooding after 1966, the so-called "Aquagranda" or "Huge (high) water (tide)". It caused more than 1 billion € of damage and, unfortunately, some deaths (Ambrosini & Clerici, 2022; Allnutt, 2021).

COVID-19 came and Italian borders closed suddenly. Flooded by huge hordes of tourists, Venice became a sort of ghost town overnight, populated by its few residents (about 50,000 people, less than a third of what it was in the 1950s). Twenty million tourists per year disappeared: the grossly underestimated income the city previously gained from these flows equalled 3 billion euros per year (Momigliamo, 2019).

The chapter aims at exploring the themes of urban reframing and revitalization by delving into the peculiarities of the city and investigating the role of higher education institutions, as actors engaged in the transformation of the social fabric they are part of, as triggers and actors of change (Svensson et al., 2012). Venice, in fact, is home to three universities (Ca' Foscari University, IUAV University, and Accademia delle Belle Arti) as well as a number of post-secondary education institutions. Students, as well as professionals working within or with these institutions, represent a multifarious set of interests: they are partly living in Venice –albeit on a temporary basis– or commuting. Finally, the recent shift towards "entrepreneurial" universities (Audretsch & Belitsky, 2022; Sánchez-Barrioluengo & Benneworth 2019) and engaged scholarship contributed to the creation and growth of novel approaches to research and training in these institutions. Based on the experience developed within a European project aimed at revitalising and rethinking cities, the chapter explores the potential of Higher Education Institution (hereafter HEI) in transforming cities, the hurdles and critical factors that might inhibit their contribution, the strategies, methods and processes that can facilitate HEIs' engagement with processes of social and economic transformations on an urban scale.

The chapter begins with a literature review, and then describes the urban challenges faced by Venice in terms of touristic destination, historical city, heritage site, lagoon city, and capital of the Veneto region. The authors embrace a collaborative approach to better understand how to revitalise the identity of Venice and trigger urban change.

BACKGROUND: REIMAGINING CITIES AS A RESPONSE TO COVID-19

Over the centuries Venice has faced countless moments of crisis and challenges (i.e. epidemics starting from the one in the 15th century, Napoleonic domination, and Aqua granda/High water) yet it has always managed to regenerate and revive itself. The question now may be if the current COVID-19 pandemic could be considered as a new opportunity for Venice to reinvent itself once again and identify a new model for sustainable tourism for the Queen of Seas, that allows a balance between residents and the presence of tourists.

Cities are material, social and cultural artefacts, wherein a multiplicity of social groups –both dwellers and external actors such as tourists– express a variety of needs and expectations. Renewing the identity of a city, thus, might be seen as an exercise in negotiation and development of novel ideas, identities and frames that bridge between different ways of looking at the urban social, cultural and economic fabric. Thus, theoretical approaches that tackle questions related to collective and collaborative innovation efforts might shed light on how complex cities like Venice might go through renewal and redefinitions.

Since Chesbrough's seminal work (2003), open innovation has a relevant role in the academic debate and nowadays it is more and more associated with innovating with users. In this chapter we look at cities as potential 'motors' and arenas for sustainable development (Rotmans et al., 2000). Venice, as other cities, has to cope with complex and persistent challenges and needs to transcend the barriers of temporal and institutional scales moving to a collaborative approach (Bai, 2007; Hamann and April, 2013): we believe universities can act as intermediary organisations and create platforms for deliberation and collaboration between diverse stakeholders as shown also by Trencher et al. (2013). Indeed, sustainable urban transformation involves promoting active collaboration among diverse stakeholders, integrating different perspectives and bodies of knowledge and expertise (McCormick et al, 2013). This is particularly true when there are ambitious goals. As we will show in the chapter, adopting collaborative approaches appears crucial to engage different key players in learning processes able to unlock new thinking and solutions. This is in line with the recent proliferation of Living Labs as experimental arenas where cities try to tackle their urban challenges: they have emerged at the intersection of open innovation and user innovation frameworks (Schuurman et al., 2016).

Frameworks and toolkits to discipline and direct open innovation efforts were developed in the past twenty years. For instance, first applied to the design process of industrial products, the Double Diamond design model has been widely applied to different contexts and it is indeed a design process model adapted from the divergence-convergence model proposed in 1996 by Hungarian-American linguist Béla H. Bánáthy. We used this model to engage several stakeholders, to clarify two main aspects:

- How to decompose the main three challenges that Venice is facing (overtourism, depopulation and protection of the lagoon) in sub-challenges that may be more accessible to tackle;
- How to apply our findings to make Venice a resident-friendly city.

The Double Diamond model, also known as the Framework for Innovation, is indeed the result of an investigation by the British Council (2005) that wanted to learn more about how people process information in order to create solutions. The model is based on a figure consisting of two diamonds: the first diamond refers to the problem space and the second to the solution space. The two diamonds represent a process of exploring an issue more widely (divergent thinking) and then taking focused action (convergent thinking). These two diamonds should help navigate through complex processes in a structured manner. The model suggests that the design process should have four phases:

- **Discover** (Divergent): this stage aims to investigate, explore and identify the stakeholder needs while simultaneously seeking to understand the complexity of the problem. It involves speaking to and spending time with people who are affected by the issues.
- **Define** (Convergent): stage that delimits the interpretations based on the Discover-stage findings, aiming to obtain a clear definition of the problem.
- **Develop** (Divergent): stage that aims to acquire various solutions, which can address the problem.
- **Deliver** (Convergent): stage that entails final testing of the solutions pursuing an improved outcome followed by feedback sessions on the process.

This model does appear to be an appropriate framework, because it helps create a structured project plan with a critical focus on problem-solving in action research, and furthermore, it is a solid way to apply Design Thinking.

Following the three steps of Design Thinking process (Brown, 2008) - Inspiration/Ideation/Prototyping-, and the recent literature on participatory urban rethinking projects based on the approach's main tenets (Medeiros, 2021), the authors have been engaging different stakeholders to assess the current scenario through different perspectives and identify the most urgent pain points about living and staying in Venice.

Our approach is also inspired by the SENSEable City Lab of the MIT (https://senseable.mit.edu/), led by Professor Carlo Ratti, that aims to create a new model for the city of the future based on the 4Ps: Partnerships, Public Sector, Private Sector, People.

Now, with this context in mind, the questions we aim to explore are: How to manage the rebirth of Venice? Which actors need to be involved? How to include all aspects of sustainability while redefining the identity of Venice? As previously stated, our basic assumption is that such challenges require a collaborative approach considering the role of the university, given its role as knowledge producer and talent attractor. The literature stream we partly refer to is the Triple Helix Model (THM), devised by Etzkowitz and Leydesdorff (2000). It implies that innovation is a circular development process grounded on reciprocal relationships and an increased hybridization among elements of different institutional settings such as universities, governments, and the industry. Carayannis and Campbell (2009) argued similarly, pointing out that innovation ecosystems are systems in which different organisations and institutions in public and private sectors (such as governments, universities, and financing organisations) collaborate and compete among each other. Throughout the last two decades, the THM model has been extended to further depict the growing interdependence with regard to innovation, and entails an innovation ecosystem driven by the systematic process of coevolution (Carayannis & Campbell, 2011), highlighting the interplay between different actors as an engine of creativity and innovation (Kretz & Sá, 2013).

Based on these frameworks and through these methods, our goal is to create and actively interact with urban communities of practice, to then develop, and implement, multidisciplinary learning interventions for university students, young researchers and lifelong learners to ideate real- life solutions to the most pressing urban challenges that Venice faces, involving the urban fabric that animates the city.

VENETIAN URBAN CHALLENGES

The United Nations (2018) estimate that 55% of the world population currently lives in urban centres. This percentage is expected to rise to 68% by 2050. With few exceptions, cities are expected to become larger and more crowded, despite the changes that the current pandemic has generated and the issues it has exacerbated. Transforming these clusters of buildings and people is now a necessity that concerns the economic, social and environmental spheres of urban life. The main challenges that Venice faces with a view to building a more sustainable community are overtourism, depopulation, and environmental and lagoon protection.

Before COVID-19, as mentioned by Momigliano (2020), hotels in and around Venice used to be overcrowded with tourists. Approximately 12.1 million mostly international guests a year used to stay in the city's accommodation structures (Italy's bureau of statistics, 2019). This number does not take in consideration day-trippers coming from cruise ships, or by train and bus tours.

The mass tourism of recent decades is a result of globalisation, of the launch of home-sharing platforms, and of the presence of cheap airfares (such as Ryanair, easyJet). Cruise ships used to bring in 1.6 million visitors per year. Eventually, the growth of emerging economies, especially Asian economies, allowed many new tourists to join the crowds travelling from Europe and North America.

During the past few years, tourism has changed the soul of the lagoon, putting the natural ecosystem at risk because of pollution and salt marshes erosion. also, grocery stores turned into souvenir shops because they were not able to sustain themselves financially any longer, with the few residents still in town and services to locals started to get more and more scarce. As an example, there is only one ferry per hour that connects the main island of Venice with the smaller islands of the lagoon. Not to mention the rising housing costs, which is another reason that contributed to pushing residents out. Prior to the pandemic, short stays used to be the rule in the city. Venice used to have Italy's highest Airbnb-to-population ratio, indeed, with more than 8,000 apartments listed on the temporary rentals' platform.

For some, Venice has the allure of a sort of Disneyland and they feel entitled to jump from bridges in the Grand Canal for example. Indeed, after a group of Belgian teenagers dangerously jumped from the Calatrava bridge into the Grand Canal, the City Council added regulations to ban taking a leap of a Venice bridge. Swimming or jumping into the canal now comes with a hefty $500+ fine.

The pandemic emergency could have been a time for residents and local leaders to rethink their city to "develop an economy that does not revolve entirely around tourism, one that would draw international investors, expand the footprint of the city's two universities, and turn its empty buildings into environmental research facilities" (Momigliano, 2020). Left indeed without its main activity, tourism, Venice started rethinking to its dwellers and to new services focused on locals. Row Venice for example, a group of women that normally teaches tourists how to row *alla veneta* (Venetian style, or standing), started delivering food for free to neighbourhoods where grocery shops are rare, to support the elderly. Following this path, in 2020, Cocai Express was launched (cocai means 'seagull' in Venetian dialect), as the very first delivery app in Venice.

However, now that 2022 has started, it looks like tourism is regaining its presence within the city in an even more aggressive way, and no lessons have been learned during the pandemic and lockdown months. This of course has an impact on the 50,000 residents who live in the historic city and on the lagoon ecosystem that is part of Venice and its identity.

In a recent research commissioned by the World Economic Forum (2021) on the concept of national brand and related strengths, it emerges first of all that Italy improved its positioning (from the sixth place of last year to the fourth), preceded only by Germany (first place), Canada (second place) and Japan (third place), and the main strengths linked to the brand Italy are Culture, Tourism, and the adaptability of Italians. These elements make Italy strong, however at least two of them (culture and tourism) can be considered as vulnerable assets for the Venetian ecosystem.

These reflections design the ground floor for the UCAMP (Urban GoodCamp), a three-year Erasmus+ KA2 project that brings together universities and citizens to co-design cities centred on their residents, students and workers. Ca' Foscari University of Venice aims at involving local stakeholders to co-design a more sustainable Venice, from a social, environmental and financial perspective, a city where people would want to move and build a life.

"The place you choose to live will help determine your success in business, in finding a life partner, and in living a fulfilling life. In fact, it may be the most important decision of your life. I believe that we are in the beginnings of a shift as fundamental as the industrial revolution was over a century ago—one that will have as dramatic an impact on how people live and work. Furthermore, it will have a dramatic impact on where they live and work." (Florida, 2008, p. 10). Venice represents a unique environment; however, at the moment, it seems to be not so resident-friendly. Also, it looks like the city is losing its own identity, meaning it is not able to properly identify what kind of tourists it wants and how to educate them in respecting the lagoon ecosystem.

That said, we began our work starting from the main challenges of Venice: overtourism, depopulation and the protection of the natural environment of the lagoon. UCAMP aims at enabling universities to explore urban challenges first-hand, creating co-design and partnership models that will involve students, citizens, workers, tourists, entrepreneurs, teachers, the public administration and cultural institutions. In this scenario, universities could act as facilitators for urban cooperation and for the co-creation of a sustainable urban ecosystem, while generating innovative learning environments for civil society, industries and local governments.

These three main challenges are linked to the fact that the more touristy Venice became, the more residents were pushed out; the fewer the residents, the more those who remained struggled to find employment outside tourism, thus reinforcing the pattern and creating a vicious circle. Approximately 25,000 Venetians are indeed now directly employed in tourism (Momigliano, 2021).

The decline of residents is linked to a dwindling number of jobs that do not involve tourism. At the same time the rising cost of food, transportation and housing are additional reasons that contribute to making life difficult for residents. Property values have indeed increased by 800% forcing young people to leave. The number of movie theatres in Venice has fallen from 20 to two. One of the largest bookstores has become a clothing store, and some business owners now charge "tourist prices" at shops and restaurants even to locals, reversing an old practice that made visitors who do not pay taxes bear a greater financial burden. In Venice, a city built on water that has a huge history in fishing, there are no fishing stores. The only place that sells nautical products is a mobile device store that dedicates some space to nautical and fishing items, to share the rent. The alternative is the online market or going to the mainland.

Talking about age, Venice seems to push away younger generations with its high costs of living. This allows only those who have the money to afford the stay in the city, indeed it seems difficult for young couples to decide to move to Venice. A paradox lies also in the fact that the city states the will of increasing the number of residents with young people, however then the Municipality approves a local policy, in 2019, that forbids children over 10 years old to play in the *Campi*, the local squares. In addition to that, there is no real space for children to play; with the reopening of the borders indeed, in summer 2021, after the first wave of the pandemic (prior to the Omicron variant), bars and restaurants are occupying all the available space on the street, making it impossible for kids to play outdoors anyway.

The risk of flooding makes life in Venice even more challenging. Who would buy a house that risks being flooded during high tides? Many houses got abandoned and those that did not, have expensive maintenance fees. More than 20 years ago, the Venice Municipal House Observatory gave to the Local

Administration a strategic solution for this: social housing. The idea was to allow people who do not meet the conditions for low income housing lists and, at the same time, cannot afford market prices, to return to live in the historic centre. However, due to bureaucratic delays, this project seems to be extremely slow in the Venetian landscape. Dreams of attracting multinational corporations, prestigious institutions and digital nomads have been discussed for years, but keep staying an open point that has not found a sustainable answer yet. At this point it is important to mention the relevance of the relationship with water that any person who lives in Venice, at least in the historical city, should have. In the Mediterranean area indeed, the lagoon of Venice represents an outstanding example of a transitional-water environment which has become more and more vulnerable as a result of irreversible anthropogenic and climate changes.

The Doge's Palace displays an 18th-century oil painting by Giambattista Tiepolo where the water god Neptune gifts a cornucopia with coins and treasure to a young lady who represents Venice. The painting celebrates the sacred bond between the city and the natural world that surrounds it. Today the relationship between Venice and Neptune has changed, because of climate change as well as the changes made by human kind to the natural ecosystem of the lagoon. This is why it is now a relationship that needs new interpretations. In this co-evoluted ecosystem where the tidal flats and salt marshes are as important as the islands, pile-dwellings, fishing villages need to be protected no less than the palazzi and churches. In fact, few places in the world are as threatened by climate change as Venice. As the city is built in the middle of the lagoon, its existence has always required maintaining a careful balance between the city and the surrounding natural environment.

However, in recent years, anthropogenic impacts and climate change have threatened to throw off the balance. Floods regularly damage lower floors of homes, restaurants and shops, and what was once a winter phenomenon is now happening more often. Experts say that climate change is likely to blame. The occurrence of exceptional high waters poses a significant threat to the protection and integrity of the Venice lagoon and its historic settlements. The increase in the frequency and levels of high tides, in addition to the phenomenon of wave motion caused by motor boats, is one of the main causes of deterioration and damage to the building structures and urban areas. This phenomenon has a significant impact on the morphology and landscape configuration of the lagoon due to the erosion of the seabed and of the salt marshes. In order to preserve the lagoon and protect its historic settlements and the whole city of Venice against flooding, several projects have been elaborated. Unfortunately, putting in place protective measures has proven difficult. These include an integrated system of public works, such as the mobile flood gates (MoSE - Experimental Electromechanical Module) to temporarily isolate the lagoon from the sea and some complementary measures capable of reducing the level of the most frequent tides in the lowest areas on the water.

The construction of the MoSE began in 2003, but it became operational only in October 2020, after the emergency of the Aqua Granda of 2019. Taking inspiration from another country with a strong relationship with water, the Netherlands, where the North Sea and its floods have been a threat for human settlement for years, we notice that tides are controlled by a complex system of floodgates with huge blocks, and other mechanisms built more than 50 years ago, after a devastating storm in 1953. These barriers are protecting the land; however, they also caused a significant biodiversity loss. This is why also the Netherlands need now to rethink their relationship with water and nature. Their experience with the barriers points out the problems related to a "walled" Venice, regarding the environmental crises the MoSE may cause. Venice's Lagoon is part of the Mediterranean Basin, and therefore linked to the climate crisis in the Mediterranean, where CNR-ISMAR has recorded an increase of 1,5 degrees Celsius of the surface temperature in the last 20/30 years. This growth in temperature due to climate change, has consequences on cities by the sea, and on Venice in particular. To avoid having a city underwater it is undoubtedly necessary to think about new models to manage the lagoon.

This is one of the reasons behind the choice of the city to run as a candidate to be one of the European cities that aim to become climatically neutral by 2030. Following this candidacy, in March 2021, alongside the Veneto Region and the Municipality of Venice, institutions such as Ca' Foscari University, IUAV, Benedetto Marcello Conservatory, Academy of Fine Arts, Cini Foundation, Confindustria Veneto and some of the main brands of the area have joined forces for the "Venice World Capital of Sustainability" project, with the goal to develop shared and integrated actions, with positive impacts in terms of sustainable development, employment growth, improvement of the living and working conditions of the population, industrial and energy transition. For example, since Venice is a city where almost everybody must walk or use public transportation, a redesign of the public transport system in a way that is compatible with the environment as well as with a sort of slow lifestyle may enhance the fact that Venice is a city that goes at a "slow pace rhythm". A question emerges here: is a city that cannot go on a rush compatible with the needs of a modern lifestyle? And this question leads to another one: Who do Venetians want to come to Venice?

METHODS AND DATA ANALYSIS

From the previous description of Venetian urban challenges, it appears relevant to embrace collaborative strategies, defined as institutionalised inter-organisational relationships that incorporate joint decision-making processes to enable short and middle term changes and transition (Evans, 2001). Inter-actor collaboration between stakeholders is crucial within a destination, because of its features of a tourism product (Fyall, Garrod, & Wang, 2012) and the required coordination and collaboration.

We adopt action research to gain new insights into urban transformation management and the role of HEI. Following McKay and Marshall's (2001) our research project distinguishes a problem-solving cycle addressing the problem of Venice, while the research cycle aimed to develop new knowledge about strategizing co-creation of a more sustainable city. The two phases unfold in parallel, allowing us to continuously transfer data and experience from the problem-solving phase to the research one.

We have reviewed and analysed urban challenges in the local ecosystem, by observing the city and the people living and transiting within Venice, developing a thorough desk research surfing through the literature available on the topic, interviewing experts, and building case studies on different local learning models that could make a pragmatic impact on the urban challenges that the city is facing. In detail, our activities include conducting 21 interviews in person or online and one consultation event (focus group) with key sixteen stakeholders in the transformation process; holding workshops with students, ideating bootcamps; and participating in the implementation of some identified interventions/actions.

We triangulated data from multiple source (Yin 2003) to validate our research: our primary data (interviews, field observations, events, workshops) are complemented by secondary data sources (desk research). We began data collection in January 2021. We collect data in parallel and to fill gaps in data, validate initial findings based on stakeholder interviews we shape events and workshops by applying the Double Diamond design model previously introduced.

STAKEHOLDER ENGAGEMENT AND COLLABORATIVE APPROACH

A sustainable Venice has indeed to evolve and probably detach from its current status of "company town" in order to become a "hub" for a community of new residents, in ways that are respectful of the city's urban and natural environment.

To achieve this, it is crucial to engage and listen to different stakeholders who are part of the city and co-design a new city model for Venice, such as:

1. **Municipality:** it has created the control room to take under control the flows of tourists within the city, and has signed an agreement with the Mayor of Florence addressed to the Prime Minister, with the goal to identify ways to educate tourists behave in Art Cities.
2. **The Hospitality Industry** that includes hotels, restaurants, bars and of course the brands part of the Travel Industry, such as cruise companies, airlines, travel agencies and also AirBnb.
3. **Local artisans**, who not only represent a continuum with the city's past of textile, lace-making and glassworks, but they could lead the way in recreating vibrant and mixed urban spaces where people can both work and live. Unfortunately, the aggressiveness of the introduction of souvenirs made in China within the Venice local marketplace is creating an issue. These souvenirs are sold for very low money, causing difficulties for Venetian artisans in terms of maintaining their factories alive and sustainable.
4. **Museums**, responsible not only for their culture share, but also for their civic role.
5. **Residents** who have chosen Venice as the city to build a life in.
6. **Students:** they are vital for the city's economy and should be protected, encouraged, supported, as part of the urban context.
7. **Nonprofits**. Some for the role they play for the Lagoon preservation, especially Venice Calls with the Clean Ups, We Are Here Venice, the Peggy Guggenheim Foundation. Others for the initiatives they develop to preserve the presence of dwellers in the city and their quality of life, especially for new generations.
8. **Media**, for the role they play in communicating the city brand.

From our interviews it emerged that universities play an important role within the urban context of Venice, mainly from three perspectives:

- as facilitators of specific processes, where this role is usually played by a local authority, a municipality or a consortium. As a third party, not an administration nor a political one, they can provide the service to activate different communities on specific projects, and, according to the transformation policies that each process or contract may require, they can activate and engage the appropriate stakeholders as *super partes* entities.
- as attractors of young talents with a high education level in the city. This is a very important role Universities already have in Venice. However, it should be supported by public policies and other institutions, to facilitate the permanence of this "kind of population" that is then supposed to grow, get married, have kids and therefore from temporary residents they can become permanent residents. With a city based mainly on tourism though, this role at the moment is very hard to be carried out without external support.

- a key actor in the real estate market. During this past year and a half of the pandemic emergency, together with the City Hall, Universities in Venice created the Study in Venice Program. The project also involves real estate companies that had no idea about how to rent empty houses and Airbnb apartments that had immediately gone out of business during the pandemic. The goal of the project is to reserve these apartments for students, building a series of concessions to get lower rents for them. The Study in Venice program wants the city to re-appropriate a part of the Venetian real estate that has been completely moved towards the tourist function and bring it back to a residential function.

Table 1. Interviewed stakeholders. Source: Own elaboration

Stakeholder Typology	interviewee	Role
Municipality	Paola Mar	Councillor for University and Promotion of the city
Museums	Luca Mizzan	Director of the Museum of Natural History of Venice.
Arts and Culture	Stefano Mondii	President Fondazione Forte Marghera
Nonprofits and resident	Giorgia Fazzini	Member of Ekos Club and Archeoclub Italy, she manages the island of Lazzaretto Nuovo that has been transformed into an ecomuseum.
Nonprofits and resident	Ludovico Dejak	Director of Innovation at the non-profit Venice Calls founded in 2018 with the mission "to give its contribution to creating and supporting new projects with all those realities that pursue a sustainable city model: economically, socially and environmentally."
Hospitality	Luca Chiais	General Manager Hotel Aquarius
Hospitality	Antholy La Salandra	Expert of Tourism with a strong knowledge related to the cruise industry and its impact on Venice.
Media	Fabio Bozzato	Italian journalist, expert in investigating cultural and social aspects of cities, and dealing with urban transformations.
Media	Rebecca Ann Huge	Journalist who writes for Forbes, she is based in Venice and originally from Scotland
HEIs and resident	Massimo Warglien	Professor of Management - Ca' Foscari University of Venice
HEIs	Maria Chiara Tosi	Professor - IUAV (Istituto Universitario di Architettura di Venezia)
HEIs	Marta De Marchi	Research Fellow - IUAV (Istituto Universitario di Architettura di Venezia)
Student	Diego Domenichelli	Student - Ca' Foscari University
Student	Nicole Favero	Student - Ca' Foscari University
Student	Sara Gavioli	Student - Ca' Foscari University
Artisans	Riccardo Bisazza	President Fornace Orsoni - Glass Making Factory
Artisans	Luciano Marson	Owner Pieces of Venice - Eco Sustainable design brand
Sport	Guglielmo Feliziani	President Basket Mestre
Sport	Massimo Zanotto	President CUS (University Sport Centre)
Resident	Gianfranco Bettin	Former Councillor for the Environment. Resident and expert of Marghera
Resident and HEIS	Francesca Checchinato	Mother and Professor of Marketing - Ca' Foscari University of Venice

COLLABORATING TO FACE URBAN CHALLENGES

Considering the context described in the previous paragraphs, we aim now to describe the adopted approach, to develop new ideas for a more sustainable Venice, where the financial, social and environmental aspects can be considered together within the urban ecosystem. In the light of the Triple Helix model and its extensions (Carayannis & Campbell, 2011) we point out the hybrid role of universities that can act as brokers to facilitate processes and policy implementation, beyond studying the context and the challenges. We can identify three stages or phases.

Based on our primary and secondary data gathered in the **Inspiration** phase, over-tourism has been confirmed to be a very urgent challenge, linked to depopulation and environmental issues on the lagoon ecosystem. In particular, engaging with Ca' Foscari University students through interviews and surveys we observed that tourism for Venice is an important resource and still a fundamental source of income, but it has also changed the lagoon ecosystem. At this stage the main question that emerged is "*Who do we want to come to Venice*?" and immediately after, the question is "*How can we develop a dialogue with these people*?".

With the 80 students of the course in Communication and Digital Marketing we have analysed how the city is communicating its brand, what is working and what can be improved. Venice seems to have a very anonymous communication approach, that targets everybody, without a specific message addressed to identified target audiences. This seems to translate to an indiscriminate invitation to visit the city without any sort of relation to the history of the city and the respect of its heritage.

With the 60 students of the Design and Innovation Management course, we applied design thinking to the concept of sustainable living in Venice. We divided the students into 12 groups. Each team was assigned to a specific borough (Sestiere) of the city, considering the historic centre, the islands of Murano, Burano and Lido, the mainland with Mestre and Marghera. In 6 weeks, each team:

- conducted an observation of the area assigned to them by visiting it
- managed to interview people who live and work there
- collected insights and identified the main challenges about what it means to live in Venice and what challenges dwellers face.
- developed ideas based on the insights gathered
- built a video (12 videos in total) describing the process they followed and the idea they would propose to make people's lives better in the city.

It is now evident that there are no longer real travellers like in the 1970sor the 1980s, who came and stayed in the city for weeks and wanted to discover Venice. Today there are tourists driven by a rush to see as much as possible in a short time. But not even during the pandemic did Venice attract "quality" tourists, because people came mainly on daily visits, from the areas of the Venetian countryside. And this only emphasised the city's lack of identity. The presence of numerous apartments on Airbnb has made the historic centre a sort of widespread hotel that, on one hand has reduced the number of residents; on the other, it has become a growing phenomenon due to the ease of earning.

This phenomenon, as well as the presence of low-cost hotels in the city, somehow cannibalises the offer of quality hospitality that has always represented a strong element in the identity of the city, just as the presence of low-priced Made in China glass souvenirs murders the offer of Murano glass factories.

Venice is a city protagonist of many movies and this, obviously, influences some of the tourists who arrive in the city and who, rather than exploring the cultural and natural beauty of the lagoon, want to see the locations where films were shot.

That said, for a city with tourist flows that are difficult to map, segment and manage like Venice, it is obvious that an overload of presences in the historic centre generates the escape of residents to the mainland, or even in other municipalities.

In the **Ideation** stage a Consultation Event (December 15th 2021) was organised pursuing two main goals:

- to open a conversation, engaging with different local stakeholders, to better identify the sub-challenges related to the three main challenges we begin our process with, which are overtourism, depopulation and protection of the lagoon ecosystem, to them move forward with more specific co-design workshops;
- to define potential synergies among projects focused on Venice, already in development within the different Departments of Ca' Foscari.

We invited 16 stakeholders from academia, local businesses, arts and culture to discuss together about the city and its challenges. What emerged is that living in Venice seems to be a privilege, but with two very different meanings:

- be residents of a unique, beautiful and one of the most exclusive cities in the world;
- but also, to be citizens of a context full of contradictions and energy-consuming.

During the pandemic, several associations worked by distributing masks and developing other kinds of support, such as grocery shopping, to residents and, by managing these activities, they realised that many areas of the historic centre of Venice had completely empty buildings. Stakeholders agreed on the fact that Venice seems to lack development strategies for public policies, as well as it seems that the city is not fully aware of its metropolitan space of Mestre and Marghera, in the mainland.

The duality Venice - Mestre, due to the uniqueness of the island, makes it difficult to perceive Venice as the historic centre of a single large city made by Venice and its mainland. Eventually, we also asked participants to the consultation event "*How can we together redesign a Venice that is sustainable from a social, environmental and economic point of view*?"

Education appears to be a very important term in this redesign process.

- Education of tourists, to make them understand that Venice is not a playground but a city and as such it must be respected;
- Education of Venetians, to make them resume and understand their own history, their ability to co-design and collaborate;
- Education regarding the relationship with the lagoon, which is an integral part of Venice and which must be known to be respected.

As further steps we are working on the ideation of Bootcamp workshops, Innovation Tours, Mentorship Programs, to brainstorm possible solutions based on the findings gathered during the Inspiration phase and the consultation event in fall 2022. We have been also building stronger internal synergies by creating links among existing academic projects that are trying to add value to the city (i.e. Venywhere and Ucamp). This may help reduce a sense of fragmentation, which seems to be a common factor related to the perception of the city by the different stakeholders we have been engaging with. Since during the consultation event of December 2021, it emerged that education is an important pillar to co-design a sustainable Venice, we think about developing innovative training experiences for those who work in the hospitality industry, so that they can be the first to educate tourists to respect the city while exploring it. We also think about creating new training experiences and developing new job opportunities within Venice to attract a new typology of residents, who are highly skilled and can make Venice a contemporary city.

In the **Prototyping, testing and validation** phase, we will develop mini boutique workshops to test possible solutions and identify which aspects work and which needs to be reassessed. This phase will be fundamental to ideally present the solutions to the Municipality and discuss with them possible implementation in the next few years.

We think that success in urban development can only be achieved through an integrated approach, where strong partnerships between local citizens, civil society, industry and various levels of governance are deemed necessary. In order to accomplish this, the role of Higher Education Institutions is crucial. They are anchor organisations that can lead transformational change in their urban environments.

FUTURE RESEARCH DIRECTIONS

The chapter presents an ideal empirical context for research efforts aimed at deepening our knowledge on two thematic areas. The first pertains social mobilisation and resource acquisition in urban transformation processes. Cities are places wherein vested interests might create inertias and difficulties in triggering change. Real estate operators, politicians referring to specific constituencies, economic actors coalescing around specific agendas and scenarios might either express the urge to subvert the status quo or the need to preserve it. Other lines of fracture might make transformation difficult to achieve, such as generational factors –youngsters obviously demand different things to a city than their older counterparts–, class and socio-economic factors –services and the access to them might differ in different socio-economic groups– and the like. Ours is not a premise that gives up on the chance to ignite change, nor a pessimistic view on the irreconcilability of diverging interests. Our experience in the investigation of these processes in Venice taught us that the desire for change is widespread across groups and constituencies; what hinders transformation is the difficulty in creating coalitions and shared frames and visions of the cities that might accommodate all the interests involved. Thus, our empirical investigation aims at documenting the mechanisms that allow some actors to legitimise a specific transformation direction and the rhetoric tools they use to actually circumvent or oppose resistances and to co-opt diverting social groups in their desired trajectory. From this point of view, we will try and test the theories stemming from studies on resource mobilisation and social movements (Davis et al., 2008; Garud & Karnoe, 2003).

A second line of research we aim at tackling is related to the actual impact of design-based methods in promoting and sustaining social change. Design thinking and design inspired methods and toolkits enjoyed a vast consensus and enthusiasm in the past 10 years as harbingers of social, organisational and economic change at all levels of our society (single firms, sectors, societies). A strengthening body of literature and a concurrent debate among practitioners and experts points to the dangers of excessive enthusiasm and the lack of concentration on the limits and boundary conditions making design thinking and similar methods effective (Kolko, 2018; Lee, 2021). Our longitudinal analysis of the follow-up to the workshops with students and the observations of the results and implications of those who will be involving stakeholders will allow us to formulate propositions to generalise a model of design thinking use for urban renovation. In particular, we will investigate the need to "translate" the tools and logics born in a corporate setting eminently for product and service innovation processes to better fit the categories and idiosyncrasies of the social fabric of a city. Second, we will delve into the shortfalls and gaps of the methods and tools of design thinking in such situations, to promote a reflection that will involve scholars and practitioners to improve the arsenal at our disposal to solve pressing urban problems.

CONCLUSION

Urban challenges can be described as complex ecosystems that require the collaboration of different stakeholders to be tackled. The chapter provided an insight into the lessons learned through the Urban Good Camp Project, as well as a summary of the participants' opinions, giving also a strong voice to students, and the collaborative and creative attitude that the design thinking methodology has the potential to inject. We explored if and how, Ca' Foscari University can reach an even stronger role in tackling pressing urban challenges in Venice, by creating and actively engaging urban communities, developing and implementing multidisciplinary learning interventions for university students, young researchers and life-long learners to design real-life solutions that can be implemented to make Venice a more sustainable urban ecosystem.

At this moment it is as if we were in front of a puzzle to build together, with some pieces that we can create from scratch, with the aim of working on the identity of Venice, to make it a resident-friendly city and, consequently, a better city also for those who come to visit it. It also seems that the duality between the island and the mainland is exacerbating. It can be described as a sort of antagonism, and can find meaning through Goffman's Front Stage and Back Stage behaviour theory (1959). In sociology, the terms "front stage" and "back stage" refer to different behaviours that people engage in everyday life, and use the metaphor of the theatre to explain social interaction. This dramaturgical perspective also emphasises the importance of the "setting," or context, in shaping the performance, the role of a person's "appearance" in social interaction, and the effect the "manner" of a person's behaviour has on the overall performance.

Running through this perspective is a recognition that social interaction is influenced by the time and place in which it occurs as well as by the "audience" present to witness it. It is also determined by the values, norms, beliefs, and common cultural practises of the social group or the place where it occurs.

According to Goffman, people engage in "front stage" behaviour when they know that others are watching. Whatever the setting of front stage behaviour, people are aware of how others perceive them and what they expect, and this knowledge tells them how to behave. It shapes not just what individuals do and say in social settings but how they dress and style themselves, the consumer items they carry around, and the manner of their behaviour (assertive, demure, pleasant, hostile, etc.) These, in turn, shape how others view them, what they expect of them, and how they behave. In the case of Venice, we may say that the front stage is represented by the historical city, a unique setting where everybody wants to go at least once in their lives. Such an important international stage is also where everybody wants to be perceived as a leader, and this brings us for example to the presence of a myriad of associations/non-profits (sometimes very small), almost all with the same goal: fight over tourism and protect the lagoon ecosystem. However, this fragmentation leads to difficulties in achieving goals because of the presence of too many initiatives, generating a vicious circle.

In the Merchant of Venice, Shakespeare wrote "I hold the world but as the world. A stage where every man must play a part" (Act I, Scene 1). Even though it is not the world, Venice actually represents a luxurious exclusive stage, where everybody aspires to be seen. It is like Venice represents a dimension that forces us to recognize that the "reality" of the action we identify ourselves in, is only a "reflection" of reality and the illusion is the master of Venice instead.

Going back to Goffman's theory, if the front stage is where people shape their behaviours according to how they want to be perceived, backstage is where they are free of any expectations and are often more relaxed and comfortable. They let their guard down and behave in ways that reflect their "true" selves. They cast off elements of their appearance required for a front stage performance. When people are backstage, they often rehearse certain behaviours or interactions and otherwise prepare for upcoming front stage performances. In the case of Venice, the back stage is represented by the mainland: Mestre and Marghera. There, since there is less need for visibility, best practises actually happen. In this ecosystem, particularly in Marghera, there is a consolidated activity of associations, focused on the urban regeneration of abandoned areas and buildings, able to create a virtuous circle, stimulating public opinion and involving citizens in the redesign of their local context.

The city we choose to settle in has an impact on the people we meet, the friends we can make, the type of job we will have access to, the prospects for professional growth and the level of salary we can aspire to, and eventually the opportunities we can create for our children. Compared to the past, today we have more freedom and possibility to choose, in terms of different travel options, the possibility of working remotely and also the financial possibilities are different, but at the same time there are tons of options to choose from, and they all have something different to offer.

According to Florida (2008), the key to making the best decision is finding a city that makes us happy and allows us to achieve our life and career goals at the same time. Therefore, the university along with the other stakeholders and the local community have to co-design a Venice that can become a city where people choose to live and build their future. University-community engagement is emerging as an important channel for social innovation (Petersen & Kruss, 2021) and the university can be the platform able to facilitate the change.

REFERENCES

Allnutt, C. (2021). Can Venice Reinvent Itself. *Financial Times*. https://www.ft.com/content/ff4c32eb-b34e-4638-a0c9-1c6656906b02

Ambrosini, F., & Cherici, S. (2022). Anthropogenic Venice: Surviving Life in a Lagoon. *Green European Journal*. https://www.greeneuropeanjournal.eu/anthropogenic-venice-surviving-life-in-a-lagoon/

Audretsch, D. B., & Belitski, M. (2022). A strategic alignment framework for the entrepreneurial university. *Industry and Innovation*, *29*(2), 285–309. doi:10.1080/13662716.2021.1941799

Avelino, F., Wittmayer, J. M., Pel, B., Weaver, P., Dumitru, A., Haxeltine, A., Kemp, R., Jørgensen, M. S., Bauler, T., Ruijsink, S., & O'Riordan, T. (2019). Transformative social innovation and (dis)empowerment. *Technological Forecasting and Social Change*, *145*, 195–206. doi:10.1016/j.techfore.2017.05.002

Bai, X. (2007). Integrating global environmental concerns into urban management. The scale and readiness arguments. *Journal of Industrial Ecology*, *11*(2), 15–29. doi:10.1162/jie.2007.1202

Bánáthy, B. H. (1996). *Designing Social Systems in a Changing World.* Plenum. doi:10.1007/978-1-4757-9981-1

Brown, T. (2008). Design thinking. *Harvard Business Review*, *86*(6), 84. PMID:18605031

Carayannis, E. G., & Campbell, D. F. (Eds.). (2009). *Knowledge creation, diffusion, and use in innovation networks and knowledge clusters. A comparative systems approach across the United States, Europe, and Asia*. Praeger.

Carayannis, E. G., & Campbell, D. F. (2011). Open Innovation Diplomacy and a 21st Century Fractal Research, Education and Innovation (FREIE) Ecosystem: Building on the Quadruple and Quintuple Helix Innovation Concepts and the "Mode 3" Knowledge Production System. *Journal of the Knowledge Economy*, *2*(3), 327–372. doi:10.100713132-011-0058-3

Chesbrough, H. (2003). *Open Innovation: The New Imperative for Creating and Profiting from Technology*. Harvard Business Press.

Davis, G. F., Morrill, C., Rao, H., & Soue, S. A. (2008). Introduction: Social movements in organizations and markets. *Administrative Science Quarterly*, *53*(3), 389–394.

Etzkowitz, H., & Leydesdorff, L. (2000). The dynamics of innovation: From National Systems and "Mode 2" to a Triple Helix of university-industry-government relations. *Research Policy*, *29*(2), 109–123. doi:10.1016/S0048-7333(99)00055-4

Evans, N. (2001). Collaborative strategy: An analysis of the changing world of international airline alliances. *Tourism Management*, *22*(3), 229–243. doi:10.1016/S0261-5177(01)00024-3

Florida, R. (2008). *Who is your city? How the Creative Economy Is Making Where to Live the Most Important Decision of Your Life*. Basic Books, Random House.

Fyall, A., Garrod, B., & Wang, Y. (2012). Destination collaboration: A critical review of theoretical approaches to a multi-dimensional phenomenon. *Journal of Destination Marketing & Management*, *1*(1–2), 10–26. doi:10.1016/j.jdmm.2012.10.002

Garud, R., & Karnøe, P. (2003). Bricolage versus breakthrough: Distributed and embedded agency in technology entrepreneurship. *Research Policy*, *32*(2), 277–300. doi:10.1016/S0048-7333(02)00100-2

Goffman, E. (1959). *The Presentation of Self in Everyday Life*. Penguin Books Ltd.

Hamann, R., & April, K. (2013). On the role and capabilities of collaborative intermediary organisations in urban sustainability transitions. *Journal of Cleaner Production*, *50*(July), 12–21. doi:10.1016/j.jclepro.2012.11.017

Italy's Bureau of Statistics. (2019). *Turismo in Italia nel 2018.* https://www.istat.it/it/archivio/232137

Kolko, J. (2018). The divisiveness of design thinking. *Interaction*, *25*(3), 28–34. doi:10.1145/3194313

Kretz, A., & Sá, C. (2013). Third Stream, Fourth Mission: Perspectives on University Engagement with Economic Relevance. *Higher Education Policy*, *26*(4), 497–506. doi:10.1057/hep.2013.32

Lee, K. (2021). Critique of Design Thinking in Organizations: Strongholds and Shortcomings of the Making Paradigm. *She Ji: The Journal of Design, Economics, and Innovation*, *7*(4), 497–515. doi:10.1016/j.sheji.2021.10.003

McCormick, K., Anderberg, S., Coenen, L., & Neij, L. (2013). Advancing Sustainable Urban Transformation. *Journal of Cleaner Production*, *50*(July), 1–11. doi:10.1016/j.jclepro.2013.01.003

McKay, J., & Marshall, P. (2001). The dual imperatives of- action research. *Information Technology & People*, *14*(1), 46–59. doi:10.1108/09593840110384771

Medeiros, E. (2021). Urban participatory planning approaches in capital cities: The Lisbon case. *European Planning Studies*, 1–18.

Momigliamo, A. (2019). The Flooding of Venice: What Tourists Need to Know. *The New York Times*. https://www.nytimes.com/2019/11/20/travel/venice-flooding.html

Momigliano, A. (2020). Venice Tourism May Never Be the Same. It Could Be Better. *The New York Times*. https://www.nytimes.com/2020/07/02/travel/venice-coronavirus-tourism.html

Petersen, I.-H., & Kruss, G. (2021). Universities as change agents in resource-poor local settings: An empirically grounded typology of engagement models. *Technological Forecasting and Social Change*, 167.

Rotmans, J., van Asselt, M., & Vellinga, P. (2000). An integrated planning tool for sustainable cities. *Environmental Impact Assessment Review*, *20*(3), 265–276. doi:10.1016/S0195-9255(00)00039-1

Sánchez-Barrioluengo, M., & Benneworth, P. (2019). Is the entrepreneurial university also regionally engaged? Analysing the influence of university's structural configuration on third mission performance. *Technological Forecasting and Social Change*, *141*, 206–218. doi:10.1016/j.techfore.2018.10.017

Schuurman, D., De Marez, L., & Ballon, P. (2016). The impact of living lab methodology on open innovation contributions and outcomes. *Technology Innovation Management Review*, *6*(1), 7–16. doi:10.22215/timreview/956

Svensson, P., Klofsten, M., & Etzkowitz, H. (2012). An entrepreneurial university strategy for renewing a declining industrial city: The Norrköping way. *European Planning Studies*, *20*(4), 505–525. doi:10.1080/09654313.2012.665616

Trencher, G., Yarime, M., & Kharrazi, A. (2013). Co-creating sustainability: Cross-sector university collaborations for driving sustainable urban transformations. *Journal of Cleaner Production*, *50*(July), 40–55. doi:10.1016/j.jclepro.2012.11.047

United Nations. (2018). *2018 revision of world urbanization prospects*. https://www.un.org/development/desa/en/news/population/2018-revision-of-world-urbanization-prospects.html

World Economic Forum. (2021) *Which 5 Countries Have The Best National Brand?* https://www.weforum.org/videos/which-5-countries-have-the-best-national-brand

ADDITIONAL READING

Calgren, L., Rauth, I., & Elmquist, M. (2016). Framing Design Thinking: The Concept in Idea and Enactment. *Creativity and Innovation Management*, *25*(1), 38–57. doi:10.1111/caim.12153

Hatchuel, A., Le Masson, P., & Weil, B. (2002). From Knowledge Management to Design-Oriented Organizations. *International Social Science Journal*, *54*(171), 25–37. doi:10.1111/1468-2451.00356

König, J., Suwala, L., & Delargy, C. (2020). Helix Models of Innovation and Sustainable Development Goals. In W. Leal Filho, A. M. Azul, L. Brandli, A. Lange Salvia, & T. Wall (Eds.), *Industry, Innovation and Infrastructure. Encyclopaedia of the UN Sustainable Development Goals* (pp. 1–15). Springer.

Lokwood, T. (2010). *Design thinking: Integrating innovation, customer experience, and brand value*. Doubleday.

Martin, R. (2009). *The Design of Business: Why Design Thinking Is the Next Competitive Advantage*. Harvard Business Press.

Paetz, P. (2014). *Disruption by Design: How to Create Products That Disrupt and Then Dominate Markets*. Apress. doi:10.1007/978-1-4302-4633-6

Von Hippel, E. (2005). *Democratizing Innovation*. The MIT Press. doi:10.7551/mitpress/2333.001.0001

Von Stamm, B. (2004). Innovation – What's Design Got to Do with It? *Design Management Review*, *15*(1), 10–19. doi:10.1111/j.1948-7169.2004.tb00145.x

KEY TERMS AND DEFINITIONS

Co-Design: A participatory approach to designing solutions, in which all members are treated as equal collaborators in the design process.

Depopulation: Substantial reduction in the population of an area.

Design Thinking: Human-centred approach to innovation, focused on understanding people's needs, generating ideas to meet their needs and developing prototypes that need to be tested with final users and adapted if necessary.

Double Diamond Methodology: Design process model developed by the British Design Council in 2005. Divided into four phases—discover, define, develop, and deliver—it is probably the best known and the most popular design process visualisation.

Facilitator: A person or institution that makes an action or process easy or easier.

Hybridization: The creation of new actors out of selective adaptation, innovation, and change.

Innovation: Process of creating new products/services or process of upgrading already existing products/services to the next level.

Overtourism: When a place of interest is visited by excessive numbers of tourists, causing undesirable effects for this place.

Urban Challenge: Main issues faced by a city and its dwellers.

Section 4
COVID-19's Touristic Economy

Chapter 14
Management of the Health Crisis on Social Media by Hoteliers in Europe

Philippe Viallon
University of Strasbourg, France

Cécile Dolbeau-Bandin
Université de Caen, France

Jérémy Picot
IUT Robert Schuman, University of Strasbourg, France

Viola Krebs
University of Strasbourg, France

ABSTRACT

The COVID-19 pandemic was a major phenomenon with strong health, political, economic, and social consequences. But the health crisis was also coupled with an equally powerful communication crisis. This double crisis has of course had an impact on the world of tourism. Since mobility has been partially authorized again in Europe, hoteliers are looking to win back their customers, especially on these same social media. What is the place of the pandemic in online communication? How can the subject of the crisis be raised without scaring customers? Are the modes of communication different depending on the category of the hotel? Based on 10 hotels of different categories taken at random in Paris, London, Berlin, and Rome, the study seeks to analyze the informational policies implemented online by individual hoteliers. The results show that the variables "country" or "category" are not significant; there are strong differences from one hotel to another, which highlight a greater or lesser capacity or interest in using these communication tools to promote the establishment.

DOI: 10.4018/978-1-6684-3369-0.ch014

INTRODUCTION

The COVID-19 pandemic was a major global phenomenon with strong health, political, economic and social consequences. But the health crisis was also coupled with an equally powerful communication crisis. Social media were the communication outlet for this informational crisis, characterized by an overabundance of contradictory information, fertile ground for a variety of scientific and conspiracy theories. Citizens had a hard time determining what was true and what was false. This double crisis has of course had an impact on the world of tourism. Since, mobility has been partially re-authorized in Europe. Moreover, this is a difficult time for hotels from an economic point of view: the general uncertainty and the rapid changes weigh on the economic situation. At the micro level, many hoteliers fear a possible closure of their hotel; at the macro level, it is the profitability of the activity (Tsionas 2021) that is scrutinized by financial investors (Sharma/Shin/Santa-María/Nicola 2020).

The objective of this research is to compare the digital communication of 40 European hotels during the health crisis. How did European hotels react during this period, which was very complicated for them? The period analyzed is from 2020 to 2022. We will not deal with the post-crisis period, which is another topic. This study aims to answer two questions: the first aims to determine whether the category of the hotel has an influence on the communication strategy implemented: the hypothesis would be that higher class hotels have more means to implement an adapted communication. The second question is whether national differences can be detected as a result of the different situations in the various countries.

BACKGROUND

State of the Art

This paper combines two main themes identified by Yang, Zhang, and Rickly (2021) in their study of tourism research and COVID-19: theme 2 "responses, strategies, and resilience" and theme 5 "technology adoption". Three successive theoretical frameworks will be considered through a up-to-date literature review. The first one will question the importance of communication and especially digital communication for tourism. Indeed, a hotel that does not have a website or a facebook page does not exist. The guesthouses in remote Laddakh have understood this. The second framework will point out the consequences of the health crisis for tourism. It will consider the place of crisis communication, in general, and more specifically in tourism, in particular, with a question that often comes up in studies: what are the expected longer term consequences of this global crisis for global tourism practices? This point is essential, not as much in order to predict how managers will react after the crisis, but rather to understand what they did during the crisis. The last part will focus on the few studies conducted on the subject of the digital strategy of hotels during the crisis.

Digital Communication and Tourism

Since 2012, more than half of all tourism-related spending in Europe was done on the Internet and the trend is growing (Thirumaran, Mohammadi, Pourabedin, Azzali, Sim 2021). The communication of hotels typically began with a web 1.0 business card type of website providing information with many advantages: almost unlimited space, updating possibilities, diversity of forms, personalization, reasonable cost, etc. In a second time (web 1.5), the websites allowed a beginning of interaction by giving the customers the possibility to book a room. Hoteliers just had to make sure that all their reservation sources were well coordinated to avoid overbooking. But the web 2.0 has changed the rules: Information circulates not only from top to bottom, but also from bottom to top and especially horizontally. Social media are the expression of these new forms of information flow. To what extent do hoteliers take this evolution into account? Do they have a forum on their website where they interact with their customers? Do they respond to criticism on facebook or Instagram? If these questions are well debated in the professional field, they are rarely the subject of scientific articles. A website is to a hotel what a shop window is to a store: it must seduce and incite people to enter. It must therefore be perfect, whether it is managed internally or by an external company.

Tourism and COVID-19

Although we are still in the health crisis of COVID-19, scientific publications on the subject are already numerous, particularly in Asia. The reactivity of researchers (Duro/Perez-Laborda /Turrion-Prats/ Fernández-Fernández 2021) and institutions (COVID-19 and tourism 2020) is strong, commensurate with the economic stakes for many countries in the world. As a matter of fact, tourism is important for the countries of the South, but also for those of the North: it represents for example 12% of GDP of Spain and 7% of France, 7% of Thailand or Morocco (UNWTO 2021). All studies agree on the extent of the impact of the crisis: all sectors of society have been affected (I 2020), of course those directly or indirectly related to tourism, but also others because of the closure of businesses, stores and administrations during the successive confinements, because of the decline in income of the population (Guibert/ Réau 2021), because of the absence of foreign currency from tourists. Among factors that have hindered tourism are the frequent changes in political decisions (Lebaron 2020), the regular arrival of new variants — modifying situations that were thought to be under control, the uncertainty about the rules of entry and exit of countries, differing regulations in the countries (Collins-Kreiner/Ram 2021, Gounder/ Cox 2021). In general, the difficulty of making forecasts that are necessary for tourism has strongly affected the sector (Foo/Chin/Tan/Phuah (2021). We have had further proof that security and stability are the two mainstays of tourism. It helps to develop attachment to a place (Limei/Ying/Qi 2021). Any crisis leads to considerations in the short term – how to overcome it? (Boto-García/Mayor 2022) – and in the long term (Lemoine 2020) – will tourism evolve and if so, in what way? Researchers agree that there is a pause in the development of tourism and that we should not expect any major changes after these special times.

Crisis communication is a field that is well known in the corporate world. It also frequently affects tourism in the case of war, terrorism, climatic or meteorological events, accidents, etc., but it is much more difficult to manage in this case, due to the multitude of actors (national and local governments, international and national companies) (Dianoux/Siadou-Martin 2020, Fabry/Zeghni 2020). Research has shown that communication is therefore even more essential: an absence of it means a lack of professionalism or leaves room for the wildest assumptions, even fake news. Finally, many researchers are wondering about the post-COVID-19 dynamics (Fabry/Mugnier 2020, Kock/Nørfelt/Josiassen/Assaf/ Tsionas 2020). The reflection is not new: mass tourism and its excesses, an alternative through a possible green, alternative, fair tourism has often been evoked in the past (Khamung/Po Siu 2021, Delaplace 2021). However, most researchers (Babii /Sanaa 2021, Moatti 2021, Rastegar/Higgins-Desbiolles/ Ruhanen 2021, Violier 2021) do not believe in a change. Instead, they rather believe that like other past crises, whether sanitary (avian flu) or political (jasmine revolution in Tunisia), tourism will start again, especially since tourists frustrated by their immobility for two years are impatient to travel again. Given this dynamic, the communication carried out today by hotels is essential (Skare/Riberio/Porada-Rochoń 2021), because it prepares these future stays.

Digital Communication in this Crisis

When physical communication is problematic, technology can provide a substitute. The use of digital social networks has strongly increased during these periods of social distancing and confinement, with both their interpersonal and intimate dimensions being used (Viallon/Gardère 2020). Some companies, in general, and tourism companies, in particular, have understood the interest they could have in being on platforms to maintain more personal links with their customers (Le/Phi 2021) or to reinforce their brand image by transmitting positive messages during a particular period (Ou/Wong 2021). If a website refers more to a rational and mercantile dimension, a facebook page or an Instagram account work more on feelings and building on relationships (Thirumaran/Mohammadi/Pourabedin/Azzali/Sim 2021). They also make it possible to address people's sensitivity and understand their representations (Wassler/Talarico 2021). Finally, they are also a marker for hotels to follow innovations, to understand the needs and desires of different generations, offering potential customers a more personalized service (Duthion/Mandou 2016).

Methodology

The principle of comparative analysis was adopted on two levels: on the one hand, it was decided to choose the four capitals of the four main European countries (Paris, London, Rome and Berlin, Geneva). The first three are also the cities that receive the most tourists in Europe. The aim was to understand whether the diversity of measures taken by the governments in each country as well as the professional practices linked to the culture of each country had an effect on the communication of hotels. In addition, ten hotels in each city were randomly selected. These are the first ten hotels that appeared in a search with the keywords "city" + "hotel" on facebook. They are hotels from 2 to 5 stars, with a overrepresentation of 4 and 5 stars. This was accepted, because the study wanted to verify if the hotels with the most financial means communicated better than the others. In addition, the prices offered are not the usual prices: most of the hotels have significantly lowered their rates, in order to quickly regain a clientele and this puts them on a par with 3-star hotels. The list is attached. We have chosen not to name the hotels in the article, as the objective is not to give good and bad points, but rather to understand the main trends. For each hotel, it was decided to analyze all of its online digital communication: the website, the facebook page, and all other social media that are used: Instagram, Twitter, LinkedIn, Pinterest and YouTube. The objective was to see if a global strategy was present and if so, how it was structured, and what role was assigned to each medium.

A triple analysis was conducted on the publications of these 40 hotels and their 130 social media postings: first, a textual content analysis: what are they talking about? Second, a semiological analysis of the photographs and videos (Ni Wayan/I Wayan 2021): who and what are they showing? And finally, a media analysis on the way each medium is managed: when is the information posted? How is it arranged? What interactions are proposed with the public, etc.). The study was conducted in January 2022, i.e., in the four countries analyzed, there is freedom of movement, and all hotels are open again. Constraints (sanitary passes, masks, distancing,...) are still present, but the populations aspire again to travel and to forget this period, which has been longer than previously expected (Xun/Jian/ Baojun/Peiwen 2021). Websites were analyzed at this time, accounts and pages on digital social networks diachronically from the beginning of the crisis in March 2020 to February 2022.

ANALYSIS

In this section, the communication of hotels dedicated to COVID-19 will be analyzed in the following way: first, city by city and then, according to the medium concerned: first the websites, then the facebook pages and finally the accounts on other social media.

Parisian Hotels

France has experienced three confinements: from March 17 to May 11, 2020, from October 30 to December 15, 2020, and from April 3 to May 3, 2021. The websites of the Parisian hotels fall into four categories:

1. three have chosen not to communicate at all on the subject,
2. two transmit very little information,
3. two others provide more complete information on several pages with detailed explanations,
4. two go even further and deal with the health problem by integrating it completely into their strategy. They aim for transparency, they seek to reassure their customers by (re)building trust. On their Facebook page, only two hotels mention the health crisis directly, three others do so implicitly. Among other social media, only Instagram is often used, the others (Twitter, YouTube, LinkedIn) are very little used and often echo Facebook. The crisis is almost absent from the publications.

London Hotels

The United Kingdom has experienced two confinements: from March 24, 2020 to mid-April, from November 5 to December 2 and from December 20 to March where the deconfinement was progressive. All but one of the London websites analyzed mention the health crisis. They are divided into two roughly equal groups: either they mention the pandemic in a minimalist way, a small cartridge in a corner of the screen, as if to meet a legal obligation, or they are much more explicit and seek to inform in detail about the measures in force and to reassure their potential tourists. On Facebook, three types of strategies have been updated: the first one consists in not communicating anymore, the closing of the hotels has probably led to an absence of activity on these specific accounts. The second strategy is to avoid the subject completely, as it seems to be disturbing. As for the third, posts are dedicated to the pandemic. All those who communicate seek to stay close to their clientele and give a good image. Nine hotels have an Instagram account, seven have a Twitter account, two have a LinkedIn account, two have a YouTube channel and one has a Pinterest account. Some accounts are either very inactive or simply copies of other accounts (Facebook or Instagram to Twitter). Overall, there is very little mention of the health crisis and the accounts mainly promote hotels.

Berlin Hotels

Germany did not experience any real containment: there were more or less strong distancing measures depending on the period and the region, Berlin was less affected. For the websites of Berlin's hotels, we can see a strong communication on the pandemic, clear information. Numerous references to government websites allow hotels to both respect the law and follow the evolution of the measures. On Facebook, four hotels do not mention the crisis, and these are often the same hotels that do not publish or publish very little information and have a strategy of avoidance; three publish some information; three others are very active with a complete strategy in the face of the pandemic and are transparent. But in all cases the number of interactions is very low. For other social media, all hotels are on Instagram. The analysis reveals a diversity of strategy that ranges from no mention and integration to the overall strategy. Twitter is little used and rarely in an original way, LinkedIn is used once as a recruitment platform, once as an echo of Facebook.

Roman Hotels

The Roman hotels have been confined for two periods: from March 9 to May 18, 2020 and from December 21 to January 6, 2021. They are in an interesting situation, as Italy was the first country to be affected by the crisis in Europe and experts have often said that the country is a textbook case for the rest of Europe. Of the ten websites, only three mention COVID-19. In Rome, the chosen hotels do not use Facebook as a privileged communication tool: with three exceptions, the pages are either no longer active or have a very low publication rate. Communication on other digital social networks is very limited: Italian hotels do not generally make communication on social media one of their priorities and consequently COVID-19 is rarely mentioned.

Geneva Hotels

From 13 March to 14 June 2020, Swiss borders were closed. On 14h June, they reopened for the Schengen zone. PCR tests, quarantine and/or vaccination passes are required thereafter, depending on countries from which passengers are traveling. Throughout the ski season of 2020/1, the French are officially not allowed to spend time in Swiss ski resorts. Continuously changing measures make it hard for tourists to have any visibility of what is permitted and the measures required for travel to happen without major issues and hiccups. Measures are constantly changing, more or less restrictive, with restaurants remaining closed –something which also affects hotels— until February 2022, when they are finally lifted progressively.

Three categories can be distinguished among Geneva hotels:

1 Hotels not mentioning anything about the pandemic,
2 Hotels providing a dedicated webpage, but no social media mentions about closures or reopening,
3 Hotels providing both dedicated information on their website and also announcements made in social media.

FUTURE RESEARCH DIRECTIONS

In this section, the use of different social media will be discussed in a transversal way.

Websites were the first tool used by hotels in the history of their digital communication. One would think that everyone would master it perfectly. But this is not the case: about a third have blurred photos, several cases of dead links have been reported. COVID-19 is either completely absent or only mentioned in a small section on the homepage, but a few hotels in different categories and countries have played the transparency card and integrated the pandemic into their strategy to show all the efforts they make for their customers.

Facebook is the "ancestor" in the digital world. It is the most ancient digital social network, still in use and it remains in the lead. It is the preferred tool for people over 40. But these are often the people who have the most money, so it is understandable that for hotels it is a preferred tool. The regularity of posts is a quality criterion: some hotels had completely stopped communicating on Facebook (they were probably closed). Others had very spaced publications. Only a few hotels have maintained the link with their community without interruption. The number of subscribers is an indicator, the number of "likes" is also an indicator, but interaction is also essential: when people write a positive or negative message, they expect a response. Whether it is on COVID-19 or another topic, this is rarely the case.

Instagram's strength [1] is in the image. Several common elements emerge. The relatively recent creation date of Instagram accounts: more than 50% of the accounts were created after 2017: the frequency of publication is low, even very low; there is no visual message, yet a priority on Instagram, centered on COVID-19; the messages are mainly informative while the platform strength is in aesthetics and originality. Hotels have many opportunities to offer a virtual journey to their future guests: their facade, their rooms, their common areas are at the heart of their communication. Others also use it to value their employees and at the same time make their hotel more human. Their restaurant --when they have one-- also offers an almost inexhaustible source of photographs: the evolution of the menus and the diversity of the dishes allows them to naturally meet the requirement of regular publication on social media. Hotels can also use images to highlight their urban environment. But only a third have made this choice. COVID-19 is almost never mentioned, which is understandable, as it is difficult to visualize and not very interesting to show.

Twitter is the microblogging platform whose strength is to allow communication in the same temporality between an event, its transformation into a message by one individual and its reception by another. If its use for a show or an event makes sense, the extent of its use by hotels was surprising. The analysis has unfortunately revealed that, in the majority of cases, it was to post copies of messages already published on Facebook or Instagram and without any interaction. An inappropriate use of Twitter which explains the poor numbers of subscribers. COVID-19 is absent from Twitter.

YouTube is a video platform the success of which is only growing: in 2022, 500 hours of videos are uploaded every minute and more than one billion hours of videos are watched on a daily basis. Tourism has a significant share in this production, with the publication of both professional and tourist videos. Only a few hotels in the analysis have a YouTube channel. Even the hotel chains that have greater means do not make it a priority. Interviews with professionals have provided us with the following insight: lack of time, skills and money. This is a pity, because to make people dream, video is the ideal tool and moreover it is very popular among the public. The epidemic is not mentioned at all, but the very limited number of videos does not allow a generalization.

CONCLUSION

At the time of writing (February 2022), we can already say that the health crisis (2020-2022) has had, like any crisis, the effect of revealing weaknesses, in this case that of the digital communication of many hotels. Almost all hotels have accounts on different social media platforms, but very few have an adapted use of them: the number of publications is often insufficient, the rhythm is irregular, there is no or little interaction with the public. The first hypothesis stating that the category of the hotels had an influence on the quality of the communication is not confirmed. Indeed, some three-star hotels communicate very professionally, while five-star hotels completely ignore basic professional communication standards. The other hypothesis, linked to the "city" variable, and therefore the country and culture, which could play a role is also not relevant either. As a matter of fact, in all countries, some hotels stand out for the quality of their communication, while others have completely neglected it. Only the hotels in Berlin stand out for their stronger communication on their websites.

In addition, the analysis of all the communication media of each hotel allowed us to verify whether or not a global strategy with respect to COVID-19 had been put in place: here again, the results are very contrasted and few hotels demonstrate that they master the specificities of social media: duplication of messages of Facebook and Instagram to Twitter or between the website and Facebook, no dedicated account when the account of the chain to which the hotel belongs does not allow for a personalized relationship, no response to the criticisms made, etc. Overall, COVID-19 is very little present while the constraints related to the pandemic are not over. Five types of strategies have been identified in all media:

1. COVID-19 does not exist. The strategy adopted is the following: the reality disturbs, it is denied. The advantage is that visitors who do not want to hear about the disease are satisfied, but it is a bit like burying your head in the sand.
2. Hoteliers give it a very small place in their communication. This strategy is that of the minimum legal coverage allowing both the minimum compliance with legal measures and the legal protection against customers who could sue for lack of information.
3. Managers play the good students, applying by the letter government instructions, showing their good will and their concern to take care of their customers.
4. Hotels integrate completely the COVID-19 communication by turning a vice into a virtue. They will strengthen their digital communication to compensate for the lack of contact in real life.
5. How to explain these fifth strategy? After two years of "stop and go", regarding the variations of the epidemic and the often contradictory speeches made by governments about it, the public is tired and aspires to forget, especially when traveling. Some hoteliers are making the choice to help their customers forget, the watchword seems to be "above all do not talk about the devil" on any medium. They have made the aversion of tourists into a theme and motto (Williams/Li/Chen/Balasz 2022). Multiple other interpretations are possible here: staff in charge of communication hay have been laid off (Zhang/Wang/Rickly 2021), thus information has not been updated; or communication is not the priority in the strategies for restarting the activity. The injunction made in 2020 by several researchers (Sharma/Shin/Santa-María/Nicolau 2021, Khalilur Rahman/Gazi Md.//Bhuiyan Miraj/Rahaman 2021, Williams/Chen/Li/Baláz 2022) to innovate, in order to manage the crisis has not always been confirmed by facts. Yet, surveys show that the issue of safety is major for tourists (Rahman/Gazi/Buhiyan/Rahaman 2021, Orîndaru/Popescu/Alexoaei/Căescu/Florescu/Orzan 2021).

But other hoteliers have made a strategic choice that seems both braver and more subtle: they have a small, discreet mention of COVID-19 on their website which –with one or two clicks— gives the interested client the opportunity to get a little or a lot of information. This corresponds to the adopting of the host culture, as defined by Wassler Philipp, Talarico Chiara (2021). This strategy also allows hotels to avoid possible legal battels. These companies facilitate the stay of their customers by communicating on practical aspects and taking the opportunity to recall all the precautions they take to protect their guests, providing them with all due respect (Bonfanti/Vigolo/Yfantidou Georgia 2021). This information about COVID-19 is confined to the website, which provides the framework for a custom navigation for the Internet user, while the platforms hardly mention the pandemic. They are playing their usual role, that of making people dream with beautiful photos, inciting them to enjoy the hotel and dream conveyed by it, creating and maintaining the community. When hotels had to shut their doors for specific periods due to COVID-19, dure to government-ordered lockdowns, most of them have stopped communicating (Siga 2020). When the links are distended by a crisis, the community becomes even more important, it must be kept alive (Blanchet 2020). Social media allow this new type of relationship, but it is important to use them well.

Finally, this study reveals that, thanks to the COVID-19 crisis, the use of social media in the hotel industry is largely susceptible to improve: the regularity of publication, the knowledge of the specificity of each platform: the right tone of the message and the place of the tourist/customer in the communication must be improved.

REFERENCES

Babii, A., & Nadem, S. (2021). *Tourism in a Post-Pandemic World.* https://www.imf.org/en/News/Articles/2021/02/24/na022521-how-to-save-travel-and-tourism-in-a-post-pandemic-world

Blanchet, C. (2020). Crise de la Covid-19 dans le secteur du tourisme. Illustration de la couverture médiatique et académique. *I2D - Information, données & documents*, (3), 101-114.

Bonfanti, A., Vigolo, V., & Yfantidou, G. (2021). The impact of the Covid-19 pandemic on customer experience design: The hotel managers' perspective. *International Journal of Hospitality Management*, *94*, 94. doi:10.1016/j.ijhm.2021.102871 PMID:34866744

Boto-García, D., & Mayor, M. (2022). Domestic tourism and the resilience of hotel demand. *Annals of Tourism Research*, *93*, 93. doi:10.1016/j.annals.2022.103352

Cao, L., Qu, Y., & Yang, Q. (2021). The formation process of tourist attachment to a destination. *Tourism Management Perspectives.* https://unctad.org/system/files/official-document/ditcinf2020d3_en.pdf

Collins-Kreiner, N., & Ram, Y. (2021). National tourism strategies during the Covid-19 pandemic. *Annals of Tourism Research*, *89*, 89. doi:10.1016/j.annals.2020.103076

Delaplace, M. (2021). Après la crise, un tourisme durable ? *L'économie politique*, *3*(91), 8-22.

Dianoux, C., & Siadou-Martin, B. (2020). Eclairages sur la communication des entreprises lors de la crise de la COVID-19: que dire et comment le dire ? *Projectics, 3*(27), 9-20.

Dufour-Coppolani, D. (2020). Une vision inclusive. *I2D - Information, données & documents*, (3), 7. https://www.cairn.info/revue-i2d-information-donnees-et- documents-2020-3-page-7.htm

Duro Juan, A., Perez-Laborda, A., Turrion-Prats, J., & Fernández-Fernández, M. (2021). Covid-19 and tourism vulnerability. *Tourism Management Perspectives*, *38*. PMID:34873568

Duthion, B., & Mandou, C. (2016). *L'innovation dans le tourisme: Culture numérique et nouveaux modes de vie*. De Boeck.

Fabry, N., & Mugnier, P. (2020). Peut-on entrevoir les contours de la nouvelle industrie touristique ? *I2D - Information, données & documents*, (3), 85-89.

Fabry, N., & Zeghni, S. (2020). Au-delà de la Covid-19: la gouvernance des territoires touristiques. *I2D - Information, données & documents*, *3*(3), 95-100.

Gounder, A., & Cox, C. (2021). Exploring the role of tourism dependency on COVID-19 induced economic shock in the Small Island Developing States. *Current Issues in Tourism*, 1–18.

Guibert, C., & Réau, B. (2021). Les travailleurs du tourisme dans la tourmente. *Economics and Politics*, 3(91), 36–46.

INSEE. (n.d.). *Activité touristique* . https://www.insee.fr/fr/statistiques/4625624?sommaire=4625628

Kasni, N. W. (2021, November). The Multimodal Forms of Tourism Promotional Discourse in the Age of COVID-19. *International Journal of Linguistics, Literature and Culture*, *7*(6), 422–440.

Khalilur, R. M., & Gazi, M. (2021, September 1). Effect of Covid-19 pandemic on tourist travel risk and management perceptions. *PLoS One*, *16*(9), e0256486. Advance online publication. doi:10.1371/journal.pone.0256486

Khamung, R., & Hsu, P. S. (2021). Striving for a new normal after the Covid-19 pandemic: Taking small businesses at Bangsaen Beach as an example. *Journal of Tourism and Cultural Change*, 1–17. doi:10.1080/14766825.2021.1987447

Kock, F., Nørfelt, A., Josiassen, A., Assaf, A. G., & Tsionas, M. G. (2020). Understanding the COVID-19 tourist psyche: The Evolutionary Tourism Paradigm. *Annals of Tourism Research*, *85*, 85. doi:10.1016/j.annals.2020.103053 PMID:32921847

Le, D., & Phi, G. (2021). Strategic responses of the hotel sector to COVID-19: Toward a refined pandemic crisis management framework. *International Journal of Hospitality Management*, *94*, 94. doi:10.1016/j.ijhm.2020.102808 PMID:34785839

Lebaron, F. (2020). Covid-19. L'anomie pandémique. *Savoir/Agir*, *2*(52), 7-9.

Lee-Peng, F., Mui-Yin, C., Kim-Leng, T., & Kit-Teng, P. (2021). The impact of COVID-19 on tourism industry in Malaysia. *Current Issues in Tourism*, *24*(19), 2735–2739. doi:10.1080/13683500.2020.1777951

Lemoine, B. (2020). Après la pandémie. *Villes en Parallèle*, *49-50*(1), 446–452. doi:10.3406/vilpa.2020.1833

Li, X., Gong, J., Gao, B., & Yuan, P. (2021). Impacts of COVID-19 on tourists' destination preferences: Evidence from China. *Annals of Tourism Research, 90*, 90. doi:10.1016/j.annals.2021.103258 PMID:34924648

Moatti, S. (2021). Les vacances d'après. *Economics and Politics, 2021*(3, n°91), 5–7.

Morand, J.-C., & Mollard, B. (2008). *Tourisme 2.0.* M21 Editions.

Nikjoo, A., & Hernández-Lara, A. B. (2021, April). The flag-bearers of change in a patriarchal Muslim society: Narratives of Iranian solo female travelers on Instagram. *Tourism Management Perspectives, 38*, 100817. doi:10.1016/j.tmp.2021.100817

Orîndaru, A., Popescu, M.-F., Alexoaei, A. P., Căescu, S.-C., Florescu, M. S., & Orzan, A. O. (2021). Tourism in a Post-COVID-19 Era: Sustainable Strategies for Industry's Recovery. *Sustainability, 2021*(13), 6781. doi:10.3390u13126781

Ou, J., & Wong, I. A. (2021). Strategic crisis response through changing message frames: A case of airline corporations. *Current Issues in Tourism, 24*(20), 2890–2904. doi:10.1080/13683500.2020.1849051

Rastegar, R., Higgins-Desbiolles, F., & Ruhanen, L. (2021). COVID-19 and a justice framework to guide tourism recovery. *Annals of Tourism Research, 91*, 2021. doi:10.1016/j.annals.2021.103161 PMID:33583982

Sharma, A., Shin, H., Santa-María, M. J., & Nicola, J. L. (2020). Hotels' COVID-19 innovation and performance. *Annals of Tourism Research, 88*, 2021.

Sigala, M. (2020). Tourism and COVID-19: Impacts and implications for advancing and resetting industry and research. *Journal of Business Research, 2020*(117), 312–321. doi:10.1016/j.jbusres.2020.06.015 PMID:32546875

Skare, M., & Riberio, S. D. (2021). Impact of COVID-19 on the travel and tourism industry. *Technological Forecasting and Social Change*, 163.

Thirumaran, K., Mohammadi, Z., Pourabedin, Z., Azzali, S., & Sim, K. (2021). COVID-19 in Singapore and New Zealand: Newspaper Zahra portrayal, crisis management. *Tourism Management Perspectives, 38*, 38. doi:10.1016/j.tmp.2021.100812

Tsionas, M. G. (2021). COVID-19 and gradual adjustment in the tourism, hospitality, and related industries. *Tourism Economics, 27*(8), 1828–1832. doi:10.1177/1354816620933039

UNWTO. (2021). *International Tourism Highlights 2020.* https://www.e-unwto.org/doi/epdf/10.18111/9789284422456

Viallon, P. (2013). La communication touristique, une triple invention. *Mondes du Tourisme, 7.*

Viallon, P., & Gardère, E. (2020). *Médias dits sociaux, médias dissociants.* De Boeck.

Violier, P. (2021). La pandémie du COVID, une parenthèse dans la mondialisation touristique. *Economics and Politics*, 3(91), 23–35.

Wassler, P., & Talarico, C. (2021). Sociocultural impacts of COVID-19: A social representations perspective. *Tourism Management Perspectives*, 38.

Williams, A. M., Chen, J. L., Li, G., & Baláž, V. (2022). Risk, uncertainty and ambiguity amid Covid-19: A multi- national analysis of international travel intentions. *Annals of Tourism Research*, *92*, 92. doi:10.1016/j.annals.2021.103346 PMID:35013630

Yang, Y., Zhang, C. X., & Rickly, J. M. (2021). A review of early COVID-19 research in tourism: Launching the Annals of Tourism Research's Curated Collection on coronavirus and tourism. *Annals of Tourism Research*, *91*, 91. doi:10.1016/j.annals.2021.103313 PMID:34611371

Zhang, C. X., Wang, L., & Rickly, J. M. (2021). Non-interaction and identity change in Covid-19 tourism. *Annals of Tourism Research*, *89*, 89. doi:10.1016/j.annals.2021.103211

KEY TERMS AND DEFINITIONS

Communication Crisis: The main rule of teaching crisis communication is the importance of preparation. No one anticipated the COVID-19 crisis. We need to learn from it for the future and train hotel managers to communicate in future health crises that experts believe will continue to spread.

Health Crisis and Hotels: The health crisis has revealed great differences in the ability of hotel managers to manage their digital communication.

Hotel Professionals: As in other areas of business, many hotel managers still have a very top-down view of information. It is not enough to adopt social media to have a social vision of information.

Social Media: Social media is a great tool to develop the e-reputation of hotels; however, hoteliers must invest in professional skills to manage it.

Tourism and COVID-19: the COVID-19 health crisis was of course a problem for the entire hotel industry, but it also revealed the weakness of the digital communication of some hotels.

Tourism Communication: Unlike marketing, which looks at the best way to make investments profitable, tourism communication studies all the elements that come into play before, during and after the stay for all the actors linked to the territory concerned.

ENDNOTE

1 These elements were obtained thanks to the work of Elisa Frécher.

APPENDIX

List of the Hotels

Paris: Hotel Mathis Paris, Drawing Hotel, Shangri-La, Hotel Molitor, Hotel du Levant, Fauchon l'hôtel, Secret de Paris, The Peninsula, Hotel Paris Lafayette, Marys Hotel.

London: Z Hotel City, Club Quarters St. Paul, Leonardo Royal London St. Paul's, One Aldwych, Club Quarters Hotel Lincoln's Innn Fields Holborn, Rosewood London, Sea Containers London, The Rookery, Hilton London Bankside.

Berlin: Radisson blu Hotel, Hotel Zoe by Amano Group, Regent Berlin, Cosmo Hotel Berlin Mitte, Sofitel Berlin Gendarmenmarkt, Hotel de Rome, a Rocco Forte Hotel, The Westin Grand Berlin, Capri by Fraser Berlin, Hilton Berlin, Courtyard by Marriott Berlin City.

Rome: Hotel Ariston, Roma Resort Termini, Best Western plus Hotel Universo, UNA Hotel Deco Roma, Golden Tulip Rome Piram, Bettojas Hotel, Hotel Nord Nuova Roma, Hotel San Remo Nuova, Palazzo Montemartini Radisson, NH Collection Roma Palazzo Cinquecento.

Chapter 15
Challenges for Urban Tourism in a Post-Pandemic World

José Dias Lopes
ISEG, University of Lisbon, Portugal & KIPT, Portugal

Ângela Lopes
KIPT, Portugal

Antónia Correia
Universidade do Algarve, Portugal & KIPT, Portugal

João Pedro Portugal
KIPT, Portugal

ABSTRACT

Urban tourism has particularities that distinguish it from other forms of tourism. The urban tourist wants to enjoy the city, its monuments, its environment, its cuisine, its events. The tourist will always be an outsider to the city, someone who will enjoy an area that is not designed specifically for tourism. But the tourist does not have to be a stranger in the city. The pandemic affected urban tourism in a more "violent" way. However, the pandemic has not affected the pull factors of cities – the cuisine, the monuments, the events, the vibe are still there waiting to be enjoyed. They will be, and city tourism will return with full vigor. This chapter argues that the quest for sustainability and change supported by a disruptive set of new technologies and the consequent skill shifts will possibly affect urban tourism intensely. These new trends have the potential to give rise to different solutions than those that have been attempted in the past. We will see if we can better reconcile cities with their tourists to the benefit of all.

DOI: 10.4018/978-1-6684-3369-0.ch015

INTRODUCTION

Since World War II, the tourism industry has been growing steadily (Kester, 2015). In 2019, the industry reaches a new record of 1,460 million in international arrivals and $1,481 billion in revenues (UNWTO, 2021a). Despite the slowdown in the global economy and the feeling of great uncertainty over Brexit, and the geopolitical and trade tensions that were felt, tourism, although impacted by this situation, persisted in a growing trend (UNWTO, 2021a). Even though this macroeconomic context and the possibility of "health scares and extreme weather events, over the long-term tourism is expected to continue to grow" (OECD, 2020, p.3).

In Portugal, tourism is widely recognized as a sector of great relevance for the country. In 2019, tourism "represented about 8.5% of the total Gross Value Added of the national economy (…) recording a higher increase than that of the national economy compared to the previous year" (Turismo de Portugal, 2021, p.4). In the last decade, the global growth trend was accompanied by Portuguese tourism, both in revenue and in overnight stays (INE, 2021; Turismo de Portugal, 2021), resulting in a very optimistic environment for the entire sector.

The competitive position of Portugal as one of the best destinations in the world is mostly related to the urban tourism infrastructure growth. Tourism infrastructures are critical to the development of urban tourism (Jamil & Puad, 2010; Lim, Zhu & Koo, 2019). The overnight stays in the biggest cities in Portugal, Lisbon and Oporto, grew between 9 and 12 percent per year in the time frame of 2014-2019.

The boom of urban tourism was suddenly, and particularly, frozen due to the pandemic. Under these tenets this chapter discusses policies, challenges, and opportunities in urban tourism on the road to excellence, demystifying the prejudice that is engrained in this concept but also bearing in mind the two major topics that COVID-19 deeply prints on a global scale: sustainability and transformation digital.

URBAN TOURISM

Urban tourism is embedded in prejudice as it is mostly considered a negative externality (Biagi, Ladu, Meleddu, & Royuela, 2020; Liu, 2013). The discussion around this topic unveils problems that urban tourism could bring, such as over-tourism, anti-tourism, touristification, and tourism gentrification (Yubero, Condeço-Melhorado, García-Hernández, & Fontes, 2021), all of which relates to the over-tourism problem.

According to UNWTO, Centre of Expertise Leisure, Tourism & Hospitality, NHTV Breda University of Applied Sciences, and NHL Stenden University of Applied Sciences (2018), over-tourism is defined as "*the impact of tourism on a destination, or parts thereof, that excessively influences perceived quality of life of citizens and/or quality of visitors experiences in a negative way*" (p.4). It is consensual that over-tourism pressures the short-term stays (Ashworth, & Page, 2011; Carvalho, Chamusca, Fernandes, & Pinto, 2019; González, 2011) and deteriorate the quality of life for city dwellers, by increasing noise levels, overcrowding, pollution, restraining access to certain public spaces and increasing the cost of living (Gonia, & Jezierska-Thöle, 2022). Nevertheless, this doesn't have to be only a negative standpoint, as most of those impacts are inflated by residents' perceptions that were educated to perceive tourists as *persona non grata* (e. g. Aall, & Koens, 2019; Koodsela, Dong, & Sukpatch, 2019). Some authors (such as García-Hernández, de la Calle-Vaquero, & Yubero, 2017; Aall, & Koens, 2019 and Yubero et al.,

2021) posited that the hyper-emphasis on tourists impacts and this tourism-phobia should be considered to mitigate the over-tourism mysticism.

Still, regarding the over-tourism concept, UNWTO *et al.* (2018) elicited four of the major myths associated with it:

Tourism congestion is not only about the number of visitors but about the capacity to manage them; Tourism congestion is commonly a localised rather than a citywide issue; Tourism congestion is not a tourism-only problem and technological or smart solutions alone are important but will not solve the issue of tourism congestion (p.5).

As most of the impacts of over-tourism rely on a lack of appropriate strategies, UNWTO et al. (2018) point out the most effective strategies to deal with it, such as: Facilitate the spatial and temporal deconcentration of visitors in and outside the city or destination; promote new itineraries and attractions – create new tourist services/products; revisit and adapt the current legislation; promote tourist demand segmentation; guaranteeing that the benefits of tourism are shared with the local community; develop new experiences for both visitors and residents; increase and improve the supply system of infrastructure and facilities in the city; communicate and engage with local stakeholders and visitors; and control and monitor the impacts of the strategies implemented.

Despite the researcher's and policymakers' recommendations, the problem of over-tourism is far from being solved. Henley (2020) wrote "Across Europe, historic cities are buckling. Mass tourism, encouraged by cash-hungry councils after the 2008 crash and fuelled by the explosion of cheap flights and online room rentals, has become a monster. The backlash, however, has begun". Some examples of over-tourism are the number of visitors in Amsterdam of about 20.3 million for a population of 854,000, Barcelona with 30 million visitors for a population of 1.6 m, Florence with 10 million for a population of 382,000 or Prague with 8 million against 1.3 million of inhabitants (Henley, 2020).

In the early 2020s, a pandemic arose with unprecedented impacts on tourism (Yang, Zhang, & Rickly, 2021; Kupi & Szemerédi, 2021; Ianioglo & Rissanen, 2020; Wen, Kozak, Yang, & Liu, 2020; Gössling, Scott, & Hall, 2020). This health issue has frozen all tourism movements, moving the cities from a situation of over-tourism to one of total emptiness, from crowds to ground zero.

TOURISM AND THE COVID-19

COVID-19 was recognized by the World Health Organization (WHO) as a public health emergency of international concern (PHEIC) on January 30, 2020, and later on March 11, 2020, as a pandemic (WHO, 2021, June 29). This pandemic has been testing the entire resilience of tourism. The impacts of COVID on tourism were extensively researched from different angles (Yang et al., 2021; Kupi & Szemerédi, 2021; Ianioglo & Rissanen, 2020; Wen et al., 2020; Gössling et al., 2020) with the same conclusion that this was the worst battle tourism sector has faced since the beginning of the history of tourism development.

According to the OECD (2018), tourism resilience has faced several challenges in the last ten years from "impacts of the global economic crisis, geopolitical uncertainty, numerous terrorist attacks, natural disasters, and other external shocks" (p. 32). All of which with impact on tourism, highlighting the vulnerability of the sector (Marques Santos, Madrid, Haegeman, & Rainoldi, 2020; Wut, Xu, & Wong, 2021; Butler, 2020). This was the case with the September 11, 2001, terrorist attack, the 2002-2003

SARS, the 2005 Indian Ocean tsunami, the 2008 and 2009 global economic crisis, the 2009 swine flu, the 2012 MERS-CoV, the 2014 Ebola, the 2015 Zika, the 2016 Dengue fever (Gössling et al., 2020; Hall, Scott, & Gössling, 2020; Higgins-Desbiolles, 2020; McKercher & Chon, 2004), and other contagious diseases that tourism seems to magnify (Richter, 2003).

In this regard, Godovykh, Pizam and Bahja (2021) state that tourism may even be the vehicle for the transmission of these diseases and that "the mobility of international tourism may introduce new infectious diseases to world populations" (p. 739). The importance of globalization and global change processes in the spread of new diseases is also mentioned by Hall et al. (2020).

However, the current COVID-19 pandemic has had a greater impact on tourism rather than the previous public health crises (Higgins-Desbiolles, 2020; Zhan, Zeng, Morrison, Liang, & Coca-Stefaniak, 2020). Higgins-Desbiolles (2020) justifies the unprecedented impacts with the velocity of the virus spread, and the lockdown that frozen completely the activity whether it be by government imposition or just because people were afraid of traveling. Furthermore, Yang et al. (2021) justify the impacts with the duration of the pandemic situation.

Although at different paces the governments ended up locking down the countries and closing the frontiers (McKercher & Chon, 2004), as this was considered one of the most highly contagious outbreaks in recent human history (Sharma, Thomas & Paul, 2021).

Restrictions on mobility and impediments on non-essential travel, border closures, lockdowns, quarantines, closure of commercial spaces, suspension of civil aviation, school closures, and mandatory telework (Marques Santos et al., 2020; da Silva Lopes, Remoaldo, Ribeiro, & Martín-Vide, 2021; Wen et al., 2020; Ivanova, Ivanov, & Ivanov, 2021; Morar, Tiba, Basarin, Vujiˇci'c, Valjarevi'c, Niemets, Gessert, Jovanovic, Drugas, Grama et al., 2021) are some of the examples of the imposed restrictions that reduced the social life and tourism to ground zero.

On the demand side, several authors suggest a change in tourism behavior (Huang, Shao, Zeng, Liu, & Li, 2021; Zheng, Luo, & Ritchie, 2021; Rončák, Scholz, & Linderová, 2021). Hall et al. (2020) argue that disease outbreaks and pandemics play a role in social and economic shifts.

Marques Santos et al. (2021) argue that "tourist behavior and choice are affected by psychological and economic factors" (p. 6). Psychological factors are related to the fear of contamination with an impact on the willingness to travel and vacation destination conditions and preferences while economic factors are associated with reduced household income as a result of unemployment or layoff. Nevertheless, tourism is one of the most affected industries (UNWTO, 2020) and the short-term impacts alone are indicative of the extent of this crisis.

In addition to the economic impacts of COVID-19, the United Nations (2020) warns of other impacts, namely, implications on the environment and climate change, and the cultural sector. In environmental terms, it is dreaded that in the context of loss of income in tourism, the most vulnerable populations may return to the exploitation of natural resources in protected areas. In culture, the impact of the pandemic was significant because the sector was temporarily closed. Abbas, Mubeen, Iorember, Raza, & Mamirkulova (2021) identify other impacts, considering that COVID-19 had consequences on tourism employment and education as "it has seriously affected job creation opportunities worldwide" (p.9). The authors argue that "Due to virtual learning and teaching, students also have to deal with training interruptions, recruitment, and unstable employment opportunities in the tourism business" (p. 9).

COVID-19 AND THE URBAN TOURISM

In the first year of the COVID-19 pandemic, 2020, the World Bank (2022) estimates that the global economy will have shrunk by approximately 3% and global poverty will have increased for the first time in a generation.

More than 90% of the countries have declined economically by 2020, in a regressive scenario worse than the ones that occurred during the two world wars, the Great Depression of the 1930s, or the global financial crisis of 2007-09 (World Bank, 2022).

Table 1 illustrates the impact of Tourism Direct Gross Domestic Product (TDGDP) (as a percentage of GDP) worldwide. The decrease in tourism contribution to the GDP from 4 percent to less than 2 percent demonstrates the huge impact of the pandemic.

The contribution decrease reflects in TDGDP, revenues, receipts, and tourists' arrivals as illustrated in table 2.

Table 1. Tourism direct gross domestic product (percentage of GDP). Source: UNWTO (2021b, p.25).

	2017	*2018*	*2019*	*2020*
World	4.0	4.0	4.0	1.8
Europe	4.2	4.2	4.3	2.2
Asia Pacific	4.3	4.3	4.4	1.8
Americas	3.4	3.4	3.4	1.6
Africa	4.1	4.3	4.3	1.6
Middle East	4.5	4.6	5.1	1.7
EU-27	4.5	4.5	4.5	2.5
G20	3.9	3.9	3.9	1.8

The drop in international tourism impacts urban tourism (Carvalho et al., 2019; Gonzalez, 2011), as they are a relevant part of urban tourism. Additionally, the uncertainty about the safety of traveling to urban places and the imposed social isolation drove tourists to choose natural places (ETC, 2021).

Table 2. The impact of COVID-19 on several tourism indicators (world). Source: UNWTO (2021b, p.25).

	2017	*2018*	*2019*	*2020*	*Variation 2019->2020*
TDGDP (USD trillion)	3.20	3.44	3.54	1.56	-55.9%
Int. tourism revenues (USD trillion)	1.56	1.69	1.72	0.64	-62.8%
Int. tourism receipts (USD trillion)	1.33	1.44	1.47	0.53	-63.9%
Int. tourist arrivals (billion)	1.34	1.41	1.47	0.40	-72.8%

Urban tourists enjoyed the cities and they want to live all at once, the gastronomy, the culture, the events, and the city in a short stay. But the pandemic put an end to all this. At least for a while. Empty cities and restrictions reduced over-tourism to emptiness to a level that raises the question will we ever be discussing over-tourism again? On the contrary and recently some authors came up with a new concept, revenge tourism, which can bring the urban tourism to what we knew it and so the concern about the over-tourism and the very needed sustainability is one of the biggest challenges of tourism in particular and to all destinations in general.

THE CHALLENGES OF URBAN TOURISM

Notwithstanding new trends in consumer behaviour and the speed of their change, we believe that two major issues will be continuously, implicitly and explicitly, printed in future policies and strategies to develop tourism whether it be in natural, rural, or urban places: Sustainability and Digital Transformation.

Sustainability

In the period before the pandemic, despite its continued growth, tourism was already facing some pressure to make the industry more aligned with more sustainable growth (OECD, 2020). Part of these pressures was to establish strategies to meet somehow the United Nations' Sustainable Development Goals (SDG). In fact, until 2020, the central issues in the development of tourism and those that would perpetuate the growth of this economic activity focused on matters such as sustainability (considering the trilogy of economic, ecological/environmental, and socio-cultural aspects) and the digital transformation, considering the technological evolution of recent years.

Moreover, in 2020, OECD also describes the goals that have been under discussion in recent times, such as "to deal effectively with overcrowding at popular destinations, spread the economic and other benefits to areas that attract fewer visitors, develop new products to expand the season, and encourage productivity, better resource use, and more stable employment" (p.3). Earlier, UNWTO&UNDP (2017) showed a significant relationship between SDGs and tourism, also demonstrating that there were SDGs at that time that stood out positively for the number of initiatives carried out by companies - 'Responsible Consumption and Production', involvement - Reduced Inequalities' and 'Sustainable Cities and Communities. The period leading up to the 2020 pandemic crisis was fruitful in the development of initiatives that sought to activate the sector's adherence to SDGs (e.g. reports like UNWTO&OAS, 2018; or initiatives such as the Berlin Declaration (Transforming Tourism, 2017); the Tourism for SDG platform -https://tourism4sdgs.org/learn/ - sponsored by the UNWTO).

In this context, the network of observatories for sustainable tourism (http://insto.unwto.org/observatories/), created in 2004 by the World Tourism Organization (henceforth UNWTO), gains special relevance.

According to data from the International Network of Sustainable Tourism Observatories (UNWTO INSTO, 2022), 31 observatories had joined the network, with Portugal being the fourth most represented country in the network with the observatories of the Azores, Alentejo, and Algarve.

The general framework that has been exposed previously evolved significantly at the last COP 26 UN Climate Change Conference held in November 2021 in Glasgow, with the approval of the so-called Glasgow Declaration, established under the auspices of the One Planet Network (https://www.oneplanetnetwork.org/programmes/sustainable-tourism/glasgow-declaration). It is expected that this declaration will have significant importance in the recovery of the sector in the post-pandemic period. Adherence to the Glasgow Declaration is voluntary and reveals the willingness of the organization that adheres to it to commit to and implements the following initiatives concerning its activity (OPSTP, 2021):

"MEASURE. Measure and disclose all travel and tourism-related emissions.
DECARBONISE. Set and deliver targets aligned with climate science to accelerate tourism's decarbonisation.
REGENERATE. Restore and protect ecosystems, supporting nature's ability to draw down carbon, as well as safeguarding biodiversity, food security, and water supply.
COLLABORATE. Share evidence of risks and solutions with all stakeholders and [...] guests, and work together to ensure [that] plans are as effective and co-ordinated as possible.
FINANCE. Ensure organisational resources and capacity are sufficient to meet objectives set out in climate plans, including the financing of training, research and implementation of effective fiscal and policy tools where appropriate to accelerate transition."

These commitments, from the Glasgow declaration, can provide guidelines for the establishment of a sustainable policy for urban tourism, for each city that is heavily impacted by tourism and to manage their crowdings. The necessary involvement to establish effective sustainable policies for tourism in cities is not an assignment/responsibility of local authorities, industry, or community leaders.

Timur and Getz (2009) present a generic model for managing Sustainable Tourism Development (STD), suggesting the intervention of three stakeholder groups - Local Authorities, the Tourism Industry, and Environmental supporters - sharing common goals concerning STD. Furthermore, the authors present an empirical study conducted with these three stakeholders' groups to assess their orientation/

Figure 1. Relevant interactions in urban tourism
Source: Adapted from Timur and Getz (2009).

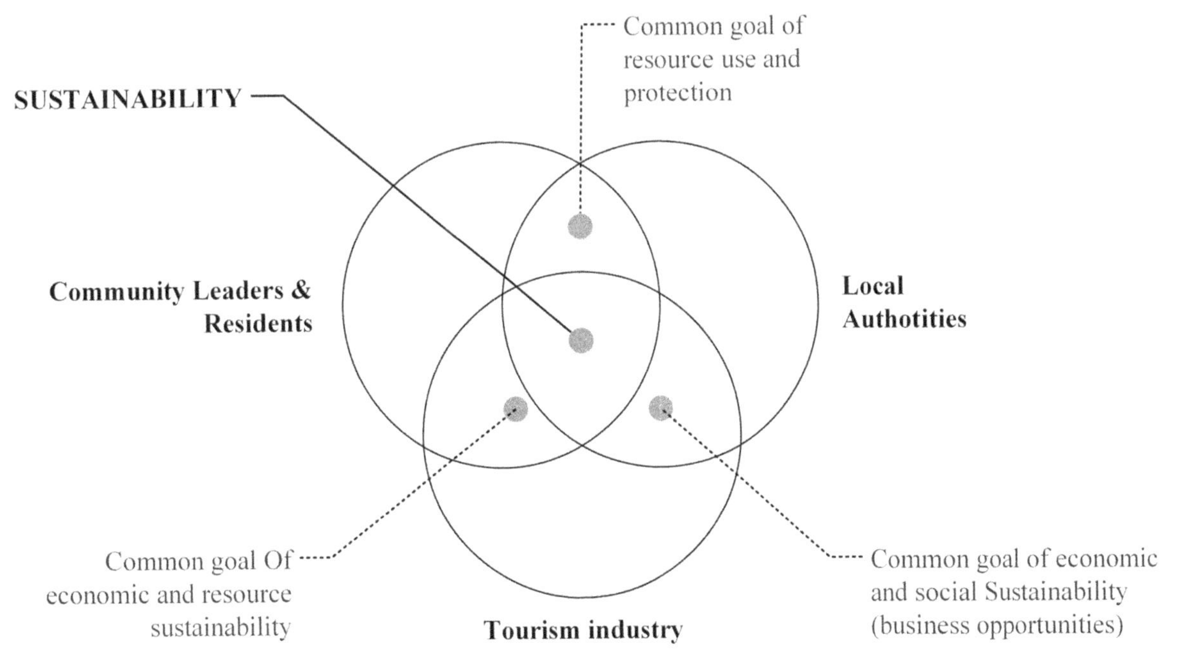

engagement with the issue of sustainability in the urban tourism context. Their conceptual model is illustrated in Figure 1 with adaptations to comprise Community Leaders & Residents instead of Environmental Supporters.

Timur & Getz (2009) emphasize the importance to mobilize stakeholders within an articulate frame. Additionally, they conclude that policymakers should consider the concerns of the stakeholders, co-creating participatory models, and sustainable networks.

These networks should measure and disseminate all relevant information on tourism-related travel and emissions. It should set and provide targets to accelerate the decarbonization of tourism, as well as establish effective policies to protect (and eventually restore) ecosystems. Further, these networks should establish models to analyze risks and anticipate solutions that could be adopted by all the tourism value chains. Last but not least it is in this frame that mobilization of resources and capacities should be developed to enhance the integration of tourism into city life.

Skills and the Digital Transition

As stated earlier, one of the biggest challenges of tourism, and for urban tourism in particular, and the one which was pretty much accelerated by this recent pandemic relies on digital transformation.

For Liberto (2013) tourism is a sector usually considered as low-skilled / low-tech labor. This view of the sector is also held by Wadley (2020), although in this case, the author considers that it only applies to mass tourism. But still, the tourism sector is continuously characterized as low-skilled and low-tech. The CEDEFOP (2020) study shows that technology skills are only the seventh most required skill in job vacancies in the tourism sector. The low technological intensity justifies the reduced importance that this type of skill has in the recruitment processes in tourism. Figure 2 illustrates the skills tourism professionals should have.

However, it is expected that the coming years will be of significant reinforcement of the technological intensity in the tourism sector. Emerging technologies - big data (Fuchs, Höpken, & Lexhagen, 2014; Li, Xu, Tang, Wang, & Li, 2018), artificial intelligence (Li, Bonn, & Ye, 2019; Samara, Magnisalis, & Peristeras, 2020; Tussyadiah, 2020), virtual reality (Guttentag2010; Tussyadiah, Wang, Jung, & tom Dieck, 2018) and augmented reality (Chung, Han, & Joun, 2015; Yung, & Khoo-Lattimore, 2019), the automation of functions interface and robotization (Tussyadiah, 2020), – will, soon, significantly shape the tourism sector.

Despite organizations' presence in online, namely through websites and/or on social media, the digital transformation goes much beyond these interactions with the consumer. In this regard, Mahraz, Benannbou and Berrado (2019) advocated that "digital transformation is a disruptive or incremental change process" (p.923), in which "new digital technologies such as social media, mobile, analytics or embedded devices" (p.923) are not only used for greater adaptation to the consumer market but also adopted to materialize "the change of organizational processes or the creation of new business models" (p.923). Furthermore, the pandemic we are experiencing today has, in a way, accelerated the consumer's dependence on the digital world (e.g. with an increase in online shopping, their online presence to contact others, etc).

Thus, the mastery of these technologies and the ability to innovate in these and with these technologies will be a critical competitive factor. For this to happen, it is necessary to reinforce the technological skills of tourism human resources, both in initial and lifelong training/learning, as the speed of the rise and development of these technologies is mind-blowing. Moreover, it is also necessary to value the tourism professions, making the sector attractive to attract and retain highly qualified professionals.

Figure 2. Tourism Professional's Skills
Source: CEDEFOP (2020).

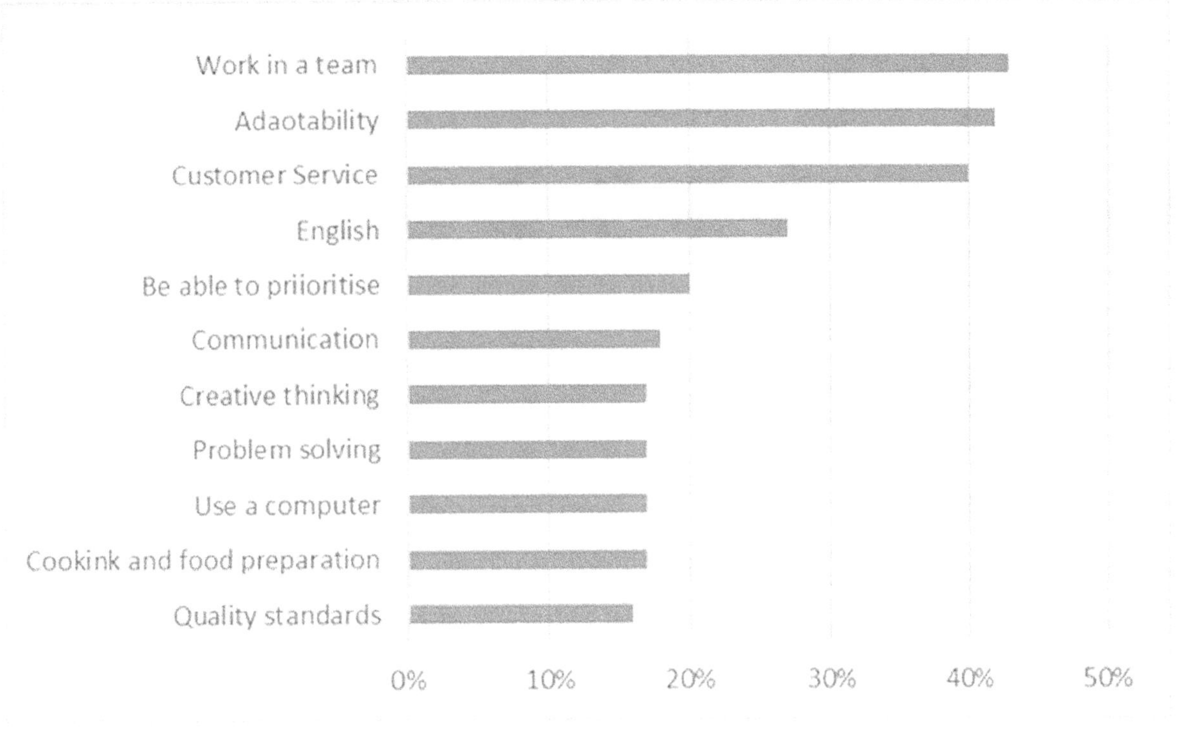

The literature on tourism in cities is also beginning to take an interest in these technological developments (see, for example, Salas-Olmedo, Moya-Gómez, García-Palomares, & Gutiérrez (2018) on "big data" or Nóbrega, Jacob, Coelho, Weber, Ribeiro, & Ferreira (2017) for "augmented reality).

This transformation of the technological nature of tourism is often referred to as smart tourism, which in the case of urban tourism is sometimes called smart tourism [in] cities (see Boletsis, & Chasanidou, 2018; Coca-Stefaniak, 2020; Gretzel, & Koo, 2021).

Regarding urban tourism, one of the aspects in which the new technologies impact the most refers to the communication with the tourists, contents, and accesses (Garcia Henche, 2018). Furthermore, skills and career profiles aligned with the digital transformation impact the restarting of urban tourism (Blöcher and Alt, 2021 and Scaltsa, Nitsiou and Georgaki, 2016).

Blöcher and Alt (2021) reflect on how these developments will impact restaurants and Scaltsa, Nitsiou and Georgaki (2016) on how tourism education will change. In both cases the symbiosis between the coming technological changes in tourism and the change in skills and career profiles and also in education for tourism professionals is evident. And this change will be particularly relevant in urban tourism.

Because of the characteristics of tourism in cities - much more based on the stay and less on traditional service - new technologies may play an important role in matching the tourist with the resident, or in cases where this has not yet proved a problem, in reducing this potential risk.

From the above, we propose the following conceptual diagram where we synthesize the evolution we anticipate for urban tourism.

Figure 3. Conceptual diagram for urban tourism evolution
Source: Authors

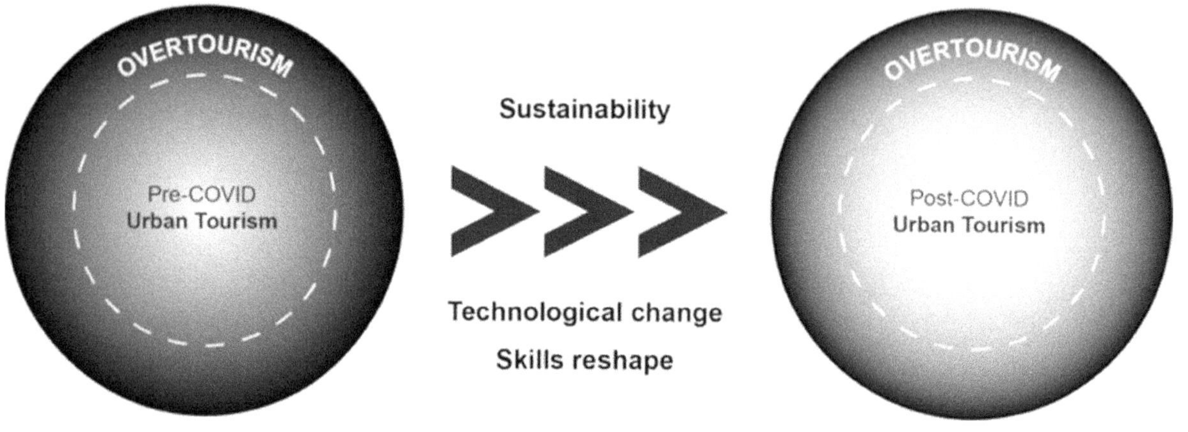

CONCLUSION

This research anticipates the main issues within urban tourism development, in a time when the focus is to resume tourism when tourists are still afraid of social contacts but desiring to revenge for these two years of contained travel. On the other hand, Covid-19 has been reinforcing major trends, giving them greater strength when it comes to consumer behaviour and tourist destinations and/or organizations. First, the issue of sustainability drives us to the relevance of eco-urban practices combined with health care systems that mitigate the fear tourists still feel in crowds. Second, the qualification and capacitation of human resources within the digital transformation are critical to the development of a new urban tourism paradigm where smart is the motto.

In our understanding, these are the challenges that post-pandemic urban tourism will bring. Some of them are basically of the same nature as the challenges of tourism in general - sustainability and the digital transition, in particular. However, the urban setting will give it a unique set of particular characteristics. The social dynamics issue is more specific to urban tourism and will require tailored responses, to which sustainability and the digital shift can make a strong contribution.

This exploratory research in its essence raises challenges urban tourism needs to take into account if the idea is to restart. Tourism is the blood of the cities, and the integration of tourism into the lifeblood of the cities is the way to keep alive the authenticity of the cities. Currently, tourists are afraid of crowds due to the pandemic situation which, although it is now more stable, is still active. Urban tourism will be restarted, but it will take time to talk about over-tourism again, which means that the government and stakeholders should activate intelligent space management models, carrying capacity models, and environmentally friendly practices which are some of the solutions to shape urban tourism. Crisis Management is another management function for tourist destinations and it will be more and more appellant in the future considering new adverse events. At the center of these solutions must be the residents and their pride in sharing their city.

This research contributes to the body of literature by offering a critical overview of the literature articulated with policy documents that brought to the discussion of this topic some practical thinking. Although the relevance of the discussion this chapter offers, this is not absent of limitations that may contribute to opening new paths of research. The models outlined in this research could be tested in different cities to understand the feasibility of the model, furthermore, the best practices in cities should be more developed to enrich the comprehension of the phenomena. Last but not least this critical overview should be repeated two years ahead to understand if the paradigm of urban tourism has changed indeed or if we are on the same page.

ACKNOWLEDGMENT

This research was, partially supported, by the FCT - Fundação para a Ciência e Tecnologia (Portugal) [UIDB/04521/2020].

REFERENCES

Aall, C., & Koens, K. (2019). The discourse on sustainable urban tourism: The need for discussing more than overtourism. *Sustainability (Switzerland), 11*(15), 4228. doi:10.3390u11154228

Abbas, J., Mubeen, R., Iorember, P. T., Raza, S., & Mamirkulova, G. (2021). Exploring the impact of COVID-19 on tourism: Transformational potential and implications for a sustainable recovery of the travel and leisure industry. *Current Research in Behavioral Sciences, 2*, 100033. doi:10.1016/j.crbeha.2021.100033

Ashworth, G., & Page, S. J. (2011). Urban tourism research: Recent progress and current paradoxes. *Tourism Management, 32*(1), 1–15. doi:10.1016/j.tourman.2010.02.002

Biagi, B., Ladu, M. G., Meleddu, M., & Royuela, V. (2020). Tourism and the city: The impact on residents' quality of life. *International Journal of Tourism Research, 22*(2), 168–181. doi:10.1002/jtr.2326

Blöcher, K., & Alt, R. (2021). AI and robotics in the european restaurant sector: Assessing potentials for process innovation in a high-contact service industry. *Electronic Markets, 31*(3), 529–551. doi:10.100712525-020-00443-2

Butler, R. (2020). Tourism–resilient but vulnerable as "the times they are a changing" in the "new normality". *Worldwide Hospitality and Tourism Themes, 12*(6), 663–670. doi:10.1108/WHATT-07-2020-0063

Carvalho, L., Chamusca, P., Fernandes, J., & Pinto, J. (2019). Gentrification in Porto: Floating city users and internationally-driven urban change. *Urban Geography, 40*(4), 565–572. doi:10.1080/02723638.2019.1585139

CEDEFOP. (2020). *Skills developments and trends in the tourism sector*. Skills intelligence. https://www.cedefop.europa.eu/en/data-insights/skills-developments-and-trends-tourism-sector

Chung, N., Han, H., & Joun, Y. (2015). Tourists' intention to visit a destination: The role of augmented reality (AR) application for a heritage site. *Computers in Human Behavior, 50*, 588–599. doi:10.1016/j.chb.2015.02.068

da Silva Lopes, H., Remoaldo, P. C., Ribeiro, V., & Martín-Vide, J. (2021). Effects of the COVID-19 Pandemic on Tourist Risk Perceptions—The Case Study of Porto. *Sustainability*, *13*(11), 6399. doi:10.3390u13116399

di Liberto, A. (2013). High skills, high growth: Is tourism an exception? *The Journal of International Trade & Economic Development*, *22*(5), 749–785. doi:10.1080/09638199.2011.603054

ETC (European Travel Commission). (2021). *European Tourism: Trends & Prospects (Q4/2020)*. Brussels, Belgium: European Travel Commission. https://etc-corporate.org/uploads/2021/02/ETC-Quarterly-Report-Q4-2020_Public-1.pdf

Fuchs, M., Höpken, W., & Lexhagen, M. (2014). Big data analytics for knowledge generation in tourism destinations - A case from sweden. *Journal of Destination Marketing & Management*, *3*(4), 198–209. doi:10.1016/j.jdmm.2014.08.002

Garcia Henche, B. (2018). Urban experiential tourism marketing: Use of social media as communication tools by the food markets of madrid. *Journal of Tourism Analysis*, *25*(1), 2–22. doi:10.1108/JTA-01-2018-0002

Godovykh, M., Pizam, A., & Bahja, F. (2021). Antecedents and outcomes of health risk perceptions in tourism, following the COVID-19 pandemic. *Tourism Review*, *76*(4), 737–748. doi:10.1108/TR-06-2020-0257

Gonia, A., & Jezierska-Thöle, A. (2022). Sustainable tourism in Cities—Nature reserves as a 'New' city space for nature-based tourism. *Sustainability (Switzerland)*, *14*(3), 1581. doi:10.3390u14031581

González, S. (2011). Bilbao and barcelona 'in motion'. how urban regeneration 'models' travel and mutate in the global flows of policy tourism. *Urban Studies (Edinburgh, Scotland)*, *48*(7), 1397–1418. doi:10.1177/0042098010374510

Gössling, S., Scott, D., & Hall, C. M. (2020). Pandemics, tourism and global change: A rapid assessment of COVID-19. *Journal of Sustainable Tourism*, *29*(1), 1–20. doi:10.1080/09669582.2020.1758708

Guttentag, D. A. (2010). Virtual reality: Applications and implications for tourism. *Tourism Management*, *31*(5), 637–651. doi:10.1016/j.tourman.2009.07.003

Hall, C. M., Scott, D., & Gössling, S. (2020). Pandemics, transformations and tourism: Be careful what you wish for. *Tourism Geographies*, *22*(3), 577–598. doi:10.1080/14616688.2020.1759131

Henley, J. (2020, January 25). *Overtourism in Europe's historic cities sparks backlash*. https://www.theguardian.com/world/2020/jan/25/overtourism-in-europe-historic-cities-sparks-backlash

Higgins-Desbiolles, F. (2020). Socialising tourism for social and ecological justice after COVID-19. *Tourism Geographies*, *22*(3), 610–623. doi:10.1080/14616688.2020.1757748

Huang, S. S., Shao, Y., Zeng, Y., Liu, X., & Li, Z. (2021). Impacts of COVID-19 on Chinese nationals' tourism preferences. *Tourism Management Perspectives*, *40*, 100895. doi:10.1016/j.tmp.2021.100895 PMID:34642624

Ianioglo, A., & Rissanen, M. (2020). Global trends and tourism development in peripheral areas. *Scandinavian Journal of Hospitality and Tourism*, *20*(5), 520–539. doi:10.1080/15022250.2020.1848620

INE. (2021). *Número de Dormidas nos estabelecimentos hoteleiros 2009-2020 em Portugal.* https://www.ine.pt/

Ivanova, M., Ivanov, I. K., & Ivanov, S. (2021). Travel behaviour after the pandemic: The case of Bulgaria. *Anatolia*, *32*(1), 1–11. doi:10.1080/13032917.2020.1818267

Jamil, J., & Puad, M. S. (2010). Challenges in managing infrastructure for urban tourism: A comparative study of three asian cities. *WIT Transactions on Ecology and the Environment*, *142*, 243–253. doi:10.2495/SW100231

Kester, J. (2015). International Tourism Trends in EU28 Member States. *Current Situation and Forecasts for 2020-2025-2030.* https://ec.europa.eu/docsroom/documents/16845?locale=pt

Koodsela, W., Dong, H., & Sukpatch, K. (2019). A holistic conceptual framework into practice-based on urban tourism toward sustainable development in Thailand. *Sustainability (Switzerland)*, *11*(24), 7152. doi:10.3390u11247152

Kupi, M., & Szemerédi, E. (2021). Impact of the COVID-19 on the Destination Choices of Hungarian Tourists: A Comparative Analysis. *Sustainability*, *13*(24), 13785. doi:10.3390u132413785

Li, J., Xu, L., Tang, L., Wang, S., & Li, L. (2018). Big data in tourism research: A literature review. *Tourism Management*, *68*, 301–323. doi:10.1016/j.tourman.2018.03.009

Li, J. J., Bonn, M. A., & Ye, B. H. (2019). Hotel employee's artificial intelligence and robotics awareness and its impact on turnover intention: The moderating roles of perceived organizational support and competitive psychological climate. *Tourism Management*, *73*, 172–181. doi:10.1016/j.tourman.2019.02.006

Lim, C., Zhu, L., & Koo, T. T. R. (2019). Urban redevelopment and tourism growth: Relationship between tourism infrastructure and international visitor flows. *International Journal of Tourism Research*, *21*(2), 187–196. doi:10.1002/jtr.2253

Liu, X. (2013). A study on the competitiveness of urban tourism destinations taking loudi city in Hunan province for example. *Information Technology Journal*, *12*(14), 2905–2912. doi:10.3923/itj.2013.2905.2912

Mahraz, M. I., Benabbou, L., & Berrado, A. (2019, July). A systematic literature review of digital transformation. *Proceedings of the International Conference on Industrial Engineering and Operations Management Toronto*, 917-931.

Marques Santos, A., Madrid, C., Haegeman, K., & Rainoldi, A. (2020). *Behavioural changes in tourism in times of Covid-19.* Publications Office of the European Union. doi:10.2760/00411

McKercher, B., & Chon, K. (2004). The over-reaction to SARS and the collapse of Asian tourism. *Annals of Tourism Research*, *31*(3), 716–719. doi:10.1016/j.annals.2003.11.002 PMID:32572281

Morar, C., Tiba, A., Basarin, B., Vujičić, M., Valjarević, A., Niemets, L., Gessert, A., Jovanovic, T., Drugas, M., Grama, V., Stupariu, M., Stoica, A., & Lukić, T. (2021). Predictors of Changes in Travel Behavior during the COVID-19 Pandemic: The Role of Tourists' Personalities. *International Journal of Environmental Research and Public Health, 18*(21), 11169. doi:10.3390/ijerph182111169 PMID:34769688

Nóbrega, R., Jacob, J., Coelho, A., Weber, J., Ribeiro, J., & Ferreira, S. (2017). *Mobile location-based augmented reality applications for urban tourism storytelling*. Paper presented at the EPCGI 2017 - 24th Encontro Portugues De Computacao Grafica e Interacao. 10.1109/EPCGI.2017.8124314

OECD. (2018). *OECD Tourism Trends and Policies 2018*. OECD Publishing. doi:10.1787/tour-2018-

OECD. (2020). *OECD Tourism Trends and Policies 2020*. OECD Publishing. doi:10.1787/6b47b985-

OPSTP (One Planet Sustainable Tourism Programme). (2021). *Glasgow Declaration: A commitment to a decade of climate action*. https://www.oneplanetnetwork.org/programmes/sustainable-tourism/glasgow-declaration

Richter, L. K. (2003). International tourism and its global public health consequences. *Journal of Travel Research, 41*(4), 340–347. doi:10.1177/0047287503041004002

Rončák, M., Scholz, P., & Linderová, I. (2021). Safety Concerns and Travel Behavior of Generation Z: Case Study from the Czech Republic. *Sustainability, 13*(23), 13439. doi:10.3390u132313439

Salas-Olmedo, M. H., Moya-Gómez, B., García-Palomares, J. C., & Gutiérrez, J. (2018). Tourists' digital footprint in cities: Comparing big data sources. *Tourism Management, 66*, 13–25. doi:10.1016/j.tourman.2017.11.001

Samara, D., Magnisalis, I., & Peristeras, V. (2020). Artificial intelligence and big data in tourism: A systematic literature review. *Journal of Hospitality and Tourism Technology, 11*(2), 343–367. doi:10.1108/JHTT-12-2018-0118

Scaltsa, M., Nitsiou, P., & Georgaki, P. (2016). Educating a new generation of tour guides for the forthcoming era, at the Aristotle University of Thessaloniki. *Tourismos, 11*(2), 240–261. doi:10.26215/tourismos.v11i2.475

Sharma, G. D., Thomas, A., & Paul, J. (2021). Reviving tourism industry post-COVID-19: A resilience-based framework. *Tourism Management Perspectives, 37*, 100786. doi:10.1016/j.tmp.2020.100786 PMID:33391988

Timur, S., & Getz, D. (2009). Sustainable tourism development: How do destination stakeholders perceive sustainable urban tourism? *Sustainable Development, 17*(4), 220–232. doi:10.1002d.384

Tourism, T. (2017). *Berlin declaration on "transforming tourism"*. http://www.transforming-tourism.org/fileadmin/baukaesten/sdg/downloads/Berlin_Declaration.pdf

Turismo de Portugal. (2021). *Reativar Turismo | Construir Futuro - Plano de ação - Plano de ação apresentado pelo Ministro de Estado, da Economia e da Transição Digital*. https://www.turismodeportugal.pt/SiteCollectionDocuments/estrategia/plano-acao-reativar-o-turismo-construir-o-futuro.pdf

Tussyadiah, I. (2020). A review of research into automation in tourism: Launching the annals of tourism research curated collection on artificial intelligence and robotics in tourism. *Annals of Tourism Research, 81*, 102883. Advance online publication. doi:10.1016/j.annals.2020.102883

Tussyadiah, I. P., Wang, D., Jung, T. H., & tom Dieck, M. C. (2018). Virtual reality, presence, and attitude change: Empirical evidence from tourism. *Tourism Management, 66*, 140–154. doi:10.1016/j.tourman.2017.12.003

United Nations. (2020). *Policy Brief: COVID-19 and transforming tourism*. https://www.un.org/sites/un2.un.org/files/sg_policy_brief_covid-19_tourism_august_2020.pdf

UNWTO, Centre of Expertise Leisure, Tourism, and Hospitality, NHTV Breda University of Applied Sciences, & NHL Stenden University of Applied Sciences. (2018). *'Overtourism'? Understanding and Managing Urban Tourism Growth beyond Perceptions, Executive Summary*. UNWTO. doi:10.18111/9789284420070

UNWTO. (2020, April 1). *Message from Madrid: Tourism and Covid-19*. https://www.unwto.org/news/madrid-tourism-covid-19

UNWTO. (2021a). *International Tourism Highlights* (2020 Edition). UNWTO., doi:10.18111/9789284422456

UNWTO. (2021b). *The Economic Contribution of Tourism and the Impact of COVID-19, preliminary version (November 2021)*. UNWTO. doi:10.18111/9789284423200

UNWTO & OAS. (2018). *Tourism and the Sustainable Development Goals – Good Practices in the Americas*. Madrid: UNWTO. doi:10.18111/9789284419685

UNWTO & UNDP. (2017). *Tourism and the Sustainable Development Goals – Journey to 2030, Highlights*. Madrid: UNWTO. doi:10.18111/9789284419340

UNWTO INSTO. (2022). *INSTO Observatories*. https://insto.unwto.org/observatories/

Wadley, D. (2020). *The City of Grace An Urban Manifesto*. Springer Nature Singapore Pte Ltd. doi:10.1007/978-981-15-1112-7

Wen, J., Kozak, M., Yang, S., & Liu, F. (2020). COVID-19: Potential effects on Chinese citizens' lifestyle and travel. *Tourism Review, 76*(1), 74–87. https://doi.org/10.1108/TR-03-2020-0110

WHO. (2021, June 29). *Listings of WHO's response to COVID-19*. https://www.who.int/news/item/29-06-2020-covidtimeline

World Bank. (2022). *World Development Report 2022: Finance for an Equitable Recovery*. Washington, DC: World Bank. doi:10.1596/978-1-4648-1730-4

Wut, T. M., Xu, J. B., & Wong, S. M. (2021). Crisis management research (1985–2020) in the hospitality and tourism industry: A review and research agenda. *Tourism Management, 85*, 104307. https://doi.org/10.1016/j.tourman.2021.104307

Yang, Y., Zhang, C. X., & Rickly, J. M. (2021). A review of early COVID-19 research in tourism: Launching the Annals of Tourism Research's Curated Collection on coronavirus and tourism. *Annals of Tourism Research*, *91*, 103313. https://doi.org/10.1016/j.annals.2021.103313

Yubero, C., Condeço-Melhorado, A. M., García-Hernández, M., & Fontes, A. C. (2021). Comparing spatial and content analysis of residents and tourists using Geotagged Social Media Data. The Historic Neighbourhood of Alfama (Lisbon), a case study. *Investigaciones Turisticas,* (22), 95-120. doi:10.14198/INTURI2021.22.5

Yung, R., & Khoo-Lattimore, C. (2019). New realities: A systematic literature review on virtual reality and augmented reality in tourism research. *Current Issues in Tourism*, *22*(17), 2056–2081. https://doi.org/10.1080/13683500.2017.1417359

Zhan, L., Zeng, X., Morrison, A. M., Liang, H., & Coca-Stefaniak, J. A. (2020). A risk perception scale for travel to a crisis epicentre: Visiting Wuhan after COVID-19. *Current Issues in Tourism*, *25*(1), 150–167. https://doi.org/10.1080/13683500.2020.1857712

Zheng, D., Luo, Q., & Ritchie, B. W. (2021). Afraid to travel after COVID-19? Self-protection, coping and resilience against pandemic 'travel fear'. *Tourism Management*, *83*, 104261. https://doi.org/10.1016/j.tourman.2020.104261

KEY TERMS AND DEFINITIONS

Challenges: Factor or a key issue of the macro environment that can negatively influence or impact the performance of a company, destination or type of tourism.

COVID-19: The disease provoked by the new coronavirus and the designation of the consequent pandemic.

Digital Transition: The adoption of new processes or business models and/or the creation of new products based on a new set of disruptive technologies such as big data, augmented reality, among others.

Over-Tourism: A situation where the carrying capacity has been exceeded which negatively impacts in different components of the tourist destination.

Sustainability: A governance and management system model that implies a balanced consumption of the different components of the territory, namely the economics, the socio cultural context and the environment, allowing the oportunity of future generations to satisfy its own needs.

Urban Tourism: A form of tourism in which the city or urban area is the specific geography where the tourist performs a set of activities and consumption based on the existing resources.

Chapter 16
Body, Exercise, and Health: Culture and Society in Portugal in the 21st Century

Manuela Hasse
University of Lisbon, Portugal

ABSTRACT

In this chapter, the authors present the current situation and the legislation produced as well as efforts done by medical doctors at the end of the 19th century in order to introduce hygiene as a natural as well as a cultural way of preserving physical health as well as mental health. The work done by doctors supported a general development of economic life as well as it saved the Portuguese society before laboratories took over our bodily management.

INTRODUCTION: CLEAN AND SAFE

Opposition between health and disease constitutes a fundamental aspect of Portuguese tourism as we know it today (Hasse, M.;1999; 2005:259-279). At the second half of the nineteenth century, the confrontation between weakness and strength it's at the core of multiple efforts done by medical professional class and by more or less favoured social class groups. By then, the confrontation seen at that time was necessary in face of limits and problems related with medicine and medical profession. These professional doctors faced a weak sanity and sparse salubrity and they pointed out the deep lack of general hygiene and, less known, the deplorable state of medical knowledge itself.

BACKGROUND: SOME PREVIOUS POSTURES

In this context, professional doctors had to intervene at the level of daily practice life with regular and systematic efforts related with the presentation and clarification made in general through their direct intervention close to diverse kinds and social levels of people as well as lectures presented to Portuguese

DOI: 10.4018/978-1-6684-3369-0.ch016

population in order to inform, explain and educate them in terms of hygiene and care. Those lectures were made at well known intellectual centres like Sociedade de Geografia de Lisboa (1875), and several kinds of associations and above all groups more open and curious about new ideas, interested and receptive to the information given. Medical doctors, amongst whom a vigorous activist intervention initiated and among whom Ricardo Jorge (1858-1939) became one of the most famous, expressed a new behaviour based on new norms and more adequate practices to promote the transformation of general population common life.

Those lectures didn't concern dispersed small villages isolated from major urban centres. It didn't concern just the rural areas, regions where people was delivered to themselves. It concerned as well and with biggest seriousness, urban centres where the active majority of the working population were gathered. As authentic health agents, medical doctors were conscious of the fact that their own social and professional legitimacy would only be assured through professional efficiency shown to put and end to disease and by the offering of ways of cure. For that, they worked with determination all over the country looking for a way to find and to introduce the proper ways to health preservation. In face of the general health state of the country, the lack of medical knowledge itself and the scientific resources available, physicians didn't hesitate to adopt an intensive public intervention. Urgency was felt in order to invert the terms of the health situation which could only be possible through a new mentality formation able to recognise the importance of hygiene – that is, prevention, action before the spread of disease – and all ways to make hygiene assured.

MAIN FOCUS: THE COVID-19 PANDEMIC, AND HEALTH FOR CITIZENS AND TOURISTS

The useful and practical lodging activity and sympathetic ways of caring from which such a significant Portuguese economic activity depends, has its own roots on a deep need felt by those social groups with a relatively comfortable income to follow a capable direction to conduct them to health problems solutions. Therefore, it's according with an opposition line between diseases that caused pain, general weakness and suffering and the proper cure conditions, well-being and health that strategic creation and development of one of the strongest economic activities started its structuring and tourism became a reality in this country.

At the core of a rich intensive legislation production concerning Covid-19 pandemic, while major laboratories all over the world and the main pharmaceutical enterprises worked hard to research in order to identify and understand the virus, as well as the scientific ways to stop and to control its effects through a vaccine, it would have been impossible for those who studied health and disease at the turn of the nineteenth century to the twentieth century to not recognize some of the ways used precisely about one hundred years ago in Europe and in particularly in Portugal. That was the time when facing deep limits of scientific knowledge, without any kind of pharmaceutical enterprises, rare laboratories established, without national or private health systems foundations, health insurance nor any kind of health agencies or health authorities found at a national level, medical doctors depending on their own and being perfectly conscious of medical science limits in itself, of un-existing pharmaceutical chemistry and the incipient biology, looked for safe sources of treatment and cure processes for other current problems (from chronicle to acute diseases) which threatened normal life especially the vulnerable youngest and compromised populations general development.

On April 2020, after the initial shock on the planet related with dissemination of Covid-19 and the subsequent measures taken which obliged the sudden closure of all the social and economic activities in Portugal, it was announced in the 1st July of 2021 that at the national space, where the *Portugal Tourism* had created the sign 'clean and safe', and it was the first country to make it available through European Tourism Covid-19 Safety Seal. In spite of fears related to the recognised virus presence at the European continent, this seal presentation implied the re-opening from north to south of Portugal of many of hotel unties and local lodgement when the safety indicator- having been created by the *European Committee for Standardization* in a straight cooperation with the European Union. For that matter, it would just be sufficient for those who had the Clean and Safe seal to get the European seal through CEN-*European Committee for Standardization site*. The European seal introduced inside an wide hotel industry universe has a non-mandatory character and assures, in tourism activities terms, that any tourism related establishment which has it and exhibits it, on their physical areas or on digital and network information follows the sanitarian and safety measures recommended by Direção Geral de Saúde (Health General Direction) to prevent Covid-19 dissemination.

The guidelines to Covid-19 prevention and control produced and diffused by this official institution for enterprises of all sectors of national activity (in conformity with European Union orientation and according with concerns diffused through World Health Organization) were continually announced and spread through public communications means through a wide information set, recommendations, orientations, measures, proceedings and informative support with the goal that this measures had to be applied to various social and organizational contexts and tourism as well, having been published and made known through Guides for institutions (like several materials and leaflets of information, orientation, health and work manuals, graphic information about AVAC systems – meaning heating, ventilation and air conditioning systems, orientation about space temperature and payment methods) a wide succession of continuing legislation publications, with consecutive Law-Decree, namely about emission, presentation and use of European Union Digital Certificate use to be shown on maritime and air traffic, inside national territory and still matters related with access to cultural, sporting, corporative or family access. There was continuing legislation about testing and infection determination, travel effects, prophylactic isolation, Government exceptional preventive measures and Ministry Council Resolutions.

At the same time, recommendations and masks mandatory use, alcohol gel and isolation were made, specific orientations were produced concerning cleaning and disinfection of environment and surfaces potentially contaminated with Sars-Cov-2 in various facilities and different establishments for public service and, as well, concerning prevention, control and surveillance proceedings inside enterprises. According with that same legislation initiatives, there were still published two more documents related with Individual Protection Equipment, under Orientation nº019/2020, from 3rd of April 2020, and Orientation nº011/2021, from September the 3rd.

About access to tourism and lodging establishment, restaurants, and similar places, gambling establishments, casinos and bars, bongos or equivalent, as well as other drinking establishments without shows and establishments with dancing areas and national continental territory rules, nothing was forgotten by authorities.

Open spaces like beaches and swimming pools were not forgotten as well. On the epidemic context by the Law-Decret nº35-A/2021 from May 18th an exceptional regime was defined for bathing beaches and particularly during summer season occupation and use. Publication of manuals with guidelines concerning beaches functioning was verified, as well as access regulation, occupation and use, and with 14th June Despacho nº 5818/2021, with some local authorities, like it was the case in Oeiras municipality, forbidding access to the walking wall that follows the sea side area and the beach. This legal document established a calculation from which it was possible to mathematically determined beaches occupation potential capacity for 2021 Summer season. In relation to the walking wall, built for health, exercise, leisure and pleasure and quickly adopted by local population – cause it opened a free access to the beach and allowed health and exercise maintenance and leisure physical activities for all ages – it was closed or its access controlled in the context of pandemic fright. It stop being a privilege place established along the sand, a place where usually open air activities like walking, run, skate, rolling skate and several other physical exercises took place and, on Winter and in safety, it was even allowed a leisure cycling practice for all the ages.

Following strictly the law, on 2021 Summer season and not putting police authorities in clear ridiculous situations for impossible access control by police agents to the beaches, as it had been seen in some places at 2020 Summer, direct acces to beaches was literally closed with heavy metallic barriers with the exception of one unique entrance and exit place – a door – in order to allow or interdict access and to determine occupation potential capacity and if it will have or not green light (in some days like Saturdays and Sundays, and in certain hours, half of the morning or sunset hours), instead of police agents it was used electric signs for beach entry permission with the respective metallic separation and occupation regulation control. On 2020 Summer season, limitation established discouraged, when they didn't forbid the use of the most popular beaches and closest to town. Confined at home by law (emergency state) or by recommendation (calamity state), submitted to mandatory isolation, it didn't take long until health problems related with mental health issues were registered and associated with direct or indirect experience with the virus and the pandemic regulations.

However, restrictive legislation didn't limit itself only to these open air places. If these were the official measures taken to the access to beaches, limitations weren't made just to access beaches and the sea. Access to public swimming poles at open air, its use and functioning were regulated from the 5th June through the Despacho nº 6134-A/2020, which made mandatory to management responsible people to assure a set of measures and rules in order to guarantee favourable conditions to avoid the infection spread by clear and strict prevention and control. Rules enforced concerned occupation ways, with social distance rigorous definition, permanency, space hygienic principles and still showers use and cleaning.

It's interesting to verify – as we will understand further – that on a severe pandemic period thermal places received were not neglected and received as well official guidelines and orientation concerning access, behaviour and care. This way, on 2020 13th was published the Orientation nº 013/2020 from Direcção Geral de Saúde, in order to establish even more careful procedure on re-opening time (usually, thermal places open to public in May staying closed from October until May) considered even more adequate to this places traditionally known by clean and hygiene establishments and facilities. The defined norms introduced a set of additional preventative measures particularly to some circumstances favourable to thermal users gathering and encounter like admission, treatments suspension and about maximum thermal users frequency, not only by period as well as by treatment area, as well as the reinforcement of all establishment hygiene procedure related with facilities, the whole equipment, hygiene protocol (before and after treatments) and even thermal users social behaviour recommendations. On this line of thought, through Despacho nº 7046/2021, the exhaustive set of measures taken, in association to Summer infection softening allowed the re-opening of aquatic parks on 2021, 15th July, that is one year later.

If the whole collective life was affected by measures taken by authorities concerning restaurants, bars and disco places, were public disenchantment, discourage and disillusion took over owners as much as clients. Those were the places for gathering and experiencing sociability lived with great intensity.

Those were sharing places, well-being, experience characteristic of warmer places in European south where moderate climate, calm nights and high temperatures invites people to go out and meet friends. Those were the life higher intensity exactly before the pandemic and, in terms of tourism, a vivid factor for visitor's interest. From there the reason for a straight set of official measures taken and recommendations made and systematically published showing the need to extreme precaution concerning infection. The high sensitivity to the major intensity and animation proper to current social life and the intense tourism before the Covid-19 pandemic time these were the places were discouragement, concern and worries were manifested with great despair by owners and represented a contrariety to usual clients. Going from total closure to restrictions and the softening of discontinuous restrictions by Summer (which didn't help) in a deep vulnerability, from ups and downs and equally impossible to manage with a positive way by their owners. In fact the access to fun places especially for Lisbon night life places, the published legislation changes according with circumstances, expectations and the risk factor weren't easy. This way, respectively between contention and restrictions liberating measures, as it is perceptible on 2020, 8th April, on 2021, 1st October with actualizations during 2021 and 2022, Direcção Geral de Saúde defined as well procedures to be followed by restaurants in general, bars and disco places with the publishing of restrictive character orientations for considering the characteristics of these establishment proper to direct and undirect contact, meaning to contagious environment and CARS-Cov-2 infection.

Concerning lodging, prevention procedure, control and vigilance were established according with Direcção Geral de Saúde orientations as it was seen on 2020, 10th March, with Orientation nº008/2020, where it was clearly specified infection prevention and control measures on hotel, lodging for tourists and still more management Orientations to Covid-19 in hotel industry in general and lodging for tourists from WEO-World Health Organization in partnership with TWO-Tourism World Organisation.

Unavoidably, public events and particularly mass events, sporting events (national and international championships and Olympic Games, for instance) cultural (music festivals, theatre, concerts, ballet and dance shows in general), cooperative (rallies, assemblies and meetings) and other (excursions, promenades, etc.) were object of official measures as well as it is clear from Orientation nº014/2021, from 31st November, from Direcção Geral de Saúde, up to date on 9th December and on 2022, 9th January while other cooperative character events were submitted to rules through 2020, 18th Despacho nº8998-C/2020 and given to public knowledge from the Minister of State, Economy and Digital Transition Cabinet, the closest to the Prime-Minister, were subject to a new legal initiative were the interpretation of principles and orientations applied to any cooperative event realization clearly established. As it had happened concerning the same sort of events established before, from the nº70-A/2020 Ministers Council Resolution, on 2020, 11th September.

Orientations and regulations extended to sports and physical activities, to touristic animation, maritime-touristic operators and leisure spaces as well as sporting facilities of all kind and gym academies, all object to specific recommendations concerning sporting infra-structures and other spaces where any kind of sporting practices including open space activities or in closed spaces might happen. Delimitation of any kind of activities was identified, interpretation and orientations were published concerning any event concerning cooperative events presented on sporting grounds, gym academies, swimming pools, sporting academies as well as dance, martial arts and similar activities, as well as massage rooms and health clubs. These sporting and physical activity facilities were all affected on a very hard way by restrictive measures of any activities. Many of these facilities stayed closed for more than one year. In face of this restrictions and interdictions, the only choice that remained was the solitary and desperate run – being this sort of physical activity allowed as long as it was done in isolation.

Finally, all the access to the country were heavily under surveillance. Not only by land borders but by air were object of all the understood restrictions and consecutive rules published consecutively in an official effort to control people circulation and to give them the necessary knowledge related with how each one travelling should behave in order to avoid and to limit virus contamination danger. As it is now well known by collective experience, a big part of these restrictive measures and orientation references were common to the general western world population and they interfered with every people life. In 2020, in a period of a few months, as soon as it was possible to realise the planetary extension of virus effects and the risks involved with Sars-Cov-2 dissemination, the entire closure and confinement political determination as well as the successive measures taken having as harsh pragmatic purpose to avoid virus dispersion and total lack of any control were taken with the understandable rush inherent to the unknown, invisible and intangible danger – but still as well as de difficulties to make sure of political capacity to control the situation through the nervous creation and consecutive publication of urgent measures underlined by the uncertainty of its own effectiveness, its acceptance and obey or even its Constitutional legitimacy.

SOLUTIONS AND RECOMMENDATIONS

If the list of a big part of these legal initiatives presented above might be fastidious, particularly because we lived it in a direct, close and recent remaining experience and due to the fact that in great part we are still enduring it, this list is necessary in order to explain its relation with other set of equivalent measures felt as necessary but slowly and hardly introduced under a form of health and security legislation which since the end of the 18th century were introduced very hardly in Portugal in order to preserve health and life in general (Crespo, J.; 1992).

In fact, the easy way of cholera and other epidemics infection dissemination, fevers and illness situations of all kinds which devastated entire villages as it still happened with cholera epidemic occurred at the middle and again by the end of the 19th century, the devastation caused by tuberculosis and syphilis, confirm resources lack, the absence of hygiene practices, the frailty of public health, bodies weak condition and the clear medical science inability to contain all the invisible evil. These measures taken in Portugal as they started to be taken in other European countries – and to which from the end of 18th century and being a fundamental part of hygiene the creation of gymnastic methods is deeply related – some tenuous measures were taken with a preventative and hygienic goal. These frail measures were in any way a significant start of something new and preparation for a greater intervention and results and in spite of their simplicity they attained all the aspects of collective life like such as hospitals, cemeteries, houses, clothes, food, amusements, schools and markets as well as animal enclosures. By then hygiene orientation measures were taken under another rhythm and under a different context. In a great part, those measures taken reached some of the main urban centre but in spite of the pace (with only a small accelerating rhythm by the end of the nineteenth century and with another acceleration given by the beginning of the 20th century) and several kind of difficulties found – due on a great measure to the very resistance nature to anything unexplored, to novelty, the attempt to find resources as well as solutions – opened the way to change.

At the second half of the 19th century, prevention as a fundamental idea didn't only became more valuable in relation to what it was before as a way to avoid debilities and disease through hygiene – while a medical concept and a concrete technics and well established practices set –, as it gave the chance to the structuring of the country propitious to tourism development. To an inner national tourism, from the second half of the nineteenth century to 1950, to a broader international tourism from the 1960s particularly with the English community finding in Algarve the perfect weather conditions, a favourite life style and the proximity to UK which in itself promote the south of the country, and lately, with clear intensity, from the second decade of the 21st century precisely during the immediate previous years of Covid-19 pandemic. Following concerns and models sketched by other nations as well in equivalent situations, it will be understandable the straight relationship between prevention methods, chronicle diseases treatment, health strengthening and tourism development in Portugal.

The idea that '*you are safe with us*' promoted by the seal Clean and Safe in consideration, it's far from just being a considerable communication initiative which importance comes out in a moment of deep world crises. In reality, this idea corresponds not only as a sign of quick adaptation ability to tourism collapse observed but in reality to an ability of reaction inherent to a long Portuguese experience in matters related to tourism and its straight relationship with health. In fact, this strength points out the Portuguese Office efforts which comes in the first place from a development built from the second half of the 19th century (in this period as it happened with several other countries) due to clear Portuguese population health needs.

Health and Hygiene

Perhaps we would a better understanding of the significance given to the growing importance and value of hygiene and sanitarian safety in the western world when we know the deepest function it has in the world order as this was established by mental frames of this side of the planet and at the core of practices inside which our collective life develops. The order of things in the world we live in western side is built under the base of life favourable principles equivalency that can be found in other societies considered on a geographic distanced perspective or distance on time societies. They are all human societies therefore obey to the same need of order of things. The western dominant world vision organizes life according with the most adequate forms according with its values, its knowledge, its ways of understanding everything that makes part of this world, the world vision e de put in place of things related with life. Perhaps it will help to considerer some old societies in India, as it is the case with the Brahman Havik studied by the English anthropologist Mary Douglas (1921-2007) (Douglas, M.; 1981), among whom she found the sacred opposition between purity and impunity established a clear and rigorous separation line between two worlds in permanent confrontation: in one side, the sacred, on the other the profane. To one corresponds purity and is favourable to life and the other opposite one corresponds to impurity, and is malignant and conducive to decadence and death. Any transgression between these two life dimensions finds itself related with danger, disease and death.

This understanding of the human universe, from which the whole human life is structured and is organizes itself determines the fact that each action people do in daily life is made with extreme care, order and safety that is, in a way that looks to assure the absence of danger, that is the opposite, meaning the presence of symbolic structures, clumsy manners which will bring unquietness, fear and danger which will *certainly* (as it is believed) will introduce fear, evil, suffering, disease and death.

This way, separation between good and evil is maintained through established religious ways, the evil spirits, impure, materialised by the outside, that is by animals and food prohibition because they are considered negative. And interiorly by impure substances like blood, pus, excretions, urine, saliva, perspiration, nasal viscosities, semen etc., body fluids that are believed to transform men in impurity and to scatter evil. According to Brahman Havik world view deep ingrained in a religious order of things, collective and individual life, a perspective from which the whole organization of the world and structuring of social life establishes itself, food, behaviour, social relationships, elementary practices of life and economic activities, everything divided and share established with extreme clarity between men and women. Once this organization was established, it's guaranteed everything linked to impurity will always be an area of woman intervention on her domestic activity related with selection and preparation of food ingredients. And this remains this way due to the fact woman and her female nature makes her more familiar with blood cycles, the reason why women can't be contaminated due to the fact it is understood she shares, through her dipper link to nature and the regular contact with blood turns her to be of the same side of referred opposition. Woman therefore had to secure the continuing established order in everything related with food preparation, that is meals, as well as with every aspects of life in relation with man and with the world. And she has always to act accordingly with this principle to preserve and protect man from any contamination source.

In this perspective the whole life is organised according with religious principles and belief systems which confirm and reinforce it, believes from which it had been established many centuries ago, with a group of integrated practices, a fundamental part of these ancestral societies culture. These kind of societies built their own system of believes, understanding and world organization and life following a traditional and religious world which didn't change, remaining unchanged and static, the same way it was understood and it had been passed to others since ever – making sure everything was delivered from the older people to the younger ones unchanged.

In the western world in general, modern resources are based on a critical thinking and inquisitive mind. It's structured under the basis of distinguished systems of believes which made it possible to introduce progress and technic by the initiative of an inquiring mind. A mind which question the sense and goals of everything, the things form. Promotion of sciences development, especially from the 17th-18th century onwards open the way to changes concerning the way of seeing the world and from that perspective, the way to organize it, to live and to understand things, to share tasks according with sexes in a totally distinctive way.

Considering the principle of impurity and purity studied by Mary Douglas, however distant it might be seen in our societies, this anthropologist considered a deep equivalency relating the same deep individual and societies fears felt towards disease and infection as an equivalent attitude, believes and feelings expressed by Braham societies studied. In western societies those fears would be related to dangers that threat life and health under one not entirely perceptible way. This is why western societies use all kinds of purification/cleaning products, from bleach to disinfectant, soap and perfumed gel, alcohol, regularly and everywhere to keep miasmas, germs, bacteria, virus and impurity – which biological, chemistry and medical sciences closer to the 20th century identified – and to keep it all away from people. In reality, our society idea of progress turned quickly its own attention, after medical doctors intervention particularly since the 19th century, to apply its progress and technical devises extensively to cleaning and hygiene arsenal.

Within the West, in fact since the 19th century, medical determined intervention stressed the need of hygiene practices adoption. Those practices concern not only the outside and public space, animals susceptible to be consumed or not, but to the internal and private world as well where unlimited cleaning and disinfection practices are long established. This set of practices were organized in order to achieve exactly the same life preservation goals. The main difference remains in the fact that instead of being built on a system of believes considered centuries ago and under religious cultural bases, it's about a system of believes based on a system of believes structured on medical knowledge, on chemistry, biochemistry, and on hygiene and clean values as sound disinfection and prevention ways from the invisible threat. This is a system of believes and values that didn't exist before, not under such a clear way, nor so extensive to every elementary area of life. The main attitude is related to hygiene adoption and to make sure the regular space, food and bodies cleaning. Attitudes and practices to guarantee hygiene and to assure the fact they don't neglect the need to follow the rules related with the need to preserve and avoid impurity, as it was the case among the Braham Havik societies, to observation, separation and control disease propagation.

Inside the western world, and particularly in Portugal, the 19th century was the period when towns and cities started to be felt as the main cause of fatigue, frailty, diseases and death. It´s under this believes bases, related with air pollution from the factories introduced after the industrial revolution and believed to be produced at the main urban center, the progress and the slow but constant migratory movement of peasant from the countries to towns, with the subsequent higher demographic concentration generated, under the anticipation of Summer, and in general from May on, Portuguese families moves towards the deep parts of the country. They were searching a more deep contact with nature. Under these seasonal migration was the believe of finding under that proximity with the natural world, far and away from polluted city air, urban conventions and straight ways related with a demanding social urbanity, a physical and nervous relief – if not the cure only possible on the re-encounter of a lost balance of the deeper functions of biological life.

That way, after Adam Smith (1723-1790) physiocracy as an economic theory had pointed out the need to return to nature and to go back to agriculture world as a form of balance and to re-establish work practices favourable to economic progress and social development, the orientation in the sense of going closer to nature continued, and it got stronger under other ways and reasons after the 19th century. After J.-J. Rousseau (1712-1778) philosophy expressed in his work *Émile* from 1760, where he expressed his understanding of the fact man is good at birth but society transforms him as a human being mainly with negative and bad feelings, in opposition to the *good savage* and the good associated with innocence and an innate purity as elements inherent to life in contact with nature, as it was well shown the (apparent) resistance of the so called savages and the peasants, people that in spite of their hardship in contact with nature became (apparently) stronger and with better resistance abilities, after Adam Smith and his economic physiocracy theory expressed on his 1776 work the *Richness of Nations*, it would be medical science to conduct circulation back to nature and the revelation of its benefit to the individual and collective health balance.

On this perspective, the group of families movement since the middle of the 19th century from Lisbon or Porto towns and other urban center, along the whole country in search – as it was declared – of a balance lost felt as necessary and indispensable to life. Those groups were constituted by families and friends whom searching, from May and close to summer season the contact with natural space, the country most inner places. This movement it's seen in the search of a closer contact with nature, its regenerative and curative potentialities, forces associated with the purity of the air and to water sources which came through the soil or rocks by the hand. It's from 1975 Ramalho Ortigão (1836-1915) work about *Banhos das Caldas e Águas Minerais*, where he indicates numerous water sources in the country, the curative indications and the different regions where they could be found. He describes as well the main occupation of the water users. This way, it was confirmed the existence of new uses concerning *vilegiatura*, a time to rest, elegant social gathering and amusement, places unkown until then. The move is indicated amongst upper class groups, media class and low media class which were part of urban population on the move to the country.

They went to *Caldas*, meaning the water places (even when waters were not always warm), travelled to places more or less inhospitable, distanced themselves from the city by car or by train and they followed a believe based on the fact they searched a mere chance to find at the country and by mineral-medical waters the lost balance, an health recovery opportunity. The waters study and waters clinical knowledge development of mineral-medical sources introduced in fact other ways of caring, through different kinds of bathing (immersion, shower, Scottish or Turkish), specific water sources treatments and controlled water ingestion, different massage technics, water injections, exercise like walking, or rowing, gymnastics, simple rest and an appropriate diet lower in spices and seen as more adequate to balance and the re-establishment on various food excesses. The regularity maintained during waters season was essential, the cure for chronical pain and suffering, impurities that might have destroyed health natural balance. In this movement throughout the country, places discovered new interests, a new life, activities unknown were introduced in order to satisfy comfort minimum needs, the levels of hygiene and cleanliness demanded.

Belief on nature power, the pure air and in particular on different kinds of water sources and water therapeutic indications without any doubt was underlined and encouraged by medical doctors guidance and control. Amongst physicians social and medical intervention, it's fundamental to remember the fact that physicians in general remained without consistent therapeutic resources. This was an important reason to consider the possibility to find and to apply innovative therapeutic solutions inside existing resources. Above anything else, mineral-medical waters curative nature represented, at that time, a prevention and controlled therapeutic way, with no risks to become an aggressive or invasive mean once its use was recommended, prescribed, controlled and under surveillance by the clinic expert on hydrology. In reality, therapeutic intervention with base on waters recommendation was presented safer due to the fact it was based on water sources from north to south, west and east serious survey and systematic researches, exhausting study of its physical characteristics like colour, odour, temperature, chemical constitution, researchers amongst whom a French chemical expert became famous by his work and teachings Charles Le Pierre (1867-1945), precise therapeutic indications. From this specific therapeutic indications and studies many articles were written and published in respected medical journals and mainly based on exhaustive survey of national mineral waters sources studied all over the country. This was the reason why once diagnostic was established clinical doctors did not hesitate to recommend those seasonal practices. Their name and work was scientifically based, their professional honour and legitimacy to take over the country sanitary situation was a simple matter of time.

More than just recommend, water contact and treatments were prescribed, technics of purification, favourable to health and with effects proved under an empirical way by local peasants as well as scientific methods which gave much confidence to medical intervention. Spa waters and spa waters science became not just one hygiene resource and medical therapeutic as they contribute enormously to the country inner places discovery and enjoyment as well as to the notorious need to be created in those places, and as soon as possible, as it happened, spa establishments with a minimum of safe and clean conditions as last but not the least adequate lodging conditions, for rest, food, leisure, pleasure and complete satisfaction of all those spa waters regular users.

In presence of the sick conditions related with infectious life in the city and thanks to medical needs associated with a still inefficient level of response from medical science to diseases in general and above all to severe epidemics or serious diseases to which solutions were felt to arrive too late, thanks to the therapeutic quality of Portuguese spa waters, the quality of the air as it was known to happen in Spain, France, Austria and other famous and prestigious European spa places, first hotels establishments appeared which characteristics advanced to the time references, the majority by private initiative and private enterprises. The first hotels were made to lodge sick people and their families under an original way of health tourism, various spaces built on which didn't took long to become famous and to promote development of places like Buçaco, an exception created by official initiative, but hotels as elegant as Vidago, Monchique, Caldelas, Monfortinho or Estoril and many others became an high reference on tourism initial times. The tourism factor introduced and developed by social needs, a response to social practises to which private initiative open the way to entrepreneurs willing to invest inside the country.

In the majority of cases, these hotel centers were the unique developed and technically civilized life capable to offer the most elaborated, modern, comfortable and refined equipment found in the European major cities. Building up those tourism initiatives was an opportunity to introduce improvements in around green places, access and spaces, parks were drawn and an all new green environment started to grow. In this growing new sites, parks and gardens were essential to match natural environment itself, but light and shade games were made with trees plantation like plane trees on the access ways, lime-tree and several others inside the parks which gave way to spa waters and to sources itself, bushes to explore adequate forms and ways, gymnastic rooms were introduced, as it was the case at Vidago, playing fields, where people could enjoy tennis, golf, volant, musical concerts in the bandstand or inside the wide open rooms adequate for a ball as well.

From the last decades of the 19th century the benefit of open air were clearly underlined and valued. While until the end of Ancient Regime (18 th century) people felt and believed the need to hide the whole body and to protect themselves with large clothes and the permanency inside the house walls had been the most adequate way to miasma and germens protection as it was believed, being invisible and to be up in the air – through which disease were spread, in association with bad smells, the bad air – cleanliness of the air indicated the possibility to usufruct the free air with confidence.

If the inside was becoming valued by the existence of spa waters with excellent beneficial and curative sources, especially in cases related with chronicle diseases, the sea side would be discovered as well and the contact with the sands and beaches something slow and progressive. At the beginning, sand beaches chosen by a few were those close to Lisbon with Cascais in first place due to the positive influence of royal family *villeggiatura,* then along the banks of the river Tagus, where people could see some bathing charts installation prepared to protect lady bathing. All along the north bank of Tagus river and in some places of the south bank, as Trafaria and Porto Brandão, bathing season start gained form. There was Paço d'Arcos after, where families concentration called the attention and made this a family kind of beach. Each year families gathered there and during Summer recent habits were full of life and playfull moments with children as well as with adults. This regular presence each Summer favoured some economic development of those small localities as it happened in Oeiras. Under such perspective, and under medical influence as well, maritime air was considered a health powerful allied, so powerful that it should be taken gradually and with moderation. The tan related to maritime air exposure, to iodine and to the sun, would stopped to be related with rural works and poverty to become related many decades after with health, a healthy life, a life outdoors, and beauty but only by the end of the 1960s. According with this set of believes, supported by medical science, sea bathing were progressively and carefully recommended due to their invigorating effects. Important to point the work of the same Ramalho Ortigão, related with this sea side kind of new life *As Praias de Portugal*, from 1976, where main Portuguese beaches were listed and described as well as the main activities groups were involved.

Sports, by its side, and by an attentive medical intervention, constituted important elements of physical invigoration and health. Sports made possible the introduction and the institutionalization of a set of innovative practices and body technics some of those would establish, for instance, the first Portuguese sporting club related with sailing practice and regattas at the river Tagus. In 1856 the Real Associação Naval de Lisboa, established by the initiative of a few Yachting men as it was the case with the Conde de Alcáçovas and one and another member of the English community living in Portugal. Sports introduced different ways of being and doing, innovative and effervescent attitudes and ways of thinking, other body technic ignored until then but demanding more exercise and repetition in order to become always better prepared and physically capable – something that bring about the pleasure about technic and attention to perfection.

These sporting body technics favoured sedentary habits, the abandonment of lazy, passive and useless habits. Transformations were introduced by the royal family to whom being linked by education with sports were decisive to sports introduction in Portugal in a way that was susceptible to cause interest, curiosity, attention from all and were decisive to the abandonment of customs linked to the fear of open air activities, and even the fear of movement and related with excessive protection from air and the outside, habits of live more conducive to disease development, frailty and rickets. In fact, sport introduction process in Portugal was made by aristocracy, considering the inner aristocratic life style were sports were a significant activity related with regular activity but with moderation and pleasure, and following horse riding, dance and fencing as part of an education and distinction as it was the case amongst European aristocratic families. In the meantime, it were medical doctors that soon started to define the limits within which sports should be practiced, advised moderation, regularity and contention. An activity that should be experienced inside leisure moments or free time of the whole working population inclined in many cases to abandon sports as a mean of development, exercise and satisfaction but instead as a way to change for a working activity, that is as a professional work, a way to achieve popularity, economic and social climbing anticipation.

Tourism, Movement and Hygiene

Another idea was taking over the country in the transition between the 19th to the 20th century. That idea would conciliate itself with medical initiatives related with health. The adverse economic crises suggested to many of the responsible people the need to find other ways to promote the country economic and social development. The need to find inside its own territory the resources and the abilities capable to the country recovery. This idea was conducive to achieve a new way of looking at the country and its resources and directed a search for income adequate to introduce a new dynamic in the population life as well as in the whole society. That concern in relation to the deep will to find a new way of looking to the country and the subsequent discourses promote the search of so far ignored national wealth, products that might represent economic and social mobilization in all the different regions. From the start, geographical diversity and the different natural resources, the beauty of natural environment are aspects immediately registered. It was registered as well climate softness, the wide diversity of the landscape and population good temperament. The guiding idea encouraged the survey of the whole architecture patrimony, cultural, journals and magazines were introduced and the main theme was monuments and history richness writing articles in periodicals already known about some picturesque national aspects and looked for ways to value the national territory in all of his aspects, geographical, mineralogical, historical cultural and meteorological features or existing diversity that offer the chance to discover the country, the mountain and the beach.

The discovery of mineral-medical waters, medical direction of those places, attributed credibility and safety to these practices as much as to the journey in itself. Spa waters represented a fact that would certainly encouraged railway need and soon after introduction. The train schedule soon introduced a especial fair to those who were going to spa places and built links to get rail ways closer to those regions. But it favoured the opening of road tart, maps introduction in order to help people orientation and as much as possible adequate signs.

It was noticed the existence with real touristic interest like rivers, lakes, bridges or archaeological roman roads, the roman vias ignored until then and covered by time achieved a new interest. The interest to reveal the nation values was also clear in the systematic study and register of each place more characteristic aspects, in the importance given to religious architecture. This one became considered as a matter of interest and continuity of religious practices that were part of the way to mineral waters places or during the period of spa waters treatments. In reality, confrontation between health and disease persisted in contact with waters, deepest attention was directed towards the feeling of being well, healthy, the frilty or recovery of energies, health and the religious feeling wasn't forgotten. It was expressed instead as a deep gratitude feeling manifested by those who leave with new strength and, above all, recharged with hope. The idea of value everything Portuguese territory offered is underlined by the revealed attitude for instance by monuments and proper local history systematic survey, the innovative publication of leaflets rich in information about the places and its saints patron and the calendar inherent to celebration and, still, small books of touristic interest of the spa waters places, the creation of guides and periodical publications rich in photographic images and diverse information about different places with touristic interest. Automóvel Clube de Portugal (ACP) creation in 1903, the distinction of having as Honorary President the King D. Carlos himself, as Honorary Vice-President the Infant D. Luís Filipe and as General Assembly Perpetuous President the Infant D. Afonso, constituted an unquestionable sign of recognition and support given by the highest responsible of this country to this association. At the same perspective, in 1906 of Sociedade de Propaganda de Portugal (SPP) (Portugal Propaganda Society) the creation revealed the growing interest in giving back to the country its own inherent value which the introduction of this new society confirmed.

CONCLUSION

Covid-19 epidemic appearance brought to our collective knowledge resources and emotions long forgotten in the 21th century society, scientific, technic and economic development became something taken for granted and the unexpected virus introduced a general attitude of disbelief at first to an attitude of fear and sometimes of panic especially in periods and places, like Italy, Spain and Portugal when death rate shown the incapacity for health services to deal and to control Covid-19 spread. Knowing this was a new virus and science hadn't yet found the vaccine to guarantee a scientific controlled situation in terms of health was urgently dwelt by experts with constant explanations about this new virus and its characteristics in order to make public a certain level of understanding about the dangers and solutions to tackle the collective and individual safety, the safe way to keep safe and sane.

All over the world different countries and different political regimes adopted a diversity of measures that went from the entire population in an absolute isolation (China), to the introduction of severe restrictions of circulation and normal activities, from work to travel (Portugal), to a variable approach according with politicians and experts (USA) or maintaining a normal life in order to get group immunity (Sweden). More important for our analyses than the adopted ways to deal with the new problem and especially its world dimension was the fact of founding at the 21th century our safe society was no longer safe at all. Fear was the dominant feeling as it had been centuries ago when science was far from any solution and finally hygiene was the solution for decades until scientific research could built and developed itself mainly in the western world. The word hygiene, and the practices related with it were substituted by the word health as science evolved positively and started to guarantee the feeling of being safe by the adoption of chemicals, medical technics, practices which seemed to have solve the general health problem with terms of old epidemics like cholera, tuberculosis or other health problems.

And yet, from one moment to another fear, disorientation, anxiety, doubt became the dominant feelings as they used to be many decades ago. In spite of the whole development the world, the developed world was confronted with the same basic feelings which had kept individuals and societies submitted to psychological, social and economic structures related with insecurity, fear and anxiety. From one day to another fear, distance, isolation, separation were the basic ways to deal with an invisible and mighty virus. Covid-19 fact revealed both collective and individual reactions equivalent to the same basic fears rooted in ancient and traditional societies where fears were kept away by religious beliefs and practices, like with the Havik, and societies having to face the unknown returned back to primitive unsafety and fear.

According with expert information urban regions with higher demographic levels and population density presented the easiest areas for the virus to spread information that conduced people who could to leave towns and move to the country side, back to a second home or renting a few rare lodgements away from major urban center.

REFERENCES

Crespo, J. (1992). *A História do Corpo*. Editorial Difel.

Douglas, M. (1991). *Pureza e Perigo. Ensaio sobre as noções de poluição e tabu*. Lisboa. Edições 70.

Hasse, M. (1998). Oeiras e o Desenvolvimento de Novos Comportamentos de Lazer. A valorização de um novo mundo: o mar, a praia e as férias, Iº Ciclo de Estudos Oeirenses. *Oeiras – A Terra e os Homens*, 265-285.

Hasse, M. (1999). *O Divertimento do Corpo. Corpo, lazer e desporto na transição do séc. XIX para o séc. XX em Portugal*. Editora Temática.

Ortigão, R. (1944). *Banhos das Caldas e Águas Minerais*. Clássica Editora.

Ramalho, O. (1966). *As Praias de Portugal*. Livraria Clássica Editora.

ADDITIONAL READING

Boyer, M. (1972). *Le Tourism*. Seuil.

Cavaco, C. (1983). *A Costa do Estoril. Esboço Geográfico*. Editorial Progresso Social e Democracia, SARL.

Corbin, A. (1988). *Le Territoire du Vide. L'Occident et le désir du rivage 1750-1840*. Aubier.

Corbin, A. (1995). *L'Avènement des Loisirs 1850-1960*. Aubier.

Elias, N. (1984). *Du temps*. Fayard.

Elias, N. (1991). *La société des individus*. Fayard.

Hasse, M. (2005). As Àguas Minero-Medicinais e o Processo de Ritualização dos Lazeres. In *Caldas da Rainha Património das Águas. A Legacy of Waters* (pp. 259–269). Caldas da Rainha. Assírio e Alvim. Câmara Municipal das Caldas da Rainha.

Henriques, J. M. (2011). *Da Riviera Portuguesa à Costa do Sol. Fundação, Desenvolvimento e Afirmação de uma Estância Turistica*. Edições Colibri. Edições Colibri. Câmara Municipal de Cascais.

Laty, D. (1996). *Histoire des Bains*. Hachette.

Wallon, A. (1981). *La Vie Quotidienne dans les Villes d'Eaux (1850-1914)*. Hachette.

KEY TERMS AND DEFINITIONS

COVID-19: Corona virus and pandemics are phenomena that have ancestors. E.g., it is possible to make equivalences between reactions to COVID-19 at the 21th century and *health fears* as some of the main reactions to disease between the 18th century and the 19th century in Portugal.

Health Fear: This process may appear as a terrifying sense of danger felt by the population, in addition to the lack of an immediate scientific solution to such problem.

Hygiene: Hygienic measures, taken by public policies and legislation, include the return to nature, as one major solution to recover health.

Medical Guidance: This strategy, including *hygiene* and *vaccines*, among other public and private measures, proved to be fundamental, and also provided some of the main conditions for tourism development.

Vaccine: Until a safe vaccine against *COVID-19* was produced, public reactions went back to the same basic *health fears* known in the past, when science was far from giving answers to serious health problems.

Section 5

Outbreak Politics, Policies, and Power Within Hospitality

Chapter 17
Workplace Harassment of Restaurant Workers

Silvena Dencheva Yordanova
Varna University of Management, Bulgaria

Hanna Stratchen
Varna University of Management, Bulgaria

ABSTRACT

Today all people become victims of harassment in one form or another. Harassment at work is a very hot topic nowadays as lot of women are objects during their work. The chapter is focused on presenting the essence of harassment and how it can take place in a kitchen. The first part discussed harassment and mobbing in the workplace. The methodology part presented a study conducted among women chefs and kitchen staff. The study was conducted through online questionnaire and included 39 respondents. Besides this, an interview with 13 women chefs has been performed. The study aims to understand if there is harassment in the kitchen and what forms it takes. It also aims to understand how women as chefs are perceived in their job positions and how they are harassed. The results show that women chefs have been victims of different forms of harassment in the kitchen.

INTRODUCTION

Today all people become victims of harassment in one or another form. Harassment at work is a very hot topic nowadays as lot of women are objects during their work. It has very bad effect on people's life as it reduces the work satisfaction, increases turnover at work.

The chapter aims at presenting what is harassment, the forms of harassment and mobbying and how it can take place in kitchen. The chapter also aims to find the link between harassment of women chefs in kitchen. It took form of a positivist approach. It includes 13 interviews with women chefs and a study among 39 kitchen workers. It aims to understand how women as chefs are perceived on their job positions in kitchen and how they are discriminated (Strachan, 2021). It presents how harassment is perceived among kitchen staff that took part in the survey.

DOI: 10.4018/978-1-6684-3369-0.ch017

The chapter starts at first with short introduction on the topic of harassment, types of harassment.

Later in the methodology part is presented the study of women chefs. A questionnaire has been distributed through online platforms to women chefs and kitchen staff. Besides an interview with 13 women chefs have been conducted to understand about what harassment is, what makes them feel bullied at their kitchen, how they perceive their roles.

Importance of the topic

The problem of harassment has become serious in the last few years. In the academic studies there has been a significant interest towards this problem and the studies show that it is quite popular problem.

For example, the study shows that over 50% of women working on academic positions and staff at universities have experienced any form of sexual discrimination at work (theguardian.com). The study above shows that it has been mainly experienced by women as they became frequently have been felt any type of harassment in the working place. This percent has been significant for women that work in academic environment compared to women working in administrative positions. In this regard, the studies show that the main type that has been reported is sexual harassment as a form of discrimination against other sex. It is supported by data of study of guardian.com. According to it, more than four-fifths of young women in the UK have been become victims of sexual harassment. It is mainly experienced by women from 18-24 years old, 86% which had felt sexual discrimination publicly, whereas 3% did not mention ever such discriminating behavior to have met. 11% of respondents prefer not to answer the question. 71% of women have felt a sexual discrimination, according to the survey (theguardian.com)

The problem of harassment can be seen from many perspectives. The legal definition provided by the Equality Act 2010 harassment defines it is as unpleasant and unwanted behavior directed towards violating an individual's respect and showing humiliation or disrespect to someone thus forming a hostile and negative for that person.

The Chartered institute of personnel and development discusses harassment as it is involving one or many people and can happen once or many times in a different form of behavior, varying from extreme forms of physical violence, to ignoring someone. The discrimination can be with or without witnesses, online or offline. Every single situation in which is a tension caused by physical or mental violence on somebody is a proof of a harassment.

One of the most studied form of harassment is the sexual harassment. In this regard, the 2018 report, the sexual harassment was considered as a form of discrimination, in term of three categories of behavior as discussed by White & Hardemo (2002).

1. **Gender harassment:** discriminating the other sex with vulgar words or impolite gestures aimed to show hostility, and exclude the other gender;
2. **Unwanted sexual attention:** use of words and gestures that are unwanted and can include assault;
3. **Sexual coercion:** all situations including undesired sexual activity that happens if you are threatened, pressured, or forced in a nonphysical way. As for sexual coercion, there are specified two types of sexual harassment in US; quid pro quo, in case a manager threatens an employee or gives a promise for a sexual favour. The second case of sexual harassment can happen in toxic work climate environment, and refers to conduct of a sexual nature, by anyone in an organization, as part of the individual's work performance. Quid pro quo harassment is existing if after it has happened once and the superiors have been informed about it but did not take any action and continue to pretend

that there is no problem. In case of such accident, the victim needs to report as she/he has felt it a humiliating, repetitive and is continuous in time as suggested by White & Hardemo (2002) .

Harassment can be also moral, when practiced by a manager or superior with higher position to an employee with lower rank in the organization. It can be practiced also by coworkers and this is the case of a horizontal harassment caused by envy, competitiveness or other motives (Grazina & Magalhaes, 2012)

The two researchers Grazina &Magalhaes, 2012 conclude that the moral harassment can also take the form of a mixed harassment if it continues over time and presents a discrimination from manager and from coworkers towards other employee and managers do nothing to solve the problem. At last, not so frequently can be met also the ascendant vertical form of discrimination when an employee discriminates or mocks a person on a higher level of the organizational structure due to incompetency of the person who is promoted to higher position as result of having personal relationships with the boss but not due to knowledge or skills (Grazina &Magalhaes, 2012).

The unwanted behavior in the workplace can take the form of mobbing too. There is a specific difference between harassment and mobbing. Mobbing can happen if aggression or sarcastic language has been used to a coworker, employee in a form of racist or sexual joke. This is case causing unfair and unethical communication on the working place towards a specific person and can be continuous process to happen often (almost every day) and over a long period (at least for six months as Leymann, 1990 connotates.

Workplace bullying includes all cases where an employee is shown bad aggressive behaviors during work by using any form of humiliation with words or threat or fear or using any form of punishment.

Mobbing can take following stages:

- Stage 1 – *the conflict is obvious as well as the conflicting parties*
- Stage 2 – *"we and the individual" principle (identified patient)*
- Stage 3 – *Victim is fired or transferred to another department or quits job*

In this regard, the leader's role is very crucial to avoid and solve the conflict on its first stage. Having in mind that people are different, their attitude to solve problem is various so the conflict competences they have differ too. If there is a potential arising conflict, the leader or the mediator must do his best to pay attention and find a reasonable solution that fits all parties.

Mobbing can exist on different levels:

Vertical – top-down. In this case, mobbing is performed by management to staff to lower levels
Horizontal – same level. It can happen on the same level, which is the most popular form in most companies. This is mobbing between coworkers, colleagues in department.
Top directed – bottom-up

Mobbing can be related to the work, person oriented or take the form of a physical intimidation.

As Einarsen et al. 2008 connotates each time when there is hostile workplace and if it happens too often for a long period of time is a signal of harassment on the workplace.

If there is imbalance between the power of two parties involved, there is mobbing as defined by Einarsen et al., 2008. This is especially true when the relationship is manager-subordinate and when the manager uses his legitimate power of the position to abuse verbally to say racist jokes or isolate other people, it can be reported as mobbing on the workplace.

The bullying relies on the interpretation of the victims that the accidents are humiliating, hostile and directed towards them on purpose as Einarsen et al., 2009 connotate.

HARASSMENT AT HOSPITALITY

In present chapter the focus is on the workplace harassment on the example of women chefs and kitchen staff.

There are different forms of harassment that have subject of study for staff in hospitality. The majority of the researchers focus on the sexual harassment. For example, White & Hardemo (2002) discuss about two studies related to hospitality industry, focused on the gender perceptions of sexual harassment in the restaurant industry, one in Hong Kong and the other in the US.

The first study, presented by White & Hardemo, 2002 displays verbal abuse including the use of vulgar language or sexual hints, sexually comments and even undesired physical touch. The results showed that women take these more serious cases of harassment than the men, according to White & Hardemo (2002).

The second reported case was about 72 hospitality students in the USA, who have responded that sexual harassment is an often met practice in restaurants. According to the data of the survey, more than 60% of restaurants have already developed instructions what to do in case of discrimination and that 42.4% of female and 21.1% of males had already experienced any form of sexually harassment at work (White & Hardemo, 2002).

In their paper, White & Hardemo (2002) studied how culture affects the perceptions of sexual discrimination in the hospitality. The two researchers studied the culture of French and Swedish in the kitchen as they included in their sample two culinary schools, one from France, the other in Sweden were selected as the data collection sites.

Working in restaurant is defined as having specific restaurant culture. In this regard, Matulewitz, 2015 presents in the restaurant a lot of sexist jokes and flirts are usual part of the work process and restaurant workers take it all as normal part of the job".

Cases of sexual harassment of restaurant workers have also been discussed by Janzen and Platy (1989). It was the case of two women waitresses being harassed sexually by a cook.

That is explained by the definition of the Supreme Court of Canada according to which sexual harassment is perceived as being a form of discrimination breaking human rights and represented by unpleasant and unwanted conduct of a sexual nature that happens in working environment or has some negative impacts on the victims (Matulewitz 2015).

In this regard, in the kitchen work environment harassment can be defined as all type of behavior towards employee such as refusal to communicate with this person or lack or misleading instructions to employee or giving too much work to one person or assign too much tasks that are beyond the competence of the employee or giving stupid tasks to do or too much critics and showing sarcasm through jokes or insulting in front of the other staff (Strachan, 2021).

Many women working in hospitality often become victims of harassment. Especially women chefs as there is a stereotype of working on chefs positions and being a male profession. The harassment in the kitchen can be verbal, physical or sexual. Working in the restaurant business is sometimes very stressful job. Due to lack of time and no time for reactions, staff becomes victims of any forms of harassment. The problem appears when such practices are perceived as normal due to stressful job conditions and intense work.

It is very popular definition of sexual harassment as undesired sexual behavior at work that is perceived by the victim as a form of a threat or too much unwanted.

The harassment also can be displayed with using unpleasant words, sexist or racial jokes towards women regarding possession of physical traits (i.e., hair, clothing, makeup) or some features of the personality (i.e., assertiveness, humor, low levels of sensitivity (Main,2021).

The difference between harassment and normal practices in kitchen has been studied by Cooper 2017 et.al 2010). Such practices taking place in kitchen including verbal aggressive behavior, unethical forms of behavior are considered as kitchen culture as the normal. (Strachan, 2021).

During interviews with chefs in UK from mid class restaurants was found that existing myths were affecting and dominant among restaurant, such as the saying that aggression in the kitchen has become a normal part of the work but should be taken as such by the kitchen staff, thus making free chefs from any personal responsibility (Einarsen, 2008). It can be explained with bad working conditions of low salaries and intense work that may create hostile work climate among kitchen staff as Einarsen, et al. explain 2008. The study of Einaster et al. 2008 concluded that in such work conditions staff can easily become a victim of harassment due to individual burnout and increased level of tiredness among restaurant workers.

To sum up, harassment can take many different forms. Most frequently met in the hospitality are verbal harassment as well as sexual. The researchers will discuss further what happens in kitchen and if women chefs and kitchen staff have experienced any form of harassment while working.

METHODOLOGY OF RESEARCH

The chapter draws attention on studying how women chefs and kitchen staff perceive the harassment. For the purpose of the study are used primary and secondary data. The study includes a combination of quantitative research methods. The questionnaire will be used to understand what the opinions of all kitchen staff are including women chefs. It also aims to provide understanding of the existing practice nowadays in professional kitchens. For these purposes, 39 women chefs were given a questionnaire to fill in (Strachan, 2021). In this regard, Strachan, 2021 has conducted interviews with 13 women chefs from all demographics. The study of Strachan, 2021 aims to understand if they have experienced harassment and what they think about it as a main factor in the work of women chefs.

The online questionnaire has been distributed online among participants by using social media platforms like Facebook, Instagram and also sent by email. The participants were 18 years old or older and participants were from both genders and were asked to answer all questions. After the first 7 questions, the questionnaire was divided for two genders on chefs positions as male chefs have to respond only to the first 7 questions concerning demographic profile they have and then on two questions related to harassment in the kitchen. On the other hand, women chefs are asked to reply many questions concerning how they perceive harassment in the (Strachan, 2021).. 13 women chefs were interviewed having professional expertise in kitchens. These women have different backgrounds and are above 18 years of age or older (Strachan, 2021).

The study aims to understand if there is any form of harassment experienced by women chefs and kitchen workers. The results are summarized below. As it becomes clear, from 39 responses given shown in table 1, only 9 of 39 respondents answered that they did not met such practices in the kitchen they have worked. (Strachan, 2021). As it can be seen from the chart below, 30 participants have declared that they have met cases of harassment, discriminatory behaviors resulting from the religion they have or based on gender differences.

For the interviews, narrative analysis has been used. The interviews are transcribed using an online platform called O'Transcribe and then analyzed by narrative analysis to focus on the opinion of every single participant (Strachan, 2021). The profile of the respondent's chef is presented below. As for the demographic profile, it includes 26 female respondents and 13 males. As for their age, 38% of the respondents are from 18 to 25 years, 28% of them are in the age group 26-34 . less is the number of people above 34. 18% of them are up to 44 years whereas only 13% are up to 54 years old. Just 3% of all are over 55 years. To sum up, the demographic profile of respondents shows they are active people.

As for educational status, almost half of them 44% are with bachelor degree followed by high education diploma- 21% and the rest are with master degree- 10% as well as associate- 10%.

Regarding the employment status of the respondents, 62% of them are employed. The number of students is significant- 38% and just 10% of all are unemployed to the moment of the survey.

Figure 1. Types of harassment experienced at kitchen by women chefs (Strachan, 2021)

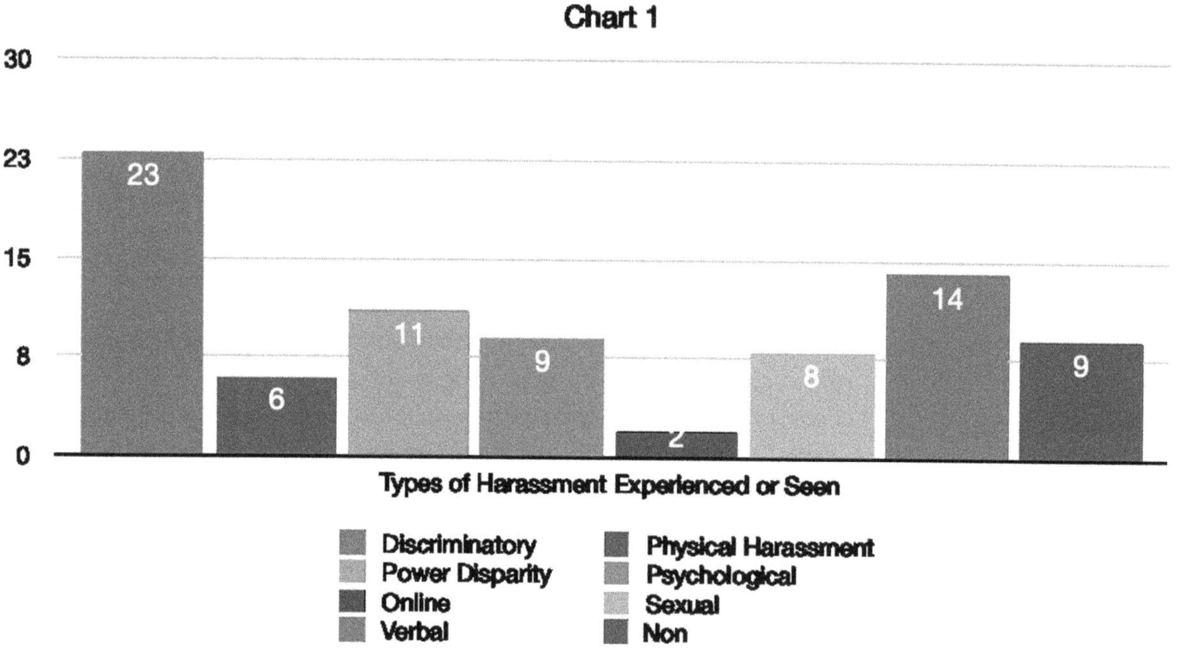

It was interesting also to understand if harassment is normal in kitchen regarding gender inequality. For this purpose, second question has been raised. It becomes clear from the chart 1, that respondents have met some form of a harassment in the working environment. The results show that harassment affects mainly women and many feel that there is gender inequality in kitchens (Strachan, 2021).

The same results were reached by Selcuk, 2018 who conclude that though women compete with men, it was hard for them and even only few of them are able to be promoted to higher positions. In other words, success is achieved harder by women. As Selcuk, 2018 states, men are hired to work and take responsibility for professional kitchens whereas women should care for the domestic kitchens.

Harassment can be felt as differences in gender. For this purpose, respondents have to answer if they see these both as barriers for women chefs and are why many women do not have the chance to take managerial positions in kitchens. The study of Strachan, 2021 proves that some professions are taken as male's and women are given the opportunity to work in pastry workstations only.

In the study of Strachan, 2021 is evident that what is not attractive in the job is the long working hours. Besides, chefs find the job stressful and hard to balance life and work. What is more, the study shows that women chefs do not receive respect on their job positions. The main factor leading to stress in the positions of restaurant industry is the intensity of work and difficulty to balance between career and life. It is supported also by the study of Selcuk, 2018 that in restaurant industry which is considered as men's territory women are feeling more times harassment. As LaPontie connotates, if women want to compete and work in the professions of chefs, they have to be be tolerant and competitive to succeed but also to take easily jokes, abasement, and sexual harassment (Selcuk,2018).

The researchers focus also to understand how many times women chefs face harassment while being at work. The results show that for some women chef's harassment is experienced by being exposed to jokes every day. This represents verbal harassment on the workplace and should not be perceived as a normal as it is the practice in the restaurant industry. For others, sexual discrimination exists as unpleasant and undesired physical touch from another person for many happened occasionally or never for some. Sexual discrimination as physical touch has been reported to happen occasionally, whereas some chefs have never met such practice in the kitchens they work. Verbal harassment is also met as using verbal jokes or unpleasant comments towards specific person regarding religion and it has happened for a few of the respondents (Strachan, 2021).

The different way women chefs and kitchen staff perceive harassment is quite normal and is explained due to individual differences they have. It is supported by the study of Nielsen and Knardahl (2015) who explain the harassment with the help of four different causal mechanisms. The first mechanism explains it that it is up to the personality to perceive it as harassment. According to the second mechanism, the so-called target-behavior mechanism, some employees provoke aggression in others with their behavior through violating expectations, underperforming, and even breach social norms (Einarsen et al.2008). The third mechanism tells that some individuals due to the personality traits they possess may interpret some behavior as negative. In other words, all employees having such personalities are more prone to experience and describe as bad incidents that happened to them during work process as harassment (Einarsen et al. 2008).

Selcuk, 2018 says that women chefs need to be having team skills, discipline and show patience to survive in restaurant business. If they cannot adapt to long working hours and are not competitive enough, it would be a problem in their career in kitchens.

The next questions are regarding discrimination at workstations. The chefs were asked if they feel restricted regarding the workstations they can work in. Most of chefs answered positively that they felt such limitations as many replied never.

The next question is treating the way chefs have felt encouraged or not to work in different stations and different positions. The majority of respondents have answered positively that it happened occasionally but for others they never felt that. When asked if because they are a woman if they were treated unfairly, most respondents said that this happened occasionally and for some it happened more often than not. The feeling of exclusion and unfair treatment of respondents confirms that harassment exists in kitchen.

Feeling of exclusion can be avoided as Selcuk 2018 suggests that women chefs should prepare themselves for the kitchen working environment on physical and mental level to be competitive and ready to balance between hard working conditions and personal life.

The next question asked concerned how many times the chefs have experienced intimidation and the majority answered that it is a frequently met practice. Whereas some of them have met it more often than occasionally. It is interesting also to know the effect of harassment on the victims if they feel safe reporting. Out of the 26 respondents when asked if the harassment has affected which departments they work on and it was divided in that for some women replied positively while for others it didn't while some could not give a definite answer (Strachan, 2021). These answers are a signal that people feel unsafe at their working place, afraid to lose their jobs especially in the difficult times of covid crisis. The study has been conducted in the period of covid crisis which also affects on the answers received of women chefs and kitchen staff part of the study.

The chefs were also asked how they are going to do in case of harassment. They were asked if they would report. Most of the respondents answered positively that they would inform their supervisors or HRM in case of discrimination on them. Few of chefs women answered no or that they couldn't provide a specific answer. It becomes clear from the answers that there is no practice of introducing workers in kitchen and chefs what are their rights and what should be done in case of harassment in the kitchen as the majority of respondents confirmed that there is such a manual or policy in the restaurant they work. At the same time, there was a big number of respondents that were not introduced to having such a policy in the restaurants they work (Strachan, 2021). Such a problem can be solved if kitchen managers and human resource managers should do their best and not hire staff with a predisposition to aggressive behavior. Prevention is the solution to harassment as Selcuk, 2018 suggests.

The study is comprehensive as described earlier as it consists also of an interview.

In the study was included also an interview with 13 women chefs from various countries having various level of work expertise and opinion. The participants showed that they know the specifics of kitchen (Strachan, 2021).

On the question if there is still discrimination in any form to women in the kitchen they all replied positively that there is such practice too often nowadays in the kitchen. Participant 12 described the kitchen as male territory, where women chefs have to work harder to prove their skills (Strachan, 2021).

The same results were confimed by Albors et al.2021 who define kitchen environment as hard to work in, especially in the very dynamic and competitive kitchens in Michelin0star restaurants.

Most of the respondents women chefs declared that sexual discrimination happens to women and Participant 3 gave an example of sexist joke used by her male coworkers repeatedly that embarrassed and harassed her many times (Strachan, 2021).

Discrimination between sexes still happens as study of Albors et al. 2021 prove with fact that male values are dominating in the kitchen as it is in army.

According to by participant 13 harassment still happens, not so frequently nowadays but there are "still underlying biases against women'' in kitchens. Most of the of women chefs believed that it should be taken seriously and this kind of mocking or verbal abuses in kitchen should not be tolerated. For others, there is no such problem as they take it not seriously but just as a fun as kitchen work is such (Strachan, 2021).

The next interview question was related to the gender discrimination. Most women chefs agreed that women have met harassment while working in the kitchen. Few of women chefs did not felt it as case of discrimination. According to participant 2, women were not taken seriously on the positions of women chefs....and that women have to prove their competency and work more and harder than men. This opinion has been confirmed by participant 13 saying that no matter what the industry is about, women must be very competitive and industrious to succeed. It becomes clear from the results, that women and men in same working conditions receive different attitude to each other. What is more, it is up to two factors mentioned by participants (11 and 3) .For them the opportunities are given to men there is still unfair treatment due to gender differences. Other chefs told that inequality exists in kitchen and even women are given more work and have to work harder (Strachan, 2021).

As Albors et al. 2021 explained the main factor that effects on the restaurant climate is the fear. If women can manage with fear, then authoritarian masculine style of management can be avoided.

On the question: Is gender a limit and how are treated women over 40 years old, most women chefs agreed on it. Participant 2 agreed that it is hard for a woman to balance between professional work and personal life. Most of the respondents summarized that if women are competent and have the ambition, they will succeed. Participant 13 said "if a woman wants to move up the ladder, she will go up the career ladder. Regarding the respect, the majority of chefs answered that women over 40 still working in kitchens are respected the same way as men, as they proved with years they deserve it. As Strachan, 2021 notes it can be explained with the level of proficiency the chefs have accumulated through years working in kitchen.

Gender discrimination exists as women cooks face many challenges while working in professional kitchens and balancing between family and career progress, which is more complicated and requiring in Michelin-starred restaurants. The main barriers that prevent women chefs from promotion in their career are family obligations, difficulties to balance work and personal life as well as fear to take responsibility in case of promotion as discussed by Albors et al.2021.

As it became clear from the studies above, harassment is a normal practice still in kitchens in different forms. As the researchers La Lopa and Gong, 2020 mention discrimination still happens in kitchens, in all forms. It has been perceived as normal as part of restaurant culture so many chefs have felt it (Strachan, 2021). Women chefs often become victims of verbal aggression such as racist or sexist jokes, being a part of the masculine kitchen culture in the kitchen they work. Gender discrimination exists in kitchens as it becomes clear from the interview. Then arises the logical question how to prevent harassment. One of the ways to deal with such practices is the role of mentors. However, there is a lack of mentors in hospitality is perceived as a main problem by Albors et al.2021.

The last question of the interview deals with: Is harassment a defining factor and should kitchens help victims?. Most of the participants in the interview shared the idea that harassment is of importance as it affects on the turnover of women chefs. They all believed that stress at work can provoke and increase the probability of discrimination, as staff is too stressed and the level of burnout is high. Many women in restaurant business find it difficult to compete as they have to work even harder in male professions as well as deal with family obligations. In this case, they consider working conditions in kitchen not so attractive and they feel not recognized or treated with respect and dignity (Strachan, 2021).

CONCLUSION

The best policy to deal with harassment in any of its forms is the prevention. It is supervisor and manager's responsibility to provide positive working climate, to inform employees about their rights and how to proceed in case of discrimination. Each worker should be introduced with the code of conduct when appointed. The rules of the company in a written or unwritten form create justice in the organization. The main responsibility of the managers is to create environment where all employees will be treated fairly and not to allow any form of discrimination to be considered as a normal part of the job as it is nowadays in professional kitchens.

The studies show that harassment is quite known in kitchens (Cooper 2017 and Patah et.al 2010), taken not seriously as fun (Webber et.al 2008 And Patah et.al 2010). It happens more often to restaurant staff and quite often as well as in different forms of harassment. Secondly, the working environment in kitchen is considered to be predominantly male. Thus the women feel discrimination quite often in a form of joke or racist comments (Strachan, 2021).

Chamberlain et al found that sexual harassment depends on the structure of the work group and its composition of males and females and also how all of them interpret harassment. The high level of uncertainty at the jobs of restaurant workers can create favorable conditions of possible harassment in the workplace as they go to work not always knowing how many working hours are they going to work. In case of uncertainty of jobs and seasonality too, the works becomes too stressful and intensified relationships between managers and employees in the restaurant busines

REFERENCES

Acker, J. (2006). Inequality Regimes: Gender, Class, and Race in Organizations. *Gender & Society*, *20*(4), 441–464. doi:10.1177/0891243206289499

Aggarwal, A., & Gupta, M. (2000). *Sexual Harassment in the Workplace* (3rd ed.). Butterworths.

Albors-Garrigos, J., Peiro Signes, A., Val Segarra-Oña, M., & Garcia-Segovia, P. (2021). Breaking the glass ceiling in haute cuisine: the role of entrepreneurship on the career expectations of female chefs. *Tourism and Hospitality Management*, *27*(3), 605–628. doi:10.20867/thm.27.3.8

Einarsen, S., Hoel, H., & Notelaers, G. (2009). Measuring exposure to bullying and harassment at work: Validity, factor structure and psychometric properties of the Negative Acts Questionnaire-Revised. *Work and Stress*, *23*(1), 24–44. doi:10.1080/02678370902815673

Equality and Humans Rights Commission. (2010). *The Equality Act: Guidance for small business: Understanding legal definitions and unlawful behaviour in the equality act.* Retrieved: 27.04.2021 from https://www.equalityhumanrights.com/sites/default/files/ea_legal_definitions_0.pdf

Friedman, G. S., & Whitman, J. Q. (2003). *The European Transinfromation of Harassment Law: Discrimination Versus Dignity* (Vol. 9). Columbia Journal of European Law.

Garrigos, J. A. (2020). Gender discrimination in haute cuisine: A systematic literature and media analysis. *International Journal of Hospitality Management*, *89*, 102569. doi:10.1016/j.ijhm.2020.102569

Garrigos, J. A., Haddaji, M., Segovia, P. G., & Signes, A. P. (2020). Gender differences in the evolution of haute cuisine chef's career. *Journal of Culinary Science & Technology*, *18*(6), 439–468. doi:10.1080/15428052.2019.1640156

Giousmpasoglou, C., Marinakou, E., & Cooper, J. (2018). "Banter, bollockings and beatings": The occupational socialisation process in Michelin-starred kitchen brigades in Great Britain and Ireland. *International Journal of Contemporary Hospitality Management, 30*(3), 1882–1902. doi:10.1108/IJCHM-01-2017-0030

Grazina, I., & Magalhaes, J. (2012). Moral harassment at work: The relationship with organizational culture. *Journal of Business and Retail Management Research, 7*(1).

Haddaji, M., Albors-Garrigós, J., & García-Segovia, P. (2017). Women chefs' experience: Kitchen barriers and success factors. *International Journal of Gastronomy and Food Science, 9*, 49–54. doi:10.1016/j.ijgfs.2017.06.004

Haddaji, M., Albors-Garrigós, J., & García-Sigovia, P. (2017). Women professional progress to chef's position: Results of an international survey. *Journal of Culinary Science & Technology*, 1–18. doi:10.1080/15428052.2017.1363109

Harris, D. A., & Giuffre, P. (2015). *Taking the heat: Women chefs and gender inequality in the professional kitchen*. Rutgers University. doi:10.36019/9780813571270

LaMontagne, Smith, Louie, Quinlan, Shoveller, & Ostry. (2009). Unwanted Sexual Advances at Work: Variations by Employment Arrangement in a Sample of Working Australians. *Australian and New Zealand Journal of Public Health, 3*(2), 173–179. PMID:19413863

Leavy, P. (2017). *Research Design. Quantave, Qualitative, Mixed Methods, Arts-Based and COmmunity Based Participatory Research Approaches*. The Guilford Press.

Lerum, K. (2004). Power and Camaraderie in Service Work. *Gender & Society, 18*(6), 756–776. doi:10.1177/0891243204269398

Lessard, H. (2007). Backlash in the Academy: The Evolution of Campus Sexual Harassment Regimes. In D. E. Chunn, S. Boyd, & H. Lessard (Eds.), Reaction and Resistance: Feminism, Law, and Social Change (pp. 164–195). Academic Press.

Lopa, J. (2020). Sexual harassment of hospitality interns. *Journal of Hospitality & Tourism Education, 32*(2), 88–101. doi:10.1080/10963758.2020.1726767

Lopa, J. (2020). Sexual harassment of hospitality interns. *Journal of Hospitality & Tourism Education, 32*(2), 88–101. doi:10.1080/10963758.2020.1726767

Main, A. (2021). Measuring Workplace Harassment Based on Gender Nonconformity. *International Journal of Business and Management Research, 9*(1), 11-19.

Matulewitz, K. (2015). Law and the construction of institutionalized sexual harassment in restaurants. Canadian Journal of Law and Society, 30, 401-419.

McGinley, C. A. (2007). Harassing Girls at the Hard Rock: Masculinities in Sexualized Environments. *University of Illinois Law Review, 27*(4), 1229–1277.

Meiser, E. T., & Pantumsinchai, P. (2021). The normalization of violence in commercial kitchens through food media. *Journal of Interpersonal Violence*. PMID:33858241

Nielsen, B., Glase, L., & Einarsen, S. (2017). Exposure to workplace harassment and the Five Factor Model of personality: A meta-analysis. *Personality and Individual Differences, 104.*

O'Brien, P. W. (2010). The careers of chef: a study of swiss expatriates working in Thailand. Bulletin of Faculty of Contemporary International Studies, 6.

Patah, M.O.R.A., Abdullah, R., Naba, M.M., Zahari, M.S.M., & Radzi, S.M. (2010). Workplace Bullying Experiences, Emotional Dissonance and Subsequent Intention To Pursue A Career In The Hospitality Industry. *Journal of Global Business and Economics, 1*(1).

Ram, Y. (2018). Hostility or hospitality? A review on violence, bullying and sexual harassment in the tourism and hospitality industry. *Current Issues in Tourism, 21*(7), 760–774. doi:10.1080/13683500.2015.1064364

Selçuk Kurtuluş, S., Kılıç, B., & Kurnaz, A. (2018). Evaluation of women chefs in professional kitchens. *Journal of Tourism and Gastronomy Studies, 6*(3), 119–132. doi:10.21325/jotags.2018.275

J. W., MS, W. C., J. A., J. T., & MS, M. M. (2008a). Sexual harassment in the hospitality industry: Perceptions of restaurant employees. *Journal of Human Resources in Hospitality & Tourism.*

Strachan, H. (2021). *Why are women still the target? How harassment in kitchens has led women to not assume or explore roles in the kitchen and even stop work completely.* Bachelor dissertation.

United Kingdom Legislation. (2010). *Equality Act 2010.* Retrieved 05.04.2021 from https://www.legislation.gov.uk/ukpga/2010/15/section/26

White, C., & Hardemo, L. (2010). Sexual harassment in kitchens. *Journal of Foodservice Business Research.*

KEY TERMS AND DEFINITIONS

Gender Inequality: Unfair treatment of other gender due to some prejudices.

Harassment: Situation in which is a tension caused by physical or mental violence on somebody.

Harassment of Kitchen Staff: Creating a hostile environment in kitchen by using verbal and non-verbal abuse towards specific person or a group pf staff.

Kitchen Culture: Specific language used by kitchen staff.

Mobbing: Hostile and unethical communication directed to a specific person that happens more than once.

Sexual Harassment: Using words or gestures showing hostility, objectification, exclusion, or second-class status about members of one gender.

Chapter 18
COVID-19 and Human Rights in a Fragile State: Guinea-Bissau

Carlos Sangreman
https://orcid.org/0000-0001-5591-6882
University of Lisbon, Portugal

Raquel Tavares Faria
https://orcid.org/0000-0002-0024-5257
University of Lisbon, Portugal

Bubacar Turé
Human Guinean Rights Ligue, Portugal

ABSTRACT

This investigation aims to study the situation of human rights during the COVID-19 pandemic in Guinea-Bissau between January 2020 and January 2022. The research organized an inquiry to families and another to market sellers about the effects of the pandemic and the measures enacted by the government and presidency to contain it. A public hearing was also organized, with various entities from the high commission to trade unions, journalists, and the public order police, for information on how each institution saw its situation and action in this period. The conclusions of the analysis of all these qualitative and quantitative data allow us to affirm that the fragility of Guinea-Bissau has such a social weight that a disease that killed fewer people than malaria, diarrhea, or tuberculosis did not overcome the problems of human rights stemming from poverty, institutional fragility, and low incomes in general.

DOI: 10.4018/978-1-6684-3369-0.ch018

INTRODUCTION

The World Health Organization (WHO) declared on January 20, 2020 "pandemic state for Coronavirus - Covid-19", and this month later, more specifically on March 11, was changed to "global pandemic state". Since then, there has been a progressive number of infected people around the world, with alarming mortality levels. This study asked a series of questions about the evolution of Human Rights (HR) over the past two years. There is an acceptance of the implicit subordination of HR to the protection of individual and public health with the support of the population and international organizations that have not used the HR benchmark for recommendations to combat the pandemic, despite the calls of the United Nations (UN): "This is not a time to neglect Human Rights; it is a time when, more than ever, Human Rights are needed to navigate this crisis in a way that will allow us, as soon as possible, to focus again on achieving equitable sustainable development and sustaining peace" (Guterres, 2020).

Even in countries with consolidated institutions and stable governments, the concern to keep the democratic regime functioning was the socio-political scenario in which packages of measures such as "states of emergency", "mandatory curfew and confinement", "states of calamity" were planned and executed. HR, as a comprehensive reference, was very little mentioned. Most people concluded, based on a probability catastrophic loss of life, that a temporary loss of freedom was justified.

The main objective of this study is to show and draw conclusions on the evolution of Guinea-Bissau in these pandemic years, regarding the measures taken to combat it and its close relationship with HR. It is essentially intended to unveil the perceptions and behaviors of the Guinean families and economic agents regarding the socio-economic impacts of the Covid-19 on HR, contributing to the design, development, and implementation of more informed public policies.

BACKGROUND: STATE OF THE ART

The methodology began with literature research in the publications of international organizations followed by articles in academics journals taking care to avoid those that only speculated about possible effects of a pandemic very little known at the time of editing. The ones published by WHO and the UN were privileged.

Documentary research focused on Guinea-Bissau, whether in official pandemic data or in books and articles by authors investigating the country. The most reliable institutional information came from the official bulletins on the epidemiological situation, with weekly frequency, produced by the High Commissioner for Covid-19. On the other hand, and regarding the violation of HR, information was collected from the Guinean League for Human Rights (LGDH).

Through this documentary research, it was concluded that families, companies, and market/street sellers could be the social components most affected by the pandemic.

From the information of the Bulletins of the High Commissioner it was concluded that cases of Covid-19 infection were mainly concentrated in the capital, Bissau, in a percentage between 70 and 85%. However, it's important to note that this concentration can also be justified by the fact that the largest and best hospital (both in terms of human resources and equipment) - Hospital Simão Mendes - is in this region, implying, from the other parts of the small country, a displacement of patients to the area, which in practical terms can translate into a location distortion of cases.

The empirical work consisted of three surveys organized in different universes: families, companies, and market/street sellers.

For the definition of the sample, it's crucial to have in mind that there are only estimates of the population since the 2009 Census. These estimates can be corrected by the voter registration data, which does not cover residents under the age of 18. Thus, and until a new census is made, we cannot consider reliable the estimates obtained 12 years later. The same is true for the distribution of the population throughout the neighborhoods of the capital. What the inquirers say is that the spaces between the neighborhoods have been filled with population coming from the interior, in some cases acquiring new names in other cases increasing existing neighborhoods.

From the point of view of the limited budget at our disposal for this research, the data collection structure was based on the establishment of 4 teams with two people, to which areas and neighborhoods were allocated.

As the most experienced inquirers always do one or two more surveys as a precautionary measure for errors that are not currently noticeable, 201 families were retained, randomly chosen with 1293 people divided by the neighborhoods. The only stratification implemented was to collect between 20-30% of families with female heads, since the 2009 Census has 22% of households at national level, with this characteristic with a difference between urban and rural. With this dimension and without reference universe it cannot be said that it is a sample with statistical significance, but it is within the number advised by the theory of sampling for investigation that does not require crossings of variables in large numbers, as is the case of a study focused on the effects of a single social phenomenon of pandemic.

For companies we have the same problem of lack of data to know the possible universe of inquiry. Thus, the inquirers made by direct observation a list of 80 companies of different sizes and with "open door" for public access. The difficulties were greater than with households or sellers because the companies had more reluctance to respond. Seventy-nine companies were interviewed, regardless of their formal or informal legal status, such as pharmacies, travel agencies, locksmiths, carpentry, mechanics, newspapers, radios, construction, commerce in general, restaurants and hotels, transport, furniture, insurance, etc...

The markets and the streets constitute a different social space from the companies when their sellers moved the most in the city and surrounding areas to be able to continue selling even at the times when measures were enacted to restrict trade and confinement of households. Thus, men and women market/street sellers were randomly surveyed in the four main markets (Central, Bandim, Caracol and Antula), collecting a total of 50 surveys.

Finally, we sought qualitative data through a public hearing where several institutions - High Commissioner for Covid-19, trade unions, professional associations, students movements, Police National Order, women associations and journalists unions -, were invited to testify about the action of their organization in the pandemic years.

The existing literature on the virus is very varied in themes and quality. But on the nexus Human Rights – Pandemic there is not enough research to be able to claim that a theoretical body has been formed on this nexus. An interview by the German philosopher Habermas to the newspaper Le Monde (Truong, Nicolas, 2020) explains the reason for this absence stating "*il nous faut agir dans le savoir explicite de notre non-savoir*".

The same is found in other authors in which uncertainty dominates the knowledge about the pandemic and on the action necessary to face a situation with multiple social and personal facets.

An author (Sousa, V, 2021) is in line with those who, without forgetting uncertainty, seek to attribute a universal dimension to the individual and social behavior resulting from the "stop of time" due to confinement at home, but only superficially deals with Human Rights. Boaventura Sousa Santos affirms his fear for democracy when he writes "confinement has such a countercurrent logic that its duration has to be limited, noting that "we act in the dark and dark there is little room for politics, especially for democratic politics" (Santos, Boaventura Sousa, 2020). Among the few authors more focused on HR, we should note (Padrão Neves, M.C., 2021) who considers the social changes caused by the pandemic so significant for HR that she argues that its formulation must be rethought, because the social pattern of response to pandemics with the isolation and exclusion of people affected has not changed throughout history (…leprosy, plague, HIV…), forgetting the evolution of Human Rights over time. For African sub Saharan countries, the literature emphasizes the fragility of structures, especially sanitary ones, and social informality in general, but the authors confess their ignorance about the reasons for the low number of cases and deaths, especially with the high numbers of victims of African-Americans in the USA to rule out the hypotheses of genetic resilience to viruses; areas of the world as or more informal in the Indian subcontinent, in large Brazilian cities or South Asia to have high incidences (Moustapha Mbow et al., 2020).

International organizations produced texts (see References) stressing above all the need to protect human rights activists from emergency policies that do not follow democratic principles, worsening inequality by reducing absolute poverty and a greater focus on vulnerable groups such as children, women, refugees, or even people with diseases other than the pandemic by decreased treatments or deterioration of their mental health. They sought that the "Principles of Syracusa", 1984, on states of emergency and freedom of movement were the standard for measures to combat the virus without unjustified attacks on HR. But in our opinion, they quickly focused more on present and future access to vaccines, inequality and poverty resulting from the decline in global economic activity and less on Human Rights.

CORE AIMS: CONTEXTUALIZATION: GUINEA-BISSAU AND THE COVID-19 PANDEMIC

The birth of the Nation State of Guinea-Bissau began with a struggle for independence. Led by Amílcar Cabral the Guineans organized in 1956 a party – the African Party for the Independence of Guinea and Cape Verde (PAIGC) – and developed a guerrilla war against Portugal, present since 1440 on the coast and, from 1936, colonizing the entire territory.

The PAIGC with Chinese, Cuban and above all the Union of Soviet Socialist Republics support in arms, training, and some direct fighters on ground, united a good part of the three dozen ethnic groups against this common enemy supported by the North Atlantic Treaty Organization. This struggle (1963-1974) was eventually at the origin of the Revolution of April 25, 1974, in Portugal, which ended the colonial regime, installed a democratic regime, and accelerated the independence of Guinea-Bissau and the other Portuguese colonies.

After independence, with the common enemy disappeared, with the main leader Amílcar Cabral assassinated in 1973, political-military conflicts and ethnic rivalries began to shake the process of state building under a single party, soon with a coup d'état in 1980, and even multi partisanship did not prevent an 11-month civil war (1998-1999), and in 2012, military officers overthrew again the Government before the second round of a presidential election.

After a transition period of two years, legislative and presidential elections were held in 2014, marking a return to constitutional order. Or at least that was foreseen. Effectively, it lasted a short time.

Disagreements between the winning party and the President of the Republic and successive presidential governments led to widespread instability that drove investors and International Aid apart.

In 2019, new presidential elections took place, with the elected candidate taking office in a hotel and not in the National Assembly while the Supreme Court was still assessing an appeal with allegations of fraud. Although the Supreme Court validated the election, the situation of political, economic, and social instability continued, with serious repercussions at the level of HR. In February 2022 an alleged failed coup triggered persecution and arrests without intervention by the judicial system.

Guinea-Bissau is a small State of West Africa, located between Senegal and Guinea-Conakry. With a population of about 2 million inhabitants, it is characterized by low levels of education, poor sanitary conditions, and a weak judicial system. Its endogenous resources are cashew nuts, which make up 96% of exports in 2020. In addition, its economic activity is supported by the sale of fishing licenses, timber, heavy sand, phosphates, and bauxite and, still in the process of initial prospecting, oil. It is estimated that the production for the internal market of rice, fruit, vegetables, breeding, pigs, and fish is sufficient to feed the population, and the food safety problems are the result of the weakness of the distribution circuits and the poor quality of the roads. The country has not developed any industrial structure.

From the point of view of digital communications, it can be said that the qualitative factor is observed only and exclusively in the capital, being nonexistent in other regions of the country, where its access is quite limited, except for mobile communications spread across the country without geographical reliefs.

The state budget depends on its implementation in more than 50% of International Aid. With a centralized post-independence economy, the structural adjustment of the 1980s led to privatization of public enterprises that were nothing more than mere transfers of public goods into the hands of a political elite in untransparent processes. With the economic fragility resulting from this instability of the institutions, it is no wonder that the economic and social indicators of human well-being place the country in the last places of the UNDP Human Development Index.

Guinea-Bissau had the first positive diagnosis for Covid-19 on March 25, 2020, and the first death on April 26.

It began by establishing an inter-ministerial commission to combat the pandemic, but soon realized that this was not the right organization for the emergency and the kind of response needed. Thus, it created in May 2020, when the country recorded 1,089 cases of Covid-19 infection, including six fatalities, a High Commission to replace that commission (Presidential Decree of 05/06/2020). To preside over the chair, Magda Robalo Correia e Silva, former WHO representative in Namibia and Ghana, was appointed, with Tumane Baldé, a physician, and Plácido Monteiro Cardoso, former director of the National Institute of Public Health, as well as other members of the board of this High Commissioner.

When there were still only 38 diagnosed cases, health authorities issued recommendations for voluntary confinement, use of masks and frequent hand washing with water mixed with bleach or soap.

Days later, on March 27, the Government declared a "state of emergency", limiting the right to free movement. That is, it was only allowed to circulate between 7 and 11 o'clock in the morning. In addition, it closed air, land, and sea borders, stopping economic activity related to foreign trade. The health system created three treatment centers for Covid-19 (one at the Simão Mendes National Hospital, another at Cumura Hospital - in the Biombo region near the capital - and another at Bor Hospital in Bissau).

Schools remained closed for twenty-three weeks, according to the United Nations Educational, Science and Culture Organization (UNESCO, 2020). This closure period, together with teacher strikes, further deteriorated the education system (and therefore the formation of human resources or capital) in the country.

In Fragile States, the possibility for populations to follow these obligations is affected by the weakness of the institutions, as if the economic and social shocks related to Covid-19 have spared no country, fragile states have been and are invariably affected more than those where institutional and family resilience is greatest.

In Guinea-Bissau the inability to provide health services mainly in rural or suburban areas, the lack of facilities, human resources, equipment, materials and medicines for the provision of health services, the insufficient central and regional management with no evaluations are the expression of the fragility based in poor governance (i.e. the institutions of the political system – including the armed forces that behave as if they were legitimately active political institutions and not only in defense of sovereignty and public order (Mendes, 2019) - can only meet the expectations of the population in a very limited way), in the absence of social protection, which results in one of the worst health systems in the world (Somalia alone was worse in 2019 according with the Global Security Health Index, 2019), where those even in cities who need treatments other than basic ones and cannot afford it are not simply treated. This weakness is also based on the economic fabric with weak operators (i.e., small, import-dependent private enterprises, scarce technical staff, little capitalized with no ability to compete with those of neigh-boring Senegal or other countries, or to meet production standards to export to more demanding markets such as the European Union).

Informality in trade and precarious labor relations as a more normal situation, complete the picture of socio-institutional weaknesses (which combines gender inequality with mid-level indicators when compared to other African countries). The presence of the threat of drug trafficking to Europe that grows or decreases with political instability, completes the picture of weaknesses.

To exemplify this fragility, we can mention the Report on Human Traffick (ENACT, 2021) which highlights the trafficking of Guinean children who go to Gambia and Senegal to study the Quran and end up forced to beg for false and corrupt teachers *"The Government of Guinea-Bissau does not meet fully the minimum standards for the elimination of trafficking and is not making significant efforts to do so, even considering the impact of the COVID-19 pandemic*".

The positive forces that promote the resilience of the country are, above all, a young demographic structure (Figure 1), a good ability to feed the population, a substantial level of micro-scale associative between family members, groups of the same generation and neighbors, and a set of civil society organizations very active on various fronts since the defense of HR, the protection of biodiversity, the fight against trafficking, the promotion of microcredit, vocational training, cultural activities, migrant-based organizations for the development of villages of origin, among many others initiatives.

The closure of the international cashew nut market due to the pandemic directly harmed the population of multiple small producers, with the decrease in revenues, and the Public Administration by reducing taxes corresponding to exports mainly to India, Bangladesh, and Vietnam.

The right to have sufficient nutrition (i.e., the right not to go hungry or malnourished) is largely a consequence of the right to pursue an activity from which income is obtained as an informal or formal employee or on its own. The measures against Covid-19 when they limited the exercise of activity, endangered that right. It is a dilemma that all governments face: to protect public health they limit economic activity and automatically reduce the incomes that people need not to go hungry.

In practice Guinea-Bissau has stipulated measures, but with breakaway points in the acquisition of food with vendors moving to markets on the outskirts of Bissau. Those who got worse were those who needed to move to work in transports, hotels, shops, restaurants, …, and could not do so. These people saw their incomes decrease and had a rough time (see analysis point of interview and survey responses).

The idea that when people are unable to obtain resources to guarantee their own food, for reasons unrelated to their will as natural or man-made disasters, the State has an obligation to guarantee this right (for instance with distributions of food) depends heavily on the existing public means and the political will of the rulers.

The capacity to produce food stuff leads to a resilience of households in their daily lives (Quételet et al., 2021) which is only overcome by armed conflicts such as the civil war of 1998/99 or by an epidemic of mass incidence (as was feared to be that of Ebola in 2014/15, or that of Covid-19 in 2020 and 2021, but they turned out both not to happen), since the country has no natural disasters such as floods, hurricanes, volcanoes, earthquakes, or devastating fires due to its flat geographical structure with larger and smaller rivers throughout the territory.

The number of cases and deaths are concentrated in Bissau which has a population in neighborhoods of very close dwellings, with little electricity, without water piped in the dwellings, cooking and drinking many families still water from untreated wells, with the largest informal sector in the country that leads to the displacement of vendors, men and women, all the space towards the markets, shops and street sales areas.

As the health services with the best equipment and more medical and nursing staff are in Bissau, the size of the country makes it easy for the populations affected by the disease, wherever they are to go to the capital in the expectation of getting better treatment.

Figure 1.
Source: World Bank estimates

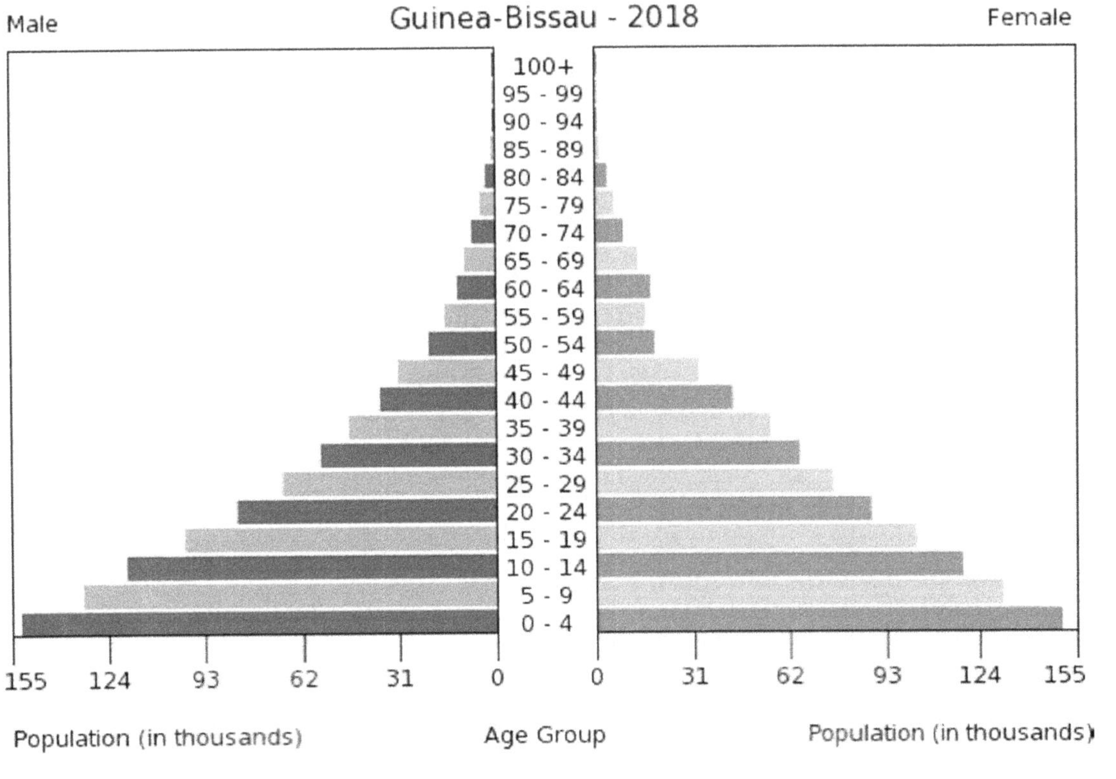

Table 1a. Cases and deaths accumulated in Guinea-Bissau October 2020 - January 2022. Source: Press releases of National Institute of Public Health of Guinea-Bissau (INASA) -24/03/2021, High Commissioner for Covid-19 - 13/06/2021 and Official bulletin about Covid-19, 2020, 2021 e 2022. - https://www.accovid-gw.org

Region	Oct/20		Mar/21		Jun/21		Oct/21		Dec/21	
Administrative	cases	deaths	Cases	deaths	cases	deaths	cases	deaths	cases	deaths
SAB-Bissau	2057	29	3059	37	3238	40	4562	79	4747	80
Biombo	141	7	7	8	164	13	275	28	324	35
Bafatá	52	0	103	8	112	12	183	17	187	17
Cacheu	42	0	67	1	73	1	227	2	249	2
Gabu	3	0	74	1	76	1	132	4	140	4
Quinara	37	0	50	0	52	2	102	2	120	2
Oio	24	0	32	0	44	0	363	6	408	6
Farim	19	0	22	0	23	0	27	0	27	10
Tombali	0	0	9	0	32	0	81	1	82	1
Bolama	1	0	1	0	1	0	2	0	2	0
Bijagós	0	0	6	0	10	0	180	2	180	2
Total	2376	36	3568	55	3825	69	6134	141	6466	149

Table 1b. Cases and deaths accumulated in Guinea-Bissau October 2020 - January 2022. Source: Press releases of National Institute of Public Health of Guinea-Bissau (INASA) -24/03/2021, High Commissioner for Covid-19 - 13/06/2021 and Official bulletin about Covid-19, 2020, 2021 e 2022. - https://www.accovid-gw.org

Region	Oct/20		Mar/21		Jun/21		Oct/21		Dec/21	
Administrative	% cases	% deaths	% cases	% deaths	% cases	% deaths	% cases	% deaths	% cases	% deaths
SAB-Bissau	86,6	80,6	85,7	67,3	84,7	58	74,4	56	73,4	53,7
Biombo	5,9	19,4	4,1	14,5	4,3	18,8	4,5	19,9	5	23,5
Bafatá	2,2	0	2,9	14,5	2,9	17,4	3	12,1	2,9	11,4
Cacheu	1,8	0	1,9	1,8	1,9	1,4	3,7	1,4	3,9	1,3
Gabu	0,1	0	2,1	1,8	2	1,4	2,2	2,8	2,2	2,7
Quinara	1,6	0	1,4	0	1,4	2,9	1,7	1,4	1,9	1,3
Oio	1	0	0,9	0	1,2	0	5,9	4,3	6,3	4
Farim	0,8	0	0,6	0	0,6	0	0,4	0	0,4	6,7
Tombali	0	0	0,3	0	0,8	0	1,3	0,7	1,3	0,7
Bolama	0	0	0	0	0	0	0	0	0	0
Bijagós	0	0	0,2	0	0,3	0	2,9	1,4	2,8	1,3
Total	100	100	100	55	100	100	100	100	100	100

However, throughout 2020 and 2021, the number of cases and deaths was advancing to other regions. the result of these factors is Table 1, where Bissau has between 85% and 73% of cases and deaths due to covid-19.

On the initiative of its President, Guinea-Bissau began by acquiring the tea to prevent and cure the symptoms of Covid-19 produced in the Republic of Madagascar (*Público* newspaper, Lisbon, April 27, 2020). There was a great distrust among Guinean health and academic staff since this tea still lacks scientific evidence, which drives the question of why betting on a compound that does not present evidence from the point of view of science.

Such tea did not give any results and quickly the country sought access to vaccines recognized by the WHO. With seven "states of emergency" or "calamity" in one-year, Guinean society has taken steps that are summarized here (with reference to Decrees 09/2020 of April and 11/2021 of March):

- *Human Right to Information:* created from its own information space on television, a weekly Epistemological Bulletin and Information issued by the Commission for Journalists with data on the global pandemic in Africa and Guinea-Bissau, also available online. This Commission also sought the help of geographically based non-governmental associations to be vehicles for information and dissemination of prevention and vaccination measures. Health professionals have established a communication network for information and sharing of international news and articles on the pandemic.
    - *Human Right of Freedom of movement*_conditional on essential issues and only in authorized time periods (usually 7-11a.m. or 2 p.m. hours), with the general use of a mask, on public transport, markets, all enclosed spaces, and schools for pupils over the age of eleven. It could only be circulated within the regions of residence, with border controls of existence of Covid-19 test, social distancing of one meter and frequent hand washing.
    - *Human Right to Health:* free tests for citizens considered vulnerable and at high risk (over 55 years or with specific pathologies); fourteen-day quarantine for people waiting for test results; border closures (except residents) in April 2020; free emergency calls in partnership with multinational Telecommunications MTN and Orange.
    - *Human Right of Free Expression:* meeting and demonstration with mask, social distancing of one meter and maximum concentration of 25 people (2 meters and 5 people in April 2020); prohibition of all gatherings, including parties and demonstrations, with more than 25 people with 2 meters away (1 meter and 5 people in April 2020). In June 2021, these indications were followed only in public acts because in the day-to-day life what was seen in Bissau was that no one wore a mask or cared about distancing.
    - *Human Right to Religious Freedom:* religious ceremonies limited to half of the assistants; funerals and traditional ceremonies with no more than 25 people, mandatory use of masks and with a social distance of one meter (10 people in April 2020, with masks and distance of 2 meters).
    - *Culture:* shows in forbidden enclosed spaces.
    - *Human Right to Freedom of labor: power for authorities to compel sectors seen as fundamental to the functioning of essential services*; prohibition of absence redundancies, as well as strikes in the sectors to combat the pandemic (April 2020); state-paid life insurance for health workers involved in combating coronavirus (April 2020).
    - *Family Protection:* fines for price speculators.

- ◦ *Human Right of Freedom of Economic Activity: take away* regime for restaurants (7-14h, April 2020) and social distance to similar open spaces; closure of markets and any kind of street sale (April 2020). In June 2021, these measures no longer existed in practice.

The High Commissioner may request movable and immovable property and make expenses without public tender in the context of combating Covid-19, and all revenue stemming from fines imposed for infringement of the measures approved reverts to that body.

The state of the health system in the face of the pandemic is highlighted in the Government Decree that on 26.09.2021 prolonged the "state of calamity" in the country until October 10, where it is stated that the level of hospitalization is above 50% of the national capacity. Such a level is a good reason for preventing the spread of the virus to a controllable level, unless the situation becomes out of control at any time. The Decree maintains the mask as a mandatory use for people over the age of 11 years on public road, enclosed spaces of public access, public passenger transport, educational establishments, and markets.

The country's framework in the global pandemic effort was made by the approval of the Covid-19 Guinea-Bissau Emergency Response Project, prepared under the World Bank Strategic Preparedness and Response Program to Covid-19, funded by the rapid instrument COVID-19, "Emergency Pandemic Financing Facility, and additional vaccine funding" adopted on October 13, 2020. It aims to prepare, respond to the Covid-19 pandemic, and strengthen public health preparedness and prevention systems in Guinea-Bissau. In addition to this project, the country sought support from partners (countries and public or private institutions) to strengthen the means of combat (equipment, protective suits, masks, gloves, hospital consumables, medicines) to the virus and vaccination.

- **Human Right to Information:** to take action to combat the pandemic, to avoid rumors, false news and to adhere to vaccination, information transmitted by health and political authorities has proved to be fundamental in all countries. Families in Bissau, heard about the pandemic mainly on radio (90%), on television (75%) and with friends and friends (80%). Newspapers and social networks were less important and there were more families not hearing any information through these channels. Market sellers responded to the same standard. Companies listened more social networks and presented radio (90%), television (92%) and social networks (84%) as preferred sources. For all respondents' newspapers are a less important source. That is, the overwhelming majority remained informed, and all families, sellers and companies reported any of the channels pointed out and there was not even one that answered that they did not hear about the virus. It can be concluded that with this indicator the Right to Information has been fulfilled.

If we refine the analysis and try to verify the quality of the information, the answers respond according to Table 2.

Sellers (58%), companies (59%) and families (48%) found it "clear", "sometimes clear sometimes confusing" (26, 29 and 39%) and a 16,11 and 13% minority "confusing and difficult to understand".

We used the criteria for quality and housing, lighting, water used for cooking and the number of daily meals in survey week to classify as poor and less poor families. As the poor responded in 17% who consider the information "confusing and difficult to understand", while for the less poor families this percentage was 11%.

Table 2. The quality of information about Covid-19 perceived. Source: Survey about Covid-19 and Human Rights in Bissau (2021)

Quality of information	Families	%	Sellers	%	Companies	%
Clear	95	48	39	58	47	59
Sometimes clear sometimes confused	77	39	13	26	23	29
Confused and hard to understand	26	13	8	16	9	11
Total	198	100	50	100	79	100

In conclusion, the investigation can state that information has not always come clearly to people who tend to have fewer resources and have less education and the means of communication, but it doesn't seem that this difference is too great, and it can be concluded that the information of health and political authorities has fulfilled its role of ensuring the respective Human Right.

FUTURE DIRECTIONS

The position of the respondents' families drew a truly clear pattern:

1. They completely agree, with approval percentages between 80 and 100%, with the measures of washing hands frequently, wearing masks and prohibiting demonstrations, parties, and ceremonies.
2. They are completely against, with percentages between 80 and 90%, measures to close markets, shops and street sales, and schools, as well as the shutdown of public transport.
3. They are divided on the measure of mandatory stay at home with 44% to agree and 58% to disagree.

This pattern is repeated with few variations in families with female and male heads or even in those with resources, poor and less poor.

This analysis can be complete by finding that only one reason to leave the house, in the period of confinement, had positive answers above 80%: buy food/drink. In other words, it can be said that there is a consensus in the society of the capital Bissau to dispense the Right to Free Expression to combat the pandemic if it takes the form of demonstrations, parties and even ceremonies, dividing itself on the right to mobility and disagreeing with measures that limit their rights to a demand for food.

They are also manifested by the Right to Education, even in a pandemic, by speaking out against the closure of schools and sending all children back to school shortly after reopening, at a percentage of 94%, making no difference between boys and girls.

Right to Health: as already mentioned the Right to Health depends on the existing system and in Guinea-Bissau this system is very fragile and insufficient for the expectations of the population in normal times and unable to respond to a crisis such as that of Covid-19. But, as we has already mentioned, the pandemic has left few dead and several infected also found low when compared to other endemic diseases and with other countries. The families answered that in 60% of the cases of disease (without specifying what disease it was about) they did not seek either hospitals, health units or traditional medicine, during this period. This percentage corresponds to the observation made by the researchers that people avoided going to treatment sites for fear of contracting Covid-19. For more when people who were sick and went to one of those places, they confirm that 2/3 (66%) had to pay for all services and an even higher number had to pay for medicines (71%).

There is no significant difference between families with more and fewer resources, as can be seen in the following Table 3:

Table 3. Consultation payments, treatments, and medicines. Source: Survey about Covid-19 and Human Rights in Bissau (2021)

Health costs for families	Poor	Less poor
Yes, I had to pay for everything	58%	65%
I did not have to pay	42%	35%
Total	100%	100%

The conclusion thus points to the that – even in the period of the pandemic – the health system has maintained one of the obstacles to the free Right to Health referred to in the country's Constitution and which consists in the payment of services in private and public by the users themselves. The investigators do not know the detail because there were people who did not pay anything and others who paid for everything. In fact, one of the participants in the Public Hearing stated that although there are exceptions to legislated and posted payments at the places of care, the health units and hospitals continue to charge costs to everyone, with the justification that they do not have a budget for the material and medicines that patients need. Other research confirms this even to prenatal consultations for pregnant women (Sangreman, et al., 2020).

Human Right to Work/Income: research sought information from the perspective of families as social agents with economic activity to maintain their income and from sellers and companies as private economic agents. The families did not choose, mostly, for having second activities during the two years of pandemic. Only 26%, that is, one in four sought to have another activity. In general, sell food items cooked or not, drinks (natural sums and ice water), charcoal and used clothing, among others.

The families with female heads needed less to resort to activities other than those who already performed (24%) against a percentage of 25% of families with male heads. Within the families that had to do so, most of them were the women who engaged in these second activities. In fact, of the total number of people in the families who reported having arranged an additional activity during the pandemic, 61% were women.

Consistent with this behavior that shows little desperation due to the decrease in activity in these years, the families answered that 84% "maintained" or "maintained with some reduction", their main activity and only half of the respondents (51%) asked for help from family, friends, or neighbors.

Families with female heads were more successful in maintaining the activity (29%) versus 18% of those who have men as heads. The opposite was the maintenance of reduced activity with 51% for the first type of households achieving it and for second type, with 63%. It may have to do with the weight of the product sales sector or the stronger informality for women's activity that allows for more flexibility and imagination in the face of measures to limit freedom of movement and market-making.

All respondents to market surveys said they had temporarily closed the activity with consequences of 76% of the business having worsened by more than half of the income, due to the lack of customers (98% of responses). To combat this decrease in activity and income, about half of the sellers' developed strategies, namely: they sought to find new products (52%), to take the sale of potential customers (54%) and resort to credit through loan (50%). However, the other half merely hoped that the limiting measures would end, not changing anything in their behavior as an economic agent.

The qualitative data from the public hearing and the answers from companies focused first on the lack of customers and secondly on the complete absence of government support for economic operators, which made the situation even more difficult.

The conclusion from these data is that the Right to Work or Activity was interrupted in some periods by the measures enacted against Covid-19, but the effect of the decrease in customers, in general, was more important than the disease. And this decrease in income in households and, therefore, a decrease in consumption is due to several factors, some derived from the pandemic, others from the fragility of the State in the face of the crisis in other countries.

The population has an opinion on the current situation after two years of pandemic based on the role of the authorities in the direction of combating it, essentially the Government and President of the Republic and the health authorities. As we can see with Tables 4 and 5, families, sellers and companies have a more positive than negative perception although the "Very well" is not significant.

Table 4. Opinion of the sellers and companies on Government/President and High Commissioner in pandemic management. Source: Survey about Covid-19 and Human Rights in Bissau (2021)

Criteria	Government and Presidency		High Commissioner Sel	
	Sellers	Companies	Sellers	Companies
Poor	32%	39%	21%	16%
More or less	32%	39%	50%	41%
Well	28%	15%	23%	32%
Very well	8%	6%	5%	11%
Total	100%		100%	

Table 5. Opinion of the families on Government/President and High Commissioner in pandemic management. Source: Survey about Covid-19 and Human Rights in Bissau (2021)

Criteria	Government and Presidency	High Commissioner for Covid
Poor	34%	21%
More or less	50%	50%
Well	14%	23%
Very well	2%	5%
Total	100%	100%

The research, articulating the previous tables and the opinion of the families on the current situation (Table 6), considers that people think they are worse off now than before, but do not think that it is the fault of the management of the pandemic by the political or health authorities. The poor and families with male heads have a more negative opinion, but the percentage values do not differ significantly.

Table 6. Family situation after two years of Covid-19 (%). Source: Survey about Covid-19 and Human Rights in Bissau (2021)

Criteria	Everyone	Men heads	Women heads	Poor	Less poor
Worst	77	80	69	85	73
The same	20	18	23	10	27
Best	1	1	2	0	0
I do not know	2	0	6	5	0
Total	100	100	100	100	100

The shortcomings of access to economic and social rights are illustrated in the published work of the Observatory of Rights (Sangreman, et al., 2020), between 2014 and 2020, and the pandemic only aggravated them.

Indeed, on the eve of the pandemic, the Human Rights Council, the Working Group on Universal Periodic evaluation, at its 35th session on January 20-31, 2020, in the Report of the United Nations High Commissioner on HR in Guinea-Bissau, stated that:

1. Widespread inequality between men and women is an obstacle to development;
2. Violations of rights against political opponents, arbitrary detentions and impositions on freedom of expression and peaceful assembly generally have no conclusive inquiries;
3. Access to justice was impossible for the majority of the population and even worse outside Bissau;
4. Justice rarely acted in cases of forced marriage, female genital mutilation, sexual or family violence, exploitation of children or drug trafficking;
5. The proportion of the budget dedicated to basic social services was lower in the countries of the Economic Community of West African States (ECOWAS);
6. For education, the country was far from achieving the SDG of universal primary education because only a quarter of primary schools guaranteed six years of compulsory schooling;
7. The health budget ranged from 3% to 7% in the last five years, without a financing strategy and with a great dependence on external funds.

In the case of the Human Right to Health the country health system is unable to cope with an increase in a new disease as evidenced by different individual authors and international organizations, giving an image of lack of resilience and failure in access to HR that is difficult to understand with the multitude of bilateral international cooperation interventions such as Cuba, Netherlands and Portugal, and multi-lateral as the UN, WB, EU and UNICEF that have existed in the sector since independence, including local training in nursing and medicine.

It is clear that the system has nothing to do with the conditions at the end of the colonial period in 1973/74 and that there are indicators with permanent positive development (except in the period of the civil war of 1998/99 with the flight of health and administrative personnel to other countries) such as infant and maternal mortality, childhood vaccination, the existence of drug sales in the regions, the number and capacity in human resources of health units and hospitals. But the expectations of the health population are also much higher, it always grows, and public policies to date do not follow developments, giving room for private arrangements of the health professionals themselves that penalize especially those who cannot afford medical care, in a practice that was generalized after the structural adjustment programs of WB and IMF started in 1987, forgetting the right to free health defended in the Guinean Constitution. The chaotic level of the entire system resulting from years of limitations, leads to an inability to deal with large unforeseen events like a pandemic.

The country is ranked in the Security Health Index, far from the rest of the sub-region (Global Security Health Index, 2019). The Human Rights Observatory shows that the number of inhabitants, per doctor, worsened between 2016 and 2019 in most regions, although the same indicator for the number of inhabitants per nurse has improved.

This situation of lack of access to HR has become even more evident with the complaints of the Guinean League for Human Rights (LGDH) of violations of political rights in this period, mainly due to the situation of permanent institutional instability in which the country lives and the inoperability of governments in that area. There have been violations, according to the LGDH, on freedom of expression, persecution of opposition party members, include deputies, and the installation of a climate of insecurity in general focusing on journalists.

The U.S. Department of State in its 2020 annual report on Guinea-Bissau (United States Department of State, 2021) considers HR violations in conditions of detention and imprisonment as significant; serious problems in the independence of justice; corruption and lack of investigation into crimes of violence against women and children, trafficking in persons, child labor and impunity in the abuses of power of the security forces, especially in arbitrary arrests, in attacks on media and media agencies, as well as in the repression of demonstrations. In the same report considers that there is no evidence of torture, murder, abduction, disappearance, or imprisonment of people from the political opposition, restrictions on the use of the Internet, censorship of cultural spectacles or academic activity. It also considers that the rights of political participation are respected, and there is no restriction for minorities or women, although no party abides by parity laws. The same document also refers to corruption linked to drug trafficking and illegal logging involving military personnel, police and members of the Government or Administration as a component of the deterioration of Humans Rights.

From these violations the climate of insecurity also has a component related to the confidence in the authorities and institutions to contain the epidemic by the populations. However, the rest are much more a consequence of the country's political situation than the impact of the pandemic.

The exposure of inequalities across the planet was one of the types of evidence of the pandemic. As international organizations have been publishing indicators on this reality for many years, the WB's report on poverty and inequalities (World Bank, 2021) notes that the pandemic, with the abrupt decline in economic activity, has broken the positive developments of poverty alleviation that had been going on for two decades, exacerbating inequalities within each country and between countries with more and fewer resources.

Figure 2. HR Violations 2020,2021

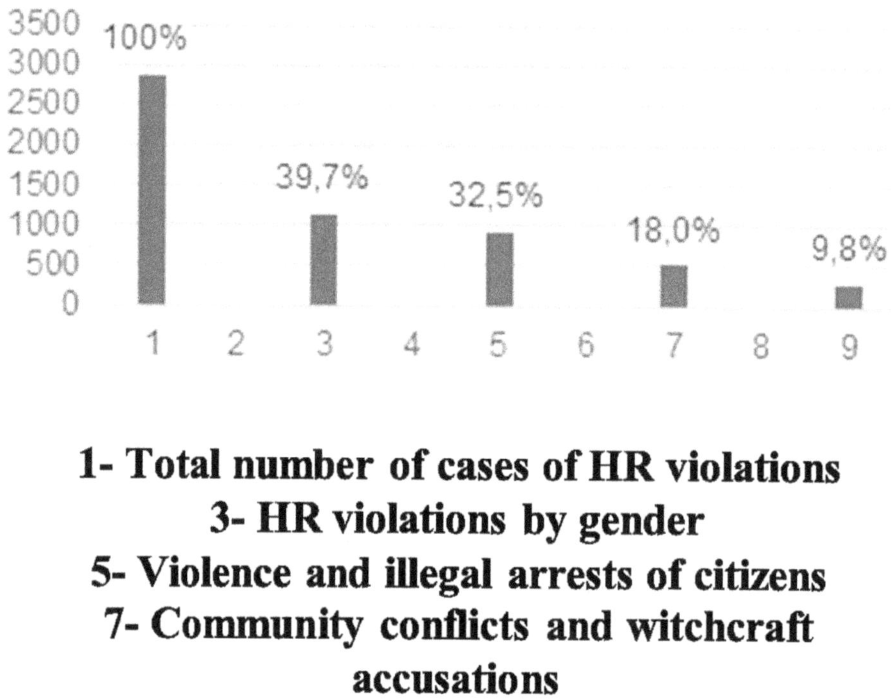

Figure 3. HR Violations by gender 2020,2021

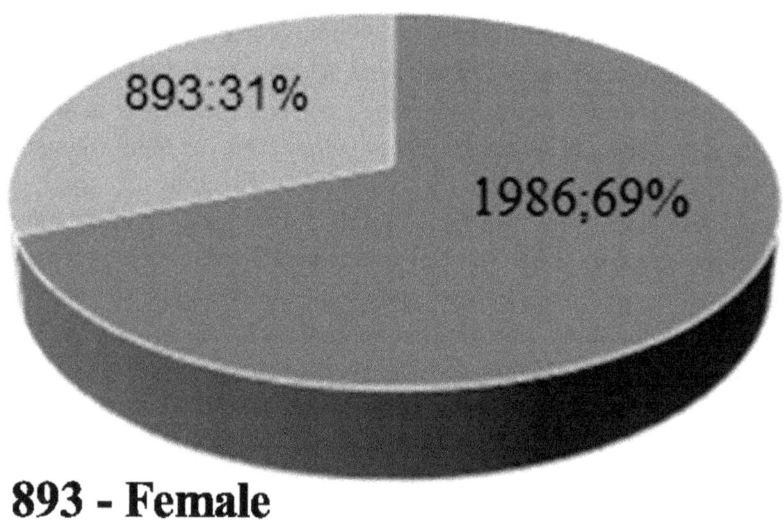

And if people already poor or close to the poverty line saw their situation aggravated, there were also people on low middle incomes, especially in urban areas, who lost their jobs or saw their activity decrease, falling into poverty situations where they had never been. Countries with fewer resources had no choice but to expect those with more resources to support them, as health systems are not resilient enough for a crisis such as the pandemic. The worsening of inequalities in the pandemic has made situations such as those caused by climate change and armed conflicts even more deadly.

For Guinea-Bissau, the data seen from the perspective of HR point to a worsening of poverty and inequalities in these pandemic years and, therefore, to a degradation of economic and social situation. The refuge in vegetable garden agriculture and subsistence fishing or for sale, in traditional health, in the work/trade of small incomes and in the choice of sons/daughters who go to school or not, the change for the worst of family food, are the compensation strategies that families develop, but that only slowly reset previous levels if they do not exist policies expressly focused on income recovery and social welfare. And the political and institutional instability of the country since independence caused by advances and setbacks in HD and the well-being of families, does not contribute anything to a rapid recovery.

CONCLUSION

The impact of the pandemic on public health and the economy has been compounded by the state's weakness in providing the population with public goods that take on constitutional HR as a greater obligation. The disintegration of the global economy due to the pandemic favored authoritarian regimes to crack down on political opponents or mere protesters and consolidate power, also stressing that in countries such as Uganda, Nigeria, Ethiopia, and Guinea-Conakry extremist groups took advantage to support the populations that their states left unaided in the face of the impoverishment that the downfall of the economy caused.

The economy is not organized from a framework of human rights. That is, the production of goods and services does not consider values such as dignity, freedom, solidarity, accountability, social justice as configurators of its organization, production, and distribution. Thus, the fight against the pandemic tends to leave the most vulnerable, the poorest, exposed to disease or the deterioration of their standard of living, as has become clear in the distribution and administration of vaccines and aid to the population affected by the stopping of economic activities. As the United Nations Educational, Scientific and Cultural Organization (UNESCO) argues: "*the Covid-19 crisis is an opportunity to rethink the hypotheses about the paradigms of development adopted by African states. The concentration of priorities on people, education, health care, social protection, and scientific research must become the basis for creating a new Africa capable of finding its own solutions to its problems at the same time as* it *secures its place on the international scene*" (UNESCO, 2020).

Current discussions around intellectual property and access to vaccines seem more like a demonstration of "I'm first", responding to internal political pressures and election cycles than a more solidarity and rational view of "global public goods". If Africa remains dependent on goodwill and the remains of vaccines, equipment, and medicines to recover and organize health systems capable of tackling this and future pandemics, the chances of sustainable structural transformation are compromised. The decision to remain in the COVAX mechanism and to help Africans have vaccine production centers in six countries on the continent (Tunisia, Egypt, Senegal, Nigeria, Kenya and South Africa) will be a clear symptom of the options to be made.

The importance that people in Guinea-Bissau attach to associative and family and help between neighbors has helped to contain the consequences of a virus even in the face of State and private sector weaknesses in the health system and the impoverishment of people who have seen their income cut by measures taken or simply by the decrease in economic activities and international cooperation.

The HR in Guinea-Bissau was more affected by the situation of institutional instability and internal security than by the pandemic itself.

However, this does not prevent us from concluding in this research, that the economy, in sectors dependent on the rest of the world, the educational system and the health system, were exposed in its weaknesses, evidencing the need to assume that without its functioning the whole of all HR in the country hardly progress.

Cases of excessive use of force by police to impose emergency measures have existed in many countries. Persecution of people from the political or civic opposition continued in authoritarian-tended countries or with a very deficient democracy. Refugee detention centers (usually overcrowded) continue to be places of HR violations. Economic, social, and cultural development were affected all over the world to stop the pandemic, with the consequent decrease in HR, even where political rights have not been achieved.

In the specific case of Guinea-Bissau, the research considers that the following consequences are clear:

By the negative: less economic activity, then slower recovery, not changing existing inequalities; continuation of some authoritarianism with intimidation of opposition figures and attitude of little respect between institutions. The strikes that degrade education have nothing to do with the pandemic, but with the government's inability to pay teachers' salaries and invest on the educational system. The fight for HR and against violations of the same, which took place during this period, is due more to the political situation than to the pandemic. The low standard of living with low access to economic and social human rights has made families and economic agents more vulnerable to the measures taken to contain the pandemic.

By the positive: strengthening of the public health system through international cooperation and with the participation of civil society - profitable and non-profitable - in the areas of health and prevention.

Addressing the response and recovery of the pandemic from the HR perspective could lead to better outcomes for all, ensuring that vulnerable countries and groups are not forgotten. Measures such as quarantines, school closures, confinement, and mobility limitation were seen as exceptional, and temporary, and very attentive to physical integrity and human dignity.

REFERENCES

ENACT. (2021). Available at: https://enactafrica.org/countries/human-trafficking

Global Security Health Index. (2019). Available at: https://www.ghsindex.org/country/guinea-bissau/

Guterres, A. (2020). *We are all in this together: Human Rights and Covid-19 response and recovery.* Available at: https://www.un.org/en/un-coronavirus-communications-team/we-are-all-together-human-rights-and-covid-19-response-and

Liga Guineense dos Direitos Humanos (several years). (n.d.). *Relatórios e comunicados.* Bissau. Available at: http://www.lgdh.org/

Mbow. (2020). COVID-19 in Africa: Dampening the storm? *Science, 369*(6504). PMID:32764055

Mendes, P. (2019). Os militares e a impunidade legítima. In C. Sangreman (Coord.), *Guiné-Bissau – Notas sobre o presente e o futuro* (pp. 129-151). Lisboa: CESA - ISEG/Universidade de Lisboa. Available at: https://www.researchgate.net/publication/337448432

Patrão Neves, M. (2021). Os desafios da COVID-19 aos Direitos Humanos. *Cadernos Ibero-Americanos de Direito Sanitário, 10*(3), 66-82 Available at: https://www.cadernos.prodisa.fiocruz.br/index.php/cadernos/article/view/825/836

Quételet, C., Bordin, G., Abreu, A., Lemi, I., & Sangreman, C. (2021). On the nature and determinants of poor households' resilience in fragility contexts. *Journal of Human Development and Capabilities*, 1–18.

Ramos, S. P. (2021). Sobre os desavessos da pandemia e o medo da anormalidade: diálogos com Boaventura de Sousa Santos. In *Sociedade em pandemia(s)* (p. 10). EDUFAL.

Sangreman, C., Barreto, N., Barros, I., & Sanhá, C. (2020). *Observando Direitos na Guiné – Bissau, educação, saúde, habitação, água, energia, justiça, igualdade de género.* Projeto Observatório de Direitos, ACEP com LGDH e CEsA, Lisboa e Bissau. Available at: https://www.repository.utl.pt/handle/10400.5/20866

Sousa, V. (2020). Tempo, espaço e Covid-19: unidade, fragmentação, dissincronia, velocidade, Einstein, Kafka, nova-normalidade e incerteza no futuro. In *C. Sales, E. Araújo & R. Costa (Orgs.), Tempo e sociedade em suspenso* (pp. 203–216). CIES/ISCTE.

Truong, N. (2020). *Jürgen Habermas: «Dans cette crise, il nous faut agir dans le savoir explicite de notre non-savoir».* Le Monde Online.

UNESCO. (2020). *Incidences socio-économiques et culturelles du Covid-19 en Afrique: quelles réponses pour l'UNESCO*? Available at: https://unesdoc.unesco.org/ark:/48223/pf0000373904

United States Department of State. (2021). *Country Reports on Human Rights Practices for 2020 Bureau of Democracy, Human Rights and Labor - Guinea-Bissau 2020 Human Rights Report.* Available at: https://www.state.gov/wp-content/uploads/2021/03/GUINEA-BISSAU-2020-HUMAN-RIGHTS-REPORT.pdf

United States Department of State. (2021). *Trafficking in persons report.* Available at: https://www.state.gov/reports/2021-trafficking-in-persons-report/

World Bank. (2021). *Poverty and shared prosperity. Reversals of fortune.* Available at: https://www.worldbank.org/en/publication/poverty-and-shared-prosperity

ADDITIONAL READING

Josephson, A., Kilic, T., & Michler, J. (2020). *Socioeconomic impacts of Covid-19 in four African countries.* Policy research working paper no. 9466. Available at: https://openknowledge.worldbank.org/bitstream/handle/10986/34733/Socioeconomic-Impacts-of-COVID-19-in-Four-African-Countries.pdf?sequence=1&isAllowed=y

KEY TERMS AND DEFINITIONS

Africa: It may refer to the entire African continent or only sub-Saharan Africa without countries where the official language is Arabic (or a derivative); the division between North, West, Central, Eastern and Southern Africa is also used. Guinea-Bissau is located in West Africa, neighboring Senegal and Guinea-Conakry.

Democracy: Political regime based on the balance of power between bodies elected by all citizens (Parliament and Presidency) or appointed by elected officials (Courts, and Government). Sovereignty belongs to citizens who delegate their representation through free elections. There are several models of democracy, but they all respect this balance. There exists an international scale between plein, imperfect, hybrid, and authoritarian regimes (The Economist Index).

Health Guinean System: Health system consisting of hospitals (central and regional) health centers of three different dimensions and basic health units in the villages. It includes an analysis laboratory, a nursing school, and a medical school. It has public, private, and religious units. The whole system lacks doctors, nurses, midwives, auxiliaries and equipment, medicines, and consumables.

Human Rights Violations: As defined by the UN are all individual (persons or organizations) or state acts that go against the 1948 Declaration of Human Rights and other documents approved by the General Assembly. These violations are reported around the world by non-governmental organizations or governments that have respect for HR in the Constitution.

Measures Anti-Pandemic: Measures taken by the authorities of countries affected by a pandemic. They are all based on the isolation of the infected and the reduction of contacts between the entire population, whether staying at home, wearing masks, or doing distance work. Vaccines are a preventive measure and not an anti-pandemic measure.

Pandemic: It is an epidemic of an infectious disease that spreads in a large geographical region, such as a continent or the entire planet. Only the WHO can declare that a disease is a pandemic. Other diseases that today would be or were pandemics: leprosy, plague, influenza, cholera, typhus, and HIV. Diseases such as Ebola would only be a be considered if they reached much larger geographical areas. Malaria is a disease widely spread all over the planet, but it is not infectious.

Poverty: The situation of people with few economic resources is absolute or relative poverty (inequality). Poverty exists in all developed countries or not and is measured by a World Bank methodology from households' budgets and access to movable and not movable goods. A poor person in a developed country has little to do with a poor person in a fragile or less developed country.

State Fragility: According to the OECD a fragile region or state has a weak capacity to carry out basic governance functions such as education, health and safety for its citizens and businesses. They are vulnerable to internal and external shocks such as economic crises or natural disasters. It can cover countries with weak institutions, armed conflict, victims of natural disaster, hunger, and extreme poverty spread. These are countries where general informality is permanent.

Section 6

The "Virtualization" of Mobile Heritage and Culture

Chapter 19
Total Tourist Experience:
From Heritage and Historical Significance of Places to New Directions for Post-Pandemic Tourism

Graciete Honrado
Instituto Politécnico de Tomar, Portugal

Ana Nata
https://orcid.org/0000-0002-7325-1754
Instituto Politécnico de Tomar, Portugal

Cecília Baptista
Technology, Restoration, and Arts Enhancement Center, Instituto Politécnico de Tomar, Portugal

Cláudia Pires da Silva
https://orcid.org/0000-0002-0803-5641
Technology, Restoration, and Arts Enhancement Center, Instituto Politécnico de Tomar, Portugal

Filipe Romão
HighSun DMC, Portugal

Luís Mota Figueira
Instituto Politécnico de Tomar, Portugal

Marta Santos Dionísio
https://orcid.org/0000-0002-4654-3399
Technology, Restoration, and Arts Enhancement Center, Instituto Politécnico de Tomar, Portugal

ABSTRACT

The aim of this chapter is to propose the concept of total tourism experience (TTE). In the current digital environment, this implies new visions on tourism research and operation. In this atmosphere, the experience of each tourist is structured according to a value chain composed of travel, accommodation, restaurant, entertainment and leisure, safety and health, access to goods, and complementary services. COVID-19 had a significant impact on the behaviour of tourism players. Heritage (natural and cultural resources) sustaining the tourism value chain is an asset worthy of special attention due to its finiteness, namely regarding the world heritage icons because its economic and social effect is revealed by each country's GDP. The pandemic destabilised the global tourism system. In this context, opportunity responses emerged. New opportunities for relaunching tourism were opened up, and thus, the TTE proposal is a timely contribution in the range of necessary, qualifying, challenging, innovative responses.

DOI: 10.4018/978-1-6684-3369-0.ch019

INTRODUCTION

Since 2015 the Tourism Laboratory (L-tour.ipt) of the Instituto Politécnico de Tomar has been studying tourism issues and testing solutions raised in the evaluation of the "state of the art" of each research line in a cooperative environment. Through completed projects and others in progress there is empirical evidence, theoretical essays and case studies of business interest.

From the Academia-Business experience to the operational innovation provided by the HighSun company, as a case study of a regional tour operator with an international vocation, the relationship established determined the affirmation of the TTE model. A TTE is a tourism model that comprises the combination of all elements and services associated with a tourist experience, since the moment that the tourist leaves home until his return (Cho, 2000).

The experimentation with a "Destination Management Company" (DMC) enhanced the specific partnership of this case. Its choice was essentially due to the mutual trust accumulated in the joint work, but also to the fact that what started as a student internship experience was enhanced to other approaches. In this line the tourism epistemology developed "as ongoing work" ensures the inclusion of this intervention model in the study concerns on the part of L-tour.ipt and consequent experimental application in the design of cultural tourism products by HighSun which positions itself in the international tourism market.

The object of study is the communicative facilitation between the person who plays the role of tourist and the person who plays the role of Tour Operator. In other words, it is intended that the objective of tourist satisfaction can be planned and provided by the Company and felt by the Tourist within the scope of his satisfaction, since the outward journey to the destination, until the home return journey. It is believed that the dissemination of the acronym TTE will increase the effectiveness, efficiency and economy of the tourist's practice established between Supply (Tour Operator) and Demand (Traveller).

The health crisis contributed to the changes in consumption trends of tourist products. Thus, the TTE also intends to generate a greater tourist literacy among consumers.

The theoretical framework supporting this process results from the literature review and the perception at each moment of the "state of the art" of the activity, but also from the critical monitoring by the company and its contact with suppliers when the acronym TTE is invoked and referenced. Do Values and Knowledge open the way to Innovation?

The main focus in this operational intervention is the profile of the Tourism Professional. Assertively planning the participation of each of the six elements of the TTE requires a prospective and updated vision on Demand trends. In this perspective, the survey to the target population also allows to understand how this opportunity was managed in the IPT's area of influence.

The tourist visitation implies travel and activities for travellers. In this atmosphere, the experience of each tourist is structured according to a value chain composed of: Travel, Accommodation, Restaurant, Entertainment and Leisure, Safety and Health, Access to Goods and Complementary Services. For these reasons, the aim of this chapter is to propose the concept of "Total Tourism Experience - TTE" developed in recent years and its operationalisation. In innovation, the use of algorithms and artificial intelligence combines with the use of a kind of "shorthand". In the current digital environment this implies new visions on tourism research and operation. Every place has historical value. This is what sustains the heritage appropriated by tourism and it is also what promotes new cultural expressions.

METHODOLOGY

In this methodology, the Tourism Laboratory is a participant observer which in this Team mobilises the Authors' collaborative knowledge because the TTE model in its essence inspires experiences, memorable moments and collaborative work in the production of tourism programmes.

The dynamics generated between People and Organisations in face of Territorial specificities animates and revives Tourism and its segmentations. Thus, aware of this framework in the recovery of the public health crisis, the energy of the historic significance transformed into narratives by Researchers and into innovative business practices, by Operators, opens up possibilities for the implementation of the TTE model. In the heritage tourism planning and local development environment, the UN Agenda 2030 is central. It should be noted that the UN 2030 Agenda for sustainable development results from the process of political intervention that, since the Brundtland Report published in 1987, aims to achieve, until the year of 2030, the 17 Sustainable Development Goals (SDGs) among the UN's member states. Companies networked with the IPT and Tourism Laboratory also contribute to this goal. Because in tourism communication, dedicated digitalisation is unavoidable, the mnemonic "TTE" as a working tool supports the development of tourism products and their operationalisation, reinforces the "academia/professionals/decision-makers/consumers" link.

In the post-covid 19 period, the structuring of the local tourist offer is as much a priority as the responses to be given by destinations in the different geographies. Economic and tourism globalisation was an opportunity to be considered in this TTE proposal.

The quantitative methodology used consists first of collecting data information through an online survey using the *Google Forms* platform, and then the *Statistical Package for Social Science* (SPSS) software, which was used for the statistical data analysis and conclusions.

The theoretical methodology comprehends various dimensions, introducing diverse, but converging themes such as Tourism, Human Resources, Environment, Culture, Economy, Statistics, Business Management, Social and Business Sciences, develops the conditions for the new labour framework of the 21st century to be understood also "glocal" i.e. global and local at the same action time.

CULTURE, LINGUISTIC AND CULTURAL CONSUMPTION IN AN INTERGENERATIONAL PERSPECTIVE

From the experience taking place in a reality where tourism audiences are increasingly cultural audiences, which the major international events are about, the issue of languages emerges. In all human activities, Communication is a form of relationship in which human actions are expressed under support models.

Acronyms have that function, decoding a whole scenario to which they allude indirectly and whose codes of more global reading the senders and receivers know and dominate. In real life there are many acronyms whose origin we do not question because we are convinced of the reading we make of that statement. The communicative processes are evolving, and we are increasingly witnessing a breakdown of the longest words in favour of "new words" as we can see in many dialogues namely between interlocutors in social networks and in the interaction dialogues cause to their players.

Experience, emotion, mood, discouragement and other states of mind are conveyed through communication. The professional language of a given sector also has its own codes. If we consider the vernacular medieval languages and the languages of certain contemporary groups, the phenomenon of cipher words is nothing new. However, the person who pronounces the acronym TTE is making his individual evocation of a set of elements that ensure a total tourist experience, from a single transmitter to all the receivers that can be reached. If we look at the Latin expression *communicare*, which means "to share something with someone else", the exchange of information means to divide, to expand.

The TTE model is, therefore, the sharing of something that in its six links are connected in a way that the first link is the Trip-and the last one is Access to Complementary Goods and Services, still covering Accommodation, Gastronomy, Entertainment and Leisure, and also Safety and Health. Tourist acculturation depends on the capacity of each communicative process to work to people whom the communicative act is intended. Tourism as a space for the production and distribution of products to be consumed by customers requires the conjugation of oral, written and gestural transmission processes and of the vehicles and strategies used for this purpose. Encrypted language is, in any productive operation, part of the guarantee of its profitability.

TOURISM AS A HUMAN EXPERIENCE DOMAIN

Theory structuring needs constructs that, in terms of epistemic epistemology as well as terminology, constitute conceptual reinforcement.

In this sense, the concept of Tourism both determines the appreciation of travel activities of those who want to get to know places other than their usual residence as well as the economic dynamics that it generates:

multisold, i.e. sold as different tourism products to different tourist segments simultaneously from which arise potential conflicts between users and by overcrowding of space Caldeira (2014:32).

The "multisold" phenomenon occurs when sales exceed product stock. In terms of tourist destinations, this phenomenon occurs when the massification of the number of travel agencies is not accompanied by a correct strategic management of tourist destinations. However, since a tourist destination is understood as a combination:

of consumer space and tourism products (Fuchs & Weiermair, 2004, p. 212)

that tourism products, in turn, can be seen as:

"the total tourism experience, comprising a combination of all service elements, from the moment you leave home until you return to it", arguably "total product is a broader concept than the destination itself" (Cho, 2000, p. 145).

This idea of combining elements of the "tourism product & tourism service" leads the players in this economic activity, under the demand for quality of their goods and services, to act in a benchmark plan whose feasibility affirmed in each destination, contribute in turn to the improvement of the value chain and its links. Thus seeing Tourism from the economic perspective means considering it as the result of value creation through a structured set of various economic activities. As a result of travelling and consequent costs and gains in related markets, Tourism is also classified as an economic and social result provided by the movement of travellers as the same type of flow is called Excursionism when performed without overnight stays at the destination.

The experiences acquired in the triad "Culture-Tourism-Management" situations inserted in a working method in which the primacy of People and the safeguarding of Tourism Values are found in an activity generated and developed according to the status of economic activity that however create impact on people's lives and therefore identifiable as a social phenomenon (Pinto (1994), Hernandéz (2002), Freire (2007), Weston et al (2019).

The concepts as applied theoretical constructs underlie the role of the manager in the cultural and tourist sector. Management includes the visions of its protagonists, guiding their missions, impregnating their activities with values that are played in the various aspects of this management and building "tourism-cultural management models". If, in Culture and in the fruition of heritage, as we know, there are reference models that are forms of management consolidated by the operationalization that each entity plans, executes and monitors, namely: the "Gulbenkian Model (https://gulbenkian.pt/)" and more models like, "CCB Model", the "Casa de Serralves Model", the "Archaeological Site of Mértola Model", the "Templar City of Tomar Model" and so on, and so on. Also in the tourism sector the management models proposed by the global strategy, under the responsibility of the UNWTO, or the national strategy under the responsibility of Turismo de Portugal, I.P. as part of the Ministry of the Economy and its Secretary of State for Tourism, include the "ERT Model - Regional Tourism Authority".

In this structuring dimension of public policies, the TTE will also come from the management capacity of public investments and their articulations with the private initiative in the attempt to bring together the irreplaceable components of what is considered a total and distinctive tourist experience: innovative, of internal reference and, of great interest, attractive for the investments and tourist flows that the country manages to attract and satisfy. Therefore, in a theoretical dimension, but which is justified under the logic of the customer experience value chain, we present our theoretical proposal that can be put into practice. The surrounding scenario also facilitates this proposal because, as France (2017:11) argues:

Technology has an impact on experiences causing frequent changes in the way they are generated, delivered and enjoyed (Binkhorst and Dekker, 2009). Because of technology, co-creation with the consumer can easily be extended beyond the moments of purchase and enjoyment, and there are opportunities for co-creation along the entire value chain (Prahalad and Ramaswamy, 2002) of a tourism experience, i.e. all the tourist-related events that occur before (planning and preparation), during (at the destination) and after enjoyment (memory) (Ek et al, 2008; Tung and Ritchie, 2011).

Therefore the value chain of a tourism experience, comprising well connected links endowed with well specified attributes and features, makes clear the need to take advantage of this experiential situation by applying it to the preparation, execution and consider the relevance of TTE we can try the methodology of mnemonics (i.e. using the TTE mental image and deducing from it the chain of links that constitute it) so that we have a tourism management guide for use by everyone. For a professional in the tourism sector, invoking these three letters has the power of immediate articulation of the six components that constitute it. These evocations under the concepts that the matrix of the tourism operation and its regulations impose, facilitates communication, a crucial part of management. In the field of tourism, France (2017:7) also argues:

The creation of user experiences is especially relevant for enterprises falling within the "primary experience sector", such as tourism (Sundbo, 2009: 441), which has experiences as its core offering.

In this perspective, experience constitutes people's motivation, desire, need and in this narrow sense makes room for emotion, for relaxation, for delight and of course for critical evaluation between what is promised in the sale of the product you choose and your perception as a consumer. Considering tourism as a "primary experience sector", tourists through the experience of each of the TTE components achieve the full enjoyment that their status as travellers provides.

Human Capital in the Tourism Sector

From a diachronic perspective and since tourism became a goods and services production environment that is tradable on the market, what was until then mainly an area of a marked social nature has inevitably gained the status of economic activity. Therefore, tourism is one of the most relevant activities in international relations that goes beyond international trade, migration and financial flows between countries. Sometimes international tourism revenues exceed those generated by exports of goods, other times they fully finance the total imports of some countries (Cunha, 1997). The situation it's continuous nowadays.

In the 1990s, society experienced significant behavioural changes with the increase in household income, the growth of economies, new technologies, transport and communications. Travel became accessible to ordinary citizens, which allowed its popularisation and consequent ease of travel. Travelling in this era is an acquired "good", present in people's daily lives. "Mass tourism" prevails, allowing tourists to choose intercontinental locations. This state of affairs also enables large hotels to expand to new countries, in order to offer their customers the same quality they have in large metropolises (Ramos & Costa, 2017).

The concept of tourism can be understood as an activity that concerns the act by which someone travels, i.e., someone who moves geographically for some reason. The human element is central here, insofar as the concept of tourism relates to people, since tourists are people and tourism products and services are provided by people, and visitors are also people. The word tourism was born from the English term tour which means "trip" or "excursion". The term "tourism" in its current sense is understood as the temporary movement of people (less than a year) to destinations other than the usual places of residence and work for a variety of purposes (leisure, health, religion, among others), excluding the purposes related to obtaining profit. We can also state that the concept of tourism encompasses all the activities developed by tourists during their trips and stays (Lourenço, 2021).

Tourism considered from both demand and supply perspectives is

a vast and varied activity which encompasses not only the movement of people and all the relationships they establish in the places visited, but also all the productions and services developed to meet their needs. It is a concept that simultaneously covers the strong competition in the markets, forcing companies to be in constant change both at a technological level and in the adaptation of the offer. And consequently, innovation and differentiation of tourism supply and demand (Cunha, 2012).

Tourism development is also strongly influenced by the level of trust in governments, namely political trust in the tourism context and general trust in institutions. The directives of the World Tourism Organization demonstrate this. Thus, it is important that government policies and strategies are seen as positive for tourism development and that it is beneficial for communities, and governments should inform about their role and responsibilities in tourism development (Pagliare, 2020). The operationalization of government policies aimed to support tourism development becomes more relevant in an era of post-pandemic recovery, to allow a fairer, more inclusive, sustainable and resilient transition. For this reason, there should be a top priority in the articulation between public policies, business actions and international cooperation as an answer to the exceptionality of the consequences resulting from the pandemic crisis.

The human resources area plays a key role in the tourism sector, taking responsibility for people management in order to adapt to constant change and simultaneously foster customer and employee satisfaction. The development of a corporate culture in this specific case of services, consistent and oriented towards the quality-of-service provision, can be the guiding element for obtaining competitive advantages (Marques, 2019). Therefore, in order to achieve business success, organisations must have trained and qualified human resources as a response to the many challenges that arise daily (Sousa, 2021).

Human capital is an important resource for the competitiveness of tourism enterprises and for the development of the tourism industry. Organisations should, therefore, reflect on the most appropriate type of management, on the best HR management practices (Pimentel, 2016). The human resources area plays a leading role in the various domains of tourism taking responsibility for people management in order to adapt to constant change and simultaneously encourage customer and employee satisfaction.

Recently, the impact of the Covid-19 global pandemic has been one of the biggest challenges facing the tourism sector at international level. Data indicate that tourism demand has declined significantly and that the negative impact on the economy, particularly on tourism, has been extremely strong, not only directly due to the cases of disease, restrictions and constraints, but also due to tourist risk perceptions (Lourenço, 2021).

Talent Recruitment and Retention in the Tourism Sector

Human resource management is understood as a series of:

policies, practices and systems that influence the behaviour, attitudes and performance of the members of the organisation towards increasing competitiveness and learning capacity in the organisation (Cunha, 2012).

In order to have a competitive advantage in terms of service quality, it is necessary to develop HRM practices enabling the recruitment and selection of competent professionals who are aware of the organisation's objectives. Thus, the degree of customer satisfaction will reflect on the quality of human contact, dependent on the motivation and personal skills of the service provider (Marques, 2019).

One of the problems faced by HRM in the tourism sector is the existence of a culture of high turnover and constant recruitment problems. Special attention to these two aspects is needed from organisations and implementation of talent management practices (Sousa, 2021).

The results indicate that human resource managers face a highly competitive market to be able to attract the best talent to perform in the labour market and, in this sense, the main difficulties of these professionals are perceived with the low number of candidates or candidates without the necessary experience for the advertised position as well as competition between organisations (Sousa, 2021).

These organisations should be concerned with HRM that leads to better performance by tourism employees, both individually and in work teams. It is important to use management strategies that lead to organisational success, particularly with regard to recruitment and selection practices, training, performance appraisal and benefits management (Lourenço, 2021).

The central idea is to understand the vision that companies have about the qualification of their employees as well as their hiring requirements (Marques, 2019).

The Human Resources departments of tourism-related companies are subject to alternate and extended schedules due to operational functions and mass hiring arising from seasonality. It is necessary to ensure that the selected candidates are effective and engaged with the organisations (Sousa, 2021).

Talent retention is a key part of talent management. In a turbulent labour market, retaining talent is a win-win situation for any organisation. There are several factors that can influence talent retention such as the need for ongoing staff development, adequate reward systems, motivational levels based on a strong commitment and the structure that allows them to raise their level of performance (Sousa, 2021).

Tourist profile has changed, leading companies to be more competitive in attracting new customers and building customer loyalty. In a competitive world, innovation is crucial particularly with regard to human capital management, since it contributes to improving the quality of services and is therefore a competitive advantage of organisations (Pimentel & Duarte, 2016).

According to the economic and sociocultural transformations experienced by the sector, and in view of the requirement for growth, competitiveness and quality, there are needs for adequacy in the tourism sector. Thus, it is professionals with specialised skills who take action when relevant situations arise. The sector also has the help of those responsible for planning and managing tourism policies as well as educational institutions and specialised consultancies, providing the sector with support in its development (Marques, 2019).

Employees in the tourism industry must possess a range of technical, personal and behavioural skills such as good interpersonal relationship skills, respect for colleagues, good ability to deal with customers and co-workers, friendliness, good conflict management, good posture, professional image and personal hygiene, ability to deal with pressure, initiative, innovation and problem-solving skills, spirit of cooperation and teamwork, sense of responsibility, job-specific technical knowledge, customer management technique, sales technique and knowledge about the market environment (Garcia, 2016).

In addition, other decisive requirements are the availability to work flexible hours, to be versatile, friendly, good communicator, sociable, to have a command of the mother tongue and foreign languages, the ability to adapt to changes, a taste for activity, the ability to work in a team and the ability to apply theoretical knowledge. Thus, it is expected that human resources to have mastery in performing tasks and well-consolidated and multidisciplinary knowledge to adequately respond to different situations and contexts (Garcia, 2016).

Human Capital Competencies in the Tourism Sector

Competencies are the range of attributes of organisations to survive in the environment such as personal skills, technology or know-how. Currently in the context of business competitiveness and uncertainty about tourism markets, tourism organisations should be concerned not only with functional skills development, but especially with the adaptation of the individual to the job and with their cognitive and behavioural skills, choosing employees with the most appropriate skills (Pimentel, 2016).

In order to become competitive, organisations and their managers must identify the essential competencies for employees and managers. A competency model is a tool that should identify the knowledge, skills and behaviours required for a better performance of employees, enhancing a better organisational performance. The competencies of the human resource manager are equally important for a successful management, since they assist in his/her role of managing individuals and teams, as well as administrative functions, namely labour relations, recruitment and selection, human resource planning, among other management functions (Lourenço, 2021).

Technological evolution is most probably the aspect that will most influence professional skill profiles in tourism. For this reason, this paradigm shift should not be ignored in the sector. Some of the professional skills needed in 2030 include:

- **Artificial Intelligence** - man-made devices that can perform certain functions without human interference;
- **IoT (Internet of Things)** - has allowed great evolution in tourism and hospitality, as the sensors and technologies that are embedded allow objects and the internet to exchange data with each other, making decisions automatically, often without human intervention;
- **Big Data** - This is also one of the keys of the current digital transformation, as it allows transforming data into useful information about users/visitors, their movements and preferences, adapting business strategies so that the offer fits the consumers' needs (Pereira, 2015).

Hotel chains, airports and airlines have invested in the area, developing actions that enable tourists to access the different services they offer. However, in addition to investment in software, a number of properly qualified professionals is needed to help installing, maintaining and supporting the use of this equipment. Emotional intelligence is a relevant factor in the social dimension of the professional-customer interaction.

Critical skills of tourism workforce, especially in the international area, include behavioural skills such as the ability to adapt to a multicultural context and to possible economic crises, a context where resilience stands out. Technical skills are also valued especially the mastery of specific qualifications related to new technologies such as computer skills.

The growing dynamics of the labour market and the notable growth in demand for work with better conditions has led to employees being less loyal to companies. An important aspect to improve in tourism is the qualification of the workforce in the sector, as well as their income, valuing them and constantly updating their conditions in line with market changes. This is the only way to reduce turnover that is so common in the sector, especially in the hospitality industry.

In this sense, the development of human resources that possess specific professional qualifications are an asset to organisations making them more competitive in the market mainly because HRM in the tourism industry promotes and develops practices that attract, retain, develop, motivate and remunerate employees (Garcia, 2016).

Regarding innovation, although top managers in the hospitality industry are doing their best to keep up with these rapid changes, they need to embrace the complexity of the concept of innovation in order to recognise the potential it brings and stimulate the development of ideas based on disruptive innovation, which can contribute to the creation of a sustainable competitive advantage. Internal communication in a company, particularly one that provides tourism services, as well as the ability of its staff to generate empathy with the brand, sometimes has a decisive advantage in clear and shared references. The use of acronyms and jargon facilitates and boosts performance (Castells, 2007).

The present research (cf section 10) shows that there is still much to be invested in the application of technology in the tourism business. Furthermore, it has been found that in what concerns the enjoyment of a TTE, variables such as Culture, Knowledge and Peace of the territories to be exploited for tourism purposes are relevant.

Globalisation, marked by the opening of borders and the rapid movement of information and people, favours a convergence of contexts. This convergence leads, in turn, to a certain approximation of practices accentuated by the instrumentalization? of HRM and the establishment of human resources information systems with common characteristics. The interest in using "best practices", the adoption of international standards, certification processes, audits and the dissemination of management models contribute to a convergence of HR practices (Alis, Horts, Chevalier, Fabi, & Peretti, 2012).

TOURISM ENVIRONMENT AND INTERACTIONS

Tourism activities develop in a variety of settings sought after for their climate, landscape and cultural features. Itinerary & Accommodation plans are not separable elements of a tourist experience which, not occurring in the lives of ordinary consumers means in the classical typology of travellers an opportunity to escape, to flee from the regular and dull life of most people.

The emergence of the so-called "digital nomads" is changing this reality but it is still in its early days. Companies and their customers are, for all intents and purposes, conjugations of human beings united by the same vision: the product built and sold and the enjoyment of that product in a certain context of space and time. In the perspective of sustainability and natural connection between economic and social life with the natural environment under pressure from the need for territorial sustainability, the experiences in this type of "new sustainable tourism destinations" where ecotourism is gaining followers are becoming benchmarks.

Certain areas find in "tourist and cultural brands" a new energy. The Vicentine Coast or the Algarvian Route, as well as the Romantic Route or the New Hand Lab in Covilhã or even the Lx Factory in Lisbon are experimental territories of culture and cultural tourism and other segments like industrial tourism, scientific tourism, etc. These new cultural landscapes join the processes of Street Arts and Performing Arts that animate urban, peri-urban and rural territories. The emergence and development of the "new rural" or "digital nomad" movement are not isolated phenomena. These phenomena result from the realignment of the concepts of work and leisure which, in the knowledge society in which we live, reorganise spaces, human behaviour and eclectic and very diverse visions. Cultural pluralism is a new avenue which cultural tourism has taken hold of and exploits with increasing quality, particularly with a view to increasing city demography as the issue of smart cities demonstrates.

TOURISM ECONOMICS AND EXPERIENCES

Tourism as an economic activity deal with scarce assets (natural and cultural resources) by enhancing them and giving them the attributes that the markets discover and consume or, on the other hand, by the demand that puts pressure on the suppliers of tourism goods and services. Research shows that whenever a tourism product is revealed and described as something new, but carrying a strong associated identity, the consequence is that it is sought to be consumed by those who, in the field of leisure travel, collect life experiences in tourist and cultural destinations. If the democratization of access to goods and services in the tourism and cultural sector is evident since the massification of travel provided such access, it is also true that we are witnessing the emergence of innovative modalities such as, for example, space travel – scarce, highly valued products in the transport market. Recent post-pandemic economic trends suggest promising *scenarios for the recovery of the world economy. However, as stated by Romão (2021:63):*

A year ago, at HighSun we were very positive about the recovery of the economy in the 4th quarter of 2020 with direct consequences on tourism. But this has not been the case. Due to the significant increase of infection rates in several regions, a second and third wave occurred, with new partial and general lockdowns all over the world.

In a perspective of economic recovery, the prospects for 2022 are more encouraging and, therefore, the current situation of shortage of by-products, for example, for the automotive industry and the production of goods dependent on electronic circuits as well as the acute problem of inflation and increases in fuel prices and the resulting cost burden, requires all tourism players to reinvent themselves in terms of work processes, remediation strategies and real-time responses. The UNWTO has constantly affirmed the values involved in testing the most appropriate measures (UNTWO, 2019)

Sustainability is therefore the central aim of the necessary rational critical thinking in an environment of permanent uncertainty (academia and its business and social surroundings) of concerted action (networks and their effectiveness, efficiency and economy) of permanent communication (consumers and age profiles, financial capacities, social status, etc.) and of sharing difficulties and possible solutions (advocacy for tourism issues and its importance in globalisation).

TOURISM TRENDS: AN ANALYSIS OF RESULTS

Sustainability is undoubtedly a trend in the tourism sector, but it is important to know what other trends are associated with the recovery of tourism in a post-pandemic era. What will be the new tourist destinations in Portugal? What will be most valued in a tourist experience? What should be the strategic positioning of national tourism? For this purpose, an online survey was carried out, guaranteeing respondents total confidentiality of the data collected and ensuring their anonymity, thus obtaining greater reliability in the answers. The statistical treatment of the data was carried out using the Statistical Package for Social Science (SPSS) software.

In addition to sociodemographic questions, the survey used for this research consists of three open-ended questions and six closed-ended questions, the latter comprising two single-ended questions, three five-point Likert-type ordinal scale questions and one yes/no dichotomous question (see Table 1).

The collected sample was composed of 51 valid answers Despite the response rate regarding the number of questionnaires sent being only 35%, the representativeness of the sample is ensured because the respondents were strategically selected and are somehow associated with tourism, whether for professional or academic reasons or as a tourist. In addition, according to the central limit theorem, the sample size guarantees the approximation to the normal distribution.

Table 1. Survey characterisation. Source: own elaboration

Data	Dimensions		Question Type	Survey Item
Measurable	Information concerning the:	• identification of the capacity in which the respondent answered the survey from among the following options: tourism professional, student in the area of tourism or tourist. If the first option was selected, the area, time and geographical region of the respondent's professional activity were identified.	Identification (single choice)	1
		• perception about the long-term permanence in a post-pandemic era of tourism habits and interests inherent to the current pandemic context.	Dichotomous Yes/ No type	2
		• relevance, in a tourism experience, of each of the following six elements: accommodation; access to complementary goods and services; entertainment and leisure; gastronomy; safety and health; travel.	Likert-type ordinal scale	3
		Transversal nature of tourist experiences with the identification of the relevance of the following seven terms: discovery; emotion; freedom; satisfaction; longing; surprise; tranquillity.	Likert-type ordinal scale	4
		• identification of factors influencing the future of Tourism.	Likert-type ordinal scale	5
		• identification of the most relevant factor in characterising Tourism trends in a post-pandemic era.	Identification (single choice)	6
Non-measurable	Identification:	• of terms not specified in question 4 that the respondent considers equally relevant in a total tourism experience.	Open-ended	4.1
		• of terms not specified in question 5 that the respondent considers equally relevant to ensure tourism development in the future.	Open-ended	5.1
		• of the main factor associated with a high quality tourist experience.	Open-ended	7

Table 2. Socio-demographic characteristics of respondents (n=51). Source: own elaboration

		Frequency
Gender	Female	68,6%
	Male	31,4%
Academic Qualifications	Secondary Education	23,5%
	TeSP/Undergraduate	11,8%
	Bachelor's degree	31,4%
	Master's degree	11,8%
	PhD	21,6%
		Anos
Age	Minimum Age	19
	Maximum Age	82
	Average age	39,9
	Standard deviation	15,8

The socio-demographic data are summarised in Table 2. It was found that more than half of the respondents are female (69.6%), the majority (31.7%) has a university degree and the age range with the highest expression (37.3%, cf. Graph 1) is between 18 and 33 years, since although the average age is approximately 40 years, there is a high dispersion of data regarding the age of respondents due to the occurrence of an upper outlier.

Figure 1. Distribution of respondents according to age grouped into classes. Source: own elaboration

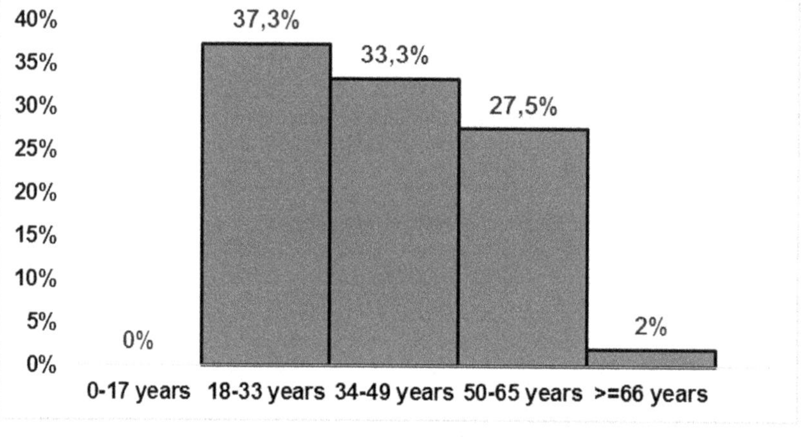

According to Figure 2, almost half (45.1%) of the respondents answered this survey as tourists, followed by students in the area of tourism who represent 31.4% of the respondents, with the remaining 23.5% being professionals or entrepreneurs in the area of tourism.

Figure 2. Distribution of respondents according to the quality by which they answered the questionnaire. Source: own elaboration

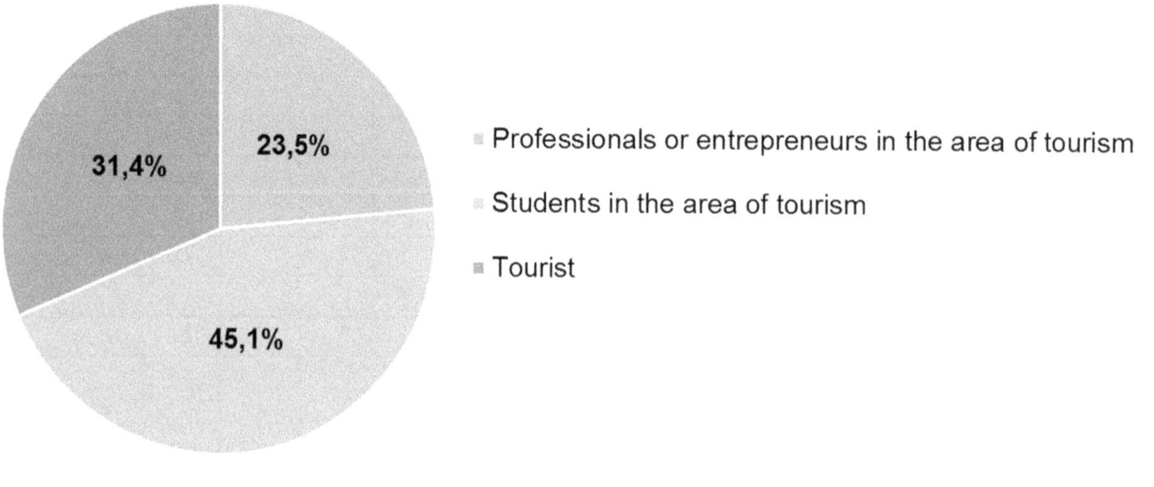

As regards the tourism professionals surveyed, as illustrated in Figure 3, the majority (41.7%) work in the hospitality industry, followed equally by ecotourism and tourism operations, both representing 16.7% of the respondents and, to a lesser extent, restaurants, tourism events and rural tourism, with only 8.3% each.

Figure 3. Distribution of respondents who are tourism professionals according to the tourism area in which they perform their professional activity. Source: own elaboration

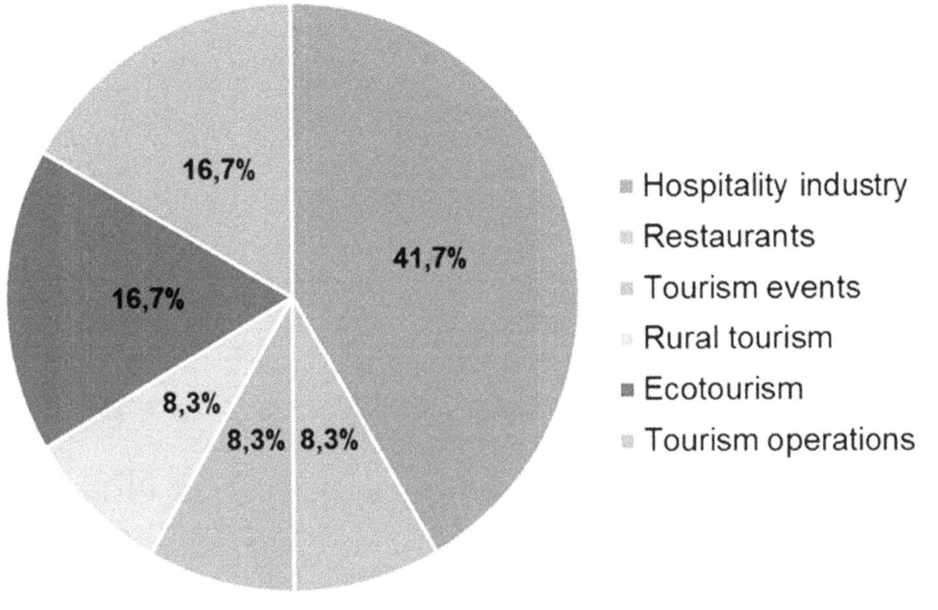

In relation to the geographical region in which the tourism professionals surveyed carry out their professional activity, the prevalence of the Central region (91.7%) was noted, with the remaining 8.3% belonging to the Alentejo (cf. Table 3). Among the respondents who are tourism professionals, the majority (75%) have been working for less than 20 years (cf. Table 3).

Table 3. Distribution of respondents who are tourism professionals according to the geographical region in which they work and according to how long they have been tourism professionals. Source: own elaboration

		N %
Geographical area of professional activity in tourism	Centre	91,7%
	Alentejo	8,3%
Time (in years) as a Tourism professional	Under 21 years old	75,0%
	21 to 40 years old	8,3%
	Over 40 years old	16,7%

We will now analyse the results in relation to the subject under study. According to Figure 4, there is a significant numerical superiority (68.6%) of agreement with the permanence, in the long term, in a post-pandemic era of the tourist habits and interests inherent to the current pandemic context. Therefore, the first conclusion to be drawn is that change is here to stay and, therefore, adaptation to new organisational paradigms inherent to the tourism sector will be determinant for the success of its recovery.

Figure 4. Distribution of respondents according to the perception about the permanence, in the long term, in a post-pandemic era of the tourism habits and interests inherent to the current pandemic context. Source: own elaboration

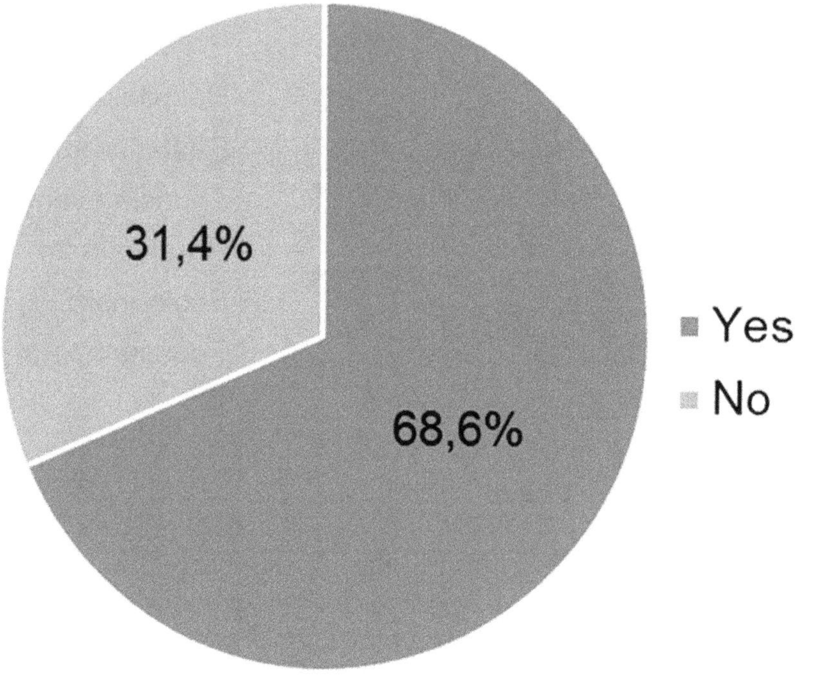

In order to assess what is most valued in a tourism experience, we analysed six inseparable items which are listed below in alphabetical order: (i) Accommodation; (ii) Access to complementary goods and services; (iii) Entertainment and leisure; (iv) Gastronomy; (v) Safety and health and (vi) Travel. As illustrated in Table 4, it was found that in all parameters, most answers are concentrated in the last two marks of the Likert scale, i.e., in the very relevant and relevant marks, which is demonstrative of the relevance of the six parameters considered in the present study, and also of the importance that they have for the respondents in a tourism experience. Only parameter (vi) Travel has an incidence of 2% and 3.9%, respectively, in the *inexistence of relevance* and in *little relevance*, there also occurring an equally residual percentage of 2% associated with little relevance in parameter (iii) Entertainment and leisure. It should also be noted that with the exception of (ii) Access to complementary goods and services and (iii) Entertainment and leisure, for which the percentage superiority lies in the relevant brand with 47.1% and 51%, respectively, all other parameters are considered very relevant. It should be noted that among those considered very relevant, there is a slight numerical superiority in (v) **Safety and health** and (i)

Accommodation, with 64.7% and 60.8% of the answers, respectively. We can therefore conclude that in order to achieve more favourable results in tourism experiences, quality must be ensured in parameters (i), (iv), (v) and (vi), with special emphasis on safety, health and accommodation. The colours shown in Table 4, as well in Table 5 and 6, represent the categories associated with the highest percentage values regarding the two first levels and the two last levels of the Likert scale.

Table 4. Crosstab created using statistical software (SPSS) with row percentages of the six parameters of the categorical variable under analysis associated with a tourism experience, according to the five point Likert ordinal scale. Source: own elaboration

	Inexistence of relevance	Little relevance	Moderately relevant	Relevant	Very relevant	
	Row N %	Row N %	Row N %	Row N %	Row N %	Total
Accommodation	0,0%	0,0%	0,0%	39,2%	60,8%	100,0%
Access to complementary goods and services	0,0%	0,0%	17,6%	47,1%	35,3%	100,0%
Entertainment and leisure	0,0%	2,0%	11,8%	51,0%	35,3%	100,0%
Gastronomy	0,0%	0,0%	13,7%	39,2%	47,1%	100,0%
Safety and health	0,0%	0,0%	7,8%	27,5%	64,7%	100,0%
Travel	2,0%	3,9%	7,8%	37,3%	49,0%	100,0%

As far as its transversal nature is concerned, Tourism allows tourists to have remarkable experiences in their lives which integrate various parameters. Since it was impossible to include all these parameters in our study, we considered seven that are shown in Table 5. It should be noted that in order not to limit the answers to the list of options presented, other parameters were collected that the respondents freely mentioned as being equally relevant in a tourist experience and which are illustrated in Figure 5. From the analysis of Table 5 it can be concluded that there is a concentration of answers in the last two marks of the Likert scale which means that most of the parameters analysed are considered very relevant or relevant for the respondents. In fact, with the exception of the parameters Homesickness and Surprise which have numerical superiority in the relevant mark with 43.1% and 51% of the answers respectively, all the remaining parameters are considered very relevant. Amongst those considered very relevant, Satisfaction stands out with 70.6% of the answers, followed by Tranquillity with 60.8% of the answers being considered very relevant.

These results indicate that, in addition to the obvious importance that satisfaction represents in a tourist experience, tranquillity and calm are increasingly valued to counteract the massification and bustle associated with the city's daily life. This conclusion was to be expected and is entirely in line with the habits that were created during the pandemic situation associated with the SARS-CoV-2 virus. It should also be noted that, albeit residually, there are two parameters which have non-zero values in the irrelevant mark with only 2% of answers which are the parameters: Emotion, Freedom and Homesickness. The parameters Homesickness and Surprise also have 9.8% and 3.9%, respectively, in the *not very relevant* mark.

Regarding the open-ended question in which the terms not specified in Table 5 have been collected, which respondents consider equally relevant in a tourist experience, it is clear in Figure 5, the emphasis associated with Safety already addressed in Table 4 and then Adventure and Nature. These trends are not recent but have been increasingly consolidated and will undoubtedly be reinforced not only in the current context, but also in a post-pandemic scenario. Indeed, safety and preference for outdoor environments, less crowded destinations that favour adventure and contact with nature, are at the basis of the demand for tourist destinations in the current context.

Table 5. Crosstab created using statistical software (SPSS) with row percentages of the seven parameters of the categorical variable under analysis associated with a tourism experience, according to the five point Likert ordinal scale. Source: own elaboration

	Inexistence of relevance	Little relevance	Moderately relevant	Relevant	Very relevant	
	Row N %	Row N %	Row N %	Row N %	Row N %	Total
Discovery	0,0%	0,0%	3,9%	45,1%	51,0%	100,0%
Emotion	2,0%	0,0%	5,9%	45,1%	47,1%	100,0%
Freedom	2,0%	0,0%	7,8%	39,2%	51,0%	100,0%
Satisfaction	0,0%	0,0%	0,0%	29,4%	70,6%	100,0%
Homesickness	2,0%	9,8%	27,5%	43,1%	17,6%	100,0%
Surprise	0,0%	3,9%	9,8%	51,0%	35,3%	100,0%
Tranquillity	0,0%	0,0%	3,9%	35,3%	60,8%	100,0%

It is therefore important to understand if these trends were intensified merely due to the pandemic situation, or if they could become a pattern and remain in the future. According to Figures 2 and 3, which will be analysed later, everything suggests that these trends should remain, at least, in a near future. Other aspects that are equally valued in a tourist experience are the concerns associated with Welfare, Health, Happiness, but also Culture and Knowledge that are also identified in Figure 5. It should be noted that, though in an occasional way, Comfort, Quality, Entertainment, Traditional and Regional cultures are also mentioned, among others.

As for the factors on which Tourism will be dependent in the future, the data in Table 6 also reveal a concentration on the last marks of the Likert scale: totally agree and agree. It should be noted that among the nine factors with the highest percentage value in full agreement, there is a slight concentration on **Financial Stability** and the existence of **Peace**, with 68.6% and 66.7% respectively. It is therefore essential that there is a government strategy that ensures the socio-economic interests of citizens and quickly achieves Sustainable Development Goal 16, which aims to promote peace, justice and effective institutions among the countries that are members of the UN. According to the data collected, this is the only way to boost Tourism to another level of growth and development. It should be noted that data analysis also allows us to conclude that 35.5% of respondents have no opinion in relation to the influence of Artificial Intelligence in the future of Tourism, which is the realisation that there is still much to invest in the application of technology in the Tourism business. In turn, only the factors related to greater investment in Education, the existence of more Free Time and again Artificial Intelligence, show non-zero values in total disagreement, although with reduced percentage values of only 2%, 2% and 3.9%, respectively.

Figure 5. Cloud of words considered relevant in a tourist experience. Source: own elaboration

Table 6. Crosstab created using statistical software (SPSS) with row percentages of the categorical variable under analysis associated with the future of tourism, according to the five point Likert ordinal scale. Source: own elaboration

	Strongly disagree	Disagree	Neither agree or disagree	Agree	Strongly agree	
	Row N %	Row N %	Row N %	Row N %	Row N %	Total
Existence of Peace	0,0%	2,0%	2,0%	29,4%	66,7%	100,0%
Existence of greater freedom	0,0%	0,0%	5,9%	58,8%	35,3%	100,0%
Existence of greater public security	0,0%	0,0%	9,8%	37,3%	52,9%	100,0%
Nonexistence (or greater control) of pandemics	0,0%	0,0%	19,6%	35,3%	45,1%	100,0%
Greater investment in Education	2,0%	0,0%	13,7%	49,0%	35,3%	100,0%
Climate change	0,0%	2,0%	9,8%	43,1%	45,1%	100,0%
Growth of extreme natural phenomena	0,0%	3,9%	11,8%	43,1%	41,2%	100,0%
Artificial Intelligence	3,9%	5,9%	35,3%	35,3%	19,6%	100,0%
Technological transformation	0,0%	2,0%	15,7%	54,9%	27,5%	100,0%
Financial stability	0,0%	0,0%	3,9%	27,5%	68,6%	100,0%
Greater participation of the State in the promotion of tourism	0,0%	3,9%	11,8%	41,2%	43,1%	100,0%
More free time	2,0%	2,0%	2,0%	37,3%	56,9%	100,0%
Health	0,0%	0,0%	2,0%	39,2%	58,8%	100,0%

Regarding the open-ended question in the survey which asks for reference to other aspects not included in Table 6, but which in the opinion of respondents are equally important in identifying the future scenario for Tourism, the content analysis of the answers illustrated in Figure 6 allows us to perceive that this future necessarily involves an unequivocal commitment to **Sustainability**. Undoubtedly, the implementation of practices that protect the environment and ecosystems is on the agenda of the world agenda and, consequently, it is also increasingly marked in the minds of citizens who, as tourists, have a greater concern in directing their choices in line with tourist offers that embody a real ecological awareness. **Hospitality** and **Accessibility** also occupy an important place in the relevance they represent for the future of Tourism, as well as **Friendliness in** customer service, **Incentives** and **Training**. Although less frequently mentioned in the answers, marketing, entrepreneurship, knowledge and training are also mentioned as relevant.

Figure 6. Cloud of words considered relevant for the future of Tourism. Source: own elaboration

When asked about the option they would choose as the most relevant in characterising Tourism trends in a post-pandemic era, the analysis of Figure 7 allows us to conclude that local cultural exchange is considered among the four trends analysed the least relevant in the future of Tourism, with only 9.8% of the answers. This is followed by wellness retreats and outdoor activities with similar percentages of 25.4% and 21.8% respectively, with a special emphasis, with almost half of the responses, on **Tourism that implements sustainable practices** which is therefore the trend considered the most relevant in the future of Tourism and which once again reinforces the conclusions drawn from the previous analysis in Figure 6.

Finally, the last open-ended question which aims to identify which features characterise a tourist experience considered to be of high quality, Figure 8 shows that **Hospitality** comes first, which had already been mentioned as relevant in the trends for the future of Tourism in Figure 6, followed by **Satisfaction**, and also an experience that transmits sensations associated with **Relaxation**. Although less frequently, emotions are also mentioned, in general, and in particular those associated with **Tranquillity,** contact with **Nature**, **Freedom**, and once again the issue of **Safety** is mentioned, among others mentioned in a timely manner, such as price, discovery, efficiency, differentiation and innovation.

Figure 7. Distribution of respondents according to the most relevant option in the characterisation of tourism trends in a post-pandemic era. Source: own elaboration

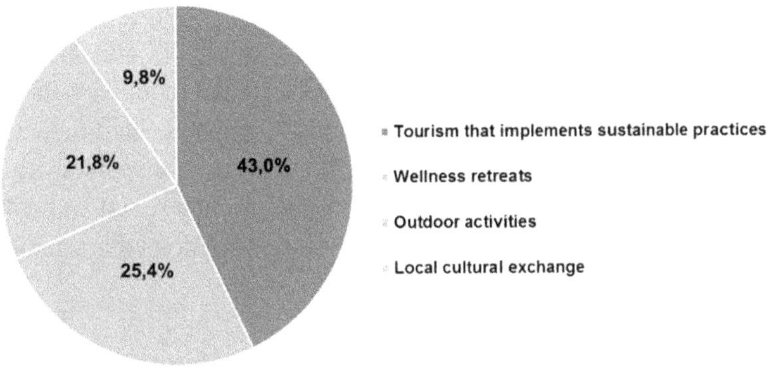

Figure 8. Cloud of words associated with a tourist experience that is considered to be of high quality. Source: own elaboration

The tourism characteristics identified in this section are essential to ensure an efficient recovery of the tourism sector, guaranteeing assertive investment by companies in this sector of activity, taking advantage of new trends as opportunities. The Figures 1, 2 and 3 were created using the free online word cloud art creator https://wordart.com/.

THE COMPANIES, THE OPPORTUNITIES, THE CONSTRAINTS, THE COMBINED SOLUTIONS

Integrated in the model proposed in the academic environment of the IPT and its Networks, the planning and execution study aiming at the implementation of the TTE model requires value-adding intervention benefiting People, their Territories and their Organisations. Companies and their suppliers are, in this context, the main beneficiaries of the operational and communicative transparency that can exist between all tourism players, which is undoubtedly an economic meeting and a social and cultural relationship.

As a result and based on this approach, we propose a range of actions that can be applied to tourism as follows:

- **The Enhancement** of endogenous resources, the attraction of exogenous resources and the management of heritage and contemporary expressions imply research, training of professionals and publics, aiming at results benefiting the tourism value chain;
- **The Management procedures** on tourist destinations and their cultural spaces/facilities bring about cross-cutting and multidisciplinary approaches and their continued use generates habits and routines sustaining and qualifying the productive rhythm, making it notorious in the global tourism market;
- **The Definition** of strategies for Demand Studies, Promotion, Programming, Creation of communicative strategy and creation and loyalty of audiences, generates audiences for the TTE model, incorporating tourist and cultural values;
- **The sense of place** derived from the Roman “genius loci” promotes community and intercultural creation and sustains the “glocal” (linking the global with the local) and is an indelible mark of every geography;
- **In cultural Globalisation**, we find avant-garde currents and mimicries such as the relocation of iconic brands (The Louvre, for example) and new management models and their “centre-outskirts” relationship promoting new experiences and, in this domain, disruptive experiences but integrable in the chain and links that make up the TTE;
- **The Territories**, considered as “social and creative experimental spaces”, are instruments for sustainable development and monitors of change (permanent phenomena with ever shorter time cycles) and coherence to increase the quality of services provided in the Visitor-Host relationship;
- **The Involvement of People** in the CBCD - Community Based Territorial Development, brings value to the Promoters, Investors, Consumers and other Beneficiaries - direct and indirect protagonists in each link of the TTE;
- **The Touristification** of territories and the crossing of the conventional offer with the online offer create new atmospheres, prospective vision on culture, new businesses and economy, jobs and infrastructures that add value to the territories, namely in attracting the “digital nomads”;
- **The Social integration** of professionals with residents and visitors promotes shared and shareable total tourism experiences (authentic-unique in distinctive and specific places) achieves results on several dimensions (social-environmental-economic-cultural);
- **The Alignment** with the relationship produced and disseminated by the pair “tacit, informal, vernacular culture - explicit, formal, academic culture” stimulates the TTE model under the aim of social, economic and cultural inclusion (aligned with the UN Agenda 2030 and other political policy) enhancing human capital;

- **The Maintenance** and continuous development of Tradition in conflict with Modernity and historical reconstitutions and manifestations of rupture involve the conventional arts and the performing arts, both seen in a broad sense and oriented towards the vision of the future, establishing virtuous links between the two fields;
- In the **constant prospective vision**, necessary as a response to permanent change and in a society eager for "disruptive novelties" associated to the fields of innovation and public, private and associative entrepreneurship, operational action under the aegis of TTE serves as a stable reference for all stakeholders involved;
- **Social Inclusion through tourism** as a central challenge of our modernity present in all its economic and social dimensions requires collaborative observation and action according to the principles of the United Nations Agenda 2030 and UNESCO Declaration and Conventions and the decisive importance of the "UNESCO CHAIRS", disseminated in many countries for incorporating the spirit of inclusive tourism practices as relevant social innovation;
- **The projection** of cultural space to the local and external surroundings through the integrated management model that TTE represents can achieve objectives of qualification of people, their territories and their organisations through the awareness resulting from the socialisation processes that tourism always promotes and on which it is dependent.

From all these points and other aspects that characterise each place and each type of territorial governance, it is clear that everything becomes dependent on the people of that place and their social organisation, Cuvelier et al (1994), Crespi (1997), UN (2020). In TTE and its practical expansion, reference to the United Nations Universal Declaration of Human Rights (1948) and consequent UNESCO publications and other international organisations is obvious. The construction of economic and social intervention models are at the basis of public policies that take into account this civilisational matrix and adapt to the realities of each country and its public administration and society. This whole issue revolves around each country's economic planning model in general and the concept of tourism in particular (Amorim et al, 2015), Guerreiro (2016), Kotler, (2017), Chatfield (2013).

CONCLUSION

The aim is to fine tune a concept that for some years has been in the concerns of this work group, reflecting on the scientific production of L-tour.ipt and on the rationale of the "Total Tourism Experience - TTE" as shown in the argumentation. The operationalisation and viability of the reflection found in the company HighSun, the market partner which allowed the first results to be revealed in this early stage of operation. As mentioned in the text, the "shorthand" associated with the acronym TTE seems to work as a facilitator of the dialogues between professionals. If TTE is evoked in the working circle that these months of discussion and refinement of the formulation represent, an important part of the challenge seems to have been achieved. If the concept represents the image of a value chain composed of the links, Travel + Accommodation + Gastronomy + Entertainment and Leisure + Safety and Health + Access to Goods and Complementary Services, the sum of all of them ensures that the experience is attainable. It speeds up the argumentation of the Offer, simplifies procedures for the Demand, streamlines the logistics of the Geographies and finally makes the work of the long and diverse chain of Operators profitable. The tourism system benefits from this reference. As the authors refer, in a perspective of full satisfaction of consumption expectations, each tourist aims to achieve the maximum number of experiences with a cost-benefit ratio in which he feels comfortable and, therefore, taking advantage of this strong reason to return to a tourist destination is a key goal of this teamwork.

In fact, enhancing tourism covering all its Players but particularly Operators responds to the desires, motivations and needs of travellers. Destinations prepared to respond to new trends using short but significantly simple languages that use abbreviations, acronyms and mnemonics, challenge us by what they represent in qualification terms. Therefore, if the preparation of Human Resources is a critical point of success in tourism, the cross-cutting nature of the tourism activity requires consistent approaches. We are sure that what we call "opportunity responses" includes innovative proposals that TTE intends to contribute to. We take the risk because we believe that TTE has the conditions to impose itself in a tourism reality that is complex but, undoubtedly, constantly changing. Therefore, by designing and building comprehensive and action structures in partnership as demonstrated we also fulfil part of the mission of higher education, particularly the polytechnic sector.

ACKNOWLEDGMENT

This work was partially supported by the Centre for Mathematics of the University of Coimbra - UIDB/00324/2020, funded by the Portuguese Government through FCT/MCTES.

REFERENCES

Alis, D., Horts, C.-H. B., Chevalier, F., Fabi, B., & Peretti, J.-M. (2012). *Gestão de Recursos Humanos - Uma Abordagem Internacional*. Edições Piaget.

Castells, M. (2007). A Era da Informação: Economia, Sociedade. *Cultura*, *2*.

Chatfield, T. (2013). *Como aproveitar ao máximo a era digital*. Alfragide: Lua de Papel.

Cho, B. (2000). Destination. In J. Jafari (Ed.), *Encyclopedia of Tourism* (pp. 144–145). Routledge.

Crespi, F. (1997). *Manual de Sociologia da Cultura*. Editorial Estampa.

Cunha, L. (1997). *Economia e Política do Turismo*. McGraw-Hill.

Cunha, L. (2012). Turismo de Portugal. Sucessos e Insucessos (1.ª ed.). Lisboa: Edições Universitárias Lusófonas.

Garcia, A. G. (2016). *Gestão de Recursos Humanos na Indústria do Turismo em Portugal:Tendências Futuras do Capital Humano: um estudo exploratório em unidades hoteleiras no Distrito de Coimbra*. Tese de Mestrado, Instituto Politécnico do Porto.

Guerreiro, J. A. (2016). *A questão da autenticidade – do objetivismo à fenomenologia do turismo*. IX Congresso Português de Sociologia, Faro: Faculdade de Economia – Universidade do Algarve. https://www.researchgate.net/publication/305279908_A_questao_da_autenticidade_-_do_objetivismo_a_fenomenologia_do_turismo

Kotler, P., Kartajaya, H., & Setiawan, I. (2017). *Marketing 4.0: Mudança do tradicional para o digital.* Conjuntura Actual Editora – Edições Almedina.

Lourenço, H. M. (2021). *A Gestão de Recursos Humanos no Turismo: o caso da hotelaria na cidade de Lisboa.* Tese de doutoramento, Turismo, Universidade de Lisboa, Instituto de Geografia e Ordenamento do Território.

Marques, A. (2019). *A Importância da Qualificação dos Recursos Humanos no Turismo: O Caso dos Hostels da Cidade do Porto.* Tese de Mestrado, Instituto Politécnico do Porto.

Pereira, I. (2015). As 10 tendências do futuro do turismo. Publituris. https://www.publituris.pt/2015/12/30/as-10-tendencias-do-futuro-do-turismo/

Pimentel, H. (2016). O papel da Gestão de Pessoas no Turismo. R-LEGO - Revista Lusófona de Economia e Gestão das Organizações, 4, 81-111.

Pimentel, H., & Duarte, I. (2016). Gestão de Recursos Humanos no Turismo. *II International Meeting of ISSOW*, 416-429.

Ramos, D. M., & Costa, C. M. (2017). Turismo: Tendências de evolução. *PRACS: Revista Eletrônica de Humanidades do Curso de Ciências Sociais da UNIFAP*, *10*, 21–33.

Romão, F. (2021). O Turismo Receptivo no "Novo Normal" – a Perspectiva da HighSun DMC. In *Turismo Mundial, Crise Sanitária e Futuro: novos tempos, novos ritmos - World Tourism, Health Crisis and Future: new times, new rhythms* (pp. 59–66). Instituto Politécnico de Tomar. http://www.cda.ipt.pt/download/Livros/novos-tempos-novos-ritmos-ebook_compressed.pdf

Sousa, J. P. (2021). *Desafios da Gestão de Recursos Humanos na Geração Millennial - Dificuldades no recrutamento e retenção de talento - Aplicação ao setor do turismo e hotelaria.* Dissertação Mestrado em Gestão de Recursos Humanos, U. Lusófona.

UN – United Nations. (2020). *The Sustainable Development Agenda.* https://www.un.org/sustainabledevelopment/development-agenda/

UNWTO - World Tourism Organization. (2019). *The Future of Work and Skills Development in Tourism – Policy Paper.* UNWTO. doi:10.18111/9789284421213

Weston, R., Guia, J., Mihalič, T., Prats, L., Blasco, D., Ferrer-Roca, N., Lawler, M., & Jarratt, D. (2019). *Research for TRAN Committee – European tourism: recent developments and future challenges.* European Parliament, Policy Department for Structural and Cohesion Policies, Brussels. https://research4committees.blog/2019/10/29/european-tourism-recent-developments-and-future-challenges

ADDITIONAL READING

Alçada, M., Lisitzin, K., & Manz, K. (Eds.). (2013). *Turismo e Património Mundial: Seleção de abordagens e experiências em gestão em sítios do Património Mundial de origem e influência Portuguesa (TOUR-WHPO) – Um projeto de cooperação do Turismo de Portugal e UNESCO World Heritage Centre*. https://centrodocumentacao.turismodeportugal.pt/Nyron/Library/Catalog/winlibimg.aspx?skey=4C6AB12C9F6540B18E66180F4CFF276D&doc=27418&img=172&save=true

Baptista, C., Honrado, M. G., Nata, A., Martins, S., & Figueira, L. M. (2021). Turismo, Economia, Sociedade – renascimento da atividade e novas realidades na empregabilidade. In *Turismo Mundial, Crise Sanitária e Futuro: novos tempos, novos ritmos - World Tourism, Health Crisis and Future: new times, new rhythms* (pp. 15–40). Instituto Politécnico de Tomar. http://www.cda.ipt.pt/download/Livros/novos-tempos-novos-ritmos-ebook_compressed.pdf

Brizard, T., Derde, W., Hamza, M., Helsen, T., Lamgen, I., Plentinckx, D., Putman, J.-L., Silberman, N., & Tency, H. (2007). *Basic Guidelines for Cultural Heritage Professionals in the Use of Information Technologies – How can ICT support cultural heritage?* The Interactive Institute AB. www.tti.se/knowhow

Cunha, L. (2017). *Turismo e Desenvolvimento – Realidades e Perspetivas*. Lidel – Edições Técnicas, Lda.

Dias, F. N. (2001). *Sistemas de Comunicação de Cultura e de Conhecimento*. Instituto Piaget.

Fernandes, A., Pereira, E., Bento, J. P., Madaleno, M., & Obaina, M. (2017). *Introdução à Economia*. Edições Sílabo, Lda.

Figueira, L. M. (2007). *Património, Identidade e Turismo Cultural: abordagem ao caso da Golegã*. Revista Superavit – Revista de Gestão e Ideias.

Figueira, L. M. (2019). *Turismo, Municípios, Cultura e Sociedade – Breve reflexão*, Instituto Politécnico de Tomar, Tomar, http://www.cda.ipt.pt

Figueira, L. M., Baptista, C., & Honrado, G. (2019). *Turismo e Cultura Turística: A função promocional dos Postos de Turismo,* Tomar: Instituto Politécnico de Tomar, Centro de Arquivo e Documentação, http://www.cda.ipt.pt/

Florida, R. (2019). *The Innovations of the Creative Class Affect a Rural Area's Fortunes*. https://www.bloomberg.com/news/articles/2019-08-15/rural-innovation-links-to-creative-class-presence

Fortuna, C. (2007). *A internacionalização da Sociologia: Notas sobre a globalização e a disciplina sociológica*, Coimbra: Oficina do CES n.º 274, abril de 2007. https://www.ces.uc.pt/publicacoes/oficina/274/274.pdf

França, C. J. F. (2017). *A Co-Criação de Experiências como Factor de Inovação no Turismo*, Dissertação de Mestrado em Economia e Gestão de Ciência, Tecnologia e Inovação, Lisboa: Lisbon School Economics and Management – Universidade de Lisboa. https://www.repository.utl.pt/bitstream/10400.5/13838/1/DM-CJFF-2017.pdf

Glup, M. (2021). *Tourism's Impacts on Local Populations* (Vol. 29). Nebraska Anthropologist. https://digitalcommons.unl.edu/cgi/viewcontent.cgi?article=1194&context=nebanthro

Gyr, U. (2010). *The History of Tourism: Structures on the Path to Modernity*. EGO – European History Online. https://d-nb.info/1020543884/34

Reichenberger, I. (2017). Digital nomads – a quest for holistic freedom in work and leisure. *Annals of Leisure Research*, 1–17. doi:10.1080/11745398.2017.1358098

Serrano, A., Gonçalves, F., & Neto, P. (2005). *Cidades e Territórios do Conhecimento – Um novo referencial para a competitividade*. Edições Sílabo.

Shaw, G., & Williams, A. M. (2004). *Tourism and Tourism Spaces*. SAGE Publications Ltd. doi:10.4135/9781446220528

Tribe, J. (2011). *The Economics, of Recretaion, Leisure and Tourism*. Elsevier. doi:10.4324/9780080890517

Turismo de Portugal, E. P. (2017). *Estratégia Turismo 2027: Liderar o Turismo do Futuro*. https://www.turismodeportugal.pt/SiteCollectionDocuments/estrategia/estrategia-turismo-2027.pdf

Urry, J. (1990). *The Tourist Gaze: Leisure and Travel in Contemporary Societies*. Sage Publications.

Zuelow, E. (2016). *A History of Modern Tourism*. Palgrave. doi:10.1007/978-0-230-36966-5

KEY TERMS AND DEFINITIONS

Attractive: Results, in tourism, in something that has interest or utility to attract travelers to a tourist destination.

Culture: The habitats, traditions, and beliefs of a country or group of people and, thus, we can consider that there is a tourist culture.

Experience: Something that someone does, sees, feels, and shares with others (or not) and integrates into their culture.

Public Administration: Management of the territory at the local, regional, and national scales through respect for the constitutional and public policies of each elected government.

Sustainable Development: The process of developing something new and causing little or no damage for the environment, society, and economy and therefore able to continue for a long time.

Territory: Land that is owned or controlled by a particular country with their public administration.

Total Tourist Experience (TTE): It results from the connection between six inseparable factors for the complete tourist experience: 1) travel; 2) accommodation; 3) restoration; 4) entertainment and leisure; 5) safety and health; 6) access to complementary goods and services.

Tourist Routes: Delineation on map of a few or middle and long distance, that tourist buying and consumes (on foot or by other means of travel).

Touristification: Planning and process of transforming a territory's endogenous resources into attractions for tourist and cultural consumption.

APPENDIX: SURVEY

This survey aims to analyse the post-pandemic trends of the components that characterize a Total Tourist Experience (TTE). The information collected in this survey is strictly confidential and the data provided will be used solely for the purpose of this research project and only aggregated results will be reported in academic publications. We greatly value your cooperation. Thank you for your time and consideration.

A. Socio-Demographic Data

1. What is your gender?
   - Female
   - Male
   - Other
2. How old are you (in years)?
3. What is the highest degree or level of school you have completed?
   - Below Secondary Education
   - Secondary Education
   - TeSP/Undergraduate
   - Bachelor's degree
   - Master's degree
   - Doctorate's degree (PhD)
4. You are answering this survey as a:
   - Professional or entrepreneur in the area of tourism
   - Student from the area of tourism
   - Tourist
   - None of the above

   4.1. If you are a tourism professional or entrepreneur, in which area do you carry out your professional activity?
   - Hospitality industry
   - Restaurants
   - Tourism events
   - Rural tourism
   - Ecotourism
   - Tourism operations
   - Another area

   4.2. If you are a tourism professional or entrepreneur, in which geographical region do you carry out your professional activity?
   - North
   - Center
   - Lisbon Metropolitan Area
   - Alentejo
   - Algarve
   - Autonomous region of the Azores

- Autonomous region of the Madeira

4.3. If you are a tourism professional or entrepreneur, how many years have you been a tourism professional?

____

B. Post-Pandemic Trends and Total Tourist Experiences

5. Do you consider that the tourist habits and interests inherent to the current pandemic context will last, in long term, in a post-pandemic era?
   - Yes
   - No
6. Indicate the relevance that, in your opinion, each of the below elements represents in a tourist experience according to the following scale: (Very Irrelevant-1; Irrelevant-2; Neutral-3; Relevant-4; Very Relevant-5)

Accommodation	**1**	**2**	**3**	**4**	**5**
Access to complementary goods and services	1	2	3	4	5
Entertainment and leisure	1	2	3	4	5
Gastronomy	1	2	3	4	5
Safety and health	1	2	3	4	5
Travel	1	2	3	4	5

7. Indicate the relevance that, in your opinion, each of the below terms represents in a tourist experience according to the following scale: (Very Irrelevant-1; Irrelevant-2; Neutral-3; Relevant-4; Very Relevant-5)

Discovery	**1**	**2**	**3**	**4**	**5**
Emotion	1	2	3	4	5
Freedom	1	2	3	4	5
Satisfaction	1	2	3	4	5
Homesickness	1	2	3	4	5
Surprise	1	2	3	4	5
Tranquillity	1	2	3	4	5

7.1. Regarding the previous question, specify other terms, up to three maximum, not mentioned in the above list and that you consider that are missing.

____

____

____

8. The future of Tourism will depend on several factors. Indicate your degree of agreement with the specified factors according to the following scale: (Strongly Disagree-1; Disagree-2; Neutral-3; Agree-4; Strongly Agree-5)

Existence of peace	1	2	3	4	5
Existence of greater freedom	1	2	3	4	5
Existence of greater public security	1	2	3	4	5
Nonexistence (or greater control) of pandemics	1	2	3	4	5
Greater investment in Education	1	2	3	4	5
Climate change	1	2	3	4	5
Growth of extreme natural phenomena	1	2	3	4	5
Artificial intelligence	1	2	3	4	5
Technological transformation	1	2	3	4	5
Financial stability	1	2	3	4	5
Greater participation of the state in the promotion of tourism	1	2	3	4	5
More free time	1	2	3	4	5
Health	1	2	3	4	5

8.1. Regarding the previous question, specify other factors, up to three maximum, not mentioned in the above list and that you consider that are missing.

____

____

____

9. Indicate the option that, in your opinion, is the most relevant in characterizing tourism trends in a post-pandemic era:
   - Tourism that implements sustainable practices
   - Wellness retreats
   - Outdoor activities
   - Local cultural exchange
   - Other
10. According to your sensitivity as a tourist, write the term that you associate with a tourist experience that you consider to be of high quality.

____

Chapter 20
Thinking Urban Heritage in Times of Pandemics:
Art Staging and Coordination

Fernando R. Contreras Medina
University of Sevilla, Spain

ABSTRACT

This chapter studies the work of the art museum exhibition during the pandemic. The confinement in citizens' homes, as a preventive measure against COVID-19, forced museums to a new virtual staging and an ordering of the works under the technical conditions of the internet. The excess of the virtualization of heritage as opposed to the rational contemplation of art is what establishes the bases of identification of the anticulture proper to capitalist consumption. Theoretical questions of art, philosophy, sociology, and politics are discussed, but only as stepping stones into the deep waters of visual culture.

INTRODUCTION: TOWARDS THE IMMATERIALITY OF THE ART STAGING

Social distancing during the pandemic also meant, on many occasions, a distancing of cultural activity from the cities. Especially for art, it meant the temporary closure of its conventional exhibition venues (museums, galleries), as well as a series of difficulties in relations with the public (Art Fund, 2020).

During the quarantine, museums had to reorganize the content generation and orient their work with a non-face-to-face public (ICOM, 2020). To do so, museums undertook digital actions that increased with social networks. This mediating activity with technologies has uncovered a recovery of artistic creativity that affects the museum's conception, visual culture, and a different perception of art in the scopic regime of virtuality (Bono, 2020; Burgos, 2021; Cultura, 2020; Ibart, 2021; Lucas, 2020; Morla 2022).

The pandemic has opened other paths to the creative imagination of art institutions. The virtual exhibition breaks with the spatial limitations of the physical exhibition. The museum space is increased, decreased, or reorganized without physical requirements; it acquires another format in a virtual network or a multiplication of cyberspaces generated by the various social networks; its extension is global, as the community of visitors and admits the online gallery connection.

DOI: 10.4018/978-1-6684-3369-0.ch020

The challenge of museums in the face of the pandemic also extended to the historical and artistic heritage of cities. Driven by the new museum concepts (open-air museum) or the museum integrated into its landscape, the cultural gaze on urban heritage also experienced the vacuum of confinement. But, on the other hand, promoting of urban heritage also avoided oblivion and the disappearance of social interest, relying on communication to show itself to the world.

The implementation of digital policies around the exhibition of art brought new challenges for these institutions and the cities themselves; inventories, catalogues, collections were digitized, as well as the face-to-face activities that are usually planned for the public, making them possible on the Internet. The UNESCO report (2020), regarding the behavior of museums in the pandemic, points out a series of digital tools based on communication platforms for their collections, social networks, the digitization of their activities (concerts, online exhibitions, or 360-degree itineraries), and even some exclusive activities created for confinement that required interaction with the public (Samaroudi, Myrsini, Rodríguez Echavarria and Perry, 2020).

The tactic of the internet outlet in the exploration of the artistic heritage of cities stopped at the idea of immateriality, and at the same time, it generated different materiality formed by computer codes, screens, interfaces, etc. (Matthews, 2021). Portable materiality that did not require the physical location that characterizes the cultural heritage inventory. In other words, the physical contracted and increased its distance from the digital, offering a useless prop and an amorphous spectacle. Technological hybridization offered a different approach to the curiosity of tourism designed from the automation of information, the mechanization of vision, and the simulacrum of action/interaction.

It is quite possible that understanding the phenomenology of this scopic regime characterized mainly by the physical closure of the spectator, who sees art through the screen, will have repercussions on its neuroesthetic conditions, the notion of plasticity of the work of art, and the limits of digital projections in the social imaginary.

Undoubtedly, a question arises in this scenario of reclusion, estrangement, and closure of the ontic coherence of art that directly connects the viewer with the materiality of the work: How does the viewer virtually delve into the spirit of the work of art? A question that arises from the evidence of aesthetic losses that make an artistic work singular, such as the breakdown of the artist's aura or the eviction of authenticity by technological mediatization (Benjamin, 2017).

The digitization of the urban artistic heritage causes a deconstruction of the classical vision of the work of art from which not only superficial technological effects arise. As a result, the artwork loses most of its empathic interests, such as the tactile exchange between the artist and the audience, the synaesthetic impetus of the work, or the internal intensity that activates the properties of art perception (light as immaterial materiality of the dynamic mechanics of the artwork).

A revisionist critique of the performance of these institutions that protect the cultural commons does not avoid thinking about the bland homogeneity that accompanies the tourist diffusion campaigns during the pandemic. The transformation of art into consumer goods that finally acquire a "trademarked article" format does not cease to worry us. The impoverishment of culture comes from the displacement of a commodified aesthetic to other cultural sectors. The artistic heritage is not exempt from this disaster, due to the commercialization and stability of art market businesses that wish to reduce tourist business risks and uncertainty during the pandemic.

Recalling the disparate character of the artistic heritage of cities (monuments, squares, open-air museums, galleries, murals, carnivals, religious rituals, etc.), the urban art experience is a kind of work in progress, intended to satisfy the Lefebvrian taste for heterotopia. Modern cities are frontier social spaces in which new revolutionary trajectories fit: "It is something different does not necessarily emerge from a conscious plan, but simply from what people do, feel, perceive, and come to articulate in their search for meaning for their everyday lives" (Harvey, 2019, p.15).

MAIN FOCUS: FROM THE IMMATERIAL TO THE UNIVERSAL IN THE NEW ART ORDER

In recent decades, the meaning of heritage has changed for citizens, as evidenced by vandalism attacks by social movements on statues representing protagonists of a history of oppression, slavery, or human suffering. Today we do not want heroes; we prefer noble ideas. Artistic activism revives structured iconophobia and an idolatrous order implemented by popular protest. At the beginning of the century, art opened a political space of consensus that renewed its function and social interest.

Looking at the artistic heritage of heterotopic groups is an opportunity for an ephemeral moment to shift to the possibility of a collective action that produces something radically different. Studying the consequences of the digitalization of art involves considering the political intangibility of the work (its virtuality), the physical alterations in the perception of social conflict, the neocolonial practices of art world institutions, the supportive behavior of the public, and even the political distraction in the face of an immaterial, energetic reality parallel to the real world and its problems. All these aspects are not interesting from a material perspective of art or aesthetic formalism. Instead, our attention is closer to the political fetishism that currently organizes cultural heritage and surrounds its memory and archive value with humanistic research. Heritage, let us not forget, is a common property that becomes a global solution without resorting to hierarchical organizational forms, since the interpretation of the artistic work opens up to human relations on a planetary scale. Ultimately, art is always a discursive arrangement of the polyphony of voices of a planetary community.

During the preventive measures of lockdown in homes in front of Covid-19, the museum loses its destiny with the disappearance of contact with the public and its significance, because the work of art is an individual and not a collective journey. The digital recreation of contemplation put an end to the deepening of the sensory experience. The meaning of art does not reach those who do not offer a specific effort required, above all, on oneself. Art is no longer an isolated refuge of consciousness. It was not during the lockdown. It did not offer a recurrent retreat for cultural illusion. The work of art languished without the strength to strike the consciousness of the multitude. From a false identity mediated by technology, the simulacrum sign that appears on the screen of digital devices is the only image that rivals another image that sought the recognition of a free expression of reality.

Artistic heritage is not only the perception of a sign from particular spaces, but also a set of outstanding narratives that organize the aesthetic, political, and idealistic weltanschauung of a community.

The revolutionary ideology of the city claimed the symbolic value associated with its artistic heritage and monuments. In this subversive ministry, the political function of the museum in the pandemic consisted of preserving the symbolic richness contained in the aesthetic dimension of the artistic work and avoiding alienation from its social destiny. An examination of each monument shows that its erection is due to its interest in history and art. For history, it is memory; for art, it is an archive. Therefore, the monument no longer remains isolated in its immediate environment. Progressively expanded, the very notion of monumental artistic heritage will refer to its physical, human, transcendental ensemble, and its symbolic consecration will come from its revelation as a world cultural asset, not only the monument itself, but also the city or the region in which it is located.

In 1889, Camillo Sitte's work (1996) entitled "Construction of cities according to their aesthetic principles, in the past and the present" aspires to also consider the framework of the built monument as a whole, i.e., the urban landscape in which the work of art is located, its aesthetic values, and contemporary artistic capabilities. The site does not conceive of "monumental objects in themselves but integrated into the fabric of their surroundings" (Recht, 2014, p. 131). Riegl (1999) also refers to such environments when he speaks of an atmosphere surrounding the monument, the Stimmung. Only the subjectivity of the narrative inscribed in the work will remain after the artistic object's conversion for its exhibition on the Internet. The political and vindicating value of memory, remembrance and the encounter with its narratives will act on the citizenry, but the other essential values of art will disappear in renouncing its ideological context.

The requirement of art institutions to subject their works to other sensory impressions concentrates the force of thought on experience, instead of sublimating itself through the aesthetics of simulation. In the virtual exhibition, an image goes beyond its status of representation; it is exhibited as a presentation by not distancing itself from sensible reality through mental operations. We live in the conviction that all possible experiences can be expressed in the language of discursive thought.

The exhibited collections and museum inventories serve as a foundation for history, and thus acquire an extraordinary character based on their sensitive specialization for the citizenry. During the pandemic, cities remain the centers of popular forces without renouncing their collective rights and the recovery and realization of the surplus value of their living space.

Reactivation of cultural heritage participated in the organization and implementation of collective and collaborative action with the citizenry (Santos & Ramalho, 2021). In the face of capitalist urbanization, the artistic heritage, being a social, political, and vital common good, has been a civil resource for union and consensus. The museums and heritage conservation institutions of cities were aware of this responsibility during confinement and reduced the uncertainty and risk of disunity that isolation and social distancing entail.

VIRTUAL STAGING AND ART MANAGEMENT IN TIMES OF PANDEMICS

During the pandemic, the circumstances of social confinement forced art to invent new imperatives of mediation and exhibition. Mediation is understandable as a general solution to the problem of consumption. Mediation and exhibition are characterizations of the function of artistic heritage. Both museums and the cultural, artistic, and historical heritage of cities illustrated by national policies for cultural development had achieved the public's general interest (Poulot, 2011).

The community's isolation revealed other particular and even idiosyncratic epistemological positions of museums. In the past, it was already known that the organization, preservation, presentation, and analysis of cultural (and natural) heritage increases its user's volume and quality of knowledge. However, there was a lack of emphasis on the phenomenon of communication and its activity oriented to the preservation of legitimate values to be shown to the public. At the most intense moment of the pandemic, museums were the social agents that ignited the resonance of knowledge and experiences. In the exhibition of objects and works that uncovered shared references, a new public scenario of political and wise aspects was consolidated simultaneously.

Between the past and the present, the virtual recovery of the city's artistic heritage stimulated the social strength of the homeland wounded by the mourning of the deaths caused by the virus. Thus, exhibitions followed one after the other in search of a new cultural policy to maintain public space. Museography ordered its art collections, urban history, the exhibition of the nature of ecomuseums, and the visibility of intangible cultural heritage. This new arrangement was based primarily on the technological possibilities of digitization (360 visits to virtual reality and video game engines in the web design) and the public's gaze through screens. The Rijksmuseum (National Museum of Amsterdam) displayed an interactive version of the work entitled "The Night Watch" (De Nachtwacht, 1642) by the artist Rembrandt van Rijn, in which the process of restoring the painting is visualized. For its part, Tate Modern (British National Museum of Modern Art in London) organized a virtual visit to an Andy Warhol exhibition and the MoMA (Museum of Modern Art in New York) offered a virtual exhibition of the photographer Dorothea Lange with a selection of texts by Lange, short videos about her work, and recorded interviews with the curator Sarah Meister.

Let us remember that in France's past, the museum that was born under the Revolution, in addition to the imperative of the catalog and the collection, was coated with a political mood, as it was part of a patriotic inventory of the country's wealth, in addition to implying democratization of knowledge and taste (Poulot, 2011, p. 19). The history of art is also the history of politics, since it imposes with its contents certain social dispositions that are even remodeled or reconstructed later by the demands of time. A successful reorientation of a current museum implies a set of approaches to the missions to be fulfilled in the community. It is not a question of inscribing the vitalization of the artistic heritage in the archaeology of collecting, but rather of contributing to the physiognomy of a country (Crouch and Lübbren, 2003). During the nineteenth century, European museums proliferated, exalting patriotic fervor. In the twentieth century, museums in the United States took on the task of representing the diversity of the multicultural, class roots (about social groups) involved in the constitution of their national identity. In the 21st century, faced with this stage of the pandemic, the new demands are predominantly social in nature, i.e., to nurture a sense of community.

At a time when the official institutions of different areas of the States (public administration, health services, security forces, etc.) demanded from their citizens the renunciation of parts of their freedom for the common good (confinement to homes, stoppage of certain commercial activities, mandatory use of masks, vaccination certificates), it was necessary to orchestrate social and political awareness in the definition of the city. The urban heritage is thus integrated into the horizon of a civil obligation of "pandemic life genres". Museums claim to be, then, an image of will in which the population can recognize its way of life linked to a territory linked to its past. The territoriality of urban heritage belongs to the logic of the community (territory, heritage, and population). Monuments, museums, and urban art reveal a pedagogical pragmatism, as their resources provide sustenance for public instruction, even above their aesthetic and sensitive function (Falk & Dierking, 2018). In these pandemic circumstances, art is considered a sociocultural factor that reconciles the human spirit with the seriousness of new civic norms.

The museum in the pandemic is the museum of mobility that tears down partitions, architectural spaces for the sake of a permanent discovery in the virtual museum scene. The virtual exhibition is a reinvention of the public purpose and community interest of the services offered by museums. The validation of the new virtual proposals of museums overcame obstacles such as their closure during confinement, the reduction of capacity during the deescalation of the number of people concerned, and restrictions on urban mobility. Recognition of the museums' work was achieved by increasing their presence on the Internet. In Spain, the newspaper La Vanguardia published the favorite types of museums highlighted by their number of virtual visits, which were art (80%), archaeological (44%), and history (32%) museums (Del Rio, 2021). In this oscillation of the public's attitude towards the new virtual arrangement of collections, catalogs, and inventories of urban heritage, the dictates of the capitalist economy and the complacency of postmodern culture are responsible.

The implosion of the virtual museum entails a multiplication, mutation, and diversification of figures more typical of a mass culture articulated by contemporary capitalism. The most severe criticism of capitalist virtualization comes precisely from the alignment of the museum institution, or the pedagogy embodied by the artistic heritage with commercial vulgarity. A malaise already expressed in the past by Clair (2011) due to the branch strategy of large museums for their expansion. This capitalist corruption contradicts the trust placed by citizens as a refuge of truth and authenticity of the past. The artistic heritage has several vocations that pose oppositions between tourism or commercial growth and its cultural, scientific, and academic legitimacy. The museology of social history, a natural heritage, or urban history ennobled didactic displays of collective cognitive adventures. The next steps were taken by offering individuals the possibility of semiotic appropriation in the collections.

Such approaches revive the intellectual capacity of social groups to maintain an original discourse on a subject and reconstruct it through artistic objects: the most "real" objects are, in turn, representations. During the pandemic, the pressure of other demands, such as the increase of tourism, the contribution to territorial planning, the reduction of cultural inequalities, or the progress in social integration, does not cease to affect the artistic heritage. Museum work and knowledge of cultural heritage affect strategic public spaces in urban renewal and the recovery of urban centers in decline due to gentrification (Rosler, 2017). Although the museum loses its attraction to the physical experience of the public through cultural mediation, it increases its social work in the community and reveals the fears raised by political claims. The timeless, the mystical, or the imaginary in art, that is, a considerable part of our artistic heritage will disappear in the face of the split of citizenship in the pandemic.

Virtualization itself participates in subjectivity in the interpretation of the works, since the museum space, like the landscape of urban heritage, obeys physical conditions that will disappear on the screen (division of the exhibition rooms, lighting, framing, the arrangement of the works, or the urban context). The installation of a work of art requires a scientific arrangement based on the classification of the works or the hermeneutic approaches of the gazes that pass through the exhibition. The materiality of heritage disappears on the screen, where there is no longer a material culture that involves the treatment, identification, and exhibition of objects as we know them. With virtualization, the physical space of the sensitive gaze becomes a set of data that generates a space for the installation of information. The new decoration of the museum environment will consist of a public sphere erected on the artifice of technology in which artistic productions are inserted in an unprecedented way. In this place, the authority of art that legitimizes the transcendence of the work is corrupted. In the digital environment, works no longer respond to what something is and what it is made of, but rather to what that work does, accepting only that form of knowledge of art.

The elaboration of a virtual public space on the networks will turn the work of art into a pretext that will be the object of discussion and commentary by connected individuals (Kist, 2020). Through digitalization, the staging of art will reconstruct the space in which private individuals meet as public and reason in public. Thus, an anthropological plot is formed that highlights the function of the exhibited objects by separating the materiality of the uses as the experience of the world and instrument of knowledge, but at the same time as social action of value. The virtualization of artistic heritage has involved successive contradictory considerations between the advantages and disadvantages of the possible "naturalization" of the gaze through the windows of the Internet, which usually respond to sociological and ideological-political models. Let us not forget that the museum or the exhibition of urban artistic heritage is a way of looking and always responds to a set of ideological norms.

The opening of a virtual place to visitors also implied a revision of the very concept of museum or heritage. The structuring of public space, even if it is on the Internet, involves discussion around a common interest and organization under a rational pedagogical program (Burnham & Kai-Kee, 2011). The different proposals offered by museums in different places during the pandemic gave rise to new approaches in a postcolonial and multicultural context. Above all, the latter was the case with museum reconstructions of historical truth. Contemporary museums exhibit their objects to provide elements that trigger social reflection through recognition, interpretation, and understanding. Postmodernity regularly questions the natural authority of museums or the social function of urban artistic heritage, such as the conservation of monuments that lose legitimacy over time under the current critical gaze of the community. The deterioration of the rectitude of art values stems from contemporary debates; it is the case denounced in 1985 by the feminist activist group Guerrilla Girls with their work entitled "Do women have to be naked to get into the Metropolitan Museum?". On this occasion, activists denounce the display of male desires in the works exhibited in museums. This was the case of the Art Gallery of Manchester, which removed from its galleries "Hylas and the Nymphs" (1896), by John William Waterhouse, a painting with young women that salivates a young man with their nudity. The pressure of social movements today, such as "Time's Up" (timesupnow.org) and #MeToo on social networks, has raised awareness of this change of attitude in conservative institutions of artistic heritage. Emerging social sensitivity is a challenge to the authenticity of the artistic heritage that is challenged by community criticism and its claims. This has sometimes involved violent vandalism of monuments or works of art by social groups, as well as the emergence of iconoclastic programs designed by institutions or social organizations that censure the amnesia of places, people, and collections in the face of the meanings of works of art.

The pandemic helped to reconfigure specific collections that lacked more committed critical attention, which is now a central point in museological reflection. Political activism sometimes helps to palliate the shortcomings of public cultural policies. Exposure to a global gaze has meant a turn in the 21st century toward museology more committed to social sensibilities and a politics of personal experience (Sandell & Janes, 2019). Both museums and the rationale for maintaining monuments in public spaces are more in line with the claim of the subject, the memory of the discriminated body, and the senses as opposed to the analytical knowledge of the past. Moreover, they will have to face mass culture and leisure consumerism that do not always understand the sense of cultural relativism and are respectful of identity politics.

During confinement in homes, the artistic heritage was staged; it was a matter of knowing where and how to present the works of art in compliance with the preventive norms regarding Covid-19, and then, with the help of technologies, we thought about how to occupy the virtual exhibition space. Then, a relevant function was granted to that contacts the virtual museum space, the work of art, and the public. The screen was the point of contact between the work of art and its social environment. The aim was to integrate the works into this virtual space to produce an aesthetic effect outside of their material space (formed by the plinth, their position in the room, their framing or pedestal, their shape, their scale, their lighting and their materiality), which would avoid the loss of the dramatization that characterizes art.

It was impossible to avoid the virtual shipwreck. The screen eliminates the historical monument's recognition, coherence, stability, and status. The advent of virtual simulation functioned as a process of transformation of art, but it also degraded the human environment, which saw the values of sensitivity attributed to the artistic heritage, especially the aesthetic ones, diminished.

In the past, the city and its artistic heritage experienced a material conversion into an object of historical knowledge due to the transformation of urban space after the industrial revolution (Choay, 1992). The ancient city then became an object of research due to the differences and in contrast to modern urbanism. The invention of virtual heritage during the pandemic is constituted against the current dominant process of urbanism and the preservation of material historical heritage. Technical progress does not favor the sacred value of the work of art but respects adequate memory by reviving particular communities' pasts through images on the screen.

THE REDESIGNIFICATION OF ARTISTIC HERITAGE IN VIRTUAL EXHIBITIONS

The space of digital networks becomes the object of comparative analysis of the aesthetic judgment. Works of art, like the object of urban artistic heritage, are reduced to digital energy that circulates on the Internet, giving body to the simulacrum. The visuality of art supplants the real object (the physical work) in its absence, and at other times the same visuality constitutes the nature of digital art. Art levels the material value of the work with digital energy in the virtualization process. This context alters the logical or conceptual categories we have of the experience of art.

The value of the experience of art diminishes when the particular disposition of the public that contemplates the work is from an Internet window. Under these conditions, the work of art has two destinies: pure destruction understood as an exercise of violence against others (the heresy of art) or violence against itself (the suicide of art). The virtualization of art drags on a different order based on the new beauty of an exhibition that invents heretical forms. During a virtual visit to an exhibition, art inverts the traditional philosophical model by presenting a visual aberration in a landscape stripped of life, as the visitor moves through a necropolis of images. In this progression towards the virtual, we witness the sequential extraction of natural life from a work of art. More than a necropolis, it could also be interpreted as a metropolis in ruins, in which we organize the pessimism of a pandemic hell, discovering a space of images in the very heart of our political behavior. The temporality of our historical life has been compromised by a virtual world that implies neither culminated destruction nor the beginning of redemption.

In this sense, the survival of the images that belong to the artistic heritage is understood because they serve as a common memory of the community; there is no source before the virtuality of the work of art, nor a subsequent horizon, only a continuous present that is updated in the very gaze of the spectator. Daily life no longer affects the contrast between the multiplicity of experiences that could be born from contact with real objects. Poor experiences or destruction of experience, finally, the semi-darkness of virtuality. The experience depends on clandestine images that escape the danger of the pandemic, useless images but transmitted to us, anonymously by the Internet, that rely on the authority of the dying museum.

Drawing on Jean Clair's (2011) term "actualization", it is possible to understand better the misery of the virtualization of masterpieces. Among the subjectivity of the powers of the museum, virtualization has disenchanted art through "a policy of brand, knowledge, and image management". While the exhibited work preserves the ephemeral symbolism of the objects it represents, the virtual gallery on the Internet banishes any significance and charm, dragged along by the immaterial economy. There is no longer any trace of its material (precise metals, shiny metals, stones), nor of its finish, its conservation, or its rarity. The aesthetics of the closed has passed to another form of subsistence for cult and culture.

- In virtual exhibitions, the artistic heritage becomes a mobile asset; national treasures are material, physical, and singular objects that can now be reproduced and distributed by digital networks in immaterial, virtual, and universal objects. A museum is no longer a place, nor are stories embodied in its objects. The artistic heritage of culture will no longer be in the network a crossroads between a certain space at a certain time.
- The monument, the museum as a temple, becomes a pilgrimage place for the tourist. Cultural objects become cult objects, which changes the ontic nature of museum collections or the artistic heritage itself. In this way, official institutions maintain the ritual of the cult of these objects by managing the flow of gazes. The virtual experience guarantees the public's right to the museum and to its popular affluence. Through these digital reconstructions, the original galleries consider the possibility of survival in material absence, increasing the rotation of flows and the rewards derived from it.
- The artistic heritage is decontextualized in the virtual exhibition when the visitor no longer has the geographical limits of the work as a reference. This is how Clair (2011: 68) recognizes it: "The work remains linked to its place and time and cannot be detached or displaced without its very meaning being ruined". The tourist visit is replaced by navigation through the Internet through huge databases that translate into a continuous visual representation. Navigation is designed to be practically useful to Internet users, offering complete information on buildings, monuments, gardens, avenues, or museums, a vision similar to documentary or street photography.
- Looking at art is an activity controlled by new technological devices. Mirzoeff (2016) explains that visual culture is now filtered by institutional powers that control all forms of data communication. The vision of art is replaced by technologically mediated screen viewing. Between the artistic creation and the viewer now stands any digital agent that decides what we can see and know through programs and screens.
- Virtualization is a form of totalization vision that adds fragments of visuality, instead of instants of eternity in a museum or in the corners of the city. Only the spirit of observation has concentrated on buildings, monuments, or masterpieces through their unity or the hierarchization with which the authorities of art arrange their value or cultural relevance (Tanga, 2021). The particular study of each work, of each figure, leads the viewer to the deconstruction of the work, the analysis of styles, and the isolation of its constituent elements.

Postmodernity gives meaning to the new meanings of art in the virtual exhibition. Simulation resembles self-referential quotations that can only be experienced under repetition. Here emerges the postmodern, which is then repeated, that follows a linear temporality and expresses itself with the recourse to the old, to construct the new, adequate, and finalized language of the museum. During the pandemic, an attempt has been made to establish emergency measures adjusted to the present that protect and preserve what was at risk of disappearing in a specific state. Heritage conservation is based on the non-mutilative reuse of the essence of the physiognomies of art. The pandemic is the premiere of a regime of circulation of objects and signs in which the authority of the institutions maintains a real political relationship with the spectator. This political relationship consists of the equal recognition of the interpretative authority of both the community and the spectator in the defiant attitude he or she maintains with the art world. In these circumstances, each viewer self-founds his or her authority from an external gaze on the screen. Finally, the culture will legitimize through a slow appropriation of the spectator's gestures in front of the art.

FUTURE RESEARCH DIRECTIONS

We could intuit that the exhibitions organized during the pandemic are projects with a more archival character of information. The archive is thought of as an open work that adopts the conceptual appearance in virtual fragments of collective and particular memory. These archives are staged in virtual exhibitions. Museums propose paths through different typologies of registers that harmonize with diverse typologies of memory. From this plurality of records, museums attempt to preserve memory through time and incorporate the objectives of their selection and the particular histories that come from the archiving process.

The spectator's interaction in the new staging of art is reduced to the impoverishment of experience. In the virtual exhibition, there is a loss of the value of the instant in which one perceives and conceives the world in terms of art, but it also implies a return to oblivion. Remembrance, the recovery of time lost in confinement, is the authentic function of the virtualization of works of art. For Guasch (2021: 251) "the proper medium of the art of the archive is digital culture or the Internet network that forces a displacement of the `archival space' linked to architecture, to `archival time' linked to the virtuality of the network". Guasch highlights a blurred logic in a culture based on digital data that makes the principles of recording in archives, as well as their histories and rituals, flexible and unstable. Digital culture is a renewal of the culture of memory and the preservation of its materials.

This scenography of art unfolds in the history/memory dialectic that questions the objects of the past and the way to activate other social, political, or cultural issues in the future. We could highlight several aspects of the innovations experienced in the staging of art and its exhibition arrangement during the pandemic:

- The flexibility of digital content also provides a means of greater learning agency. Through greater choice in decisions about what area to work on and when the digitization of the museum generates a remarkable variation from its formative function. Digital is also credited with breaking away from the linear narratives of "normal" museum learning in the physical space, facilitating a different kind of experience with artistic heritage (Noehrer, Gilmore, Jay, et al., 2021).
- The digital revolution altered the meaning of cultural work in both theory and practice. During the pandemic, it became more apparent that the functions of art organizations and managers needed to adapt and reorient their work to the possibilities of digital tools. During the Covid-19 crisis, art world administrators became aware of the impact, potential, and value of digital collections. They realized the importance of the presentation of art assets and the preservation of institutional cultural capital for the future. The presentation of art in the pandemic brought about a new reality with an enlightened horizon of ethics and aesthetics of solidarity, correctness, and altruism within museum institutions. The activist and vindicative stories of civil society found their space on the Internet platforms. The museums offered collections and exhibitions oriented to particular narratives of both the public and artists. With digital instrumentation, exhibitions were organized for co-creation in virtual spaces that allowed the inclusion of private emotion and "intimacy" in a narrative of collective memory (Sesic & Stefanovic, 2021).
- The pandemic was a global phenomenon. Additionally, the news chain became global on social networks, requiring museums to look for new opportunities and generate global solutions. The International Council of Museums (ICOM) adopted an old slogan "Museums have no borders, they have a network" dating back to 1946 on the initiative of Unesco in Paris. According to Akyol's study (2020), museums used global networks to maintain cultural awareness of the public

and to invent, in turn, cooperation programs that developed among themselves according to their needs.

- Museums ordered their educational activities following participatory creative pedagogy guidelines (Noble, 2021). In this way, museums and the virtual exhibition of artistic heritage were able to function as a collaborative space to construct stories between the institution, the artist and the public, represent narratives, and establish conversations with diverse subjects, stimulating discussion and intellectual debate. Participatory creative practices provided inclusive encounters with objects in neutral spaces. In these activities, historical injustices and inequalities could be acknowledged, opening collections and heritage sites to other individual and group sensibilities. Openness to collaboration and participation of individuals in cultural production meant offering other services that enriched the sense of representation and identity. Museums provided cultural capital when the contents had relevance and meaning related to the participants' own lives. The new cultural services facilitated a process of exploration, reflection, and discussion that resembled conversations that might have taken place during a museum visit.

The perception of the works and objects was different in the virtual tours. Kaplan (2021) recognizes that crucial aspects of experiencing art are not available when the viewer is in front of the screen. The dimension of the artwork is already a factor that provokes reactions in the visitor's body and disappears in the virtual tour. The work of art is often perceived as the size of one's own body. So other psychophysical factors are at work in aesthetic contemplation that no longer exist on the screen: the position of the work at eye level, the lighting, its placement on the wall or a pedestal. Looking, as Kaplan admits, implies locomotion. The movement of the viewer in front of the work is meaning forming. Proximity to the work allows the viewer to focus on details or, on the contrary, to cut out the images by omitting the same details.

CONCLUSION

In conclusion, technologies have played an essential role in managing art and urban artistic heritage. Virtual tours have contributed to art education. The democratization of access to collections, exhibitions, and displays through the Internet has opened up a common space for the community in which critical thinking, participatory reflection, and debate on art and culture have become possible. Digital staging has provided hermeneutic material that invites visitors to a cognitive experience of the objects. The interactive order of the information on the screens organized the educational program of the art more comfortably. However, digital staging has simultaneously revealed the fragility of aesthetic activity. The will of art in the pandemic era no longer seems to be bound by exclusivities, nor to expose reticence towards the monuments of particular epochs or civilizations. In the new digital scene, the aesthetic sensibility seems to be able to welcome all the global manifestations of art through the planetary diffusion of the museum. The inflation of artistic heritage and the fetishism of art brands (or the signatures of famous artists) are also visual pathologies of a narcissistic composure that museums return in the Covid-19 era.

REFERENCES

Akyol, P. K. (2020). Museum Activities During and After the Covid-19 Pandemic. *Milli Folklor, 127*, 72–86.

Art Fund. (2020). *COVID-19 impact. Museum sector research findings. Summary report*. https://www.artfund.org/assets/downloads/art-fund-covid19-research-report-final.pdf

Benjamin, W. (2017). *La obra de arte en la era de su reproducibilidad técnica*. La Moderna Ediciones.

Birringer, J. (2015). The New Digital Materialism. *PAJ: A Journal of Performance and Art, 37*(3), 102-110. https://www.jstor.org/stable/10.2307/26935691

Burnham, R., & Kai-Kee, E. (2011). *Teaching in the Art Museum*. J. Paul Getty Museum.

Choay, F. (1992). *Alegoría del patrimonio*. Gustavo Gili.

Clair, J. (2011). *Malestar en los museos*. Trea.

Crouch, D., & Lübbren, N. (2003). *Visual culture and tourism*. Berg.

Didi-Huberman, G. (2017). *Supervivencia de las luciérnagas*. Abada editores.

Falk, J. H., & Dierking, L. D. (2018). *Learning from Museums*. Roman and Littlefield.

Guasch, A. M. (2021). *Ensayos críticos sobre arte y pensamiento*. Akal.

Harvey, D. (2019). *Ciudades rebeldes. Del derecho de la ciudad a la revolución urbana*. Akal.

ICOM. (2020). *Cómo comunicarse a distancia con su público. Consejo Internacional de Museos*. https://icom.museum/es/COVID-19/recursos/como-comunicarse-a-distancia-con-su-publico/

Kaplan, D. (2021). "What Does It Matter Where My Body Happens to Be?" A Personal View of Online Art Museum Tours. *Journal of Museum Education, 46*(4), 531–546. doi:10.1080/10598650.2021.1987768

Kist, C. (2020). Museums, challenging heritage, and social media during COVID-19. *Museum & Society, 18*(3), 345–348. doi:10.29311/mas.v18i3.3539

Magliacani, M., & Sorrentino, D. (2021). Reinterpreting museums' intended experience during the COVID-19 pandemic: Insights from Italian University Museums. *Museum Management and Curatorship*. Advance online publication. doi:10.1080/09647775.2021.1954984

Matthews, M. (2021). Expanding from the small screen – arts practice for affective digital presence. *International Journal of Art & Design Education, 40*(3), 492–507. doi:10.1111/jade.12359

Mirzoeff, N. (2016). *Cómo ver el mundo: una nueva introducción a la cultura visual*. Paidós.

Mondzain, M.-J. (2015). *Homo spectator*. Orfeu Negro.

Noble, K. (2021). Challenges and Opportunities: Creative Approaches to Museum and Gallery Learning during the Pandemic. *International Journal of Art & Design Education, 40*(4), 676–689. doi:10.1111/jade.12380

Noehrer, L., Gilmore, A., Jay, C., & Yehudi, Y. (2021). The impact of COVID-19 on digital data practices in museums and art galleries in the UK and the US. *Humanit Soc. Science Communication*, *8*(1), 236. doi:10.105741599-021-00921-8

Poulot, D. (2011). *Museo y museología*. Abada Editores.

Recht, R. (2014). *Pensar el patrimonio. Escenificación y ordenación del arte*. Abada Editores.

Riegl, A. (1999). *El culto moderno a los monumentos: caracteres y origen*. Visor.

Rosler, M. (2017). *Clase cultural: arte y gentrificación*. Caja negra.

Samaroudi, M., Echavarria, R., & Perry, L. (2020). Heritage in Lockdown: Digital Provision of Memory Institutions in the UK and US of America During the COVID-19 Pandemic. *Museum Management and Curatorship*, *35*(4), 337–361. doi:10.1080/09647775.2020.1810483

Sandell, R., & Janes, R. R. (2019). *Museum Activism*. Routledge.

Santos, T. H., & Ramalho, G. (2021). Escuta em tempos de pandemia: Participação em museus a partir da experiência do Museu da Imigração do Estado de São Paulo. *Simbiótica. Revista Eletrônica*, *8*(2), 92–114. doi:10.47456imbitica.v8i2.36380

Sesic, M. D. & Stefanovic, M. (2021). The Ethics and Aesthetics of Museum Practices in Serbia During the Pandemic and The Digital Turn. *IPSI BGD Transactions on Internet Research*, *17*(2), 33-44.

Sitte, C. (1996). *L'Art de bâtir les villes: l'urbanisme selon ses fondements artistiques*. Seuil.

Tanga, M. (2021). Let's Imagine a New Museum Staff Structure. *Journal of Conservation & Museum Studies*, *19*(1), 7. doi:10.5334/jcms.197

UNESCO. (2020). *Museums around the world in the face of COVID-19*. https://unesdoc.unesco.org/ark:/48223/pf0000373530

KEY TERMS AND DEFINITIONS

Confinement (of Citizens' Homes): The governments of some countries adopted this preventive measure against COVID-19.

Cyberspace: Space generated by digital networks (internet) for communication.

Gentrification: Process of transformation of a deteriorated urban space with the rehabilitation causing an increase in rents.

Immaterial (Artwork): It refers to a work of art that can only be perceived by sight or hearing.

Mechanization (of Vision): Use of machines in vision processes.

Mediatization (Technological): Phenomenon resulting from the intervention of communication and information technology in human practices.

Virtualization: Process of transformation of reality and its objects into digital information.

Section 7
Confined Travels on Art Worlds and Museums

Chapter 21
Impacts, Resilience, and Creativity in Cultural Tourism and Leisure in a Time of Pandemic:
Presential and Virtual Visits to Lisbon Museums

José Cunha Barros
Tourism Laboratory, Polytechnic Institute of Tomar, Portugal

ABSTRACT

Museums were one of the cultural sectors most negatively affected by COVID-19. The restrictions introduced in the crisis period had multiple negative effects in the case of people's mobility and tourist travel. However, the temporary closure of museums also created opportunities. Museums have given greater relevance to the use of digital technology and virtual networks by increasing new forms of virtual communication with the public in the access to collections as well as the flexibility of activities. The proactivity and creative responses in moments of crisis, although of museums' initiative benefited from strategies promoted by cultural agents with contributions from the community itself. The crisis was also an opportunity to discuss and reflect on desirable changes for the museum of the future. The predictable socio-demographic changes in the public who frequent museums and the role that virtual communication will play were some of the concerns assumed by museums. The research focused on a restricted number of museums in the city of Lisbon, national museums.

DOI: 10.4018/978-1-6684-3369-0.ch021

INTRODUCTION

Heritage assets are resources that facilitate the affirmation of culture, being preserved and musealised at national, regional and local level. They provide access and enjoyment to the public, which is intended to be as diverse as possible, in order to democratise culture. Their activity is carried out through the dissemination of their collections, and through interpretation and exhibition, attracting and inviting the public to visit. They have a greater expression in urban contexts and are manifested above all by the material and symbolic value of the museum collections and by the experiences they offer, often constituting, in the case of tourism, one of the main reasons for trips and visits.

Heritage is not a static concept and has undergone not only conceptual changes, but also a widening of its scope. There has also been a significant change in the relationship of subjects with heritage. It should also be noted that society, including inhabitants and tourists, has an active role in financing, safeguarding and even in the very definition of what is understood as heritage. Authors such as Jacquot (2018) understand that heritage cannot be removed from its everyday environment for conservation or safeguarding purposes, also considering that heritage values are not opposed to the use of values (cf. Jacquot, 2018: 7).

According to the study, museums are an important part of cultural heritage and are places where real objects are displayed and communicated, and spaces where digital technologies have been used more frequently, being configured as facilitating resources for communication and interaction with real and virtual audiences (Roque, 2020).

The attendance of Portuguese museums has had an increasing trend, following the same evolution observed at European and international level (Fig.1). However, from 2020, a temporary interruption or reduction of cultural activities was observed, with emphasis on museum visits. Such negative impacts were caused by the pandemic outbreak of Covid-19, whose effects, transversal, affected all economic and cultural sectors. As museological activity has a global character and, knowing that museum attendance, as in the Portuguese case, depends significantly on tourist flows and visits, this was one of the main factors that led to a significant reduction in museum admissions, as well as internal circulation. The reduction or absence of public was not limited to traditional museums.

However, it is known that even before the pandemic crisis, museums were no longer characterized as mere repositories of pieces, assuming, on the contrary, a dynamic, proactive and creative character, providing their public with new experiences, namely through the use of new communication strategies and vehicles, as in the case of the increased use of new technologies. It is also worth mentioning that, even before the pandemic crisis, Portuguese policy makers, especially those responsible for cultural activities, such as cultural agents, were already reflecting on the function of museums and, at present, in an even greater way, on the museums of the future.

Traditional museums themselves have been diversifying the forms of communication with the public, namely through the articulation of face-to-face narratives, as well as, of other narratives that favour digital resources that often allow attracting more public, and facilitate subsequent physical visits.

RESEARCH THEME

The central question of this research is the analysis of the relevance of new technologies in the communication process between museums and their public, as a facilitator element of access to new publics through virtual networks. For this purpose, five national museums were selected in the city of Lisbon which present characteristics that fit into the universe of study that was intended to be analysed. As for the national museums' publics, the socio-demographic characterization of the publics was taken into account. The classification of the public was based on the criteria adopted in the in-depth and cross-sectional study on the museum public, promoted by the Directorate-General for Cultural Heritage (DGPC, 2015). As there is no available data on the public who are users of virtual networks, we considered, for the definition of their profile, the criteria applied to users who make physical visits to museums.

For the purposes of the study, a set of national museums in the city of Lisbon was selected, considering, besides the new technologies, the diversity of the dominant collections. In the definition of the object of study, two museums that present the arts as their main subject were included, i.e., the National Museum of Contemporary Art (MNAC) and the National Museum of Ancient Art (MNAA); the National Museum of Archaeology (MNA), the National Museum of Costume (MNT), and the National Tile Museum (MNAz) are also included. The two art museums cover a long period of time ranging from the 12th century to the 19th century, in the case of the MNAA, and from the 19th century (1850) to the present, in the case of the MNAC. Besides the two art museums, the object of study also included a set of national museums in the city of Lisbon which cover very diverse areas of the museological domain. Acronyms have been clarified in the "end notes", as well as for some of the concept presented in the text1.

Regarding visits to these museums in the period before the pandemic crisis, there was an increase and, in some cases, a tendency for the number of the respective publics to stabilise (Fig.1).

The pandemic crisis had a very negative impact on the cultural sector and, in particular, on museums. According to the DGPC, national museums, monuments and palaces recorded a 68, 7% drop in visitors in the first half of 2021, compared to the same period in 2020. With the progressive easing of health restrictions, these sectors have seen a gradual recovery in visitor numbers.

Physical visits to museums often occur within the framework of tourism activities both for national and foreign visitors, and the visit to museums is conditioned by a number of factors such as, for example, the time available which necessary affects the quality of visit.

Figure 1. Source DGPC.

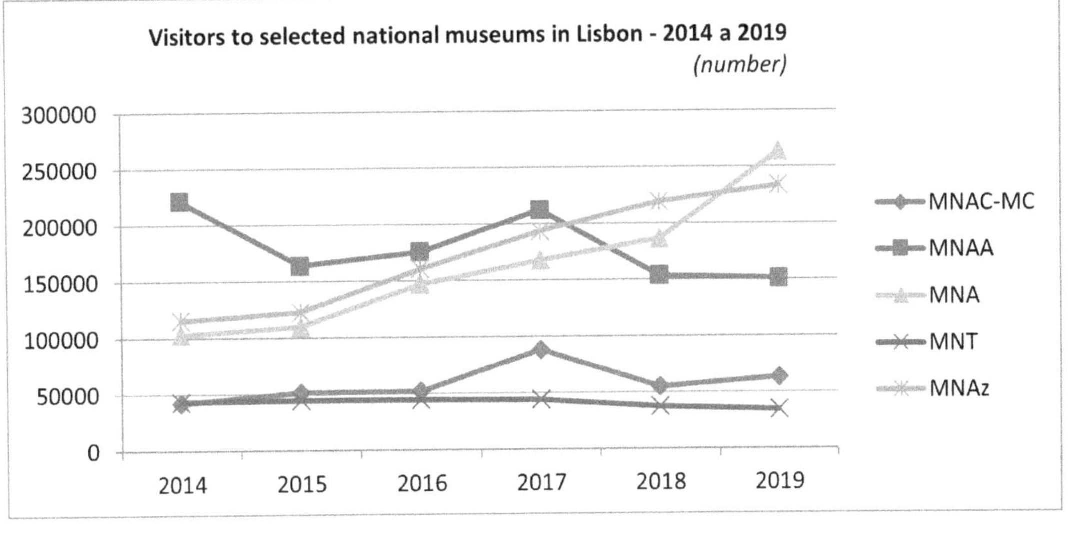

In the case of the Study of Audiences of National Museums carried out in 2015, geographical proximity was one of the criteria used as a factor in assessing the attractiveness of the museums under analysis2.

There are also issues related to the type of visit, the routes, the knowledge and interpretation of the objects.

Therefore, visits to museums are conditioned by a number of factors and, in addition to interest and motivations, the time available is also one of the main conditioning factors, reflected in the choice of the type of visit and the routes to adopt, affecting the knowledge and interpretation of objects and collections. These factors essentially affect the quality of the visit.

Besides the time available for the visit, often scarce, there is also the issue of the visitor's relationship with the exhibited works. In this case, John Berger calls attention to the visitor's incapacity to concentrate when faced with the large number of works on display, considering that they often feel "overwhelmed" (Berger, 2018:106).

In addition to "time" and "creativity", there is a set of tools that can contribute to the quality of the visit and the experience. According to Ferreira (2016: p. 80) the extrinsic and intrinsic facilitating factors should be considered. Among the former we find the "quality of experiences and knowledge", as well as the "physical, social, relational, temporal contexts", among others; in the case of intrinsic elements we find "motivation", "analytical ability", "imaginative ability", among others.

There are indeed a number of factors that can positively or negatively condition the quality of experiences resulting from museum visits. Essentially, the aim is to provide visitors with participatory experiences that stimulate "imaginative and creative thinking" and "critical thinking".3 Besides "time" and "creativity" there is a set of tools that can contribute to the quality of the visit and the experience. According to Ferreira (2016: p. 80) the extrinsic and intrinsic facilitating factors should be considered. Among the former we find the "quality of experiences and knowledge", as well as the "physical, social, relational, temporal contexts", among others; in the case of intrinsic elements we find "motivation", "analytical ability", "imaginative ability", among others.

The quality of the visit to museums is also dependent on the contribution of mediation instruments or agents. According to Ferreira (2015: 115) mediation is done through a series of cultural agents, called mediation instruments or agents. The mediation process allows questioning the object contributing to the construction of meanings. The mediation agents are responsible for the approximation between visitors and the object of the visit4. It is also intended to facilitate the visitor's interpretation and provide a participatory experience. The referred author highlights a set of mediation instruments/agents: people (in the case of guided tours), machines or texts, thematic contents, tables, websites, podcasts, blogs, catalogues, among others.

The visit to museums is made by very diversified audiences so the quality of the view depends on factors that condition the visitor's dialogical relationship with the museum, and there are, in addition to the facilitating role of mediation, personal sensory and cognitive skills of sensory visitors that should be pointed out, as well as other constraints and attributes.

Figure 2 presents an analysis of the geographical proximity of visitors to the museums under study. The graph presents the same configuration for all the museums under analysis although there are some differences regarding the frequencies. The dominant public is composed by foreign tourists, with particular relevance for the MNA-AC and, although the MNT also has foreign tourists as its dominant public, it is the institution which presents the lowest frequency. Regarding the Portuguese public it should be noted that the museums under analysis are visited by a very significant number of local publics, especially local visitors.

Figure 2. Source DGPC/CIES-IUL, EPMN, 2015

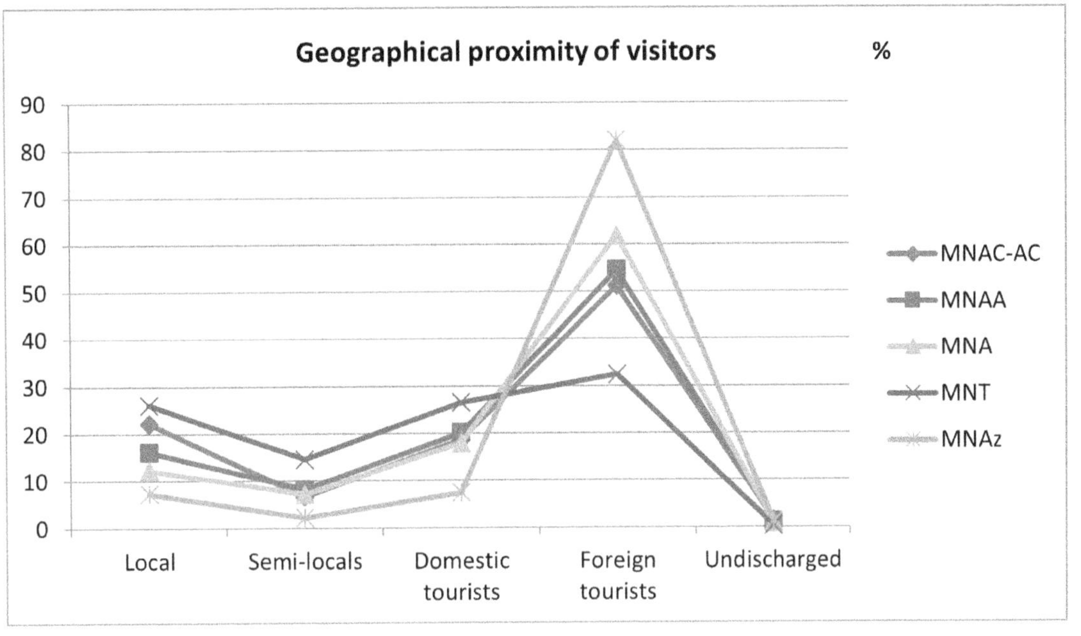

RESEARCH OBJECTIVES

The study constituted a reflection within the scope of world culture, i.e., according to Lipovetsky and Serroy, about the expansion of communication, information and mediation (Lipovetsky and Serroy, 2010, p. 14), having as general objective not only the analysis of the impacts caused by the pandemic in the museums' public, considering tourists and other visitors, but also the creative actions in pandemic period promoted by each of the museums and, also, the joint initiatives, in network, with the contribution of their addressees and other cultural agents and the community itself.

The research is based on the following theoretical foundation: the impacts of crises on museum activities are activating factors of self-assessment as they prompt the promotion of creative and innovative activities that contribute to the acceleration of change.

As a specific objective, the author sought to assess whether virtual visits can come to integrate the strategic objectives of the museums under analysis, due to the importance they assumed in the crisis period and, subsequently, and can contribute to the qualification of the face-to-face visit and also to the affirmation of the public attending virtual networks.

The author also established as specific objective the analysis of the community participation, case of friends and volunteers, in the programming and parallel activities promoted by the museums under study, with emphasis on the initiatives carried out with recourse to new technological means, by face-to-face and virtual ways.

METHODOLOGY

In the research, the author adopted the case study as a research method. Within the scope of the case study, he opted for an analytical approach taking into consideration the unit of analysis and the context. The museological activity, considering the particular case of the national museums of the city of Lisbon, constitutes a well delimited but complex "phenomenon" that was intended to describe, explore and explain.

Based on a set of national museums in the city of Lisbon and, in a critical sample (of convenience), the study established, as an object of analysis, the museums' publics, with emphasis on tourists5, on the existing technological resources, on virtual visits in pandemic times, and on the importance that virtual visits assumed in pandemic times and in the future.

Besides the publics, the research aims to verify if the existing technological resources in the museums were important to face the critical periods raised by the pandemic and, if the digital communication strategies used raised the increase of the virtual visits, both in the periods of closure and in those in which the physical visits were very limited.

The author considers that the use of a non-probabilistic sample does not as a rule allow great generalisations regarding the results obtained. However, he understood that in this type of research it is possible to find valid results.

Inter-methods triangulation was carried out by combining several different research strategies that were applied to the same empirical unit.

The work started with bibliographic and documentary research, based on studies about museums' technological resources, audiences and virtual visits. When accessing the museums selected as research object, participatory observation was privileged, as well as the realization of (semi-structured) interviews with people in charge of the museums under analysis. Since the selected museums provide virtual visits to access the collections in their permanent exhibitions, virtual visits were also made to each of the museums.

In addition to the documentary and bibliographical research, and the empirical work carried out in the museums, the "Study of Audiences of National Museums", promoted in 2015 by the Directorate-General of Cultural Heritage, was of significant importance to better contextualise the institutions under analysis and the socio-demographic characteristics of their audiences.

The study had as main instrumental resource the questionnaire survey, having as object the public of fourteen museums of the Portuguese mainland, under the DGPC tutelage, selected for this purpose, with collections of different types. The five museums included in this study participated in the survey. It was considered, for the purposes of the study, the national and foreign public aged fifteen or over. The sample had a probabilistic character, being representative of the universe of each museum included in the project. The samples considered in the five national museums in this study were: 1142 questionnaires at MNAC-MC; 1010 at MNAA; 1817 at MNA; 731 at MNT; and 1194 at MNAz. The field study began in late 2014 and lasted until the end of 2015, with the results published in 2018 and 2019.

NETWORK SOCIETIES AND VIRTUAL COMMUNITIES

The confinement raised the involvement of the international community through the production of a multiplicity of texts that help us to better frame the issue of the impacts of Covid-19 at an international level and the responses and the promotion of initiatives aimed at the 'construction' of societies more able to face crises and more sustainable. This was the case of the United Nations Educational, Scientific and

Cultural Organisations (UNESCO) which, with the contribution of its member countries, has published a series of reports and guiding texts related to the pandemic crisis.

Among these documents, the report published by UNESCO in 2021 on the issue of museums entitled "Museums Around the World. In the Face of Covid-19" and which was fundamentally based on the results of a questionnaire survey covering 87 member states.

Within the aspects pointed out by UNESCO regarding museological activity, it should be noted the asymmetric distribution of museums in the various regions of the world.

According to the aforementioned UNESCO report, 61% of museums are located in Western Europe and North America; 18% in Asia-Pacific; 11% in Eastern Europe; 8% in Latin America; 0.8% in Africa; and 0.7% in the Arab States.

The 2021 report follows on from a similar document produced by UNESCO in 2020, thus following the emergence of COVID-19.

It should be noted that UNESCO attaches significant importance to museums and attaches particular relevance to the use of new technologies. As a matter of fact, already in the Recommendation dated 20 November 2015, concerning the protection and promotion of museums and collections, as well as their diversity and their role in society, particular emphasis was given to the growing influence of new technologies and their potential for the promotion of museums.

The aforementioned 2015 recommendation follows other UNESCO guiding measures, such as the "Recommendation Concerning the Most Effective Means of Rendering Museums Accessible to Everyone", which deserves special emphasis for its emphasis on museums as intellectual and cultural centres in communities, on audiences whose museum attendance should be accessible and encouraged, especially considering the working classes, with particular relevance also being given to the educational function of museums.

With reference to 2020 it should be noted that around 90% of the total number of museums (estimated at 104000), i.e. more than 85000 establishments, were closed in that year for more or less brief periods. However, even those museums that were not closed saw their activities negatively affected by the fact that there was a very significant reduction in the mobility of people, particularly in the case of tourism.

It was also found that 83% of the Member States that responded to the survey adopted, in 2020, very differentiated total or partial closure measures, covering periods ranging from less than one month to one year. With regard to the segmentation of the measures taken in 2020 by the national museums of the member states that participated in the survey, it was found that 29 countries opted for total closure, while 43 countries opted for partial closure, with 14 countries not closing their museums and one country not mentioning museums. This closure registered an average value of 150 days for the year 2020. In addition to the national museums, the survey also included the other museums in the member countries which, in the vast majority, chose to close down in 2020 (71%). In 2021, as the pandemic crisis evolved, museums had to observe new restrictions, with closure measures that affected 43% of the museums in the member countries that responded to the survey.

The confinement served to make virtual reality and the Internet come to be considered, by an increasing number of people, as an integral part of everyday life. In this case, Manuel Castels stated in a reference work of his that "the internet is an extension of life as it is in all its dimensions and modalities" (Castels, 2004:147). Also, according to Castels, "social activity in all its diversity has appropriated the internet, although this appropriation has specific effects on this social activity ..." (Castels, 2004: 147).

Transnational Hyperculture

There is an asymmetric distribution of heritage and museum resources on the Portuguese mainland, with the coastal areas presenting a more significant concentration in terms of resources and attendance.

It should also be noted that, on the coast, the Metropolitan Area of Lisbon, although it does not have the largest number of museums, is the continental region whose museums have the highest affluence, especially the number of foreign visitors (Single Table).

The pandemic did not allow Portuguese museums, especially in critical periods, to promote activities based on the physical presence of the public. So their managers attributed greater importance to virtual communication, in cyberspace, registering not only a more intense use of virtual communication, but also a significant increase in the respective national and foreign audiences and, consequently, a more significant number of accesses to museums using the Internet. It was also proved that, in periods when they were not available for physical visits, the museums under analysis dedicated more time to the production and publication of online contents, being possible to maintain the connection to the public and, by diversifying the offer, to raise online visits of visitors, even by non-regular visitors.

As mentioned, the pandemic crisis caused negative effects and the culture sector was one of those that faced more difficulties. The mobility of people was significantly restricted and, in the international chain, Portugal was no exception. In the case of tourism there was a sharp drop. INE statistics show that, in the period between 2019 and 2020, there was a very significant drop in tourism demand translated into decreases of -74.9% abroad and -13.4% in the national market.

The (single) table presents global data regarding the attendance of Portuguese museums, whose results already reflect the effects of the pandemic crisis.

The effects were necessarily differentiated given the museums' geographic distribution, the indicators regarding demand, and the fact that public museums have a greater capacity to respond to crisis situations. As regards the geographical distribution, it should be noted that most museums are located on the continent, totalling 400 museums, with the other 36 museums in the Autonomous Regions of Madeira and Azores.

Within the mainland, the northern regions of the country are those with the most museums, totalling 237 (59.2%). The Metropolitan Area of Lisbon (AML), which is the macro-region where the museums under analysis are located, has a total of 87 museums, corresponding to 21.7% of the total. However, although the aforementioned region does not have a very significant number of museums, it nevertheless has a dominant position in terms of attraction capacity and developed activity. It is the region that generates almost half of the visits to museums that take place in the continent, i.e. 48.9%.

The very significant attendance of the museums in the AML is necessarily associated with the tourist activity promoted in the region, since it is known that the physical attendance of museums is made by tourists. In the case of the Portuguese mainland, the regions of Lisbon and the Algarve are those with the most significant overnight stays, according to data from INE. In 2019, from January to November, they recorded overnight stays of around 17 and 20 million, respectively. In the same period of 2020 the number of overnight stays in the two regions was around 3 million, registering in 2022 a trend towards recovery.

In the case of the Portuguese, this means that the number of nights spent in tourist accommodation in Portugal fell very significantly, with a very significant drop in the number of nights spent in the country as a whole.

In the critical phase of the pandemic there was a great resilience on the part of Portuguese museums. In this case, the number of museums that had to definitively close their activity was very limited, even considering those of a private nature (in general with less capacity to respond to crises6.

Although we have analysed a restricted set of museums in the city of Lisbon, national museums, the research was conducted taking into account that they, as cultural resources, are part of a broader framework that Lipovetsky and Serroy called hyperculture-transnational or, world culture (Lipovetsy and Serroy, 2010).

From an anthropological point of view, Lipovetsky and Juvin consider that world-culture means “a new lived relationship with distance, an intensification of the consciousness of the world as a planetary phenomenon, as totality and unity” (Lipovetsky and Juvin, 2011: 20).

It will also be important, to pay particular attention to the future, considering not only the great unpredictability associated with modern and postmodern societies, as well as the fact that the idea of future is alien to most models of culture, as Appadurai points out (2013:226). In this case, Appadurai considers that “we need to build an understanding of the future through the study of the interrelations between the three human concerns that shape the future as a cultural fact, i.e. as a form of difference: imagination, anticipation and aspiration” (2013:359).

Table 1. Museum visitors by regions (NUTS II) 2019.

geographic range	Number of museums	Visitors of which are			Visitors	
		Total	Inserted in school goups	Foreighers	Free admission	Temporary exhibitions
Portugal	436	19777691	2011659	10342761	5900297	12134568
Mainland	400	19273085	1932396	10103457	5733250	11824497
North	115	6002853	613753	2976824	2041654	4167428
Center	122	2260630	357507	612021	920154	1614148
Metropolitan área of Lisbon	87	9425594	814059	5717479	2203061	5095443
Alentejo	60	956253	106676	392268	227560	659972
Algarve	16	627755	40401	404865	340821	287506
AR Açores	18	291472	47231	125886	122563	226553
AR Madeira	18	213134	32032	113418	44484	83518

Source: INE - Museum survey, 2019 (own elaboration)

The pandemic made it impossible for Portuguese museums to promote, in certain periods, activities based on face-to-face audience criteria, so an increased importance was given to virtual communication, in cyberspace, through the increase of virtual visits to Portuguese and foreign visitors. In this case, even in periods of closure, it was possible to maintain the connection with the public and, through the diversification of the offer, to encourage visits by non-regular customers.

Still according to Lipovetsky and Juvin, cinema was one of the main determining factors for the affirmation of world culture at the beginning of the 20th century. They associate the emergence of the "modern world culture" to the affirmation of cinema, with emphasis on American films whose exportation allowed, among others, not only to be seen by audiences from all continents, but also to use a language of easy diffusion, which allowed mass consumption, and the simultaneity in the diffusion of images (Lipovetsky and Juvin, 2011: 61).

They also consider that it was television, in the second half of the twentieth century, that imposed "the universe of the direct image, the instantaneous, the exclusivity, the insignificant. to the same information at the same time". Television changed the world, the political world, political communication, advertising, leisure, the world of culture..." (Lipovetsky and Juvin, 2011: 61).

It is also important to note that in the new reality, "global space-time is further reinforced by the great risks and catastrophes that accompany hypermodernity and ignore the borders of nations..." (Lipovetsky and Serroy, 2010: 22).

Hyperculture creates a new form of tourism. This is the hypercultural tourist, so called by Han who, as he mentions, "moves from one here to another here (Han, 2019:76). It is also worth noting that Han considers that globalisation creates more than a global here, in that it delocalises and distances the there (Han, 2019:79).

In the hypermodern era, culture has acquired greater cultural importance and centrality character unlike in the previous phase of modernity (Lipovetsky and Serroy, 2010: 31).

Technological Resources

The use of digital technologies in museums is a fairly recent reality and with use in restricted areas. However, it is a resource that intends to become an integral and essential part of the museum of the future, facilitating the insertion of museums in digital networks, that is, in the virtual society. In this case, Maria Isabel Roque, based on Andreacola, considers that "digital technology began to be used in the museum in a more expressive and consistent way since the last decade of the last century, but very limited to the development of inventory databases and collection management" (Roque, 2018: 21).

New technologies can make different contributions to museum visits. Applications on visitors' mobile devices are one of the resources that have been used in physical visits, as an alternative to guided tours. In turn, online visits (virtual visits) to the permanent exhibitions offered by museums can complement the "real" visits or even become an alternative, especially for those who, for economic reasons or physical distance, cannot access museums.

The five museums under analysis, all national museums, integrated the set of 22 institutions participating in the project "Portugal, Art and Heritage", which resulted from a partnership between "Google Arts & Culture" and the Ministry of Culture, through the Directorate-General of Cultural Heritage. The launch of the said project took place on 15 January 2019.

From the date in reference, the possibility of direct connection to access the collections of museums, palaces and national monuments under the responsibility of the Directorate General of Cultural Heritage became available through the Internet.

The Internet is not only a resource that facilitates access to national museums. It is also worth noting the use of the Internet as a means of information frequently consulted before visiting museums, as is the case of visitors living in Portugal.

One of the platforms that allows online access and visits to museums is "Matriz-net". It is also an important resource that facilitates virtual visits, but with different characteristics and purposes from those provided by the "Google Arts & Culture" platform. According to information from the DGPC, this technological resource is an online collective catalogue that covers a wide range of cultural assets, including 34 museums that are members of the Portuguese Museum Network, including the museums included in this research. In addition, this catalogue, which has been online since 2002, is a vehicle for visiting over 100,000 selected cultural assets. Access can be guided and the search can be made by museum, theme, author, exhibition and date.

SOLUTIONS AND RECOMMENDATIONS

The restrictions raised by the pandemic had very significant impacts on museums, as was the case in the city of Lisbon. There were periods of temporary closure of museums, reduction of activities and limitation of accesses, teleworking and "mirror-working", among other restrictions. However, the national museums in Lisbon continued their activities, despite being made up of reduced teams and, as mentioned by the author, virtual communication was the process that allowed and facilitated contact with the public and access to new audiences.

The first confinement resulted from a measure enacted by the government to halt the spread of Covid-19 in the period between 14 March and 17 May 2020.

The second period of confinement (with closure of museums) took place in the period between 15 January and 9 April 2021.

With reference to 2019 (INE data), and in an analysis by regions, it can be seen that a significant majority of visitors to national museums, around 4/5, preferred to attend museums located in the Lisbon Metropolitan Area (47.7%). This confirms the asymmetry observed in the Portuguese territory. As for the other regions of the Portuguese mainland, the distribution of visitors was as follows: in the North region (30.4%); in the Centre region (11.4%); in the Alentejo region (4.8%); in the Algarve region (3.2%); in the insular (autonomous) regions of Madeira and Azores (2.6% in total). Considering foreign visitors, it can be seen that more than 5/6 of the visitors attended the museums located in the A.M. of Lisbon (55.3%), while the North region registered a less significant attendance (28.8%).

In addition to the aforementioned regional asymmetry, it should be noted that even the museums located in the Lisbon Metropolitan Area show a lower frequency in some periods of the year. The available monthly statistical data allow us to see not only that the frequency of the museums under analysis present large variations throughout the year, but also that they are coincident in all museums under analysis (Fig. 3).

Since virtual visits to museums assume an increased importance, it is increasingly important to collect and process statistics on virtual visits. The systematic collection of statistics on virtual visits will allow museums not only to have a better perception of the users of the various digital resources, but also to promote and manage this communication vehicle in the most appropriate way.

New technologies allow museums to diversify their offer and address a wider public and even make museums accessible to those who are not regular museum goers. In agreement with other authors, Isabel Roque when analysing the museological object in the museum and in a virtual environment considers that the object, in the museum, keeps its unique and original character which gives it authenticity. In a virtual environment, the object exists through its virtual reproduction, but the information associated with it and the interaction that allows, contribute to the perception of a genuine experience and confer greater authenticity to the representation (Roque, 2018:28).

When talking about museological object one has to refer, necessarily, to the subject. In this case, the actual or potential visitor. In this case, Roque considers that museums have been giving greater relevance to the user who, as he says, presents a diversity of characteristics, referring in particular to their skills, interests and expectations (Roque, 2018:24).

Figure 3. Source: DGPC/DMC (Own elaboration)

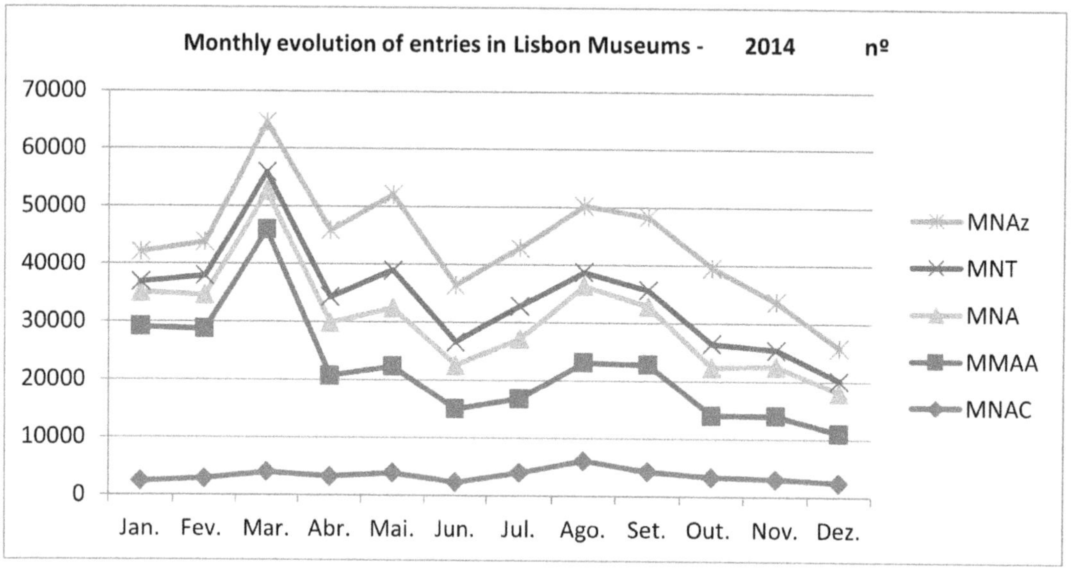

INE's statistics also allow us another type of analysis on museum attendance and on the impacts caused by the pandemic on Portuguese culture, with particular reference to museum activity. This analysis is based on absolute values. According to the National Statistics Institute (INE), Portuguese museums recorded a total of 19.8 million visitors in 2019, which represented an increase of 283.6 thousand visitors compared to 2018.

In relation to foreign visitors, tourists predominate, whose number reached the figure of 10.3 million (an increase of 1.1 million compared to the previous year), and who represented a percentage of 52.3% of the total number of visitors to museums. However, 2019 corresponded to the end of a cycle of growth in the visitation of Portuguese museums and, in particular, of the city of Lisbon. In 2020 occurred the pandemic crisis caused by COVID-19, which also had negative impacts on the operation of museums and their attendance (Fig.4). The graph shows the presence of the Portuguese public (P) and also foreigners, registered as others (0).

Regarding the museums under analysis, based on the data on "Audiences of National Museums" it was confirmed that there is a predominance of female visitors (57.7%) while male visitors were less expressive (41.7%).

The frequency by age of the Portuguese and foreign visitors presents some generic characteristics that can be pointed out. Considering the visitors in general (Portuguese and foreigners), the central value is occupied by the visitors from 35 to 44 years old. The extreme classes, i.e. 15-24 and 65 and over are the ones that present a lower frequency.

The aforementioned results suggest that the educational activities promoted by museums and those which are promoted with the elderly, the senior class, as their preferred clientele, are of the greatest interest for museum programming. In the case of the elderly, more and more relevance is given to active ageing, and museums can play a very important role. Anyway, considering specifically the Portuguese publics of the museums under study, visitors classified as belonging to the "local" and "semi-local" categories presented a very significant weight when compared to the total number of museum visitors.

Figure 4. Source: DGPC/CIES - IUL, EPMN, 2015 (own elaboration)

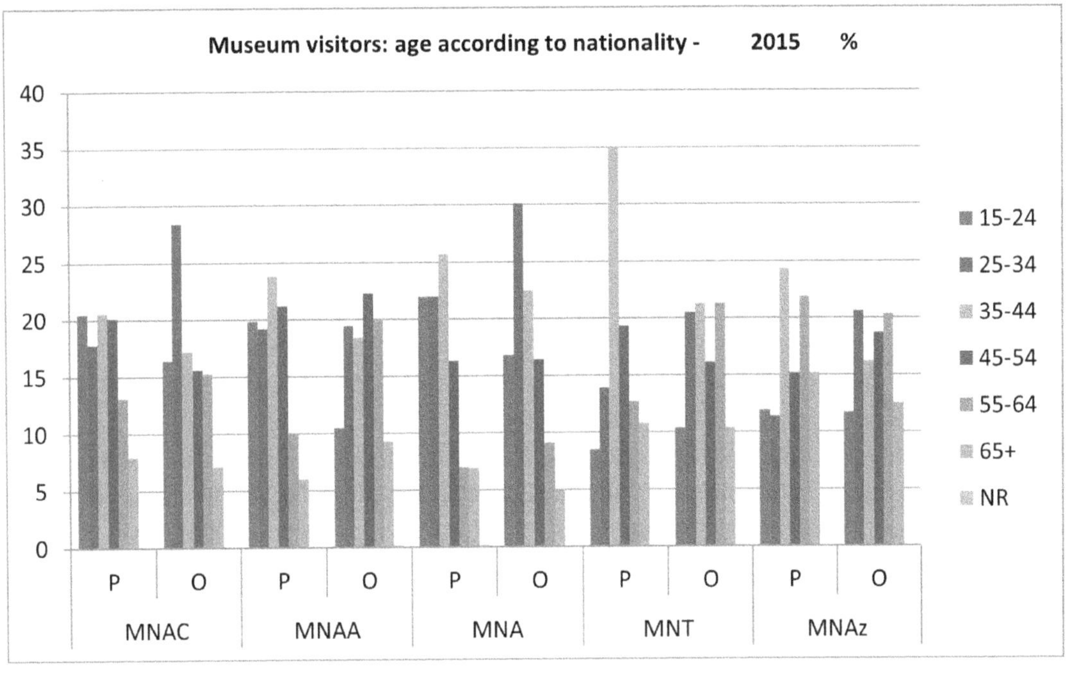

In addition to the museums' own means for online visits to their collections, two platforms deserve to be highlighted: the two online platforms already mentioned. As already mentioned, they allow, in one case, access to the collections of museological spaces and national monuments and, in the other, access to the museums' online collection catalogue, namely those under the responsibility of the Directorate-General of Cultural Heritage and the Regional Directorates of Culture. The first case is Google Arts & Culture and the second is the Matriz.Net platform.

Confinement in the cultural sector was not limited to museums. In fact, museums, monuments and palaces were closed between 14 March and 17 May 2020, after the first closure, and from 15 January to 9 April 2021. During these periods, particularly in relation to museums, there was an increase in contacts by museums and also in online visits by the public. Museums increased the maintenance and use of their respective platforms, such as Facebook.

In the case of the National Museum of Contemporary Art - Museu do Chiado (MNAC-MC), before the two confinements, they carried out very few online activities. With the pandemic and the confinement, they devoted special attention to contact with the public through virtual platforms. In particular, a very significant increase in the number of followers on Facebook was observed. Based on the data shared by the museum, it can be observed that, in the first confinement, the number of followers on Facebook went from around 5,000 to 25,000 and, in the second, from around 25,000 to 130,000. Through its Youtube channel, it also had an exponential growth reaching a total of 45,000 people worldwide.

The National Museum of Costume (MNT), like the other national museums analysed, took advantage of the two years with greater incidence of crisis to dedicate more time to the publication and production of online contents, also responding to a greater demand from the public and the greater interest shown by the promotion of the digitalisation of Museums and collections. In this case, it was confirmed that the Museum, as a result of greater visibility and the respective online offer, recorded an increase in visitors and virtual interaction/communication on Facebook and Instagram, and it is known that many visitors used the links published to access the pages of Google, Matrix and the Museum's website.

In the case of the Google Arts & Culture Platform, it was confirmed that the Museum in question has 12 online exhibitions available (in Portuguese and English), 551 photographs and catalogue cards of pieces and 2 Street View visits. During the year 2021, according to information provided by the platform, more than 10,000 visitors from different parts of the world were recorded, with particular reference to Europe, America and the United States of America, visitors who used tablet, mobile phone and computer to access, directly through the Google page of the Museum or, alternatively, through the link published on Facebook.

In the case of the National Tile Museum (MNAz), one of the online activities carried out in the wake of the pandemic crisis was the virtual guided tour for groups, by appointment, accompanied by a Museum technician. This initiative is mainly aimed at people who cannot visit the Museum, particularly when they live in more distant places.

The author considers that virtual visits can not only raise the interest of museum audiences, but also reach new audiences and complement the visits made by Portuguese and foreign publics, which in both cases are characterised by a very short duration, i.e. classified as being predominantly very quick (11.6% and 13.8% respectively) and quick (41.9% and 44.8% respectively).

The impacts caused by the pandemic have led many cultural actors to promote a variety of reflections and publish studies aimed not only at assessing the impacts on the cultural sector, but also at finding models and strategies for dealing with unforeseen situations in the future. The International Council of Museums for Portugal (ICOM-Portugal) was one of the public interest entities dedicated to issues related to international policies for Museums that promoted some initiatives.

In particular, reference should be made to the questionnaire promoted by that institution under the title "Museums at the time of Covid 19", carried out in 2020, applied during the period between 24 April and 18 June, which covered 38 museological institutions, among which 54% of participants were not members of ICOM: 35% in the Lisbon area; 32.4% in the North; 13.5% in the Centre; 11.0% in the South; and 8.0% in the Madeira Archipelago without responses for the Azores Archipelago. The survey covered museums under the jurisdiction of the State and others that do not depend on the State (private, associative, or others).

Among the survey's results, it should be noted that 74.8% of museums closed completely during the period and no activities were carried out in their physical space. With regard to new technologies, it was confirmed that museums paid special attention to virtual communication with their public. In this context, 89% changed their communication strategy and invested significantly in digital communication; 85% through Facebook, 69.7% through their website, 45.5% using Instagram and 42.4% publishing on Youtube.

The content that was made available was mostly created on the occasion, with 24% resorting to pre-existing material, combined with others that already existed. This reinforcement in communication was essentially based on internal resources (85%), with only 15% of the cases resorting to external professionals. Also in relation to new technologies, it was mentioned that the massive use of digital content in the period of confinement accelerated the use of this resource, while recognizing the need to integrate these changes in the daily life of museums.

During the periods when measures were implemented to reduce the possibility of contagion by restricting the concentration of people in enclosed spaces, measures based on the assessment of the carrying capacity for the areas in question.

The issue of carrying capacity and sustainability is a problem that has been discussed and that museums will have to address and solve in the future. In the case of actual visits, one of the main indicators of museums at present is based on the number of people who attend them. However, when it comes to audiences, the massification of museum attendance is very questionable. In the work of Bourdieu and Darbel, already mentioned, on the public of European museums, considered even today as a work of reference, it was proved that both the public of the museums under study, members of the upper and middle classes, were more interested in visiting museums with few people, with frequencies of 70% and 60% respectively (Bourdieu & Darbel:232).

As for the lower classes, the dominant frequency was between those who preferred visits with many people and those who preferred visits with few people, the frequency being 39% in each case (Bourdieu & Darbel:232).

It was also found that museum audiences currently continue to consider that the occupation of museum spaces by fewer people contributes to the quality of visits. However, the issue of revenue must be part of the equation.

One of the dimensions of the research was the characterization of virtual and face-to-face museum visits, considering, in both cases, situations in which the visit takes place with recourse to mediation. Without minimising the role and importance of mediation in museum visits, Gielen considers that museum visitors are, particularly with regard to the historical dimension, active interpreters. He states that museum visitors "observe by means of a viewing grid or schema that he has acquired through education or socialization" (Gielen, 2009:91). Indeed, to this end, he considers that active skills manifest themselves first, and visitors then passively enjoy the pleasure of sight.

The mentioned author identifies a set of elements that characterize the visit: "the relation between subject and object, between the viwer, reader, listener and the museum display, consists of a double synchronous movement of active passion and passive action" (Gielen, 2009:92). In the development of the research the author concluded for the complementarity between face-to-face visits and virtual visits promoted by museums, by the existence of common elements, considering however that their main characterizing elements are diverse.

The increase in the use of new technologies and virtual visits to museums was triggered by the emergence of the pandemic crisis raised by Covid-19. However, considering the national museums under analysis, supervised by DGPC, the author concluded that actions increased as a response to the crisis, based on reflections promoted previously and guidelines considered essential in museums of the future. In this case, the study carried out by the "Museus in the Future Project Group" was very important in the actions implemented in the crisis period, even having a "premonitory" character of what was to come. In fact, the project itself, which was initially going to focus only on museums, also covered, by decision of the supervising minister, the palaces and monuments.

It was pointed out and recognised that museological and heritage entities are not static and fixed realities, but have a great capacity to adapt to changes and crises. In the context of the long period in which they operate, museums build the future. The project and interaction were structured along several axes. They are, considered as a unifying axis and, networks and partnerships, as well as digital transformation, in this case integrating the two transversal axes.

The conception of the museum of the future is not consensual and is open. Indicative of this is the proposal presented by ICOM at the 25th General Conference held from 1 to 7 September 2019 in Kyoto, Japan, to replace the definition of the museum, which was not consensual and was not approved. According to Scheiner, the aforementioned definition, while taking the traditional museum as its starting point, did not make explicit what is to be understood by a museum, merely presenting a set of qualities and attributes intended to integrate the museum (Teresa Scheiner, 2020: 51).

For Scheiner, the contingency of the pandemic "also demonstrated the agility of many museums in re-articulating their presence in the digital environment, leaving materiality in second place (even if temporarily), to make their collections available in the form of journeys viewable on social media" (Teresa Scheiner, 2020:54-55).

CONCLUSION

The international networks had a very significant importance during the pandemic period due to the multiplication of international meetings, as was the case of the cultural sector, often online, the publication of a multiplicity of reports, not only of an evaluative nature, but also the presentation of guidelines aimed at achieving sustainable future objectives, a greater ability to face crisis situations.

The study showed that digital technology was very useful during the pandemic period and was used more frequently in all museum operations with very positive impacts.

The experience of national museums, as was the case of those under study, in the use of new technology in physical visits by the public, such as mobile devices, was very useful in providing information also in virtual visits.

In the case of the museums under study, the author concluded that hyperculture and virtual networks assumed a growing importance during the pandemic crisis, becoming the main way of keeping virtual contacts with their publics and allowing access to new publics.

Scheduled activities were reactivated due to the resumption of travel, the revitalisation of tourism and economic activity in general. Museums acquired new skills as a result of the experience gained and virtual communication now plays an important role, not only as a complementary activity to physical visits, but also as a factor that can generate interest and motivation for subsequent physical visits to museums.

National museums in the city of Lisbon showed great resilience during the pandemic crisis, even in critical periods caused by the significant increase in contagions. This was due to the fact that these public museums depend on the state budget for their operation. Therefore, these museums have a higher degree of economic autonomy compared to private museums. In order to maintain a regular activity, these museums depend more significantly on revenue.

Mediation is a customary resource in the actual guided visit of the public to museums. During the pandemic crisis, the use of mediation in virtual visits became part of the methodologies used by museums, which also started to promote virtual guided tours in groups. This practice facilitates public access to the collections and allows promoting creative strategies to create new audiences.

The museums under review have a very close connection to the community and in addition to providing guided group and autonomous physical visits, they have explored, directly or indirectly, virtual communication, through virtual guided tours, online visits to the collections through platforms such as Google Arts & Culture, Matrix-net, also giving volunteers and friends the opportunity to participate in their programmed activities in real and virtual ways.

The pandemic crisis allowed and provided opportunities for museums to diversify and make their activities more flexible by acquiring new skills to promote virtual communication, to attract new audiences and to deal with unpredictable situations.

The virtual visit to museums has assumed a growing importance, this communication process being a complement to the physical visit, a form of enrichment and information and even an alternative to the physical visit and a pretext to navigate in hyper cultural space.

Hyperculturality has led to the emergence of a new form of tourist, the hypercultural tourist who, by moving in the hypercultural space, will tend to have increasing access to cultural goods and, in particular, to museums.

FUTURE RESEARCH DIRECTIONS

With globalization, societies have become increasingly intercultural, which often implies a tendency for the loss or decline of cultural references, i.e. identity values. Museums play a fundamental role in allowing the public, regardless of age, to encounter and participate in experiences that stimulate critical thinking and the construction of meaning. Museums can be considered as cultural resources and also as identity resources.

As a rule, mediation facilitates the visitor's relationship with the museological object. Therefore, it is through mediation agents (people, machines, texts) that the visitor integrates and contextualizes the museum pieces and objects. New technologies allow any individual, regardless of where they are, as long as they have the necessary technological resources, to access a very significant number of museums worldwide, visit the collections and enjoy creative cultural experiences. Although museums are considered as a privileged object of study, they still constitute an area of work that should be explored.

The issue of the relationship and interdependence between virtual and real visits to museums is one of the dimensions to be explored. Virtual visits can not only enable museum visits for people who will never be able to make physical visits, but also create an increased interest for physical visits. In turn, as physical visits to museums often lead to very limited experiences, not very enriching or not at all, it is considered that virtual visits, subsequent to physical visits, may complement the initial experience, even allowing the visitor to dwell on some particularities that have been considered relevant and of particular interest.

Museums also often show great resilience and the ability to adapt to change, namely in the face of crisis situations. New technologies constitute an important resource which has been privileged. These issues are of interest to be explored. Statistical studies on museum visits are also of interest to study. In this case, everything indicates that there is a deficit of research on virtual visits.

REFERENCES

Appadurai, A. (2013). *Condition de l'Homme Global*. Payot.

Berger, J. (2018). *Modos de Ver*. Antígona.

Boletim ICOM-Portugal. (2020). Os museus de portas fechadas. *Os museus em tempo de Covid 19, 2*(14), 64-69.

Boudieu, P., & Darbel, A. (1992). *El amor al arte. Los museos europeos y su público*. Paidós.

Burke, P. (2017). *Testemunha Ocular. O Uso de Imagens como Evidência Histórica*. Editora Unesp.

Carvalho, A. & Matos, A. (2018). Museum Professionals in a Digital World: Insigths from a Case Study in Portugal. *Museum International, 70*(277-278), 34-47. htpps:// doi:10.1111/muse.12191

Castells, M. (2002). *A Sociedade em Rede* (Vol. I). Fundação Calouste Gulbenkian.

Castells, M. (2004). *A Galáxia Internet. Reflexões sobre Internet, Negócios e Sociedade*. Fundação Calouste Gulbenkian.

Eidelman, J., & Céroux, B. (2009). La gratuité dans les musées et monuments en France: quelques Indicateurs de mobilisation des visiteurs. In *Culture études. Ministère de la Culture – DEPS*. CAIRN INFO.

Ferreira, I. (2016). *Criatividade nos Museus. Espaços Entre e Elementos de Mediação*. Edição Caleidoscópio.

France Europan Commission. (2020). *Strategic Foresigt Report*. https://ec.europa.eu

Gastenhaus, A. (1997). *Minds in Motion- Using Museum to Expand Creative Thinking*. Caddo Gap Press. https://www.cairn.info/revue-culture-etudes-2009-2-page-1.htm

Gielen, P. (2009). The Murmuring of the Artistic Multitude: Global Art, Memory and Post-Fordism. Antennae, 3.

Han, B.-C. (2019). *Hiperculturalidade. Cultura e globalização*. Vozes.

INE. (2019). Visitantes dos Museus por região (NUTS II). Análise de resultados do Inquérito aos Museus e do Património Cultural Imóvel. *Estatísticas da Cultura 2019.*

Jacquot, S. (2018). Élargissement de la sphére patrimoniale. De la diversification des lieux de patrimoine à l`engagement populaire. *Espaces,* 340.

Lipovetsy, G., & Juvin, H. (2011). *O Ocidentes Mundializado. Controvérsia sobre a Cultura Planetária*. Lisboa: Edições 70.

Lipovetsy, G., & Serroy, J. (2010). *A Cultura Mundo. Resposta a uma Sociedade Desorientada.* Lisboa: Edições 70.

NEMO. (2020). *Survey on the Impacto of the COVID-19 Situation on Museums in Europe: Final Findings and Recommendations*. https://bit.ly/2X5NOI3

Roque, M. I. (2018). *O dilema do museu: apenas real ou também virtual? In Património Cultural e transformação Digital. Povos e Culturas 22. CEPCEP*. Universidade Católica Editora.

Rosa, H. (2014). *Aliénation et accélération. Vers une théorie critique de la modernité tardive*. La Découverte.

Scheiner, T. (2020, January/July). Museologia, hipercultura, hipertextualidade: Reflexões sobre o Museu do Século 21. *Museologia & Interdisciplinaridade.*, *9*(17), 46–63. doi:10.26512/museologia.v9i17.31592

Silva, S. (2013). *Turismo interno. Uma Visão Integrada*. Lidel.

Vidal, L. (2020). *Les Hommes Lents. Résister à la modernité XV – XX Siècle*. Flammarion.

ADDITIONAL READING

Anjos, M. (2017). *Contraditorio. Arte, Globalização e Pertencimento*. Cobogó.

Bauman, Z. (2017). *Retrotopía*. Paidós.

Becker, H. S. (2010). *Mundos da Arte*. Livros Horizonte.

Braga, P. (2021). *Arte contemporânea: modos de usar*. Editora Elefante.

Canclini, N. G. (2004). *Diferentes, Desiguales y Desconectados*. Editorial Gedisa.

Crary, J. (2017). *Técnicas do Observador*. Orfeu Negro.

Larsen, L. B. (2014). *Networks*. Whitechapel Gallery & Mit Press.

Lockood, A., & Medlick, S. (2001). *Tourism and Hospitality in the 21st*. CenturyButterworth Heinemann.

Marques, D. (2018). *Realidade Aumentada em Exposições de Museus. Experiências dos Utilizadores*. Edição Caleidoscópio. doi:10.30618/978-989-658-563-1

Santos, M. S. (2012). *Memória Coletiva e Teoria Social*. Universidade de Coimbra. doi:10.14195/978-989-26-1135-8

Silva, J. M. (2008). *Les technologies de l'imaginaire. Médias et culture à l'ère de la communication totale*. Table Ronde.

Virilio, P. (1989). *La Maquina de Vision*. Catedra.

KEY TERMS AND DEFINITIONS

Creativity: Is a personal or collective aptitude beyond routines and habits, facilitating change by generating ideas, discovery, and innovation.

Critical Thinking: Critical thinking is a reflexive way of thinking, based on analysis and consideration of facts, with great relevance in the digital age.

Digital Resilience: Digital resilience relies on new technologies and contributes, especially in crisis and threat scenarios, to finding innovative and creative solutions.

Digital Technology: Are resources that contribute to visitors' experiences and, regardless of distance, provide virtual visits.

Guided Tour: In the guided visit to a museum, real or virtual, the mediator helps the visitor or participant to interpret and interact with the object.

Hypercultural Tourism: Hyperculturality produces the so-called hypercultural tourist who moves in a symmetrical space without the here and there.

Mediation: Mediation through cultural instruments and people can help the visitor to become an active observer and builder of meaning.

Participation: Museums are nowadays cultural institutions in which the public and the community are considered as partners, as active interlocutors.

Participatory Experience: Implies the involvement of the person (visitor), having a more or less active character, the museum visit should be participatory.

Virtual Visit: In this case, the "visits," such as to museums, take place in a virtual environment, using tablets, computers, or other mobile devices.

ENDNOTES

1 Museum acronyms as well as the acronym of the Directorate-General for Cultural Heritage (DGPC), the authority responsible for national museums, are presented according to the Portuguese designations. The same criterion was used for the Lisbon Metropolitan Area (AML).

2 The variable in reference (geographical proximity) aims to assess the attractiveness of museum audiences, according to their place of residence, considering local and semi-local audiences and national and foreign tourists. The classification used, as referred, was based on the criteria used by Eidelman and Céroux (2009) in a work published within the Ministry of Culture (DEPS) of France. According to these criteria, "local" publics are considered to be those living in the municipality of the reference museum; "semi-local" publics are those living in the neighbouring municipalities of the reference museum; domestic tourists and also foreign tourists are considered to be visitors/participants of the reference museums. In this research we also crossed nationalities and places of residence, considering both Portuguese emigrants (living abroad) and immigrants, i.e., foreigners with habitual residence in Portugal.

3 According to Inês Ferreira (2016: p. 79), creativity can play a key role in crisis situations. The referred author points out that "it has become fundamental in adapting to change and for the success... personal and professional of individuals" (Cf. Ferreira, 2016:79).

4 It is important to distinguish "objects" from "things" that are not objects. Appadurai (2013:323) considers that objects are things that humans introduce into the realm of social life

5 Tourists are the main public of the museums under analysis, and cultural tourism deserves particular relevance. In the chapter title a distinction was made between tourism and leisure. However, as pointed out by many authors, such as Sancho Silva (Silva, 2013:3), tourism and leisure, not being synonymous, present, however, a very significant area of convergence both in the public and private sector perspective. In this case, it is not justified to replicate the aforementioned distinction in the body text due to the fact that tourists who frequent museums do so predominantly during holidays

and as a leisure practice. In the case of the other categories of museum audiences the necessary disaggregation was made, such as "visitors", "visitors/participants", and "residents".

6 The issue of resilience in organisations is also of concern to European institutions, as was the case with the European Commission. The *2020 Strategic Foresight Report* puts foward resilience as a new compass for EU policies. Resilience is defined by the Report as the ability not only to withstand and cope with challenges but also to undergo transition, in sustainable fair, and democratic manner. In the Report, one of the dimensions that is particularly relevant is digital resilience. In this case, it is considered that "the crisis has accelerated hyper-connectivity and the integratior of new technologies affecting the human condition and the way we live. Strategic foresight can help us antecipate how key emerging technologies could develop, their impact on all spheres of life, and ways to seize upcoming opportunities" (Cf. *2020 Strategic Foresight Report European Commission*).

APPENDIX: TEACHING NOTES: CASE STUDY

Chapter based on a case study:

TITLE: Impacts, resilience and creativity in cultural tourism and leisure in a time of pandemic. Presential and virtual visits to Lisbon museums

José da Cunha Barros

L-tour.ipt researcher-correspondent, PT

1. Questions and Answers

1.1. What is the overall problem presented in this case ?

The overall research problem was the analysis of the importance that digital communication assumed in museums during the pandemic crisis with particular reference to public access to museum collections and objects.

1.2. What are the factors affecting the problem(s) related to this case ?

The study took into account the fact that the pandemic is a global problem that has led to an increase in solidarity networks and responses at global, regional and local level, with new technologies and virtual communication being a very important resource for the continued operation of museums and communication with the public, namely through virtual visits.

1.3. Discuss managerial, organizational, and technological issues and resources related to this case.

The museums under study, which are under the responsibility of the Directorate-General of Cultural Heritage, are an integral part of the Portuguese Museum Network and, in addition to their own digital resources, they integrated the group of museological institutions whose collections are already available online through the Google Arts & Culture Platform and the Matriz-net. Therefore, the five museums under study, in addition to the in-person visits, already had their own online communication vehicles and in partnership which allowed them to communicate with the public and provide access to their respective collections.

1.4. What role do different players (decision-makers) play in the overall planning, implementation and management of the information technology applications ?

Under the supervision of the Directorate General of Culture, it was the directors of the museums under analysis, with the collaboration of the members of the respective teams, in the case of the technicians, who defined the strategies considered most appropriate to respond to the pandemic crisis and by increasing the use of new Technologies.

1.5. What are the possible alternatives and pros and cons of each alternative facing the organizatin in dealing with the problem(s) related to the case ?

The study showed that the institutions under analysis established their crisis response strategies in a coordinated manner, taking into account the particularities of their collections, the type of public and the resources available.

1.6. What are some of of emerging Technologies that should be considered in solving the problem(s) related to the case ?

The reinforcement of the use of digital technologies in all museum operations was perhaps the most evident aspect, as well as the realisation of virtual guided tours in groups.

1.7. What is the final solution that can be recommended to the management of the organization described in the case ?

The study showed that the museums in question, which are national and public in nature, intend to capture new audiences, increase community activities and diversify their activities, based on the experience they have gained and the increased use of new technologies. The Report "Museums in the Future", which resulted from a governmental initiative, in particular the Ministry of Culture, has also included Monuments and Palaces and has made it possible to determine a significant set of recommendations and proposals that have been very important and followed by the target institutions, dependent on the General Directorate of Heritage and the Regional Directorate of Culture. In addition to management models, issues such as networks and partnerships, digital transformation, collections management, and audiences and mediation were competences that were analysed and the subject of guiding standards.

2. Epilogue and Lessons Learned

The research allowed the author to get a sense of the importance of new technologies not only nowadays but also for the new demands, needs and motivations of the public in the present and in the future.

3. List of Additional Sources

Anjos, M. (2017). *Contraditório. Arte, Globalização e Pertencimento*, Rio de Janeiro: Cobogó

Braga, P. (2021). *Arte contemporânea: modos de usar*. São Paulo: Elefante

Carvalho, A. (2016). *Museus e Diversidade Cultural. Da Representação aos Públicos*. Caleidoscópio: Lisboa

Cuetos, M.P.G. (2009). *Humilde condición. El património cultural y la conservación de su autenticidad*. Gijón: Trea

Heinich, N. (2011). *Le Triple Jeu de L`Art Contemporain*. Éditions de Minuit

Munari, B. (2006). *Design e ComunicaçãoVisual*. Artes e Comunicação. Lisboa: Edições 70

Silva, J.M. (2008). *Les Technologies de l`imaginaire. Medias et culture à l`ère de la communication totale*. Paris: La Table Ronde

Chapter 22
Portuguese Public Art Primary Itineraries:
A Map of Discoveries

José Guilherme Abreu
Universidade Católica Portuguesa, Portugal

ABSTRACT

Since public art works are normally city collections, park installations, or landscape interventions, in order to achieve a rich, diverse, and representative sample, it is necessary to hit the road, to ramble in the park, and to explore the city. So, the main assumption is that public art value should also be judged by the consideration of its environment context, its social substratum, and the means used for its discovery. A literary feeling is required for its encounter. A journalistic curiosity is required for its inquiry. Travelling by car enables these aims, for it remains a reliable way of getting to know a region or a country. Besides that, on a period of pandemic or post-pandemic times, travelling by car is a defensive form of travelling for it avoids the agglomerations of airports and railway stations.

INTRODUCTION

The text "Portuguese Public Art Primary Itineraries" is an attempt to build an itinerancy model, providing certificated data about both historical and contemporary Portuguese public art collections, in order to allow a well informed and judiciously selected touristic discovery.

First of all, it is useful to note that the concept of itinerancy postulated here is defined by the idea of a journey by car. Driving is a defensive way of traveling during a pandemic crisis, for it avoids the large transport interfaces, such as airports, train stations, and port terminals, while providing a deeper immersion in the territory to be discovered, its people and cultural substrate.

Beyond the attempt at grasping the diverse nature of public art, the proposed itinerancy is also defined by the background institutions that assist cultural itinerancy, as extensions and complement of the journey. Each itinerary has a backup institution: museum, academy, gallery, library, etc., in this sense

DOI: 10.4018/978-1-6684-3369-0.ch022

adopting and adapting the aims and methods of the Council of Europe's Program for Cultural Routes (Berti et al, 2015).

Using our references as a new starting point, the driver may discover other public art pieces and programs, relying on the backup support and materials the correspondent background institutions provide, such as smart phones interactive and multilanguage guides, booklets and maps.

Since it is conceived as a journey by car, the structure proposed is based on the Portuguese road map itineraries, and in order to provide easy and comfortable trips, the roads that are recommended are highways and main national roads, linking major cities, small towns and even recondite rural villages and landscape spots, where relevant public art pieces may be discovered.

Because it is not possible to bring together in a single chapter all "Portuguese Public Art Primary Itineraries," a sample will be presented: the northwestern primary itinerary, travelling between the border of Galicia and Minho, and back, using IP_1 as descending itinerary and IC_1 as the ascending one, entering Portugal via Valença do Minho and leaving Portugal via Vila Nova de Cerveira's bridge over the Minho River.

TRAVELING BY CAR: STATE OF THE ART

This journey to Portugal is a story. The story of a traveler inside the journey he made, the story of a journey transporting a traveler, the story of both travel and traveler united in a desired fusion of the seer and the seen, a not always peaceful meeting of subjectivities and objectivities. Soon: shock and adequacy, recognition and discovery, confirmation and surprise. (Saramago, 1985, p. 5)

Travelling by car has indeed its own mystic flavor, as literature shows. In 1907, the notable French novelist and art critic Octave Mirbeau (1848-1917) published *La 628-E8*, whose title referred to the number plate of his four cylinders 1904 *Charron*, which was the central character of Mirebau's journal of his 1905 audacious journey through France, Belgium, Holland and Germany. In 1908 a luxury edition was published, gathering more than a hundred Chinese-ink drawings by French painter Pierre Bonnard (1867-1947).

Its narrative begins as follows:

Here, then, is the journal of a motor tour through parts of France, Belgium, Holland and Germany; and, more especially, through a small part of myself.

Yet is it a journal? Is it a tour?

Perhaps it is a simple tissue of daydreams, memories, impressions and stories that have nothing to do with the countries visited, but simply brought to mind by the recollection of a face, a passing landscape, a half-heard voice singing or weeping on the wind. But am I certain that I really heard such voice or genuinely saw the face which brought back so many memories, both joyous and sad; and that I truly glimpsed the landscape which inspired the rough poetry of the following pages? […]

Motoring has such a bewildering effect that one cannot be sure. The motor car is all whim and fancy; it engenders incoherence, forgetfulness. You set out for Bordeaux and that evening you are in Lille. But

Lille or Bordeaux, Florence or Berlin, Budapest, Madrid, Montpellier or Pontarlier – what difference does it make? (Mirbeau, 1907, p. 1-2)

These predictive existential interrogations emphasize the impact that motorcars had on annulling distances and dissolving borders, including those which were supposed to establish the frontier between the inner and outer experiences of the self, thus blurring perceptions and reflections, sensations and feelings.

Curiously, in his trip, Mirbeau describes statues in a humorous way, as it is the case of the *Manneken-Piss*, in Brussels, but also monuments, in a critical way, as is it the case of the *Palais de Justice*, also in Brussels:

I went to the *Palais de Justice*, where they piled up haphazardly, as much as they could, memories of monuments on monuments of memories, only to end up with a monument of incredible ugliness. They stacked Assyrian on Gothic, Gothic on Tibetan, Tibetan on Louis XVI, Louis XVI on Papuan ... It's so ugly, it becomes beautiful. (Mirbeau, 1907, p. 68)

La 628-E8, by Octave Mirbeau, may therefore be seen as the first literary reference to an attempt to conceive a roadmap for public art discovery.

We think that discovering public art by car routes still remains valid, for it was recently used by the Portuguese architect Manuel Graça Dias, as a method for discovering Portuguese architecture, as the author explains in his essay:

At that time, I was doing a radio program for TSF, "Driving through the city", now in its third consecutive year. I interviewed architects, according to what was seen during a car trip, in which I was driving, through streets and sectors of the city that, in general, guests indicated me. The main challenge of the model was to make listeners understand what was passing through our gaze in motion. I have good memories of that work, the way the sounds were later assembled, at night, in the TSF studios, with Pedro Vieira, trying to keep the sense of what had been said, plus the noises of the car, the city, the traffic; the music that we made coincide with that of a hypothetical car radio, scoring or establishing the connections between the various exposed ideas. (Dias, 2014, p. 15)

Once again it is the qualities of the lived experience of seeing, describing, analyzing and discussing that gives to such a methodology some kind of phenomenological flavor, which normally escapes other more formal attempts to catch the real.

So, instead of showing the pieces, providing the data and even their specific curiosities, our purpose is to look for a meaning whose extraction relies not only on objective information, but, above all, on the entanglement of both objective and subjective approaches, since the meaning of the work of art cannot be extracted only from the object, nor only from the observer, even if the observer is an expert.

The entanglement of inner and outer aspects is also some kind of journey, for there is a close connection between the voyager and the tourist, in the sense that both share a similar gaze: the gaze that observes both inner and outer dimensions of the real.

MAIN FOCUS: STRUCTURAL AXES

On the 26th of September 1985, the Portuguese Government approved the National Road Plan (*Plano Rodoviário Nacional*), nine "principal itineraries" (*itinerários principais*) created, in order to "*serve as a support base for the entire national road network and ensure the connection between urban centers with supra-district influence and between these with the main ports, airports and borders*." (DR, 1985:3207).

Conceived as a network of Portugal's main road connections, the nine principal itineraries provide a proper means for the discovery of Portuguese public art to those coming from abroad, as they link the main ports, airports and borders.

Nine *Principal Itineraries* (IP in Portuguese) were defined, linking the northern and southern parts of the country, (IP_1, IP_2 and IP_3) and the seacoast to the interior (IP_4, IP_5, PI_6, PI_7, PI_8 and PI_9), while twenty-four *Complementary Itineraries* (IC in Portuguese) were defined, establishing relevant articulations between the *Principal Itineraries*, and providing a second range network of main road communications covering the Portuguese continental territory.

As we have assumed that the means of transportation used in our journey would be the automobile, we needed to consider the possible routes, while also establish that the criterion to be followed would be, not a geographical one, but essentially a cultural one.

Portugal's National Road Map provides a pattern for the spatial distribution of the several structural elements we need to consider in order to establish, and to expand, a few discovery routes for Portuguese public art collections.

SOLUTIONS AND RECOMMENDATIONS

Santiago de Compostela Declaration approved by the Council of Europe on its 23rd of October 1987 meeting, launched the European Council Program for European Cultural Routes, and remains now a heritage reference of excellence – historical, cultural and artistic – for strong and reliable cultural routes.

The European Institute of Cultural Routes (EICR) is the central agency that manages this program, and since 1998 it is located in Luxembourg, under an agreement established between the Council of Europe and the Grand Duchy of Luxembourg. EICR organizes seminars for Cultural Routes operators, assists the Enlarged Partial Agreement (EPA), as well as the candidates for certification, with the evaluation and certification cycle, providing advice and support to candidate networks and certified routes.

Let us begin by the definition of what a Council of Europe's Cultural Route is meant to be:

Cultural Route: a cultural, educational heritage and tourism co-operation project aiming at the development and promotion of an itinerary or a series of itineraries based on a historic route, a cultural concept, figure or phenomenon with a transnational importance and significance for the understanding and respect of common European values.

From this definition, a Cultural Route is to be understood not in the restricted sense of physical pathways. "Cultural Route" is used in a more conceptual and general sense, expressing a network of sites or geographical areas sharing a theme (Berti et al, 2015, p. 15).

As we can see, Public Art is an adequate and pertinent theme for organizing a cultural route, as it closely relates cultural and natural heritage, pedagogy and tourism, while also reflecting on phenomena of transnational importance, such as the 19th century positivist cult of the great men and the great historical deeds, or the memories of totalitarianism, chauvinism and genocide, during the European most repressive dictatorships. It is documented by memorials and monuments that have been erected in the most sinister Nazi Concentration Camps, as well as in the center of major European cities. Two main themes, ranging from an ecological theme, such as artistic installations in the natural environment, to participative artistic urban projects, or temporary public space installations or interventions.

Besides the concept of cultural routes, it is useful to refer to another definition that is crucial for the implementation of the kind of cultural routes that seek deeper and more ambitious objectives than mere anonymous mass entertainment.

We mean the concept of Cultural Route operator:

Cultural Route operator: an organization or a grouping of organizations legally registered in one or several of the Council of Europe member States, or a public institution, which carries the legal, financial and moral responsibility for the management and functioning of a cultural route and represents the route vis-à-vis the Council of Europe. (BERTI et al, 2015, p. 177).

This is indeed a crucial element, for it is because there are institutions involved in the organization of the cultural routes that these routes overcome the initial objectives of mere leisure and entertainment.

In fact, it is because we rely on such an institutional and operational background that we will be able to set up a few complementary tools, aiming to provide reliable and adequate content that may be useful for illustrating the historical context and artistic problematics involved in each specific public art ensemble. These operators may be national or local institutions and community associations or partners, closely related to one or several aspects or problematics involved in a specific ensemble, or only a single piece, as we will see further on.

Portuguese primary public art itineraries will therefore be conceived under this conceptual umbrella, and we would be very pleased if our contribution could be seen as the first step towards the definition of a future wider and more elaborated European Public Art Route.

PORTUGUESE PUBLIC ART ROUTES

First of all, we have to point out that *Portuguese primary public art itineraries* are not directly developed from the Principal Itineraries of the Portuguese National Road Plan. That is the reason why we call the proposed public art itineraries "primary", instead of "principal".

Nevertheless, it is clear that the primary public art itineraries we propose are based on the principal itineraries of the national road plan, since the principal itineraries have the Portuguese border as their initial and final points, which is crucial for our project. Portuguese public art itineraries intend to not remain closed inside the country, but rather to establish articulations and connections to Spanish public art ensembles, therefore intending to become Iberian.

It is also useful to remark that, for our project, public art itineraries are not the same as public art cultural routes, since a public art cultural route should include and/or combine several public art itineraries.

In order to propose any public art itinerary, first we have to identify the most notable public art pieces, hierarchized in four different series: centers, ensembles, nuclei and single works, defined as follows:

A *center* is a large and most representative series of public art pieces that entails a wide chronological range, while keeping a certain artistic consistency due to the effect of a specific art school or a distinguished tradition, whose value attracts students and artists from abroad. An *ensemble* is a smaller and less coherent collection that entails a medium chronological range, while conveying heritage value, often beyond strict artistic excellence.

A *nucleus* is a specific collection of pieces that entails a narrow chronological range, while keeping a high coherence as it reflects the implementation of a specific public art program, art in architecture project or public space art curatorship.

An *ensemble* is a less coherent collection that entails a medium chronological range, while conveying heritage value, often beyond strict artistic excellence.

Finally, a *single piece* is a specific public artwork that stands out its context, because of the excellence of its aesthetic, symbolic or narrative content, and often because all these features are present.

Table 1 synthetizes the referred groups and displays their quantities by categories and by itinerary.

Table 1. Map of the most representative Portuguese public art collections by category and by itinerary. Source: own elaboration

Itineraries / Categories	IP_1	IP_2	IP_3	IP_4	IP_5	IP_6	IP_7	IP_8	IP_9	**Total**
Centers	2	-	-	-	-	1	-	-	-	3
Ensembles	2	3	2	-	1	-	3	-	1	12
Nuclei	7	5	1	3	-	2	3	-	-	21
Single pieces	7	4	1	3	2	6	1	2	2	28
Total	18	12	4	6	3	9	7	2	3	**64**

Table 2 shows, by category and by case, the several institutions – Museums, Academies, Foundations, Administration, and Archives – that should be partners and/or supporters for the implementation of the Portuguese public art primary itineraries.

*Table 2. Map of institutional partnership/support by Center. *See list of acronyms. Source: own elaboration*

Center / Institution	**Porto**	**Caldas da Rainha**	**Lisboa**
Museums*	MNSR; MMIPO; MACS; CMTL; MIEC; CAAC; MBC; CMAS	MAD; MBF; MJF; MLA; MJM; CMSR; CMHD; MSF	MNAC; MCG-CAM; ML; MARL; MNR; MMCP; MNA;
Academies*	FBAUP; UCP; UM	IPL-ESAD.CR;	FBAUL; FCSH;
Foundations*	FEAA; FAMS; FJR; FBVNC	FHD; FASM;	FCB; FMS; FCCB; FEDP
Administration*	**DRCN**; CMPrt; CMST; CMPds; CMVC; CMVNC;	**DRCC**; CMCR; CMA; CMO;	**DGPC**; CML; CMA;
Archives*	ADP; AMP	ADL; ADC	ADL

The point is that the mentioned centers are, not only the poles that gather most information and documentation about Portuguese public art, but they also function, each one of them, as regional focus of its own region: Porto, for the northern part of the country; Caldas da Rainha, for the central part of the country, and Lisbon, for the southern part of the country. We think that the background support provided by these institutions will be enough, at least for launching the program, by assuring its first operative tasks and assisting and assessing its initial results.

STATIONS OF PORTUGUESE PUBLIC ART PRIMARY ITINERARIES

Further criteria, beyond the adoption of the general aims strived by the Cultural Routes of the Council of Europe have not yet been discussed.

However, in order to achieve those aims, it is necessary that our program may integrate, in a coherent and cohesive way, cultural, pedagogical and symbolic dimensions, thus gathering scientific expertise, didactic methodology, spiritual approach and feeling.

Hence, because it is not possible to develop all the Portuguese public art primary itineraries in this text, we will explore a specific segment of them, focusing on the northern part of the country.

1st Station: Valença Do Minho Border Post Building

Based on IP_1, and IC_1 (E_1 and A_{28} in Google maps) in its northern section, limited on the North by the Minho River and on the south by the Douro River, the first public art primary itinerary is designed for tourists entering Portugal by the Minho region, in its northwestern border at Valença and having Porto as destination. In its descending course the route will follow IP_1 (E_1 on the map), and then follow IC_1 (A_{28} on the map), for returning. This itinerary is used both by Spanish tourists coming from Galicia, as well as by Portuguese tourists travelling to Galicia, since the connection, if not the empathy, between Minho and Galicia is very well established, due to deep and strong historical and cultural reasons. In fact, the origin of the Portuguese language is the Galician-Portuguese dialect, first spoken in the territory formed by both provinces, later on divided and turned apart for political convenience, as Agostinho da Silva mourns:

... they left Galicia to their fate; the Portuguese knights betrayed the girls who, in the cantigas d'amigo, cried for their absence and longed for their pilgrimages. [...]

Galicia should never have been abandoned; if we had to die, we had to die along with her. Portugal is guilty for Rosalía's tears, and every emigrant who does not return to Galicia, accuses him. (Silva, 2000:249-250)

Our public art tour has its first station in the border post that appears after passing the old international iron bridge that crosses the Minho River border.

The image above shows the first building that appears to the driver arriving to Portugal from Spain, in Valença do Minho. It is a polychromatic mosaic by the distinguished painter Júlio Resende, which is implanted over the northern wall of the border post, while slightly detached, forming an entrance, as the building is also a remarkable work of architecture. Its project, by the modernist architect José Carlos Loureiro, dates from 1956, and can be now retrieved at Fundação Marques da Silva (FMS): a proposed partner for the present Routes.

Figure 1. Júlio Resende, Coastline, 1959, mosaic, border post, Valença do Minho.
Source: JG Abreu

Actually, this mural by Resende is a recent discovery because it had been first attributed to the painter José Gastão Seixas (1926-1982). In fact, it was only during its 2015 restauration that its cleaning revealed Júlio Resende's signature, which can be read, depicted in light green, near the bottom of the composition.

The composition is displayed vertically, thus counteracting the horizontal depicted landscape, which represents a maritime theme, with an elegant feminine figure playing lute by the seaside, in the upper part of the scene. The sea waves are marked in different shades of blue, and we can see fish crossing them, horizontally. A narrow brownish bar seems to represent the beach sand, covered with starfish, shells and fish nets, while depicted in several bars, a vertical garden shows plants, flowers and seaweed, mixing up figurative and abstract forms, as if "weaving" some kind of external and exquisite tapestry.

However, Valença do Minho is not located on the coast, nor even nearby, for the distance to the seaside is about 30 km (18 miles). Maybe in the artist's mind was the romantic idea of depicting Portugal as a "*Garden of Europe planted by the seaside*", as the Portuguese poet Thomás Ribeiro wrote in his poem *Portugal*, first published, in 1863, in his book *D. Jayme*.

Near the Valença do Minho border post, appearing on the opposite side of the building to those that leave the country, stands the bronze sculpture, *Rhythm of Spring*, signed "*Arlindo Rocha, 1961*", which is, the first abstract sculpture implanted in the Portuguese public space (Oliveira, 2011, p. 70).

Figure 2. Arlindo Rocha, Rhythm of Spring, 1959-61, bronze, border post, Valença do Minho.
Source: JG Abreu

Its present location is not the original one. Initially, the piece was implanted near the South wall of the border post, inside a flowerbed full of agapanthus, where the vegetal-like forms of the sculpture established a curious dialogue with the surrounding plants.

As in Júlio Resende's mosaic, we observe here an interplay between figuration and abstraction, for the dynamic forms of the sculpture remind us of the movement of leaves and branches. However, the piece intends not to capture the floral movement, but instead it intends to capture the invisible energy, the eidetic dynamism of life or, as Henri Bergson said, its *elan vital*, i.e., its rhythm. That is the reason why those abstract forms depict an odd geometry. They dematerialize movement, as if reducing it to a vortex.

Finally, it is necessary to say that initially the sculpture had a ceramic red application in its frontal top, being the presence of such a form as unexpected as the flowers on the top of a certain type of cactus seem to be.

In brief, these first public art pieces are notable artworks. They assert the centrality of Oporto's academy, showing how its centrifugal influence can be clearly felt in the perfect integration of the three arts: architecture, painting and sculpture.

Hence, in an almost surprisingly way, we recognize here an exercise of modernism, echoing the 6th and the 7th CIAM recommendations, concerning the synthesis of the arts (Giedion, 1951), as an alternative to rhetorical monumentality, in a time deeply conditioned by the traditionalist/nationalist credo that prevailed in Portuguese official aesthetics.

2nd Station: Devesa Park – Vila Nova De Famalicão

After the first station, which served as a gate entry to Portuguese public art, we drive south along highway A_3 to Vila Nova de Famalicão, in the district of Braga. During the trip, we can see the mountainous relief covered with pine trees and listen to music, while driving.

Our destiny is *Parque da Devesa*, an urban park opened to the public since 2012. Projected by the architect Noé Diniz, the construction of Parque da Devesa was the result of a partnership between Câmara Municipal de Vila Nova de Famalicão, CESPU, CITEVE, AML and ADRAVE, as well as other Complementary Actions: a mix of public and private investments, involved in maximizing the intervention and its effects.

Located near the city center, *Parque da Devesa* provides an area of twenty-seven hectares (66.7 acres) for recreation, as well as cultural and pedagogic activities.

Our 2nd station gives us the opportunity for a pleasant walk in a green park that has many distinct charms, from archeologic pieces and reconstitutions to permanent and temporary exhibitions, both indoors and open-air, as well as a cultural program for its amphitheater.

That is not all. Besides its environmental concerns, Parque da Devesa develops its own public art program, whose central commitment is ecology. So, coherently, it deals mainly with trees, something which remains its major singularity.

In 2012, Noé Dinis created "Painting Trees", a group of five dead poplars painted in vivid colors. In 2013, an artist from Mozambique, Gonçalo Mabunda, was invited to create two pieces "Verticality" and "*Surge et Ambula*" (Stand up and walk), whose vertical presence dialogues with the surrounding trees. In 2017, the landscape architect Gonçalo Nunes de Andrade authored "Memories in the Park", a spatial piece whose purpose was to fix the memory of the spatial presence of an old dead cork oak.

But it is the piece by Isaque Pinheiro, *Knock Down a Tree*, created in 2016, that surprisingly was the ablest in dealing with the sad and sinister presence of dead trees in a joyful park, and succeeded in modifying it completely.

Isaque Pinheiro fixed the memory of a centenary oak tree, measuring thirty meters high and weighing seven tons, and succeeded in transforming its trunk into a sculpture: a giant hinge. The huge tree is now displayed in a reclined position, bringing the branches of his dead top close to the public. The dead tree seems to be no longer dead, for it is no longer a tree. It is an object-like surreal organic thing. In a way, it became a fossil. Its form keeps the image of a living thing, but its nature, as if being fossilized, acquired a quite distinct presence: the abstract presence of a document.

Figure 3. Isaque Pinheiro (esc.), Knock Down a Tree, 2016, Parque da Devesa, Vila Nova de Famalicão.
Source: JG Abreu

3rd Station: Museu Internacional De Escultura Contemporânea – Santo Tirso

After the second station, the driver returns to highway A_3, heading now to Santo Tirso, in the Oporto district. Santo Tirso is an old historic town. Its peopling was the consequence of the foundation in the territory of S. Bento monastery by D. Unisco Godiniz and her husband Abunazar Lovesendes, primer overlord of Maia, whose first installation goes back to the 10th century, being therefore older than Portugal (Correia, 2009).

However, instead of seeking Santo Tirso's history and heritage, our purpose is decidedly focused on contemporaneity, that is, on the pacific cohabitation between contemporaneity and heritage. This is perhaps the best way to express the singular character of Santo Tirso. In fact, Santo Tirso is the hometown of *Museu Internacional de Escultura Contemporânea de Santo Triso* (MIEC): an important project launched by the sculptor Alberto Carneiro (1937-2017) with the institutional, operational and funding support of the local Municipality. This is a paradigmatic case of the intertwining of heritage and contemporanity, as the MIEC is a modernist buiding next to a 18th century hospedary of S. Bento monastery.

Established by an agreement between Alberto Carneiro and Santo Tirso City Hall, the implementation of this project took almost thirty years, between 1991 and 2020, and began with the organization of bienal international sculpture symposia, reaching now the quite impressive number of fifty seven pieces, all of them created by notable national and foreign sculptors.

The MIEC is a mandatory presence in any public art route worldwide. In MIEC's website (http://miec.cm-stirso.pt/en/collection/) it is possible to see the images of its collection, ordered from the most recent piece, *Tirsa*, by the German sculptor Robert Schad (1953-), dedicated in 2020, to the oldest one, Alberto Carneiro's *Água sobre a Terra*, dedicated in 1990. They are a quick portrait of the collection.

MIEC's sculptural collection is spread throughout Santo Tirso's public spaces, not in a continuous way, as if establishing a track – that is, for instance, the case of Paredes Public Art Circuit – but instead creating distinct concentrated areas, in different parts of the city, both central and peripheric.

Obviously, it is not possible to describe, analyse and comment on each one of the fiftyseven pieces. Therefore, in order to spare time and space, we will add a sample, involving only the initial nucleus, the one that was the cradle of MIEC, as Alberto Carneiro explains:

Everything began in 1987, when Mayor Joaquim Couto invited me to conceive a sculpture to install at one of Santo Tirso plazas. The sculpture "Water upon the Earth", erected in Plaza Camilo Castelo Branco, was dedicated in 1990, and soon after another sculpture was implanted there too, "The boat, the lune and the mountain", in order to equilibrate the spaces of the plaza. By that time, in an informal conversation, I told Dr. Joaquim Couto that an important museum of contemporary sculpture could be created in the town, through the organization of ten biennial symposia, during a period of twenty years, inviting both Portuguese and foreign artists. (Carneiro, 2015, p. 15)

Figure 4. Alberto Carneiro, Water upon the Earth, 1990, Granite, Pç Camilo Castelo Branco, Santo Tirso.
Source: JG Abreu

It is useful to notice that when Alberto Carneiro proposed the organization of an international symposium in Santo Tirso to Mayor Joaquim Couto, the artist had already participated in the international symposium *Forma Viva*, when in its 1986 edition he created, in a premonitory way, the sculpture "*Tree, flower and fruit*", at Kostanjevica, Slovenia.

Tree, flower and fruit and *Water upon the Earth*, beyond the radical contrast of their materials, are indeed related: they similarly evoke the descending flux of life, facing the ascending irruption of matter, being its genetic power, most clearly stated in Santo Tirso's piece.

Water upon the Earth is more than just a sculpture. It is a symbolic cosmic event. As the author says, it is a mandala. Let us evoke his own words:

Water runs over the Earth's surface and shapes its movements in recognizing matter. Water and stone: the river and the mountain. The stone that unfolds itself in the vibrations of water. Grooves of life over the earth for the anamnesis of the body. The elements left their marks on the surface of the stones. The water runs from the mountain and seeks the horizontality of the land. Liquid scintillations that make the stone fluid and appeal to the mobilities of the body. Mandalic space where water seeks the center. Cosmos of the unity of the body and the universe (Carneiro, 2001, p. 118).

Figure 5. Alberto Carneiro, The boat, the lune and the mountain, 1990, Granite rocks, Pç CCB, Santo Tirso
Source: JG Abreu

These unequivocal hermetic correlations become even clearer if we consider the relationship between *Water upon the Earth* and *The Mountain, the Boat and the Lune*, the other piece formed by three elements: installation becomes constellation! The boat, or the cultural endeavor, the mountain, or the natural ground support, the lune, or the poetic ethereal evasion, the earth, or the cosmic capsule, the water, or the flux of life, as the author says:

Time and the elements have shaped these stones. They are mountains. They are formed by the passing of time in the recognition of a body on the earth. Chosen as identifications of the mountain that interiorizes them, of the moon that reveals the night and of the boat that seeks the mystery, these stones propitiate nature. Between the hands they are the grain of sand and the infinity. They reveal the body in this intimacy with the simple things that give us aesthetic enjoyment. Archetypal consciousness that takes us to the dawn of time as a future path of non-being. They are added in a triangle on the spiral that the earth draws. They are the call to the sacred that the body carries in its cosmos (Carneiro, 2001, p. 119).

This installation/constellation was the cradle of the MIEC project. Metaphorically, we can say that they constitute the DNA of the project. The main assumptions of the future works are already there: the dominant spatial character of the works, instead of a strict minimalist objectual and/or literal presence; the ecological dimension and/or the relationship with the landscape; the subtle and/or discreet, but close, dialogue with the viewer; the spiritual or meta-narrative/meditative conveyed sensitivity; and so on …

Now the driver has to choose if he wants to step down from this station and dive in MIEC's collection, or if he prefers to follow our itinerary, heading to Oporto. Maybe he will take a break, and walk along Santo Tirso's public space, in order to get acquainted with its public art collection, with the aid of a smartphone app (http://www.miec.mobinteg.com/).

4th Station: Oporto Public Art Map

Considered as a center, Oporto plays a prominent role in northern public art, being the corollary of such a condition the existence of a large and diversified public art collection, from historical statuary to contemporary installations.

Oporto's public art has been the object of academic dissertations, conferences and studies, and as the corollary of its relevance it was recently assumed as a first range area for the Municipality's cultural policy.

That was the impacting action by the late cultural city councilor Paulo Cunha e Silva (1962-2015), who in 2015 launched Oporto's public art program, with the dedication of the mural, "*To the City*", by the architect-artist Fernando Lanhas (1923-2012), who alone with Arlindo Rocha was one of Portuguese abstract art pioneers.

While affected by the premature death of Cunha e Silva, Oporto's public art program remains one of the cultural priorities of the Municipality. In 2017, a *Map of Oporto's Public Art* was published, gathering not only the pieces already installed in the city by the public art program launched by Cunha e Silva, but also other representative works, from historical pieces to contemporary artistic production, thus establishing the central status of Oporto's public art.

Oporto's public art map presents a record of about two hundred sculptures and murals, from which fifty pieces are detached, distributed over five thematic routes of ten pieces each, as follows:

We suggest, for starting point, these five thematic itineraries which help to tell Oporto's narratives: the Historical Route, covering some of the most important moments of the almost two thousand years of this burg; the Route of Water, vital element to any life and privileged means of Oporto's expansion;

the Route of Letters, which honors the great writers who sang and dreamed us; the Route of Fine-Arts Academy which formed the majority of the signalized artists; and Route of Contemporary Art which let us discover some of the most relevant artistic languages of our time, and projects towards the future the creative capability of Oporto's inhabitants. (CMP, 2017)

Hence, the departing point of Oporto's route cannot be other than the Map of Oporto's Public Art. So, we shall pick up two pieces of each route, analyze them, and discuss their relevance.

From the Historical Route, we choose the *Monumento aos Heróis da Guerra Peninsular*, whose initial project was authored by the architect Marques da Silva and the sculptor Alves de Sousa, 1909, but was finished by Henrique Moreira and Sousa Caldas, in 1952.

From the Route of the Water, we choose *Monumento à Grei*, by Lagoa Henriques, 1960, and *She Moves*, by Janet Echelman.

From the Route of Letters, we choose *Rosalía de Castro*, by Barata Feyo, 1954, and *Abel Salazar*, by Hélder Carvalho, 2009.

Figure 6. Marques da Silva (arch) & Alves Sousa (sculp), Monument to the Heroes of Peninsular War, 1909-51, Porto.
Source: D Morais

From the Route of Fine-Arts Academy, we insert the iconic statue *Juventude*, by Henrique Moreira, 1929, and choose *O Guardador do Sol*, by José Rodrigues, 1963, because we do not understand the selection of the rehabilitation of Old City Hall as a public art piece.

From the Contemporary Art Route, we choose *Pirâmide*, by Zulmiro de Carvalho, 1985, and *O meu sangue é o teu sangue*, by Rui Chafes, 2015.

The Monument to the Heroes of the Peninsular War is the most representative achievement of the Portuguese positivist monumental model, honoring in an impressive way great men and great historical deeds. Besides its specific artistic value, it is also notable as a heritage testimony of the late 19th century and early 20th century monumentality: the golden age of the commemorative monument (Reyero, 1999). In this sense, it dialogues with other similar monuments such as Barcelona's *Monument a Colom*, the Parisian *Colonne de Juillet*, London's *Nelson Column* or Berlin's *Siegessäule*.

Selected after a national competition, the winning prize had "People and Troops" as title, alluding to the alliance between the population and the army that led to the victory against the Napoleonic invasions of Portugal, in early 19th century.

In Abreu's study of Oporto's public sculpture, the monument is described as follows:

The monument is a Doric granite column with considerable entasis, surmounted by a bronze sculptural group of a lion dominating an eagle, representing respectively the Luso-Britannic and the Napoleonic armies. On the stem, near the granite base, are sculpted in low relief the figures of four Generals: Silveira, Bernardim Freire, Champallimaud and Ebben, heroes of the Portuguese army, who dominate, without direct participation, the fierce struggle that takes place on the flanks of the pedestal, where, in an impressive composition involving soldiers and popular forces, allegorical figures and real people cohabit, blended in the profusion of the bronze. On the East and West sides, in symbolic opposition, the allegory of Victory, alluding to heroism, incites rebellion, raising the national flag and the flame of insurrection, while the figure of a crying Mother about to be swallowed by the waters, representing sacrifice, alludes to Oporto's tragedy of the bridge of boats. On the front of the column, the date of MDCCCVIII and on the back of the column, that of MDCCCIX, refer to the 1st and 2nd invasions, emphasizing the local and regional character of the event that is intended to be commemorated. (Abreu, 2012, p. 116-117)

While its conception follows the architectural model of the commemorative column, it is nevertheless the sculptural image of a lion dominating an eagle in a most unlikely way that creates the allegoric key that solves the most crucial symbolic/narrative problem brought by this commemorative monument.

Other monuments to the *Peninsular War* use the symbolism of the lion and the eagle, both in Portugal and Spain. In Lisbon, the eagle is on the monument's top, while the lion is on the base, facing North. In *Monumento a Agustina de Aragon*, by Mariano Benliure, in Zaragoza, 1908, the lion is attacking the eagle. In *Monumento à los Sítios*, by Enrique Higuero, 1910, Astorga, the lion dominates the eagle. In Oporto, the lion is on the top of a column, attacking the eagle.

We can see how any public art route reveals international connotations. In statuary that becomes common, for the models tend to be a lot more rigid and redundant, as well as the intentions tend to become limited and rhetorical.

So, late 19th century statuary, instead of standing for strict aesthetic or plastic excellency, more adequately stands for its associated cultural values and social uses, displaying its vocation to work as places of memory (Nora, 1986).

The other historical route piece we have chosen is the equestrian statue of Vímara Peres: Vimara Peres was a vassal of the Asturian King Afonso III and a medieval Christian knight, who in 868 commanded the *aprisio* of *Portus Cale*, which was the name of the ancient settlements of Oporto and Gaia that later would give the name to Portugal.

In 1968, on the celebration of the 11th century of the foundation of *Portucale County*, and of the time when Vímara Peres was named first Count, an equestrian statue was commissioned to the sculptor Salvador Barata Feyo, to be implanted near Oporto's Romanic Cathedral, inside the old defensive wall of Penaventosa.

Figure 7. Carlos Ramos (arch) & Barata Feyo (sculp), Vímara Peres, 1968, bronze, Cal. de Vandoma, Porto.
Source: D Morais

The statue represents Vímara Peres as a medieval knight, with his iron helmet on. Firmly standing still on the territory he had just occupied, the knight wears a round plate *hauberk* and holds a round Celtic shield in his left hand, while raising high a medieval Christian banner with his right arm, towards the Romanesque cathedral, not yet built at that time.

This equestrian statue is a good example of the balance between classicism and modernism, which Barata Feyo strived for. The statuary tradition is patent both in the Roman equestrian model of the figurative group and the solid plinth with rusticated masonry, by the architect Carlos Ramos. The figurative group establishes and assures the classical irreprehensible unity and consistency of the knight and the horse. The high and robust plinth enhances monumentality, if not sublimity, to the work. Sculptural modernity is patent, on the one hand, in the unusual pose of the horse and the knight, both depicted in a suspended and most expectant immobility – as if profiled for a joust – and on the other hand in the synthetism and simplicity that shapes the figure of the knight, as well as in the subtle geometrization of the anatomy of the horse, the rough texture of the bronze, and its dark patina.

The statue was dedicated on the 16th of June 1968, during an official ceremony in which took part the President of the Republic (AHMP, F-P/CMP/10/485), thus assuming the symbolic and political role these achievements had for the Portuguese *Estado Novo*.

Figure 8. Lagoa Henriques, Monument to the Grei, 1960, bronze, Jardim António Cálem, Porto.
Source: D Morais

From the Route of the Water, we choose *Monument to the Grei*, by Lagoa Henriques, 1960. This is a relevant piece of Oporto's public art, not only for its plastic features, but also for the context in which this piece was produced: the commemoration of the 5th Centenary of the death of Prince Henry the Navigator, in 1460.

The initiative for the creation of this monument belonged to a local delegation of the Executive Commission for the celebration of the 5th centenary of Prince Henry's death, which under the presidency of the Dean of Oporto University, Professor Amândio Tavares, decided to promote the creation of a local monument, with the aim of "*implanting a memory dedicated to the effort of a Pleiad of sailors, naval carpenters, caulkers, tireless workers who gave their valuable contribution to the discoveries, specifically in the equipment of the fleet to Ceuta*" (Abreu, 2012, p. 295).

The monument is implanted at public reach on a lower pedestal, composed of a sculptural exercise combining figuration and abstraction. Conceived as a sculptural group, the piece includes three figures: one passive, standing up, and two active, crouched, working on a wood vessel, suggested by the stylized curved line that rises from the prow to form a sail. The standing figure observes, stiffly, the horizon line in an expectant and attentive attitude. At his side, a second figure, alluding to naval carpenters, is cutting the wood for the boats. In the rear side, a third figure carves the meat of a dead ox, already stripped of viscera, alluding to the provision of supplies by Oporto's populations for the expedition to Ceuta. The fact that the figures do not show any historic costume gives to the ensemble a subtle timeless cut, not connoting it with any specific period, thus introducing an unexpected modernity.

Besides the sculptural group, the monument includes an exercise on landscape architecture, presenting a granite map showing the maritime routes launched by the Portuguese navigations, lawn spaces, park benches, and two bronze cannons and a flag-waist, in order to grant a martial character to the precinct.

Commissioned by the urban regeneration Polis Program for the waterfront of Matosinhos and Oporto, the intended public artwork should fulfill the following: 1) Create a sculptural symbol that would unify the waterfront redevelopment of the two cities at their border 2) Be visible from 1 km in all directions 3) Not block out the view of the Atlantic Ocean 4) Withstand high winds, sun, harsh salt air, and pollution 5) Require minimal maintenance.

The other piece from the Route of the Water is the sculpture She Changes, by the North American artist Janet Echelman: a huge suspended structure, made of painted steel, cables and tenara fiber net, which is a material used by NASA for spacecraft engineering.

Named "Anemone" by the public, this sculpture is formed by a red iron ring suspended in the air by six high cables and six lower ones stressed from three white steel pillars. About forty cables link the circle of the ring to its center. From the outer circle of the ring, a red and white net forming three distinct curved levels drops smoothly, lowering from the periphery to the interior, and then surmounting to the very center of the ring, thus suggesting the organic form of an anemone.

Being the very first aerial net sculpture conceived by Janet Echelman, *She Changes* was recognized by Americans for the Arts Year in Review – Public Art Network (PAN) as a most compelling work of the year 2005: https://www.americansforthearts.org/2005-public-art-network-year-in-review-she-changes.

While keeping its sophistication, Echelman wanted the work to be accessible to a large audience, and so she turned to the history of the site. She established a link between the old fishing village and the industrial zone it became in the latest decades. In this sense, the work echoed the surrounding smokestacks with their red and white-striped patterns, as well as the masts of the Portuguese vessels and the fishing nets that local fishermen and their wives sewed.

Figure 9. Janet Echelman, She Changes, 2005, steel and tenara net, Pç Cidade de S. Salvador, Matosinhos-Porto.
Source: D Morais

From the Route of Letters, we choose the statue of the poetess from Galicia Rosalía de Castro, about whom we have already spoken. Considered the founder of modern Galician literature, the implantation of the statue of Rosalía de Castro was the corollary of the creation of the new Plaza of Galicia, opening a new axis of Oporto's urban expansion towards west.

The dedication of the statue was a significant event that gathered Rosalía de Castro's daughter – Dª Gala Murguía – and local authorities from Oporto (Abreu, 2012, p. 251).

The Poetess is carved in a reclining pose with a contemplative gaze, next to a water mirror. Her face, slightly raised, looks up. On the head, stylized grooves representing the waves of the hair, stand out, harshly carved. The trunk, twisted on its base, is covered with a sober garment that blends with the body itself. The left arm, following the trunk, rests on a pillow, the right arm, bent at the elbow, rests on the lap, holding a book. Over the crossed legs, fine drapes adapt to the body and its support, developing forms that become autonomous, according to a geometric composition that lends a modern plasticity to the modelling. On the feet, the sandals denote the classic feature of the representation, also reinforced by the configuration of the base, in the form of a stylized triclinium.

The work refuses the canonical model of the pedestrian statue and states the primacy of formal research instead of the naturalistic transcription of the figure, displaying a plastic interpretation. In terms of composition, the work uses an intermediate formula between the seated figure, used in the statue of Abraham Lincoln, and the reclining figure, frequently used by Henry Moore (1898-1986). In turn, stylization is present throughout the composition, including the plinth that here acquires the personality of a sculptural object, whose tiny elevation brings the figure closer to the audience, while assuring a certain horizontal distance, established by the presence of the water.

Figure 10. Barata Feyo, Rosalía de Castro, 1954, Granite, Praça da Galiza, Porto.
Source: D Morais

Figure 11. Hélder de Carvalho (sculp) & Morais Soares (arch), Abel Salazar, 2009, bronze, Jardim do Carregal, Porto.
Source: D Morais

The seated statue of Abel Salazar (1889-1949) is a homage to a gifted personality. He was a physician, researcher, lecturer, essayist, painter, sculptor, Freemason and, last but not least, a tough opponent to Estado Novo dictatorship. He graduated in medicine and surgery in *Escola Médico Cirúrgica* do Porto, in 1915, with his doctoral thesis earning him the highest rating. In 1918 he was titled full professor, beginning his career as both lecturer and researcher, after the creation of his own laboratory of histology and cytology, whose results granted him international acknowledgement.

In spite of his scientific and cultural prestige, in 1935 Abel Salazar was expelled from Oporto University, due to political disputes with the Government, thereafter, focusing his activity on art – drawing, engraving, painting, sculpture – and writing – science, philosophy, aesthetics and travel literature.

His strong personality, determination, and dominant presence are well expressed in this statue. Alluding to the cathedra he first earned, Abel Salazar is represented seated in a firm and steady way, with a facial expression of disobedience and revolt. While static, his pose, attitude and expression denote a powerful will and a strong conviction, which are enhanced by the rough texture of the bronze, the thickness and angular form of the clothes, and of course of the scale, whose inflation exceeds natural size, while its dimensions seem minor due to the seated positioning of the figure.

Concerning the Route of Fine-Arts School, as we have already stated, we decided to replace the rehabilitation of the Old City House building, with the iconic marble fountain and statue *Juventude*, by Henrique Moreira, 1929.

The construction of this decorative fountain is inseparable from the renovation that the city council wanted to induce, reflecting the city's economic growth.

One of the instruments of this affirmation was the creation of new squares and, above all, the opening of impressive avenues that adequately framed the equipment and services that reflected modernity: Banks, Administrative Corporations, Commercial Firms, Offices, Newspapers, etc.

With these objectives in mind, the City Council launched two distinct competitions for the creation of decorative sculptural motifs that could fill the void of the large esplanade of the recently opened Avenue of the Allied Nations.

The winning project for the first element is formed by a prismatic pedestal with profusely faceted edges displaying an exquisite exercise of art-deco geometric stylization, surmounted by a seated marble female nude, depicting a smiling young woman with her head gracefully tilted to the right. Her feet are resting on one of four bronze heads, alluding to the four seasons of the year. These four bronze heads are placed at half height on the central mainstay, from which flows the stream of water that feeds the middle and lower reservoirs of the fountain.

With a bold design and beautiful decorative effect, harmoniously fitting into the urban axis of the Avenue of the Allied Nations, this work was re-baptized with the popular designation of Naked Girl. It became an *ex-libris* of the new image of the city, enhanced by the sophisticated quality of the pedestal's design by the architect Manoel Marques which stripped of allegorical or rhetorical allusions, adequately matched the grace of the statue.

Youth, by Henrique Moreira, has some kind of an avatar in Barcelona. It is the statue also named Youth by Josep Clarà, 1928, (*Joventut*, in Catalan) which was implanted in Paseig de Gracia, in 1929, just like in Oporto, *Juventude*, was.

The second piece of the route of the Fine Arts School, is *O Guardador do Sol* (The Guardian of the Sun), by the sculptor José Rodrigues, 1963. In fact, it was the final work as a student, being for that reason implanted in the Garden of the old Oporto's Fine Arts School, now Faculty of Fine Arts of Oporto University.

Figure 12. Henrique Moreira (sculp) & Manoel Marques (arch), Youth, 1929, marble, Av. dos Aliados, Porto.
Source: D Morais

In *Guardian of the Sun*, a few pairs of oppositions, underline a complex set of dualities, such as the teluric tension of his musculature and the ascendent lightness of the pose that makes the figure to rise, seeming to float above the ground.

However, beyond contrasts and oppositions, beyond tension and distension, the statue shows an unmistakable elevation that makes the figure appear suspended, not even resting the feet on the ground, raising high in the air like a Maasai warrior, during the *adumu* ritual dance, proudly and vigorously wielding his shield and spear.

Figure 13. José Rodrigues, The Guardian of the Sun, 1963, bronze, Gardens of FBAUP.
Source JG Abreu

The image of the *Guardian of the Sun*, instead of conveying the iconography of Western sculpture, comes from the Masaai people of Southern Africa, and their rites, who were one of the first peoples of Black Africa whose culture and rituals were better known and described.

But the *Guardian of the Sun* is not an ethnological sculpture. Above all, it is a solar figure, and as such, triumphant. So, it was also, if not mainly, a manifest. A manifest against the Colonial War that ignited the Portuguese African Colonies, being to the author's generation an inevitable destiny, as it was to José Rodrigues, himself, two years before he created the *Guardian of the Sun*, which gave him the top classification.

Finally, the Contemporary Art Route has many notable works, both by national and foreign artists. Because the foreign artists are well-known artists, such as Richard Serra, Claes Oldenburg, Minoru Niizuma or Juan Muñoz, we choose two not yet mentioned Portuguese artists, one from Oporto, the other from Lisbon.

Figure 14. Zulmiro de Carvalho, Pyramid, 1985, granite, Garden Paulo Valada, Porto.
Source: D Morais

This sculpture is a monumental structure composed of two rectangular triangles, built with cyclopean blocs of non-trimmed granite, overlapping each other without cement, and offering a crack between them. The implantation of both parts forms an obtuse angle, suggesting a pyramidal volume.

In addition to its size, the work is characterized, on the one hand, by the roughness of the cyclopean granite blocks, ripped out by mechanical processes and explosive charges at the quarry. It is also powerfully characterized by an image of monumental ancestry, which is suggested not by a straight transcription, but by a kind of abbreviation, which eliminates redundancy and refines the final result, reducing the elements of its own decoding to a minimum.

This piece was created during the second international symposium of stone sculpture, which gathered sixteen sculptors both Portuguese and worldwide, during the month of July 1985. The first symposium had been organized, in 1981, in Évora. The Oporto symposium used only Portuguese stone and was registered on film by the prestigious film director Manoel de Oliveira, with the assistance of his son Manuel Casimiro, from which we add a few frames.

The last piece of the Contemporary Art Route is the sculpture *O Meu Sangue é o vosso Sangue* (My Blood is your Blood), by Rui Chafes, dedicated on the 15th of July 2015, when *Museu da Misericórdia do Porto* (MMIPO), opened to the public.

Figure 15. Rui Chafes, My blood is your blood, 2015, painted iron, MMIPO, Rua das Flores, Porto (inside/outside).
Source: Own elaboration, based on MMIPO and JG Abreu

This specific context is relevant, for the sculpture is composed of three distinct segments: one located inside the museum and two located outside, a lower one and an upper one, as if piercing the wall as the extension of the first one located inside.

In fact, as the title of the piece suggests, the sculpture intends to establish a link between the 16th century painting *Fons Vitae*, by an anonymous painter of the Flemish school, displayed in the same room where the inner segment of the sculpture is, as the following images show.

The sculpture "My blood is your blood" is therefore an artwork about connectiveness. It aims not only to connect the inner and outer spaces of MMIPO, but also to connect contemporary art and renaissance art, being the connection established by a vein which shares the common immanent nature of blood, both sacred and carnal.

In this black iron sculpture, the metallic matter displays organic forms: something between a tendon or a vein, or a combination of both. The black color normally connoted with death, appears as a support of life. But it is also true that the negativity it symbolizes may suggest a pure enunciation – the purest of all – of its radical refusal of any possible formulation, as it is the sole single way of avoiding what would necessarily become a depreciation of the sacred nature of all blood, the blood of all.

With Rui Chafes's piece, we close our itinerary within Oporto's Public Art Map, leaving behind much to see and yet much more to discover. The driver has to make now a choice, and decide if he is willing to resume our itinerary, or if he prefers to stay and navigate in Oporto's Public Art.

If the driver chooses to stay, he can go to the nearby Porto Welcome Center and pick up an *Oporto's Public Art Map*, to guide him on his journey. Instead, if he chooses to go on, before picking up his car, let me remind him of José Saramago's words in his *Travel to Portugal*, when expressing his feelings towards the city, he dedicates his farewell to an ancient public art piece:

When the traveler is about to leave, he will return to Pelican Fountain, he will look at those angry women who tied to the stone are defiantly, he will know that there is a secret there, that no one will come to explain to him, that is what he will take from Porto, a hard mystery of gloomy streets and earth-colored houses, all of this as fascinating as the lights that turns on at dusk on the slopes, the city joined to a river they call Goldy. (Saramago, 1985, p. 77)

5th Station: King D. Sebastião – Esposende

The statue of D. Sebastião, by Lagoa Henriques, 1973, enters our itinerary as a single piece, mainly because, in our opinion, it represents the last of the "statuary statues".

In another place, is written that "statues are political statements" (Abreu, 2020), and in this case we think that the statue of D. Sebastião, by Lagoa Henriques, was the very last enouncement of the "national-historicist" model for *Estado Novo* official statuary.

In fact, the dedication of this statue occurred in 1973, the last complete year of *Estado Novo* dictatorship, for most of year 1974 was already spent in democracy, as the Carnation Revolution took place on the 25th of April.

By that time, Esposende was a modest village of Minho whose population was about twenty-five thousand people, and whose motor of development was its growing use as resort during the summer and on the weekends by second housing dwellers.

A few years later, in that same avenue, the statue of King Sebastião was dedicated, celebrating the designation of Esposende as an autonomous village, in 1572.

The statue of King Sebastião, while being above all an intelligent exercise of plastic art, it is in no lesser degree a remarkable narrative piece.

In the statue, King D. Sebastião appears depicted, both figure and pedestal, surrounded by clouds, in an allusion to the myth of the return, in a foggy morning, of the king who disappeared during the battle of Ksar el-Kebir, in northern Morocco. Besides, the King is represented uncrowned, with the left half part of the trunk naked, while the right part of the trunk, as well as the neck and the lower part of the body, are dressed in an Augsburg field armor (https://www.metmuseum.org/art/collection/search/23084), reflecting the iconography of the portrait of King D. Sebastião by portuguese Renaissance painter Cristóvão de Morais (http://www.matriznet.dgpc.pt/MatrizNet/Objectos/ObjectosConsultar.aspx?IdReg=247677&Lang=EN).

While dressed to fight, the king holds no sword, has bare naked legs and feet, and appears floating over the clouds. Both figure and pedestal are made in bronze, which gives a great visual coherence to the monument.

The image of the King has an oneiric quality. The king is depicted as a contradictory person, partially bellicose, partially vulnerable, whose impossible synthesis aims to suggest his idealistic persona. With this statue, Lagoa Henriques took the statuary model to its limits, reducing the respect to natural resemblance and narrative to its minimum.

Figure 16. Lagoa Henriques, King D Sebastião, 1973, bronze, Av. Arantes e Oliveira, Esposende.
Source JG Abreu

It is curious to note that also in 1973 another statue of D. Sebastião, by the sculptor João Cultileiro was dedicated in the southern town of Lagos, Algarve, which is considered the first modern Portuguese statue. While very different, both share the same naïve figuration of the young paradoxical soldier-king.

6th Station: Vila Nova De Cerveira – Village Of The Arts

Our last station will be the most artistic. Vila Nova de Cerveira's *Bienal Internacional de Arte* (BIA) is the oldest and largest Peninsular Biennale art event. Located in a small rural Community in the border between Minho and Galicia, it is the sole Portuguese artistic program that has chosen as its *ex-libris* a modern public sculpture.

Launched in 1978, as the *V Encontros Internacionais de Arte – I Bienal Internacional de Arte* by the painter Jaime Isidoro and the local Municipality (Silva, 1978).

Once again, our itinerary will only consider a selection of two permanent pieces implanted in the urban space, whose criterion is the oldest and the newest, and two pieces implanted in the natural landscape environment, whose criterion is one in the mountain and the other by the river.

The first piece to be implanted was the sculpture Granite-Rhythm, by Clara Menéres, 1978, produced during the first Biennale.

Granite-Rhythm is a sculpture willing to establish a relationship between contemporary art and the traditional elements of rural life. The use of granite mainstays and iron bars echoes the application of agricultural materials and structures, while the curved line and the increasing distance variation from the curved line to the straight line traced by the mainstays, suggest abstract form and arithmetic progression (rhythm), adequately merging minimalism with ancestral imagery.

Figure 17. Clara Menéres, Granite-Rhythm, 1978, granite and iron, Vila Nova de Cerveira.
Source: JG Abreu

The next piece is *River: water and blood*: a neon light sculpture by Silvestre Pestana, implanted next to the Biennale Pavilion, in 2020. However, it is not its first installation. The same piece was first implanted in a larger lake, also in Vila Nova de Cerveira (Caeiro, 2015, p. 192), when Silvestre Pestana wan the Great Prize *Bienal de Cerveira 2003*, and was invited to create a work for Cerveira's public space.

Figure 18. Silvestre Pestana, River: water and blood, 2003-20, Biennale Pavilion, Vila Nova de Cerveira.
Source: Patrick Esteves / FBAC

Because of the construction of the new Court of Law, the work had to be removed and was finally re-implanted in a new location, for the 21st edition of the Biennale.

As a neon light piece, it is better to see it at twilight. It is a neon light colored pillar, implanted on the center of a small lake. The previous implantation, closer to the river, had in fact a more lyrical landscape integration, while the present implantation appears more urban and architectonic, if not institutional. Nevertheless, the contrast of the red warm color with the blue cold one, its reflection on the water surface and in the granite border of the lake, contaminates the environment with color, suggesting the spatial quality of Dan Flavin's work, amplified here by the fact that it is an outdoor installation.

The most scenic piece of all is the *Cerf*, by José Rodrigues, 1984. This black painted iron sculpture, suggesting a geometrized Cerf, became Vila Nova de Cerveira's ex-libris.

Made of several cropped and welded sheets, bars and tubes of iron, forming a silhouette like image without volume, the sculpture has a powerful presence at the top of Serra da Gávea, dominating the landscape with its monumental scale. Implanted on four high concrete blocks, the Cerf is depicted with its head turned up (as a wolf howling), and marks a dynamic presence in the bucolic (the river valley), but also wild (the surrounding mountains) landscape.

The brutalism of the high concrete blocks contrasts with the lightness of the constructivist-like sculptural technique, which discards any eventual reading of romantic sublimity or banal poetic lyricism.

Figure 19. José Rodrigues, Cerf, 1984, painted iron, Cerf belvedere, Serra da Gávea, Vila Nova de Cerveira.
Source: JG Abreu

Tear in the soil, by Gabriela Abergaria, 2018, is implanted on Vila Nova de Cerveira waterfront, next to River Minho. The work shares with other sculptures the same amazing environment, while assuming, unlike the other pieces, its nature of an earth work intervention, maybe as a strategy to underline its outsider status.

In fact, *Tear in the soil* has no relationship with the *International Biennale of Vila Nova de Cerveira*. Its origin is *Desencaminharte*, a recent public art program promoted by CIM – *Comunidade Intermunicipal do Minho* (Minho's Inter-municipal Community), whose scale of action englobes the entire District of Viana do Castelo, having its 2018 edition spread ten public art pieces across the territory of Alto Minho.

It is a discrete intervention: a snaking line groove carved in the soil and covered with granite stone, forming a path which refers to the water line of the river, during a flood. The groove is carved forty centimeters lower, and the granite stones are reused and minimally worked, intended as a reminder of traditional rural constructive techniques.

With this artistic intervention on River Minho's banks, and not far away from the modern bridge that crosses the border between Minho and Galicia, ends this first primary itinerary around Portuguese northern public art collections, and the driver returns home, or if he has the will and the energy to continue, maybe he could use this last piece as an impulse for the discovery of the other pieces of the program *Desencaminharte*, using the full and accurate indications given by its website: https://www.desencaminharte.altominho.pt/.

Figure 20. Gabriela Albergaria, Tear in the soil, 2018, granite, Parque de Lazer do Castelinho, Vila Nova de Cerveira.
Source: JG Abreu

But if the driver prefers to take a break before any further explorations, he could watch online *Desencaminharte 2018* film (online here: https://vimeo.com/350316280), or quietly scroll its book (online here: https://issuu.com/desencaminharte2018/docs/des_issuu).

FUTURE RESEARCH DIRECTIONS

Further developments should include other itineraries, already mentioned on Table 1, as well as the related institutional background support mentioned on Table 2.

Moreover, it would be pertinent to organize trans-European public art routes with the required strategies and means, since it is possible to establish meaningful relationships with public art collections of other European countries echoing similar typologies, techniques, artistic affinities and even common memorial content.

CONCLUSION

The main conclusion asserts the high touristic potential of both historical and contemporary Portuguese public art collections. Although the present study deals only with a single artistic center, focusing a restricted geographical area, it stands for the diversity and richness of the analyzed works.

It seems also clear that no adequate public art routes can be organized without a proper institutional background support, since much benefit may be achieved, if the driver, stimulated by the already given examples, is free to make his own interpretation about the specific pieces or ensembles he discovered, helped by its background institutional support.

In this sense, public art car itineraries seem to provide a deep and genuine contact not only with public art collections, but also with its surrounding natural environment, social substratum and cultural frame, while assuring a defensive way of travelling.

ACKNOWLEDGMENT

I would like to thank Dr Laura Castro, Director of *Direção Regional da Cultura do Norte* for her commitment to our effort in cataloguing public art collections of the North of Portugal and for organizing public art cultural routes.

I would also like to express, in a most personal way, my gratitude to Dr Maria Georgina Abreu, PhD in English Culture, for her kind proofreading of the English text and all the benefit that her suggestions brought to its fluency and clarity.

REFERENCES

Abreu, J. G. (2012). *A Escultura no Espaço Público do Porto. Classificação e Interpretação*. Universidade Católica Editora.

Abreu, J. G. (2016). Tempos Românticos. Ou o Romantismo como Recorrência nas Artes. In *Atas do II Congresso O Porto Romântico*. CITAR.

Abreu, J. G. (2020). Letter on Statuary and Power. Statues as Political Statements. Cadernos de Arte Pública/Publica Art Journal, (2), 6-33.

Almeida, B. (2007). *Alberto Carneiro. Lição das Coisas*. Campo das Letras.

Berti, E. (2015). *Cultural Routes Management: From Theory to Practice*. Council of Europe Publishing.

Burnshaw, S. (1995). *The Poem Itself: 150 of the Finest Modern Poets in the Original Languages*. University of Arkansas Press.

Caeiro, M. (2014). *A Arte na Cidade. História Contemporânea*. Círculo de Leitores.

Carneiro, A. (2007). *Das Notas para um Diário e outros Textos. Antologia*. Assírio & Alvim.

Castro, R. (1909). *Obras Completas de Rosalía de Castro II. Cantares Gallegos*. Librería de los Sucesores de Hernando.

Chafes, R. (2016). *O Meu Sangue é o Vosso Sangue*. Santa Casa da Misericórdia do Porto.

Correia, F. C. (2009). *O mosteiro de Santo Tirso, de 978 a 1588: Estudo*. CMST.

Dias, M. G. (2014). *Ao Volante pela Cidade. Dez Entrevistas de Arquitetura*. Relógio d'Água.

Giedion, S. (1951). *A decade of new architecture*. Girsberger.

Lecea, I. (2009). *Barcelona. Art Public*. Ajuntament de Barcelona i Àmbit Serveis Editorials.

Mirbeau, O. (1907). *La 628-E8*. Bibliothèque Charpentier.

Moreira, A., & Carneiro, A. (2015). *Museu Internacional de Escultura Contemporânea de Santo Tirso 1990-2015*. CMST.

Nora, P. (1984) Les Lieux de Mémoire. Gallimard.

Oliveira, A.L. (2011). *Esculturas de Arlindo. Quando a escultura passou a ser apenas escultura.* Tese de Mestrado, FBAUL.

Pacheco, E. (2001). *Alteração das Acessibilidades e Dinâmicas Territoriais na Região Norte: expectativas, intervenções e resultantes*. Tese de mestrado. FLUP.

Reyero, C. (1999). *La Escultura Conmemorativa En España: La* Edad De Oro *Del Monumento Publico, 1820-1914*. Cátedra Caduernos Arte.

Ribeiro, T. (1863). Sociedade Typographica Franco-Portuguesa. Academic Press.

Rincón García, W. (2013). Escultura del siglo XIX en Zaragoza. De la imagen devocional al monumento conmemorativo. In M. C. Lacarra Ducay (Ed.), *Arte del siglo XIX* (pp. 13–79). Instituto Fernando el Católico.

Salazar, A. (2013). *The European Crisis*. Orfeu.

Saramago, J. (1985). *Viagem a Portugal*. Editorial Caminho.

Silva, A. (2000). *Ensaios sobre Cultura e Literatura Portuguesa e Brasileira I*. Âncora Editora.

Silva, M. (2018). *Bienal de Cerveira (1978-2007). Memória e singularidade*. PhD thesis, FBAUP

KEY TERMS AND DEFINITIONS

Center: This is a large and most representative series of public art pieces that entails a wide chronological range, while keeping a certain artistic consistency due to the effect of a specific art school or a distinguished tradition, whose value attracts students and artists from abroad.

Cultural Route: A cultural, educational heritage and tourism co-operation project aiming at the development and promotion of an itinerary or a series of itineraries based on a historic route, a cultural concept, figure or phenomenon with a transnational importance and significance for the understanding and respect of common European values.

Cultural Route Operator: An organization or a grouping of organizations legally registered in one or several of the Council of Europe member States, or a public institution, which carries the legal, financial, and moral responsibility for the management and functioning of a cultural route and represents the route *vis-à-vis* the Council of Europe.

Ensemble: It is a less coherent collection that entails a medium chronological range, while conveying heritage value, often beyond strict artistic excellence.

Nucleus: This means a specific collection of pieces that entails a narrow chronological range, while keeping a high coherence as it reflects the implementation of a specific public art program, art in architecture project or public space art curatorship.

Portuguese Primary Public Art Itineraries: This process is conceived under a conceptual umbrella that includes the *Cultural Route* and *Cultural Route operator* projects, in order to establish a first step towards the definition of a future wider and more elaborated European Public Art Route.

Single Piece: It signifies a specific public artwork that stands out its context, because of the excellence of its aesthetic, symbolic or narrative content, and often because all these features are present.

APPENDIX: ACRONYMS

ADL- Arquivo Distrital de Lisboa
ADC- Arquivo Distrital de Coimbra
ADP- Arquivo Distrital do Porto
AMP- Arquivo Municipal do Porto
CAAC- Centro de Artes Alberto Carneiro
CITAR- Centro de Investigação em Ciência e Tecnologia das Artes
CMAlc- Câmara Municipal de Alcobaça;
CMAlm- Câmara Municipal de Almada
CMCR- Câmara Municipal de Caldas da Rainha
CML- Câmara Municipal de Lisboa;
CMO- Câmara Municipal de Óbidos;
CMPrt. Câmara Municipal do Porto
CMPds- Câmara Municipal de Paredes;
CMST- Câmara Municipal de Santo Tirso
CMVC- Câmara Municipal de Viana do Castelo
CMVNC- Câmara Municipal de Vila Nova de Cerveira
CMAS- Casa Museu Abel Salazar
CMHD- Casa Museu Humberto Delgado
CMSR- Casa Museu São Rafael
CMTL- Casa Museu Teixeira Lopes
DRCN- Direção Regional da Cultura do Norte;
DRCC- Direção Regional da Cultura do Centro
DGPC- Direção Geral do Património Cultural
FBAUL- Faculdade de Belas-Artes da Universidade de Lisboa
FBAUP- Faculdade de Belas-Artes da Universidade do Porto
FCSH- Faculdade de Ciências Sociais e Humanas
FAMS- Fundação Arquiteto Marques da Silva
FBVNC- Fundação Bienal de Vila Nova de Cerveira
FEAA- Fundação Engenheiro António de Almeida
FJR- Fundação Júlio Resende
FHD- Fundação Humberto Delgado
FASM- Fundação Aristides Sousa Mendes
FCB- Fundação Calouste Gulbenkian
FMS- Fundação Mário Soares
FCCB- Fundação Centro Cultural de Belém
FEDP- Undação Eletricidade de Portugal
IPL-ESAD.CR- Instituto Politécnico de Leiria-Escola Superior de Arte e Design das Caldas da Rainha
MAD- Museu António Duarte
MBF- Museu Barata Feyo
MJF- Museu João Fragoso

MLA- Museu Leopoldo de Almeida
MJM- Museu José Malhoa
MNAC- Museu Nacional de Arte Contemporânea do Chiado
MCG-CAM- Museu Calouste Gulbenkian. Centro de Arte Moderna
MNAA- Museu Nacional de Arte Antiga
MNSR- Museu Nacional de Soares dos Reis
MMIPO- Museu da Misericórdia do Porto
MASC- Museu Amadeo de Souza-Cardoso
MIEC- Museu Internacional de Escultura Contemporânea de Santo Triso
UCP- Universidade Católica Portuguesa
UM- Universidade do Minho

Chapter 23
From Tourist Destination to Tabernacle

Regina Frank
Independent Researcher, Portugal

ABSTRACT

The author describes her personal experience and insights on Portugal as a German artist who has lived and worked worldwide. She observes the changes from her first arrival during the European Football Championship in 2004, the experience of the WebSummit and various exhibitions in Portuguese Museums, up to the COVID-19 confinement and reopening in 2022. She tells the story of 18 years of development of the country, its tourism and economy, the spiritual path, and also the "zen exercises" the community has and had to offer: getting lost in Lisbon, finding faith in Fatima, sanctuaries in Santarem, and tabernacles in Tomar, as Portugal became her literal port of the grail (Port-u-gal). From children to education, poetry, prose, science, food, art, and religion, she paints a picture from the view of a foreigner mingling with the Portuguese and the ex-pat community. Isolated in the countryside during COVID-19, she connected online with Portuguese artists and created new alliances.

INTRODUCTION

Exhibiting and performing mainly in the capitals of the USA, Asia, and Europe, Regina Frank always carried with her Fernando Pessoa's (1982) Book of Disquiet. She had neither spoken a word of Portuguese, nor had she been to Portugal. After her artist in residencies in Japan and Taiwan, the Gulbenkian Foundation funded her new project in Portugal called Dreamweaving. (see Figure 8)

After numerous museum exhibitions, traveling only for professional reasons, she landed in Portugal for her first holiday in twenty years. Portugal then became her favorite destination - a place where she found refuge between projects enjoying the nature, the culture, and the food it had to offer. Watching the country grow, she observed changes in all areas: from arts to soccer to zen practice. Between 2004 and 2007, as a tourist, she got lost in Lisbon — later, as a resident, she watched Lisbon gentrification.

DOI: 10.4018/978-1-6684-3369-0.ch023

Torn between seeing a town overwhelmed by tourism, its unique culture partially destroyed, and appreciating the new opportunities for artists, she developed strategies to find the positive in all change. In Portugal, she connected with an international crowd during the WebSummit, sat with thousands of Portuguese people at the Pavilhão do Conhecimento just 56 days before the lockdown with her installation iLAND. Then COVID19 seemed to change everything. Watching the Urban desire for culture crumble, she found a stronger calling for the virtual and viral in the arts and environmental sciences.

She took time for magical meetings every year in the mountains of Monchique, celebrating ancient Vedic ceremonies. In 2017, among thousands of people at the 100 Years of Fatima celebration, she reconnected to her Catholic cool. Long durational performances at MAAT coincided with a Festival of Light organized by the Jesuit community, deepened her faith. Walking the Ways of St James, along the shore on her path to Porto, tracing the Templars in Tomar, and savoring the sacred week in Braga, she connected to the spiritual tourism of this Port-o-Gral, (Harbor of the Grail).

During the COVID19 confinement, she celebrated solitude, living in the Portuguese countryside, breathing deeply near the churches of Santarém and the Tagus River. While her daughter zoomed and zapped, broadening her horizon online, Regina Frank returned to her roots in a routine of writing, painting, sculpting, filming, editing, cooking, and working with plants. She used the opportunity to build a new body of work based on Silenced Science and her Silenced Sides, later shown at the Museum of Natural History and Science and the CPS (Centro Português de Seriegrafia). With a group of professional artists, she cofounded SOS Arte PT, spending many hours in online conferences. Here she made more meaningful friendships during the COVID than at the superficial openings she previously tended to avoid anyway. She created several virtual exhibitions paired with poetry, pearls, text, and textile.

In the text, she describes the zen exercises the Portuguese culture offers: situations in traffic, public offices, and supermarkets, and her solution and salvation. She watched in awe as conformity became a code for coexistence and tolerance as a key to transformation, admiring the Portuguese organization of the vaccination.

This text features Regina's transformation of seeing Portugal from the perspective of a frequently-visited tourist destination to eventually becoming her home and a tabernacle and vessel for ritual structure and spiritual connection - a sanctuary.

In the text she describes the "zen exercises" that the Portuguese culture has to offer: situations in traffic to public offices and supermarkets, and her solution and salvation. She watched in awe as conformity became code for coexistence and tolerance as a key to transformation, admiring the Portuguese organization of the vaccination.

This text features Regina's transformation of seeing Portugal from the perspective of a frequently-visited tourist destination, to eventually becoming her home and a tabernacle and vessel for ritual structure and spiritual connection, a sanctuary.

Home was my AdDress

Until 42, I lived minimalistic, being homeless and home-filled at the same time, squatting buildings, sleeping at the beach, under a bridge, in an open field, in a bus and boats, being a guest, cooking for people in exchange for a room, sleeping in fancy hotel's presidential suits or in cockroach filled cellars. Being a guest professor in faculty housing or an artist in residency, I was flexible and often in awe of what the universe provided. Like this, I was able to spend time in many cultures on all continents. I was creating work, selling, donating, or recycling it for the next creative cycle. My home was my AdDress.

Making Art was like breathing air. This was the case until the age of 42, when I arrived in Portugal. My pencil was my best friend. Of all the countries I have lived in I love the mix of Germany and Portugal. Especially now, we move so freely in space and time through social media, zoom, meditation, mediation. Travel seems almost old-fashioned -- still, it is wonderful and full of wonder to perform in public IRL (in Real Life) and to feel the rising of energy in a mesmerized audience.

Figure 1. iLAND performance at São Pedro Beach in Portugal, 2013
Photo: António Pedro Ferreira

From Art to HeArt

When I published my first book, The Artist is Present, in 1999, I had no idea about Portugal. And yet the first publisher (Lusitania Being Online, Net Subjectivity (1996)), who put me on their book cover in 1994, was Portuguese.

Once I settled in Portugal and Germany, I changed my motto to The HeArt is Present. The way I changed from Artist (from the Artist is Present 1989-2010) to HeArt, my nomadic life ended, and I grew roots. During an annual vedic ceremony, I connected all the dots in my life, and arrived where the cruci-fixation of time and space vanished. I noticed that HeArt is where all begins and all ends. In the Heart all comes together, and in the end all is about love.

Before my arrival, I had been living and performing mainly in the USA, Japan, China, and the capitals of Europe. I had it all figured out. Kunst Kind und Karriere, the three KKK do not come together. If you want to do art (Kunst), it has to consume you, nothing can be more important than that, and once you have a career (Karriere) and/or kids (Kinder), they will take away from that committment. Without speaking Portuguese and never having seen Portugal, I went to sleep and woke up with the work of the poet Fernando Pessoa. For more than a decade, his *Book of Disquiet* was my source of calm and inspiration - in the subway it was my fast track to light, and after my daily dive into life, I dove into his words. When I lost my German copy in New York, I replaced it with an English one, only to find it even more compelling. I could not resist the temptation of his texts. Later in Paris, he sounded new in French again, and finally, I learned some Japanese through him by pointing to some paragraphs - a performance in itself. This man's soul seemed so strangely familiar, especially when I realized that he had created different personalities with different birth dates and lived according to their astrological charts.

Numerous Portuguese writers, even the highly appreciated and widely published, were often unable to live in their profession. Unless independently wealthy, they were forced to take a job as a translator, editor, or even something vastly different. Fernando Pessoa earned his daily bread as a bookkeeper — one of the greatest Portuguese poets and writers. Little did he know that now people make millions with special editions of his books, the house where he lived is now a museum with thousands of tourists stamping through, admiring the box that delivered the quotes for the great *Book of Disquiet*. What a pity that we rarely write like this. Leave the editing to the professionals, to the historians! Like Pessoa: let your mind wander, capture the thought in flight, without weighing words, balanced by the soul!

Food, Fear, and Fuss

The Portuguese have a thing for slow food without making much of a fuss about it. Most workers go out for lunch, which consists of soup, a main dish, dessert, and a glass of wine or beer, and of course a coffee. The kids have lunch at school or Kindergarten, and a sandwich just doesn't do it. These menus usually cost around 3-15 Euros depending on the location. Between 12h and 14h people flock to their nearest restaurant and enjoy a meal together, which is a lovely tradition. In New York, this common European part of life is like a myth: wine for lunch seems to be the ultimate luxury.

In the performance "Fast Food Slow Motion," (see Figure 2) I ate a vegan burger in a fast-food restaurant, taking 45 minutes. The result was a time-lapse movie of 70 seconds. The reaction of the audience was priceless.

Eating flexitarian, avoiding meat wherever I was, practically on a vegan diet (vegan wasn't even a word yet) didn't seem to be possible when I first came to Portugal. In Japan, I could survive on soba noodles, soups, vegetables, rice balls filled with salty plums, and occasionally I ate sushi. In Taiwan and in India it was incredibly easy because most of the food is vegetarian and tasty, offered in a buffet style, and the meat is usually separate. In Germany, it wasn't as easy, but it was possible. But in Portugal, there was hardly anything vegetarian on the menu in 2007. So, for the first couple of weeks, I lived on fries, bread with olive oil, boring boiled vegetables, lettuce, tomato, and olives. Then I learned not to be so attached to a special diet and experimented and tasted everything.

Needless to say, I was in paradise when this changed and vegan restaurants started to pop up due to increasing demand. Along with the global concern for greater sustainability, the Portuguese youth started to create great projects, with Alliança Animal and Escola Vegana among them. At the Pavilhão do Conhecimento (see Figure 3 - iLAND Silent Science) in 2020 we collaborated to create a wonderful buffet with vegan versions of traditional Portuguese food. To get a vegan option included in the school menu was a fight in another dimension.

Figure 2 . Fast Food Slow Motion Performance in a Fast Food Restaurant, 2017
Photo: Peter Treffler

Figure 3. iLAND Silent Science at Pavilhão do Conhecimento 2020
Photo: Ana Cristina Marques

It was also difficult to get fresh organic ingredients locally, so every Saturday I went to the market in nearby villages to buy local organic produce. In some ways the Portuguese had never really adopted large-scale factory farming, so it was also not so crucial to buy organic produce. The local stores had fresh produce that was much tastier than many of the organic vegetables I had eaten in other countries. Before this, I had to cook or learn to digest animal products, and I learned that the Portuguese have great fish dishes like Caldeirada, and meat dishes like Cozido à Portuguesa. The vegetables in those stews taste like heaven and thankfully my boyfriend ate the rest….

With the confinements of COVID19, a lot of the Portuguese habits changed. Restaurants had to close, some never reopened, and many opened for delivery. Apps like "Too Good To Go" (an app that connects customers to restaurants and stores that have unsold food surplus to reduce food waste) were extremely helpful, but apps like UberEats and Glovo nearly destroyed some of the restaurants. Places I used to frequent gave me a discount if I came to pick up the food myself and paid directly, as they claimed to be owed thousands of Euros by UberEats. Some stopped working with the apps altogether, and none of this was very transparent to the consumer. Soup kitchens struggled to survive and keep up the good work of feeding the poor. Some locations were locked down at short notice, often leaving the restaurant owners stranded, having shopped for the weekend and invested in fresh produce. This discouraged some places from investing in the next weekend, and to focus on their "to go" business instead, resulting in the wait staff being furloughed for extended periods.

In Portugal, the restaurants that remained operational seemed exemplary in terms of hygiene - nowhere else in the world have I seen waiters clean and wipe every chair and table, put cutlery in paper bags, and have the menus available only via QR code. Hand sanitizer was available at every entrance and often on every table, or someone was standing at the door to greet you with a spray. Carousels were wiped down after every ride, and in stores, the conveyor belts and credit card terminals were cleaned after every client passed through.

The fact that you had to show a vaccination pass to enter restaurants may have contributed to the high vaccination rate in Portugal. In Germany, the anti-vaxxers seem to protest by eating outside in the freezing cold.

I started to cook all meals at home and prepared them in restaurant style, enjoying the creative process during the pandemic, growing vegetables and herbs in spirals, created herbal and mushroom dresses (see Figure 4). It was a form of painting with food, combining flavors and smells.

Now you feel the celebration of the slow food (in Portugal a tradition that no one called that way). The Portuguese always had mostly slow food. Of course now in light of tourism unfortunately a lot of fast food restaurants opened, taking a lot of business from the original local places. The Portuguese versions of fast food are Bifana, Prego (a nail), Sopa da Pedra nails nutrition - a healthy snack (if you eat meat). In general, since the death toll of COVID sank low, Restaurants are crowded again, and people enjoy both, the dishes and the company. As tourism is picking up, people seem to be more appreciative of the culture and the luxuries we have. The food in Portugal is generally amazing and besides the beautiful landscapes probably one of the strongest attractions for tourism. The simple Tasca to a fancy restaurant has a lot to offer: an excellent quality price balance, freshness, taste with great super friendly service.

Figure 4. Mushroom Dress, first version, Garden of the Palace in Dreieich 2012
Photo: Walter Rogger

All for One, One for All

I was never much into what the Americans called soccer. Though my football feelings are still mixed, Portugal opened that door a few cracks for me. In Portugal, you can hardly find a way around it. I have to admit: I was more of a soccer-sucks- kind-of-girl, until I came here and some German ex-pats invited me to watch a game. Germany playing against Portugal in 2004 was a big event. So sipping Sangria, I noticed Ronaldo's incredibly joyful and elegant dance with the ball, his creative and surprising rollick, suddenly somehow shooting the sphere into the goal. Watching the German weaving a web, he almost felt like a soloist in the concert - so beautiful - but also agonizing! When I shifted my gaze to the Portuguese team, it often stood forlorn, unemployed, alone, and hardly ever all one. Watching the Portuguese win, the people celebrating in the streets, dancing on top of their cars. It was delightful.

Later I watched this Ronaldo Syndrom in schools, universities, at work, and in arts. Group work is being done so often by one alone, information rarely shared, competition is fostered, and work not evenly distributed. How often did I had to watch my loved ones sweat to make their team look good? In my German, American, and Japanese groups, we would divide the tasks equally, which could sometimes take a while. Then report to each other frequently, exchanging information, trying to find the most efficient way to reach our aim, supporting and balancing each other, and learning. While we shared our thoughts in a safe space we created excitement, wild and crazy dreams, deeper knowledge gained through our focus on a small area and through the group exchanging questions, ideas. But in Portugal, it seems that everyone has to reinvent the wheel: This can be ingenious, but it is exhausting. One person ends up taking control, and the creative potential of a group, through the safe exchange of ideas and dreams, vanishes into thin air.

While the individual genius is supported, teamwork is not learned in the schools or dangerously neglected. What other cultures seem to find essential from early on, sharing, playing together, and helping each other to achieve greatness as a team, seems to be lacking in Portugal. Groups of artists working together and promoting each others work, shared studio spaces inviting curators are seldom, while solo stars like Joana Vasconcelos and Pedro Cabrita Reis have large teams of employees behind them and factory size spaces sweeping up nearly every commission. Pedro Cabrita Reis, an avid collector of mainly Portuguese young artists sold his collection to the EDP Museum, thus helping and promoting other artists but the real group spirit like Group Zero, the Neue Wilde, Group Material, Guerrilla Girls, just to name a few, seems to be inexistent in Portugal. During the Pandemic, SOS Arte PT was founded to promote and help artists through the pandemic and beyond. For some of the projects, we worked day and night to make this a refreshing community with tremendous success from early on. The talks and openings had up to 5000 views on Facebook, inspiring the change and evoking discussion and dialogue. Within a few months, their social media had a following of more than 3000 people and more numerous active supporters with lawyers, managers, and politicians joining the cause. Due to the diligent dedication of the director Margarida Sardinha and a few active founding members an online exhibition with 117 participants could be realized, which culminated in an exhibition and comprehensive catalog featuring all the works at Maia in Porto carefully curated by Antonio Cerveira Pinto and the director of MAIA. The mood was celebratory when we finally could come together IRL again and met first time in person. It occurred to me that I had spent more than 60 hours on the phone and in video conferences with a person without ever meeting live. What a strange feeling to finally hug or have a drink together!

A professional website could be established, designed by photographer and designer João Bacelar. There were numerous well-prepared and strategized meetings with the PSD (Social Democratic Party), the PS (Socialist Party) parliamentary groups, the XII Culture and Communication Commission, the Councillor for Culture of the Mafra Municipality, City Council of Lisbon, Civil Society Forum of the UNESCO Convention 2005 (Convention on the Protection and Promotion of the Diversity of Cultural Expressions), the PLMJ Foundation, Rock in Rio and more.

The established collaborations and connections resulted in a change: the approval of the New Statute for Culture Professionals with the Ministry of Culture, (for the first time officially recognizing that making Art is a profession (something that most other European countries take for granted). It inspired the law that 1% of the public building's budget would go towards art. SOS ARTE PT took also to heart the Promotion of the Diversity of Cultural Expressions), where it participated in the elaboration of the recommendations to be presented to the States Parties. On social media, SOS Arte PT started a Campaign to raise awareness and defend Artist's rights.

Tiago Duarte moderated monthly online conversations with international artists. Countless press releases resulted in numerous articles in the print media and online newspapers. There were plans and clear concepts and designs for an online magazine, an online gallery, an emergency fund, micro credits, etc. SOS ARTE PT developed big plans for exhibition projects in museums and public spaces, a concept for an O2 award, and started various fundraising activities, applying at DGArtes, and RAAML (Lisbon City Hall). Unfortunately, the grant proposals were not accepted (even though they passed until the last round) and the group, tired after two years of voluntary work, had to focus on their paid projects.

You could see that good teamwork can result in something great. If there are too few active, it is not sustainable. Ronaldo has learnt how to team-play, and it is now a delight to watch him juggle the ball and the team. Just picture one additional percent of the soccer budget going into culture: how many changes could be implemented? Soccer lives of the moment, but aren't we profiting immensely from the previously created art and culture? Picture Portugal without its churches, buildings, paintings — a country without its cultural heritage, without its artists, its contemporary design, inventions, creative contributions, and the changes this brings to our lives.

Clothing and Colors

After a trip to New York with far too much luggage in 1992, I had scaled down and was living out of a small carry-on for 14 years. The rest of my large dresses that were art pieces, traveled alone or were in permanent museum collections. With so few pieces of clothing, my palette was black for many reasons: it's easy to fit in wherever, all items match, it's non-distracting when you compose colorful installations, and spots and stains are not very visible or they brush out quickly. In Portugal I found a pink raincoat with a floral lining. The idea of wearing pink for me was like eating pork knuckle is for a vegan. But the price was so tempting, that I thought I might as well keep it as a joke. However, this coat became a steady companion here in Portugal as I was pregnant, and with Sofia, color came into my life.

Figure 5. Mother Mandala, Permanent Installation, International Forest Art Path in Darmstadt, 2020 made from donated dresses, sculpture, stones, mosaic
Photo: Peter Treffler

There were inspiring charities in Portugal, and one of my favorites was Labels for Less, a charity event organized by the WRVS Women's Royal Voluntary Service. Every spring and fall, women would donate their hardly-worn wardrobes, and all the proceeds would go to a local charity. Here I was able to expand my wardrobe from three pieces of black clothing to a closet filled with color. And so I decided to turn this into an unpublished performance. For seven years I bought exclusively from charities. During the summer, I would wear a different colorful dress every day, and in winter, I would wear black and white.

The Portuguese designers have a unique style that combines comfort with smart ideas and good-quality fabrics. The few relatively small brands aren't exporting too much, guaranteeing exclusivity. It is refreshing that most women wear comfortable flats or sneakers, and seeing women teetering in the back-breaking heels like in New York offices is almost as rare as a white peacock sighting.

Also, clothing in Portugal can be from the previous seasons. Women wear good quality and use the same items of clothing for several seasons. Their style is modest, practical and simple, yet elegant. It is beautiful without being a strain on the body or on their health. Men mostly wear classical suits, although the tie can speak volumes and sometimes looks quite historical — whoever cares to do so, can always donate a tie to the unique restaurant Adega das Gravatas (Ao entrares nesta casa deixa aqui a tua gravata. O vinho que se bebe aqui faz a vida mais barata-- When you enter this house leave your tie here. The wine you drink here makes life cheaper). Sweatpants, athleisure, and streetwear are styles that seem to be more common now. Never before have I seen so many leggings worn with comfortable, wide, short dresses, under which you can hide a multitude of sins. Second-hand shops, which were commonplace and extremely cheap, started to close down during the COVID19 confinement. That, or they became more expensive and less frequently visited. Shopping malls emptied and what used to be the Portuguese way to spend the weekend (shopping and eating lunch at the mall) was replaced by hiking and taking walks in nature - leaving the shopping malls that often overfilled, far less crowded.

Figure 6. Hermes' Mistress, Spiral Wacoal Art Center, Tokyo, 1996
Photo: Masao Yamamoto

Performance - Transformance

Performance art is a bit neglected in Portuguese galleries and museums. Some artists, like Helena Almeida, create interesting performance artwork in the studio and then transfer it to film and photos. Theatre has a tradition and is widely accepted, while performance has a slightly negative reputation. When I first arrived, I was advised not to mention that I am also a performance artist - the few performance festivals I saw here in Lisbon made me understand why. At my first performance in Lisbon I found an extremely attentive, hungry audience. I invited some of my dancer and actor friends to wear dresses I created in the 90s. In slow motion we showed the piece Slowdown Runway. The piece was such a success that we presented the project at the London Artfair. The audience was enthousiastic as the artwork became alive and clusters of people framed the moving artwork. Today, online conversations and videos are seen by more people than we could ever have dreamed of before the pandemic, and more institutions have switched to social media and other internet applications. However, live performances are an entirely different experience: the energy, the dialogue with the people, the space, the small, the tension through spontaneous interaction, the acoustic can hardly be conveyed by an online platform.

I had started an unpublished performance series in 1986 at the age of 21, being moved by the German artist Joseph Beuys, who had just died after planting 7000 oaks for the documenta7. I decided I would do private performative statements in his honor: Wearing black for 21 years, mourning the man and celebrating his activism and work. (If someone gifted me a color, I would wear that as an exception.) Only during an active performance installation did I choose the color of the piece. For example, in Hermes' Mistress (see Figure 6 Hermes' Mistress) my clothes were all red - red coat, dress, gloves, shoes. In this performance-installation I surfed the internet and embroidered the information that touched me with letter beads on the gown, slowing down time and materializing information, herby questioning its value.

Three years later, in 1989, I decided to start an unpublished performance that was supposed to last 33 years until 22.02.2022: When people asked for my contact details, I would give them the slogan The Artist is Present, on a business card, with no phone number, nothing else. Most people pocketed the card without looking at it, only to realize later that it was useless for getting in touch with me (search engines on the Internet didn't exist yet). I would keep all of their contacts for 33 years until 2022, and then handwrite them a letter. If they receive it and still have the card in their possession, they could exchange it for one of my art pieces, as a gift. The value of the simple business card would become a voucher for a present in the future. I had no idea at that time that handwritten letters would become almost as rare as the floppy disks I would have to put into my Atari to start writing on my first computer.

I published this performance on 22.02.2022 at 20:02, the date and time of which is both a palindrome and an ambigram. In 2022, I will release many of my unpublished pieces, which, in itself, is a social media performance.

In 2003, I performed Whiteness in Decay, (see Figure 7) where I broke my body free from a plaster cast. I was surrounded by fruits and vegetables, also shelled in white gypsum to be broken, cut, and shared with the audience. For the first time I shared the stage with the audience, allowing them to touch and paint the porcelain doll that was becoming human. In Taiwan, I shared hugs; in China, cakes and cookies; in France, dough; and I even let people get under my dress – ideas that are impossible to execute now, after COVID19.

Since COVID19, like many artists, I have ventured more into virtual and video performances, AR (augmented reality) pieces, and NFTs. But I have also expanded my painting practice, text, and tapestry creation. The live performances are still significant, but documentation became a crucial element in my process.

Figure 7. Whiteness in Decay, San Diego Museum of Art, San Diego, USA, Permanent Collection donated by Paulo Coelho, 2003 Video stills

Dreamweaving and Sofia

When I first landed in Portugal during the European Football Championship, I was developing the piece iLAND. While spending my 40th birthday sharing silence at Cabo da Roca, it occurred to me that when we take a picture, it is to tame the overwhelming power of nature. To break the breathtaking beauty of the sun dancing with the wind and water smashing against the rocks, seeping into the sand, thus creating the perfect sound and surroundings for our souls to surrender. The human ego in us takes a picture - a last desperate act, to define the distance between the body and the beyond. (This connection to nature here in Portugal grew stronger during COVID. The previously nearly empty areas, where one used to be able to walk without encountering anyone, were suddenly crowded. The promenades and beaches along the river and ocean, when not locked, resembled a procession.)

From then on, Portugal was my "tourist" destination whenever I could find a few weeks between commissions. It was the only place I ever went for holidays without a scheduled work project in sight. I taught a master class at Bauhaus Dessau and then went straight to Lisbon. The feeling was priceless, landing at Portela Airport, the view of the Tejo glittering golden in the sun, the seven hills, the green grid of the Avenidas and public parks - it was like opening the door to heaven. My routine during those months was attached to water and the views: walking at the shore, writing at an empty coffee shop with a view to the ocean, at the house weaving a website. At lunchtime, I ate and enjoyed watching, exploring, and expressing. At low tide, taking a long walk in absolute solitude, from Pedra do Sal up to Parede and back. Then fetching my computer, I wrote for a few hours until business people arrived, chatting animatedly.

Figure 8. Dreamweaving, Egyptian Code, funded by Gulbenkian Foundation, 2007/

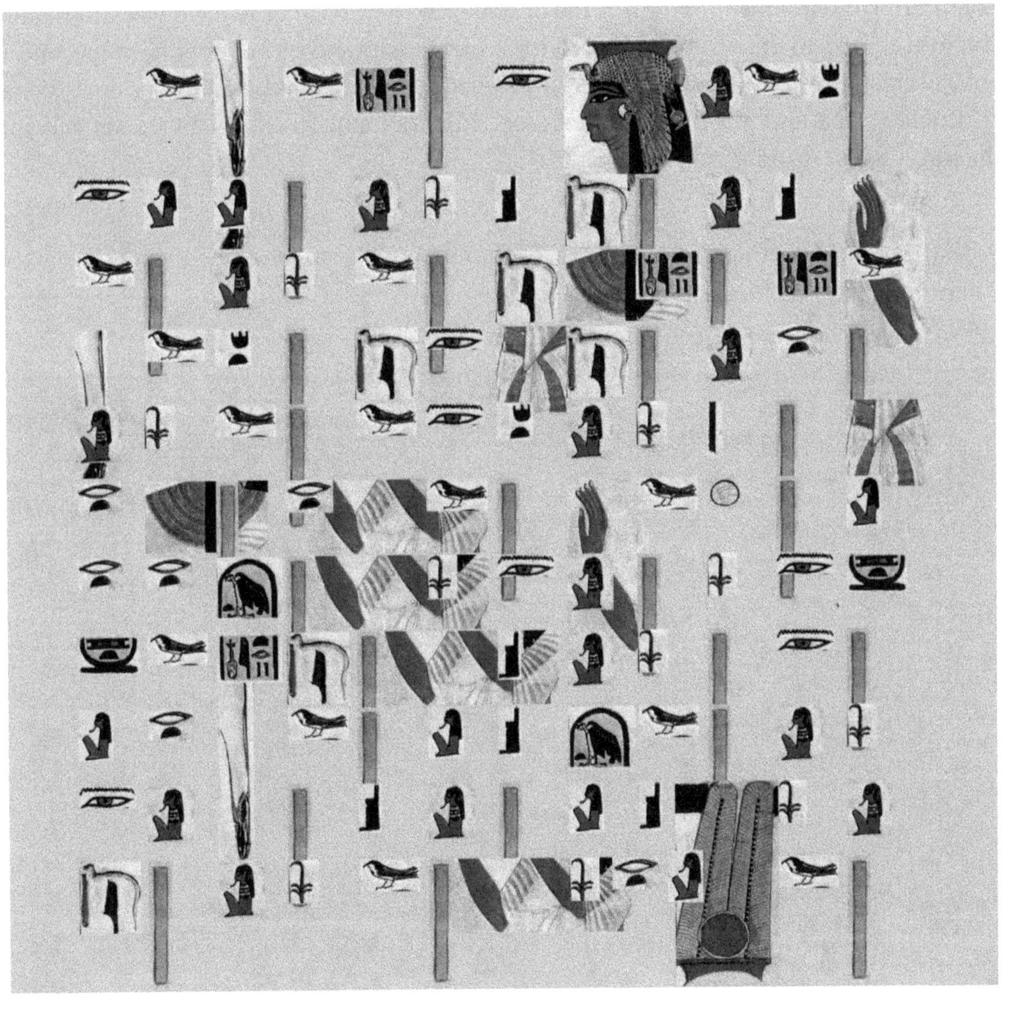

Between 2006 and 2007, I worked on a piece for the Gulbenkian Foundation called Dreamweaving (see Figure 8). Tapestries made from text, letters transformed into images and sounds, and those pictures, woven by a jacquard loom, honored the connection between the inspiration for the computer (punchcards) and the marriage between 0 and 1. Celebrating the absence and the presence of threads, I wove text into textile. I tried to find a weaver, which should have been easy as Portugal is famous for that. By the time I received the quote from the Portuguese weaver, the German company had already sent samples for approval and delivered all seven of the tapestries.

Kids have Great Spaces or Just Space

When you want to know the Portuguese culture, raising a child is the best way In Germany, children have lots of beautiful spaces where they can play. In Portugal, children have lots of space. In every restaurant, children are treated with respect and admiration. They are welcome everywhere and receive lots of love and attention. There is a menu for the little ones, but everyone shares anyway - kids eat everything from early on: olives, snails, eel, sushi, chili, curry - you name it. My daughter was no exception, and she grew up in this nurturing environment - going to the beach every day, once or twice a week to a museum, to the library, to the park, to the playground, and for a picnic with moms and tots from the International Women of Portugal group. When in Germany I traveled with Sofia and made solar dress installations and solar benches - she also got her solar dress (see Figure 9 and 10- Solar dress) but she could also play in the forest Kindergarden….

Figure 9. Solar Dress, Dress made for my daughter to play while I constructed the installation, International Forest Art Path in Darmstadt, 2008
Photo: Holger Hertling

In Portugal, kids wore an apron with white and blue stripes, which for miraculous reasons, never seemed to need washing. Only the German Kindergarten in Portugal seemed to let the preschoolers play in the rain and the dirt. But once in the Portuguese system the kids stayed clean, inside the house or outside on tarred surfaces. Very little creativity or contact with the soil was allowed. But in summer, all go to the beach, in big groups. All with an obligatory hat in an easy to spot color, they play in the sand and puddle in the small pools that the high tide leaves behind. It is a beautiful experience for a tourist, from 9-11 at the beach, observing those toddlers enjoy before the heat and the masses make the beaches less bearable.

Kids were always welcome - the adults and adolescents are great with kids. While kids in Germany are usually supposed to be in bed by eight at the latest, and dinner invitations often exclude children, in Portugal they could stay up and are always welcome. The adult friends would leave their living room to smoke, while the children were present as if it was the most normal thing to do. My child could profit from the open milk bar of my breasts and slept in a sling while I sang, danced, worked, or spoke, and friends remained friendly, all enjoying the company of the baby, toddler, and later the teenager she was.

We would get lost in Lisbon or meet twice a week with the IWP (International Women of Portugal), and I became Playgroup Captain. The museums offered special activities, and had storytellers and theatre events for youngsters. Once a year, the city of Cascais celebrated children´s day on the first of June, with free activities such as jumping castles, big wheels, climbing, sliding, and sweets - we just loved it. In August, the Festas do Mar, a seven-day open-air concert and festival, was a safe place to bring your children and hoist them onto your shoulders. Everyone sang along, and the mood was spectacular. I had never been in such a crowd, let alone felt safe, and suddenly I was enjoying myself with my precious child and felt protected in a community of peaceful music fans.

In 2020 and 2021 those events were canceled, and the playgrounds closed. The youngsters had to retreat from public spaces to small personal playdates, stay home, and were not even allowed in restaurants anymore. However, schools did a fantastic job of making the online lessons almost more interesting than the in-person classes. The first confinement classes were on TV, and only one hour was hosted online every day. This didn't do the trick, as it ended up amounting to barely more than a "hello". Since the TV classes thematically served two educational levels at once, like 5th and 6th year for example, it was boring or ineffective for the 6th year, as it was repetitive. During the second confinement, the classes were exemplary and incredibly interesting, and the homework was so well-designed and project-oriented that even the public-school students felt encouraged and got creative. My friends with children in German schools were not so lucky. In Germany youngsters learn a lot from group work, but Portugal encourages individuals to shine. But there are rewards for socially responsible behavior, such as a Peace Prize or a Citizenship Prize, which was one of the greatest presents my daughter brought home one day.

The HeArt is Present

It was here, on the Portuguese shores, as I was watching ships waiting in the glimmering water, that I heard that Marina Abramoviç titled her retrospective at MOMA with the title I had been using for 21 years: The Artist is Present. Taking this as a Zen exercise, I hoped she experienced that same bliss through the Artist's presence. Although I was advised to sue her, I decided to sew, and sow and reap. Those who love everything they receive, receive everything they love. That night, my beautiful daughter saw me in distress and said for the first time: "Mom, I love you." It was here and then, that I realized: "in the end, everything is about love." I found my new motto, The HeArt is Present.

Figure 10. Solar Dress, photo while I constructed the installation, International Forest Art Path in Darmstadt, 2008
Photo: Holger Hertling

For me, the Artist was the perfect name for that nameless entity that inspires us to create, evolve, and shine. Also calling it LORT… Love+art: as if I was melting feelings, I found, that the real presence is an absent one, that the artist can best be present when the person disappears into the background. It is the ability to surrender to the gift of every moment - the present of the now. Through deep meditation, I received confirmation that now was the time to understand once and for all that we are all one.

Living in Portugal brought me deep peace. I was now living in a country that had had a revolution without killings. It was the only country in the world that could completely overthrow a government by sticking carnations into the barrels of their guns. Carnations, a symbol for reincarnation — flower power. Portugal was a power that would create the first global village long before anyone could imagine the Internet. A country that was able to expand by intermingling. Such a small country with such great power —- owning half of the world at some point in history.

Globalization and collaboration in a positive sense were sped up by the lockdown: with researchers collaborating, people solidarizing, and seekers meditating together. Via zoom and social media virtual exhibitions, videos and dialogues were created.

Zen Exercises

Portugal is full of Zen exercises for me. In no culture have I seen that you could be at the supermarket check-out, remember that you had forgotten something, and then have the cashier waiting for you until you had the sack of potatoes to add to your groceries. All this while people are patiently queuing and waiting for their turn. This happened until I was finally just able to laugh about it and appreciate the luxury of it. Traffic lights are a frequent invitation to meditation - it seems traffic lights in Portugal are programmed to maximize the waste of time and petrol. I never understood why I rarely got a green wave if I went at the legal speed. It seemed that, even if I were moving at the speed of light, I would inevitable pass through several red lights. Later it dawned on me: What if I started to fold my hands while waiting? Breathing, practicing awareness, surrounded by the metal shield of my car. At the green light, I would restart, savoring each moment as if it were a welcome break - inspiration, expiration.

Running a red traffic light at breakneck speed seems to be a common sport. The typical scene: it says 50 and "Velocidade Controlada" (speed control). The car next to me speeds up, passes me on the right side, the traffic light turns red, and the car shoots ahead - At the next traffic light, I see the car again.

In India, I prayed many times in the back seat while the drivers blew their horns and passed through what seemed like bottlenecks. I waited in a bus for a holy cow to move, but I have never seen a woman blocking a road for several minutes to pick up her dry cleaning. Naturally, as a foreigner, I wasn't tempted to lose my temper but somehow the locals were very understanding.

Public offices, such as Finanças (the tax office) and the Loja de Cidadão (a one-stop shop for all things administrative) is another. There is no one in line, two people are working, and somehow they manage to make you wait at least ten minutes. I must admit that when I need to go to an office like this in Germany, I am always shocked when I am attended to at the precise time of my appointment — not one minute before, but also not one minute after. I never understood how they perform such magic. In Portugal, you had better bring a good book, a good app, or the ability to meditate or to medi-date (speak on spiritual stuff with strangers).

Magically missing a paper? How many times was that paper lacking, the one with that signature in this particular place? To get this paper, I had to arrange a meeting. To do that, I had to wait twenty minutes, and when it was my turn, they told me that the appointment for that paper, can only be scheduled in another building. After finally finding the other edifice, (it was impossible to call) I found out, that appointments can only be made on Mondays, and it was Tuesday!!! Is this Kafka`s castle, or is this a Koan?

Wilting Words and Writing Wealth

JK Rowling started Harry Potter in a coffee shop in Portugal. The café - and even the restaurant at the beach - were fabulous offices. It used to be an immense privilege unknown in any other country: with just an espresso, you could sit and write for hours, and no one would bother you. This is one of the great gifts of Portugal that I never wanted to abuse. The waiters treated me well - the food was consistently delicious. It is not usual to leave a tip, but I felt if I rented an office, I would pay more and be less inspired, and consequently was perhaps a little too generous.

In the mountains of Monchique, I put together my second book, The Art is Present, a book of black and white drawings inspired by the star-filled night skies. Later at the café on the rocks of São Pedro, I composed my third book, The HeArt is Present, featuring my performances since the 90s. When I first arrived in São Pedro, I thought of the quote: "Upon this rock, I will build my temple." The dirt lot with the heavenly view had no coffee shop yet, but plenty of parking, and it was a perfect place to bring work. Later they built the lovely café: Pedra do Sal.

On the last day of my 49th year, I performed at the Centro de Ambiente, a stunning museum/ seminar space with an amazing view of the sunset over the sea. This performance was the last with the first version of the iLAND tapestry, before my gallery sold it to a famous young actress a few months later. This celebration was special as I had 49 guests sharing food and drinks, and at dinner later that night, I turned 50.

iLAND is a tapestry with aerial images featuring environmental issues. The performance is a slow-motion meditation designed to reflect on these problems and consider possible solutions. I did another performance directly on São Pedro Beach (see Figure 1- iLAND São Pedro) for a film and an audience that would be there by chance. I sat until the water of the rising tide touched my iLAND dress, sharing bread and butter with the people, and some words with the beach audience in bikinis and bermuda shorts. The people pretended not to notice me at first. Only when some people started talking to me, did strangers come closer and ask interesting questions. The resulting film was a smashing success, inspiring me for countless appearances in other countries, in Venice, Helsinki, Séte, and Guangzhou.

In 2018 I connected with around 20.000 computer geeks during the WebSummit, and I sat with thousands of people at the Pavilhão do Conhecimento just 33 days before the lockdown. It had taken me 11 years to show in a notable museum in Portugal. It suprised me that it took so long, because I had received my first commission for a solo project in the window of the New Museum of Contemporary Art while I was still finishing my Master's in Fine Arts at the Berlin University of the Arts in 1993. (see Figure 11- L'Adieu).

Figure 11. L'Adieu – Pearls Before Gods, 1993 Window of the New Museum of Contemporary Art, New York

Performance installation (duration: 28 days): pearls, bowls, silk dress, dress-maker's mannequin, bread, flowers, gold-leaf and seaweed tiles, surveillance camera and monitors, LED-display. Size: variable, min. 1m x 2m x 1,80m

Meanwhile: Focus on my Masterpiece

I trusted, when the time was right, that projects would materialize. In the meantime, I focused on my masterpiece: Sofia. Imagine creating a piece without making any decisions, choosing and placing the eyes, the heart, the intestines - all at the right time, and functioning perfectly. Here she was: such a joy. For the first seven years, I took her with me to every installation, performance, and exhibition. She was wonderful, attentive, and secure, and played the art field with such innocent charm and diligence that she achieved perfection in her way.

We played daily for several hours, in German or in English, depending on which dolls or figures we would play with. Although she'd slip in a few Portuguese words as well. Once, we played with the dollhouse. Suddenly, she put a chair to bed. "Oh, I am so tired!" She exclaimed: "The whole day people have been sitting on me."

Once I told her she could sing that song by heart, she answered: "Oh mom, I have to sing from the heart and not by heart!"

One day she asked me to sing an O-apple. "What is an O apple?"

"OO-orange," she laughed.

"How can you sing an orange, my dear?"

"Ah no! An O pair."

"Ah! You want to sing an Opera!" I sang the aria of the Queen of the Night.

"Yes! Yes!" she chimed in, just as I remembered that in Portuguese, Péra from Opera also means pear.

The lockdown gave us a chance to connect on a different level. Those hours of playtime were gone, but we found ways to draw and paint, and we made movies and projects in the garden. We built the Mother Mandala together, and she painted the lines of PTS Pandemic. Her writing inspired me to write more as well, and when she had to write a poem for school, I wrote one too.

Faith, Fatima and Fascination

100 years of Fatima happened during my tenth year in Portugal. I had broken my bond with Catholicism when I was 14, after a long conversation with a priest where I realized that my path would have to lead me to become an atheist. As a child, I was as faithful as I could be, never missing a Sunday mass and looking forward to it, preparing myself for it with prayer and the proper clothing. Once a month, on a Saturday, I loved to goto the nearby Beuron Monastery with my mother, to hear the Gregorian chanting at the Vesper. This was after going for a walk in the beautiful valley surrounding the church. The Church was my way of partying - it was a feast for the eyes and ears. I cried during my first communion because ultimately, I could be one with Christ. Even during all my later travels as an artist and an atheist, I visited every temple and church that I came across, spending at least a few minutes, though often a few hours, in the silence of these beautiful art installations.

So when I came to Portugal, naturally the churches were at the top of my list. I spent a week in Santarem and explored every single church in awe. These striking examples of architecture, with their unique blue tiles or azuleijos, mixed in with goldenen ornaments and dark paintings - I couldn't stop looking, admiring, meditating, celebrating, enjoying. While most of the museums left me quite cold, the churches took my breath away, and in Tomar, I felt at home. I somehow understood that Port Grail, the harbor of the Holy Grail, would be here and that the templars, hunted everywhere else, found peace and shelter in my beloved town, Tomar.

A year later in 2007, I accompanied two Hindu Priests to Fátima. I was appalled. It was one of the very few places of worship that didn't work for me. I passed by countless shops selling merchandise of "the miracle", and I saw Jesus overthrowing the temples in anger and hissing: "My house shall be a house of prayer, but ye have made it a den of thieves." Utterly unimpressed, I made it to the sanctuary, where I watched an old woman from afar. She was dressed in black and slowly approached the sanctuary on her knees. Deeply immersed in her prayer, she transformed me with her devotion. I couldn't take my eyes off her until she reached the end and lit a candle. I cried a while with her, deeply touched by her tears. I tried to remain unnoticed, sharing her grief as if it were the world's misery. I knew then that Fátima is made through the people, manifesting belief.

Ten years later, on the 13th of May, I woke up early and touched my phone. I must have posted something the day before because my phone went straight to Facebook. In shock, I see the Pope in a video, as if directly speaking to me, inviting me to Fátima. As if on autopilot, I went to Blabla Car and immediately got myself a convenient lift from Santos (Saints) to Fátima. The driver turned out to be a filmmaker on Faith, traveling from one religious or spiritual ceremony to the next and documenting it. Two Canadian priests joined us, and we had a wonderful conversation.

Upon arrival, we each went our separate ways. Entering the large courtyard in Fátima at the nearest entrance from where we parked, I got pushed toward a building. The Pope was far away, but the crowd was dense. At some point, I found a pillar and a little bit of shade, and without really being able to orient myself, I sat down. After what could have been several minutes or several hours, I suddenly realized that the Pope was within touching distance from me. At that moment, I still had one eye closed and the other open, but then right eye flew open and our eyes met, as we were in another world. People had started to kneel or sit, and I realized that I had been leaning against one of the sanctuaries' pillars. I stayed there for the coming night and the next day. I was utterly unprepared, without water or food, just with my clothes, but somehow unable to move. I sat as if on ecstasy, tripping through the prayers in different languages. At some point, someone offered me some water, someone else a sandwich - I was far from hungry or thirsty, but I ate and drank obediently.

That night, a beautiful woman shared her blanket with me. I said, "But you need your blanket." She told me she preferred to share it with me. Someone put a pillow under my body, which felt as soft as a cloud. I felt so much love I could hardly stand it. I sporadically just burst into tears, which I couldn't tell whether they came from pleasure or pain. Probably both. The next day, someone offered me a sandwich — the father of a beautiful family, who asked me if I wanted a lift back to Lisbon. We drove together to Santos, and I took the train back to São Pedro. To my left, the sea was dipped in pinkish blues. The evening sun played its best game of light and color on its smooth surface.

I would never have imagined that I would reconnect to my Catholic cool through 100 years of Fátima, the one place I liked the least. But again it was the Portuguese people that made that place magical. The architecture was just the vessel, the surrounding walls and edifices just markers, an empty space for transformation.

A few weeks later, on the 6th and 7th of June 2017, the artist Ana Pérez-Quiroga invited me to stay in her stunning installation at APQ-HOME at MAAT. I invited the filmmaker Frederic Daudon to stay there with me. On the second night, we stepped out to get some air, when the museum closed to the public. We were very surprised that the spacious courtyard was filled with people. We got invited to participate in a meditation on light. This was the continuation of a blessing on a different level: we marched along the river and after delved into the stunning installation of diverse street lamps: shining strong and dimming down, alternating at the speed of breath (see Figure 12- Sacré Coeur Collider at MAAT).

Figure 12. Sacré Coeur Collider tapestry in the installation overlooking the exhibition at MAAT by Carlos Garaicoa: Yo Nunca He Sido Surrealista Hasta el Día de Hoy (I Have Never Been Surreal Until Today) 2017
Photo: Frédéric Daudon

Later, on the roof of the MAAT, the mass was celebrated, encircled by candles, streetlights, and dimly lit homes. Prayers, accompanied by birdsong, the sound of traffic, and the interim silences weaved a carpet of trust. I wrapped myself in my tapestry, the Sacré Coeur Collider (see Figure 12- Sacré Coeur Collider). The blanket, features a marriage of the sciences and the sacred, between the Large Hadron Collider of Cern and the window of creation at the Notre Dame de Paris, both showing a version of the beginning of time as we think we know it.

The Portuguese Museums, the new MAAT and WebSummit

In 2006, I explored all the museums in Portugal, and both Public and Private collections in a VIP tour organized by the Art Fair. I was committed to writing about the Portuguese art world for American magazines. My suggestions to write about the Pedro Cabrita Reis exhibition at Gulbenkian, and about Joana Vasconcelos, who was not yet as internationally known as she is now, were accepted by Sculpture Magazine.

I loved the churches all over Portugal, the aquarium in Lisbon, and the Pavilhão do Conhecimento, but the other museums failed to touch me. So when I got lost in Lisbon, I usually squandered my time in the streets of Alfama, rested in the Sé, overlooked the seven hills from various viewpoints, and left the museums after browsing briefly.

Figure 13. iLAND in Sickness and Health at MAAT Museum of Art Architecture and Technology Exhibition View Eco-Visionaries 2018
Photo: Regina Frank

When the MAAT opened, it was in a nearly empty building - an impressive space. In 2018 the director of the MAAT (Museum of Art, Architecture and Technology), Pedro Gadanho (who was formerly at MOMA in New York and thus knew my work), came for a gallery visit and lunch at Flores do Cabo. It was the first lunch with a Portuguese curator I had had in seven years. As a result of our brainstorming, he invited me to show in his exhibition, EcoVisionaries, for which he commissioned a new iLAND. (see Figure 13- iLAND, MAAT)

From being featured on the cover of the monthly magazine of the MAAT, I forged a path into other shows. Antonio Cerveira Pinto, who I had met in person on a train to the MAAT, invited me to install iLAND at WebSummit. The first few hours were a shock: The thousands of young people at this overcrowded event were more interested in staring at their phones and computers, and used my cork-stage, on which iLAND was laid out, as a platform to sit and put their heavy backpacks to rest. I had to find a connection and create a subtle separation between my installation and their need for places to sit. After all, a fair is not a museum: my carpet-like tapestry looked so soft and comfy, inviting enough to lay down on. I had three rolls of colored thread in my suitcase, making the perfect fine-spun fence that cocooned the sculpture, and made it complete. Amazing what a bit of thin yarn can do! People now sat down around it, watching my slow-motion performance. I did this for 77 minutes each day, juxtaposed against fast-forward films, recordings of the previous appearances, just like we only see the consequences of our actions to the environment when we fast forward and look into the future.

Figure 14. iLAND in Sickness and Health at WebSummit 2018
Photo: António Pedro Ferreira

From the WebSummit (see Figure 14- iLAND at WebSummit), the piece went straight to the next event at the MUHNAC, the museum of natural history and science. I packed the sculpture at WebSummit by myself, sat it in the front seat of my car, fastened the seatbelt, drove over to the museum in darkness, and installed it there in the dim but spacious hall of the museum. I installed blinking lights underneath that added a vibrant glitter through the weavings. We had a virtual garden growing in augmented reality on the piece, on your cell phone. Next to the tapestry stood a monitor with videos from previous performances.

Great conversations with young people inspired me to contact the science museum, the Pavilhão do Conhecimento, across the street from the WebSummit. A few months later, just before COVID 19 struck us, around 21 thousand kids and adults saw the installation iLAND Silent Science (see Figure 15 - iLAND at Pavilhão do Conhecimento). There were panic buttons shedding light on the environmental problem, shown through aerial images on the tapestry dress. QR codes led to a website discussing environmental issues. They were structured into three categories, namely air, soil, and water, and showed projects and solutions, as well as short descriptions of the problem. The website is still active, and to this day, I receive emails with suggestions and solutions.

We had a vegan buffet, inspired by the traditional Portuguese cuisine - all vegan versions of hearty meat or sausage dishes. It was such a hit that people complained that the food was too tasty to be vegan and suspected us to have cheated. The former Minister of Environmental Affairs, Carlos Pimenta, who was a visionary politician and a patient and passionate advocate for sustainability in energy production and consumption, gave the opening speech. "Meat should be only for events like Christmas and Easter, maybe once in a while, but the kind of consumption we seek, meat or fish with every meal, can't go on," were some of his inspiring words. Don Quixote fought against windmills. Carlos Pimento fought for windmills and other renewable energies. Little did we know that this would be the last party for a long time.

Our next vegan opening lunch for my exhibition at Flores do Cabo was canceled because of COVID rising. Earth without Art would be just eh! We had to close before it opened. I installed the work, made a film, and took it down a week later. We showed the exhibition only on video, on social media.

Figure 15. iLAND Silent Science at Pavilhão do Conhecimento 2020
Photo: Regina Frank

Lockdown, PTS Pandemic, Mother Mandala

My father-in-law, Gaspar Barreira, would have loved to see the piece at the Pavilhão do Conhecimento (Pavilion of Knowledge), but he could not live to see it. He left us his sanctuary in a valley near Santarém, next to the Tagus river. As the world went into lockdown, we had the unique opportunity to enclose ourselves in a beautiful garden on an ancient Quinta. We cleaned and cleared, and made a daily pilgrimage to the recycling bin and back. My daily ritual was long meditations, listening to podcasts. I cooked two meals a day, enriched with garden-grown vegetables and herbs from my greenhouse, where I also painted. I sat on the riverbank and watched the sun playing with the water, bushels of algae floating slowly South, surrounded by the glittering reflections of the blue sky. My social life had stopped, but I didn't miss anything, except singing with others, at some point. A few long and regular conversations with my oldest and most enduring friends and some new ones brightened my day.

Figure 16. PTS Pandemic at CCB, Centro Cultural Belem, Um Todos, 2021
Photo: Regina Frank

Once a month, we filled a shopping cart with food which we diligently disinfected before stacking them in perfect order in the fridge and freezer. For the first time, I stayed in Portugal for eleven consecutive months. We bought a small pool, which I used as a think tank. I floated in it religiously every day for an hour or more, sometimes sipping a gin tonic with red pepper. It was paradise on earth. Our friendly neighbors gave us vegetables and eggs from their garden, which we exchanged with an inherited ham that we did not dare to cut or consume. Every second day, the baker delivered fresh bread into a fabric bag that we hung outside the gate. Sometimes weeks passed without talking to anyone, even while walking through the empty streets, as if awaiting a storm. I learned to focus on peoples' eyes, seeing beyond the masks, on the natural Portuguese beauty, inside and out.

Then I got invited to have an artist in residency in the Darmstadt Forest in Germany, with a big studio and an apartment, to create Mother Mandala and PTS Pandemic in the studio. I researched statistics related to deaths, infections, and the progress of the virus spreading on Our World In Data, and through other news agencies. I paired the countries with articles and observations, focusing on the environmental and political impact of the disease.

For the first public performance after the COVID 19 Lockdown in Germany and Portugal (see Figure 16- PTS Pandemic), I created a white dress featuring the COVID development curves of 77 selected countries, from Afghanistan to Zimbabwe, according to the Corona Virus Pandemic Data Explorer at ourworldindata.org. In contrast to the cold, printed lines of the data development graphs, I painted the lines by hand, following the movement of the graphics as closely as possible. My duktus symbolically interfered with the curve, leaving space for the emotional quality of the line. Combined with headlines and short bits of news, the piece captured the political climate, focusing on diverse aspects of the crisis. While I was painting this dress in the spacious studio at night, or my daughter stayed and traced the lines during the day, I worked in the forest on Mother Mandala. The first layer of the Mandala consists of donated, colored clothes enriched with tiny treasures. The personal stories behind the dresses are in digital form on the Internet. Ultimately, each segment links to the history of the garments, their previous owners, and their contemporary story. Mother Mandala celebrates the origin and the correlation to nature, its healing powers, and the complex connections of the Mandala, the communities on the Internet, and the underground "communication networks" in the forest through mushrooms and roots.

The Darmstadt audience participated generously with donations of clothes. Every day during construction, groups of people passed by, talking to me, asking questions, or just watching curiously how the piece was growing. Many donors of dresses came to see for themselves where their donations ended up in the artwork. Some fans of Forest Art (Waldkunst) donated very personal things, such as the dress they wore at their wedding, or a grandmother's favorite self-made jacket. We brought up our daughter to appreciate used clothes and toys, as they hold the energy from previous playtimes and happy moments. But it struck me that most people thought it was something that only poor people would do in Portugal. For me, to consume secondhand things is an environmental decision, as every toy comes packaged in plastic, every new piece of clothing comes full of chemicals that need to be washed, further polluting our waters. Consequently, also for my child, I prefer preloved products as a healthier, more sustainable choice.

Mother Mandala honors the stories behind the clothing, the memories attached to the fabric of life (see Figure 5 - Mother Mandala). It is a statement for sustainability, but also an identity and its expression through clothing that lives, shines, and decorates different women for several generations. While most dresses nowadays have a lifespan of being worn 0-50 times, I remember my mother going to the tailor every season and being astonished when her tailor managed to make an outfit after only one fitting. I would receive a matching dress from the leftover fabric, and sometimes even my doll got one stitched together with all the tiny pieces, if they weren't needed to dress the buttons. Almost nothing was discarded, and everything was worn on numerous occasions and then changed, mended, or transformed into another dress.

Thousands of textile workers didn't get paid during the pandemic. Their work became redundant and the multi-billion dollar clothing companies took no responsibility for compensating them. A few influencers helped with the #PayUp campaign, but only after hundreds had to starve.

Code for Coexistence

The pandemic became exceedingly polarizing, and most of us have friends in both camps, from pro-vaccination and fear, to declaring Covid a hoax. I felt like the experience of the pandemic in Portugal was much more relaxed, because this constant fight between the camps was practically non-existent. I talked to my German friends about this atmosphere, with all of the museums and restaurants closed while most people ran into nature, as this was the only thing you could do besides work. The Germans love their museums, Kunstverein, their culture. There is a museum or a cultural center in almost every town, almost to the same extent that every city has its bank and bus stations. I think art is mainly an urban desire If you are surrounded by nature, art becomes less important. The dialogue of art with nature is so much more enriching. (see Figure 17- iLAND at Felsenmeer)

In Portugal, there are sports centers everywhere. With those infrastructures empty, the military used them to provide vaccinations in the largest possible space. Mobile walls provided separations, and a fabulously organized process could begin. My husband received three phone calls before he went for his vaccination, because he was in one of the first groups to profit from protection. In my fifties, I could register myself for a free vaccination when it was my age group's turn. I received an SMS, to which I replied with my National Health Service number. The atmosphere at all the pavilions (I went to four different ones) was generous and peaceful, and the nurses and staff were impeccable. A day before my second shot, I received an SMS with the time slot. Because I hadn't responded, they called me. What was incredible, was that I rarely waited more than ten minutes in line.

Later, my daughter wanted to follow the lead of her classmates, who all opted for vaccination. I wasn't sure, as children were rarely symptomatic, but I didn't refuse. It made our life much easier. I was prepared not to sit inside of a restaurant for another year. At the vaccination center, the silent eye-to-eye communication was refreshing and felt full of loving-kindness. I felt that conformity and tolerance, as well as acceptance, could become a key for transformation. One might argue that the people just took the path of least resistance. But maybe the innate capacity for peace in Portugal, the ability to trust, the acceptance of race, culture, religion, language, and circumstance is unique in the world, and is a keystone for unification and love.

As hard as the Covid19 was for many of us, it seemed at least to pause a conflict, delay wars, and create some peace and unity among most of us. For many, it was an accelerator to tend to the most significant issues in life. As we face another major conflict in Europe, my heart goes out to the people that suffer the most from it. In 1991 the Gulf War triggered such fear and sadness in many of us: while people were bombed and terrorized - we were bombarded with images of terror. It was the first time that visual information could travel so fast and many of us were shocked to their bones. A hundred years ago, young soldiers (practically boys) had no idea what they would experience at the front a few days before being sent to die (see Figure 18- Sans Souci). Today the clearer idea we seem to have, due to the overflow of images, does not seem to touch us anymore. When will it be enough? Hopefully, before cutting the last tree, polluting the last river, and breathing sickening air, we will realize that wealth is not in bank accounts, and we cannot eat our money!

"Terrorism is the war of the poor, and war is the terrorism of the rich," Peter Ustinov said, but it's the poor that suffer the consequences in any case.

Being like a Tool

Honestly I feel a bit like a feather shed by a bird, blown by the wind, dipped in ink, being used to write sentences, I have no knowing of until I read them, a vehicle being driven to places I can't imagine and a vessel filled and emptied in cycles. Most of my decisions come from visions, or dreams, if they don't rapidly transform into reality I let go easily, only to breathe a next adventure. Inspiration leads to expiration almost as inevitable as the next wave crushes to the shore, a bit like a drop of water that falls into the ocean only to rise later into a cloud and fall again as a snow flake … So given this gift of the present I share thankfully and feel myself floating along. Whenever I try to intervene I mess up….so I have learned to let go and go with the flow.

The ego doesn't like it much, but the GO in Ego goes along with it. Let'(s) go!

Figure 17. iLAND in the Rocky Sea (Felsenmeer), 2021
Photo: Samira Jamali

Figure 18. Sans Souci dress with soldiers from the second world war in front of Sans Souci- 1991 beginning of the 2nd Gulf war.
Photo: Stephan Wiesner

REFERENCES

Pessoa, F. (1982). *The Book of Disquiet*. Penguin Classics.

ADDITIONAL READING

Frank, R. (2016b). *The Art Is Present. Drawings and Paintings*. Kreisgalerie Schloßmuseum, Meßkirch, Freier Berliner Verlag.

Frank, R. (1999). *The Artist is Present*. Senate for Cultural Affairs and Science.

Frank, R. (2016a). *The HeArt Is Present. Installation and Performance*. Kreisgalerie Schloßmuseum, Meßkirch, Freier Berliner Verlag.

KEY TERMS AND DEFINITIONS

Artist Is Present/The Heart Is Present: It was Regina Frank's Motto from 1989 until 2011. When Marina Abramoviç adapted this title for her exhibition at MOMA Regina Frank took this as an invitation for change and created The HeArt is Present. This motto or title refers to the omnipresence of the true Artist within all of us. The power of divine creation through the body being solely a tool, vessel, or receiver. The Artist or HeArt is impersonal and does not refer to a specific body-mind organism but humankind as a whole, experiencing true inspiration and guidance through the Artist´s presence.

Artists in Pandemic: Often reoriented their creative practice to cope with the lockdown. Many augmented their social media presence and participated in exhibitions on-line. Some invested in NFTs, performed from their home on camera, or created digital artworks.

Cultural Tourism: Is a type of tourism activity in which the visitor's essential motivation is to learn, discover, experience and consume the tangible and intangible cultural attractions/products in a tourism destination.

Internet Art: Is a form of new media art distributed via the Internet. This form of art circumvents the traditional dominance of the physical gallery and museum system. In many cases, the viewer is drawn into some kind of interaction with the work of art.

MAAT, Lisbon: Is the Museum of Art, Architecture and Technology, an international institution dedicated to fostering critical discourse and creative practice that inspire new understandings of the historical present and an empowering engagement with the common future. The MAAT has shown numerous well-known artists and has commissioned many outstanding artworks that were highly technological, inventive, and visionary.

Performance Art: Is an artwork or exhibition created through actions executed by the artist or other participants. It may be witnessed live or through documentation, spontaneously developed or written, and is traditionally presented to a public in a fine art context.

PT: It was an association created during the first lockdown to help artists by fighting for artists rights, meeting with political parties, creating exhibitions, awards, a magazine, a podcast, fundraising, studios, and microcredit.

Rituals: Are regulated, often stereotyped procedures that serve the communication of people with each other and often the communication with spirits and divine forces. Every culture has their specific rituals. There are ancient rituals, vedic, shamanistic rituals, magic rites, and new rituals, such as challenges that are performed in front of a camera and posted on social media to connect to peers.

Slow-Food/Organic Food: It means a reorientation towards conscious eating with economically and ecologically justifiable food contents. The healthy nutrition of the people comes back to the foreground, which is a sensual experience of pleasure.

Spiritual Tourism: Is an act of traveling to visit spiritual places such as (1) mosques, churches, and temples and (2) natural environments such as forests, rivers, oceans, lake, spiritual gardens, wildlife parks for birds and animals, botanical gardens, or (3) teachers, priests, monasteries, ashrams, retreat centers.

Zen Exercises: Are challenges, usually created by circumstances or people. The exercise is to stay with a zen (calm, peaceful) mind and not to take it personally but instead let go and act with loving kindness, taking this as a reminder to be in this world but not from this world, as each situation is an opportunity to learn, give and forgive.

Section 8

Itineraries on Education for Deconfinement

Chapter 24
Virtual Tourism as a Pedagogical Practice:
Adaptation of Curricular Content During and After a Pandemic

Sérgio Rafael Eliseu
Instituto Superior de Lisboa e Vale do Tejo, Portugal

Marta Noronha Sousa
Instituto Superior de Lisboa e Vale do Tejo, Portugal

Ângela Berlinde
https://orcid.org/0000-0001-8597-5648
Instituto Superior de Lisboa e Vale do Tejo, Portugal

Mihaela Bogdan
Instituto Superior de Lisboa e Vale do Tejo, Portugal

ABSTRACT

Post-pandemic anxiety and potential new outbreaks of this and other diseases could be a push, started long ago with the explosion of low-cost mobile devices, for immersive technologies, such as augmented reality, virtual reality, and mixed reality, to assert themselves as credible solutions in contrast to traditional ways of knowing the world. Effectively, these technologies already offer safe alternative ways to "travel" through continuous and uninterrupted interactive experiences for travelers who do not want to give up their own private space, safety, and health. However, this possibility brings with it new questions and challenges, mainly in terms of the development of these experiences, more specifically in the academic and pedagogical field. The authors analyze a project that articulates, in a transdisciplinary way, contents of several curricular units attended by Erasmus students around a VR experience of one of the most emblematic places in the region, the famous Castro of Monte Mozinho.

DOI: 10.4018/978-1-6684-3369-0.ch024

INTRODUCTION

Immersive technologies, such as augmented reality (AR), virtual reality (VR) and mixed reality (MR), are increasingly asserting themselves as credible alternatives to traditional ways of knowing the world. Post-pandemic anxiety (Covid-19) and potential new outbreaks of this and other diseases have contributed to an increase in this tendency, in a process that started long ago, with the explosion of low-cost mobile devices. Effectively, these technologies already offer safe substitutes to "travel", through continuous and uninterrupted interactive experiences for travelers who do not want to give up their personal space, safety, and health. However, this possibility carries with it new questions and challenges, mainly in terms of the development of these experiences, more specifically, in the academic field. Old and new design principles need to be articulated to create a digital travel experience in VR, AR or MR that represents a positive perception of a tourism destination.

In the context of the course on Production of Interactive Content and Multimedia at the Higher Institute of Educational Sciences of Douro – Penafiel (ISCE Douro), articulating with the research objectives of the Multimedia and Arts Research Group (NIAM – renamed recently to CI-ISCE) of the Department of Arts and Multimedia (DAM) in partnership with the Institute for Research in Design, Media and Culture (ID+), the authors seek precisely to explore, develop and stimulate tourism projects in VR, AR and MR in three axes:

- Local (i.e. Castro of Monte Mozinho – Penafiel – focusing research on the development of 360 degrees VR films, writing, game design, digital design, 3D modelling, augmented reality and a variety of social art and multimedia design practices, along interdisciplinary collaborations linked to aspects of science, technology and archaeology);
- Regional / National (i.e. Rabelo boats, focusing research on the construction of a VR and AR digital archive around it's build - the Rabelo boat is a Portuguese vessel – typical of the Douro River – which traditionally carried wine barrels from Alto Douro, where the vineyards are located, to Vila Nova de Gaia – Porto, where the wine is stored and later marketed and shipped to other countries);
- International (International VR tours in partnership with external institutions, such as the Federal University of Ceará – Quixadá, among others).

In this chapter the authors will focus on the first axis, specifically, on a project which is currently in progress (TPRV - Tourism and Heritage in Virtual Reality). Its transdisciplinary articulation regards contents from several of the curricular units attended by Erasmus students and takes form in a VR experience for one of the most emblematic landmarks of the northern region, the famous Castro of Monte Mozinho. It is a Roman Castro village that though founded in the 1st century AD, has a wide chronology of occupation, reaching as far as the 5th century.

Background

Virtual tourism, in a broader sense of the concept, is a way of experiencing a place without actually going there (vide e.g., Guttentag, 2009; Hernández Beltrán, 2019). Possibly it will not be wrong to say that as long as there is a transmission of knowledge between human beings it is possible to travel without physically going to a given place. Indeed, who has not travelled to different places using only their imagination through stories told, books, photos, films or music? This is the concept established by the philosopher, sociologist and researcher in communication and information science, Pierre Lévy, who states that "the world of meaning is the true virtual world, we can say, it is a world that begins with language, it is not a world that starts with computers"; for the author, we live "in this abstract world, in this virtual world of meaning", since the beginning of humanity and "what computers do is automatically manipulate the signs of language" (Lévy, 2013).

As Hernández Beltrán (2019) states, tourism has evolved over the years to not only include the usual activities related to relaxation and rest, as the public starts to be more and more demanding, and requiring more specific activities, pertaining to their changing tastes (p. 13/4). The increasing influence of information and communication technologies has provided new information sites for planning, talking about and even experiencing tourism. VR technologies, in particular, provide a vast number of new applications for tourism (e.g., Guttentag, 2010, p. 637).

So, when the authors talk about virtual tourism, they necessarily evoke the context created by the introduction of digital technologies in our lives, more specifically since the internet was created and it entered our homes. Initially by the exponential sharing of texts, images, and videos, but with a totally different expression from 2007, when Google Street View appeared. An interactive metaphor of the world map that allowed people to start being able to take "virtual tours" through city streets and from any part of the globe. Google Street View also allowed users to experience famous landmarks from any angle and time of day - even those that are closed to the public, like the Eiffel Tower at night. A navigation that has become greatly streamlined with its introduction into Google Earth and, more recently, a deeply immersive experience with Google Earth VR.

Figure 1. Google Earth + Street View

Figure 2. Preview of the sales for VR equipment

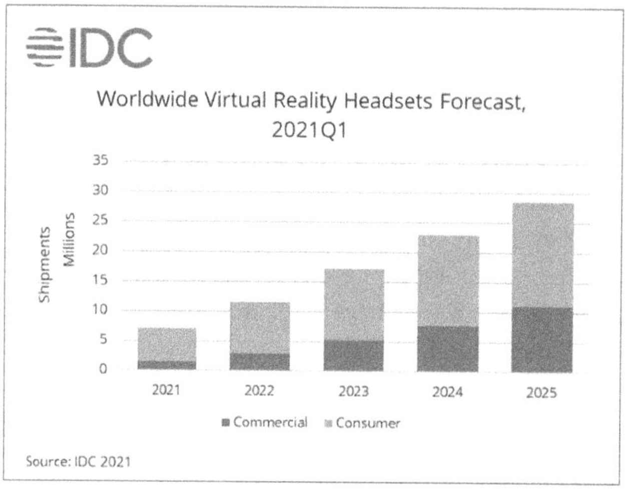

It is precisely with VR technology that, thanks to the massive introduction of low-cost, high-quality equipment on the market (e.g., oculus quest 2), an increasing number of people can access a new form of travel experience. Experiences until then were only available in theme parks (Disneyland Paris, France, and Isla Mágica, Spain, for example) and subject to long queues. Numerous applications now allow people to travel all over the world without risking their personal safety or leaving their home country. Some of the many benefits of virtual tourism are that: it is cheaper than traveling in person; more environmentally friendly, as it does not require visitors to use resources to reach the destination; more realistic as it provides an immersive audiovisual experience; and safer, as it removes the risks of transport. Virtual tourism is increasingly an accessible and safe alternative for travelers who do not want to give up their own private space, safety and health. This is particularly important in a pandemic or post-pandemic scenario. It is, indeed, a credible way to experience another place without actually being there, looking through a digital device or using VR equipment.

For Gomes and Araújo (2012), virtual tourism is a subdivision of post-tourism, in which the artificial environment is formed by VR inserted in cyberspace and chronological time is seen in a more flexible way, as technology is able to offer a timeless field.

However, for this to be possible, for virtual tourism it is important to digitize reality or, alternatively, to create new three-dimensional universes. Nevertheless, the focus of our analysis is the promotion of existing (or formerly existing) sites. In this sense, this approach seems technically to be performed in four ways: photogrammetry, 3d laser Scan, 360º Video (stereoscopic and mono) and 3d Modelling.

Photogrammetry itself is not new (derived from the Greek: light, description, and measurements), it is defined as the applied science, technique, and art of extracting from metric photographs the shape, dimensions and position of the objects contained therein (vide e.g., Kucukkaya, 2004). This process, not only for the speed of execution and high quality of results, as well as for the reduced cost of implementation and technological apparatus, when compared to the systems mentioned above it is far better: multiple high quality photographic records from various angles and the creation of a textured base that facilitates the computation of the images used. Several commercial and free softwares allow access to impressive quality results (Colmap, Meshroom, MicMac, Multi-view Environment, OpenMVG, Regard3D, VisualSFM, among the free ones, and the following stand out in the commercial aspect: 3DF Zephyr, Autodesk ReCap, Agisoft Metashape, Bentley ContextCapture, Correlator3D, DroneDeploy, Elcovision 10, iWitnessPro, IMAGINE Photogrammetry, Photomodeler, Pix4Dmapper, RealityCapture, SOCET GXP, Trimble Info, WebODM). In terms of hardware, although there is photographic recording equipment specifically prepared for this purpose, a common DSLR camera with 12 megapixels or more (either Canon, Nikon, Sony, etc.), allows good results. Or even using the cell phone itself, although the higher the resolution of the camera, the better the quality of the photogrammetry.

Many examples of projects using this method can be mentioned. The authors highlight, just as an example, some works by "raiznewmedia", which through the "othersight" app allows an immersive visit to the church of Santes Creus, Tarragona, to a museum in Madrid by the painter Joaquin Sorolla, as well as to an alley of Tokyo and a famous street in Havana (Neptuno Street). Or the "Brink Traveler" app more dedicated to great North American natural settings such as Horseshoe Bend, White Pocket, Mount Morrison, Mount Whitney, Alabama Hills, The Wave, Arches National Park, Death Valley National Park, among many others.

In terms of 3D laser scanners, they use a high-precision scanning method that allows the automatic capture of large amounts of data in a short time. Despite the speed of the process, it implies the use of quite expensive equipment (it is true that a low-quality portable 3D scanner can cost less than €200 - the XYZprinting 3D scanner, for example) which can reach more than €30,000 for advanced, state-of-the-art 3D scanners (example of Artec Leo 3D Scanner). Photogrammetry also allows you to cover a large amount of space quickly, albeit with a low degree of accuracy when compared to 3D laser scanning. Laser scanners process data using a laser beam aimed at the surface of objects and then convert that data into a point cloud that is processed in tools like Autodesk ReCap, which converts it into a 3D model or 2D drawing. For this reason, it is on the most used methods in architectural projects or topographic surveys.

The fastest and easiest way to build virtual tours is, without a doubt, the use of 360° cameras. However, these only allow the experimentation of the VR image in 3 axes which can create the feeling of motion sickness, as the image follows the movement of the head. In this field, there are many solutions on the market, ranging from €250 (for example, the GoPro Fusion, with 5k resolution - it is not stereoscopic) to more than €5,000 (the case of the high-end camera INSTA360 PRO2 + FARSIGHT - 360 Professional 8K stereoscopic VR). This type of solution is widely used in the construction of virtual tours in the real estate sector due to the ease of development and browser integration (possibility of viewing in 360°, navigating with a mouse and thus dispensing with VR equipment). Given the ease of acquiring and processing content, most virtual visits available use this technique. The "wander" app is a good example, as it combines a kind of mixture of street view, 360° photos, with 360° videos from all over the world. But with the possibility of integrating this format of videos on YouTube, users and institutions will exponentially add content to this platform. Integrated systems such as the one from the company "Matterport", also allow the rapid creation of interactive tours simulating a three-dimensional environment.

3D modelling is certainly the most expensive of the methods mentioned here and is frequently used by the film and video game industry. There are some approaches in the sense of virtual visits to this type of scenarios (as is the case of the studios Feliz and Paul – which allow you to visit Jurassic World with their app, feeling the presence and proximity of dinosaurs). However, given our focus on existing sites, the authors will not extend this approach further.

Figure 3. Virtual visit to The National Museum of Computing with Matterport system

VR may not yet replace traditional trips, but it offers visits that can work as a preview, encouraging a real visit, without the existence of a carbon footprint or also avoiding the disappointments created by idyllic postcards. Indeed, there are already tourists who prefer to visit tourist sites virtually before visiting them physically (López, yagual e aquino, 2016, p. 64-65). VR trips can, on the other hand, address issues related to accessibility, taking parts of the world to people who are physically unable to visit these places or, for reasons of human damage, giving access to sensitive content that real visits would compromise. This type of travel is currently one of the most anticipated technological developments to greatly impact the tourism industry" (Tussyadiah, et al., 2018, p. 140).

Another aspect that is worth mentioning comes with AR applications, making it possible to transform any place into a virtual tourism destination. For example, StraatWorld lets you visualize what it would be like to walk down the busiest shopping street in Amsterdam, as well as hundreds of other public places around the world. This is, however, an analysis that will be pursued further in future projects.

Teaching and Local Community

The role of higher education is, first and foremost, of course, to train its students to practice a given profession. However, its function does not stop there. An educational institution (of higher education or otherwise) does not exist in a vacuum. It exists to serve the community in which it is inserted and exists only insofar as it serves it, a function without which it finds itself useless. Furthermore, the objective is to form citizens that are integrated in the context that will welcome them when they finish their studies, aware of their history, prepared for the concrete challenges they will encounter there and able to provide innovative and creative solutions to these challenges.

ISCE Douro, in particular, sees protected in its statutes, with regard to its scientific, cultural and pedagogical project, the mission of: "serving and promoting the intercultural community according to a humanist and humanizing perspective, preparing global citizens and critical thinkers", capable of dealing with the challenges of a world "in permanent change", as well as of "actively collaborating with the local and regional living forces with which they interact directly, namely at the level of the Douro region and the Tâmega and Sousa sub-region (...), in an effort to continuously improve its products and services", through the training and qualification of future professionals, "based on the economic and sociocultural reality and taking into account the potential in the domains of historical, rural, cultural and touristic heritage" (Ministério da Educação e da Ciência, 2015, p. 2575/6). In other words, ISCE Douro's mission is also to train professionals and develop projects that promote the development of the region, namely in terms of tourism and heritage.

In this sense, it is the tradition of the Department of Arts and Multimedia to participate in initiatives and encourage students to develop, even before entering the job market, within the curricular units of the Multimedia courses they attend, works inspired by this mission. Some works focus on raising the awareness of the students and external population to problems occurring in the local, regional, national, or even international community, such as pollution and anthropogenic climate change, consumerism and materialism, mental illness (including depression, an increasingly frequent problem among the young population), among others. The 2021 Christmas project, for example, aimed to alert the community to "the other side of Christmas": to the loneliness of those who spend this time away from their family, especially in times of Covid-19, to food and resources waste, to the increasingly rapid obsolescence of electronic equipment and the consequent electronic waste produced. This project, entitled Sharing Empowers, is part of the Penafiel - Christmas City municipal project, and consists of an urban art installation, exhibited on a city street, which also featured, this year, street animation through multimedia students' performances. These projects, however, have so far not had a VR or AR dimension, although this may happen in the future.

Other projects seek not only to raise awareness of the population, but also to present multimedia solutions to promote the region, its tangible and intangible heritage and tourism. The best example is certainly the participation of ISCE Douro in the 5th edition of the Virtual Archaeology Marathon and Symposium. Organized in 2019 by the Municipality of Penafiel, through the Municipal Museum, in partnership with ISCE Douro, the Museum Friends Association and other entities, this initiative aimed to "appreciate the Castro of Monte Mozinho through the production of graphic content that helps the interpretation of this unique archaeological site with a long historiographical tradition" (Municipality of Penafiel, 2019). The 5th Virtual Archaeology Marathon, focused on the Castro of Monte Mozinho, gathered at ISCE Douro a team of renowned experts from the Iberian Peninsula to develop a VR model of the Castro, based on the three-dimensional digitization of the available finds (buildings and artefacts), which virtually reconstructed the elements that remained until today, but also carried out a projection of the buildings and human practices supposed to have taken place at the time of its occupation and full activity. This team tried to rescue not only the spaces, constructions, and material artefacts, but also the human practices of this Castro community. As Teresa Soeiro, the project's scientific coordinator, said some time before, it is as important to collect information about the material culture of ancient peoples as about their habitat, their ways of life and occupation in this material space (Soeiro, 2008).

Figure 4. Plan of the VR model of the Monte Mozinho Castro, developed at the 5th Marathon of Virtual Archaeology

Inspired by the success of this project, a group of students later developed a project in Augmented Reality about the Castro of Monte Mozinho. A physical Styrofoam model of the Castro made it possible, using an App that was created for the effect, to visualize this landmark of Castro culture in the Sousa Valley. In 2020/2021, the project was extended to the Municipality of Penafiel, and entitled "ARP Model City - Augmented Reality as a tool for the dissemination of Heritage". In an interdisciplinary logic, the project was developed within the scope of the curricular units of Interactive Media Lab, Creative Entrepreneurship, Usability Assessment, and Seminar, of the degree in Production of Interactive and Multimedia Content. In this project, a vinyl model represented the map of the municipality and the new App created allowed not only the search for information about the main tourist landmarks in the municipality, as well as the visualization in AR, when pointing with a smartphone at the landmarks displayed in the installation, of these emblematic spaces and buildings in three-dimensional projections (see figure 5).

The ARP project was presented in June 2021 to the community during the BETA exhibition, an annual exhibition organized by the Department of Arts and Multimedia to showcase the best works done by students throughout the year. Figure 9 represents the moment when the installation is being demonstrated to the President of ISCE Douro and to two representatives of local institutions, namely the director of the Penafiel Municipal Library and the secretary-general of the Penafiel Business Association. Both visitants recognized the interest and attractiveness that these types of applications can represent for the promotion of tourism, local history, and culture.

Projects like these constitute an innovative and diversified way of presenting to the community and, thus, promoting the heritage and culture of a region, both in an educational, cultural, and touristic context. Applications in AR and VR have the advantage of being particularly appealing to young people, who are the main recipients of education, but who often do not exhibit a particular interest in matters related to the past. Being keen on trying innovative devices, media, and applications, VR and AR in particular, young people may be stimulated to learn more about their heritage and culture.

Figure 5. ARP Model City app interface (with the options "AR view" and "list of monuments") and AR visualization of the Sameiro Sanctuary

Figure 6. Presentation of the ARP Model City project in Beta 2021

In the authors' opinion, this type of project is also important because it allows the community to show the type of results that can be generated in a Multimedia course. For Guttentag (2019), "many relevant ICT developments are not made directly for the sake of tourism, so tourism researchers and professionals may not be fully aware of the developments and, therefore, are unprepared to adopt and adapt to the new technologies" (p. 637). If higher education institutions are able to show the potentialities of these technological innovations to these professionals, through these kinds of projects and their exhibition inside and outside the school walls, they will be fulfilling their mission towards the community they serve. Furthermore, Multimedia still seems to be, for many people, a vague, poorly defined concept, which may motivate the relatively low demand for courses in these areas, despite the high employability rate they represent. This has been an increasing trend, even more so since the beginning of the pandemic since companies have had to adapt and to create a stronger presence in the virtual market. Whatever the evolution may be in the future, it is certain that technology will be more and more present in every daily activity. Training professionals capable of facing these challenges is, therefore, a major step to the promote local, regional, and even national development.

Monte Mozinho – Virtual Tour

Old and new design principles need to be articulated in order to create a digital travel experience in VR, AR or MR that constitutes a positive perception of a touristic destination. What role can educational institutions play in this emerging paradigm? In the context of the Interactive and Multimedia Content Production course at the Higher Institute of Educational Sciences of Douro – Penafiel (ISCE Douro), articulating with the research objectives of the Multimedia and Arts Research Group (NIAM) of the Department of Arts and Multimedia (DAM) in partnership with the Institute for Research in Design, Media and Culture (ID+), the authors seek precisely to explore, develop and stimulate local tourism projects in VR and AR, more specifically, in the region of Penafiel. The research methods were characterized by the use of artistic and design practice and its multiple materials, with a view to building an academic discourse that is manifest in works presented publicly, as well as in the practices carried out. The results are shared with the general public, as in the present case, as a democratization factor in access to knowledge. An example of this is the multiple exhibitions held inside and outside the institute, as well as the applications developed and published.

As the authors mention in the introduction, the VR universe brings with it new questions and challenges, especially with regard to the development of these experiences, specifically in academia. The first challenge that comes to mind is how best to use these technologies for all types of students. This is because some may find these immersive technologies more immersive than others or may be more inclined to learn from a two-dimensional screen than from a fully immersive environment. The second challenge is whether traditional educational institutions will replace their current teaching methods with VR and AR technology. However, VR and AR have not been around long enough (in a general way in teaching) for educators to study them extensively to see if they are viable alternatives to traditional methods. Another concern is that VR and AR can still be expensive, so these technologies may not be accessible to many people if they become mainstream teaching tools. Fortunately, at ISCE Douro, the focus on VR technology has been consolidating for some years, with different types of VR equipment, so the Covid-19 Pandemic, at least in the Art and Multimedia Department, was seen as an opportunity to deepen its exploitation even further, given the imposed context of social distancing.

The development of content that goes beyond 360º video capture comes into the equation, involving video and sound editing, 3D modelling and animation, design of interfaces with menus suitable for the devices to be used, production of scripts and storyboards. Contents that, despite the development of VR experiences, consider several factors to take into account, which are not unfamiliar with the usual usability objectives common to other multimedia products. It is therefore very important for us to understand that the development of interfaces with good usability is only possible through the adoption of an iterative cycle centred on the user. In this process, the analysis of needs, the development of solutions, prototyping, evaluation of prototyping, etc., stand out, the results of which will be visible in factors that measure learning, effectiveness, efficiency, and error prevention.

The Castro de Monte Mozinho Virtual Tour is a 3D and 360º experience aimed at history and culture enthusiasts. The application allows users to relax and absorb a day in the ruins of an ancient Roman Castro. Fortified, the real space is presented with two lines of walls with about 22 hectares and presents several urban reformulations, being possible to observe various types of construction, from nuclei of courtyard houses of the castro tradition, with circular compartments and vestibule, to the complex square or rectangular Roman dwellings. Located in the parishes of Oldrões and Galegos, in the municipality of Penafiel, it enjoys wide horizons, which extend to the Marão Moutain to the East and to Montemuro Mountain to the South, sloping over the depression where the Cavalum / Camba river flows.

Figure 7. Splashscreen of the project

The execution of the project used the "Unity3d" platform, which was configured to provide a virtual space. The result was an environment responsive to physical controls as well as user positioning, to which support for VR glasses (Oculus Quest) was added. The pedagogical methodology used in the development of this project was predominantly active, based on demonstration. However, given the need to introduce students to basic notions and characteristics, such as, for example, narrative structures and sequences, the expository method was simultaneously adopted. This was accompanied by other pedagogical techniques, such as participatory exposition, questioning, group work, project, and simulation oriented to critical analysis through the presentation to colleagues of the tasks proposed by the teachers. The students involved participated in the work from conceptualization to collection and treatment of contents, to the final execution, having carried out several previous studies, where the authors highlight the execution of models and animated three-dimensional models that allowed the development of skills in 3D modelling programs.

In this project, the user can be transported to different times of the same place, experiencing new situations. Here, the storytelling game dispenses with the storyteller by letting the audience become the storyteller, the traveler. In many ways, VR storytelling/travel maintains the same principles as its traditional counterpart. However, in this case, the exploration does not end as a simple VR trip, 360º film, it can be simultaneously an interactive time travel in a space where various activities would take place, such as games, assemblies, market, among others.

Figure 8. Development environment in unity 3d

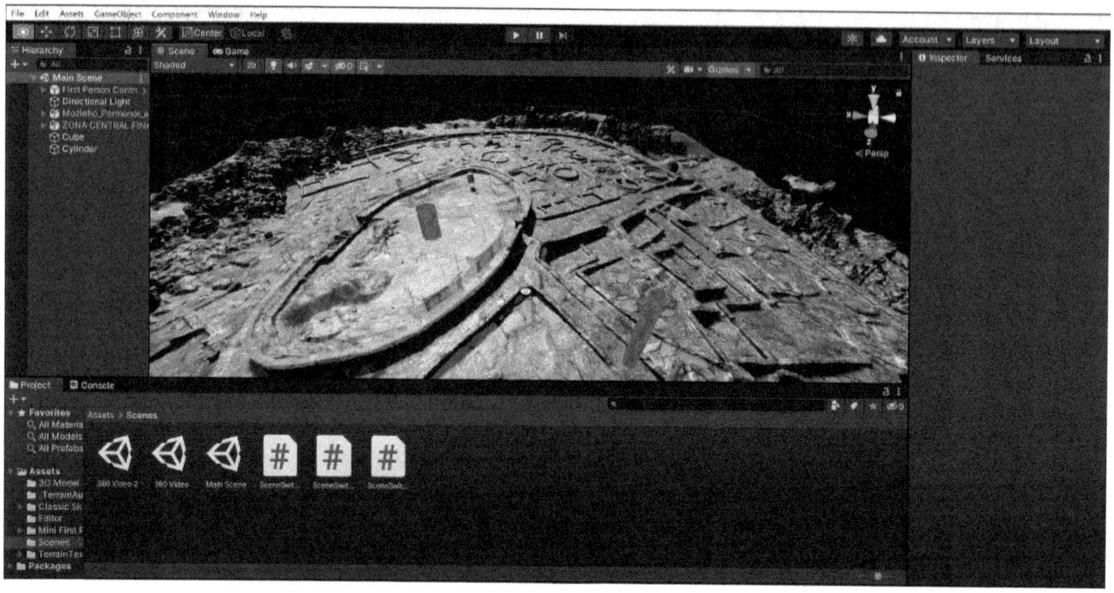

This digital experience, which aims at a real context, takes place inside and outside the educational institution. Allowing us to promote a culture of sharing between teachers and students, as well as a greater understanding of the space and community in which the educational institution operates. For this reason, a consequent reflection is expected on how projects of this nature can also act as catalysts for prosocial cooperation involving partnerships with commercial actors in the community in which they operate.

One of the goals of the project is also, as was already stated, to draw attention to local cultural heritage. It is essential that we understand our past to understand our present and build the future. There is a massive shift to the digital world, and then we must take advantage of it not only to create content for entertainment purposes but also for educational purposes.

Castro de Monte Mozinho: Virtual Tour - allows you to simultaneously explore what life was like in Roman times, while making a distinct parallel with the present. The contrast between past and present is intended to be similar through a series of zones of interaction that allow the user to switch between visual elements of the present and the past in the same zone of perception. In order to introduce this dimension of "time travel", a photogrammetric model (captured by drone) of the Castro of Monte Mozinho was used, which was the subject of a process of historical reconstruction during the V symposium on Virtual Archaeology (promoted by the company Archeo3D and the Penafiel Municipal Museum), which took place in 2019 at ISCE Douro, in Penafiel, as was already mentioned. On this model, objects with interactivity in the form of cylinders were integrated that allow the user to be transported to the current 360º images. Although this project was designed to be experienced in VR, aiming for a more immersive experience for users, in addition to the application, a small 360º video was also created with various perspectives within the Castro of Monte Mozinho that allows users to spend "an afternoon relaxing" on site. The tools used to create this project were Unity 3D, a 360º camera (Insta360 One X2), Blender modelling software, and Adobe Premiere as video editing software.

Figure 9. Photogrammetric models and interactive artifacts

Figure 10. video 360º

PHOTOGRAPHY AND THE NEW MODELS OF VISUALITY

The authors believe that this study allows a broadening of the understanding of the construction of the tourist experience and its transgressions presented in the industrialized and computerized society. In this sense, the image has been an important tool for the tourist experience, as it portrays in the contemporary world the relationship with the imaginary of the visited destination, since when composing this landscape cut of a place, it provides users with subjective ideas that, in turn, attribute meaning to the location.[1]

When it comes to photography, there is something magical that seems to last: 180 years have passed since Louis Daguerre's daguerreotype was patented in Paris and the spell seems to last. Despite the rampant and indiscriminate media proliferation of the image in the digital age, photography maintains, to this day, a kind of inevitable magical attraction.

It is true that in its early days the photographic image emerged associated with the concepts of objectivity and transparency, having been a powerful tool for the evolution of tourism, being the testimony and concrete proof of a trip, as well as its meanings, representations, and emotions. With its invention in the 19th century, photography would forever change the way of seeing and feeling the world, through accessing images to places never imagined before. The 20th century, on the other hand, is deeply marked by transformations, mutations, and reconfigurations. New technologies have substantially changed the way we relate, how we have access to information and the flow of ideas. Reality and truth are no longer hegemonic terms and concepts and, perhaps for this reason, photography has revised its objectivity aspirations, approaching the world, from simulation, fiction or staging.

Today, there are as many ways to produce images, as there are ways to share, propagate, manipulate, and use them. Within this infinity of possibilities, it becomes increasingly essential to create new spaces both in academia and in society to analyze what happens with today's images, and how we want to work with them and relate them to deal with uncertainty and with their truth codes. Photography, more than ever, is integrated into the reality experienced at the same moment, living with it, accompanying it and, at times, surpassing it. To such an extent that it led the photographer and essayist Joan Fontcuberta to state in his book "La Furia de las Imagens" (2016) that photography had died and that we would be in the post-photographic era. We can speak in a strictly academic sense of the end of photography, understood as the hegemonic account of photography as a continuation of painting or as the mimesis of the world. Admittedly, we may have been a little naive in trusting the idea that photography always tells the absolute and unquestionable truth. However, the multiplicity of technologies that we are forced to deal with contribute to the opening of a wonderful world of infinite possibilities.

It is impossible not to connect our present to the Internet and immersive technologies, where a notion emerges that shapes visuality in our societies, involved in a multiplicity of applications and visual experiences. Today we have an interesting picture of the world corrupted, sublimated, altered, recreated, fragmented, appropriated. This appropriation mixes photography with Internet software, sculpture, painting, or installations, and this is what current research focuses on, adapting to the concepts that represent today, as we reflect on new ways of traveling, associated with new ways to know the world. And this new reality, understood in its intertextual dimension, provokes a modification of the perception of time and space, appropriation, mixture and hybridism.

In fact, when we reflect on digital tourism, the definition and experience of concepts such as time, memory or space are inevitably changeable according to society or the individual. It is true that since the era of digital photography and its democratization through cameras incorporated into phones and mobile devices, there has been a disproportionate increase in the production of images and photography has become a form of commodity. The multiplication and mediation of screens make everything perfect and polished in a different perception of time and make it difficult to reflect on what we see and how we see it. The imperfection of the word against the visible makes it easier, for example, to photograph a sunset instead of describing it to a friend who is unable to contemplate it. When the artist Penelope Umbrico appropriated photographs of the sunset from a set of images on Flickr, it allowed her to create a reflection on the way we relate to images, both individually and collectively.

Figure 11. Penelope Umbrico 541,795 Sunsets of the series - Sunsets from Flickr

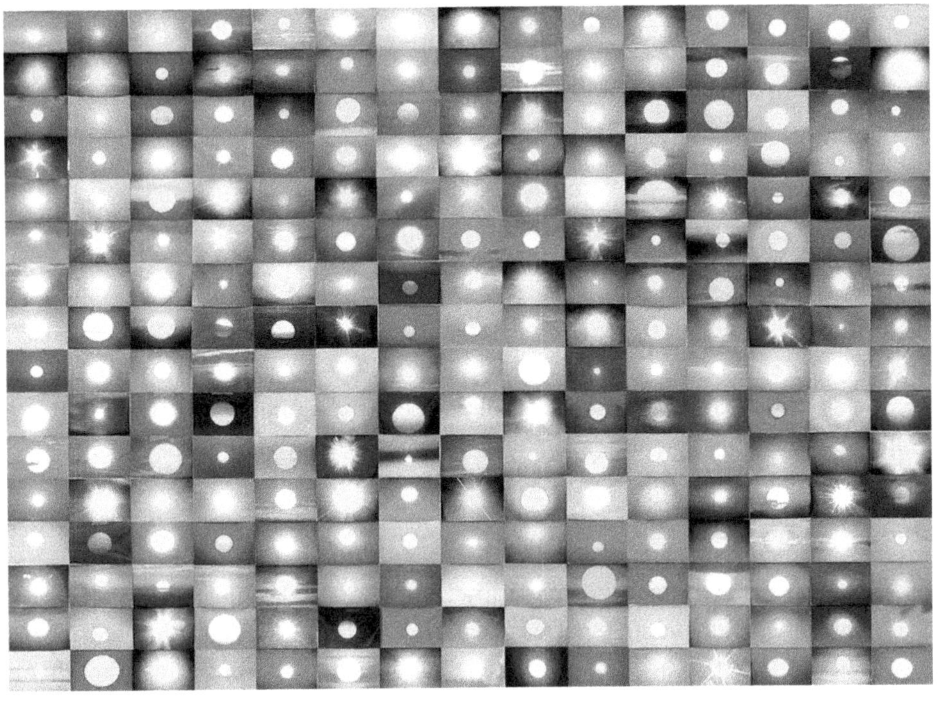

It is precisely when current visual artists appropriate this multiplicity of images that saturate the veins of the Internet to generate their discourse and narration, becoming active agents and not passive spectators, when the stronghold of the exercise of aesthetic reflection is observed. A good example of this is Joan Fontcuberta's Landscape without Memory project, in which he uses software developed by the United States Air Force, translating two-dimensional cartographic information into a simulated three-dimensional image. The software allows the user to build photorealistic models based on information digitized from two-dimensional sources. The result gives the user the illusion of navigating in three dimensions that were previously viewed only as a flat image. With this widely available freeware, he created two series that constitute Landscapes without Memory. Instead of loading maps in the software, Fontcuberta inserts landscapes by artists such as Van Gogh, Cézanne, and other painters. The contours and tones of these painted landscapes are transformed into mountains, rivers, valleys, and three-dimensional clouds. The software reads the new virtual landscapes and what Fontcuberta calls post-landscape. In this way, it creates a "nobody" land between the real and the virtual, between truth and illusion.

Figure 12. Joan Fontcuberta, Landscapes without Memory, 2002-2005

In these post-landscape, post-memory worlds developed from computer software, secondary memories are present – of the original paintings, displaced and reinterpreted by the software and of the place interpreted by the original artist – that form a simulated memory. Landscapes by Cézanne, Gainsborough, Dali, Hiroshige, and Watkins, among other great names in the history of art, are all altered by the software chosen by Fontcuberta in so many generic, smooth and predictable postcards through precise and mechanical arrangements of grey, green, yellow and blue. These photographic landscapes lead to the relentless, banal, and undemanding repetition of pictorial clichés. Perhaps this is what Fontcuberta's landscapes are all about, banal computer-generated landscapes that reflect the impossibility of fulfilling your own visual rhetoric, of bridging the gap between desire and experience. The project does not place so much the history of painting as the origin of nature, but makes visible both the defining terms, nature and landscape, reality and representation.

These questions are equally important to reflect on the notion of space. If in the old paradigm, the notion of space was conditioned by a single point of view, a concrete space, and the existence of an ordered horizon line, today the perception is of a global, multiple, and fragmented space, creating profound transformations in the way we experience travel. It is a fact that the gaze had already become mobile and mechanized with the invention of photography, but new technologies allow the gaze of the distant observer to become increasingly global and omniscient, to the point of becoming at the same time "massively intrusive, microscopic and macroscopic" (Steyerl, 2016).

Chalkroom, Laurie Anderson's virtual installation that allows the viewer to enter a world of words and chalk drawings as they move through 3D space, is another exercise in technological appropriation that allows us to reflect on the new three-dimensional spaces of representation, capable of generating a multiplication and delinearization of horizons and perspectives, breaking with the teleological point of view from top to bottom, of domination and control.

Our sense of spatial and temporal orientation has been changing drastically in recent years, motivated by new technologies, anticipating a new space of transition to new visual paradigms. If we accept the multiplication and delinearization of horizons and perspectives, the new tools of vision can also serve to express, and even change, the contemporary conditions of this transition.

The authors do not know what the future will be like and what travel and tourism will be like in the third millennium, but the authors know for sure that the 21st century will have been, ravaged by a pandemic that will have shaped it.

TESTIMONY OF THE STUDENT'S PERSONAL EXPERIENCE

One of the main challenges that virtual reality faces lies on the comprehension of the elements that need to face transition and adaptation from the real world into a completely digitalized environment. Virtual Reality is a powerful and complex tool that has a great potential in exploring different approaches to our world. At the moment, the general opinion on this technology is strongly connected to the concept of entertainment. A great majority of people that were unfamiliar with this type of technology and who never came close to it, seem to believe that its main purpose is to serve the game community, but, in reality, I came to learn that it comes to a much more than that.

As part of the Interactive and Multimedia Content Production course, I worked on the development of the project "*Castro de Monte Mozinho: Virtual Tour*". My main task was understanding and researching the particularities that immerse the users in the VR world, and then documenting visually Castro de Monte Mozinho in such a manner that delivers the end result in a form that can be experienced trough VR. My approach started with experimenting VR in the first place. I had very few opportunities to experience VR, and the ones that I did, mainly focused on gaming. Therefore, it was essential to me that I understand the digital media that I am going to use. I was provided from ISCE Douro with an Oculus Rift S which I later used to experiment and immerse myself in projects similar as concept to mine, such as 360 movies or virtual tours. At the same time, this allowed me to experience both the undoubtable advantages of this technology, as well as its flaws (to me, and many other users, motion sickness was one of the most unfavourable inconveniences of VR). I understood the immersive power this technology has, and how 2D almost completely changes in a VR world. „*Castro de Monte Mozinho: Virtual Tour*" 360 short clip that immerses the users in a relaxing afternoon at Castro de Monte Mozinho can be experienced both as a 2D film where you simply pan around and curiously explore the surroundings from your laptop/phone, but when you experience the same thing in VR, the experience takes a rather unexpected turn. Your focus on elements that you first overlooked trough 2D, such as the movement of trees or the birds that cross the sky, or details carved by time into the stones surrounding you. I came to understand that making a 360 movie needs to adapt to different human senses and that previous knowledge of cinematography rules do not apply as expected. You can no longer constrain your viewer to look at what you want him to see.

On top of the aspects mentioned above, as a foreign student, everything I experienced at Castro de Monte Mozinho was unfamiliar. I remember that while walking to the main site, the first thing that drowned me into the atmosphere of the place was the surrounding nature. Most of my visual comprehension of nature comes from The Carpathian Mountains. At Castro de Monte Mozinho I was enchanted by the way tall trees are hiding timeworn stones structures that imagination tries to integrate in a certain scenario. I believe that my attitude and feelings towards the place was later integrated in the concept. The project was designed in a way so the users get to experience a mix of an immersive relaxing afternoon at Castro de Monte Mozinho while also allowing you to curiously explore the surroundings. It immerses you into a moment of contemplation and introspection while daydreaming about what life in the Roman times must have looked like.

All things considered, as a result of working in the development of this project, I came to visualize the world through the prism of a new digital media and to sense concepts, ideas, and the world differently. It is now clearer to me that VR technologies are, slowly but surely, going to be part of our daily lives in in the twinkling of an eye, and even if at the moment the potential of this technology is mostly overlooked by the general population, I believe it has the potential to change the world. It's just a matter of time.

CONCLUSION

In sharing this experience, the authors hope to contribute to the discussion concerning the research and development processes of virtual tourism in the academic context, as well as in terms of the implementation of similar strategies in higher education and beyond, pursuing the configuration of novel solutions for a pathway that seems to assert itself further with each passing day in the universe of digital worlds.

Indeed, the transition from the second decade of the 21st century to the third brought about profound economic and social changes. As the technologies of VR, AR and RM establish themselves as valid, effective, and competitive options, this transition causes a considerable transformation of the relationship between the individual and knowledge itself, affirming digital memories in place of the fading of traditional forms and concepts of travel in face of impositions such as those which arose from the Covid-19 pandemic. As a consequence, where one will certainly discover that the components of immersion still need to go a long way, one may also already be witnessing a time of paradigm transition, where new cultural centers stripped of physical existence are affirmed, where knowledge and a free-thinking spirit that will allow us to travel and collect digital artefacts that cross traditional forms of visual arts, architecture, history, geology, as well as uses and customs in various temporal dimensions, is accumulating.

Despite the complexity of this project, as well as the sensitivity of the theme that guided it, the main objective of developing students' skills – in terms of the processes of design, planning, development, and prototyping of interactive multimedia systems and applications – was always present. This was achieved by not only focusing on the result, but especially on monitoring it's different stages, namely, in terms of sound and video acquisition and treatment, animation and interaction, prototyping, validation and implementation. To this end, knowledge development strategies were simultaneously implemented in terms of programming fundamentals, as a tool used in the practices of expression and visual communication, within the scope of the creation and articulation of interactive visual dynamics.

ACKNOWLEDGMENT

This research was supported by ERASMUS+ Study Mobility Grant.

REFERENCES

Ascape, V. R. (n.d.). Retrieved August 18, 2022, from https://ascape.com/

BRINK XR - See the world through Virtual Travel Experiences. (n.d.). Retrieved August 18, 2022, from https://www.brinkxr.com/

Fontcuberta, J. (2016). *La furia de las imágenes: Notas sobre la postfotografía.* Galaxia Gutenberg.

Gomes, D., & Araújo, M. (2012). Oferta Turística Virtual: Un estudio del metaverso. *Estudios y Perspectivas en Turism. Volumen, 21*, 876–903.

Google. (n.d.). *Cardboard - apps on Google Play*. Google. Retrieved August 18, 2022, from https://play.google.com/store/apps/details?id=com.google.samples.apps.cardboarddemo&hl=en_US+%E2%80%93+Google+Cardboard

Guttentag, D. A. (2010). Virtual reality: Applications and implications for tourism. *Tourism Management, 31*(5), 637–651. doi:10.1016/j.tourman.2009.07.003

Han, B. (2012). *A sociedade da transparência.* Herder.

Hernández, A. (2019). *Turismo virtual y experiência turística: el caso de la zona de monumentos arqueológicos de Teotihuacán.* Tesina da Licenciatura em Turismo, Universidad Autónoma del Estado de México.

Home. (n.d.). Retrieved August 18, 2022, from https://www.raiznewmedia.com/

Kucukkaya, A. G. (2004). Photogrammetry and remote sensing in archeology. *Journal of Quantitative Spectroscopy & Radiative Transfer, 88*(1-3), 83–88. doi:10.1016/j.jqsrt.2003.12.030

Lévy, P. (2013). *O que é virtual? 2013. 1 vídeo (2,54min).* Publicado pelo canal Fronteiras do pensamento em 25 maio 2013. Disponível em: https://www.youtube.com/watch?v=sMyokl6YJ5U&feature=emb_logo

López, L., Yagual, Patricia, C., & Aquino, J. (2016). El uso de la realidade virtual como herramienta tecnológica para fomentar el turismo em la península de Santa Elena. *3C Tecnología, 5*(3), 53 – 67.

Ministério da Educação e da Ciência. (2015, May 18). Portaria nº 135/2015: Estatutos do Instituto Superior de Ciências Educativas do Douro. *Diário da República.* https://www.iscedouro.pt/Files/Pages/12/estatutos_iscedouro.pdf

Municipality of Penafiel. (2019, May 8). *5º Simpósio Internacional de Arqueologia Virtual.* https://www.cm-penafiel.pt/5-o-simposio-internacional-de-arqueologia-virtual/

Que Ver e Fazer, O. (n.d.). Retrieved August 18, 2022, from https://www.rotadoromanico.com/pt/experiencias/o-que-ver-e-fazer/

Soeiro, T. (2008). A Arqueologia Histórica no Município de Penafiel. *Oppidum Atas do I Encontro de Arqueologia das Terras de Sousa*, 195-209.

Steyerl, H. (2012). *Os Condenados da Tela.* Caja Negra editora.

Tussyadiah, P., Wang, D., & Jung, T. M. C. (2018) Virtual reality, presence, and attitude change: Empirical evidence from tourism. *Tourism Management, 66,* 140-154. Disponível em: https://www.sciencedirect.com/science/article/abs/pii/S0261517717302662?via%3Dihub

visit - castro of Monte Mozinho. (2022, January 20). *YouTube*. Retrieved August 18, 2022, from https://youtu.be/33jZ6tGZCyo

Visitar Penafiel. (n.d.). Retrieved August 18, 2022, from https://www.cm-penafiel.pt/visitar-penafiel/

YouTube. (n.d.). *Home - youtube VR*. YouTube. Retrieved August 18, 2022, from https://vr.youtube.com/

ADDITIONAL READING

Eliseu, S., & Sousa, N. (2021). Stylized Artistic animation for virtual reality – Boundaries between the virtual and the physical. *Conferência Internacional Avanca Cinema, 2021*, 14–20.

KEY TERMS AND DEFINITIONS

Augmented Reality: A technology that superimposes a computer-generated image on a user's view of the real world, thus providing a composite view.

Cyberspace: The notional environment in which communication over computer networks occurs.

Heritage: Full range of our inherited traditions, monuments, objects, and culture. Most important, it is the range of contemporary activities, meanings, and behaviors that we draw from them. Heritage includes, but is much more than preserving, excavating, displaying, or restoring a collection of old things.

Mixed Reality: Blend of physical and digital worlds, unlocking natural and intuitive 3D human, computer, and environmental interactions. This new reality is based on advancements in computer vision, graphical processing, display technologies, input systems, and cloud computing.

Photogrammetry: The use of photography in surveying and mapping to ascertain measurements between objects.

Tourism: Social, cultural, and economic phenomenon which entails the movement of people to countries or places outside their usual environment for personal or business/professional purposes.

Virtual Reality: A computer-generated simulation of a three-dimensional image or environment that can be interacted with in a seemingly real or physical way by a person using special electronic equipment, such as a helmet with a screen inside or gloves fitted with sensors.

ENDNOTE

1 The imaginary of Paris, for example, makes Paris what it is. This is a historical construction, but also the result of an atmosphere and, therefore, an aura that continues to produce new images. These imaginaries lead people to construct certain meanings and (re)significations about places and people.

Chapter 25
The Cordel Pandemic:
The Daily Life of Students From the Interior of Paraíba, Brazil Told in Verse

Filomena Maria Moita
Universidade Estadual da Paraiba, Brazil

Leandro Mário Lucas
https://orcid.org/0000-0001-9627-4951
Universidade Estadual da Paraiba, Brazil

ABSTRACT

The COVID-19 pandemic established a transmutation, which, although initially temporary in appearance, produced dramatic changes in the secular way of teaching and learning in methodologies and resources. It also caused profound social, emotional, and educational impacts on the academic community in general. From this perspective, in this chapter, the authors aim to identify the impacts caused by the pandemic on daily life and revealed by students in their storytelling in verse, more specifically, in cordel. For this, they used as methodology the assumptions of content analysis, in the expectation of having an in-depth understanding of the content presented by the students and making inferences, in order to achieve the intended purpose. As a result, they found in the verses impacts associated with three categories of daily life, which they created from the data processing—sanitary, personal, and social—and they found that cordel literature, by the very essence of capturing the images of life as it is, proved to be an adequate way to capture the students' voices about the topic in question.

INTRODUCTION

The Covid-19 pandemic established a transmutation, which, although initially temporary in appearance, produced dramatic changes in the secular way of teaching and learning, in methodologies and resources, as well as caused profound social, emotional, and educational impacts on the academic community in general. As a result, teachers and students undewent an abrupt adaptation to a remote teaching process, or, in some cases, to a hybrid-mediated one via Digital Technologies (DT).

DOI: 10.4018/978-1-6684-3369-0.ch025

In this context, there were many challenges faced by the school community because of the adaptations imposed by remote teaching. However, studies that highlight the voices of students are still scarce, silenced, or do not give due emphasis to the main actors in education, whether in times of pandemic or not.

It was in the quest to fill this gap by focusing on capturing the "everyday lives" of students, during the most critical phase of the pandemic and in this moment of returning to hybrid classes, that the chapter we entitled The Cordel Pandemic: The Daily Life of Students from the Interior of Paraíba - Brazil Told in Verse is structured. For this, we asked the students in the 3rd grade of High School from a full-time public school in the interior of Paraíba to produce *cordéis* about the Covid-19 pandemic.

Regarding the structure of this text, it is composed of an introduction, the theoretical framework, which briefly discusses the pandemic and its impacts on teaching, *Cordel* Literature and the assumptions of content analysis, which it is based on Bardin's Theory (1977). There is also a section dedicated to methodology and another to the processing of results, inferences and interpretations, in addition to our final reflections.

From this perspective, we seek an answer to the following research question: how was the pandemic and its impacts on our daily life revealed by students in the verses of their *cordéis*? Thus, this text aims to identify the impacts caused by the pandemic on daily life revealed by students in their storytelling in verse.

THEORETICAL FRAMEWORK

In this section, we present the theoretical framework that supported our study, which is composed of the following items: some aspects of the pandemic and its impacts on Remote Learning; *Cordel* Literature; and Content Analysis, which is based on the assumptions of Bardin's Theory (1977). 1,2

The Pandemic and its Impacts on Remote Learning

The Covid-19 pandemic has brought about many changes in the traditional ways in which people live, are, and relate to each other. Although the transformations it caused took many forms, in general, they happened in order to transform the way people do their daily activities in person, in a remote way, and its variations, which include totally synchronous or asynchronous forms, as well as the hybridization of tasks, partly done online through digital technologies (DT).

In this scenario, "perhaps because it is an eminently social act, education was one of the most affected and disfigured segments of its traditional ways of materialization" (Double blind review), which motivated many studies on Remote Learning and its implications. in education and reflections on its possible impacts in the post-pandemic period.

For Kuklinski and Cobo (2020), the education that had to be implemented in the pandemic period redesigned the traditional face-to-face educational model towards virtuality, and actors and educational institutions resistant to the use of DT had to appropriate digital culture. This movement, in the view of these authors, forces the most resistant people to take ownership of technologies much more so that they can carry out remote learning experiences, becoming a central aspect for teacher training processes, which represents a great opportunity for teachers to become designers of learning experiences and stop being mere disseminators of content.

Still on the pandemic, Arruda (2020, p. 257) asserts that this moment we are experiencing is a historic milestone for education "because it generated the unprecedented situation of 90% of the student population being isolated throughout the world", imposing enormous challenges to public policies implemented worldwide to serve students remotely, through DT.

For this author, due to the lack of face-to-face sharing, social isolation has deconstructed the way in which teaching and learning are seen socially and revealed the deficiencies of educational training and structural policies in several countries around the world, which cannot do without knowledge and practices linked to the use and appropriation of digital technologies, otherwise these instruments will be used only as a palliative, without much effect on the acquisition of knowledge.

Moreira & Schlemmer (2020) state that Remote Teaching has transposed face-to-face and physical teaching in its courses, curriculum, methodologies and pedagogical practices to digital and network media. In this perspective, the teaching process is centered on content and, although there is a geographical distance, synchronous classes are privileged, following the principles of face-to-face classes, with predominantly bidirectional communication, as in one (teacher) for many. (students), focusing on the information transmitted or on their ways of transmitting

For Hodges et. al. (2020), Remote Learning is not only a basic approach to traditional instruction, but also a way of reflecting and mapping modes, methods and media to meet the needs and limitations of contexts that require rapidly changing resources. Due to its emergency nature, it stripped itself of previous organization, systematization and planning and, in this way, differs from already established models, such as Distance Education, considering that practically the same resources are used. It therefore involves fully remote or hybrid emergency solutions to meet the circumstances of the crisis caused by Covid-19 in education.

Alongside studies that emphasize the already measurable impacts of the covid-19 pandemic on education, there are others that, without neglecting them, in an exercise of inseparability, weave reflections on the unfolding of this moment in post-pandemic education. In this sense, (Double blind review) pointed out a diversity of practices carried out through Digital Information and Communication Technologies (DICT), during Remote Teaching, partly aligning with the *modus operandi* of traditional teaching and partly seeking to overcome it.

In the first moment, there is only a process of digitalization of the traditional, in which, "despite all the potential of the resources that are available to teachers, it seems [...] they continue to teach predominantly in a traditional way. They just changed from analog to digital form" (Hodges et al., p. 9). However, the study of these authors reveals a tendency towards originality and pedagogical innovation, through resources such as games, gamification platforms, social networks, google resources, among others, which, as a whole, provided teachers and students with new ways of learning. teach and learn with the potential to impact post-pandemic education.

Along these lines, Cani et. al. (2020) assert that the pandemic accelerated the process of integrating ICT with education, and Covid-19 demanded a reinvention of the school mediated, primarily, by these technologies. In our view, this reinvention must extrapolate issues related to the means by which teaching and learning are carried out and reach didactic-pedagogical aspects, on penalty of having the already presented digitalization of the traditional, because, despite its potential, the use of DT, as a means of reproducing traditional teaching practices, is not capable of arousing engagement, motivation and, above all, allowing the student to be the subject of their own learning.

Notwithstanding these issues, using DT as a means, but *Cordel* Literature as a pedagogical tool, we implemented a pedagogical practice that aimed to work on Portuguese language skills and identify the impacts of the pandemic on the daily lives of students. Next, we address aspects of this literature.

The Cordel: A Translation of Everyday Life into Verses

Cordel is a type of literature whose origin is not Brazilian, given that it arrived in Brazil brought by Portuguese settlers around the 16th century. Corroborating with us, the website of *Fundação Casa de Rui Barbosa* [1] states that this literature was characterized, initially, in Brazil, by the reproduction of oral literature from the Iberian Peninsula and, later, incorporated facts from the Northeastern sociocultural environment itself. At that time, it became one of the main sources of entertainment and information for the northeastern population and assumed functions very similar to those performed by encyclopedias and newspapers.

From the 19th century to the present day, *cordel* has been characterized as a literary genre that captures and describes the heart of the Northeast region, in a popular, oral and rhythmic language, revealing the Northeastern culture in its myths and realities, problems and solutions, stories and imaginations, standing out for its great cultural, historical, social and, consequently, pedagogical value.

It should be noted, however, that, despite the richness of its signs and symbols and its sociocultural value, *cordel* still does not occupy a prominent place in the dominant conceptions of school and literature. Corroborating our thinking and citing the former president of the Brazilian Academy of Cordel Literature (ABLC), Gonçalo Ferreira da Silva, De Sousa and Passos (2018) state that this entry appeared in 1881, in the Contemporary Dictionary of the Portuguese Language, by Carlos Aulete, in which it is referred to as literature of little or no literary value. This description is still available in the Online Dictionary of Portuguese, in the definition of literature. [2]

The *Fundação da Casa de Rui Barbosa* website affirms the existence of two great generations of *cordelistas* who established this literary style based on orality, with a metrified, rhymed, sung and playful structure, intended for the collectivity, whose majority of the target audience was formed of illiterates and semi-illiterates. However, for the aforementioned foundation, its informal and popular bias, devalued in traditional literature, does not make Cordel Literature a less valuable literary expression, as it is considered a very significant poetic expression by internationally recognized writers, such as Ariano Suassuna, Carlos Drummond de Andrade, Jorge Amado, Guimarães Rosa, Mário de Andrade and João Cabral de Melo Neto, among others.

There are countless renowned *cordel* writers. One of these who needs to be highlighted is Patativa do Assaré, from Ceará, whose improvised poetry was such a natural gift that he once stated that everywhere he looked, he could see a verse stirring (Assaré,1978). This motto, the theme around which an entire poem is developed, reveals the view that exists in literature in terms of describing everyday life.

Another influencial writer is Manoel Monteiro, from Pernambuco, a member of the Brazilian Academy of Cordel Literature, whose works are, at the same time, exponents and faces of this literature. In this sense, we emphasize 'Games at school. Electronic games: good or bad?', ' Cordel of the Conscientious Consumer, 'The Power of Plants in the Cure of Diseases' and 'The Diabetic Guide' (Montenegro, 2014).

The poet from Paraíba, born in the city of Pombal, in 1865, Leandro Gomes de Barros, is considered the greatest exponent of Cordel Literature. Author of works such as 'The Testament of the Dog' and 'The Horse which Defecated Money', which inspired Ariano Suassuna to write *Auto da Compadecida* (Bandeira, 2020), he was "capable of translating, in verses and rhymes, social issues and policies which originated from readings of the world in its social context", transforming information taken from the observation of everyday scenes and public events into satire (Montenegro, 2014, p. 28).

In terms of construction, Cordel Literature must abide by certain rules and can be determined in various types such as sextet, septet, decastitch, *martelo agalopado e galope à beira mar* according to the *A Corda* website. [3]

The sexteto is a *cordel* structure with six-line stanzas, with seven poetic syllables, and the septet, with seven-line stanzas. These structures are the most common and typical of *cordel* writers. The decastitch has stanzas with ten lines of seven poetic syllables, while the *martelo agalopado* contains stanzas of ten lines with ten poetic syllables in each line and the *galope à beira mar* is formed of stanzas with ten lines of eleven poetic syllables, obeying the same tenth rhyming rules, with the last stanza ending with "mar". In the table below, we present examples of *cordel* stanzas of each of these structures.

Table 1. Cordel Literature Structure.

Sextet	English equivalent
1º Eu/ vou/ con/tar/ uma/ his/tó/ria [no rhyme] 2º De um pavão misterioso [rhyme] 3º Que levantou vôo na Grécia [no rhyme] 4º Com um rapaz corajoso [rhyme] 5º Raptando uma condessa [no rhyme] 6º Filha de um conde orgulhoso. [rhyme]	1º I will tell you a story [no rhyme] 2º Of a mysterious peacock [rhyme] 3º That took flight in Greece [no rhyme] 4º With a brave young man [rhyme] 5º Kidnapping a girl [no rhyme] 6º Daughter of a proud count. [rhyme]
Septet	**English equivalent**
1º Já falei de saudade [no rhyme] 2º Tristeza e ingratidão [1st rhyme] 3º De amor e de prazer [no rhyme] 4º E cantei de emoção [1st rhyme] 5º Quero agora cantar [2nd rhyme] 6º E também quero falar [2nd rhyme] 7º Das coisas do meu sertão [1st rhyme]	1º I've already spoken about memories [no rhyme] 2º Sadness and ingratitude [1st rhyme] 3º Love and peace[no rhyme] 4º And sung about emotion [1st rhyme] 5º Now I want to sing [2nd rhyme] 6º And also speak [2nd rhyme] 7º Of my Northeast [1st rhyme]
Decastitch	**English equivalent**
1º As obras da Natureza [1st rhyme] 2º São de tanta perfeição, [2nd rhyme] 3º Que a nossa imaginação [2nd rhyme] 4º Não pinta tanta grandeza! [1st rhyme] 5º Para imitar a beleza [1st rhyme] 6º Das nuvens com suas cores, [3rd rhyme] 7º Se desmanchando em louvores [3rd rhyme] 8º De um manto adamascado [4th rhyme] 9º O artista, com cuidado, [4th rhyme] 10º Da arte aplica os primores [3rd rhyme]	1º The works of Nature [1st rhyme] 2º Are so perfect, [2nd rhyme] 3º That our imagination [2nd rhyme] 4º Can't match its greatness [1st rhyme] 5º To mimic such beauty[1st rhyme] 6º From clouds and its colors, [3rd rhyme] 7º Melting in praise [3rd rhyme] 8º From a apricotish mantle [4th rhyme] 9º The artist, carefully,, [4th rhyme] 10º From Art they use its essence [3rd rhyme]
Martelo agalopado	**English equivalent**
A Se/**nho**/ra/ dos/ **Tú**/mu/los/ ob/**ser**va [10 poetic syllables. Strong syllables in bold] O vaivém da tacanha mocidade, Que despreza a virtude e a verdade E dos vícios se mostra fiel serva, Porém nada no mundo se conserva: Sendo a vida infindo movimento, É a Morte um novo nascimento A inveja é o túmulo dos vivos — O herói repudia esses cativos, Galopando o Cavalo Pensamento.	The lady from the tomb watches us [10 poetic syllables. Strong syllables in bold] The come and go of feeble youth That despises virtue and truth And is loyal to vices But nothing in this world remains Being life a short movement, And Death a rebirth The tomb is envied by the living— The hero despises these matters Riding the Thoughts Horse

continues on following page

Table 1. Continued

Galope à beira mar	English equivalent
Can/**tan**/do/ Ga/**lo**/pe/ nin/**guém**/ me/ hu/**mi**lha, [sílabas fortes em negrito] Pois tudo que existe no mar aproveito, Na ilha, no cabo, península, estreito, Estreito, península, no cabo, na ilha, No barco, na proa, em bússola e milha! Medindo a distância eu vou viajar, Não quero, da rota, jamais me afastar, Porque me afastando o destino sai torto; Confio em Deus pra avistar o meu porto, Cantando Galope na beira do mar!	Singing "Galope" nobody beats me, Since everything in the sea is a cue, On the island, the cape, the península, the strait, Strait, península, the cape, on the island., On the ship, the prow, in the compass, the mile! Measuring the distance, I'll travel I don't ever want to divert my route Since If I do the destiny won't come right I believe in God to show me my harbor Singing "Galope" on the shore

Source: A Corda website

We can, therefore, affirm that, in its structure, the cordéis are covered with certain rigor, technique and method, as elaborate as the other types of texts in erudite literature.

It is worth emphasizing that the last three structures presented – the decastitch, the martelo agalopado and the galope à beira mar – are frequently used by *repentistas*, whose singing is not considered *cordel* literature by a group of *cordelistas*, which leads some scholars to affirm that this distinction is anchored much more in power relations than in the distinctions they have between themselves (De Sousa & Passos, 2018).

All the characteristics and elements that constitute the *cordéis* are covered by historical-cultural and social factors that make them, from a pedagogical point of view, very rich. Naturally, its educational potential is intrinsically linked to the study of Language and Literature, having the Brazilian Northeast as its context.

This literature, as stated by Ribas and Da Silva Malafaia (2021, p. 65), is capable of absorbing erudite and popular culture, supposedly irreconcilable, and transforming itself into a kind of "popular classic" poem. In the view of these authors, their educational potential must be urgently absorbed by the school, from the point of view of literary literacy, as an artistic element, media and staging.

In this sense, the pedagogical use of Cordel Literature is in line with a recent educational paradigm of valuing popular knowledge in completeness with academic knowledge, typical of pedagogical aspects that have assumed critical and post-criticism. In our view, ultimately, this use goes beyond pedagogical issues and plays a social role, that of highlighting, in the school context, the highly discriminated way of communicating of the Northeastern people, which contributes to overcoming historical prejudices.

It should be noted, however, that, due to its own transcultural nature, the potential of Cordel Literature is far from being limited to the area of languages. S. Lima (2013) asserts that this literature is an important tool to be used in teaching and learning, as it provides memorization via metrics, rhythm and playfulness and develops more general skills, such as creativity, artistic sensitivity and aesthetic enjoyment, at the same time as it favors cultural plurality and the transversality of knowledge, because themes such as the environment, ethics, health and others have always been addressed in *cordéis*.

We also highlight the studies by J. Lima (2013), who talks about the use of this literature as a way to popularize Science; that of Fonsêca (2009), who discusses the use of this resource for teaching Geography; and Machado, de Moraes Almeida e de Paula (2021), which aims to present *cordel* literature as an alternative tool for the teaching-learning process in Chemistry.

As literature that captures the social pulse, the *cordéis* also portrayed the Covid-19 pandemic. When analyzing the content of four articles in 2020, Santos, Lopes and Souza (2021) concluded that the information and guidance on covid-19 disseminated through this popular literature are effective means of health education, because they translate scientific information. in a more accessible language, along the lines of Folkcommunication, the area of Communication founded by Luiz Beltrão, from Pernambuco, in the 60s.

Considering the pedagogical dimensions of *Cordel* Literature that we present, we asked students from a public school in the interior of Paraíba to produce *cordéis*, in order to develop the Portuguese Language skill H4: to identify the theme of a text. However, due to the objective of this article, we are interested in analyzing the content presented by the students in their productions. For this reason, in the next section, we will approach the methodology of Content Analysis.

Laurence Bardin's Content Analysis

One of the most complex tasks of scientific investigations is the data analysis phase, since it is in it that the researcher, supported by the theoretical framework and their ability to interpret and produce arguments, manages to achieve the objective they want to achieve and see what the research data say in its depth, beyond what is already perceived in its surface.

When the data to be analyzed are submerged in texts produced by third parties, Content Analysis presents itself as a possible path, because it provides a set of methodological techniques that apply to various discourses, clarifying their messages and their implications and enabling the unveiling of the hidden and the latent, retained by any message (Bardin, 1977). From this perspective, it is defined as

[...] a set of communication analysis techniques aimed at obtaining, through systematic and objective procedures of description of the content of the messages, indicators (quantitative or not), that allow the inference of knowledge related to the conditions of production/reception (inferred variables) of these messages (Bardin, 1977, p. 42)

Content analysis is applicable to all forms of communication, both those with linguistic codes and those with non-linguistic semiotic codes, but carrying meanings, such as postures, gestures and different behaviors. Its purpose is to infer knowledge about the sender of the message, through quantitative and/or qualitative indicators, which materialize in three phases, namely: description, inference, and interpretation. In the description phase, the characteristics of the text are enumerated; in inference, the logical deduction of the data is made, which makes it possible to move on to interpretation, which, in turn, is the meaning given to the characteristics described, in order to articulate the semantic or linguistic structures of the texts with psychological or sociological issues of whomever produced them.

The object of content analysis is the word or sign in its meaning to understand the discourses of its senders or the environment in which they are located. Therefore, its phases are organized around three chronological poles: 1) pre-analysis; 2) exploration of the material; 3) and the treatment of results, inference and interpretation (Bardin, 1977).

Pre-analysis is the moment when the research corpus is produced and the material to be investigated is selected, with the objective of systematizing and making the initial ideas operational through the choice of material to be analyzed, the formulation of hypotheses (or not) and objectives, and the elaboration of indicators that will support the final interpretation. One of the most important moments of this phase is the free-floating reading, which serves for the researcher to remove the first units of meaning and build the sample to be used in the research, observing the rules of exhaustiveness, representativeness, homogeneity and pertinence.

The exhaustiveness rule establishes that the definition of the research sample must take into account all the elements of its corpus; representativeness is associated with the need for the sample to be representative in relation to the total number of documents previously selected; the sample will be homogeneous if it obeys precise selection criteria and does not present components that are too unique in relation to them; and the pertinence determines that the documents retained must be adequate sources of information and correspond to the objective that raises the analysis.

For Bardin (1977), when creating categories, the question and objectives of the investigation must be observed. This can be done in charts that list and group the thematic axes in tables, through differentiation or regrouping, obeying the principles of 'Mutual Exclusion', 'Homogeneity', 'Relevance', 'Objectivity and Fidelity' and 'Productivity'.

The interpretation of the data stems from the aforementioned moment and requires a resumption of the theoretical framework to support the analyzes and acquire meaning. Bardin (1977) considers that, with the advent of the computer, analysis and interpretation can be done more quickly. In order to reveal the impacts caused by the pandemic on the students' daily lives, through the cordéis they produced, we detail below each of the steps we carried out and present the sample that served as the basis for the treatment of the results, the inferences and the interpretation.

METHODOLOGY

Before detailing how we carry out the stages of content analysis, it is worth recalling the idea that this is a data analysis methodology that can be seen from the perspective of qualitative research in Education (Mendes & Miskulin, 2017). In this sense, it is the qualitative approach that underlies the findings of our study (Bogdan & Biklen, 1994).

The subjects of our study were students from the 3rd Grade of High School at a public school in Paraíba, in the year of 2021. This group consisted of 29 students, aged between 16 and 18, 24 of which were considered active. Of these, 14 chose to remain in remote learning and only 10 chose blended learning.

Obeying the norms of ethics in the research, it was requested the completion and signature of the Informed Consent (IC) by the adults or legal guardians, in which they chose a nickname to identify the authorship of their *cordéis*.

The production of *cordéis* by students entered our pedagogical planning for two initial reasons: an institutional one, because the school curriculum of the State Schools of Paraíba requires that all Common Core Curriculum components and the diversified part work on Portuguese Language skills; and another investigative, since we have always been interested in hearing the voices of students about the impacts of the pandemic on their daily lives.

From the intersection of these two initial motivations came the need to find a way for these voices to be heard. Thus, considering that *Cordel* Literature is able to capture information drawn from observation, everyday scenes, and public events (Montenegro, 2014, p. 28), we asked them to write *cordéis* based on the theme 'Pandemic in Cordel'.

In the description of this activity, posted on Google Classroom, on November 26, 2021, we told the students that the aforementioned theme was not necessarily the title of the cordel to be created, that the activity was not mandatory, and that we intended to work on the fourth skill presented in the Portuguese Language State Curriculum- Identify the theme of a text. Therefore, it was clarified that the titles of the *cordéis* would be freely chosen. However, the students did not know, until that moment, that their productions would be used in this article. We made it clear that *cordéis* should address aspects of the pandemic that impacted their daily lives. From our point of view, the conditions presented for this activity ensured freedom for students to produce content that represented the impact of the pandemic on their lives.

Table 2. Constitution criteria from analyzed sample.

Exclusion Criteria	Association with Bardin's rules
Productions that did not fit in Cordel	Homogeneity rule: in this case, any other text that did not fit in cordel made the sample heterogeneous.
Cordéis that did not address the topic presented.	Pertinence rule: in this case, cordéis that did not address the topic 'The pandemic in cordel' were not in line with our research objective.
Cordéis that contained false social or personal information or statements.	Pertinence rule: in this case, the *cordéis* were not adequate sources of information.

Source: Devised by the authors based on Bardin (1977

On December 14, 2021, the return period ended, so we obtained the production of ten cordéis with authors of both sexes. From that moment on, we started the pre-analysis phase of the material. Our study opted for an exploratory perspective, which does not require the formulation of *a priori* hypotheses. Bardin (1977, p. 98) asserts that "in fact, hypotheses are not always established during pre-analysis" so that it is possible to start from the texts themselves to understand their connections according to a deductive process, and adds that formulating hypotheses a priori, sometimes, consists in previously mastering dimensions and directions of the study that, in the last analysis, work in the process. In this sense, we seek to answer the following research question: how was the pandemic and its impacts on everyday life revealed by students in the verses of their cordéis? Thus, this text aims to identify the impacts caused by the pandemic on daily life and revealed by students in their storytelling.

The ten *cordéis* produced by the students constituted the corpus of our research. It is worth remembering that not all material in the corpus is a valid sample for analysis, as they must obey certain rules, according to certain criteria (Bardin, 1977). Before doing the free-floating reading, we established the some exclusion criteria to constitute the sample that we would analyze (see Table 2):

After performing the free-floating reading by applying these filters, our sample consisted of four cordéis whose authors are female, which are presented below [4]:

This sample of *cordéis*, in our view, encompasses the other two rules established by Bardin (1977) for the constitution of the sample, namely: the rule of exhaustiveness, since we analyzed all ten cordéis produced to reach this sample; and the representativeness rule, since we are using 40% of the *cordéis* produced (40%), a percentage that, from a statistical point of view, can be considered representative. Furthermore, from our preliminary reading, it is possible to affirm that the present sample is also representative in terms of content.

Table 3. Cordéis sample of our research.

The (un)glorious pandemic	The pain that makes you stronger
This year is coming to an end Many things took place Even a thing called coronavírus Showed itself up around us Letting everybody down With all the deaths to date After some many months The vaccine finally came A shot of hope For those who lived Without the vaccine Death prevails Since many passed away About this vaccine thing Some rumors came up We would turn into alligators So they were saying Even our president Feeble minded, I'd say The rumor did confirm But our people are wise We saw what happened Scientists doing tests The rumor faded away Using sanitizer and masks Society being ready The vaccination took place Through these verses I want to thank The attention you gave me I could clarify That the vaccine saves This is the statement That here I want to say Author:Elinha	Neighbors, friends and relatives In a memory from the past Important people Who left a message This virus kills people You better be careful From good moments and hugs What's left is a memory Of relatives and friends To whom we were loyal But today they are remembered By the whole society What's left for us are Fast and recent losses Grief and pain That linger endlessly An unforgettable time It'll be remembered forever The warriors who survived Will never forget the past Because they fought bravely So as not to be erased From people's memories Who fell into despair A loss in our lives For many is crucial We have to learn What's fundamental To awaken love In the spiritual world Author: Nya

The Pandemic in Cordel	Here I am
And suddenly the news come The one that comes from the other side of the world A virus that attacks the city And leaves it devastated Everybody gets nervous Others think it is bullshit But deep down they are afraid At school everybody is happy To watch TV only There's nothing better No tests, class or tasks But today they miss it Because of this calamity For 2 years they haven't had the June parties Around the continent Everybody is desperate They are stocking food And breaking into supermarkets They were in quarantine Instead of two weeks Gosh, everybody was fooled. The long-waited vaccine has come Scientists, we thank you all Doctors and nurses Since, because of them, we'll win So as not to end this in silence Stop being foolish Get your vaccine, wear masks, and the rest we'll get it Author: Anne	It was by March When everything happened Something unexpected A virus came up Affecting the whole nation Chaos took place The internet helped Not to fall behind But so much effort, shoot, it was too much Today I feel prepared Since I see myself ahead. Towards a future that waits for me One year has passed And now here I am Saying goodbye to the school That taught me so much And in spite of the obstacles My fingerprint there will stay And to end this I say Great was my evolution Our name? Resilience Since in agony we adapted ourselves That's why I say it one more time JRDT, my eternal gratitude! Author: Emy

Source: Authors' files

We emphasize that some previous answers to our research question could already be given based on reading the *cordéis*. However, as stated by Bardin (1977), one of the main functions of content analysis is to go beyond the obvious and uncover what is veiled, including what is not directly exposed in the content.

The next stage that we will present is the exploration of the material, in which the systematic administration of what was established in the pre-analysis phase is carried out and the registration and/or context units, themes or thematic axes are defined. In our particular case, we use the word and its meanings as registration units. Therefore, all the words in the text must be taken into account and analyzed for their frequency or their most recurrent meanings (Bardin, 1977).

Manually counting the frequency of the words contained in the four cordéis was too demanding, which led us to use the WordClouds.com website [5], which, while creating word clouds from texts in different formats, it makes it possible to count customized characters for free.

In the first count, the four *cordéis* produced by the students totaled 240 words. However, as our emphasis is on meaning, on words associated with the students' daily lives and with the potential to form themes and categories of analysis, we excluded those that were only formed by a letter, and some, with onle a syllable. Therefore, we excluded words such as '*that*', among others, and 228 words remained. Of these, those with absolute frequencies greater than 1 are shown in the table below.

As it can be seen, due to the exhaustion that would cause us, we omitted, in the above table 137(228-91), words that only appeared once. Thus, considering that, in the universe of all significant words, those that appeared isolated and could communicate something important to us, we made a word cloud to present them.

Figure 1. Cloud of words that were not repeated in the cordéis
Source: Devised by the authors

Naturally, as we knew that these words were not repeated in the cordéis, the character of the image above is descriptive. However, from the intersection of the words contained in Table 4 and/or in Figure 1, it is possible to constitute themes from the grouping of words, either repeated or not, considering the meaning assumed by them in the context of the cordéis. For this, starting from the words with repetitions, we analyzed all 228 words that we considered significant, in order to identify the following themes:

Table 4. Absolute and relative frequencies of words that were repeated in the verses of the cordéis.

Repeated words	Repetitions per word	Absolute and relative frequency of repetitions per set of words
world, since	6	12 (5,3%)
Vaccine	5	5 (2,2%)
appeared, I am, today, my, news, past, so, virus	3	24 (10,5%)
Happened, friends, year, scientists, thing, here, desperate, because, pain, school, got, people, memories, there, better, many, mask, pandemic, happened, prepared, restablish, recollection, society	2	50 (21,9%)
Total	**16**	**91 (39,9%)**

Source: Devised by the authors

Table 5. Themes identified in the cordéis.

Themes	Associated words
Vaccination	vaccine, dose, scientist(s),nurses, doctors, saves
Pandemic	happened, appeared, world, virus, pandemic, society, surroundingsm city, continente, coronavirus, nation
Pandemic effects	Since, calamity, chãos, thing, because, school, got, people, many, happened, death, deaths, loss, losses
Communication	rumor, news, Internet, alligator, president, rumor
Emotional state	I am, desperate, pain, memories, recollection, affecting, agony, nervous, broken, down, silenced, hope, strong, grief, fear, so, resilient, here
Protection measures	scientists, sanitizer, mask, tests
Social relations	friends, school, people, parentes, hugs, supermarket, neighbors, love
Overcoming	prepared, adptated, bravely, evolution, survivals, survived, won, warriors
Time	years, days, eternally, future, endlessly, months, month, pass, quarantine, fortnight, past, time, there

Source: Devised by the authors

After exploring the material, we now move on to the next phase of our content analysis - that of processing the results, inferring and interpreting the data. Due to its importance, we will give it special emphasis, approaching it in the form of a section.

DATA PROCESSING AND INTERPRETATION

According to Bardin (1977), at this stage, the raw data must become "talkative", meaningful, natural, in line with the objectives outlined. Charts, graphs and tables can be used to make inferences well.

From this perspective, we were interested in identifying the presence of the themes in Table 2 in the all of the four cordéis. To do so, we put them in a single google document and then used the Ctrl+F command to search for the simultaneous presence in at least two cordéis at least one of the words associated with the aforementioned themes in Table 2. This operation served to confirm the contextual meanings of the words previously analyzed. After this confirmation, we produced the table below.

Table 6. Simultaneous presence of the themes in Table 2 in the cordéis.

Themes	Cordéis			
	The (un)glorious pandemic	**The Pandemic in Cordel**	**The pain that makes you stronger**	**Here I am**
Vaccianation	x	x		
Pandemic	x	x	x	x
Pandemic effects	x	x	x	x
Communication	x	x		x
Emotional state	x	x	x	x
Protection measures	x	x	x	x
Social relations	x	x	x	x
Overcoming	x	x	x	x
Time (past, presente, future)	x	x	x	x

Source: Devised by the authors

In Table 6, it can be inferred that the themes approached are dense in the four cordéis, with the exception of 'Vaccination' and 'Communication'. This allows us to deduce that the impacts of the pandemic on their daily lives are associated with these aspects. In this perspective, without losing sight of the objective of our research, we categorized the themes into possible daily lives, according to their most relevant elements in the health, personal and social aspects.

Table 7. Most affected daily life categories by the pandemic revealed in the cordéis.

Daily life categories	Theme
Sanitary	Vaccination
	Protection measures
Personal	Emotional state
	Overcoming
	Time
Social	Pandemic
	Social relations
	Pademic effects
	Communication

Source: Devised by the authors

Making a counterpoint of Table 7 with the verses of the cordéis, we found that 'vaccine' and 'protection measures' were present in the *cordéis* from a health perspective, as it can be identified in the following verses:

Box 1.

After some many months *The vaccine finally came* *A shot of hope* *For those who lived* *Without the vaccine* *Death prevails* *Since many passed away*	*The long-waited vaccine has come* *Scientists, we thank you all* *Doctors and nurses* *Since, because of them, we'll win* *So as not to end this in silence* *Stop being foolish* *Get your vaccine, wear masks, and the rest we'll get it*

In these excerpts, the topic of vaccination was established with a positive perspective, since, in the first, we observed a direct relationship between the lack of vaccine and the number of deaths and its arrival as a feeling of hope. In the second, there is a direct relationship between the vaccine and Science, while recognizing the work of health professionals at this delicate moment and the importance of wearing a mask, despite the denialism that revolved around these themes on the part of some speeches.

Discourses of denial were criticized in the *cordéis*, shedding light to a critical view of the students about fake news that emerged, as shown in the excerpts below:

Box 2.

About this vaccine thing *Some rumors came up* *We would turn into alligators* *So they were saying* *Even our president* *Feeble minded, I'd say* *The rumor did confirm*	*But our people are wise* *We saw what happened* *Scientists doing tests* *The rumor faded away* *Using sanitizer and masks* *Society being ready* *The vaccination took place*

These excerpts resemble a satire that emphasize the negative effects of vaccines and some misinformation which have been disseminated in order to contest their benefits. There is also explicit mention of scientists, once again confirming the relationship between vaccination and Science.

Attitudes that reflect personal issues in their emotional states, feelings of overcoming and reflections on the past, present and/or future time can be seen in the following excerpts:

Box 3.

Around the continent *Everybody is desperate* *They are stocking food* *And breaking into supermarkets* *They were in quarantine* *Instead of two weeks* *Gosh, everybody was fooled.*	*The internet helped* *Not to fall behind* *But so much effort, shoot, it was too much* *Today I feel prepared* *Since I see myself ahead.* *Towards a future that waits for me*	*Neighbors, friends and relatives* *In a memory from the past* *Important people* *Who left a message* *This virus kills people* *You better be careful*

We perceive, therefore, that the notion of time is emphasized: the duration of the pandemic, in the left stanza; of the lessons it brought and its impacts on the future, in the middle stanza; or the time that preceded it, in a kind of nostalgia, with explicit mention of personal losses (stanza on the right).

With regard to the fact that a feeling of greater preparation for the future is observed, based on the experiences during the pandemic, in the central stanza, this feeling was identified by the study by (Double blind review) in the voices of some teachers, who said they had acquired a lot of pedagogical learning during this period, especially through DT, due to what Kuklinski and Cobo (2021) called educational redesign during the pandemic.

Finally, topics such as the pandemic in general and its effects and social relations, such as deaths, were predominantly treated from a collective point of view.

Box 4.

From good moments and hugs What's left is a memory Of relatives and friends To whom we were loyal But today they are remembered By the whole society	*Neighbors, friends and relatives In a memory from the past Important people Who left a message This virus kills people You better be careful*	*And suddenly the news come The one that comes from the other side of the world A virus that attacks the city And leaves it devastated Everybody gets nervous Others think it is bullshit But deep down they are afraid*	*It was by March When everything happened Something unexpected A virus came up Affecting the whole nation Chaos took place*

In the first stanza on the left, death is referred to as a ubiquitous social feeling, which allows us to infer that, despite the personal impacts, it has been perceived as a consequence of broader social issues. This, in a way, is contemplated in the stanza that has as its opening line "Neighbors, friends and relatives" as it brings a collective idea about death, which is maintained in the phrases "everybody gets nervous" and "affecting nation", in the stanzas on the right.

However, considering the whole *cordel* production, were the impacts of the pandemic revealed by students from a health, social or personal perspective? Crossing the data from Table 7 with Table 8, we came to an answer. For this, we classified as strong (S), medium (M) or weak (W) the presence of the themes in the cordéis according to Table 7, respectively, for cases in which they appear in four or three cordéis, in two and in just one. For example, according to the aforementioned table, the theme vaccination was present in two cordéis; thus, it will be classified in the table below as medium presence (M). Soon after, we will quantify the strong presence (S), in percentage, of the daily life categories.

We can, therefore, infer that the daily impacts caused by the pandemic revealed by the students in the cordéis are more of a personal and social nature and materialize the intertwinings that we know exist between these daily dimensions, since social experiences profoundly impact the personal ones and vice versa.

Having Table 7 as a starting point and analyzing it vertically, it is possible to identify the degree of absorption of these themes by each of the four *cordéis*.

The *cordéis* "The (un)godly pandemic" and "Pandemic in cordel" covered all of the topics, and the one "The pain that makes you stronger" only 77%, as it did not cover the topics of vaccination and communication. The *cordel* "Here I am" did not include vaccination. In terms of category, we can say that the first two are more directed towards the social, while the last two towards the personal.

Table 8. Intensity of the presence of daily categories in the cordéis.

Daily life categories	Themes/Presence in the cordéis	Strong presence percentage		
		S	M	W
Sanitary	Vaccination (S) Protection measures (S)	50%	50%	0%
Personal	Emotional state (S) Overcoming (S) Time (S)	100%	0%	0%
Social	Pandemic (S) Social relations (S) Pandemic effects (S) Communication (S)	100%	0%	0%

Source: Devised by the authors

Table 9. Absorption of the themes by the four cordéis.

Cordel	Theme absorption percentage
The (un) godly pandemic	100%
The pandemic in cordel	100%
The pain that makes you stronger	77%
Here I am	89%

Source: Devised by the authors

As a means of capturing the voices of the students, the potential of *cordéis* to absorb the informal language was evident in phrases such as "Everybody gets *nervous* (aperreia)", "Stop being *foolish* (abestada)", "*gosh*" and "*shoot*" (oxe e vixi) typical of the Northeastern people. Furthermore, it has the potential to satirize situations, criticize everyday and scientific events in an easy-to-understand language (Montenegro, 2014).

The structures of the *cordéis* did not always obey the metric and poetic rules of this type of literature. However, content analysis is more concerned with the message than with the structure of the text (Bardin, 1977). Therefore, we can say that this fact does not imply an inferior quality of the messages described, treated and analyzed here.

FINAL REFLECTIONS

Our quest to find answers to the question 'How was the pandemic and its impacts on daily life revealed by students in the verses of their *cordéis*?' led us to analyze the content of the *cordéis* produced by them in order to identify that such impacts were predominant in three categories – sanitary, personal and social – with emphasis on the social and personal aspects over sanitary ones.

From a social point of view, it is possible to identify in the *cordéis* a collective understanding of the problems caused by the pandemic in everyday life. In this sense, in verses from 'The blessed pandemic', such as "a certain coronavirus", "the people were devastated" and "everyone with their heads down" with the deaths that occurred, leads us to identify an understanding of death as a consequence of a collective engine that did not work well at this moment in which a society that is aware of its obligations to their peers. Thus, it can be said that there was a strong impact on students' critical awareness of the collective effects of the pandemic.

In this perspective, some of the problems in this collective engine were satirized by turning into rumors news propagated by some official channels that "We would turn into alligators" if we took the vaccine. At the same time, there is an "Ode" to the vaccine, to Science and to health professionals that, by the way, deserves this recognition. This"Ode" is very well exposed in these verses from 'The pandemic in string': "Scientists, we thank you so much" and "Doctors and nurses were essential", "Since, because of them, we will win". For this, according to this *cordel*, it is still necessary to wear masks and vaccinate, underlining important health issues to reduce the impacts caused by the pandemic.

The impacts from a personal perspective are presented in a lyrical 'I' that suffers and mourns the death of "Friends, neighbors and relatives", as can be seen in the *cordel* 'The pain that makes you stronger'. However, it is clear that the suffering caused by the Covid-19 pandemic was also a factor of self-overcoming, as there is in it a narrative that presents the survivors of this humanitarian tragedy as subjects who do not weaken and are strengthened in by the difficulties faced in this period. This narrative is evident in excerpts from the same *cordel*, such as "The surviving warriors", who will never forget the past due to the brave struggles they had so as not to be erased (killed) and overcome the "Fast and recent losses" and the "Feeling of pain and grief".

These impacts in the personal sphere are presented, therefore, as something that not only left scars, but also a process that provided learning. This view is evident in the *cordel* 'Here I am', because it reveals a poetic 'I' that considers the effort made to "not stay in the rear" as something that left people more prepared for a future that awaits them, very much along the lines pointed out by Lucas & Moita (2021) regarding the opinion of some teachers about the learning they acquired at that time and its possible implications for post-pandemic education.

Other relevant impacts on *cordeis* are associated with the element of time, mainly because of the nostalgia for what could have been done in the past and it was no longer possible to do it due to the restrictions of social isolation, the time of establishment of the pandemic, or hope of its end and expectations for the future. These impacts are revealed in verses such as "They were in quarantine", "Instead of a fortnight", "In the memory of a past" and "For a future that awaits me", and can be associated with many diagnosed cases of depression and anxiety at the school where the authors of the *cordéis* studied.

We emphasize that the research data revealed that Cordel Literature proved to be a potential pedagogical resource to capture the images of everyday life in its pulse and describe them with a popular, easy-to-understand language, at the same time that it presented us a refined poetic structure, combining the facets of the erudite and the popular, revealing that these ways of building knowledge need to walk side by side.

There is also an emphasis on the element of time, either because of the nostalgia for what could be done in the past and cannot be done today, due to the restrictions of social isolation, or the deaths it caused; or even due to the continuation of the pandemic or the hope of its end and expectations for the future. This emphasis is revealed in verses such as "Were in quarantine" "Instead of two weeks" "In a memory from the past", "Since I see myself ahead" and "Towards a future that waits for me".

We come to the end of this article with an invitation to other researchers to engage in new studies, since, even though *cordel* is a pedagogical resource of unique potential and richness, the research revealed a lack of publications at different educational levels, both nationally and internationally. This points to the emergence of research and publications that give visibility and importance to the contribution of knowledge and popular language in general, as it happens with scientific knowledge and language.

ACKNOWLEDGMENT

To the authors of the cordéis The (un)godly pandemic, the Pandemic in cordel, The pain that makes you stronger and Here I am, identified here respectively by the nicknames Elinha, Anne, Nya and Emy, we express our sincere gratitude to their important contribution since, without them, this research would not be possible.

REFERENCES

Arruda, E. P. (2020). Emergency Remote Education elements to Brazilian public education policies during Covid-19 [Educação remota emergencial: elementos para políticas públicas na educação brasileira em tempos de Covid-19]. Distance Learning Journal, 7(1), 257-275.

Assaré, P. D. (1978). *Sing there that I sing here* [Canta aí que eu canto aqui]. Voices.

Bandeira, M. L. B. (2020, Novembro 19). *Cordelista Day: who was Leandro Gomes de Barros?* [Dia do Cordelista: quem foi Leandro Gomes de Barros?]. Itaú Cultural. Retrieved October 15, 2021, from https://www.itaucultural.org.br/secoes/entrevista/cordelista-leandro-gomes-barros

Bardin, L. (1977). *Content Analysis* [Análise de conteúdo]. Lisbon: Editions 70.

Bogdan, R., & Biklen, S. (1994). *Qualitative Investigation in Education: an introduction to the theory and methods* [Investigação qualitativa em educação: uma introdução à teoria e aos métodos]. Porto Publisher.

Cani, J. B., Sandrini, E. G. C., Soares, G. M., & Scalzer, K. (2020). Education and Covid-19: The art of reinventing the school by mediating learning "primarily" through DITC [Educação e covid-19: a arte de reinventar a escola mediando a aprendizagem "prioritariamente" pelas TDIC]. *IFES Science Journal, 6*(1), 23–39.

de Oliveira, E., Ens, R. T., Andrade, D. B. F., & de Mussis, C. R. (2003). Content Analysis and Research in Education [Análise de Conteúdo e Pesquisa na Área da Educação]. *Educational Dialogue Journal, 4*(9), 1–17.

de Souza, L. R. D. S., & Passos, O. V. D. O. A. (2018). Cordel Literature: A pedagogical too [Literatura De Cordel: Um recurso pedagógico]. *FASETE Scientific Journal, 75.*

Fonseca, F. V. L., & Fonsêca, K. S. B. (2009). Contributions from Cordel Literature to Cartography teaching [Contribuições da Literatura de Cordel para o ensino de Cartografia]. *Geography, 17*(2), 123–132.

Hodges, C.B., Moore, S., Lockee, B.B., Trust, T., & Bond, M.A. (2020). *The difference between emergency remote teaching and online learning*. Academic Press.

Kuklinski, H. P., & Cobo, C. (2020). *Expanding the university beyond emergency remote teaching: ideas for a post-pandemic hybrid model* [Expandir la universidad más allá de la enseñanza remota de emergencia: Ideas hacia un modelo híbrido post-pandemia] (1st ed., Vol. 1). Outliers School. https://outliersschool.net/wp-content/uploads/2020/05/Expandir_la_universidad.pdf

Lima, J. M., de Sousa, J. M., & Germano, U. M. G. (2011). Cordel Literature as a means of science popularization: an intervention in Physics teaching [Os PCN e as potencialidades didático-pedagógicas do cordel].VIII ENPEC.

Lima, T. (2013). The national curriculum and the Cordel teaching and learning potential [Os PCN e as potencialidades didático-pedagógicas do cordel]. *Acta Scientiarum. Education*, *35*(1), 133–139.

Lucas, L., & Moita, F. M. G. S. (2020). Emergency distance education: The impacts in the pedagogical practices during Covid-19 [Ensino Remoto Emergencial (ERE): impactos na prática pedagógica durante a Covid-19]. *Techonology Research in Education Journal*, *6*, e143320–e143320.

Machado, L. J., de Moraes Almeida, I., & de Paula, L. B. (2021). Cordel Literature as a means of facilitating the teaching and learning process of chemistry [Literatura de cordel como recurso facilitador do processo ensino-aprendizagem em química]. *Brazilian Journal of Development*, *7*(9), 86407–86424.

Mendes, R. M., & Miskulin, R. G. S. (2017). Content analysis as a methodology [A análise de conteúdo como uma metodologia]. *Research Notes*, *47*, 1044–1066.

Montenegro, M. D. S. M. (2014). *Manoel Monteiros and the multifaceted Cordel texts* [Manoel Monteiro e as várias faces do texto de cordel] [Thesis]. Federal University of Paraíba, João Pessoa.

Moreira, J. A., & Schlemmer, E. (2020). For a new concept and paradigm of online digital education [Por um novo conceito e paradigma de educação digital onlife]. *UFG Journal, 20*(26).

Ribas, M. C. C., & da Silva Malafaia, R. (2021). Cordel Literature and Edication: an interartistic mosaic [Literatura de Cordel e Educação: um mosaico interartístico]. *PÓS: Arts Postgraduate Program Journal from EBA/UFMG, 11*(21), 61-89.

Rui Barbosa House Foundation. (n.d.). *Cordel: popular Literature in verse* [Cordel: Literatura popular em verso]. Retrieved October 15, 2021, from http://antigo.casaruibarbosa.gov.br/cordel/apresentacao.html#

Santos, P. P. P. D. O., Lopes, B. B. G., & de Souza, N. (2021). Covid-19 and cordel literature: health education through folkcommunication [Covid-19 e literatura de cordel: educação em saúde pela via da folkcomunicação]. *International Journal of Folkcommunication, 19*(43), 247-265.

KEY TERMS AND DEFINITIONS

Abestada: A silly person, a goof.

Cordel: Literary genre of Portuguese origin, which was established in the Northeast Region of Brazil and assumed its own characteristics associated with the Northeastern context. It is characterized by an informal and popular language. Initially, it was exposed in leaflets hanging on ropes or similar objects, whose specificity gave rise to the term Cordel Literature.

Cordelista: A person who writes *cordéis.*

Martelo Agalopado: Kind of cordel that has as its main characteristic a structure in which the stanzas are composed of ten lines containing ten poetic syllables, having the stress falling on the third, sixth and tenth poetic syllables.

Galope à Beira-Mar: Kind of cordel that has as its main characteristic a structure in which the stanzas are composed of ten lines containing eleven poetic syllables, having the stress falling on the second, fifth, eighth and eleventh poetic syllables; plus, the last stanza finishes with the word "mar."

Oxe: Abbreviation of Oxente, which conveys the idea of astonishment, surprise, admiration, or fright.

Repentista: Term associated with the expression de repente, which means "unforeseen," "unexpected," and "sudden." It referes to singers who make verses and songs in the form of improvisation, performing alone or challenging another singer.

Vixe: Although its meaning varies according to the context in which it is being used, it generally denotes surprise or fright, which can sometimes be replaced by the expression Virgem Maria (Virgin Mary).

ENDNOTES

1 This is one of the most comprehensive websites about Cordel Literature in the world. It has a collection of approximately 9,000 cordel leaflets, all available for online consultation, by index, author, title, subject, place of publication, publisher/typography and date. Of this set, around 2,340 leaflets are available in digital versions, with their original and variant versions. Available at: <http://antigo.casaruibarbosa.gov.br/cordel/acervo.html>. Accessed on 13 Oct. of 2021.

2 Availbale at: <https://www.dicio.com.br/literatura/>. Accessed on 31 Jan. of 2022.

3 Available at: < http://acorda.net.br/estrutura-do-cordelprojetoacorda/ >. Accessed on 20 Jan. of 2022.

4 As mentioned before, the names of the authors of the *cordéis* are fictitious.

[5v]Available at: <https://www.wordclouds.com/>. Accessed on 28 Jan. 2022.

Section 9

Contaminated Cinema Destinations

Chapter 26
To Identify the Evolution of Tourism Practices in a Sensitive Way Through Documentary

Natacha Cyrulnik
https://orcid.org/0000-0003-0530-1331
Aix-Marseille University, France

ABSTRACT

As an extension of the audiovisual work carried out for 15 years in the cities of the South of France, the tourist sites are filmed in turn as part of a new documentary series in order to try to better understand, through their narration, the daily life in these easily caricatured territories. While this series "Getting Out of From the Postcard" aimed to bear witness to the tourists who took precedence over these sites, the pandemic reoriented it by (re)presenting daily life in these places, which are both tourists and inhabitants.

INTRODUCTION

In the wake of fifteen years of audiovisual work in cities in the south of France, tourist sites are now being filmed[1] in turn as part of a new documentary series entitled "Getting out of the postcard"[2] in order to try to better understand, through their cinematographic presentation, everyday life in these so easily caricatured territories (Cyrulnik, 2020). While this series was intended to bear witness to the tourists who were taking over these sites, the pandemic and the changes in use that it brought about reoriented it by (re)presenting[3] (Bougnoux, 2006, p.53) the daily life in these places, practiced by both tourists and inhabitants. It deals with ordinary places that become touristy, thus making them extraordinary. Based on the behavior of the inhabitants and tourists, who are more or less intermingled, the director and researcher seeks to capture the atmosphere of a place in order to bear witness to the complexity of the situation (Morin, 1990). It is not a question of 'inhabiting' the space insofar as we are more in the public space than in the home (Paquot, 2021), but rather of giving an account of the atmosphere that emanates from a place through the uses of all those who practice it by a sensitive approach (Sumartojo & Pink, 2019). Documentary allows this.

DOI: 10.4018/978-1-6684-3369-0.ch026

At a time when the world-place is being manufactured, in the sense of a universal site created by and for tourism, the fact of taking a cinematographic look at a long-established tourist site, a heritage site, contributes to the risk of its being transformed. The uses of an urban space bear witness to the culture in which they are embedded. But they are sometimes modified by an invasion of tourists who change their practices or become part of them. Tourists sometimes move around on sites that they themselves contribute to altering through their behavior. The living space is important to us, with its practices, its uses, even its consumption... Public space is modified. Its perception too. This is what the documentaries in this series propose to question in turn, going so far as to finally engage the viewer in this reflection.

While tourists come to a site for its experiential value, the researcher proposes through this series to make another experience, this time cinematic. " Getting out of the postcard " is offered as a case study to initially question the media forms for approaching tourist sites. Inspired by the media power of the postcard, with all that it conveys to tourists, this hybrid work between sociology, travelogue and documentary film, bears witness to the mediation devices that organize this relationship between people and nature. With the creative documentary, we are in an anthropological and artistic approach to tourist sites, with a method in a comprehensive approach that affirms a way of addressing these territories and bearing witness to them. A method is defined that allows us to nuance the representations of this territory, and also to better understand the ways in which it is practiced, by tourists as well as by the inhabitants.

The creative documentary forces the narration of a territory in images and sounds: whether through the attitudes of tourists (Augé, 1997) who pass quickly to take their snapshot, or travelers (Paquot, 2014) who stay a little longer to try to soak up an "atmosphere"[4] (Cyrulnik, 2021). This construction of a filmic discourse about a tourist site is based on the relationship that people establish with the place and its inhabitants according to the time they spend there. The creative documentary participates in this immersion insofar as the viewer can experience it in turn.

With Covid 19, tourism practices have evolved, and the first two films in the series "Getting out of the postcard", shot at the beginning and then in the middle of the pandemic, have had to adapt to this event and change the way they film and tell the story of these trips. This is what we will see.

The analysis becomes geocultural and interdisciplinary: from film studies to anthropology, via information and communication sciences associated with the artistic dimension. Various disciplinary approaches are thus solicited in this series through the knowledge it allows us to acquire on the uses of these sites.

BACKGROUND ABOUT PUBLIC SPACE(S)

It is a question of apprehending public space from both an urban (Paquot, 2009) and media (Habermas, 1978) point of view in order to try to better understand it through film (Niney, 2000, 2002). The creative documentary offers both a device (Agamben, 2007) and the possibility of sharing an experience through art (Dewey, 1934) which favors a sensitive approach to these tourist sites which are often the subject of a selfie or a photo taken in the axis of the monument without taking the time to see the life that takes place there on a daily basis. This is what this series aims to do, by making us feel the atmosphere of the place.

Urban Space

Marc Augé (1997) tries to understand the city through the memory of those who live in it:

Memory and history are combined in the city. Each of the city's inhabitants has his or her own relationship with the monuments, which themselves bear witness to a deeper and more collective history. In this sense, the urban journey of each individual is a way of appropriating history through the city (p.135).

Do the inhabitants break with their environment when they feel that it is invaded or denatured by tourists? Is it the tourists who are in rupture because they come to look for something other than their own daily life? Maria Gravari Barbas (2005) talks about the common rules of life on a heritage site which contribute to the affirmation of a strong territorial identity. The public space asserts itself physically and existentially as well.

But "crossing the history of places and the lived experience of those who invest them [...] is not done without clashes and conflicts" and the daily life of the place is then only a series of adjustments of current practices not only to the materiality of the place, but also to its symbolism when it is ideologies of inhabiting that presided over its construction (Le Corbusier's La Cité radieuse for example) (Brochot, 2013, para. 3).

This is the question through this series. She continues:

Living in a "heritage place", a fortiori when it is globally recognized as such, also makes it possible to establish and reinforce one's social legitimacy all the more so as the symbolic power of the places refers to a glorious history (royal palaces...). From then on, the division between the private, the collective and the public can prove problematic and the "heritage place" emptied of part of its meaning if the connection cannot be established between these three orders (Brochot, 2013, para 4).

In this way, we measure the extent to which the narratives composed about a public space participate in its definition. An element of subjectivity is asserted.

Art goes in this direction, and the creative documentary offers a filmic narrative of a specific situation.

Art objects are expressive and that is why they are a language. Better still, they are languages. For each of the arts has its own medium and this medium is particularly suited to a form of communication. Each medium communicates something that cannot be expressed as well or as completely in another language. The needs of everyday life have given greater practical importance to one mode of communication, speech (Dewey, 1934, p. 188).

This is the logic behind documentary. It therefore seems suited to this understanding of a place.

Jürgen Habermas (1978) describes a public space in a participatory logic that is at once interactional, deliberative and political, associating the power of the media with it according to communicational considerations. Thierry Paquot (2009) in turn proposes a more physical and geographical point of view, in a philosophical and urban approach. But he also considers the newspaper as a means of making a place such as a neighborhood collective, by giving it a social and public dimension (Paquot, 2009, p. 92). The media power, and therefore the communicative power, of a newspaper allows us to explore the relationship between public space and territory. The connections, exchanges at different levels, the sharing, circulation and relationships that are created are at the heart of the definition of all public spaces, in the plural as well as the singular. Indeed, Bernard Miège (2010) speaks of a 'confused approach' between 'public space' in the singular, which is more political, and the more urban 'public spaces'. The latter refers to squares, streets and green spaces, but it also refers to places of encounter and spatial interaction. In any case, the political dimension is apparent in the civic commitment that the territory implies, placing people at the heart of the city. It is in fact affirmed insofar as the documentary bears witness to the lives of the inhabitants. But the documentary as a form of narrative of a territory is also invested in the urban approach it proposes through what is filmed, what is articulated, what is highlighted, etc. (Cyrulnik, 2017, pp.153 - 154).

Media Space

The media participate (therefore) in the structuring of the public space beyond the forms and contents of the discourse and information they disseminate... The 'gaps' in the media are therefore opportunities for making alternative discourses visible. In his historical approach, Habermas considers that the irruptions of the media, invading the public space, causes its decline. The links between territories and media emphasize that the latter are likely to participate in a dynamic of discussion and argumentation (Gadras & Palliart, 2013, p. 31).

The documentary series 'Getting out of the postcard' stands at the intersection of these urban and media public spaces, positioning itself more precisely on the side of tourist sites and documentary.

Tourists go to a site because of what it evokes before they go, in relation to literature or art. The representations of these sites arouse the desire of tourists to come and experience them. The images influence the cinematographic representation that can be made of them, as well as the artistic one that is already constructed beforehand. "Marcel Proust and Thomas Mann are part of Venice. Many Italian cities will always have, for those who visit them, something Stendhalian" (Augé, 1997, p. 131). Poets and novelists interact with the space of these cities. These territories are apprehended in a subjective and social way at the same time. Cinema also participates in this. A town like Arrezzo in Tuscany, for example, has installed explanatory panels in all the squares where Roberto Benigni's "La vita é bella" (1997, 116') was filmed, thus playing on this link between cinema, town and tourism. Just like René Clément's "Plein soleil" (1960, 112') in Procida, or more recently the series "El hierro" in the Canary Islands, and so many others now. Cinema thus acts on the representation of the territory. It gives a representation of it before going there, which can influence the way of perceiving it. It is then thought of as a new tourist asset. Media representations of the site therefore influence tourists in both directions. Within the framework of the series "Getting out of the postcard", the creative documentary is proposed as another way of representing these sites in order to get away from the clichés. Far from the promotional film, it aims to approach them in a subjective and ethical way in order to bear witness to reality as accurately as possible. However, the anthropological and artistic approach claimed had to be adjusted to the tourist upheaval caused by the pandemic, as we shall see later.

"Poetry" comes from the Greek "poïen": "to do", "to create". The creative action is valid for the filmmaker as well as for the spectator or visitor. The city and its imagination shape an urban fabric and a social space that is recounted throughout the world (Augé, 1997: 134). Thus, the multitude of screens that can overlap in St. Mark's Square in Venice, for example, or in millions of other emblematic places of tourism, come to constitute themselves another image of the city, between the imaginary and the virtual. This global and intangible vision modifies a more local approach to these tourist territories. Yves Winkin (2001) even comes to speak of the "enchantment" of the place for the tourist:

one could suggest that if the euphoria is related to interaction, and limited like this in time and space, the enchantment would refer to places and landscapes created with the intention of inducing a state of euphoria in those who frequent them (pp. 215 - 216).

The imagination linked to tourism offers a way to fictionalize the city. In this context, territorial storytelling is a particular facet that tends to go towards territorial marketing. This time, it is another way of looking at a territory, closer to the economic and tourist dimensions, which also influence its development. Storytelling sets up "narrative gears, whereby individuals are led to identify with models and conform to protocols" (Salmon, 2007, pp. 16-17). Political communication champions or marketing strategists use them to catalogue a precise territorial vision to be put forward for development. However, the critical mind develops with the writing or reading of any story. A narrative responsibility, dear to Paul Ricoeur (1983), is affirmed for the author as well as for the reader or spectator who decides whether or not to believe in it. This form of territorial storytelling has been a little too common; however, it reveals communicative realities for understanding a territory. Roland Barthes (1957/2014) already denounced the alienating dimension of media stories. Web 2.0, and the participatory situations it multiplies, as new media forms, would tend to enhance emancipation, while at the same time, this approach to mass media inhibits. It is then a question of trying to interpret and distance a narrative in order to humanize and democratize the public narrative space, even if it becomes fragmented. In this case, it values diversity through the different discourses that can be made of it, the series "getting out of the postcard" proposing precisely a form of media alternative.

The model of imposing a narrative imposing its vision of the world, as conceived by Salmon, would now be outdated, as the multiple circulating narratives can be captured in diverse environments by users used to interact with the fragments of narratives they collect in a fragmented way on multimedia media (Lits, 2015, p. 36).

Territorial storytelling, if it is a way of reciting space in a very coded and chaotic way at the same time, raises the question of the most appropriate narrative format according to the territories and what one can (wants) to say about it.

Gérard Genette (1966) defined the notion of narrative as "the representation of an event or series of events, real or fictional, by means of language, and more particularly written language" (p.152). In line with this territorial storytelling approach,

we propose not to see a break in territorial communication strategies by moving from discourse to narrative, but rather to consider the narrative as a marker of the recomposition of the communities' discourse. Indeed, Gérard Genette points out that while these two terms are theoretically opposed (discourse is characterized by being attached to a transmitter, while in the narrative, no one speaks, so the reader does not wonder who speaks to receive the meaning), in most literary texts, this dichotomy is not observed. The insertion of narrative elements into the discourse does not, however, make it possible to remove any reference to a speaker, who remains present in the background' (Genette 1966, p.161, quoted in Le Corf, 2015, p. 148).

The story about a territory is rather transformed into a discourse through the words of the inhabitants. While storytelling has valued a form of territorial narrative, it finally appears that it is the fact that it is carried by a person in a documentary that makes it possible to better embody the described space. Interactions and the affirmation of point of view determine the territorial approach to them. The more specific example of the tourist who travels supports this. Storytelling is a technique used to promote this economic windfall on which local authorities rely to increase their budgets. The definition of who comes as a tourist rather than a true traveler (Paquot, 2014) contributes to the evolution of the territory in terms of development, as well as in terms of social, psychological, economic and heritage representation, etc. (Cyrulnik, 2020, pp. 236 - 237). The fact of taking time in a public space and seeing the daily life that takes place there as a traveler rather than going through it quickly as a tourist, will allow us to understand it better, to integrate into it and perhaps thus to participate in its development in a more sustainable way. It is therefore a question of choosing who tells the story of the site in order to understand it differently.

Public spaces, heritage, arts and cultural diversity are highlighted when the researcher wants to enhance the brand image of a city. The perspective of urban marketing is clearly displayed to place the inhabitant in the same position as the tourist as a consumer. And when it comes to art, it is thought of in terms of a cultural product. This precision is noted here because it exposes the different ways of approaching a territory through art. Those who work on the creativity of a city and its image rely on values at the same time as they create them. "Showing talent and tolerance', 'being inventive', 'being open to difference" (Pagés, 2013, p.155), are some of the expressions that can be used to praise the assets of a creative city. It would encourage "practicing the arts as a member of a community (whether artistic, technical or commercial)" (Pagés, 2013, p.155). The inhabitant then becomes a consumer-actor of his own territory (Pagès, 2013, p.157). Does he join the tourist in his behavior through this consumption process?

Jürgen Habermas (1978) already established a direct link between the media and the public space. Bernard Miège (2010) updates this: 'The state of the media is considered *ipso facto* as an evaluation of the Public Space and the debates that take place there' (p. 115). The documentary, in an almost omnipresent media environment (at least for tourists via their mobile phones), therefore proposes an alternative device. The fact of positioning oneself with a camera or a phone, entering a cinema, or speaking in public during the debate that follows, extends this notion of public space in another way.

THE METHODS OF DOCUMENTARY

In line with the visual methods particularly identified by the Anglo-Saxons, the researcher proposes here an audio-visual method, where documentary meets visual anthropology (Cyrulnik, 2022). By capturing images and sounds from reality, it is possible to construct a filmic representation that bears witness to the site as much as it affirms a point of view, even if it is not a question of the director as author putting himself forward too much, but rather that he reports on a situation.

The Evolution of Representations of a Territory through Documentary

The history of documentaries about a territory bears witness to the different devices that emerge according to social and societal developments, technical performance and related methods. The history of the representation of territories through documentary reveals different devices that are invented as techniques become more democratic (especially with the advent of digital technology), and as they are invented and reinvented. The history of the way in which a territory is approached with a camera thus bears witness to the link between technique, context, culture and society. The researcher is dealing with an approach that is at once technical, aesthetic and rhetorical (Soulez, 2011).

However, the geographical approach to documentary cinema has been a promising one since the early days of cinema with operators from the Lumière brothers (such as Mesguich or Promio) or Pathé, who travelled the world to bring back images that were the first to bear witness to these distant lands in such a realistic way. The famous first tracking shot invented on a gondola in Venice by Promio, or the first films (of travel as well as westerns) recorded on the front of locomotives, which became a genre in their own right and were recorded in the catalogues of most American production companies, bear witness to this need to explore the world as a crazy adventure that would become possible. The cinema, the locomotive, the journey around the world testifies to the "progress" of this period. The colonial period in which cinema was invented also favored this.

These first films oscillated between documentaries, news and reports, and combined observations, reflections and research on the peoples and cultures filmed. With the remote territories, it is the peoples who are discovered, including the Westerners who position themselves in front of these new social, cultural and cinematographic representations.

The first operators no doubt sincerely believed in the fidelity of their images, indisputable reflections of reality, which led them to immerse themselves in the countries they crossed. What they did not yet know was that unbeknownst to them, their images are worked on from the inside, and that memory and the imaginary disturb the objectivity of the gaze. The 'new look', so passionately sought after, is hardly possible unless one discovers a totally unknown world without prior iconic or literary references, and, increasingly as film art advances, fed by film memories (Gauthier, 2010, p. 30).

The approach becomes increasingly rhetorical.

Around the 1920s, the exotic and colonial documentary emerged as a clearly identifiable genre (Gauthier, 2004, p. 68). After many years of technical advances, the form of documentary cinema asserts itself in order to distinguish itself from television reportage, and even comes to determine more precisely what creative documentary can be today: it combines a capture of reality with a true point of view on the world. The latter becomes an asset for understanding tourist sites in the context of this series. It offers itself as a new means of disseminating information about these sites. It offers a media alternative.

In this logic,

Just as rationality has not emancipated us from nature, the direct access to nature that we claim to owe to modernity - transport or laboratories, the 'democratization' of the tourism market or the organization of scientific work - have multiplied the mediations that separate us from it (Babou, 2009, p. 102).

The socio-technical devices (Cyrulnik & Zénouda, 2014) that the researcher puts in place must remain situations of mediation that maintain this link between people and his territory. It is not a question of one taking precedence over the other, but rather of measuring their possible interactions. This cinematographic reflection becomes ecological.

A Documentary Method

There are therefore different media approaches to these sites: that of the tourist (selfie, camera, home movies, etc.), that of the advertiser who promotes the site, that of the journalist who reports on it and that of the documentary filmmaker. They have technical and narrative consequences. While the tourist often takes selfies with his mobile phone or photographs with his camera, or even family films (Odin, 1995) with his phone or a small camera, the idea for this documentary series is to use a camera with a high image quality and some synchronous sound (a Lumix GH5) to blend in with the crowd, and above all to take the time to capture the atmosphere of these places. An extra microphone is needed to articulate a reflection on these sites in images and sounds.

Thus, by blending in with the tourists with this equipment, the director and researcher immerses herself among the tourists, captures the daily life of the inhabitants of these exceptional sites, and relates a sensitive point of view through short films of about ten minutes to bear witness to life in these places in a different way. This series aims to denounce certain invasive practices that make tourists victims in a certain way, to propose an alternative to caricatures in terms of images of the sites, but also to nuance excessive behavior, etc. In general, it seeks to have the most accurate approach possible to the site, in a sensitive and ethical way, because the question of the aesthetics of a documentary is linked to its ethical approach to a field. For the documentary there is an unspoken ethical contract between the director who assures that everything shown is from reality, and the viewer who decides to believe it (Lioult 2004). The sensitive approach through images and sounds captured on the site to relate its atmosphere implies a form of commitment.

These technical considerations participate in the form of narrative to be constructed and show that the form is consistent with the substance. The choice of an appropriate socio-technical device to capture a reality implies a form for the film. This one thus depends on the subject to be addressed. The device set up goes hand in hand with the method implemented to capture a reality and relate it, represent it.

The documentary allows us to acquire a better knowledge of the filmed situation, in order to better understand it (Niney, 2000, 2002). It is a question of proposing to experience reality with the help of a camera. The cinematographic relationship to reality is questioned.

A method emerges in a comprehensive approach insofar as it aims to "make explicit the meaning that the objective world of realities has for us (all men) in our (shareable) experience" (Mucchielli and Paillé, 2005: 14), by making us experience realities in order to construct representations which are filmic here. The process of immersion is put in place as soon as the material is chosen to blend in with the crowd. Thus, the images and sounds captured are as close as possible to the tourists and inhabitants filmed. Their behaviors, their actions, relate a daily life as close as possible in an attempt to be as accurate as possible. But a global view of the site is also important. These documentaries alternate between close-ups and long shots, which help to create a filmic experience of the site through the representation that is made of it. The documentary film is itself an attempt to understand the mode of creation of a representation of the world. It becomes a medium to define a reality.

This method is also political insofar as these documentary films bear witness to a daily life that few users take the time to see, and that it is relative to the place of people in these places. The sharing of the sensitive (Rancière, 2000) then engages the spectator until the moment of the debate after the projection. It is thus a new apprehension of these sites often stigmatized or captured quickly, to acquire a true knowledge by the means of the experience (Dewey, 1934) that the spectator will make of it (Cyrulnik, 2015).

RESULTS AND SUGGESTIONS

While the approach asserts itself as being artistic and sensitive at the same time, an anthropological knowledge is developing over the course of the documentaries made (Sumartojo & Pink, 2019). For the moment, only two films have been made. The pandemic also interferes with the production of these films insofar as the local authorities that usually subsidize these artistic approaches in France have been blocked because they had to commit sums of money to support other associations (notably sports) that had suffered too much from the repeated confinements. Some cities wished to engage in the valorization of their territory in an original way through these documentaries, but they could not do it because of the consequences linked to Covid 19. Finally, only the city of Marseille, which had already supported a first film on the 'vieux port' before the pandemic broke out, continued its support to allow the realization of a second one on the small fishing port of the 'vallon des Auffes'[5].

However, the time elapsed between these two productions testifies to the evolution of tourism practices in connection with this pandemic. The series had to be slightly reoriented to adapt to new tourism practices. It is thus on these realizations that the director - researcher can rely to testify at the same time of what the pandemic changed in terms of tourist practices and of documentary methods which adapt to it.

The Evolution of Tourism in Marseille

First of all, it should be noted that tourism is relatively new in Marseille. In this respect, the pandemic did not completely disrupt it. But it is a city that is rather frightening through its media representations, especially in the television reports on the northern neighborhoods that make the headlines against a backdrop of drugs and settling of scores. But the nomination as European Capital of Culture in 2013, and the birth on this occasion of the emblematic building of the MUCEM (Musée des Civilisations de l'Europe et de la Méditerranée) have changed the image of Marseille. Since this event, many tourists have come to this city where Provence, the Mediterranean and the whole world meet through the origins of its inhabitants. These "ugly, dirty and nasty" cities, as Thomazeau (2009) puts it, which are Marseille, Liverpool, Bremen, Naples and Gdansk, "have made their 'bad' reputation a rallying point and a means of asserting an identity and giving themselves visibility (Gravari- Barbas, 2013)" (Condevaux et al. 2016, p.10).

The people of Marseille have therefore discovered new practices and have more or less adapted to them. Urban tourism in Marseille is changing the way people look at the city. Thus, the inhabitants of certain districts are now beginning to be bothered by the noise of wheeled suitcases, as are the inhabitants of Venice who forbid their use at certain times of the night. This reflects the fact that the influx of tourists is increasing, and that many apartments in certain neighborhoods are being converted into one-time rentals for tourists. The city is changed, as is the case, for example, in the 'Vallon des Auffes' where the small fishermen's huts that make up the charm of the site are reinvested in small one-story dwellings with bay windows and gray facades that no longer have much to do with their original vernacular and colorful side. This architectural evolution testifies to the expansion of the number of tourists, who come to divert what they came for in the first place.

However, documentaries relate all of this and allow the viewer to experience it and gain knowledge about the state of the site. "Documentary" comes from the Latin "documentum" which translates according to the dictionary called the Robert from "example, model, lesson, teaching" to "demonstration"

(Niney, 2009: 13). The pedagogical character of the documentary raises awareness of this situation while helping to acquire knowledge about this particular territory.

Evolution of Tourism Practices and Films with the Pandemic

Tourism and sites have changed. Other practices and other uses are being put in place. This method of approaching a site through documentary itself is changed by Covid 19. The first film[6] shot on the 'vieux port' shows tourists coming to a site where fishermen have been working to sell their fish for a long time. It shows how the meeting between tourists and inhabitants of Marseille is organized on the 'vieux port': the fish sellers come to play with their accent and their gouaille to sell their fish a little more expensive to the "estrangers", said in Provençal. A tacit agreement is put in place. The game between the "welcoming" and the "welcomed" is not really stereotyped, it is nuanced. This raises the question of the articulation of the local culture with a culture that tends to become globalized.

In the second film[7], later on and as Covid 19 entered a new way of life for the second year, the tourists came to another site but in the same way as the inhabitants of Marseille this time. That is to say, the locals felt the need to get together for a drink in the charming little port of the Vallon des Auffes in Marseille to resume that social life they had missed so much during the confinements; and the tourists who came from further afield did the same. While in the first film there is a game between the tourists who discover the site and the locals who are amused by their presence by selling their fish and accusing their accent or their price, in the second they mingle from the start. In the first, the people were presented as two identified parties who did not have the same practice of the territory (working and fishing or buying boat tickets and fish), in the second they mixed much more because they wanted and needed the same things: to share pleasant moments. Is it the succession of these periods of confinement that comes to unite people in need of a social life? Is it the activities of the sites that bring this about? In any case, practices that might have seemed to be opposed to each other have come to resemble each other, to merge. In the Vallon des Auffes, people fish, they swim, they come to visit by bike or with a suitcase to go to a 'cabanon' rented by AirBnB, and we all end up having a drink by the water. The implementation of this series has thus evolved from a desire to present everyday life despite the clichéd uses of tourists, to a testimony of the everyday life of tourist sites, quite simply. What many tourism theorists call "authentic experiences" takes shape.

The inhabitants are the masters of the game, but they accept the tourists. The challenge lies in the fact that the arrival of tourists does not disrupt the life of the territory, which is what they came for. The set of fishermen's shacks transformed into AirBnBs changes both the architecture that responds more to the expectations of people from the North with large windows and buildings painted in gray to represent a certain standing, while the windows were small to protect from the heat in summer and the colors of the buildings very varied recalling Italy which is not so far. The symbolic reproduction of urban forms more or less idealized and conceived at their origin in strict adequacy with their "social content" is transformed. This raises the question of legitimacy in the appropriation of a place, each one considering that this heritage place belongs to him, at least symbolically. Are these interactions conflicting in their practice? For now, in Marseille, the answer is no. And these first two films testify to the fact that they influence each other and intermingle, but without overt conflict. However, the price of a square meter in the vallon des Auffes district has particularly risen since the flood of Parisians leaving the capital after the first confinement to come and settle in a city more in touch with nature, more precisely the sea here.

This is a source of resentment for the inhabitants of Marseille who are finding it increasingly difficult to find housing in this neighborhood (and in other neighborhoods as well). This migration is played out more on a national level than on an international level with foreign tourists. It is nonetheless the consequence of the pandemic and the restrictive measures that have been put in place, particularly in Paris where apartments are often small because of the price of rents and where even the parks were closed during the first lockdown. Here it is more the Parisians than the tourists who cause problems. It is no longer really a question of tourism, but of local establishment within the framework of a residence (primary or secondary, the latter still giving rise to controversy while the people of Marseille feel "dispossessed" as Mickaël Labbé (2019) puts it). The investment of a tourist site generates a form of gentrification here. The well-to-do, rather young population (characterized by the term Parisian "bobo", coming from the contraction of "bourgeois" and "bohemian" corresponding to a way of living with money but in connection with a certain relationship to nature, to situate it quickly), changes the habits of daily life in this small fishing port.

Also, the elected officials feel challenged and even come to imagine ways to channel the tourists. The first example that has been set up in Marseille is on the side of the calanques (classified and known as "of Cassis") where the beaches were full of tourists last summer and where regulations will be put in place to regulate this next year. These measures to ban tourists in the calanques in order to regulate them are reminiscent of those on the Amalfi coast for example where they are no longer allowed to use cars but can take advantage of the shuttle systems put in place a few years ago to go from fishing village to fishing village. This form of warning against tourism practices is therefore now thought of by local authorities to propose alternatives to mass tourism practices. The issue is becoming more and more about getting out of tourism (Condevaux et al. 2016), while knowing that its economic activity matters. Policies are therefore trying to intervene in these interactions between tourists and inhabitants of Marseille in order to ensure that the reasons that make tourists want to come and test a certain quality of life on a daily basis are maintained. This is the ambiguity of the situation that tourism generates in many places in the world. However, this envious outside look contributes to the value of the place. This recognition of the Other is constitutive of these territorial representations. But, with this mass tourism, should we speak of a "path of recognition" as Ricoeur (2004) does, or rather of a "struggle for recognition" as Honneth (2010) does? The investment by tourists of a place comes to raise philosophical questions. This is precisely one of the ambitions of the creative documentary, since it wishes to provoke reflection in the viewer who discovers a particular situation on the screen.

In both films, the economic stake for the residents is impacted by the presence of tourists. But since local work is not at all at an industrial level, the adaptation is rather easy. The risk, however, lies in the fact that the fishermen disappear little by little. They are well present and adapted to tourists in the 'vieux port'; they are less and less numerous in the 'Vallon des Auffes'. The latter sell their fish to the prestigious restaurants that are established there and that the tourists reserve. This working-class memory through fishing is preserved but for how long? The economic model is becoming more and more fragile. And the social cohesion, the community spirit, and the territorial identity of the small port of 'Vallon des Auffes' is destabilized, while it is precisely a sincere commitment that the tourist comes to seek!

This articulation between the local and the global is more and more present. The ecological dimension, even in its method, in its approach, is asserting itself. It is linked to the method that documentary film offers (Cyrulnik, 2018a). Do these situations resonate on a global level as well? Undoubtedly, since they raise questions related to tourism that can be heard in Barcelona as well as in Thailand, for example.

But here the pandemic more precisely impacts more at the national level with a caricatured cleavage towards Parisians that soccer exacerbates in particular, than at the international level for the moment.

The initiative for this documentary series was transformed by Covid 19. The idea of this series was born before the pandemic with precise objectives that aimed to report on the place that tourists take more and more on these sites. It was a question of taking more time to live with the inhabitants, and not only to come to make a selfie and to leave. It becomes the witness of the daily life of these sites. While this relationship to time is fundamental to understanding a site, the form of documentary film corresponds exactly to this temporal approach in comparison to the reportage that is too often quickly shot to be broadcast on the air that same evening. Here again, the form matches the content. Taking time is also an ecological approach.

CONCLUSION

At a time when UNESCO intends to give greater prominence to "cultural landscapes" and the European Landscape Convention reaffirms that "landscape [...] represents a fundamental component of Europe's cultural and natural heritage", the researcher proposes to transpose the upheavals caused by tourists to heritage sites on this urban and human scale to Marseille. The whole city is affected, with its atmosphere and its way of life on a daily basis. Thus, tourism "off the beaten track" is to be deconstructed. These new tourist offers are multiplying around the world and encountering the practices and imaginations of tourists, which jointly tend to renew the tourist system (its practices, actors, spaces, temporalities and forms) (Gravari-Barbas & Delaplace, 2015).

The goal of this series is to change the way of seeing a tourist site through documentary. It also testifies to another way of acting as a tourist: by mingling more with the locals, by adopting their way of life, as if to go towards an "authentic" experience. These films become witnesses of what the pandemic is changing in tourism practices, even if they were already underway. They are nuanced. These documentaries are not in themselves an object of study that changes the pandemic itself, but their method of apprehending the place had to adapt to what the pandemic shook up. They are a representation of it.

Territorial representations are thus nuanced for a better understanding of these sites. From a more economic point of view, the brand image or the storytelling that is done evolves. The storytelling and formatting of the documentary thus offers new filmic and territorial representations. While Covid 19 is changing tourism practices, and attractiveness is no longer the main economic claim of territorial communities, this series aims to testify and participate, by nuancing these tourism practices, to a greater coherence and territorial and social cohesion.

The documentary approach is becoming more and more sociological, thanks to its artistic approach; but also to all that it can arouse as a sensitive approach that makes it a suitable media object to approach these sites in a more comprehensive way. Experience through art (Dewey, 1934) is proposed as a strength of the documentary method, which is increasingly asserting itself as moving towards a filmic ecology (Cyrulnik, 2018a) insofar as it is a whole ecosystem that is being put in place.

At the same time, by depicting the atmosphere of the place in this way (Sumartojo & Pink, 2019), a form of imagination is constructed on the basis of this captured reality. This series aims to "transform(s) the material of everyday experience into something else - that re-enchants the city and creates mutual attachment between the city and the tourists" (Wynn, 2010: 150). This element of imagination or suggestion is important in the representation of these sites. These ordinary sites do become extraordinary. But *in contrast to these experiences that value the 'mundane,' the touring of nearby spaces, especially metropolitan ones, can rely on an exoticization of the latter, the showcasing of a cultural 'otherness' presented as great as that discovered on the other side of the world (Condevaux et al. 2016, p. 19).*

The fishermen of Marseille become exotic and they themselves come to play with the Other.

"The tourism of ordinary places and the diversification of tourist enclaves would participate in a de-differentiation of the tourist and non-tourist spheres, of the everyday and the off-daily, characteristic of the new post-tourist "era"" (Condevaux et al, 2016, p. 19). Thus, the realization of these original postcards in other cities (such as Martigues for example, which should constitute the next site to be filmed in this series) tends towards this other way of considering tourism, far from any exoticism and as close as possible to everyday life. Even the tourist offices think of communicating differently about their so beautiful sites. The creative documentary is not a marketing object, but it offers (and even more so with this collection) a more human and sensitive approach to these places.

The continuous enlargement of the tourist ecumene as it has been set up since the beginnings of tourism is manifested by the integration of more and more distant places, socially and geographically, in the tourist offer and demand. Can we say that these are trends that suggest that the phenomenon of tourism is being overtaken by the emergence of intermediate situations that are neither tourist nor totally local? Are the places that concentrate tourist practices off the beaten track post-tourist places (overtaking of tourism by other forms of practices, which are no longer tourist) or hyper-tourist places (extreme sophistication of the tourist phenomenon, individuation, segmentation, disintermediation)? (Gravari-Barbas & Delaplace, 2015, para. 48).

The experience that begins with this series rather testifies to an adequacy between the tourist and the inhabitant that are even more intertwined since the pandemic. But at the rate these practices are evolving, perhaps the next film that will help "Getting out of the postcard" will relate yet other uses. In any case, whether it is *post* or *hyper*, there is no doubt that tourism, which had already begun to evolve, will be further shaken up by this pandemic, which is giving rise to another way of traveling, and will thus provoke yet other ways of experiencing these sites.

REFERENCES

Agamben, G. (2007). *Qu'est-ce qu'un dispositif?* [What is a device?]. Rivages Poches, Petite bibliothèque.

Augé, M. (1997). L'impossible voyage, le tourisme et ses images [The impossible journey, tourism and its images]. Rivages Poches, Petite bibliothèque.

Babou, I. (2009). Disposer de la nature, enjeux environnementaux en Patagonie argentine [Disposing of nature, environmental issues in Argentine Patagonia]. In Sociologie et environnement. L'harmattan.

Barthes, R. (2014). *Mythologies* [Mythologies]. Seuil. (Original work published 1957)

Bougnoux, D. (2006). *La crise de la représentation* [The crisis of representation]. La Découverte.

Brochot, A. (2013). Maria Gravari-Barbas (dir.) 2005, Habiter le patrimoine: Enjeux, approches, vécu [Living in heritage: Issues, approaches, experiences]. *Strates*, *14*. doi:10.4000/strates.6713

Condevaux, A., Djament-Tran, G., Gravari-Barbas, M. (2016). Avant et après le(s) tourismes(s). Trajectoires des lieux et rôles des acteurs du tourisme « hors des sentiers battus ». Une analyse bibliographique [Before and after tourism(s). Trajectories of the places and roles of the actors of "off the beaten track" tourism. A bibliographical analysis]. *Via*, *9*. doi:10.4000/viatourism.409

Cyrulnik, N. (2015). Le documentaire, un espace de liberté pour une nouvelle communauté [The documentary, a space of freedom for a new community]. In G. Gramaccia (Ed.), *Revue Française des Sciences de l'Information et de la Communication* (Vol. 7). doi:10.4000/rfsic.1744

Cyrulnik, N. (2017). Représenter le territoire, filmer la cité [Representing the territory, filming the city]. Habilitation à Diriger des Recherches, Alain Kiyindou garant, Université de Bordeaux - Montaigne.

Cyrulnik, N. (2018a). Le documentaire, vers une écologie filmique [The documentary, towards a filmic ecology]. In D. Urrutiaguer (Ed.), *Des expériences artistiques au prisme du développement durable, Revue Registres de l'Institut de Recherches en études théâtrales, Hors-série n°5. Presses Sorbonne Nouvelle.*

Cyrulnik, N. (2020). Documentary narrative for new understanding of a stigmatized public space. *Revista Lusófona de Estudos Culturais / Lusophone Journal of Cultural Studies*, *7*(1). https://rlec.pt/article/view/2121/2678

Cyrulnik, N. (2021). Questionner l'atmosphère au cinéma pour expérimenter l'interdisciplinarité, une méthode audiovisuelle [Questioning the atmosphere in cinema to experiment interdisciplinarity, an audiovisual method]. *Revue Française des Sciences de l'Information et de la Communication*, *22*. Advance online publication. doi:10.4000/rfsic.10632

Cyrulnik, N. (2022). Les méthodes audio-visuelles qu'offre le cinéma documentaire [The audio-visual methods offered by documentary film]. In Méthodes visuelles en sciences humaines et sociales: de la captation à la diffusion des résultats: données, méthodes, résultats, valorisation. ISTE Open Science.

Cyrulnik, N., & Zénouda, H. (2014). La place des dispositifs socio-techniques d'information et de communication (Distic) dans les différentes situations de recherches [The place of socio-technical information and communication devices (Distic) in different research situations]. Cahiers de la Sfsic, 10, 133-224.

Dewey, J. (2005). L'art comme expérience [Art as experience]. Gallimard. (Original publication 1934)

Gadras, S., & Pailliart, I. (2013). Les territoires et les médias dans la conception de l'espace public [The territories and the media in the conception of public space]. In Médias et territoires, l'espace public entre communication et imaginaire territorial. Presses universitaires du Septentrion.

Gauthier, G. (2004). *Un siècle de documentaristes français* [A century of French documentary filmmakers]. Armand Colin.

Gauthier, G. (2010). Géographie sentimentale du documentaire [Sentimental geography of the documentary]. In Champs visuels. l'Harmattan.

Genette, G. (1966). Frontières du récit [Boundaries of the narrative]. Communications, 8, 161.

Gravari-Barbas, M. (2005). *Habiter le patrimoine: Enjeux, approches, vécu* [Living in heritage: Issues, approaches, experiences]. Presses Universitaires de Rennes. doi:10.4000/books.pur.2208

Gravari-Barbas, M. (2013). *Aménager la ville par la culture et le tourisme* [Developing the city through culture and tourism]. Le Moniteur.

Gravari-Barbas, M., & Delaplace, M. (2015). Le tourisme urbain « hors des sentiers battus » [Urban tourism "off the beaten track"]. *Téoros*, *34*, 1-2. https://journals.openedition.org/teoros/2790

Habermas, J. (1978). *L'espace public, Archéologie de la publicité comme dimension constitutive de la société bourgeoise* [The public space, Archaeology of advertising as a constitutive dimension of bourgeois society]. Payot.

Honneth, A. (2010). La lutte pour la reconnaissance [The struggle for recognition]. In Passages. Cerf.

Labbé, M. (2019). *Reprendre place, contre l'architecture du mépris* [Reclaiming a place, against the architecture of contempt]. Payot.

Le Corf, J.-B. (2015). Le récit métropolitain, composante de la communication touristique. Les cas de Marseille et d'Istanbul face au label « capitale européenne de la culture [The metropolitan narrative, a component of tourism communication. The cases of Marseille and Istanbul in the face of the "European Capital of Culture" label]. In N. Cyrulnik & V. Meyer (Eds.), *Synergies Mondes Méditerranées* (Vol. 5, p. 148). Retrieved from https://gerflint.fr/Base/MondeMed5/le_corf.pdf

Lioult, J.L. (2004). *A l'enseigne du réel, penser le documentaire* [In the sign of reality, thinking the documentary]. Publications de l'Université de Provence.

Lits, M. (2015). Storytelling: réévaluation d'un succès editorial [Storytelling: reassessing an editorial success]. In Le storytelling, succès des histoires, histoire d'un succès. L'Harmattan.

Miège, B. (2010). L'espace public contemporain [The contemporary public space]. PUG.

Morin, E. (1990). *Introduction à la pensée complexe* [Introduction to complex thinking]. Le Seuil.

Niney, F. (2000). *L'épreuve du réel à l'écran. Essai sur le principe de réalité documentaire* [The test of reality on the screen. Essay on the principle of documentary reality]. De Boeck Université.

Niney, F. (2002). La poétique documentaire comme forme de connaissance [Documentary poetics as a form of knowledge]. *Etats généraux du film documentaire*. http://www.lussasdoc.com/etatsgeneraux/2002/sem_poetique.php4

Niney, F. (2009). *Le documentaire et ses faux-semblants*. Klincksieck.

Odin, R. (1995). *Le Film de famille. Usage privé, Usage public* [The Family Film. Private use, Public use]. Méridiens Klincksieck.

Pagès, D. (2013). Passages de l'image: d'Extramuros aux cahiers de la Métropole, l'imporbable figure de l'habitant métropolitain? [Qualitative analysis in the humanities and social sciences humanities and social sciences public]. In Médias et territoires, l'espace public entre communication et imaginaire territorial. Presses universitaires du Septentrion.

Paillé, P., & Mucchielli, A. (2005). *L'analyse qualitative en sciences humaines et sociales public* [Qualitative analysis in the humanities and social sciences humanities and social sciences public]. A. Colin.

Paquot, T. (2009). L'espace public [The public space]. La découverte.

Paquot, T. (2014). *Le voyage contre le tourisme* [Travel versus tourism]. Eerotopia.

Paquot, T. (2021). Demeures terrestres, enquête vagabonde sur l'habiter [Demeures terrestres, vagabond investigation on living]. Terre urbaine, l'esprit des villes collection.

Rancière, J. (2010). *Le partage du sensible, esthétique et politique* [The sharing of the sensitive, aesthetic and political]. La Fabrique éditions. (Original publication 2000)

Ricœur, P. (1983). *Temps et récit* [Time and narrative]. Le Seuil.

Ricoeur, P. (2004). Parcours de la reconnaissance [Recognition course]. Gallimard.

Salmon, C. (2007). *Storytelling, la machine à fabriquer des histoires et à formater des esprits* [Storytelling, the machine for making stories and formatting minds]. Poche.

Soulez, G. (2011). *Quand le film nous parle, Rhétorique, Cinéma, Télévision* [When film speaks to us, Rhetoric, Cinema, Television]. PUF, Coll. Lignes d'art.

Sumartojo, S., & Pink, S. (2019). *Atmospheres and the experiential world, theory and methods - ambiances, atmospheres and sensory experiences of spaces*. Routledge.

Thomazeau, F. (2009). Les cités mal aimées se rebiffent [The unloved cities rebel]. *Le Monde*. https://www.lemonde.fr/europe/article/2009/12/31/les-cites-mal-aimees-se-rebiffent_1286303_3214.html

Winkin, Y. (2001). *Anthropologie de la communication – de la théorie au terrain* [Anthropology of communication - from theory to field]. Le Seuil.

Wynn, J. (2010). City Tour Guides: Urban Alchemists at Work. *City & Community*, *9*(2), 145–163. doi:10.1111/j.1540-6040.2010.01322.x

KEY TERMS AND DEFINITIONS

Audio-Visual Method: Inspired by the visual methods of the Anglo-Saxons, it is a question here of thinking of the audiovisual in general as a medium which implies a method to approach a situation.

Documentary: The creative documentary is a cinematographic genre that allows us to acquire knowledge of a situation while affirming the artistic and anthropological point of view of the director.

Pandemic: The pandemic caused by the covid 19 crisis has changed tourism practices and shaken up the world.

Public Space: The place is in a participatory logic that is at once interactional, deliberative and political, associating the power of the media with it according to communicational considerations.

Publics Spaces: A more physical and geographical point of view of the place, in a philosophical and urban approach.

Territory: At the same time geographical, economic, anthropological, social and political, the territory is also thought in the relation that people maintain with the public space, going until questioning the territoriality.

Tourist: The one who travels just to say he has been to a place while the traveler takes more time to live in the place.

ENDNOTES

1 By the researcher, who was already a documentary filmmaker before and turned it into a research project.

2 This documentary series is supported for the moment by the city of Marseille, but it aims to open up to other territories; more information on http://www.lacompagniedesembruns.com/index.php/docu-realisations/sortir-de-la-carte-postale

3 Daniel Bougnoux explains in "La crise de la représentation" (2006, p.53) that it is the prefix "re" that is important in this last word, because with it a distance, the need for a more or less phenomenological reformulation, takes shape and anchors the presented object a little more in a relation.

4 "AtmosphèreS" corresponds to a structuring research programme that I carry out within the UMR 7061 Prism (Perception, representations, Image, Sound, Music) and which aims to question what the atmosphere can be from different disciplinary points of view, in order to understand the interdisciplinarity within our laboratory. This documentary series of short films of about ten minutes entitled "Getting out of the postcard" is part of this reflection.

5 A third is already planned for 2022, probably on the site of the Mucem mentioned below.

6 To see on http://www.lacompagniedesembruns.com/index.php/docu-realisations/sortir-de-la-carte-postale/l-ombirere-marseille

7 To see on http://www.lacompagniedesembruns.com/index.php/docu-realisations/sortir-de-la-carte-postale/le-vallon-des-auffes

APPENDIX: FILMOGRAPHY

- *La vita é bella,* R. Benigni (1997, 116')
- *Plein soleil*, R. Clément (1960, 112')
- Series « *El hierro* » 2020.

Chapter 27
Film Festivals in Contamination:
Streaming Strategies in Brazil

Jane de Almeida
https://orcid.org/0000-0003-1520-0729
Mackenzie University, Brazil

Cicero Inacio da Silva
Federal University of São Paulo, Brazil

Alfredo Suppia
State University of Campinas, Brazil

Davi Marques Camargo de Mello
São Judas Tadeu University, Brazil

ABSTRACT

Film festivals are venues with public screenings that hold a great potential for meetings, tourism, and economy. This chapter analyzes the alternatives found by some international and Brazilian film festivals in the face of the COVID-19 pandemic for the future horizon of festivals and the possible impacts of this new scenario on the audiovisual system. The chapter also presents the formats chosen by several festivals around the world, organizing their strategies into five correlated models: traditionalist, hybrid, mimetic, online, and nostalgic strategies. Such organization also originates from the research on technological forms found to soften, replace, or even expand the audience of the festivals (i.e., the development of their own streaming platforms, employment of existing platforms, and development of additional tools) to the streaming services. In the face of a digital panorama, with streaming platforms screening and distributing films, questions on the new relations between film festivals and digital platforms and their prospects arise.

DOI: 10.4018/978-1-6684-3369-0.ch027

INTRODUCTION

Film festivals represent a sector of great circulation of people and ideas, with a direct positive impact on the economy, culture, and tourism of any country. With a market difficult to be evaluated due to the fact that it includes not only the event itself – but a large and complex circuit of exchanges encompassing publications, schools and universities, hotels, and transportation networks –, film festivals have been hit by the Covid-19 pandemic in a very direct and particular way.

The closing of movie theaters almost all over the world suggested that the festivals were destined to fail or to postpone their editions. But that was not exactly what happened. Due to the ease provided by streaming and online resources, some festivals quickly retraced strategies and faced the challenge posed by the pandemic offering online versions still in 2020, as in Europe with the Oberhausen German festival (https://www.kurzfilmtage.de/en/) and the Visions du Réel Swiss festival (https://www.visionsdureel.ch/en/). (Mello, 2020).

From that period on, festivals reinvented themselves and, in many cases, expanded their audience by means of creative streaming proposals, displaying films in a broad and international manner – besides counting on the presence, though virtual, of more filmmakers and thinkers. Some festivals ended up resorting to platforms such as Festival Scope to house their films, as in the Thessaloniki International Film Festival (TIFF - https://www.filmfestival.gr/en/festivals-en/tiff-en), the French festival of short films Clermont-Ferrand (https://clermont-filmfest.org/en/) or the Locarno festival (https://www.locarnofestival.ch/LFF/home), in Switzerland. The platform houses so-called "independent" festivals and even allows a community to create their own online film festival.

The aim of this chapter is to create an inventory of case studies regarding film festivals at the age of the Covid-19 pandemic. In other words, the film festivals' organizational and structural response to Covid-19 will be scrutinized in order to gauge the impact of their respective changes. Some festival curators and organizers were specially interviewed for this chapter, to provide us with more detailed information on the most successful strategies adopted, what has been changed temporarily, and what changes have come to stay. In addition, the impact of those changes on the broader scope of the cities' tourism and local economies will be considered as much as possible. As mentioned above, film festivals trigger a chain reaction that affects a complex circuit in terms of the cities' economy and infrastructure. Let us take an example from Brazil: a loss of about R$ 1.5 million in a period equivalent to one month. This is the reality of Tiradentes, a historic city in the state of Minas Gerais that is more than 300 years old and has 99% of its economy focused on tourism and culture. This loss is estimated by Henrique Rohrmann, Tiradentes's Secretary of Tourism, Culture, Sport and Leisure, and Rodrigo Mendonça, the city's Finance Secretary (Prata, 2021).

There will be no space to look further into the economic impact of the Covid-19 pandemic on the cities' tourism revolving around film festivals, but this is an issue to be developed in future works. However, the present chapter will provide an inventory of the film festivals' renewed strategies in response to the pandemic. This inventory of case studies may offer an introduction to further developments in terms of the future of film curation, film festivals' organization and arthouse film fruition. It can be added to similar research efforts, as in the methodology developed by Salti (2020), and Hanzlík and Mazierska (2021). Based on a bibliography on the subject, special interviews with curators and organizers, and

a research on Brazilian and international festival websites during the pandemic, the authors developed a methodology structured as a qualitative typology of film festivals' strategies, as follows: (1) traditionalist strategies, (2) hybrid strategies; (3) mimetic strategies; (4) fully online strategies, described in detail over this chapter.

A NEW SCENARIO FOR FILM FESTIVALS – WITH A SERIES OF PLOT-TWISTS

Despite some quick responses and renewed strategies set forth promptly, the alternatives and online screening have not been simple during the Covid-19 pandemic times. A renowned festival, such as Sundance (https://festival.sundance.org/), was totally held on streaming in 2021, with online film exhibitions and public participation. For 2022, curators envisioned a hybrid festival, with both live and online events, resorting to the lessons learned in the prior year. But the emergence of the omicron variant in the beginning of 2022 made Sundance return to the online mode.

The year 2020 started with countries closing their borders to foreign visitors around March, after renowned and well-established festivals such as Sundance, Film Festival Rotterdam (https://iffr.com/en) or Berlin Film Festival (https://www.berlinale.de/en/home.html) had been held in January or February. In the case of the Cannes Festival (https://www.festival-cannes.com/en/), which usually happens in May, the pandemic rendered its realization impossible, even though they were able to hold a special three-day version in October. Venice Festival (https://www.labiennale.org/en/cinema) – the oldest of the traditional festivals – occurred in September, as usual, without a wide presence of international guests and more circumscribed to the participation of Europeans, following strict health safety protocols, as the use of masks and distanced seats. The realization of the festival was considered a "miracle" by the press at the time, mainly because it was held in Italy, a country severely shaken by Covid-19. (Romney, 2021). There were no streaming initiatives or proposals for virtual broadcasting. The festival was held in its traditional format, although the "specter" of the virus was hovering over it, according to its director Alberto Barbera. (Roxborough, 2020)

Even considering the relative success of the Venice Festival, the 2020 festivals calendar was extremely shaken, with several cancellations and delays. Besides Cannes, Tribeca Film Festival (https://tribecafilm.com/) and San Francisco Film Festival (SFFilm) (https://sffilm.org/) canceled their events in their traditional March to early June dates. Tribeca organized a series of sessions in New York stadiums drive-ins in July. Other festivals, with more time to organize themselves, made proposals connected to the pandemic, as was the case of the Locarno festival, which is held in August. However, once recognized the difficulties caused by the pandemic, Locarno live festival was canceled, but created a support fund for films whose making was compromised by the sanitary crisis. It is The Films After Tomorrow, arising from the value of the prizes that were to be offered by the 2020 Festival. Another initiative by Locarno's organization was to promote the teaching position named "Locarno Film Festival Professor for the Future of Cinema and Audiovisual Arts" at the University of Italian-speaking Switzerland. Moreover, the films were released in streaming sessions to the Swiss audiences as well as audiences from some other countries, aiming to solve a complex question, namely international copyrights, because films on streaming go beyond the original place of screening. Therefore, from July on, some festivals were able to adapt in time to hold their online versions, outdoor projections, and meetings with filmmakers through digital platforms. This is the case of the Shanghai International Film Festival (SIFF) (https://www.siff.com/english/) which, after delaying its edition for a while, carried out a still large version with online ticket sales, sold out in a few hours. The Toronto International Film Festival (TIFF) (https://www.tiff.net/) was also held in a hybrid format, with online and live events, even featuring a virtual red carpet.

In Brazil, the pandemic did not affect some festivals held before the March 2020 turmoil, such as the 23ª Mostra de Cinema de Tiradentes (https://mostratiradentes.com.br/). However, the expected exhibitions throughout the state of São Paulo were altered. Its edition in the city of São Paulo, scheduled to happen in March and April, was postponed to October, when an online version was presented. The traditional Festival de Gramado (https://www.festivaldegramado.net/) adopted a multi-platform system, broadcast by a cable TV channel, Canal Brasil and a streaming channel belonging to Globo TV network. The 25° Festival Internacional de Documentários É Tudo Verdade (It's all True) (http://etudoverdade.com.br/en/home/), an important event hosted in São Paulo and Rio de Janeiro, also had its screenings restructured and divided into two phases. Between March and April, more than 60 hours of non-competitive exhibitions were shown, in the online format, on the websites of Itaú Cultural (https://www.itauculturalplay.com.br/) and Spcine Play (https://www.spcineplay.com.br/). Subsequently, between September and October 2020, the screening of the competitive part of the show was displayed on the Looke platform, on specific days and times.

Traditional festivals which are part of the Brazilian São Paulo state calendar, such as the Festival Internacional de Curtas-Metragens de São Paulo (Kinoforum) (https://www.kinoforum.org.br/) and the Mostra Internacional de Cinema de São Paulo (https://45.mostra.org/) needed to adapt themselves to the "new normal". While Kinoforum established a partnership with the virtual platform InnSaei.tv, the Mostra Internacional de Cinema de São Paulo adopted a hybrid format, in-person and online. At the Mostra de São Paulo, around 198 titles could be viewed on three institutional Brazilian platforms: Mostra Play in partnership with the Festival Scope system, Sesc Digital (https://sesc.digital/) and Spcine Play. Some titles were shown at two outdoor cinemas as well: Belas Artes Drive-In and CineSesc Drive-in. However, there was an in-person closing and award ceremony at Parque Ibirapuera's Bike-in, a sort of bicycles drive-in. With free tickets to the public, the award ceremony was broadcast by YouTube.

In the case of the Festival do Rio, besides the pandemic, the festival claimed that funding difficulties led to its cancellation. The pandemic year coincides with a year particularly impaired by federal laws of cultural incentive. The same happened to the Festival de Brasília, the oldest Brazilian festival. However, after many controversies involving the Secretaria da Cultura (Federal Culture Department), the Festival de Brasília was held in a completely digital format, in December 2020. All those interested in it could watch the film selection on the cable TV channel Canal Brasil and via the Globoplay streaming platform.

Regional Brazilian festivals like Curta Taquary (https://curtataquary.com.br/), Curta Caicó (https://www.curtacaico.com.br/), Guarufantástico (https://www.guarufantastico.com/), and Cine Tamoio (https://cinetamoiofestival.com.br/), among others, were one of the first to experiment with the virtual format, for they migrated very quickly to the online format. Initially, the exhibitions used the same links provided by the filmmakers and producers, at the very moment of the registration of their films. Because of this, there was no rental of specific platforms for hosting the short films linked to the festivals websites from addresses directly hosted on Vimeo or YouTube players.

The 17th edition of the Fantaspoa – Festival Internacional de Cinema Fantástico de Porto Alegre (www.fantaspoa.com), main showcase dedicated to fantastic genre cinema in Latin America, offered 138 films selected on the Darkflix (www.darkflix.com.br/) platform, Brazilian streaming service specialized in horror, fantasy, and science fiction films. As explained by the organizers of the event, online exhibitions allowed more people to discover Fantaspoa. Around 80 thousand people from all over Brazil watched the films, a number of viewers six times higher than that of the 2019 edition, held in an in-person format (Fantaspoa, 2021).

In 2021, all the experience acquired in 2020 yielded solutions to festival organizers and better adaptation to the new scenario on the part of the public, added to the hopes of mass vaccination. Berlin International Film Festival, known as the Berlinale, which had not suffered the impact of the pandemic in the previous year, was affected by a new wave of the pandemic in Europe. Its planning was modified many times and it was eventually held in parts, in a first event called Industry Event, carried out online in March, followed by the Summer Special in June, when the films previously awarded were exhibited to the general public. The Summer Special had live sessions, in the traditional manner, but outdoors, respecting health protocols, with some online meetings. Venice Festival proceeded with its traditional in-person format, while Cannes postponed the event to July, holding it in the traditional format as well. That is to say: apart from the need for health protocols and a certain apprehension with new infection waves, the bigger and more traditional festivals were held in a format somehow similar to the period before the pandemic. On the other hand, smaller festivals, with lower visibility, seem to have benefited from the digital technologies and the online format more effectively.

In the first semester of 2021, most of the Brazilian festivals maintained the online format for their activities. It is believed that there has been a fatigue of the format because of the accumulation of screenings so close to one another. This was a consequence of the heavy promotion on social networks and the new policies for funding local events, such as the Lei Aldir Blanc, an emergency fund especially created during the pandemic that allowed cultural events to happen online, producing a certain proliferation of events. With tight deadlines for their implementation, festivals were impaired in their public reach, as there was not a strategy to promote the events or form an audience. Since the previous year, the proliferation of festivals paved the way for the confusion between dates and events which competed for the viewers attention in Brazil. Conversely, festivals focusing on gender and racial identity themes selected by the Aldir Blanc fund, were gradually discovered or rediscovered. Mostra Lugar de Mulher é no Cinema (https://mostramulhernocinema.com.br/) and Semana do Audiovisual Negro (https://www.semanadecinemanegro.com.br/) are examples of that, among other less well known festivals such as Rainbow: Festival de Cinema e Cultura da Diversidade (https://www.forrainbow.com.br/) and Curta-Se (https://curtase.org.br/) Festival Iberoamericano de Cinema de Sergipe.

Inaugurating the festival calendar in Brazil this year, the 24th edition of the Mostra de Cinema de Tiradentes kept all its screenings online, from the exhibition of short and full-length films to the debates. In January, 114 films on competitive and theme shows were exhibited, besides 22 meetings with 102 guests at the 24th Seminário do Cinema Brasileiro (Sacramento, 2021). É Tudo Verdade also displayed 69 films online, based on the experience acquired the year before. The competing films were again shown by the Looke platform, with daily screenings. Retrospectives and parallel shows happened with free access during the period of the event through several platforms like Spcine Play, Em Casa com Sesc, Sesc 24 de Maio's YouTube channel, and Itaú Cultural's website. In the São Paulo state capital city, Kinoforum proceeded with exhibitions on virtual platforms maintaining the partnerships with InnSaei.tv, Spcine Play and Sesc Digital.

FILM FESTIVALS: A CITY, AN EVENT AND THEN…STREAMING

With the turmoil created by Covid-19, it is worth asking how the innovative solutions that resulted in experiences that will remain in post-pandemic festivals appeared. It can be noted that most festivals decided to use streaming online systems in a more or less intense way, an operation that, apparently, has the power to increase the audience, democratize accesses, and promote productions that would not be visible otherwise. In this regard, it is possible to speculate on possible transformations of the "festival" system caused by streaming, understood in a broader manner, not only as a distribution operator, but also as an independent medium, capable of producing language.

The history of film festivals begins with the Venice Festival in 1932, an international event with an annual calendar and the presence of important persons of the cinema milieu of the time. This period, as De Valck recalls, is marked by the end of the transition from silent to spoken films (De Valck, 2006, p. 29). Since then, European film festivals have been historically related to questions of geopolitics and the so-called art films.

The next festival to be created was Cannes Festival, in 1939, as a reaction against the fascism prevailing at the Venice Festival. It was organized from a joint effort by Americans, the British, and the French (De Valk, 2006, p. 29 et passim). The success of festivals outside of the industrial circuit of Hollywood cinema "(…) is related to the interest by niche film with artistic value and/or social relevance" (De Valck, 2008, p. 18). In 1933 FIAPF (Fédération Internationale des Associations de Producteurs de Films) was established and during the first Berlin festival in 1951, a decision to control the "inflation of awards" was made, accrediting competitive festivals (Strandgaard Pedersen & Mazza, 2011; de Valck, 2006). At the time, Venice (1932), Cannes (1939) and Berlim (1951) were immediately accredited and, later on, Mar Del Plata (Argentina, 1954), Shanghai (China, 1993), Moscow (Russia, 1939-1959), Karlovy Vary (Czechoslovakia, 1946), Locarno (Switzerland, 1946), San Sebastian (Spain, 1953), Tokyo (Japan, 1985) and Cairo (Egypt, 1976) (De Valck, 2006; Strandgaard Pedersen & Mazza, 2011).

The list of European festivals highlights the fact that they are events profoundly connected to the cities, which grants them the status of privileged meeting places and even labels of touristic recognition as cool cities (Ooi & Strandgaard Pedersen, 2010). This aspect denotes a sort of relationship between the festival and the place where it takes place which certainly affects different types of cultural and economic sectors. There are no festivals accredited by FIAPF in Brazil, even though lists with over 40 competitive annual festivals can be found. Some of them, older and more recognized, also carry city names, such as the Festival de Brasília, the oldest in Brazil, established in 1965, besides Festival de Gramado (1973), Mostra Internacional de Cinema de São Paulo (1977), Festival do Rio (1984), Festival de Cuiabá (1992), Festival de Recife (1997), and the Mostra de Tiradentes (1998), among others.

In 2020, Locarno Festival offered 400 tickets to the Brazilian audience through the Festival Scope platform. Each production could open their exhibition to the countries which interested them, according to their distribution strategies. As a result, many films could be viewed before their commercial distribution and in places they might never reach. Thus, the "city" effect seems to grow, with Locarno Festival reaching an audience that perhaps had never heard of the Swiss town. This potential phenomenon will certainly be assessed by marketing and culture professionals as well, thinking about its effect regarding recognition by the locals.

Another key element of festivals is their format as "event", which preserves the hic et nunc as a unique real time happening, such as a theatrical or musical performance (Harbord, 2009, p.44). So, the shift of the festivals from the traditional format to a streaming status might impact this feature. Yet, the charm of the event status seems to prevail, regardless of a remote or online mode. It is worth noting that, even though the festivals began considering digitalization back in the 1990's, there is no evidence that they were prepared for such fast and radical changes (De Valk, 2006).

As everyone knows, streaming films is a pre-pandemic event. Nevertheless, it appears that it was during the pandemic that the festivals plucked the courage to adopt models in which the film's premiere could occur outside of the then sacred movie theater. In the traditional model, the film is "baptized" by the festival and only then is "cleared" to circulate in other means like movie theaters, television networks, the once physical mediums such as DVD's, and now through database platforms that provide the streaming of films.

The appearance of online film platforms such as Amazon and Netflix, or YouTube and Vimeo, among others, was deemed important for the consolidation of a sector of the distribution chain. However, due to the pandemic, those platforms also became fundamental as a means of distribution and final exhibition. In a Benjaminian conceptual operation, the online exhibition yields a gain of public and access, but the loss of a supposed "aura", as cinema itself has once done (Benjamin, 2008), of the already mentioned hic et nunc characteristic of the festivals.

It is noted as well that the hybrid models of festivals, with online films on platforms and "live style" debates with filmmakers and participants, in many cases were organized in a mimetic way in relation to the traditional in-person festival. A bit like the popularization of the internet in the 1990's, when a number of sites and blogs emulated the written page, and like many non-linear editing software and applications which emulate a Moviola, the festivals sought to build, in a remote and virtual environment, the same sensation of the "traditional" event. Film screenings in a programmatic form, ticket sales, talks with filmmakers on big screens, prizes announced at the end, among other procedures or protocols, aimed to preserve pre-pandemic standards.

On the other hand, it seems clear that, with online platforms, the festivals were able to reach a much broader and international audience. In the case of the documentary festival É Tudo Verdade (It's all True), held since 1996, its curator Amir Labaki affirms that an unprecedented public of "over 116 thousand viewers" was achieved (Labaki, 2021) in 2020, first year of the pandemic. The festival announced its program to the press and the following day the World Health Organization (WHO) declared the Covid-19 outbreak a global pandemic. With all the promotional material already printed, the festival had to reorganize itself very quickly and facing a lot of uncertainties given the exiguous time between the scheduled date and the agreements to use the space. Therefore, the first measure was to postpone the event. At times, its organizers considered carrying out live events, but decided to have only an online award ceremony while displaying the films on a virtual platform (Labaki, 2021). É Tudo Verdade (It's all True) is exemplary of a festival which had the diffusion of its audiovisual works amplified by the online platforms.

Eduardo Santana, cultural manager and creator of Cinefantasy (https://cinefantasy.com.br/), describes the same increase in audience. He explains that the festival's change to the online format was thought of as soon as the pandemic broke out, still in March 2020. Santana notes that the last "pre-pandemic" edition of CineFantasy, held in Museu da Imagem e do Som de São Paulo (MIS-SP), had an audience of approximately 4,500 people. In contrast, the subsequent edition (2020), the first online and during the pandemic, had 70,000 views. On the other hand, Santana highlights that "the biggest difference between a festival in a movie theater and an online one is the quality of the exhibition. While in the movie theater, the projection can take place in maximum quality, with a file of 20, 30, 40Gb, for example, online film festivals work with files of 4, at most 8Gb. It loses in image quality and also in sound". (Almeida, J., da Silva, C., Suppia, A., Mello, D., 2021a).

Even considering the democratization of films through streaming, it is also not evident that less promoted films have managed to reach a larger audience. In this regard, it is important to remember that the ocean of films offered by platforms and diverse online systems may even dispel viewers. The executive director of Festival do Rio (http://www.festivaldorio.com.br/), Ilda Santiago, according to an interview conducted by the authors, understands that "having in-person screenings as well as online, in my opinion, requires a much more sophisticated selection work. In online mode, I do not know who the public is, it is not a geographically defined "place", where I envision the audience I can attract. It is like programming an "indefinite" room of an audience with whom I will never be in direct touch with, and whose reaction I will only know by internet. She finds this exciting, though she doubts most festivals will be able to manage it. She adds that "I speak without any arrogance because this doubt haunts me every day. But one thing is certain: possibilities have changed, productions will have to change their prospects of being selected and how they will be seen. Some films will be better welcomed in an online environment. Others will have a better chance to seek the audience in the movie theaters." (Almeida, J., da Silva, C., Suppia, A., Mello, D., 2022b).

Even though it is not clear that there has been democratization and there are doubts about the quality of films shown, some new practices have provoked profound changes in the ingrained systems of festivals. Even the practices that intended to perpetuate such systems ended up leading to changes in the structure of cinema and/or festivals. This brings us back to the consequences of the Spanish Flu in 1918 as the trigger of a new phase of cinema as industry. Authors who addressed the "early cinema" affirm that the pandemic "shook the industry to its very foundation" (Hampton apud Koszarski, 2005, p. 467). Although some authors attribute the first great mishap suffered by independent producers to World War I, it was the flu outbreak that led to the closing of movie theaters throughout the USA in that time. "No New 'Movies' Till Influenza Ends," was the headline of The New York Times in 1918, recalls the columnist Richard Brody (Brody, 2020), based on his reading of "History of the American Film Industry" (1931), by Benjamin Hampton. The recovery of cinema was very hard in terms of production and distribution and, when movie theaters reopened, a few months after having been closed by public decrees, the audiences did not feel safe. Exhibitors had great difficulty masking viewers and the orchestra accompanying the films (Koszarski, 2005). Countless small movie theaters closed, which facilitated their acquisition by Adolph Zukor, a crucial figure in the consolidation of classic Hollywood cinema. Zukor merged his Famous Player studio with Paramount and reshaped the authorial and independent system through the so-called "vertical integration", that produces the film from beginning to end, from the script to exhibition and promotion by means of a production chain centralized in one company (or studio).

In Brazil, while movie theaters were closed in São Paulo, Belo Horizonte and Curitiba for a few months, in Rio de Janeiro, the epicenter of the 1918 pandemic, newspapers of the time suggest that there was no shutting down, but only recommendations against agglomerations. (Jornal do Brasil, 1918)

Prompted by this relation between the Spanish Flu and the reshaping of the cinema production chain, Brody articulates the question on how the new Covid-19 pandemic would affect the studios willing to reduce the exhibition time of their products in conventional movie theaters, collecting part of this sum and offering their films on digital platforms. Reflecting on this relation, today we can notice the ascendance of some centralizing players in terms of controlling distribution networks, such as Netflix or Amazon (Brody, 2020).

Despite the bad joke that the coronavirus would have been sponsored by Netflix, a somewhat "transmigration" process was already happening. As a result of the pandemic or not, Paramount already owns its film platform. What interests us most here, however, is to reflect on an equivalent process within the framework of film festivals: the effects of the pandemic, the accelerated the streaming processes and produce new audiovisual conglomerates would be, in turn, affecting festivals?

It is understood that festivals, even encouraging new creators, are not capable of, nor are aimed at, producing a substantial number of films. They boost a creative network by disseminating a certain perspective on films, greatly affecting the distribution chain. However, instead of democratizing their films and discussing their selection and awarding criteria openly, in a true intellectual debate, would the festivals be colluding with the most powerful streaming platforms, only aiming to reach an even larger international audience, virtually inaccessible via their traditional modes? In this process, because of a commercial model of monthly fees, would films pre-conceived for the platforms be the ones adopted by the festivals, causing an emptying of the idea of "festival" as shaped throughout the XX century? In other words, would the selection made based on festival curators' criteria (be it biased or not) conducted by the conveniences of the platforms – which even assume continuous "curatorship", such as small festivals?

This article does not answer such questions but tries to better comprehend this situation and its potential transformations already underway, taking into account the technological systems adopted by some international and Brazilian film festivals.

CHALLENGES OF STREAMING

Some of the traditional film festivals that migrated to online or hybrid formats, such as Berlin and Locarno, decided to employ closed streaming systems, through which the participants could acquire tickets, receive a code or password, and have access to selected films. In contrast, during the most difficult period of the pandemic, other festivals chose to completely open their online screenings for the people registered, so that anyone could watch the film sessions on the streaming platform they chose.

Festivals like CPH:DOCS, from Copenhagen, had to create inventive solutions to overcome the limitations imposed by the pandemic. According to Salti "Copenhagen's CPH:DOX team collaborated with the professional platform Festival Scope, online streaming service Shift72, social-media platform Facebook, and online video-conferencing tool Zoom." (Salti, 2020, p. 89). That demonstrated the need to act and interact in several spaces of production and exhibition.

The festivals that chose to sell tickets for specific shows of each film, through integrations with online collection banking platforms integrated to the issuing of the tickets, such as the Mostra Internacional de São Paulo, had to prepare themselves to manage software and their compatibility with more sophisticated streaming systems.

It means taking into consideration that it was necessary to include collection, separation of the online shows by room, creation of links to the exhibition of films individually, user's login validation to access the show and clearance upon confirmation of the payment systems. For instance, the SXSW Festival (https://www.sxsw.com/) made a partnership with Amazon Prime Video in 2021, so that the platform's subscribers could have access to some contents promoted during the event – but not all, because there still is some resistance from the distribution market concerning film premieres via streaming, in addition to overwhelming criticism against the entry of the streaming big techs in traditional festivals (Salti, 2020, p. 90). This production and digital organization structure of the festivals led the "live view" film streaming enterprises to a new level of demands and complexity.

Before the Covid-19 pandemic few festivals paid attention to film streaming platforms and did not even care about having their own virtual spaces. With the global closing of movie theaters these platforms were no longer a mere option and became, in many cases, the only way the festivals could survive. In addition, the global closing of movie theaters and consequently, of festivals, caused the appearance of countless online registration platforms specifically for festivals, with various participation tools, many of them added after or the migration of the festivals to the online format.

The main tools were developed by companies in the segment of video streaming which, on account of the need to adapt to the new scenario, ended up creating tools and additional software layers to promote the festivals. Digital entertainment companies and social medias, such as Festivee and Eventive were able to include other tools beyond the traditional purchase of tickets, "room entry", and film fruition, i.e. the inclusion of specialized audio chats for participation in the lobbies of the movie theaters, where viewers can "talk" with one another and exchange ideas before the beginning of the streaming of the film. Other functionalities were the inclusion of written chats, lobbies with the festival's films trailers and advertisements, and a program of debates in several internal screens being conveyed while the viewer awaits the beginning of the show, among other functionalities.

There is a considerable variety of digital systems that nowadays enable us to host, manage and administer, estimate and control the entry and exit of viewers, besides the film screening licenses. The main hosting, streaming and festival management systems and platforms, according to the Film Festival Alliance (https://filmfestivalalliance.org/), are: Film Festival Flix, AgileLink VOD, Cinando Video, CineSend Screener Rooms, Elevent, Eventive, Eventival, Festivee, Festival Scope, Filmbot, Film Festival Plus, Its A Short, Seed&Spark, ScreeningRoom, Endavo, and Qudio. In Brazil, the platforms InnSaei.tv, Mostra Play (of the Mostra Internacional de Cinema de São Paulo), and Itaú Cultural Play stand out.

Festival management platforms are experiencing an exponential growth in terms of the number of registered events, almost all of them seeking solutions for the migration of their editions. In several festivals a decision was made to try the hybrid format, understood as an event in which part of the audience, organizers and filmmakers was physically present in the venue of the event, but the vast majority of the audience was watching it remotely. This participation format demanded the integration of technologies which were part of the reality of the live streaming world, combined with aspects originating from social networks. That is to say, platforms like Film Festival Flix, AgileLink VOD, Cinando Video, CineSend Screener Rooms, Elevent, Eventive, Eventival, Festivee, Festival Scope etc. selected by the festivals to develop hybrid formats had to create conditions in terms of software to align the demands of film directors and organizers on aspects such as guaranteeing the quality of the image on not so adequate internet bands, geolocation of the user with tags on screening permits per country, digital rights management (DRM) tools in countries where there was no license for screening the film, blocking systems of VPN (virtual private lan's), which allow one user to hide their physical internet address and access the films even from a location where the film is unlicensed.

Moreover, as it was necessary to include elements of social media to exhibition spaces, such as systems to share the status of the viewer, image exchange tools, audios and even short videos, audio spatialization and audio chat systems, other resources that were mixed with the experience of online participation in various film festivals. Another tool which was necessary to develop was the creation of a mock movie theater in terms of viewers entry and exit management – as already mentioned, festivals that sought to follow the "physical" format of film screening and charged for the tickets, with scheduled time and date, where the film, even on streaming, started and ended at a scheduled time and place.

Because of some legal distribution issues, there was the need to develop layers of systems to manage the access to the streaming, through controlling the login of each viewer who had paid for their ticket, or ticket package, as at the Mostra Internacional de Cinema de São Paulo, among others. The ticket paid by the viewer could not only grant access to films, but also to debates and meetings with filmmakers, actors and actresses, among other attractions that festivals normally offer, such as courses, workshops, lectures and/or seminars.

For example, in the lectures, seminars, and workshops sessions, the StreamYard platform was employed by Mostra de Cinema de Tiradentes, with the tool linked to the YouTube streaming system, where the videos of these activities remain hosted until today. It is worth noting that part of the audiovisual memory of the festival will be stored and, at the same time, may be monetized by an international video hosting website, in this case YouTube.

At the Mostra Internacional de Cinema de São Paulo, in the two years of pandemic, a streaming system with ticket sales for virtual shows, with the offer of packages or individual tickets, was employed. This format required the development of digital systems to control the access to the streaming platform. In this case, it was employed their own platform developed for the Mostra via VOD (video on demand). In 2021, particularly, some films of the Mostra de São Paulo were screened on platforms such as Itaú Cultural Play and Sesc Digital as well. The films could be viewed in shows that, despite being on streaming, were scheduled to start and finish at specific times, in a very programmatic model.

Monica Trigo, president of FantLatam (Aliança Latino-americana de Festivais de Cinema Fantástico - https://www.fantlatam.com/), and co-director of the annual festival CineFantasy (Festival Internacional de Cinema Fantástico), now in its 12th edition, stresses that "the challenges for the migration of festivals to online or hybrid formats have to do with questions related to the security of the platforms and the debate about copyrights" (Almeida, J., da Silva, C., Suppia, A., Mello, D., 2021c) which is a real issue to deal with. The Festival Cine Fantasy employs the InnSaei.tv online platform for its streaming editions.

Festival de Filmes Científicos (FeFiCi - www.fefici.org), going to its fourth edition, held the two previous ones employing the YouTube and Vimeo platforms, both for the submission and exhibition of short films dedicated to science and technology. For the award sessions and lectures StreamYard, in 2020, and V4H (Video for Health) platform in 2021, were employed, with both being connected to the YouTube streaming system. The 2022 edition will be held in a hybrid format, with some shows happening on streaming, while others should be held in-person.

In turn, the É Tudo Verdade (It's all True) festival employed YouTube for the maintenance of the online collection of events, in the sessions of lectures and in the seminars. As mentioned before,the festival employed Looke platform, Sesc Digital and SPCine Play's system. The film's exhibition format was phased in terms of days and number of views to the streaming systems, ers. For some films and events, the festival had to limit the number of participants. At the opening and closing, as well as for the domestic full-length films competition, the number of viewers was limited to 2,000 and, for the foreign films shows, a limit of 1,000 viewers was established.

The concerns about the limitation of public demonstrate the necessity to think of strategies which include more sophisticated technological methods for something that is today being debated in the sphere of copyrights, because it can be inferred that the fact of a film festival restraining the visualization of a film by a user does not preclude the same user from sharing their screen on some other videoconferencing or social media platform and watching the film with friends by means of this sharing. At least not technologically.

In this sense, limiting the number of viewers in digital movie theaters when films are screened via streaming seems a naïve attempt compared with the real problem regarding the distribution of films in festivals. That is, the relation between the number of viewers and broadband capacity, and the preservation of sound and image quality of the film screened via streaming, would effectively be limitations to the number of viewers? It is believed that one of the real advantages of festivals in the digital/streaming format lies in their democratization of access. Restricting the public, all the more the paying public, appears to be something that is in contradiction with one of the main advantages of network cinema. On the other hand, as remembered by Trigo, film festivals and shows have always been the most prominent showcases for independent audiovisual production. Producers usually preserve a new work for release at festivals with originality and still compete for awards. Trigo also points out, "numerous international productions have not accepted to show their works in the online format." (Almeida, J., da Silva, C., Suppia, A. Mello, D., 2022c).

Furthermore, the perpetuation of an ideal of cinema premiere ends up jeopardized when the festivals system and its commercial developments are challenged by a totally new and unpredicted conjuncture.

FESTIVAL STRATEGIES

The article's focus is to analyze the alternatives found by some film festivals in the face of the Covid-19 pandemic and also the implications of these alternatives for the future horizon of these same festivals and the possible impacts of this new scenario in the form of audiovisual content.

After introducing a panorama of the more traditional festivals' positioning and their alternatives in the face of Covid, this article presents a panorama of Brazilian festivals, depicting and commenting on some of the strategies deployed by them. Based on research in catalogs and online platforms, interviews with curators and festival organizers, the article reflects on the technological strategies and forms found to soften, replace or even expand the audience of festivals, such as the development of their own streaming platforms, use of existing platforms, development of additional tools to streaming services, implementation of virtual "rooms" for joint sessions, shared access, use of voice, development of forms of expanded access to debates with directors, screenwriters, actors and actresses, among other professionals of the sector.

Based on this panorama, it is understood that streaming practices happened on different levels and, from them on, the perspectives of their application in future festivals may be reflected upon. Investigating the strategies employed by the festivals, it is noted a division into five distinct modalities. Each modality was given a name and is characterized by the application of one or more strategies related to holding events in pandemic times. They are: (1) traditionalist strategies, or those which just implemented health protocols, keeping the event in-person; (2) hybrid strategies: they combine the traditionalist strategies and the online strategies, in which part of the festival is held in a conventional manner and part migrated to virtual environments; (3) mimetic strategies, are those in which the use of the streaming technology seeks to emulate the conventional experience of going to the movie theaters; (4) fully online strategies, when there are no in-person meetings; and finally (5) nostalgic strategies, which retrieve or revisit older models. Here, one element that may be present in more than one of the festival models described above is added, the participation of the audience through innumerous online strategies.

The traditionalist strategies are the ones adopted by festivals that did not renounce their status of a public event and even having to resort to delays or cancellations, maintained the in-person, in loco event. These events were held live and in a traditional manner, with the appropriate health measures. This was the case of the Venice Festival in 2020 and 2021 and Cannes Festival in 2021. In Brazil, in 2021, Mostra Internacional de Cinema de São Paulo and Festival do Rio made live presentations as well, with awards. However, they also promoted online screenings of films on exhibition platforms, following a traditionalist strategy, yet hybrid.

Hybrid strategies are those which offered part of their screenings in the in-person format and part in a remote/online format. Sometimes these hybrid systems, in their online format, preserve more traditional features, possibly also becoming mimetic in that version. Hybrid festivals with online and in-person systems were the most common in 2021, when there were a number of people vaccinated and an exhaustion of the fully online systems was perceived.

Mimetic strategies, in turn, are those whose events imitate live festivals, with the staging of award ceremonies, screenings that prescribe the length of the film with live shows, a virtual red carpet and, instead of obtaining the advantages of online systems, they adapt film screenings to the insurmountable limits of the in-person. The Brazilian case that most resembles the mimetic strategy is, perhaps, the Festival de Gramado, that made its award session with presenters, trophy giving ceremony and participants in an online platform session. It should be made clear that a festival may be of the mimetic type without being restrictive.

The openly online festivals are the ones which employ their own or an external platform to exhibit their selections of films and all sessions related to the event. They may vary from more open versions, with a free audience, to more programmatic versions, with some or all times prescheduled, ticket sales etc. However, they explicitly assume the features of online systems – such as the availability of films for longer periods (a few days or one week), use of open and free platforms, debates with authors on channels where the audience may ask questions, workshops, round tables –, aesthetically bearing a resemblance to the database models, yet preserving the curatorial direction, equally stated/made explicit.

Festivals of the nostalgic type are those which included in their events the rescue of models, practices, rituals, or technologies that are older or have fallen into disuse, such as the drive-in cinema – revisited, now, just to guarantee the maintenance of social distancing among viewers inside their vehicles, amid successive lockdowns or mobility restraints imposed by governments and city halls all over the world. Drive-ins were implemented by independent initiatives, but at the option of some festivals, such as Tribeca, as well. In Brazil some initiatives were explored by the Mostra Internacional de Cinema de São Paulo in the bike-in experiment, by Cinefantasy, and by the short films festival, Curta-Se. An important element of the "nostalgic" model (whose most obvious example is perhaps the drive-in) was to ensure an in-person ritual of public participation, as the buzz inherent to festival meetings could no longer happen – at least not at certain times.

The models described here are subject to amalgam and/or superposition. For instance, a given festival may release their films throughout all the period it is being held, without audience restraints, and still perform opening and closing ceremonies which follow a "mimetic" pattern. This strategy was quite common in the festivals investigated here, such as CineFantasy or Tiradentes. Mimetism does not exclude the deeper and broader exploration of the potentialities of streaming or digital platforms.

However, the model's classifications serve to discern the tendencies of each strategy and think of how they cut across the festivals. They may also serve as operators to comprehend the new organizations of the festivals, noticing those which tend to assume streaming as a screening element in future festivals following the relief or the end of the Covid pandemic. In other words: judging by the findings of this investigation, the most systematic option or prevalence of a given strategy in relation to the others may be indicative of how festival organizers understand the scenario, be it in a more transitory way – i.e., employing digital tools as a "cap" or temporary plan –, be it in a more integrated and friendly way of permanent changes.

Regardless of the strategies adopted, interactivity is a factor that imposes itself on practically all the festivals investigated here, that is to say, online participation, more or less open to the public, through chats, commentaries, employment of platforms such as Discord etc. The interactivity factor may have compensated or, at least, emulated, to some degree, the buzz inherent to festival meetings, now prevented from happening on account of sanitary restrictions. Possibly, the interactivity factor emerged as a form of "ritual" so far absent in festivals, embracing, for example, the multi-screen phenomenon.

CONCLUSION

This brief set of case studies aimed to describe and comment on the main strategies adopted by international film festivals, but above all the Brazilian ones, in a moment of profound crisis in the cultural and artistic sector, particularly in the cinema market in its facet dependent on public spaces and social congregation.

Cinema stories are prodigal in stressing the cinema's character of social ritual, its potential to agglutinate people. Even if, since the first movie theater, other versions of this art medium were able to dispense with social meetings, it is true that going to the movie theater or theater, purchasing a ticket, talking in the lobby (before and after the show), followed by a likely group conversation, in a bar or restaurant, all those rites were part of the most celebrated and remarkable cinema experience in the memory. And, in this respect, festivals emerged already in the first decades of the XX century as "hubs" of the most integrated cinema experience, be it for artists/filmmakers, producers, entrepreneurs in the audiovisual sector or, to a large scale, the general public attending them. There is no festival without an audience. Besides all this, the festivals were consolidated, throughout the XX century, as the most reputable institutions with respect to the recognition of the film as a work of art – keeping this status of "guarantor" of the art of cinema to date.

For behold, we are living through one of the greatest pandemics in the history of humankind, with the enactment of a state of emergency by the World Health Organization in the beginning of 2020 and, immediately thereafter, all forms of social agglomeration started to pose a risk of death to people. Among the most affected economic and artistic sectors are theater and cinema, for whom the presence of an audience, paying or not, has always been so important. In the case of film festivals, it is fair to affirm that the formation of an audience, with the consequent agglomeration of people, has always been essential. In parallel with this tragedy of global proportions which is the Covid-19 pandemic, the system of on demand films or cinema streaming had been expanding at a fast pace since at least the mid-2000's. Pandemic and streaming became almost binomial for the cinema industry, and a solution for the survival of the festivals. As seen above, the Spanish Flu in 1918 had already impacted cinema in a way that seems to be forgotten today (Hampton apud Koszarski, 2005). The Influenza virus changed habits and had an influence on the way cinema is consumed. The Covid-19 pandemic, around one century later, yields new changes, perhaps broader and deeper, although against a radically different backdrop: a virus in analog time does not cause the same reactions as a virus in digital times. Today we live the digital paradigm, and it is in this scenario that film festivals responded to the contingencies imposed by the new pandemic.

As the case studies were conducted, the strategies adopted by some international (e.g., Berlin and Venice) and domestic (e.g. Mostra de São Paulo and É Tudo Verdade) festivals were investigated. This investigation led us to organize these strategies, arriving at five strategies or modalities which were created and/or adopted separately or jointly by the festivals studied. It is worth remembering that none of these strategies, profiles, models, or modalities of festivals can be seen as watertight. Although some festivals have invested all their resources in one strategy or profile or have privileged some strategy or profile at the expense of the others, all the strategies listed here and temporarily qualified must be understood as instruments subject to contamination, parallelism, and superposition. For example, festivals with hybrid strategies, that is, with part of the screenings online and part in-person, may also assume mimetic strategies (e.g. charging for the tickets and scheduling show times) and/or nostalgic (e.g. adopting screenings in public places or drive-ins).

Even though it is not yet possible to reliably determine the future of film festivals in Brazil and in the world, both in terms of form and content, it may be supposed, based on some of the strategies investigated, how certain festivals envision their post-pandemic future. Undoubtedly, there are modulations or differences among festival curators and organizers who bet in the transitory nature of the moment and the brief return to the traditional modes, while others already see themselves in a significantly different future, in which the words festival and streaming will maybe be indissociable, with the addition of new technologies capable of expanding even further the reach, the immersion, the interactivity and the experience of festivals.

ACKNOWLEDGMENT

This research was supported by CAPES (Coordination for the Improvement of Higher Education Personnel) [grant number: 88881.478202/2020-01] and the Mackpesquisa fund.

REFERENCES

Almeida, J., da Silva, C., Suppia, A. Mello, D. (Interviewers) & Santana, E. (Interviewee). (2021a, Out 10). *Eduardo Santana interview for the authors by e-mail.*

Almeida, J., da Silva, C., Suppia, A. Mello, D. (Interviewers) & Santiago, I. (Interviewee). (2022b, Jan 12). *Ilda Santiago interview for the authors by e-mail.*

Almeida, J., da Silva, C., Suppia, A. Mello, D. (Interviewers) & Trigo, M. (Interviewee). (2021c, Sep 14). *Monica Trigo interview for the authors by e-mail.*

Benjamin, W. (2008). *The Work of Art in the Age of Its Technological Reproducibility*. Belknap Press. doi:10.2307/j.ctv1nzfgns.6

Brody, R. (2020, March 17). Lessons for the Movie Industry from the 1918 Influenza Pandemic. *New Yorker*. https://www.newyorker.com/culture/the-front-row/lessons-for-the-movie-industry-from-the-1918-influenza-pandemic

Changsong, W., Kerry, L., & Marta, R. F. (2021). Film distribution by video streaming platforms across Southeast Asia during COVID-19. *Media Culture & Society*, *43*(8), 1542–1552. doi:10.1177/01634437211045350

De Valck, M. (2008). Screening the Future of Film Festivals? A long tale of convergence and digitization. *Film International*, *6*(4), 15–23. doi:10.1386/fiin.6.4.15

de Walck, M. (2007). *Film festivals: history and theory of a European phenomenon that became a global network*. Amsterdam University Press. doi:10.1017/9789048506729

Fantaspoa – Festival Internacional de Cinema Fantástico de Porto Alegre. (2021). *Fantaspoa bate recorde de público e divulga seus premiados*. Fantaspoa. https://www.fantaspoa.com/2021/noticias/1/109/fantaspoa-bate-recorde-de-publico-e-divulga-seus-premiados

Hampton, B. (2005). A History of the Movies. *Film History, 17*(4), 466–485. doi:10.2979/FIL.2005.17.4.466

Hanzlík, J., & Mazierska, E. (2021). Eastern European film festivals: streaming through the covid-19 pandemic. *Studies in Eastern European Cinema*. . doi:10.1080/2040350X.2021.1964218

Harbord, J. (2009). Film Festivals as Time-Event. In D. Iordanova & R. Rhyne (Eds.), Film Festival Yearbook 1: The Festival Circuit (pp. 40–46). Academic Press.

Labaki, A. (n.d.). *É tudo verdade*. Festival É Tudo Verdade. http://etudoverdade.com.br/br/pag/apresentacao

Mello, D. M. C. (2020). O cinema durante a Covid e a memória ressignificada no experimento Odradek [Cindema during Covid and re-signified Memory in the Odradek experiment]. *Revista Z Cultural, 10*. http://revistazcultural.pacc.ufrj.br/wp-content/uploads/2020/10/O-cinema-durante-A-COVID-e-a-mem%C3%B3ria-ressignificada-no-Experimento-Odradek-%E2%80%93-Revista-Z-Cultural.pdf

Ooi, C., & Strandgaard Pedersen, J. (2010). City branding and film festivals: Re-evaluating stakeholder's relations. *Place Branding and Public Diplomacy, 6*(4), 316–332. doi:10.1057/pb.2010.30

Prata, T. (2020, May 6). Caos silencioso: acostumada a sediar grandes festivais, Tiradentes sofre com prejuízos e a pandemia [Silent Chaos: Sued to hosting major festivals, Tirsadentes suffers from losses and the Pandemic]. *Hoje em Dia*. https://www.hojeemdia.com.br/entretenimento/caos-silencioso-acostumada-a-sediar-grandes-festivais-tiradentes-sofre-com-prejuizos-e-a-pandemia-1.785625

Romney, J. (2020, Sep 12). Venice film festival 2020 roundup – against all the odds, a triumph. *The Guardian*. https://www.theguardian.com/film/2020/sep/12/venice-film-festival-2020-review-human-voice-almodovar-tilda-swinton-vanessa-kirby-cate-blanchett-nomadland-new-order

Roxborough, S. (2020, Jul 28). Venice Boss Alberto Barbera on Hosting the First Post-Lockdown Festival. *The Hollywood Reporter*. https://www.hollywoodreporter.com/movies/movie-news/venice-film-festival-boss-alberto-barbera-hosting-first-post-lockdown-fest-1304607/

Sacramento, J. P. (2021, Jan 14). *24ª Mostra Tiradentes abre calendário audiovisual brasileiro e homenageia a cineasta Paula Gaitán. Mais vertentes* [Mostra Tiradentes opens the Brazilian Audiovisual Calendar and pays tribute to Filmmaker Paula Gaiatn. More Strands]. https://www.maisvertentes.com.br/noticia/1985/24-mostra-tiradentes-abre-calendario-audiovisual-brasileiro-e-homenageia-a-cineasta-paula-gaitan

Salti, R. (2020). Do Not Go Gentle into That Good Night: Film Festivals, Pandemic, Aftermath. *Film Quarterly, 74*(1), 88–96. doi:10.1525/fq.2020.74.1.88

Strandgaard Pedersen, J., & Mazza, C. (2011). International Film Festivals: For the Benefit of Whom. *Culture Unbound, 3*(2), 139–165. doi:10.3384/cu.2000.1525.113139

Watercuter, A. (2022, Jan 22). Film Festivals Are Evolving for the Better. *Wired Magazine*. https://www.wired.com/story/film-festivals-hybrid/

KEY TERMS AND DEFINITIONS

Digital Festivals Platforms: Computational systems specifically developed for film festivals, granting access to specific festival models' attendance as rooms to watch film premières, data analytics of film attendance, rooms with ticket sales for streaming of films during a certain period of time reproducing the traditional film festival environment.

Film Festivals Analytics: A model of cultural and data algorithmic analysis of viewers participation in film festivals based on attendance, geolocalization, online digital presence in order to grant film festivals organizers the analysis of the festival's impact on the local economy, measuring concepts as soft power of the city, touristic impact of the digital presence of the festival and the creation of an online digital presence of the Festival in social media and alike.

On-Line Film Festivals: Film festivals around the globe that adopted the online participation of viewers by using the streaming computational methodology. In some cases, the festivals adopted the traditional audience participation, i.e., online tickets to enter a digital room to watch a film première or a feature film in a specific time and date. Some festivals adopted the model of an open streaming of their films schedule during the time of the venue and in other cases film festivals decided to publish their selection open to the public.

Streaming: A form of online film distribution based on a global computational distributed grid media environment to be viewed on any digital gadget with screen such as cell phones, tablets, or computers. A streaming film can be also projected on theaters in real-time.

Section 10
New Media for Post-Viral Flows

Chapter 28
Extended Reality as the Immersive Environment Conjunction of Augmented Reality, Virtual Reality, and Mixed Reality:
A Phenomenological Response Facing Modern Culture

Andres R. Montenegro
Purdue University, Fort Wayne, USA

ABSTRACT

Extended reality environments will exceed virtual reality and mixed reality environments and will be understood as a new level among several of the metaverse modalities that will become technology standards on the internet of the next future. Extended reality will address as well how the user's interface will experience dramatic changes in the way users manipulate digital 3D content. Interactive objects will adopt a holographic status. The use of natural tactile interactions determines the availability of our body as an integral interface that will obtain responses in several levels of perceptions, one of which is haptics. To create the meaningful experience of our body and the possible responses obtained from a spatial interaction or interface inside this metaverse, the content and its design must achieve a fully immersive dimension. In this crucial crossroad of body and interface interactions, phenomenological experiences will be vital to understand contemporary problematics and cultural changes.

DOI: 10.4018/978-1-6684-3369-0.ch028

INTRODUCTION

Today the emerging technologies are capturing widespread attention in terms of new digital gadget profusion. Those gadgets clearly carry a mission encapsulated by the manufacturers to engage consumer attention and expand the number of social media applications today's users are familiar with. In fact, the most relevant component of the mobile communication technology is its multimedia range of applications. Users' widespread familiarity with video, audio, and text messaging is rooted in their needs to address a complex or simple chain of submissions through an app. However, the evolution of cellphones, tablets, or mobile gadgets that contain practical applications is modestly showing the great potential for a big leap in the way users could perform the dissemination of a new level of exchanges: the built-in camera.

The built-in cameras in cellphones, tablets, and smart glasses not only capture reality, but they are also able to measure the reality and create a rendition of it with clear and accurate coordinates. Emerging technologies are not appearing today as an overnight development. They have been around for many years. Scanning done through a camera started 30 years ago when the first experiences of overlapping digital objects on top of what the camera was perceiving revealed a germinal but compelling prospect — the ability to mix the digital and the real world. The camera therefore was the electronic eye that permitted the scanning of environments. Certainly, the visual outcome of this very appealing technological act was somehow familiar from television graphics: All the time, heads-up displays, or associated animations, acted visibly and were an everyday companion of talking heads, while the flow of information was bolstered whenever the weather or news needed to display information on top of the host or presenters.

The presence of a camera in every kind of device was routinely adopted by users as the intrinsic appliance that captured glimpses of mundane existence. Nevertheless, the capability of a camera-eye was evident for the combination of digital content and real time video for content creators, artists, and scientists. Alongside the banal use by consumers, researchers were experimenting with the overlaying of 3D models and specific routines of real time recording. The groundwork of overlapping digital objects on top of reality is the real time video. The quality of built-in cameras has evolved in whirling speed, allowing the standard of hyper-high definition or what is called 4K (horizontal resolution of 4096 pixels). The Augmented Reality term was by then the best denomination for an emerging technology based on camera registration.

Perhaps the term augmentation in the beginning represented a long-term prospect in which way the capability to push the boundaries of an object registration will take place. Semantically, the term augmentation is eloquent. The Augmented Reality technology will be able to provide a continuous expansion of boundaries in terms of what the camera device is perceiving and what the digital layout can reach or installing within the camera mapping. In this dimension, artists, videographers, scientists, engineers, architects, and communicators confer with each other at this common crossroad: the content visualization.

This exercise of visualization not only displays immediate information, but it also installs digitally a system of visual inferences related to the real objects that are perceived by the camera. In fact, this system performs a meta-capability of visualization with a strong flavor of what semiotics practiced before the advent of hypertext. Little by little, the attainment of what augmentation meant for new users or experts happened around more friendly interfaces or apps rather than experimental or shrunk communities developing isolated experiences.

Background

Thanks to the evolution of mobile technology and its predictable widespread embrace by end users, the availability of apps that demonstrated simple AR experiences was clearly visible. Perhaps, it is still today problematic to make a more scholarly definition of Augmented Reality due to its heavy dependencies on technology. However, the most attainable definition of Augmented Reality relies upon its immediate analogy, putting a digital object on top of a real one (a physical one) and making it visible when the digital content is trigger by a marker, QR code, or a target image.

At this point, understanding of the concept requires another important component associated to the analogy of the one object on top of the other: the indispensable attachment of the QR code, marker, target, or anchor that enables the visualization of the digital content. Some definitions vary about this attachment, depending on how the action or triggering unfolds. A quick acknowledgment materializes better around the more familiar metaphor of the glued object, operating as the action of the camera that sees a marker and triggers the visibility of the digital glued object.

Another modality for how the paradigm of "the one object on top of another" turns up is when users can get access to a 3D library, text, pictures, or other multimedia object, and, using an app interface, literally drop the object on a specific point the camera points out. In this case, users are casting or setting anchors through the camera, and the digital content is planted in the world that the camera is seeing. In this sense, the interfacing component is ever present; it is indispensable to have a display between the real world and the user to perform the digital overlaying. In this case, it is quite self-explanatory to experience a simple app that performs the dropping action.

Art, science, education, entertainment, and cross-pollinated disciplines require a common ground to make sense of the utility of such a technological achievement represented by Augmented Reality. Any relevant contribution Augmented Reality could attain in any scholarly or professional realm depends on the paradigm embedded in it. Is there a real need for such a visualization? Is the answer multifaceted? Or basically is it a novelty available just for entertainment or a technological curiosity used in the huge sea of memes in social media?

Any discipline can think about how content should be elaborated in levels of visualization and how this layering could implement complexity through a sequenced triggering. The abstract correlation of levels of visualization could appear overwhelming if it responds to just a mere mental exercise. If it remains only in-the-mind's-eye task, it will not have a concrete outlet to display a logical structure. Thereupon, visualization operates like the quintessential paradigm that allows the analogy of an object glued, or on top of, another to make sense.

Visualization can have an organized hyper-hierarchy of levels to permit a discipline to establish a philosophical domain of principles. In a more common way, a systemic structure would be enough to develop a hierarchy of thoughts, not needing an emerging technology to produce an accurate or a coherent logical flow. It is in this larger sense when a meta dimension is needed to support how its immersive correlation would expand the reality.

The common ground for any discipline is the opportunity for expansion in terms of metadata, hyper hierarchy, or instant responses about complex and abstract problems. In aesthetics, for example, the appreciation of an object always has been problematic if this exercise relies upon "empathy" or "abstraction." Certainly, the upper level of thought in consequence will be the abstract conquest of the observed object. In this aspect, in art such an aesthetic visualization through Augmented Reality will expose a deeper or more intricated meaning transformed into an object-essence in front of the critical, experienced, or novel eye of a user.

Here is the common ground that emerges for every discipline, the possibility of experience or acknowledgement through levels of visualization. The most important component in this approach, perception, or mere registration of the reality through a digital or electronic device, is the data generated. In fact, this generated data is no longer an analog registration like the previous technologies based on video tape. The camera can measure the dimension of the space or the environment in such a way that is no longer determined by a flat visualization. The data generated allow the perceived space for a layout in depth, which is the way digital objects will be planted, whether they have volume or flat information in a planar basis. Even the pictures or digital movies are available for this change of paradigm.

Figure 1. Camera showing a spatial mapping
Credit source: Microsoft HoloLens. Spatial mapping. Mixed Reality documentation

The flat aspect of the information codified in a picture will no longer be 2D. It will be planar, because the object that supports its presence in the space configurated, the plane, has a volumetric identity inside the environment. In the distinctiveness of this emerging technology, which is the way a camera maps the environment, a common situation arises in terms of how users convey the interaction the camera can provide, regarding the screen. The screen is the final window users have available in front of them to interface the information. The problematic paradox in this situation is the way a screen determines the interfacing. Mobile devices, computers, tablets, and other devices expose the digital information through a flat screen.

The flat screen is not a limitation, perhaps explained by the user's dexterity to manipulate the interface and perhaps the intuitive notion of that flat space, (still a 3D depth can operate the Z axis). The problem is perceptual limitation of that flat interface. That perceptual limitation creates a spontaneous user's behavior, that has assumed a widespread and universal familiarity nowadays. Today nobody can question the inability of a flat screen to render a 3D environment, even if the tv monitor or a device screen uses stereo cameras, or a modality based on it.

The problem is the inability to turn that flat screen into an immersive experience or navigation. The immersion in a situation is the new paradigm. A situation could be rendered in an idea, environment, a scenario, a diorama, or a metaphor. Every object regardless of its nature could be carried out within an immersive paradigm. However, to start or inaugurate a transition from the perceptual status of a flat rendition to a mapped or spatial rendition would require a change of attitude from the technology creators, hardware manufacturers, and, finally, end users. The immersive paradigm is not yet a familiar interface experience to anyone.

The immersive experience is determined by the perception of the real space combined with the mapped virtual space obtained by the camera registration. In simple words, the device camera scanned the real space and created a tridimensional map with all the objects surrounding or present in that environment. A major problem in this change of paradigm is the users' notion of what immersive means or how they get familiar with the new interfacing. One of the consequences of this problematic relationship is the novelty of certain devices that incorporate an immersive notion of the augmented reality technology. Those devices are the smart glasses.

Smart glasses incorporate the space mapping through the camera registration and even more interfacing as well, including the ability to use device controllers to navigate in the mapped space (Naemura, et al., 2002, pp 70-71). In this aspect, smart glasses lay the foundation for just the use of users' hands recognized by the smart glass as the interfacing controller. For such a big leap in using interfaces and controllers, this is a great feat within the emerging technologies' evolution. Manufacturers and engineers are conceiving these new devices in the right dimension by choosing the use of our body as the interface. Perhaps this incorporated notion is still incipient.

THE PROBLEM

Users need abundant training in learning a new usability archetype. However, a widespread or massive use produces a sort of natural incorporation of a technology handling; in plain terms, a culture of use is acquired. It is very important not to forget how today, on an everyday basis and in a ubiquitous fashion, the common sight of people digitizing smartphones in a very dizzying speed surprises us with how they have achieved such dexterity. The answer is that it might fall under the category of a cultural impact and involve the necessity of being wired, since smart phones contain very important applications for communication.

Is the key the desire of the user to be inside a network which requires learning the mechanics and the intuitive adoption of hand gestures that help to manipulate the smartphone interface? A problem here remains at a very clear section of the smartphone, the screen. Still screens provide a solid framework for the interfacing, but the flat surface confuses the concept of interfacing for a further evolution of the same. The users are still tied to a paging concept of any information overlaid on a screen display.

If the users deem efficient and expeditious to trigger a link or grab a link with a single tactile pressure, the action remains ephemeral or very automatic. There is no time to imagine or assess a further interaction. For a technology commentator this judgment could seem irrelevant, but, in a strict definition, the concept of a document manipulation is the page, that virtually turns, and shows another one in an instantaneous appearance depending on the speed of the available internet service provider or data roaming setting in a mobile device being used. The virtual paging of layouts on a screen is conservative and almost reactionary if we think about the conventional and traditional act of handling a document.

Still a document is determined by a format in terms of pixels dimensions on a screen. The fundamental roadblock to a cultural shift in what could be a usability revolution is the resistance of manipulating a document or information in terms of other than just a 2D pixel dimension. The screen archetype or the standard use of what we need to read in a document remains prevalent and almost perennial, being almost impossible to dismantle if we recalled hundreds of years of editorial and text evolution.

A second archetype is the electronic book. It keeps the format, translating its standard mechanical action of navigating pages with equal analogic procedure as if the user manipulates a physical book. Too many branches in this tradition surface through a heads-up display or screen interface: the reading action, the paging turning, the treatment of how information is visually scanned by our sight, even eye persistence in just browsing the information.

However, the problematic conundrum posed by Augmented Reality as emerging technology not only is troubling users but also is basically rushing the technological optimization of manufacturers in relation of how a new paradigm in the document visualization will takes place without persistent glitching.

Several systems have been announced and hyped since the introduction of mobile communication. One of them is the smart glass introduced by Google in early 2013 that caused certain criticism in terms of prospective concerns rather than immediate effectiveness testing of the system. Since then, several prototypes have been introduced by Epson and Microsoft mostly for experimental purposes or for specific research methods.

However, the reception by society in general and in the technologically oriented literature has been even more problematic, adding a sterile borderline discussion in terms of how this technology could generate potential disruptions in the user's everyday behavior. Not even fully established or released as a massive and adopted technology, Augmented Reality could reveal certain concerns and preoccupations; in fact, the inheritance of user's behaviors from mobile technology is what causes a wrong appraisal about changing a paradigm in the way a document or information will be visualized.

Augmented Reality applications based on QR coding, anchoring digital objects or overlaying effects on what the camera device is registering have been around for quite some time in the universe of applications provided by the Apple Store or Google Play. Some of them just expose a superficial or banal aspect of what could be a powerful and relevant archetype for the visualization of a document.

The principle of using an augmentation experience is just the overlaying component, but this use falls short in terms of what potential the document would expose if the digital object could involve more usability modalities. In consequence for an emerging technology such us Augmented Reality, the problems are not only installed just in the user's world, or the manufacturers realm, but are in the very center of the cultural debate, oriented towards the area of art, science, communication, humanities, and computing disciplines.

In these worlds, or realms, of knowledge, the transformation or adoptions of paradigms should happen. The slow embracing of revolutionary notions of computing is caused by the traditional exercise of merging what is information and knowledge. In this case there is an almost gestalt drama occurring in the background of a discipline adopting an emerging technology, the inability to distinguish what is the value of it in terms of fundamental basis, or how just the mere component of functional use impedes the ability to see the radical transition towards a change in paradigm use.

The problem stated in this analysis finally configures the three roadblocks impeding an evolution towards a change of paradigm from users and the technology world: the prevalence of the screen archetype for visualization, the traditional usability of a conceived digital document (object), and the inability to relate the immersive notion of reality to a digital object or document.

MAIN FOCUS

The prospects to resolve this problematic conundrum lies on a very rich range of experimental acts and interdisciplinary exchanges between researchers, artists, philosophers, scientists, and technology innovators. However, it is important to refer to concrete events that augurate the advent of a possible revolution. Facebook recently changed its strategy to redefine its approach to the universal use of social media. The name changes of its main brand, *Meta*, then therefore carries an embedded concept in the brand itself, the *Metaverse*.

The concept was announced with plentitude of messages and articles during 2021; however, it responded to a broader definition that includes social media, e-commerce, and a cultural acknowledgement of an immersive integration of everyday communication. Certain commentators argue that the *Metaverse* in its very fundamental essence is already present, through video games such as *The World of Warcraft* and others, offering persistent and compelling virtual worlds running in virtual reality headsets. In these headsets, users can perform activities familiar from the real world to simulate a parallel reality. In strict terms, these experiences could apply to the gamification of any discipline interested in performing simulation or installing a hypothesis in a fictional world.

The quintessential condition of a virtual world is its immersive length, which introduces users in a fictional domain, where objects, scenarios, and avatars play different roles — even triggering actions to generate responses from users. The immersive paradigm and its spatial coordinates redefined in a virtual world are very powerful in terms of enveloping users, who are playing strapped in a VR headset. In a video game containing a virtual world, the user experiences the immersion thanks to the nature of the headset visualization, which is the stereo vision generated by cameras embedded in the headset as sensors. In this aspect there is a very smart approach in how immersion would look rendered in a headset device.

The virtual world and the generated objects are digitally stemming through 3D modeling and animation processes, and they can respond to physically simulated properties. In addition, though, in these virtual worlds Artificial Intelligence is systematically incorporating generative algorithms that can spawn instances and clones with independent and autonomous behaviors without the intervention of a user or designer. In the fascinating subject of a virtual world inside a video game, users endure a compulsory immersion that enables instantaneously a very decisive definition of what would happen on a major scale, like the Metaverse concept that attempts to champion the replacement of hyperspace.

The intended replacement of the hyperspace concept is not an overnight occurrence in the computing world. It is the consequence of the prevalence of emerging technological evolution and of how new systems incorporate better ways to interact with the computing dominion. Years of improving computer graphics development led to graphics displays on screens with astonishing quality and resolution to present a digitally generated object. The consequence of better processed graphics implies the maturation of the concept of a virtual world, in ways never seen before.

It is important not to forget that the literary and fictional notions of virtual worlds within an implicit universe have existed for more than 40 years. Since moviegoers went to theaters to experience a movie adapted from Stephen King's *The Lawnmower Man* in 1992 (Leonard, 1992, 1:48:00), people have had a clear picture of what Virtual Reality was about. Unfortunately, at the same time users and technology aficionados were excited and enthusiastic about the new world that computing was offering. The popular culture was then more informed about the implementation of a virtual world through powerful computer graphics, processors, and — more importantly — the way a person got access to that virtual world through a bulky headset.

But this essential ensemble of components was still within an incipient cognizance worldwide of what was the core tenet of Virtual Reality. Years of quasi-deception came after, once the computer world was not able to really bring what the movie was showcasing to market as a transforming and groundbreaking emerging technology. Doubtless, the massification of personal computers at the end of the '70s, and '80s helped to disseminate a very strong culture of operating systems interaction through graphical interfaces, and the scope of this activity experienced a global reach. On a parallel basis, the video game world started a great expansion in a very revolutionary way: making special devices available for users that were able to run just the video game program with proprietary operating systems.

This proprietary availability did not preclude the personal computer version of the same video game, however; at the time, the evolution of these programs was based on the quality of graphics and the power of the console built-in processor to run the dynamics of the game. An almost endless list of video game titles was very popular at the time, but their relevance in the assessment of what they contribute to cement what would be the paradigm of a virtual world really relied on what 3D graphics offered in terms of processing speed and framerate performances. It might be necessary to make a separate study of how video games were critical in the anticipation of what a compelling virtual world appearance should look like and what components were required to understand 3D graphics in terms of geometry and polygonal principles.

Basically, for a video game to run properly, these polygons and compelling geometry required a game engine written specially to configure the parameters of scenarios, models, illumination, and, more importantly, the physics simulation so required for the user's interaction with what was known in those years as game levels. Those game levels were in fact 3D scenes containing very low polygonal density in terms of model details.

It is not necessary to describe a whole historical sequence of how video games evolved through game engine capability. It is important to mention how the limitations of graphical processor units to initially manage a whole scenario with sophisticated 3D models would be a very daunting challenge for any attempt to create a realistic rendition of nature. Not only nature was required to be rendered in terms of realism, but also the characters and virtual personas involved in the video game story required some level of graphically accurate aspect.

Video games were experiencing transformations, and video game designers and artists were obtaining important attention from a very loyal universe of aficionados and gamers. In the beginning, the "storyline" was more important than the realistic rendering of any scenario and, more importantly, the capability of the graphic processor in managing the whole game with speed and efficiency in terms of framerate, which is the way a player runs the first person or third person camera seen through the screen or the game's window interface.

The prevalence of best performance instead of graphic realism was an expected property in a video game title. However, better graphical processors were releasing important updates and new versions of the video cards required by the game, so that in a generation, graphics were climbing in sophistication, and with a persistent unfolding the polygonal aspect of models and scenarios attained progressive realism in the graphics. In 1992, not a device could run the VR version of The *Lawnmower Man.* In fact, the movie's cinematic effects were produced by the emerging concept of Computer-Generated Imaging,

CGI, nowadays very ubiquitous once a theatrical release is offered. The film industry was in position of working with effects and hugely expensive hardware available at the time.

The Lawnmower Man movie exposed a wide range of video compositing techniques, together with very powerful computer processing potency capable of rendering very realistic 3D graphics, in a modality known as batch rendering, which technically was the development frame by frame of synthetic scenes containing richly texturized polygonal surfaces, modeled with heightened realism and illumination. So, the cinematic version of a virtual world was within the realm of a linear narrative, with props and diegetic machinery to furnish the virtual world paradigm.

The experiments using virtual reality headsets at the time were incipient but affected by multiple problems associated with motion-tracking sickness, because of the headset framerate latency, updating the frames rendered by the processor and the built-in camera for the first person, plus the inability to handle the uncomfortable bulky headset, and the short tolerance for an extended time interacting with the system. In 1985, Jaron Lanier and Thomas Zimmermann, creators of the term of Virtual Reality, supported their endeavor through the creation of VPL Research Inc. for the development of efficient devices like the Data Glove, the HMD (Head Mounted Device), and the Audio Sphere. The VR patents and concept were absorbed by Sun Microsystems to create software that could build and implement 3D worlds in a personal computer.

The time lapse between late 1980s and the mid 2010s exposed a gap of almost 30 years and revealed the computing industry determination to finally make a massive effort to ripen the technology, fixing the motion sickness originated by the camera tracking and the physical design of the head mounted device required for a more comfortable experience (Cummings. J, Bailenson, J., 2016).

It is worth noting during the past decades how the technology and computing world announced several breakthroughs in the form of permanent building blocks, paving the way for reaching a maturity in visualizing virtual worlds: the massification of computing power, high-speed Internet, plenty of resources to develop and create games for independent creators, and the more modular systems that permit a home-made VR experience. The rife criticism from popular technology magazines and the press in general all the previous decades had truth to it about the raw status of VR to render virtual worlds through a meaningful involvement, but these breakthroughs did improve outcomes due to ongoing experimentation.

Unquestionably and paradoxically the several or ephemeral technological breakthroughs laid concrete ground for improvement of HMD, interfacing, motion track sickness, and frame rate latency. Since the creation of Morton Helig's *Sensorama* in 1956, Ivan Sutherland *Ultimate Display* in 1965, *The Sword of Damocles in 1968,* Myron Krueger's interactive *VIDEOPLACE in 1975,* the *SEGA VR* in 1994, and the Palmer Luckey *Oculus in 2012,* there is a clear unfolding of an almost natural process as a testimony of progress for a subject that today deserve the denomination of a discipline.

There are concerns absolutely related to a superficial evaluation of a discipline or domain, as affected photography in the nineteenth century. It is the same polemic and unfounded blueprint that aggrieved technological feats in the past. Mostly the polemics and criticism are sourced from commentators instead of scholars. It is the superficial notion of something that literally triggers precipitated judgments. In this case, the precipitated judgments are that a clear responsibility of the computing world is evident, due to the urgency to compensate or provide quick assurance of the value of investment, and evident or naïve prospects about VR or virtual worlds have accompanied the breakthrough campaign promotions.

Figure 2. Myron Krueger. VIDEOPLACE
Rights and credits: Source: http://www.flong.com/texts/essays/essay_pose
Myron Krueger's VIDEOPLACE installation, 1974

One of the most persistent misconceptions is the assumption of reproduction of real experience virtually as if it would act as a substitution in the end. Certain commentators hasten dismissal statements, some of them with reason, about the inability of VR of replacing the real experience. Thomas Trescack in his article *'The main problem with virtual reality? It's almost as humdrum as real life'*, states: *Why would we prefer a humdrum virtual experience to a real one? No one needs a virtual Toyota. We need to give users good reasons to leave their reality behind and immerse themselves in a new one* (Trescack, 2019). Once again, the judgment is assertive, but it is just in the peripheral domain of criticism, and the concern replicates an uncanny aesthetic overtone, the mimicking of reality as a mistaken principle. A sort of déjà vu is surfacing every time intellectuals and the public in general judge the efforts of artists, scientists, or creators in grasping the reality in terms of implication of a given process.

Undeniably, the Virtual Reality concept implies a relationship with reality in the end as an impersonation rather than a new perceptual approach. In this dimension of the widespread and interdisciplinary work it is so important for any participant in content creation or technology development to define properly the core entity of it. At this point of the discussion, not only Virtual Reality alone as a breakthrough occupies attention for judgment, but also parallel experiences like Augmented Reality configure definitions and paradigms in the former the *replacement*, and for the latter the *overlapping* as agents manipulating the real world containing the presence of objects as virtual entities (Thibault, et al., 2019). These two notions helped to break the ill-flavored mindset of just mimicking superficially the real-world events, objectual and situational. Efforts have been made to resolve this sterile trend in public opinion. Technology has implemented a way out for this misconception addressing the same conundrum of agents conversing with reality, the expansion or extension of it, under the denomination of XR (Extended Reality).

Extended Reality is the conjunction of three paradigms combined in a unique entity of immersive nature, the result of overlaying digital objects on top of real ones, the immersive scenario that could respond to the classic boundary of a HMD, and the holographic presence of interactive models blended with classic gaming controllers. In this powerful juxtaposition, the uneasiness of frustrated surmise surrounding the VR topic as a mere replacement or mimicking of the real world is re-rendered by the expansion of the immersive perception beyond any current or future visualization technology. It is true that the media criticizes scornfully users wearing bulky headsets and with disdain also emphasizes the little interest or disillusion of audiences as opposed to the initial awe experienced for the first time using a relatively light VR modular system.

It is inaccurate to predict the decline of a technology in this case, there being available strong and evident evolution, just for the mere fact that content developers have not made a killer app yet to captivate the entertainment industry audiences. Extended Reality certainly owns several modalities of immersion, that a captive audience would not be exclusively the main target for obtaining metrics of commercial gratification or revenues (Marr, 2020).

Extended Reality could cover diverse endeavors of visualization, like the classic manipulation of planar documents to the tactile and haptic real time simulated and expanded version of the 3D render of the same 2D document. In this dimension the 3D document will be animated in real time, and what real time implies is the graphic acceleration of the 3D content with frame rate efficiency and perhaps extreme realism (Thibault, et al, 2019). However, realism is not the precondition for a successful immersive interaction. In this domain, the successful immersive interaction, the realism is just a visual component, but not the main paradigm of effective rendition. Different immersive interaction rendition styles could be implemented and constructed.

Digital artists and animators can bestow enthralling styles and novel graphical concepts, necessary to respond to a wide and well-grounded audience's reach. The Stephen King novel, *The Lawnmower Man* (King, 1975), or Neal Stephenson's book *Snow Crash* (Stephenson,1992), initially fascinating audiences in the early Nineties through amazing batch render imaging, materializes today through the power of real time render and a profusion of high poly count geometry (Leonard, 1992, 1:48:00). High poly count geometry is the nucleus for a persuasive implementation of better haptics, or simulated kinesthetics, which it configures as covenants of a phenomenological experience where the body will start blending with the immersive XR environment (Suh, Prophet, 2018). 3D objects will produce a responsive dialog with the users, like the pioneering Myron Krueger's visionary *VIDEOPLACE* (Krueger, 1974), or Jeffrey Shaw's *Legible City* (Shaw, 1989). These two artists lay the foundations of what is today the reality of XR. Certainly, the disdain to VR technology as a declining one is just the consequence of a clearly myopic account.

Extended Reality will require the vivid experience of the "space," regarding its core component the "environment" (Wehbe, et al, 2016). When an environment is set, the place will adopt familiar settings, where users or audiences will be exposed to collective impressions. Each participant in this environment interaction will obtain individual insights about how the experience affected their senses (Perkis, A., Timmerer, C., et al., 2020). When Second Life was introduced in the early 2000s, it promised to be a game changer in every field whether academic, educational or a in more general scope. However, the successful introduction ended up years later in a decline, or lack of interest.

Figure 3. Jeffrey Shaw. Legible City installation. Nagoya Japan
Credit source: ADA, Archive of Digital Art

A whole study could be dedicated to understanding the reasons of this curious process of hype/decline of an emerging technology. However, a very evident nature of an emerging technology turned down is the anxious strategy by the technology developers to speed up its widespread adoption. This effort basically aims to change user's behaviors overnight, and through this process essential components like "usability" and "effective interfacing" miss out on exhaustive testing.

When an immersive environment adopts a "familiar" surrounding, it impacts the user's behavior in a well-defined usability skills gradation (Rubio-Tamayo, J., et al., 2017). Usability skills pertain to this "familiar" environment, and those will convey the quick and instant interaction expected. A very interesting and compelling example of a possible familiar "environment" is an art installation. This setting defines an environment and invite users to intervene or question the piece's components, if there is a visible number of interactive objects included in it.

SOLUTIONS AND RECOMMENDATIONS

The art installation example is a clear and circumscribed situation inside a gallery or a museum. However, if this model system is transferred to a public space, it wouldn't have to attract only museum goers. Every passerby could be invited to trespass the boundaries of the installation, to discover its situational content. Many artists today work together with programmers, digital artists, and designers to collaborate in a common endeavor to turn the art installation interactive.

If the *Metaverse* future incarnates in just being wired or connected to a private Head Mounted Device or to an ephemeral killer app, there will be few opportunities to exert a cultural transformation in general audiences. During the Covid 19 pandemic, hyperspace responded quite efficiently to the emergency, allowing almost the entire world's population to perform distance communication, education, household chores, and job duties. However, the Internet will continue providing its mission in the ways users expect in a post pandemic world, since its massification in the mid-Nineties.

Internet has become a ubiquitous technology and rapidly evolved into a better universe of applications thanks to high-speed bandwidth. But, essentially, if the paradigm of the *Metaverse* will rely only on the classical framework of screen interaction for the opportunity to shed a widespread insight of what "immersion" means, the decline of a very powerful and amazing emerging technology will be affected by the Second Life online world fate.

Universal "culture" will have a very important forum to expose the ways global communication is expected in the *Metaverse*. The computing world must start a very productive dialog with universal culture to introduce the XR emerging technology as the core component of immersion, perhaps designing better and light HMD. The *Microsoft HoloLens* device is a clear exponent of an efficient device to introduce immersive interactivity, owning hand gestures, voice command, and holographic renditions of 3D objects to be manipulated within a mapped "space," determined by its sensor's cameras. However, still the field of view is determined by a window view, which is not allowing the full immersion in as space. Seen by the glass modality, or more exactly 6 DOF (6 Degrees of Freedom), which is the headset immersive simulation, the window view is still circumscribed inside the headset.

Unquestionably, the full immersion is on the works, for rapid development and the concept of wearing a light glasses HMD is the next step to furnish full immersion. Microsoft has developed voice commands to transform and modify objects inside the "mapped environment." The digital objects mixing concrete surfaces seen through the HoloLens can be detected as physical presences. In the same way, our bodies are interacting or colliding with the same mapped surfaces the HoloLens has perceived and set.

Holograms open a wide range of choices. In many fields the digital visualization is a real necessity for the understanding of systems, machinery, or abstract notions contained in the emerging technologies language (Ravenscraft, 2021). The holographic rendition in the Microsoft HoloLens system is still based on the HMD interfacing immersion; however, it is the closest advent to what the body interfacing would require interacting with digital conceived objects.

The computing world is engaged in several technological endeavors to bring out the 3D holographic presence, so required to avoid the glasses interfacing. The realm of visualization also demands more HMD, or glasses free experiences, to visualize the presence of 3D content. An example is the concept of Cue Holography (Carter, 2017), based on how eye perception will capture the presence of the hologram in relation with the surface rendering the object. There is an interesting notion of screen interfacing to obtain or attain the presence of holograms (Zhaoui, L, et al., 2014).

The company *Looking Glass (*https://lookingglassfactory.com/4k) has introduced the holographic 3D presence through their holographic screens to render interactive 3D objects in real time. The interfacing is in the screen, but one component of the whole experience is missing, the immersion. Looking Glass is basically a display that renders cues on a bulky screen to produce the desired stereographic construction of the 3D object. The novelty in this system is the real redemption of the screen paradigm, but in a different dimension.

Looking Glass expand the presence phenomena beyond the screen, and the users interacting with the objects involve the body in this action. Hand gestures grasp the 3D object in the screen. The users stand up in front of the screen bulky surface, then, by executing an almost choreographic performance using body action, the simulation happens as the user see how the object is modified in the screen.

Figure 4. XR simulation of the Wedding Chamber. Artech Conference 2021
Source: Interactive hologram through Microsoft HoloLens.
Wedding Chamber Project. Montenegro – Ushenko, 2021

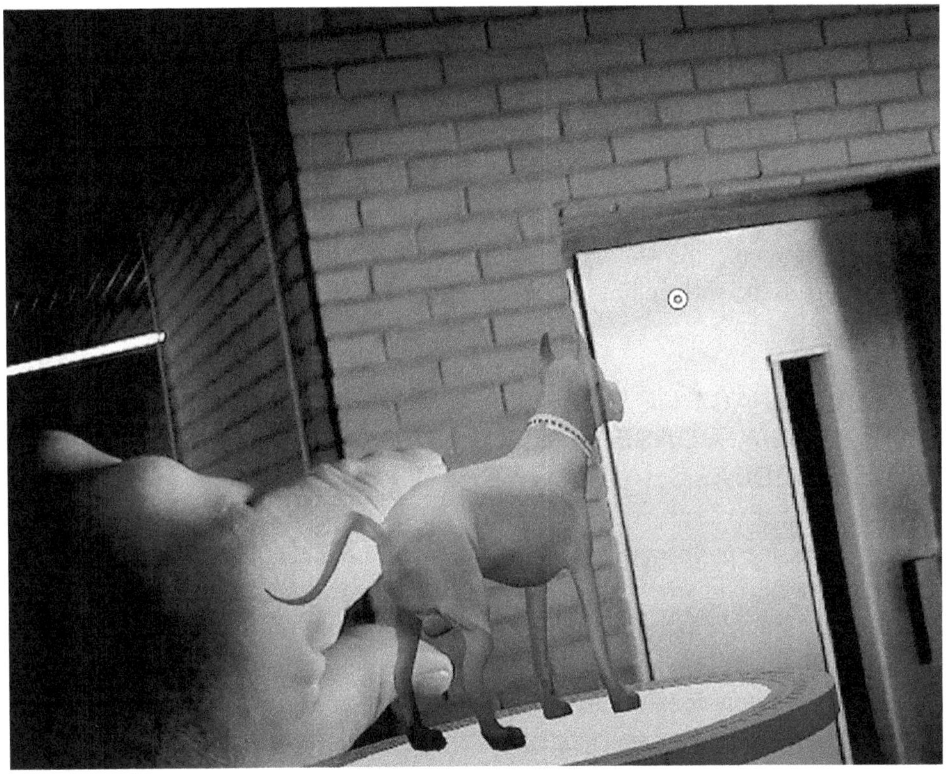

Figure 5. Looking Glass holographic screen
Image credit: Looking Glass Corporation. Looking Glass 4K holographic screen device

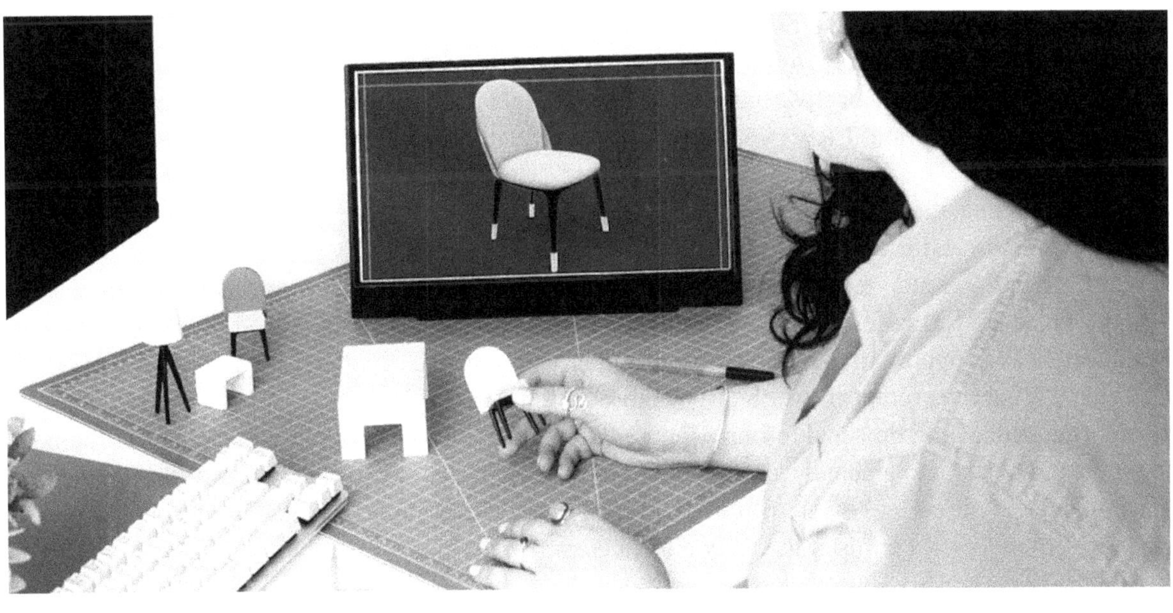

This time, there is no tactile action or physical contact with screens; the screen is just the placeholder of the 3D holographic object. Hands gestures, body stances, and the effort to telecommand the 3D holographic presence of the screen content makes this experience phenomenologically unique. The *Metaverse* will integrate a variety of ways of visualizing immersive experiences and not only determined by the control of autocratic technological companies. The universe of XR reality is determined by the way society will intuitively perceive the next future of what something we are calling the *Metaverse* will represent.

The necessary blending of what is real and what is digital relies upon the *functionality* paradigm. Society wants this progression for a real meaning and purpose, and beyond just the mere entertaining realm. All indications are that, due to the profusion of stable and incipient companies' gadgets, presenting or pitching their developments, the *Metaverse* experience will be diverse and conveyed by a wide range of devices, glasses, headsets, surface holographic screens, and holographic projections.

THE STRATEGIC IMPACT OF EMERGING TECHNOLOGIES ON URBAN TOURISM, A CASE STUDY BASED ON ANDREA MANTEGNA'S WEDDING CHAMBER FRESCO SIMULATION

The novelty and impressive performative features of emerging technologies observed in the realm of visualization are overwhelming, likewise its attainment process is at initial stages. However, users can utilize it soon as visualization, in a practical and functional applicability, therefore is a promising prospect. Several fields and disciplines of societal developments are pursuing the implementation of "immersion" as a central paradigm not only for the sake of experiencing remote or virtual simulations, but for the meaningful "phenomenological" acknowledgement of a given subject.

Museums and galleries are in the top of the list of institutions that would obtain direct benefit from "immersive" virtual experiences, however, standards procedures, scholarly methodologies, and technological normalization are still dangling around undefined, but clearly visible in terms of usability (due to an incipient cultural widespread), waiting for structural systematization. Universities and higher education institutions are also in a normalization process to integrate "virtual" distance education, mostly extort by the sudden Covid-19 disruption. Educators rushed in a frantic race to shift the "presential" modality to the "distance" one overnight. However, the time allocated to prepare and develop content was contingent to a couple of components mostly needed to perform the telematic delivery of content, the web camera, and the internet connection. These two components immediately reminded us that the "virtual meetings" were just emergency means to barely enact the face-to-face interaction. Several "phenomena" emerged during those virtual meetings, "a pseudo presence," a "debilitated engagement," and a "customary way" to comply scantily with content attainment.

None of these events really performed a full immersion or body interaction, however, they were acknowledged as "virtual meetings." Despite the contingency Covid-19 unleashed, it was not grasped by educators as an opportunity to implement meaningful immersive experiences (Perkis, A., Timmerer, C., et al., 2020). The prompt answer bides by the lack of available platforms to provide immediate support for the demanding time and resources required by the development of an immersive environment for content delivery. Undoubtedly, an immersive environment that involves virtual worlds, augmented 3D content, and mixed reality interactions, requires years of study to decisively obtain comprehensive responses from users in terms of usability effectiveness.

Urban tourism as an arising discipline provides the precise opportunity to thrive "meaningful immersive" experiences to understand and implement the phenomenological nature of emerging technologies in virtual visualization. In this case, it is not the conventional content development of the discipline the one is in action, it is the "presence" reenactment of the multiple experiences provided by the "physical place" that urban tourism would showcase. Covid-19 in the end was a catalyst for important insights related with "usability" in terms of virtualization, however, it is important to mention that emerging technologies for immersive environments were having their own challenges in terms of reliable device's usability before the Covid-19 irruption. Those challenges were related with specific and precise usability performances, the use of controllers to navigate in the immersive space, and the required presence by using hands and body as agents of interaction. At this point, a handful of online sites (Acadecraft.com, eonreality.com, immersed.com among others) offer immersive experiences, but it is compulsory the availability of hardware and systems to enable the necessary presence that full immersion depends upon. In this matter, a physical location, and the internet availability conjugate the paradigm for the befitting meaningful experience demanded by full immersion to consistently simulate another distant place without being there.

Google 3D offers an interesting way to navigate an explore determined locations that users desire to visit, before initiate the trip to a touristic destination. The online application does not need special devices, or headsets to perform credible navigations, and it is a very helpful resource to understand mostly the way travelers would find landmarks and points of orientation in a particular geographic site or urban venue. Strictly Google 3D concedes users to walk through, but scarcely be in contact with a particular simulation or a more complex presence experience.

The Steam portal published in 2018 the virtual reality reconstruction of the palace of Versailles, which is the simulation and the immersive experience of one of the most popular touristic destinations in France. It is compelling to perform this experience with required equipment components, headset, and a gaming computer to properly run the necessary frame rate demanded in terms of 6 degree of freedom, to efficiently teleport or navigate within this virtual world.

The fascinating aspect of the app inhabit in the photogrammetry technology used to map and render the immersive space of the palace in 3D real time. Also, the navigation is guided by an interactive map attached to the users' controllers. The controllers are two devices required to navigate inside the virtual world, performing the "teleporting" action, which is the way navigations occur in terms of "strides" to freely experience the presence inside the virtual architecture. Also, the controllers contain buttons that enable the grabbing and triggering actions, to effectively interact with 3D objects, references embedded in the walls, and each piece of artistic interest immured in the palace rooms.

This app reveals the promising future of how urban tourism will spell the virtual presence of visitors from a distant place to the target destination. However, the Palace of Versailles app, still is based on the notion of a traditional tethering device-computer and is in the format of a stand-alone program that can be downloaded from the Steam or Oculus stores. Perhaps, this app presents an innovative manner of making a virtual visiting, or experience attainable, to recreate the urban exploration from a touristic standpoint.

As described before in this chapter, the art installation will become the concrete paragon that will enact the remote presence of a real environment as a simulation of a touristic case. This case study is based on an Extended Reality project started at the Art and Design Department, at Purdue University in 2015. The main researchers and creators of this project are Associate Professor Andres Montenegro, and Professor Audrey Ushenko. "The Pictorial and Virtual Simulation of Andrea Mantegna's fresco The Wedding Chamber" is a Mixed Reality installation artwork that recently has been presented in the

Artech Conference on October 2021 in Aveiro Portugal. The installation is based on VR, AR and MR modalities of immersion and interactivity.

Within the immersive experience, the rendered imagery narrative, based on the original frescos, are structured in each of the walls of the virtual immersive room. The 2D and 3D animations embedded in each wall panel offer a vivid depiction of the Gonzaga's family worldly surroundings and emblematic symbolic narrative. The virtual immersive environment implements the manipulation of 2D and 3D animated objects that trigger from the room's walls stories, creating an interactive intervention of the original fresco narrative to render the immersive content as an emotional and cultural experience (Beck, D., et al., 2020).

Figure 6. The Wedding Chamber, AKA. Camera Picta
Image credit: Andres Montenegro

The art installation's fundamental endeavor is to reenact Andrea Mantegna's influence over the Renaissance's naturalistic rendition of the space. Mantegna's works continued to inspire in the next centuries different schools of painting in Europe. His use of the concept of "Trompe L'oeil," inaugurates a proto-technological system that expands the architectural space of the fresco's technique to illusionistic dimensions. Recent studies have discovered that some of his frescos depict a more complex representation inspired by writers from the classical world such as Lucian of Samosata among others, seeking to plunge readers and viewers in a more literary narrative and visual perceptions commonly known as Ekphrasis. Frescoes masterpieces followed a high-reaching desire to play with illusion as a device, to expand their pictorial narrative beyond their formal physical boundaries. That is the case expressed in the series of frescoes located at The Palazzo Ducale di Mantova. Italy.

Figure7. Palazzo Duccale, Mantua
Image credit: Andres Montenegro

The location of this universal masterpiece executed between 1465 and 1474, constitutes a cultural touristic destination, and its importance atop the close-by surrounding museums in terms of artistic feat regarding the urban and historical layout. Once the visitors arrive to the monument, the notion of "awe" becomes patent due to the immense number of artistic masterpieces around, and it becomes almost impossible to remain focused on the specific spot represented by the "Camera Picta," or "Wedding Chamber" fresco. However, there is a sequential route of visitation that is needed, and the "stay" within the room, demands from the viewers a compulsory concentration to appreciate the art.

Mantua or Mantova, is one the most visited cities in the Lombardy region, and its attractive is based on many art treasures located in distinct points of the city. The physical location of the "Wedding Chamber" is on the second floor of the Saint Giorgio Castle, which also communicates with the *Palazzo Ducale*.

The chamber's dimensions are modest in comparison with other rooms in the same castle and in the *Palazzo Ducale,* those containing works of arts, frescoes, architectural features, and illusionistic designs, rendering volumetric effects on their ceilings as artistic devices with a central point of visual interest which is the oculus.

Andrea Mantegna's masterpiece inaugurates the use of the "Oculus" as a compositional device, that sets an illusionistic circular window on the ceiling. The oculus reveals an opening that incorporates the ceiling section of the room as part of the whole fresco narrative. The phenomena resulting from this design is the "di sotto in sù" (seen from below), which represents in a painted vaulted ceiling an open sky, depicting characters glancing down. Andrea Mantegna playfully organizes this effect and suggests a perception that goes beyond the mere effect of illusion, articulating complex and opposed vantage points. The visiting time is limited to 5 minutes per each group of visitors, and clearly this time constraint, curtails the viewer's ability to appreciate with detention the fresco's technique, the physical space, and the illusionistic components depicted in the vaulted ceiling. For this reason, the life-size experience of being there is ephemeral, perhaps, visitors can return anytime, there will be incremental additions to the original appreciation.

Figure 8: Saint Giorgio Castle, Mantua
Image credit: Andres Montenegro

The *Wedding Chamber* has suffered damage overtime, although has had successful restorations to protect it from decay and deterioration, reestablishing its magnificent compositions and vivid colors. However, what makes exciting this life-size immersion, or the real and concrete presence in the room, is the details that conform not only the fresco's composition, but the architectural structure of the room that Andrea Mantegna integrates and blends with illusionistic painted mastery. Public opinion, unlike specialist or scholars, might argue, that none, or any virtual, or remote reenactment of these kind of masterpieces would replace the original experience of being there.

Then the primary goal of the project to address these concerns, required that the researchers and artists conducting the simulation implementation of the Wedding Chamber, needed to visit the real location, not only to collect concrete technical details, historical references, and institutional support, but for how the "presence" condition would be simulated through a remote way. This endeavor differs from the mere fact of creating a library resource, or a catalog description for consultation, it rather is the implementation of a vivid cognizance of the original life-size of the work of art.

This project began before Covid-19 breakout, and certain principles were defined beforehand respect the way users can share or interact with the art installation environment. The original implementation of the Wedding Chamber's remote simulation through extended reality was that the space installation that could be experienced by just two users simultaneously to allow an experimental social presence (Blaskovitch, J., et al., 2009). This meant that two headsets, and a couple of pairs of controllers could be set. Fortunately, the original requirements of use were not against sanitary guidelines dictated at the time.

The whole creative/research endeavor was developed utilizing VRTK (Virtual Reality Tool Kit), MRTK (Mixed Reality Tool Kit), and implemented in Unity 3D. The hardware included 2 Alienware stations equipped with Nvidia 2070 video cards, desktop computer Core i7 with Nvidia 1060 video card. The VR system implemented was HTC VIVE Pro, HTC VIVE Plus, and HoloLens 1.

The project was first presented in Spring 2020 and Fall 2022, experiencing two different states of mind, the one that was not concerned about wearing a headset (pre-pandemic) and the one that needed to oversee use frequency of headsets and controllers by sanitizing those after each session of use. Another important component tested in those two times, was the willingness of users to engage in such experience, and how this sometimes was a real challenge for anybody that was wholly unfamiliar with Extended reality gadgets.

Observation of these variables shed light on certain insights about helping to turn usability and accessibility familiar and expedite for the immersion and the navigation functionalities of the virtual world.

The main goal was to create and implement the desired “presence” as the main driver to engage the users through an intuitive use of controllers to teleport or navigate in the room (McCreery, 2013). The room size is square shape about 8.1 by 8.1 mt (26.5 by 26.5 feet), and it can be turned in a walkthrough. Unlike 360 degrees reconstructions based on stitched images or photogrammetric scans, the virtual reconstructed room needed to follow a real scale in the XR immersion. When the physical simulation of the fresco’s room occurred, following the original measures, the “presence” reenactment is set.

Figure 9. Installation physical environment
Image credit: Andres Montenegro.
Wedding Chamber Project. Montenegro- Ushenko, 2021

However, after this initial stage, comes the navigation to experience the simulated space, and in consequence, the users will elucidate the space notions by teleporting or getting closer to the walls to appreciate the fresco's superior technique. For that reason, a virtual world that commits to simulate a specific subject, must rely on the quality of computer models, but moreover in the images that expose the fresco's narrative.

Figure 10. Immersive chamber. Snapshot from the headset view camera
Image credit: Andres Montenegro
Wedding Chamber Project. Montenegro- Ushenko, 2021

Any imperfection will spoil or ruin the "presence" momentum. The very expected outcomes, the presence and interaction with objects within the virtual room, must enable the user's engagement to acknowledge a "meaningful" live through. The expected outcomes, the virtual simulation, and the presence, must be able to dispel the common assumption that the original life size of the art piece would not be replaced or poorly mimicked (McCreery, 2013). The compromise of virtual simulations must go beyond the mere classification of "virtual visit", due to the generic or inaccurate classification of this eventual term.

In a generic definition, virtual tourism could include brochures, maps, apps, locations, Goggle maps 3D, 4K videos and vivid pictures of the target destination. This documentation's cluster certainly helps to collect and process data about the desired place of visitation, but in the end does not implement the vivid reenactment required for a remote immersive experience. An essential goal is to basically rely on virtual simulations to implement the closest life-size recreation, fundamentally when the target touristic place suffers forms of contingent disruptions such as wars, earthquakes, pandemics, political and institutional instability, or maintenance negligence.

In this immersive environment, the visitors, viewers, or users will expand the preconceived notion they have about the target site, by getting an interactive involvement with the historical subjacent content (Cummings, J., Bailenson, J., 2016). Each viewer can explore or walk inside the installation with no overlapping collisions or conflicts with other objects located in the physical place of the playing area. The "playing area" is a fundamental component of an immersive space. Essentially the playing area, is the physical boundaries of a determined immersive space defined by the XR or virtual reality system.

Figure 11. Interactive deconstruction of the immersive content area
Image credit: Andres Montenegro
Wedding Chamber Project. Montenegro- Ushenko, 2021

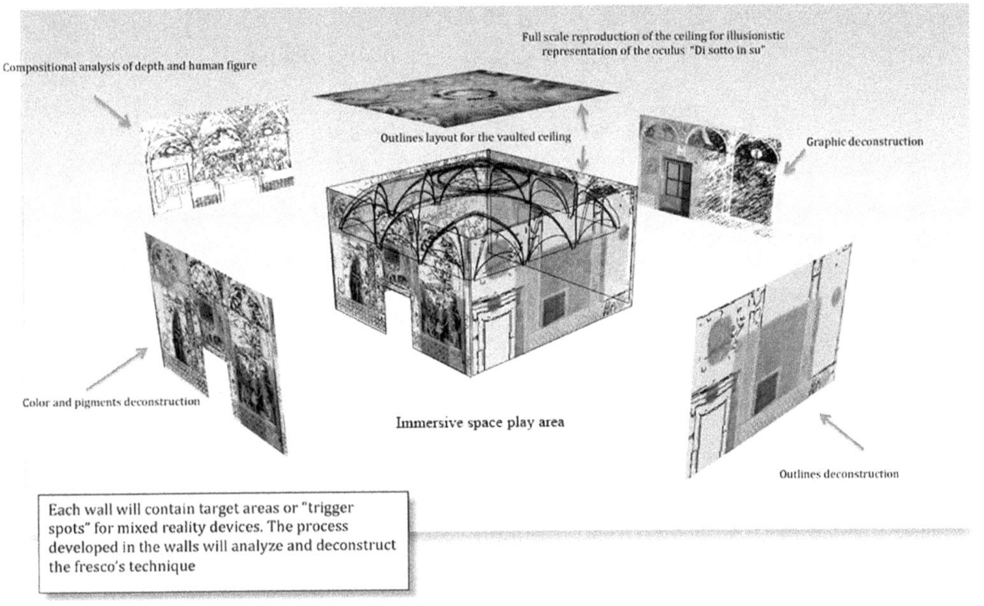

There are two conditions expected to engage in an immersive experience, the stationary viewer location, and the room's playing area. In the stationary modality, the users just perform the immersive experience in mostly the way a gamer play in front of a screen. This traditional usability does not differ from what is already well recognized as a user's computing behavior, seated in a workstation, and interfacing the content with keyboard, mouse, or a gaming controller, i.e., joystick, pedal, steering wheel, or a proprietary gaming controller. The user just performs stationary if not limited movements, all the gaming dexterity relies on what the interfacing demands in terms of how the controller's device manipulation takes place (Wolf, et al., 2021). There is not a physical area of engagement.

In a room playing area, the users can move in a determined perimeter that normally is set around 6 by 6 mt (19.6 by 19.6 feet). This physical boundary also allows stationary modality, but the novel "phenomena" is the enactment of what is being sensed inside the headset motion tracked camera (Turner, et al., 2016) and in the physical boundary where the play area takes place. This modality is fully met in the virtual simulation of the "Wedding Chamber" (8.1 by 8.1 mt), the boundaries are set using the life-size walls of the fresco as well, then the users can walk toward the walls and appreciate the surface simulation of the fresco implemented with high resolution imaging, normal maps subsurface, and lightmapping shading.

Figure 12. Simulated virtual space
Image credit: Andres Montenegro.
Wedding Chamber Project. Montenegro- Ushenko, 2021

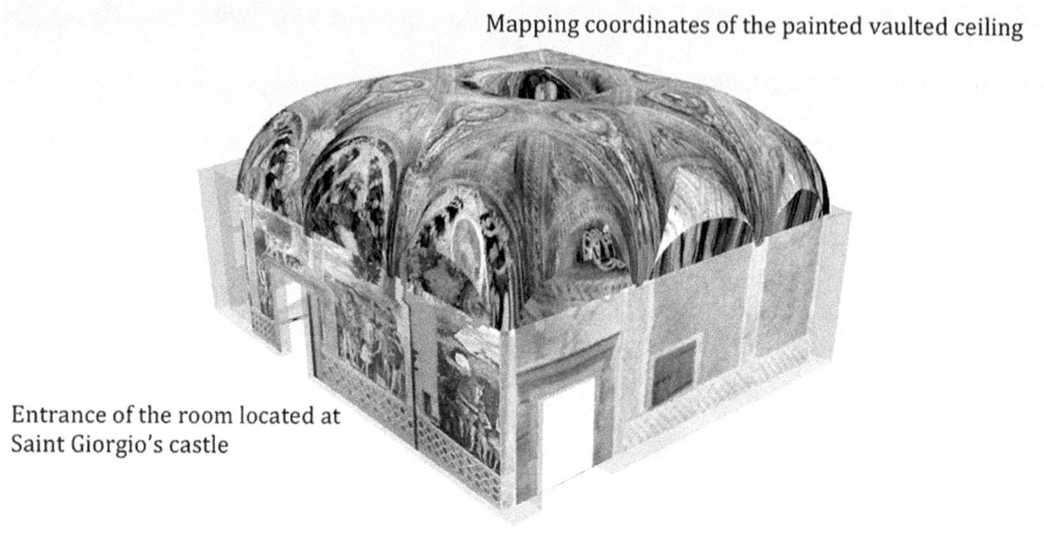

The walk-in is fundamental to appreciate and experiment the spatial "presence" feature (Hruby, F., et al., 2018), undoubtedly the more accurate way to feel the fresco's room space. The "imaging" component works as a critical asset in an immersive environment due to the way the viewer's virtual mobility, approach the immersive environment objects by visually perceive them (Li, Ch, et al., 2021).

If the texturing of the walls rendering the fresco is not set in a clean resolution, the proximity will spoil the sense of presence, revealing pixelation, jpg artifacts, or a common flaw in gaming real time graphics, the aliasing, i.e., in computer graphics, the jagged saw-toothed lines originated by a low-resolution rendering. Therefore, when the immersive action takes place, the "walking" is something the users trigger intuitively, acknowledging a walkable virtual world, to scrutinize every graphical component contained on it.

Extended Reality grants the answer for the criticism of "total detachment" that eventually users would suffer by full immersion. The walkable environment is protected by the virtual boundaries, the system sets whenever the users leave the immersive space, then the headset camera sensors activate immediately the physical surroundings visible through the headset. This enables the mixed rendering of real and 3D content as well.

The immersive environment is a system that not only deploys the computer graphics components as virtual 3D objects, but also renders a cluster of media that can be reorganized in the same fashion 3D objects can be sensed. Animations in 2D or 3D can be deployed with vivid quality, videos in 4k resolution, audio with full integration in the space, and interactivity based on the media content are able to transform the "presence" experience into a cinematic one (Mateer, J, 2017). The virtual simulation of the *Wedding Chamber* of Andrea Mantegna currently implements most of these components, to recreate the meaningful distant experience of "being there" (Coelho, et al., 2006).

Figure 13. Fresco's wall simulation view from the headset
Image credit, Andres Montenegro

The important value of the XR piece "Pictorial and Virtual Simulation of Andrea Mantegna's Fresco the Wedding Chamber Trough a Mixed Reality Installation," is the constant update of technological resources that can be implemented for more interactions within the physical and virtual space of the installation in terms of future research directions associated with "phenomena" derived from current and future emerging technologies. Moreover, its esteemed versatility serves to multiple disciplines and fields of studies to exemplify the life-size portrayal of distant "places" in terms of a historical, artistic, scholarly, and at length as an Urban Touristic endeavor.

FUTURE RESEARCH DIRECTIONS

There will be a very intersecting crossroad in the *Metaverse's* huge content offering, and at this moment, a single platform would not be able to monopolizes the paradigm embodiment. The interesting consequence of this technological diversity around emerging technologies' propositions is the constant preoccupation for improving the presence of 3D digital content in terms of cinematic experience (Mateer, J, 2017). Another important component is the 3D video rendering not being exclusively 360-degree projections or stereograms to be seen with low-tech glasses. This modality renders the video through a real time scanning of the scene, and the immediate outcome is constructed by cloud points materializing the objects with depth and 3D surfaces. Clearly, the scanning is the real time video modeling of the physical environment, and viewers or users can walk inside and feel the presence of objects as well.

The complex and diverse scenario of technological approaches to expand the reality under different modalities could appear overwhelming to say the least. However, the core component remains; the craving for an intuitive and instant rapport with unfamiliar "objects" to turn them into familiar entities. Fortunately, to understand this complex, and sometimes confusedly overwhelming, information ocean, the necessity of having commentators to pin-point the latest development is compulsory. These commentators, down to a vernacular judgment on the scope of a technological advance and even from the superficial standpoint of their apparently perfunctory assessments, contribute to an emerging technology better settling into society. The *immersion* in extended reality, through a diverse hub of channels' interactions, will embody the ingress of audiences such as designers, artists, and engineers (Du, J., et al, 2017), into a world of knowledge of multiple layers of information and perceptions.

The expectation of a quick functionality of Extended Reality could appear quite compulsory; however, content creators must have a clear comprehension about how the content must display meta information. The assured "presence" of users and objects within the immersive environment is not only the required equation for a meaningful experience, but also the multiple levels of the reality scanned, which must be displayed and exposed through the intuitive and phenomenological bodily approach of viewers and users (Van Roof, M, et al., 2016).

An effective or a mechanical expectation of what Extended Reality could contribute with, within the *Metaverse,* could lead to rapid judgments thanks to indetermined or misguided experiences. In this domain, the *effectiveness* of the experience will depend exclusively in the way cogent "usability" would be implemented by content creators.

It is worth noting that the interfacing interaction within an immersive environment is still tied to classic controllers' manipulation: joysticks, keyboards, pedals, steering wheels, or tactile devices commands. However, a progressive detachment from devices' inputs will be a fact that will include the body (Van Roof, M, et al., 2016).

Perhaps, audiences' reluctance in the adoption of new interfacing modalities will slow down the widespread use of human computing interaction through the body. Despite this reluctance, the fascinating prospect of working out the 3D content using *haptics* to produce the elaboration and manipulation of multiple levels of expanded reality through digital content is a condition basically for a paradigm change (Coelho, C, et al., 2006). Otherwise, the attachment to traditional interfacing will be preeminent.

CONCLUSION

For users in the world of education, science, art, communication, and disciplines in general, the rapport with Extended Reality as a method to produce knowledge is fundamental. However, the *Wedding Chamber's* simulation results study, containing meaningful evidence of widespread adoption, would be helpful for other domains, practices, or disciplines. For social sciences or humanities, communicating or delivering content based on evident or immediate responsive users' behaviors, thanks to encapsulated knowledge or information in 3D content, would turn the immersive experience in something that will increase the integration of our body as an interactive agent.

The discussion and understanding of the technological aspects, trends, achievements, testing and subtlety of emerging technologies is unavoidable, due to its impact on society. Mobile technology has had significant repercussions in society in general. In hindsight, mobile communication, modifies substantially the cultural landscape in society, and some of the modifications were not anticipated as facts before their outbreak.

Mobile communication has taken over all aspects of our daily agenda, from the administration of the user's pass codes, to the tracking of the simultaneous device's users manage. In some ways, it has been invasive, and, with the justification of security, all the ubiquitous technological agents work behind the scenes to control the complex traffic of information. From airfare booking, domestic chores, professional tasks, and leisure time reservations, as general examples, users are instantly informed about choices, therefore the association with immersive experiences will turn inevitable.

Cultural changes are determined by the way society interfaces with technology. The outcome of an emerging technology is its widespread use, and even more important the way their complex inner systems are masked with friendly interfacing. The origins of the WYSIWYG (What You See is What You Get) interface were based on the concept of the PC massification, and this powerful acronym represented the perennial principle of what we expect today from every emerging technology.

Visual structure is the core component of rapid, efficient, and usable systems, and the coherent match of icon and information, is the quick universal identification of its omnipresence. For that reason, technology occupies a global norm to basically standardize and dictate usability protocols.

Usability protocols within an immersive environment elaborated through Extended Reality (the conjunction of Augmented, Virtual, and Mixed Reality) are being implemented, tested, or studied at this point, in a very experimental manner. However, there will be a more complex scenario of findings once the experimental conventions of Extended Reality enter the realm of global cultural adoption. Therefore, the anticipated prediction of any emerging technological breakthrough under the assumption of ephemeral or episodic user's perceptive fluctuations is nevertheless dubious.

All indicates that Extended Reality once is universally recognized as the agent that generates the immersive environment paradigm, it will cover important and prominent colloquies about its advent beyond its germinal wonderment. Then it will come the natural familiarity of its interfacing, considering the users' body as a physical countenance. This materialization will cast presence, within a cultural understanding of what the "presence" means in terms of a phenomenological definition. The phenomenological responses generated from our individual and social bodies (Coelho, C., et al., 2006) will reverberate in benefits for a global and universal language associated to the dynamic relationship within an Extended Reality immersive environment, and the individual's presence in terms of digitally constructed 3D entities.

In time of crisis or facing the current Covid-19 events and consequences, an Extended Reality experience will offer several "phenomenological" approaches to information or knowledge fundamentally functional for several disciplines, or fields of studies such as Urban tourism, simulating sensing experiences, for the required vivid reproduction of distant places on a life-size basis with a meaningful handling of the digital 3D content.

Digital objects within an immersive environment will work as vehicles of expanded apperception surveying the physical and visible world, generating new levels of interpretation of reality and novel manners to produce responses in conjunction with the body not only to mimic or simulate physical routines but to perform unique methods of "human consciousness" or "self-awareness" to grasp better our cultural systems and singular world.

REFERENCES

Beck, D. (2020, August). Finding the Gaps about Uses of Immersive Learning Environments: A Survey of Surveys. *Journal of Universal Computer Science*, *26*(8), 1056-1057.

Blascovich. (2002, November). Immersive Virtual Environment Technology as a Methodological Tool for Social Psychology. *Psychological Inquiry*, *13*(2), 110-111.

Carter, J. (2016, September 17) Holography: VR Without Glasses. *E-Magazine*. https://emag.directindustry.com/holography-vr-without-glasses/

Coelho, C. (2006). *Media Presence and Inner Presence: The Sense of Presence in Virtual Reality Technologies. In From Communication to Presence: Cognition, Emotions and Culture towards the Ultimate Communicative Experience. Festschrift in honor of Luigi Anolli*. IOS Press.

Cummings, J., & Bailenson, J. (2015, May). How Immersive Is Enough? A Meta-Analysis of the Effect of Immersive Technology on User Presence. *Media Psychology*, *19*(2), 272–309. doi:10.1080/15213269.2015.1015740

Du, J. (2017, October). Zero latency: Real-time synchronization of BIM data in virtual reality for collaborative decision-making. *Automation in Construction, 85*, 59-60. doi:10.1016/j.autcon.2017.10.009

Hruby, H. (2019, July) Geovisualization with immersive virtual environments in theory and practice. *International Journal of Digital Earth, 12*(2), 123-136.

King, S. (1975). The Lawnmower Man. *Cavalier.*

Krueger, M. (1977). Responsive environments. In *Proceedings of the June 13-16, 1977, national computer conference. AFIPS '77*. Dallas, TX: ACM. 10.1145/1499402.1499476

Leonard, B. (Director). (1992). *The Lawnmower Man* [Movie]. Allied Vision.

Li, C. (2021, December). Synthesizing Scene-Aware Virtual Reality Teleport Graphs. *ACM Trans. Graph*., *40*(6).

Marr, B. (2020, December). The Future of Virtual Reality. *Forbes*. https://www.forbes.com/sites/bernardmarr/2020/12/18/the-future-of-virtual-reality-vr/?sh=dc008d127be8

Mateer, J. (2017, May). Directing for Cinematic Virtual Reality: How the traditional film director's craft applies to immersive environments and notions of presence. *Journal of Media Practice*, *18*(1), 19–20. doi:10.1080/14682753.2017.1305838

McCreery, M., Schrader, P. G., Krach, S. K., & Boone, R. (2013, March). A sense of self: The role of presence in virtual environments. *Journal of Computers in Human Behavior*, *29*(4), 1635–1640. doi:10.1016/j.chb.2013.02.002

Montenegro, A., & Ushenko, A. (2021, October) Installation Pictorial Representation of the Wedding Chamber. In *Artech 2021 ACM Conference proceedings* (Vol. 1, pp. 726-729). ACM.

Naemura, T. (2002). Real-Time Video Based Modeling and Rendering of 3D Scenes. *IEEE Computer Graphics and Applications Journal*, 70-71.

Perkis, A., & Timmerer, C. (2020, May). QUALINET White Paper on Definitions of Immersive Media Experience (IMEx). *European Network on Quality of Experience in Multimedia Systems and Services, 14th QUALINET Meeting*. https://arxiv.org/abs/2007.07032

Ravenscraft, E. (2021, November 25). What is the Metaverse Exactly? *Wired*. https://www.wired.com/story/what-is-the-metaverse/

Rubio-Tamayo, J. (2017, September). Immersive Environments and Virtual Reality: Systematic Review and Advances in Communication, Interaction and Simulation. *Multimodal Technologies and Interact, 1*(21), 9-20.

Shaw, J. (1989). *The Legible City*. Interactive Installation, Interface and Computer Monitor Projection.

Stephenson, N. (1992). *Snow Crash*. Bantam Books.

Suh, S., & Prophet, J. (2018, April). The state of immersive technology research: A literature analysis. *Computers in Human Behavior Journal,* 77-90.

Thibault, L. (2019, November). Is it Real? Measuring the Effect of Resolution, Latency, Frame rate and Jitter on the Presence of Virtual Entities. *ISS'19,* 10–13.

Trescack, T. (2019, November 17). The main problem with virtual reality? It's almost as humdrum as real life. *The Conversation*. https://theconversation.com/the-main-problem-with-virtual-reality-its-almost-as-humdrum-as-real-life-126761

Turner, C., Hutabarat, W., Oyekan, J., & Tiwari, A. (2016, December). Discrete Event Simulation and Virtual Reality Use in Industry: New Opportunities and Future Trends. *IEEE Transactions on Human-Machine Systems*, *46*(6), 888–889. doi:10.1109/THMS.2016.2596099

Van roof, M. (2016). Late-Breaking Work: Games & Playful Interaction. *CHI 2016*, 1990-1991.

Wehbe, R. (2016, November) Above Water: Extending the Pay Space for Health. *ISS Conference*, 499. 10.1145/2992154.2996882

Wolf, D. (2021, June). *Augmenting Teleportation in Virtual Reality with Discreet Rotation Angles.* arXiv:2016.04257v1 [cs. HC].

Zhaohui, L. (2014, October 29). Embedding perspective cue in holographic projection display by virtual variable-focal-length lenses. *Proceedings Optoelectronic Imaging and Multimedia Technology III. SPIE/COS Photonics Asia.*

KEY TERMS AND DEFINITIONS

Emerging Technologies: Are technologies evolving from known developments, and whose potentials were largely unrealized until they were improved or better implemented. In the field of visualization or virtual visualization, these are represented by Augmented Reality, and Virtual Reality.

Frame Rate: Frame Rate/FPS - Frames Per Second. The frequency at which frames in a television picture, film, or video sequence are displayed. In Virtual Reality and in video games is the frequency of the computation of the frames in real time rendering. The standard frame rate for VR is 90 fps (frames per seconds).

Immersive Environments: Immersive environments are digitally mediated learning environments designed to engage users in an artificially created make-believe "world."

Interface: A system, or program that enable users to effectively communicate with a computing process. In computer graphics, is the friendly user aspect of visual elements to understand and control a complex computation.

Latency: Is the delay between action and reaction. Having low latency is crucial when using a Head-mounted Display for VR or AR. As you move your head, your HMD should display the resulting changes immediately.

Metaverse: Commonly accepted definition of a version of the Internet through 3D models, and virtual worlds. Perhaps a relatively novel concept, it implies a wider universe of communities, platforms, and innovative ways of telepresence and communication through immersive environments.

Paradigm: A framework that defines a common methodology, theoretical approaches, and practical outcomes, that has been widely accepted and adopted by different disciplines and fields of study.

Presence: Is referred to the feeling of acknowledging in a perceptive and cognitive basis the spatial awareness of an individual in a virtual realm, reacting to stimuli if he/she was in the real world.

Real-Time Rendering: A field of computer graphics that study and develops modalities of reproducing images in high speed originated from 3D models, and 3D animations.

Chapter 29
Urban Tales Visible With Augmented Reality:
A Street Exhibition in Campanhã (Porto)

Pedro Miguel Azevedo Rocha
Universidade do Minho, Portugal

ABSTRACT

With the pandemic lockdowns, society was confronted with its social relations and everyday routines and how it relates with the conformity of life. To understand how we may evolve from this experience, past and present history (knowledge) comes to fruition, and its importance becomes paramount. So, an alternative way of information availability and relation gives ground to experiment with new ways of story consciousness, of true and unbiased knowledge about anything. This chapter presents a form of augment community search for purpose and consciousness through current and future technology. In the case of augmented reality, and how information its exhibited to the public, it brings a lively dimension as if information becomes commonly present in this reality, as past meets the now, as history is alive in the present. This chapter shows an example of how, in a context of an urban exhibition, a life tales' exhibition in an emptied neighborhood of Porto city, augmented reality might connect and socially relate people, either locals or tourists, with interesting information.

INTRODUCTION

Human beings are, by its inherent biological drive, a social being. It wants to connect, to dialogue, to communicate its thoughts, afflictions and desires. Current digital technologies came to offer a boost to such pursuit, and the demand satisfied such requirements. So, even in a pandemic situation where one would be restrained into its home walled confinement, existing digital social networks gave permission for such autopoietic lively communicative desire, and most of all, the desire to know. Information runs wild, and it's there for the taking, in search for the wishful comprehension of it. Therefore, due, but not only, to the pandemic and consequential lockdowns, society was confronted with the transformation of social relations status quo and everyday routines, and escapades (tourism wise).

DOI: 10.4018/978-1-6684-3369-0.ch029

To understand how we possibly evolve from this experience, past and present History (knowledge) comes into fruition, and its importance becomes paramount. An alternative way of information availability and relation gives ground to experience new ways of information consciousness, of true and unbiased knowledge about anything, for instance, as it's the topic nowadays, about pandemic news and social rights in this context, or some way to augment society search of purpose and consciousness through current and future technology. Such is the case of Augmented Reality (AR), and how information it's exhibited to the public, as an enhanced perspective of the real physical world. AR brings an undisclosed dimension of reality, through the use of digital visual elements, or other sensory stimuli by technologic means. In this, other information would become present in our reality, as past histories meet the present, building a more informative embodied space-time dimension. The author will describe an example of how, in the context of an urban exhibition of a community's social stories - an exhibition in the space of a demolished neighbourhood of Campanhã, Porto, Portugal (see Figure 1.) -, AR offers an unveiled space of information, and socially relate people, through everyday life's stories, in a safe and social distant manner in a pandemic context. All this context in a smart city environment (Casado-Aranda, 2021), can be seen this way, where an advanced technological development in AR may be a solution, and produce a social space where before there was none, remembering Lefebvre (1991) *Production of Space*, and Foucault's (1984) *Heterotopias*.

Figure 1. S. Vicente de Paulo seen from above
Source: image taken from Google Maps

As a viewpoint, Augmented Reality can be considered a type of digital nature transforming into a mediator of worlds, increasing the perception of this reality, of its environment in real time, within the medium that surrounds us. AR permeates smartphone technology in conjunction with camera, video, and positioning system, with visual information superimposed, into seen and felt stimuli of the place's reality where we participate as social beings. With AR it's offered a see-through information in radiographic surfaces of reality, where reality assume a degree of transparency. Since 1985 the access to goggles and gloves, that interact with virtual reality, established a gateway to an experience of our senses that would go beyond our imagination. We literally were linked to a world that could be seen with the aid of special glasses (eg, Google Glass, Facebook Oculus VR, Microsoft HoloLens). With promise of design creativity, several technological companies (eg, Garmin, Atheer Labs, Space Glasses, Ydreams, Total Immersion, Metaio, Layar) invested in new instruments and applications that would mediate and serve the frontier between the real world and the digital information world.

This street social experience, that took place on April, 2017, on how a lost community (the neighbourhood of "S. Vicente de Paulo", Campanhã, Porto) became again visible, was part of an European project called "Reclaim the Future – Nomadic Carnivals for Change" (570397-CREA-1-2016-1-SE-CULT-COOP1) -, translated into Portuguese as "Reclaim the Future / Exige o Futuro" -, involving five European theater companies, "Visões Úteis" (Portugal), "Teatermaskinen" (Sweden), "Rural Nations" (Scotland), "Dirty Deal Teatro" (Latvia) and "Compagnie des Mers du Nord" (France), with the support of Porto City Hall's European project "Creative Europe", and its "Cultura em Expansão", an annual cultural and artistic promotion program.

São Vicente de Paulo, built in 1947 (see Figure 2.), and nowadays a small urban garden, was bulldozered (see Figure 3.), around 2007, supposedly for real estate interests, by City Hall's decision, relocating and scattering its small community to other places. Through this augmented reality experience, and pinpointing local previous families' homes, it was possible to bring its past and everyday stories to present. In memory of those days, showing, in some way, that past will always be present in that territory, because one day that territory was dimensionally imprinted with life, in a timeline where the Past, in the public recollection of those people, will never be forgotten. The construct of such a connection was one of the main objectives for which those theater companies joined forces, to create a nomadic carnival for change, a carnival that opened for a strong interaction with the local communities.

This European project created a meeting place where local experiences and stories blended with international works. They proposed that "Reclaim the future" would become the creation of a common ground where stories of Europe could be voiced and shared. Together with local communities they would create carnivals celebrating the local perspectives and joined them together to share the experience of contemporary Europe. Carnivals form parallel realities, as so through AR, where one can see our lives on a second level, giving tools to reflect on ourselves and the world we live in. Building places where cultures and voices meet in a common and diverse narrative of this Europe in which we all co-exist. Advocating citizens as wanting to be part of the building of our future, based on co-creation, cooperation and participation. Promoting collaboration through an international network joining artists, civil society, researchers, stakeholders. And specifically, those whose voices are seldom or ever heard. In this case, of S. Vicente de Paulo, silenced by relocation and community separation. Carnivals are the celebration of diversity as well as a space where you can escape reality for a while and build your own narrative.

Through a process of social cartography, for a split moment, those personal stories regained a new life. Those tales were gathered, and the personal, the most intimate stories, were made public, opened in a breach in the fabric of time and space by a technology called Augmented Reality.

Figure 2. S. Vicente de Paulo seen from above, 2007
Source: image taken from Google Maps

Figure 3. S. Vicente de Paulo seen from above, 2013
Source: image taken from Google Maps

SOCIAL CARTOGRAPHY

A Brief Description

Urban territories are a social construct defined, and continuously redefined, based on the meanings and use of their inhabitants. In a way, the territory becomes us, our relations with the natural and social world. Social Cartography (Liebman & Paulston, 1994; Robinson, 2002; Gottlieb, 2009; Lobatón, 2009) allows a collective exercise of recognition of the socio-territorial environment through the construction of maps. Through this exercise of social relations, it is possible to obtain the knowledge that allows its participants to have a better understanding of their reality. When used to build knowledge collectively, it becomes a community approach to its geographic, social, economic, historical and cultural space. The construction of this knowledge is achieved through a collective mapping, which triggers communication processes between the participants and highlights the different types of knowledge that are mixed to achieve a collective image of the territory. This social mapping is presented as a conceptual and methodological proposal that allows approaching the territory and building a comprehensive knowledge of it. It is a planning tool, and a possible social transformation tool, which allows the construction of knowledge through participation and social relationships, opening up to its own transformation.

Social cartography (Liebman & Paulston, 1994; Paulston, 1996; Acselrad, 2008; Diez Tetamanti (Ed.), 2012; Diez Tetamanti (Ed.), 2014) will allow us to gain awareness of the space we inhabit. It becomes a representation of the paths we travel, the places where we become habitual, the spaces we occupy and shape, individually and collectively. They express images of our own personal and emotional space, assuming and mirroring forms of what and how we are individually, collectively resulting in a map that identifies a region or universe of interactions. They are representations of spaces that we fill.

Cartography means creating maps. A social cartography is a type of map of a territory with legends and symbols created by those who live there. The territory becomes completely original, referring only to that area. Those who make social cartography can pinpoint, for example, a place that has a remarkable history for those who live in such a location. This type of map does not necessarily mark the limits of streets, corners and intersections, but the internal codes that arise from the presence of passers-by. The narratives and testimonies of the participants also build the material. In a way, social cartography will be able to say more about a place than any map by the standards of geography. As González's (2012) social cartography expressed by "mapping the neighbourhood we build territory".

Forms of cartographic representations are important tools to analyse and understand how the relationship between space and society is established. However, certain classical mapping methodologies do not consider relevantly the social aspects of space, that is, they present themselves as produced representations that emphasize the physical, political and economic elements of space. The inhabitants of these spaces, of these communities, contribute to the cartographic construction, specifying their cultural and social way of life, their ethnic and collective identity and the delimitation of a territory that is not visible by conventional methods.

Possibilities that are elaborated from social mapping (through conflict maps, network maps or resource maps), intend to propose an instrument at the service of citizens. At the same time, they enable the intervention and guidance of social actors and any other professionals committed to social transformation. By giving back to the constituents of these communities the power to map their territories, this approach places them in the role of protagonists of their own identity. These maps reflect the understanding of these people about their own territory and the relationship of their culture with that space. According to

Andreotti (2016) the "maps produced, in the Paulston tradition of social cartography, can be understood as metaphorical devices that trigger provisional visual images of the difference between and within intellectual communities", where the resulting maps "can help make visible how problems and solutions are situated between and within particular social perspectives and locations".

Nowadays, there are in some countries, groups that invest into the production of such cartographies. For instance, the Iconoclasistas, a duo formed by Julia Risler and Pablo Ares in May 2006, from Argentina, that began as a social communication laboratory from which they produced graphics (posters, publications, cartographies, etc.) and urban interventions. Currently, its activity unfolds in three dimensions of knowledge and practices: artistic (poetics of production and graphic devices), political (territorial activism and institutional drift) and academic (critical pedagogies and participatory research). From 2008, they developed collective mapping workshops at various places in the world, stimulating critical perception of the territories and promoting the production of collective meanings. In 2013 they published "Manual de mapeo colectivo", where they systematized and shared methodologies, resources and dynamics for the self-organization of workshops, exercises, tables and interventions of collective mapping and development of collaborative research processes.

A mapping process, which includes the production of symbology, the description of captions, the choice of a scale and layers of information, may have the potential to produce criticism and collective actions, because: (1) the participants have the possibility to develop their own language to describe their reality, and the production of terms and definitions that reflect their values, (2) shared personal experiences allow groups to be able to analyse patterns and identify experiences collectives, and (3) the role of institutions and the extent of their political power in guiding the shaping of collective experiences may become more evident. These maps are made by the community, together with the research, in a participatory planning process. It is a collective process of discovery and knowledge construction. Not only that, social cartography is applied according to the conceptual bases of participatory action research (IAP).

Although there has already been some research activity in this area in South American countries since the 1980s, in Portugal social cartography only began to gain expression mainly around 2006, as a tool used in research. For instance, in 2009, with Ana Vale Pereira and Joana Lúcio through the project "Cartografia social do Porto" ["Porto's Social Cartography"] (which had no continuity or conclusion, due to the research lack of resources), reducing itself to Porto's neighbourhood of Massarelos (Lúcio, 2011), and Patrícia Ribeiro (2014) with a project ("Cartografia social: (Re)Conhecer perceções; Mapear territórios.") about the city of Espinho.

Methodology Overview

Social cartography process contemplates a participatory mapping of the past, present and/or future, as a strategy for visualizing social and community points of view. Thus, presenting a way to better understand what happened, what is happening and what is about to happen in a given territory. According to Bastidas (2009), There is no base model for the use/application of this methodology (Social Cartography) that can be applied regardless of the context, however, there are general procedures that guide mapping activities. Normally, this process includes three phases: (1) diagnosis, (2) construction of maps, and (3) interpretation of the information gathered on the maps. The mapping activity helps participants understand the complex interactions between context, structure, actors and goals of community members and other similarly interested external parties. The methodology proposes maps of the past, present and future that respond to the dynamic nature of conflicts. The conceptual bases of this intervention process are outlined by Participatory Action Research (Bergold & Thomas, 2012; Kemmis & McTaggart, 2005; MacDonald, 2012):

a) In Investigation (I) of Social Cartography, the community is a participant in the research, bringing its knowledge and experiences. Here it is considered that the maps suit and promote the culture of storytellers, and that the collective construction of maps materializes the individual and collective local memory. Bringing to the public a vision of the creation and transformation of the local territory, present only in the minds and daily lives of its inhabitants.
b) Action (A) indicates that through knowledge of (a) reality one can act on it, and, in many situations, its validity originates from and can be proven in and through action. It is about knowing reality in order to transform it and not investigating it just for the purpose of knowing it. The fact of creating a collective moment of sharing and cartographic exhibition of information will provide that the individual notion about the territory (or cohabitation/circulation space) of each participant-inhabitant is increased or transformed in face of the resulting collective vision.
c) Participation (P) is considered a continuous process of social construction that brings together knowledge, experience and proposals for development transformations. The participation by characterizing it as active, organized, efficient and decisive, it must be carried out within the scope of dialogue and based on strategy. Community participation must be expressed in and throughout the investigation process.

The systematization of this methodology should be a fundamental element to learn the reality under study and allow the possibility of a transformation, so the systematization will allow the measurement of such knowledge, data and practices for sustainable social development. The method contemplates the elaboration of three types of maps: conflict maps, resource maps and network maps.

As such:

a) Preparation of conflict maps: represent situations in question through images, icons, texts, narratives, etc.
b) Production of maps of real and potential resources: these represent the material or human resources that are considered as potential elements.
c) Elaboration of network maps: it is defined as a representation of relationships between actors, which, according to their connections with regard to a conflict and their positions in relation to it, allows us to jointly articulate an intervention in the face of conflict.

The process in each execution has the following phases: diagnosis phase, cartographic production phase, and devolution phase.

Other sources of social cartography suggest the elaboration of 4 maps. Starting with a geographic and political map of the location of the local population, in addition to their territories, resources and production activities. Thus, obtaining an economic-ecological map. On another map, the productive infrastructure, services, that is, roads, paths, bridges, schools, churches, energy networks, parks, etc. Third, a map of the population's social and cultural relations would be produced, identifying actors and possible scenarios. Finally, the fourth map would represent the conflicts, risks, vulnerabilities and potentialities. These maps can be made according to three major types of maps:

a) Maps of the Past: The configuration of this historical map is important for the recognition of the changes that took place, thus collecting the collective memory of its inhabitants. This exercise allows characterizing the territory.
b) Maps of the Present: allow presenting the current situation of the place under study. By contrasting with the previous one, it will be possible to understand the evolution of the community.
c) Maps of the Future: Also called "dream map", they reflect how the inhabitants see the community in the future, or how they would like to see it.

In general, there are fundamental principles that are specific to the IAP (Kemmis et al., 2014; Kindon et al., 2007; McIntyre, 2008): (i) a collective commitment to investigate an issue or problem, (ii) a desire to engage in collective and self-reflection to obtain an explanation of the subject under investigation, (iii) a joint decision to engage in individual and/or collective actions that lead to a useful solution that benefits the people involved, and (iv) the creation of alliances between researchers and participants in the planning, implementation and dissemination of the research process.

These objectives are achieved through a cyclical process of exploration, knowledge construction and action at different times throughout the investigation process. IAP is seen as a recursive process that involves a spiral of adaptive steps that include asking a specific issue, reflecting on and investigating an issue, developing an action plan, and implementing and refining an established plan.

The reality may be that many people who live in communities where IAP projects take place do not participate in the actual research process. Some people are unable to commit to an IAP project because of work, family, and other responsibilities. Others choose not to participate due to opposition to a project idea or the direction they think the project might take. Still other people are unaware that a project is in progress and may only hear about it during development. However, regardless of the reasons why people do not participate in an IAP project, the focus of the project must be to provide opportunities for local people to develop strategies and raise resources to change their environments/spaces for the better. Thus, as many people as possible should be invited to participate, to voice their concerns, and to inquire about the development of the IAP process as it evolves. (McIntyre, 2008)

AUGMENTED REALITY AND INFORMATION TECHNOLOGY

Augmented Reality presents as more than virtual, by its juxtaposition to the real world, as if our mental information about a given subject assumes a real representation out of mind momentarily turning real space into an extension of itself. Augmented reality may be considered another type of life, directly or indirectly, whose elements are sensory stimuli generated by computers in the form of images, sounds, images or positioning data which are linked to different types of information layers. AR assumes as a mediator of worlds, its own reality, increasing the perception of reality in real time and in a meaningful semantic context with the world that surrounds us, within a mental "computer", a space where you can interact with and manipulate requests and static actions in overlapping graphics to the real world (Rocha, 2015a).

In an article by Drew Bartkiewicz (2013), he writes that AR is increasingly possible to implement and to blend in our day-to-day due to an "overload of big data bombarding the human mind, in work and in our lives." Since 1995 the data have been increasing "being produced, processed, and aggregated at such volumes" (id.) whose "average company doubles its data every year," (id.) however people have not followed this increase not duplicate the "cognitive and critical thinking capacities at the same rate." (id.) We arrived in 2013, a turning point for high-volume databases, it was believed that this would be the "trigger that drives a massive adoption to Augmented Reality, where the display of data and the capture of content are delivered in more natural and hands-free methods." (id.)

Augmented reality, for now, makes use of smartphone technology (and such), in conjunction with its camera and screen for information visualization superimposed in real image. But to become one with a real environment it needs projection technology. One of the first videos broadcast massively in which we saw a coexistence with the projection in real space was the *Star Wars - The Return of the Jedi* hologram in 1983, where we could observe in the center of a conference room a colorful transparent planetary simulation in proportional size and its relationship with the surrounding elements. The author infers that was one of the moments of the viewer's mind transformation beginning, and the population's future imagination of what would come in the next half century in terms of how information would interact or would be perceived. There was another one before in a previous movie, *Star Wars - A New Hope* in 1977, but compared to the 1983's projection it's limited to a faint resolution 3D image of a character communicating through 'live' recording, nevertheless it was in those days that creativity gave way to be materialized through the cinematic screen magic, bringing new worlds to the imaginary of society.

According to an article, by researcher, of Management Information Systems, Ahmed Elmorshidy, published in 2010, about the observation of changing the world through the influence of the emerging holographic projection technology, it declared that this would represent a "new wave in the future of technology and communications, in different application of technology, dramatically affecting areas of life, including business, education, telecommunications and health." This technology, rapidly growing, was expected to change "the way we see things in the new era." (Elmorshidy, 2010) Today it is a fact that we are in immediate contact anywhere, and through a touch of a screen we can decide on almost any action or business remotely. We are living in a kind of a new world dreamed or imagined by someone, or by many, supposedly in a collective consciousness that expresses in what we are and what we do, described by de Kerckhove (1998).

Augmented reality it's, in general, an evolution of virtual reality extra space needs on the real world. However, each person knows his own real, and reality, and no other truly understand the vision of the first if she doesn't reveal it, since the reasoning it's peculiar from person to person and the way they look, design and understand their own judgments, it's what may build us the sense that contributes to the formation of its unique existence in this incomprehensible whole. The reality, or the real, the natural real (Husserl's real), in addition to palpable sensory and sensitivities, is everything that makes up the stability of individual reason, beyond that, the reason is lost and the sense of consciousness sinks for lack of gravitational force of some referential. Nevertheless, taking into account the underlying disagreements that come from the habit of assuming that the individual perception of a given knowledge is the same, one can establish a common basis of understanding of reality taking two given observations: divisibility of reality in elementary pieces (signs), possibly modular, and the perception of a dimensional transformation (given for instance by Kant's sensitivity of time and space).

We see, and immediately we believe in the realism of the seen. A healthy person rarely questions what the vision, or any other sense, hands to her to interpret and process. We see, and immediately we accept the realism of the seen. The brain is eager to fill it with information, affecting everything that curls inside the eyeball as an observer actively passive transparent to its reader role. The perception of the essence of each element, observed in a process in which knowledge is apparently given to us, and its implicit contextualization in a space that includes it, it is received with open arms as a longtime friend.

According to Edmund Husserl, the "giving itself up, whether it manifests something of simple representation or something of true existence, something of real or something of ideal, something possible or something impossible, it is always a giving up onto the knowledge phenomenon, onto the thought phenomenon in the broadest sense of the word." (Husserl, 2008, p.104) The innocence of a newborn that looks at the moon for the first time is staggering in the sense of mostly not being surprised by the huge pebble hanging out there, so he appears to see everything as normal, as present; sees and accepts the phenomenon as it is.

As stated by Kant (2001, p.79), "it is completely unknown to us the nature of the objects themselves." The primordial nature, of its not perceived and timeless state, we do not materially possess its pure *percept*, and in for the mental shape its pure *construct*. We perceive their existence completing it with the realization of being real for its allocation and expressed by its equivalent in thought in a *construct*. As discussed by Paul Carus, writer and German-American philosopher, the "sensory impressions are data, they exist before ideas, being these constructions that are produced from sensory impressions. Sensory impressions are facts, but the ideas are from deductive nature; they are (to use the excellent term of Conway Lloyd Morgan [English psychologist and ethologist]) *constructs*" (Carus, 1892; Rocha, 2015b) Through perception we transform the knowledge of reality by changing the perspective depth of objects, both from the side of details as from the context of its finite totality in the space to which they belong.

INFORMATION SEEN IN RADIOGRAPHIC SURFACES OF REALITY

The see-through, or transparent, screens although there is already here at least a decade, only around 2012 have begun to have a bet by companies from the IT branch, such as, for example, the "Smart Window" Samsung, exhibited at the CES (Consumer Electronics Show) 2012, due to its attractive appearance specifically the effective use of applications for augmented reality. While smart devices observed the real world indirectly from a camera, which transmits the image and subsequent composition with other information on the device's screen, in case of transparent screens the space around us it's observed directly which it's only interrupted by a radiographic "film" introducing graphic information in real time in conjunction with what you see.

Perhaps the first manifestations of this technology have been before, shy or sporadically fruit of creativity, but the most important have been the use in the animated film *Ghost in the Shell* in 1995, from there arise rhythmically in other films, such as *Minority Report* (2002), *Paycheck* (2003), *James Bond – Quantum Solace* (2004), *Iron Man* (2010), *Avatar* (2010). And recently in many others, as the cinematic technology progresses. Or included in a vision of the near future given by Microsoft in the project "Microsoft: Productivity Future Vision" of 2008 or through the experimental innovation project of the company TAT in 2010, the future of screen high-tech with stretchable screens, transparent screens and e-ink displays.

The technology application of transparent screens/glasses in conjunction with augmented reality concept, which currently it's slowly massifying, it's related to smart glasses (goggles with access to the internet or in contact with local devices, or *cloudy* servers). Its origins go back a few decades ago. In 1985, Lanier, a pioneer in the field of virtual reality (a term for which he is credited to have popularized), left Atari to found the VPL Research, Inc., together with Thomas Zimmerman (former researcher at IBM), the first company to provide goggles and gloves interacting with virtual reality. This way we would have at our homes our own visionary device of visionary view, all this in an accessible way.

Garmin, a previously American company, is since 2010 a Swiss company that develops GPS applications for the consumer in the field of aviation and marine technologies. With this experience and the concept HUD (Heads-Up Display), with enough use in the front area of fighter planes cockpits, they took the initiative, in 2014, to apply it to car windshields in terms of information and entertainment. Facebook (now Meta) has purchased for $2 billion in 2014, the Oculus VR, a company founded in California in 2012 that created a head-mounted display, a helmet that one experience the visual interaction with a virtual reality; On other hand, Microsoft has submitted in April 2013 a patent for augmented reality glasses, called Microsoft HoloLens. In 2016, as mobile technology advanced and became more involved with most aspects of daily life, the mobile APP Pokémon Go provided a perfect example of the triggering to the widespread use of this technology. It sent people into streets and parks, onto beaches and trespassing properties in the week it was released. Space seemed not to have limits, a situation that led people losing the sense of the reality where they were. However, those, and other devices from other companies, are in process of being experimented by the public in general, testing financial effort against its pragmatism, waiting for the confidence in dealing with the handling of a world that physically does not exist, which it's invisible for who observes it from the outside. However, current past pandemic situation put some shackles, and other forms of restrains, removing that freedom of expansion, and movement, that the pan piper mobile was leading with AR as its compass.

GESTURAL MOVEMENTS IN REACHING TO THE INVISIBLE

Augmented reality with information lies in space, in the atmosphere that surrounds us, where the medium is its own message. When this symbolism takes mainly over our visual perception, immediately we establish it as an integral part of the element we observe in a sense of belonging implied in the depth of the place that information appears. And so, we see it as such, connoting a world that doesn't exist physically, but it is accepted as normal in the same way as mental imagery, only this time this imaginary we see it "really" in a space before us, that only we see it. Digital information takes over and becomes "form" in an existence which hitherto did not exist, thus becoming the latest "species" of the natural world, but still is only a "spirit" of what can possibly be. Interaction with this information is produced through physical form like if you're really there, which is performed by speech or gestural movements with its own language. The movie *Minority Report*, of 2002, did the chores of the house and introduced the topic publicly in the media, and from here on there would be no turning back – life wanted to imitate art. Director Steven Spielberg to the design and discussion of producing this movie, brought together a team of thinkers and consultants, to build the image of an imagined real world 50 years into the future. Among them was Jaron Lanier, already mentioned, but in particular John Underkoffler.

John Underkoffler designed the computer interfaces that appear in the movie "Minority Report" which are notoriously gestural based systems. However, they do more than simply pointing, it allows users to move their hands to interact with the images produced by computers. Underkoffler led the team that produced this interface, called "G-speak Spatial Operating Environment", allowing applications to be developed and implemented on multiple screens and multiple devices in the resolution and processing of large databases, providing effective collaboration, transforming the action of a user (or more) on a screen (or more) in a fully interactive and shared experience, to combine data presentation features and analysis and to allow real-time insight and intelligence development. His company, Oblong, was founded in 2006 to transfer the "G-speak" to the real world, and with the aim of creating the next generation of computer interfaces. In February 2010, Underkoffler gave a presentation at TED Talks, called "Pointing to the future of UI", which was received with great interest. It began thus concretely the interaction of the human being (or user interaction – UI) with the spatiality of the information world around him, by that moment long was the time of the article "A hand gesture interface device" published by Thomas Zimmerman and Jaron Lanier, among others, in 1987.

Nowadays, gestural interaction it's limited to screen interaction, where the fingertips hit these windows, and do not float in mid-air, yet. Multi-touch technology on screens and several international companies are investing in this field due to the use and existence of smartphones, tablets and other "tables" of the same application. A Portuguese case is represented by Edigma, a technology company founded in 2000, under the guidance of Miguel Peixoto de Oliveira, specializes in developing hardware, software and interactive systems based on multi-touch and gesture recognition. Edigma through Displax develops a multi-touch layer, adding more features and use of new materials, and through Unedged develops sensory experiences by touch via a tailored software to all Interactive hardware, having received several awards in past years.

Although with all these devices, to complete the actual interaction with the emptiness of that available information, there's still a need to give the "sense of touch" to these projections making it possible to handle them. With this in mind a team of Japanese scientists at the University of Tokyo, conducted by Hiroyuki Shinoda, created touchable holograms. His technology consists of software that uses ultrasonic waves to create the feeling of pressure in his hand that touch the projected hologram. Its applicability was produced thinking of light switches, devices that could not interact directly or in hospital environments to prevent contamination by touch. Only later, in 2012, Professor Shinoda was invited to TEDxTomsk to present his work: Super Haptics. Our pace in this technology is slow but we, humans, are also the result of a long path.

TALES IN DIGITAL WINDOWS ON PUBLIC STREETS

Is Campanhã a crossing or a passage? Or any neighbourhood, for that matter? A rest room or an incubator of lives? What makes people connect with the place? What imaginary or identity (institutional, expressive, symbolic) moves them from this place or disables them from movement? The notion of where you are, and how this place is portrayed, may be part of the inhabitant's imagination. This place may be correlated with the imaginary of himself, his expectations and desires, with his vision of himself and others, and eventually with his intimate space (Hall, 1990), personally as an individual, and publicly as a citizen. The visual scale of the urban form, in which it cohabits, may transcend the capacity of his senses to perceive its scale. With technologies, the idea of territory, or at least the limits of such space, extend and enter the morphology of the digital. With layered information, or alternative realities, whether virtual or augmented, created by technologies, man becomes mobile without physically leaving the place. The images of these realities refer man to a meaningless world, absent of interpretation, where those images and words may fill his imagination. But if it is immobile, it visibly does not advance, giving the sensation of non-production, so apparently there will be an incongruity of not having meaning.

The whole conjunction of their active presence in these places creates a kind of ecosystem, a social space, a human ecology. As Ericksen (1980, p.15) pointed out, in the words of Amos Hawley, that Human Ecology would be the "descriptive study of the adjustment of the human population to the conditions of their respective physical environment". This ecology would be expressed by the description of the space to which a community adjusts and establishes itself according to a given format, thus being able to present this movement, and occupation, through the joint elaboration of a cartography. In this mapping, it will be produced a sense of how they see things, a look at the place they inhabit, where they live, where they move, where they have communion through speak and touch, a cartography that expresses the possibilities of being (or becoming) of the territory that is inhabited, a cartography that speaks their tales, and revisit once more their lives.

The morphology, and that of the inhabitant's mind, may cease to be intelligible from the point at which this individual is proposed to create the graphic expression of that space. A space that portrays itself (a personal representation) and exists only in its understanding. People, and their institutions, when located on certain terrestrial surfaces act, position themselves, direct themselves, based on their desires and emotions, added by the meaning they attribute to the places. It is questioned, then, how people connect – because, for Maffesoli (1987), the place makes a connection –, and identifies with each other, in specific places, how they socialize through the places, through the spaces – what makes them there? Does the geographic place (characteristics, surroundings, neighbourhood, topographical predisposition) attract or define its inhabitants? Does the predisposition of the territory provide the place for the necessary transformation of, or for whom, lives there? These are some question marks that arise from participatory research, and for a moment, from participating in the living of someone's life. This intrusion, with these neighbours' permission, offers, through their stories, scope into the significance of the meaningful lives they breed on that ground. And by the assistance of augmented reality, and pinpointing some local places, it was possible to bring its past and everyday stories to present, in memory of those days, showing that past will always be present in that territory, because one day that territory was dimensionally imprinted with life, in a timeline where past, in the public consciousness of those people, never forgets.

As already stated, this work followed the European project "Reclaim the Future – Nomadic Carnivals for Change", involving five European collectivities, "Visões Úteis" (Portugal), "Teatermaskinen" (Sweden), "Rural Nations" (Scotland), "Dirty Deal Theater" (Latvia) and "Compagnie des Mers du Nord" (France), grounding on the theme of 'carnival'. It emerged as an interaction with local communities, creating a meeting point where local experiences and stories mingled with international works, to share the experience of contemporary Europe. Stories were collected by Márcia Andrade (2018), from 13 former neighbours of S. Vicente de Paulo. Subsequently the present author, based on these texts, produced the exhibited material (audio videos opened by AR – see Figure 4.). These dialogues were used for making the cartography, as the "Past Map", using an APP named Zappar. Zappar was, from the apps available at the time, the most interesting choice, free to use, and had a desktop studio integrated on the product, that gave the possibility of the production of the augmented element that the author wanted to make. Zappar technology was born out of the University of Cambridge, and has been powering AR campaigns since 2011. Its mission is to "democratise augmented reality".

Because, near S. Vicente de Paulo, exists the Elementary School of Corujeira, to where most children of S. Vicente, and possibly their parents, attended many years ago, "Visões Úteis" arranged to do some activities with some current school classes within the now empty compound. Inês de Carvalho, "Visões Úteis" mediator, asked them to draw and paint over several photos, previously taken from that now urban garden, their vision of how that place should be, or what to build, or to what purpose should it be used for. So, this way they help create the cartography for the "Future Map". Some selected paintings were digitalized and incorporated into the AR system (see Figure 5.). Children would represent new houses, not empty ones; gardens, hospital, new playgrounds, new roads, nature in communion with people, sunny days, football fields, a place to ride horses, an aquapark, fireworks, pools, ballet schools, hotels, libraries, overall, life happening in contrast with its quietness.

Figure 4. iPhone AR view of a video opened by its interaction (at S. Vicente de Paulo)
Source: photo taken by the author on site

Figure 5. iPhone AR screenshot of an opened painting disclosed by its interaction (at S. Vicente de Paulo)
Source: mobile screen photo taken by the author on site

Within the carnival "festival", S. Vicente de Paulo was the starting point of the carnival procession. This procession would represent a timeline - past, present, future -, where the location of the non-existent neighbourhood of S. Vicente de Paulo would play the point of the past, and that of the future will be assumed by what will be the new cultural space of Campanhã, the old and big Municipal Slaughterhouse. As they say, nothing is lost, everything is transformed. Few days before the procession event, with the help of the other European companies, at least 40 signboards (made by IEFP, a national Institute for Employment and Vocational Training), were scattered on the empty streets of paving stone (see Figure 6 and 7); 20 for Future Maps, and other 20 for Past Maps. On those signboards were placed the generated images that would establish the connection with the mobile app, through a sign made in the Zappar studio. This unique sign had the role of permitting the connection, as a representamen, embodying as the carrier for the message, or desired element, wanted to be experienced, as a form of an image, or a video with the recorded tale. In this case, the "sign", was an image composition of a symbol that related with the story theme, and the Zappar generated code (see Figure 7 and 8). In the end, the author made a map (see Figure 9) to identify the places and the stories. In the map it's identified the stories' locations with a number and a coloured dot to distinguish from the dots signalling the places where, randomly, were put the children's paintings of "tomorrow". The author also made a bigger printing of the map to place it at the carnival procession starting point, and also marking the entrance to S. Vicente de Paulo.

Figure 6. Signboard erected (at S. Vicente de Paulo)
Source: photo taken by the author on site

Figure 7. Closure photo of a signboard example of an AR sign image of a story (at S. Vicente de Paulo)
Source: photo taken by the author on site

Figure 8. Zappar AR code example
Source: image made, by the author, at Zappar online design platform (Zapworks)

Figure 9. Map of the exhibition
Source: Own elaboration

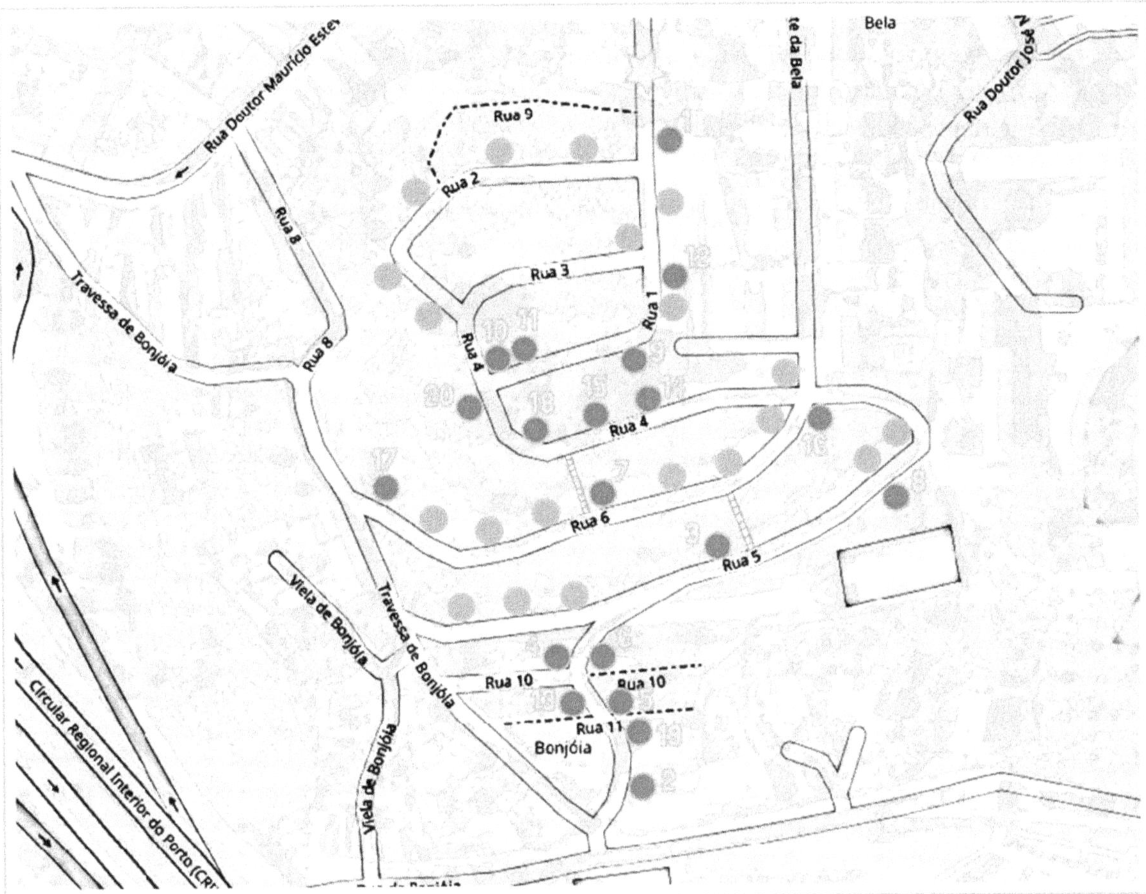

The following list represent the selected recorded stories (indicated on the legend of the map – see Figure 10), that would be present on the field of S. Vicente de Paulo:

Story 01: The Fiscal clerk (1st Street, House 1): Told by Helder and Ângelo;
Story 02: Last resident's soup pot (13th Street, House 18): Told by Carla Carvalho;
Story 03: Surviving Dog (5th Street. Hood story): Told by Elisa;
Story 04: Hanging black coat (10th Street, House 15): Told by Fátima;
Story 05: Christmas Tree (11th Street, House 29): Told by Sãozinha;
Story 06: Bunk + 2 days and would die (10th Street, House 15): Told by Fátima;
Story 07: The neighbourhood's first bride (6th Street): Told by Alice, Lito's mother;
Story 08: Sung Whistle and Football Field (at the Football Field): Told by Ernesto;
Story 09: My tree (Tree in front of 1st street): Told by Ernesto;
Story 10: Fado of my neighbourhood, fado of love (Hood story): Told by Amélia;
Story 11: Smell of mint (1st Street): Told and sung by Amélia;
Story 12: Iron (At the end of 1st Street): Told by Amélia;

Story 13: Mother who counted the cars from the balcony (11th Street): Told by Sãozinha;
Story 14: Lollipops (4th Street): Told by Ângelo and Helder;
Story 15: Slide (4th Street): Told by Ângelo and Helder;
Story 16: Viewpoint and stone bench (Next to 5th Street): Told by Ângelo and Helder;
Story 17: Chicken's ass and lollipop (Between 4th and 6th Street): Told by Ângelo and Helder;
Story 18: The first television (4th Street): Told by Ângelo and Helder;
Story 19: "São João" (Beginning of 11th street): Told by Sãozinha and Fátima;
Story 20: Yellow shoes (4th Street): Told by Ângelo and Helder.

Figure 10. Legend of the map
Source: Own Elaboration

– histórias –

01 | Fiscal
02 | Panela da sopa da última moradora
03 | Cão sobrevivente
04 | Casaco preto dependurado
05 | Árvore de Natal
06 | Beliche + 2 dias e morria
07 | A primeira noiva do bairro
08 | Assobio cantado e campo de futebol
09 | A minha árvore
10 | Fado do meu bairro, fado do amor
11 | Cheiro a hortelã
12 | Coradouro e ferro de passar a ferro
13 | Mãe que contava os carros da varanda
14 | Pirolitos. Caramileira
15 | Escorrega e Fundilhos
16 | Miradouro e banco de pedra
17 | Cu da galinha e pirolito
18 | A primeira televisão
19 | São João
20 | Sapatos amarelos

On Figure 11 the author shows examples of four AR composite images, to interact with the audio videos of the told stories, for the construction of the Past Maps. And on Figure 12. the author shows examples of children's collage and painting, for the construction of the Future Maps.

These were some of the stories, very personal stories. Memories when they were leaving, the last resident's soup pot, a surviving dog that still lingers around the streets, a black coat that was hanging right before a demolition, a still mounted Christmas tree taken to the new home. Memories of the first television for all hood, and of the first bride. Memories of how life was back then, of how things were done and overcome. Those personal and intimate memories were made public, for all who wanted to know and remember. Now the memories aren't just of them, they become of the place, in the same space and time, they become memories of a bigger crowd. A collective that became conscious of a story that increased their own story, individually, and their life's memory got bigger than what life has to offer. Because they, who left, gave it all, sounds of laughter and sounds of cries, and still they kept everything, as a silent nomad that dwells from hood to hood.

Figure 11. Examples of AR interactive symbology
Source: set of images made, by the author, at Zappar online design platform (Zapworks)

Figure 12. Examples of children's works seen through the mobile
Source: images made by various children from Corujeira Primary School, Porto

ENTERING AN AR POSSIBILITY

Augmented Reality has something different to offer, but it has to have an easy-to-use mediating device, either way it will be another hype Game Boy just for the need of a momentary distraction. During its few days of use in S. Vicente de Paulo, constructed images, seen on the mobile, fought against weather and light conditions. If it was wet, one could not properly interact with the respective picture; if it had too much sun, it would reflect back the light and the device would not be able to read the AR code. It is an interaction driven act that comes from outside, not from within, like Pokémon Go. And most of the interested public, older than the average enthusiastic of these technologies, were not equipped with mobiles that stood to its minimum requirements usage. The few people that tried it out of curiosity, agreed that it was interesting, and that local social History now was not lost. Maybe other parts of Porto's smart city have more to offer, through and by the people's tales that give life to those places. They also are, and became, part of the city's History, and not a mere name on a birth certificate book or on a tombstone ledger. Life is in-between.

Remembering what Husserl once said, in "perception, perceived things must immediately be given. Here's the thing before my eyes that they perceive; I see it and I grab it. But perception is simply a living of my subject, the subject that perceives it." (Husserl, 2008, p.40) At the end, after having the experience of augmented interaction, about what one has seen, or heard, it all comes to how anyone personally perceives it. Targeting this form of AR utilization, it all sums to how it satisfies the customer, as a local citizen or tourist. How this experience fills his expectations and delivers something more that could not be represented in any other way. AR it's a construction of visual hyperlinks to an information dimension that lives hidden in digital interpretation, superimposed to the real world. In the end, AR can offer a disclosure of an information dimension embedded in the digital world, that can be available in the feeling of the real world.

CONCLUSION

Regarding the situation of this innovation applied to tourism's recovery strategies, and how a pandemic environment might impact its industry, Abbas et al. (2021), gives a study where "innovation and change can contribute to the tourism industry's revival to the next normal" (Cuomo, 2021). Not forgetting that "technological innovation requires changes in e-tourism work, including history, reflection, openness, equality, diversity". (Gretzel, 2020), in the case of AR, lately appeared multiple studies regarding tourism re-invention and overcoming of this pandemic context (Cranmer, 2020; Loureiro, 2020; Serravalle, 2019; Shen, 2022). Thus, putting in perspective the adoption of AR, value and its multiple applications, and other competing digital realities, as in the case of museums (Andrade, 2016). This article presented an experiment where the author tested the usage of AR in an urban environment. The event in S. Vicente de Paulo became a scenario of how AR may be applied outside, or even inside as in a museum exhibition, collecting tourists or locals' attention to its capabilities and uses.

In this case, its usage did not have enough time to be experienced by a larger number of users - a couple of days were not sufficient -, failed to address a more extended study on its applications. Mainly to stick to the initial purpose of this European project, to its time limitation and report obligations, and due to lack of appropriate marketing to promote it on a broader vision to its public tryout. And, moreover, because a person had to download the Zappar APP to unlock the interaction, many of the former S. Vicente inhabitants apparently did not interact with it, nor in the day of the carnival the people who attended gave feedback enough to sustain a broader conclusion. The few that use it, gave positive feedback. They could listen to the story, in the middle of the empty lot, lost in the silence of the field, a story that came out from the sign-image. And invisible paintings from childish future visions that were there all the time. In the end, it became a practical rehearsal of an AR form of application in an urban scenario, where we could observe the openness of an eager public to expose their feelings through a digital world. A public that felt that its life memories had meaning, and should not be put aside and forgotten. Cities' tales have a memory imprint taken from citizens' life, and it is hidden in their minds. Their memories are also urban territories of the city's map, and through Augmented Reality they were added to the local landscape.

REFERENCES

Abbas, J. (2021). Exploring the impact of COVID-19 on tourism: transformational potential and implications for a sustainable recovery of the travel and leisure industry. *Current Research in Behavioral Sciences, 2.*

Acselrad, H. (2008). *Cartografias Sociais e Território.* IPPUR/UFRJ.

Andrade, M. (2018). *"Ficou tudo ao Deus dará.": S. Vicente de Paulo. Histórias e memórias de um bairro demolido na cidade do Porto.* Edições Afrontamento.

Andrade, P. (2016). *Comunicação Pública da Arte: o caso dos museus de Arte locais e globais.* Caleidoscópio.

Andreotti, V. O., Stein, S., Pashby, K., & Nicolson, M. (2016). Social cartographies as performative devices in research on higher education. *Higher Education Research & Development, 35*(1), 84–99. doi:10.1080/07294360.2015.1125857

Bartkiewicz, D. (2013). *APIs in Your Eyes with the Augmented Reality Wave of Bigger Data.* Available at http://drewbartkiewicz.sys-con.com/node/2569342

Bastidas, E. P., & Gonzales, C. A. (2009). Social Cartography as a Tool for Conflict Analysis and Resolution: The Experience of the Afro-Colombian Communities of Robles. *Peace and Conflict Studies, 15*(2). Advance online publication. doi:10.46743/1082-7307/2009.1094

Bergold, J., & Thomas, S. (2012). Participatory Research Methods: A Methodological Approach in Motion. *Historical Social Research (Köln), 37*(4), 191–222.

Diez Tetamanti, I. J. M. (Ed.). (2012). Cartografía Social: investigaciones e intervención desde las ciencias sociales: métodos y experiencias de aplicación. Universitaria de la Patagonia.

Carus, P. (1892). What does anschauung mean? *The Monist, 2*(4), 527-532. Available at https://www.jstor.org/stable/27897002

Casado-Aranda, L. (2021). Tourism research after the COVID-19 outbreak: Insights for more sustainable, local and smart cities. *Sustainable Cities and Society, 73.*

Cranmer, E. E. (2020). Exploring the value of augmented reality for tourism. *Tourism Management Perspectives, 35.*

Cuomo, M. T. (2021). Digital transformation and tourist experience co-design: Big social data for planning cultural tourism. *Technological Forecasting and Social Change, 162.*

de Kerckhove, (1998). *Inteligência conectiva: a emergência da cibersociedade* (A. Viseu, Trans.). Lisboa: Fundação para a Divulgação das Tecnologias de Informação.

Elmorshidy, A. (2010, May). Holographic Projection Technology: The World is Changing. *Journal of Telecommunication.*, *2*(2), 201. https://arxiv.org/ftp/arxiv/papers/1006/1006.0846.pdf

Ericksen, E. G. (1980). *The Territorial Experience: Human Ecology as Symbolic Interaction.* U. T. Press. doi:10.7560/780385

Foucault, M. (1984). Of Other Spaces, Heterotopias. *Architecture, Mouvement, Continuité*, *5*, 46–49.

González, J., Miguel, M., Rosso, I., Toledo López, A., & Toledo López, V. (2016). Mapeando el barrio construimos territorio. Experiencia de cartografía social en Villa Aguirre, Tandil. *Revista Másquedós*, *1*(1), 61–71.

Gottlieb, E. E. (2009). "Somewhere better than this place/nowhere better than this place": The lifemap of Rolland G. Paulston (1929–2006). *Prospects*, *39*(1), 91–101. doi:10.100711125-009-9110-5

Gretzel, U., Fuchs, M., Baggio, R., Hoepken, W., Law, R., Neidhardt, J., Pesonen, J., Zanker, M., & Xiang, Z. (2020). e-Tourism beyond COVID-19: A call for transformative research. *Information Technology & Tourism*, *22*(2), 187–203. doi:10.100740558-020-00181-3

Diez Tetamanti, I. J. M. (Ed.). (2014). Hacia una Geografía Comunitaria: Abordajes desde la cartografía social y los sistemas de información geográfica. EDUPA.

Hall, E. T. (1990). *The Hidden Dimension.* Anchor Books Editions.

Husserl, E. (2008). A Ideia da Fenomenologia. Edições 70, Lda.

Kant, I. (2001). Crítica da Razão Pura (M. P. d. Santos & A. F. Morujão, Trans., 5th ed.). Lisboa: Fundação Calouste Gulbenkian.

Kemmis, S., & McTaggart, R. (2005). Participatory action research: communicative action and the public sphere. In N. K. Denzin & Y. S. Lincoln (Eds.), *The Sage Handbook of Qualitative Research* (pp. 559–603). Sage Publications.

Kemmis, S., McTaggart, R., & Nixon, R. (2014). *The Action Research Planner: Doing Critical Participatory Action Research.* Springer. doi:10.1007/978-981-4560-67-2

Kindon, S., Pain, R., & Kesby, M. (2007). *Participatory Action Research Approaches and Methods: Connection People, Participation and Place.* Routledge. doi:10.4324/9780203933671

Lefebvre, H. (1991). The Production of Space (Nicholson-Smith, Trans.). Blackwell Publishing.

Liebman, M., & Paulston, R. G. (1994). Social Cartography: A new methodology for comparative studies. *Compare: A Journal of Comparative Education*, *24*(3), 233–245. doi:10.1080/0305792940240304

Lobatón, S. B. (2009). Reflexiones sobre Sistemas de Información Geográfica Participativos (SIGP) y cartografía social. *Cuadernos de Geografía. Revista Colombiana de Geografía*, (18), 9–23. doi:10.15446/rcdg.n18.12798

Loureiro, S. M. C. (2020). 20 years of research on virtual reality and augmented reality in tourism context: A text-mining approach. *Tourism Management, 77.*

Lúcio, J. (2011). *A Mediação Sócio-Educativa na Construção da Cidade Educadora: O projeto "Massarelos, Freguesia Educadora"* (PhD essay). Universidade do Porto.

Lúcio, J., & Pereira, A. V. (2009). *Cartografia social do Porto.* Paper presented at the I Fórum de Investigação em Ciências da Educação, Lisboa.

MacDonald, C. (2012). Understanding Participatory Action Research: A Qualitative Research Methodology Option. *Canadian Journal of Action Research, 13*(2), 34–50. doi:10.33524/cjar.v13i2.37

Maffesoli, M. (1987). *O conhecimento do quotidiano* (J. Lamy, Trans.). Vega.

McIntyre, A. (2008). *Participatory Action Research.* Sage Publications. doi:10.4135/9781483385679

Paulston, R. G. (1996). *Social Cartography: Mapping Ways of Seeing Social and Educational Change.* Garland Publishing.

Ribeiro, P. O. (2014). *Cartografia social: (Re)Conhecer perceções; Mapear territórios.* Paper presented at the 1st Research Methodology School – What needs to be done? Coimbra University.

Robinson, W. I. (2002). Remapping Development in Light of Globalisation: From a Territorial to a Social Cartography. *Third World Quarterly, 23*(6), 1047–1071. doi:10.1080/0143659022000036658

Rocha, P. M. A. (2015a). *O Constructo da informação no espaço.* 9º Congresso da Associação Portuguesa de Ciências da Comunicação (SOPCOM) – Comunicação e transformações Sociais. Coimbra.

Rocha, P. M. A. (2015b). *Augmented Reality - The Augmented Construct of Communication.* International Conference on Communication and Light, 2-4 November 2015, Universidade do Minho, Braga, Portugal.

Serravalle, F. (2019). Augmented reality in the tourism industry: A multi-stakeholder analysis of museums. *Tourism Management Perspectives, 32.*

Shen, S. (2022). Exploring the factors influencing the adoption and usage of Augmented Reality and Virtual Reality applications in tourism education within the context of COVID-19 pandemic. *Journal of Hospitality, Leisure, Sport & Tourism Education.*

Zimmerman, T. G., Lanier, J., Blanchard, C., Bryson, S., & Harvill, Y. (1987). A hand gesture interface device. *SIGCHI Bull., 17*(SI), 189–192. doi:10.1145/30851.275628

KEY TERMS AND DEFINITIONS

Augmented Reality: Offers an enhanced perspective of the physical world. On a larger sense, by using digital visual gadgets (adaptive googles, any processor-based technology with a screen and a camera), its produced by a superimposing computer-generated image on the user's view of the felted and seen world. Thus, giving a composite perspective of our dimensional nature and its environment.

Human Ecology: Given as a descriptive study of the adjustment of a human population to the conditions of their respective physical environment, being them as an individual, or collectively as a social group, configured by their interactions, and their relation with the occupied territory.

Maps of the Future: Also named "dream maps", they reflect how the inhabitants see the community in the future, or how they would like to see it. They allow to co-create a probable outcome of how the territory could be better fitted to the needs of those living it.

Maps of the Past: The configuration of this historical map is important for the recognition of the changes that took place, thus collecting the collective memory of its inhabitants. The production of these maps allows characterizing the territory.

Neighborhood Revival: A moment in time and space, where neighbors cooperate and participate on remembering and sharing everyday situations and stories of their lives. Co-creating a stage where such memories are transmitted to younger generations, thus making that not be forgotten.

Participatory Action Research: A qualitative research methodology that involves researchers and participants collaborating to understand social issues and promoting probable actions to bring about a certain social change for the better.

Social Cartography: A collective exercise of recognition of one's socio-territorial environment, with legends and symbols, created mainly by those who live there. Through this exercise of social relations, it is possible to obtain the knowledge that allows its participants to have a better understanding of their reality. Building knowledge collectively, it becomes a community approach to its geographic, social, economic, historical, and cultural space. The construction of this knowledge is achieved through a collective mapping, shared with the general public.

Chapter 30

Street Art, Intersectional Feminism, and Digital Media-Art:

Report on the Cyberperformative Artefact "Make Me Up!"

Juliana Wexel
https://orcid.org/0000-0001-5830-5729
University of Algarve, Portugal

Mirian Estela Nogueira Tavares
https://orcid.org/0000-0002-9622-6527
University of Algarve, Portugal

ABSTRACT

The chapter is dedicated to demonstrating artivist aesthetic discourses produced in converging relations between urban space, street art, intersectional feminism (gender, class, ethnicity), and digital technologies in a post-pandemic context. The research results derived from three stages: a survey of state-of-the-art study on relations between street art, feminist artivism, and digital media art; production and application of autoethnographic data in the creation of digital artistic artefact; and curatorship of video-installation. The analysis focuses on the case study of the post-digital art artefact Make me up! an immersive and cyberperformative experience that connects augmented reality (AR) technology, street art, Instagrammism and Selfiecity. The digital artefact Make me up! was launched during the "10th International Conference on Digital and Interactive Arts - ARTECH 2021: Hybrid Praxis: Art, Sustainability & Technology" in the historical city of Aveiro, Portugal, also known as the "Portuguese Venice."

DOI: 10.4018/978-1-6684-3369-0.ch030

INTRODUCTION

Since the emergence of the pandemic in 2020, the experiences in the arts in the digital field have been intensified in a progressive and exponential way, including the intersections between public space and digital creations. One can list numerous practices that have been adapted with the use of digital and telepresence technological resources, especially those focused on performing and scenic arts, and made available "through digital platforms, exhibited via streaming, or on social networks such as *Facebook* and *Instagram*" (Wexel, 2021, p. 43). Discussions of the kind that are commonly part of the scenario between the relationship of digital technologies in the construction of new aesthetics and languages in the performing arts, "such as cyber-performance, have obtained even greater relevance after the pandemic, new points of view and necessary confrontations regarding the insertion of the artistic panorama in digital languages." (Wexel, 2021, p.43), as well as their displacements concerning street art, urban art or public art. The chapter presents the case study of a digital artefact of cyber-performative and immersive character, of post-digital concept, developed in 2021, in a post-pandemic context. The artwork *Make me up!* was developed from the resource of augmented reality (AR), using street art creations circumscribed in aesthetic discourses that present a feminist and intersectional character. The discussion also derives from the production of aesthetic discourses from the artefact's cyberperformative (Jamieson, 2008) character. Make me up! is based on their relationship with the epistemological power of artistic images (Zielinski, 2017), coupled with the condition that images in contemporaneity cross bodies (Baitello, 2017) and the interpretation of the body itself as a place of combat and disputes (Bento, 2017), along with the idea that the space of the city is also a body (Baitello, 2017). Based on the idea of the "all-screen" (Lipovetsky & Gilles, 2010) and transmedia language (Scolari, 2013), the *Spark AR Studio* platform resource was used in the creation of augmented reality and facial recognition filters, modelled from 12 works produced by feminist artivists operating from Portugal, Brazil and Italy. For the design of the artefact *Make me up!*, we invested in an autoethnographic and ethnographic (Foster, 1996) research methodology for artistic creation (Candy, 2006). The journey is part of the research for the doctoral degree in digital media art, originally entitled *Gender, urban art and media art: the aesthetic discourse of arti(vi)stas in an autogynographic map* and currently under the new title *Vulva art, art(vi)smo and digital media art: Imagetic contributions in new feminist aesthetic discourses for a culture of equity*. The research results derived from the instancing (Marcos, 2016) of 3 stages: a survey of a State of the Art Study on street art, feminism and digital media art; production of an autoethnographic map of works produced by feminist artists, application of the ethnographic data in the creation and curation of the *AR* artefact *Make me up!* and, finally, in the exhibition of the artefact in an art installation at *ARTECH 2021 - Hybrid Praxis - Art, Sustainability & Technology - the 10th International Conference on Digital and Interactive Arts* - held by the University of Aveiro. It is understood that this product is part of the scientific exercise of contributing to a transversal and interdisciplinary debate between digital media-art applied to urban artistic interventions, intersectional feminism and artivism, besides new visions in the field of post-pandemic arts.

BACKGROUND: STATE OF THE ART

Nascimento et al. (2019) replicate Solnit (2016) when she mentions that for men, it has always been easier to walk down the street than for women. The discussion is on "Is walking for all? A Latin American women's approach to drifting and flâneries in contemporaneity". In the study, the main question is whether "walking has a gender?". She highlights that "Among the tacit orders of the city, questions of gender and sexuality establish times and contexts in which a given body can or cannot occupy the street (...) what delimits a woman's mobility through public space is the fear she feels when walking through it." (Nascimento et al., 2019, p. 6). Nascimento (2019) contextualize that even in Western societies, "women were customarily punished and intimated for experiencing the simplest of freedoms, that of going out for a walk because their walking and, indeed, their very existence was inevitably and continuously sexualized in societies concerned with controlling female sexuality" (Solnit, 2016, p. 387).

Based on this, it is also worth the considerations made by Karen A. Franck & Paxson (1989) in an article discussing the issue of "woman's" freedom regarding the use of public spaces, "woman" this contemporary in western countries, in industrialized countries, who highlighted that the use of public space should be seen in the context of her daily life and experiences. Alongside them, dozens of researchers began to discuss the occupation by women in spaces that were previously "naturally" perceived as predominantly male.

In a survey on artistic interventions in urban space by women artists during the emergence of the new social movements between the 1960s and 1970s, Dawsey (2008) points out that "Specifically, this development concerns the emergence of an explicitly spatialized form of feminist art production: sculptural, photographic, conceptual and performance practices that pointedly engaged the spatial parameters of the city." For the researcher, "many women artists began to imagine their feminist responses to space-to imagine the production of new kinds of spaces." (p. 21).

In exploring the intersection between gender studies and creativity in urban art, Parisi (2015) was interested in focusing on the observer's perception of the work of women street art artists. When it comes to Portugal, Valente (2016) proposes a comparative analysis "between the Portuguese case and the international context (...) and the current place of the female artist in contemporary art" supported by a "historical chronology focusing on the 20th and 21st centuries" (p. 4). The researcher outlines a profile of street art artists in the Portuguese country, also focusing on expression through *graffiti*. She points out that "in Portugal, this manifestation is mostly represented by a male audience: the rivalries between the masculine and feminine genders manifest themselves, above all, in the representation of the woman as a sex symbol and object of desire" (p. 4), but that "with the approach of women to urban artistic expressions we witness an amalgam of discourses and narratives, whose potentialities are growing (...) and with the appearance of new vanguards and manifestations, conflicts are replaced by inclusion and union to exalt the diversity of aesthetic languages." (Valente, 2016, p. 4). The growing number of studies on the occupation of street art and its relationship with gender studies, feminism and digital technologies focus relative emphasis on case studies about their use in resistance and struggles for social rights, inclusion policies and human and environmental rights. An example is Ruiter's (2015) study Imaging Egypt's political transition in (post)- revolutionary street art, which offers some concepts on the interrelations between social media and *graffiti*, the latter being elected as a means of political engagement and expression by young women in Cairo.

More than relevance, there is, above all, a need for researchers and artists who invest in producing scientific, artistic and cultural knowledge in the imbricated field of digital technologies and digital media with application in arts, even more so when it comes to access and appropriation of these devices by women artivists and the perspective of feminist studies in the production of aesthetic discourses with potential for social transformation. Many of the approaches to which scientific research is concerned deal directly with the dialogue between art and cities, with an emphasis on the relationship with public space, architecture, with the transformation of the city, but in a very incipient manner in terms of digital technologies, their uses and the impact on emancipatory processes of gender and non-binary visions (Bento, 2016). Group organization, aesthetic choices, thematic content, political engagement actions and the relationship with the theme and digital technologies and new media still receive incipient attention, especially when electing a cut based on feminist epistemologies. About the younger populations, we highlight research that focuses on the reality of black youth, the practices of deconstruction of racism and racial discrimination, the role of black feminism in the resistance movement, the production of new meanings and discourses and the construction of new pedagogies from the use of digital technologies, artistic practices and new media, such as Restorying as political action: authoring resistance through youth media arts, by researchers Stornaiuolo & Thomas (2018), from the University of Pennsylvania. This article discusses the realities experienced by young people who grow and learn in an atmosphere permeated with religious intolerance, racial discrimination, misogyny and xenophobia and highlights the importance of schools providing access to digital media and offering students opportunities to fight against social injustices, tools for resistance to such discourses and systematic practices of oppression and exclusion.

In the Brazilian context, there are relevant studies on the work of women artists in electronic and digital arts. One example is the historiographic research conducted by Osthoff (2010), which lists and analyses the contrast of the play between presence and absence, physical and virtual spaces in the performances and installations of Artemis Moroni, Simone Michelin, Bia Medeiros and Diana Domingues, the works of Tânia Fraga, Rejane Spitz, Suzete Venturelli, Giselle Beiguelman and Patrícia Canetti. These women artists "explore virtual worlds, design, art and online activism, respectively" but who "do not all act in public art" but who, however, in terms of Brazil, "is a reference when it comes to the feminine universe in the creation of new technologies and part of electronic and digital art." (p. 89). Regarding studies related to digital technologies and the public, there are some mentions about "female" users of electronic games, especially young Brazilian women, as in Anne-Marie Schleiner's research, Transactional Play: Piracy, Urban Art, and Mobile Games, especially in the section that deals with The Free-To-Play time in Brazil: Localized Mobile and Casual Games.

About her Intervention in the urban context with computer-mapped projection, Venturelli (2016) reveals that "in the urban interventions, co-authored with Bia Medeiros, videos and photographs were the means of communication and questioning about the use of the female body as an object in the media" (p. 2002). Moreover, to the artist, "in the urban interventions, performance was the medium, where the body should reinvent itself as strategic of the action in the social context and its environment." (p. 2003). Nowadays,

Venturelli (2016) elaborates on urban interventions mediated by technology with "computer projections, mapped projections, and interactive projections, enabling something new, until then, in the dynamics that materialize in the process of collaborative creation, of social participation, even at a distance." (p. 2003).

Carvalho (2016), who developed the research entitled "Locative media-art and dynamic mappings in the understanding of urban flows: the contribution of the mobile device in the activity of walking as artistic practice", demonstrated in his doctoral studies how one can, "from the locative media- art and through the development of new mappings, structure the analysis of urban spaces based on locative media processes and the production of multidimensional cartography." (p. 154). The researcher also mentions the relationship of artistic interventions with digital technologies: "E.g., Burke (2007) addresses the theme of interactive urban artistic interventions, which make use of technology and location-based informational networks" (Carvalho, 2016, p. 36). In this context, "locative media-art uses mobile devices allowing to read the cities from the referred psychogeographic maps associated with collaborative cartographies" (Carvalho, 2016, p. 36), thus promoting a hybridization not only of the physical and digital space, but also in the totality of relationships and connections that are established between both.

MAIN FOCUS OF CHAPTER

In an updated discussion for inclusion and access policy discourses in the development of intelligent cities, "Mainstreaming gender equality in smart cities", Nesti (2019), a researcher at the University of Padova, Italy, highlights grounds that unveil deficits in the gender equity treatments in this debate. In this sense, discrimination arising from implementing policies at a local level has been intensely fought by international organizations and also discussed by the literature, especially after the 1990s. "For example, the UN Habitat program specifically addressed the importance of adopting anti-discriminatory practices to empower women and girls in urban life." (UN Habitat, 2000; UN Habitat, 2010; Nesti, 2019, 290).

The concept of *Make me up!* privileges intersectionality theory (Lauretis, 1976) in feminist studies, which draws on the intersection of ethnicity, class, and gender in the analysis of socio-cultural phenomena, including digital culture. It is based on the premise that, when it comes to feminist artivism, the private sphere is also public in the face of a historical and patriarchal association that "women" belong to the domestic/private space and "men" to the public space. A type of approach which took place from the 1970s onwards and which is present in studies such as those of Franck & Paxson, (1989), who state that "From early times until today and in societies throughout the world, women have been associated with and even restricted to the private space of the dwelling whereas men have had freer rein to frequent public spaces." (p.122).

It is proposed, from the genesis of the digital artefact *Make me up!*, to contribute to the objective number five of the UN Agenda 2030, of promoting gender equity in the world, and in particular, to the indicator that aims to expand the use of essential technologies, information, and communication to promote female empowerment. *Make me up!* is a research project in post-digital artistic practice (Manovich, 2017), of immersive, interactive, performative character (Dixon, 2009) and which consists in the use of an Augmented Reality facial recognition *app* as an interface for the transmediation of street artworks, originally produced by feminist artists, into digital filters that can be directly accessed via smartphone.

The semiotic concept of the computational artefact is inspired by the prop of the mask, present in ancestral rites, for example, the performative ritual and mimetic character of the *ditirambo*, a procession led by the representation of Menades and Satyrs enacting the mythical vicissitudes of the Greek god Dionysus (Modeo, 2018). This ancestral rite "refers to the possible anthropological origin of part of the artistic manifestations of public performance in the Western world [...] a choral song, danced, with flute accompaniment, in homage to the god Dionysus", a myth originating in Asia-Main." (Belleza, 1966, p. 14; Wexel, 2012, p. 14). The ritual was continued in Greece alongside the " phallic chants of a procreation cult, which later merged with the mimes or satirical representations, used mainly in Sicione, and which were also associated with the festivities of the Dionysian cult itself." (Belleza, 1966, p. 14; Wexel, 2012, p. 14). It is noted in Wexel (2012), that "both Belleza, Lesky and the German philosopher Friederich Nietzsche draw on the concept, stemming from Aristotle's Poetics, that tragedy has its origins in the dithyramb." (p.13).

Since immemorial periods, together with the use of make-up and body painting, masks have acted as identity elements for ancestral and original peoples and are central to the aesthetic discourse of their artistic and ritualistic expressions. Depending on the contexts in which they are used, masks and make-up become instruments that allow the amplification of the expression of identity or even offer the possibility of new representations, being both gesture and performative act. It is considered that both the makeup and the mask go beyond the representation of ornament or decorative adornment: they are symbols of belonging to cultural, social, and temporal contexts and have their maintenance in carnival traditions and in contemporary performative actions. Just as masks and make-up are ancestral devices, rock art manifests the first visual expressions on surfaces. The research published by archaeologist Dean Snow of Pennsylvania State University in 2013 about the relationship between gender and cave artists suggests that at least 75% of the pictorial hands of the Palaeolithic period, discovered in caves in Spain and France, would have been produced by women.

The title *Make me up!* also refers to the various uses of that phrasal verb:

- In the first sense, it denotes the act of being made up or of self-make-up.
- In the second sense, it connotes the act of "inventing" something from an external action or internal motivation.
- In the third plan, it alludes to the idea of receiving "encouragement" from some external agent, which is understood as an imperative quite adapted to the moment of generalized uncertainty due to the global pandemic crisis.

In this sense, the artefact is also an allegory for the daily and contingent use of the prophylactic mask.

Figure 1. In the collage, on the left, the project logo; in the centre and on the right, the Brazilian actress Aline Tanaã cyberperforms the filter of the street artwork "Toda Forma de Amor" (2019), by @RafaMon, interacting with a material mask in plaster, critical conceptual element of the conception of the project.
Source: Make me up!

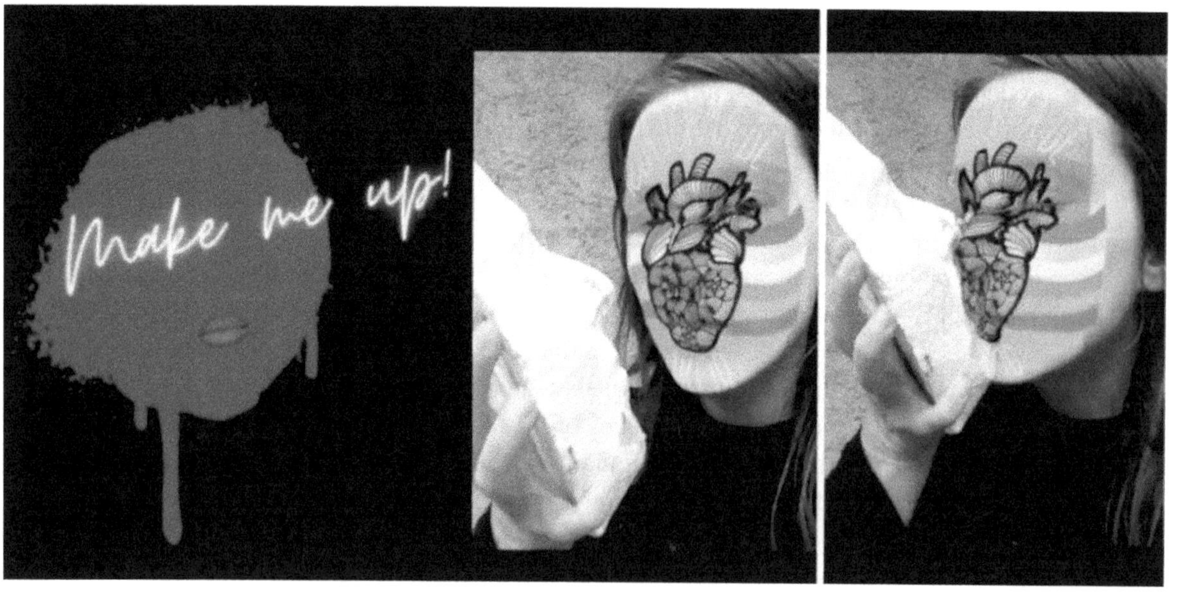

DEVELOPMENTS AND RESULTS

The choice of artists and street artworks that composed the research sample derived from an autoethnographic and ethnographic survey (Foster, 1996) based on the creative journey of the first author of this article in Brazil, Portugal and Italy. To systematize the production of the digital artefact, also with a view to the production of qualitative data and the organization of production processes, it was decided to instantiate (Marcos, 2016) three stages of the development of artistic research: the first stage, that of the ethnographic data survey; the second stage of selection of a sample of works that dialogued with each other from themes that recur in the artists' work and aesthetic statements and, finally, the third stage: the application of research data in the elaboration of the final version of the artefact *Make me up!* The first stage of data collection was used to obtain a stratum of characters and potential works for the beginning of the ethnographic phase. In this phase, 62 artists and collectives were identified in Brazil, 39 in Portugal and 29 in

Italy, all artists working in the urban space of those countries. When analyzing the thematic set of the works of these 130 artists in question, themes that recur and permeate from the perspective mentioned above of intersectional feminism (ethnicity, class and gender) stand out: black feminism, domestic violence and femicide, misogyny, sexuality and vulva art, LGBTQIA+ rights, including the fight against homophobia and transphobia, valuing indigenous peoples, environmental sustainability, decolonial thinking.

For the second phase of the project, with a view to the production of a video-installation exclusive to the *10th International Conference on Digital and Interactive Arts - ARTECH 2021: Hybrid Praxis: Art, Sustainability & Technology*, in Aveiro, Portugal, a selection of street artworks in the modalities *graffiti* and paste-up was chosen as a basis for the modelling of digital masks focused on the themes of sustainability and defense of indigenous peoples, black feminism, LGBTQI+ rights, sexuality and vulva art. The images are authored by artists who have their base constituted in the autoethnographic axis delimited in the scope of the research, but that, however, extend their productions in the public space of several cities in the world. Among the artists who collaborated with the assignment of using images for the first cyber-performative version of the artefact, *Make me up!* on the profile *@makemeup.artproject* on *Instagram* is the Italian architect Mariella Gentile, @Marioky, with consolidated works both in Italy and Portugal; the Brazilian visual artists Rafa Monteiro @RafaMon and Carla Felizardo, @Negra_Graffiti, besides the Brazilian visual artist Carine Panigaz, the latter creator of the project @ Vulvabell and working from Lisbon, Portugal.

In *Make me up!* the digital mask is used as a digital make-up, analogous to the physicalized gesture of "putting on" a "second skin" and crossing it (Giannetti, 1998). By making use of the technological resource of augmented reality (*AR*) and facial recognition, the public incorporates the street artworks and assumes the "fourth skin of the city" in its transmedial virtual skin (Baitello, 2017) through a cyberperformance (Jamieson, 2008). In this case, augmented reality serves as a resource to enable the interface between the digital image and the face of the interactor, which becomes new support for the art previously generated in the public space. In terms of choice of technology, a correlation of this with urban media art, "a still emerging and rapidly developing domain in public art, constantly transforming itself and forging new relationships with technology, architecture, design, science, biology, urban development, film, music, performance, and other disciplines, was privileged." (Toft, 2016, p. 52). From transmedia creation, a possible field of production of new senses is generated through its new uses in digital and networked support. Based on an idea of all-screen (Lipovetsky & Serroy, 2010), we used the *Spark AR Studio* application, an augmented reality platform for *Mac & Windows* that allows the user to create *AR* effects for mobile cameras through sketches and other tools similar to Photoshop. The resource of the *Spark AR Studio* platform was used in two stages of creation and curation of the artefact: first, for the modelling of the digital filters based on the original work of street art and, in the second plan, for the availability of the final filter on the network. Before its viability, the filters went through a careful evaluation from the policy of use of the *Spark AR Studio* platform. The first version of the digital artefact was composed of 12 filters for selfie photographs developed and made available via Instagram, where the user could try them with no limit of time of use or number of accesses, perform them, register them in photographs and screenshots, download the results and, still, share their experiences on the network. The only requirement to access the performative artefact would be to have an active profile on *Instagram*.

Figure 2. On the left, visual artist Rafa Monteiro @RafaMon poses in panorama next to the graffiti !Yara"(2018), created on a gable in the cove of Botafogo, Rio de Janeiro, Brazil, in one of the most famous tourist cards in the world, with Sugarloaf Mountain in the background. On the right, Albanian architect Dorina Pllumbi shares via Instagram her experience (Selfiecity) with the work transmediated into a digital filter !Yara" via Spark AR Studio.
Source: Make me up!

The assignment and sharing of *Make me up!* filters on *Instagram* follow the concept of Instagrammism (Manovich, 2017) and Selfiecity (Manovich, 2019). Since its creation, this social network has differentiated itself by offering editing tools integrated into the app itself and, with this, stimulates the production of aesthetic experiences and narratives based on photography and self-representation in digital visual culture and, more recently, by the introduction of the aesthetics of filters in *AR*. An important detail about the digital curatorship of the Make Me Up! masks on the profile @makemeup.artproject on Instagram is that they are titled respectively with the names of the works and the profile of their creators. The goal is to potentiate from the conceptual dissemination of the project and its original creations to the exchange between the audience and artists besides reinforcing the artivist character of the computational artefact and facilitating feedback on its reception aesthetics. Following the example of the *collage* in *Figure 2*, the Albanian architect Dorina Pllumbi performs (Selficiety) the filter *Yara*, authored by the Brazilian artist @RafaMon, and shares the performative experience on her profile @dor_pllumbi. The original street art piece represents the myth of the mermaid *Yara* in Brazilian indigenous folklore, a mythological figure corresponding to *Iemanjá* in Afro- Brazilian culture, both protectors of the waters and seas. It was graffitied on a gable in Botafogo cove, located among the main emblematic landscapes of Rio de Janeiro, one of the most visited tourist cities in the world. A few months before the conception of the computational artefact *Make me up!*, the artist @RafaMon was represented in a lawsuit in favor of the maintenance of the graffiti Yara (2018), as it was in danger of being eliminated at the request of the building's condominium to make way for an architectural restructuring. The idea of opting for *Yara*'s filter mask in the first version of the artefact centers on the idea of an activist discourse of engagement in the networks for the maintenance of the *graffiti* through the hashtag *#YaraLives*.

The art educator and *graffiti* artist Carla Felizardo, or @Negra_Graffiti, is part of @AmoCrew, a *graffiti* and arts collective formed by black and peripheral women, especially in regions of the North Zone of Rio de Janeiro, and works under the perspective of themes focused on the agendas involving black feminism. In June 2020, during the pandemic crisis, the artist created the work *Butterflower*, at the Escola Municipal Escultor Leão Velloso, in the Pavuna neighbourhood, in Rio de Janeiro's North Zone, one of the oldest places in the city, founded during the imperial (and slave) period, when the city was the capital and seat of the Portuguese Crown in Brazil. The work has a connotation of blossoming amid chaos in an oppressive society, where black women, among minorities, are the most affected by the patriarchal chain of racial and social inequalities.

Similar to the work "Politizar as novas tecnologias" by the sociologist Laymert Garcia dos Santos, the debate between the borders and relations between technology and art, mentions the opening of a myriad of possibilities to the problematizing intervention of art: "the critique of the new forms of domination based on gender, class, race or nationality such as, for example, imperialist wars, genocides, terrorism, international migration and xenophobia, and above all, the critique of universal surveillance and predatory globalization, not to mention the problems of the *spectacularization* of life and, above all, environmental degradation." (Machado, 2007, p. 56). Baitello presents an alternative approach for the image sciences, and that serves as a theoretical basis for the constitution of the artistic artefact in question, from the perspective of the body, "this complex other that ends up, many times, renegade to a second plan, or even ignored completely". The author starts from the bioanthropological premise that the body asks for body (Baitello Jr., 2017) and reflects on the image and the body from authors Hans Belting and Dietmar Kamper. One can consider that "The aesthetics of physicality, as well as of high technology, computers, internet and video, can become tools and an environment for an awakening of social and political interest" (Lehmann, 2013, p. 865) and of constructions of new languages. (Wexel, 2021, p. 45).

In *Make me up!* the digital body of the filter also requires a physicalized body to complete its realization and in the impossibility of contact with the street space also in the pandemic cycle, the body of the city is transmedia in the body of the viewer from the resource of augmented reality. Thus, the relationship between the body of the viewer-screen-body of the city is central to its creation. The first official experience of the fruition of the masks was the constitution of a video-art homonymous to the artefact, made from the curation of performances by users in Portugal, Italy, Ireland, Albania, Finland, Estonia and Brazil. Its script was based on three premises: the behavior of use of the users, the repercussion of the performative interaction of the public and, finally, its aesthetic reception. From the results of the performances on the users' "virtual skin" and the narrative potentiation of gestures, video art also fulfils the function of demonstrating, *a priori* and together with the hybrid concept of the artefact, the polyphonic possibility (Bakhtin, 2003) of the performative discourses produced by its use and that enhanced the original aesthetic discourse of street artworks. *Figures 4, 5, 6, 7*, and *8* are a cutout of the 12 creations that make up the first version of the digital artefact at @makemeup.artproject.

Figure 3. In this collage, at left, the art educator @Negra_Graffiti shows Butterflower (2020) in a post in her Instagram profile, in the north zone of Rio de Janeiro, already in a post-pandemic period; at the centre, the Butterflower filter approved by the Spark AR Studio use policy in Instagram (in the Portuguese language); at right, the art critic, curator and professor of the Academia di Belle Arti G. Carrara-Bergamo, the Italian Claudio Musso, performs the same work in AR at his home in Bologna.
Source: Make me up!

Our Lady of Aparecida (Nossa Senhora Aparecida) is the Christian patron saint of Brazil. In Afro-Brazilian religious syncretism, it is a figure that represents *Oxum*, the goddess of beauty in Candomblé and Umbanda. According to visual artist Rafa Monteiro's concept, publicly expressed on social networks through his profile @RafaMon on *Instagram*, the aesthetic and activist discourse of the work communicates an appeal to the heavens, a "pray for us", due to the worsening of the health crisis in the biennium 2020-2021 due to the absence of a pandemic containment policy, the denialism and genocidal strategy orchestrated by the president of the republic, Jair Bolsonaro, and his peers in the Brazilian executive. In *Figure 4*, the actor Lucas Asseituno, instead of applying the filter as a mask of his own, chooses to use another support in his performance: he transfers the filter in *AR* to the face of the image of the singer and composer Chico Buarque de Holanda on the cover of the LP "Feijoada Completa" (1978), a classic of Brazilian popular music (MPB) and which contains one of the famous songs and anthem of resistance to the Brazilian military dictatorship since the coup of 1964, entitled *Cálice*. Chico Buarque is an artist who has a close relationship with the history of left-wing political struggles in Brazil based on the Partido dos Trabalhadores (PT), is directly opposed to the neo-fascist actions of the extreme right-wing government of Brazil and expresses support for former President Luis Inácio Lula da Silva to a possible new presidential race and possible re-election in 2022. On the right in the *collage*, the Brazilian therapist Neiva Rodrigues assumes the religious aspect of the figure, dresses up as *Our Lady of Aparecida*

and "embodies" the character, applying the filter on the right side of the face and generating a kind of "double" in the aesthetic representation of the work in a *cyber- performance.*

Figure 4. On the left, Lucas Asseituno projects the "Aparecida" filter onto the face of singer-songwriter Chico Buarque; in the centre, the original artwork in digital design and, on the right, Brazilian therapist Neiva Rodrigues cyber-performs in the representation of the saint.
Source: Make me up!

In *Orgulho* (2020) (*Pride*), the artist uses some colours of the LGBTQIA+ flag and inserts the lettering of the title in work. In its AR version, the artivist discourse of Pride is reinforced in the performers' actions, as can be seen in the collage in *Figure 5*. Both put themselves bodily to reproduce and "mirror" the gesture with a clenched fist, symbolizing the struggle for civil and human rights to LGBTQIA+ diversity. The original *graffiti* was reproduced by the artist and projected in urban media art in the technique of video mapping, in the historic and tourist district of Santa Teresa, in Rio de Janeiro, Brazil, in a post-pandemic period. Since the contingency cycle of lockdown and social isolation, the resourceful use of the practice of digital visual projection has been enhanced by artists worldwide who, like Rafa Monteiro, use the city as artistic support.

Following the example of the @Vulvabell project, numerous artistic projects circumscribed in the fourth feminist wave, which began around 2012 and is characterized by a focus on the empowerment of women, the use of internet tools, and intersectionality, focus on the issue of knowledge of the vulva's anatomy, especially the clitoris. In this action, the bell is a direct analogy to the stimulation of the clitoris, the central element of the production of *jouissance* in bodies with vulva and only fully described in the annals of contemporary medicine only in 1998 due to the investigations of Hellen O'Connell, urologist and professor of the Department of Urology and Surgery of the University of Melbourne, Australia. The artivism of the @Vulvabell project takes place through the aestheticization of mini vulvar sculptures, *graffiti* and paste-up drawings of the anatomy of the clitoris, which have been stylized in public spaces in dozens of countries across Latin America and Europe since 2017. The mini sculptures are interactive: the clitoris is represented by a rattle, an element that instigates the viewers to interact with this kind of symbolic "bell". Furthermore, through the stop motion animations disseminated in the networks and the posting of the vulvar mini- sculptures in public spaces, the artist invites the viewers and passers-by to interact with the aestheticized anatomy of the vulva. For the project of the artistic artefact *Make me up!* four images of the project *Vulvabell* were used, two referring to the shape of the vulva: the first, in which the genitalia is a girl and another allusive to the anthropological myth of the toothed vagina, besides other two in which the anatomy of the clitoris is personified and applied through the technique of paste-up in spaces already inserted in the circuit of street art in Lisbon, Portugal. The use of these works in the constitution of the first version of the artefact also serves to develop a concept about vulva art, an object of investigation and problematization for the next stage of research of the doctoral research in progress, which integrates the elaboration of *Make me up!*.

For the presentation of the digital artefact *Make me up!* in the space of *ARTECH 2021 - 10th International Conference of Digital and Interactive Arts*, held in the historic city of Aveiro, Portugal, it was privileged the concepts of interactivity and immersiveness. The curatorial proposal of the artefact at *ARTECH 2021* united the creation of the facial recognition digital masks application resource via *Spark AR Studio* for sharing on *Instagram*, the exhibition of the result of its uses through the recording and sharing of the experiments along with the artistic and audiovisual intervention of the video-performance. To offer access to the artefact for free experimentation purposes, *QR Codes* were affixed next to the image corresponding to each of the street artworks allowing redirection to the profile @makemeup.artproject on *Instagram* via smartphones. The curatorship of both resources, audiovisual exhibition and *QR Code*, privileged the public's experience in loco: in the installation, the video performance fulfilled a narrative and support function to the interaction offered by the masks and its singular experience.

Figure 5. On the left, Brazilian journalist Leonardo Kluck performs the gesture in front of the Coliseum in Rome; in the centre, an image reproduced in the artist @RafaMon's Instagram profile of the artistic intervention via visual projection; on the right, cyberperformance by visual artist Rafa Monteiro, author of Pride (2020). In the cyberperformances, the repeated gesture denotes the collective engagement to the cause in a representative act similar to the original symbol.
Source: Make me up!

Figure 6. On the left, visual artist Carine Panigaz in the process of creating a representation of the personified vulva; in the centre, the finished work; on the right, journalist from Rio de Janeiro Ana Beatriz Petrini potentiates the aesthetic discourse of the work in AR through another device: a plaster statue inspired by the painting La nascita di Venere, (1482-1485), by Sandro Boticcelli. In her gesture, the journalist "reveals" what in the Italian painter's original work is hidden by the goddess's hair.
Source: Make me up!

Figure 7. In this collage, the Argentinian tourist technician Julieta Barlaro evokes the gesture of the taboo subject of masturbation and clitoral stimulation from the paste-up inspired filter in the centre, applied in public space in Lisbon, Portugal; on the right, for the video-performance (available on the @makemeup.artproject profile on Instagram as well), Brazilian actress Aline Tanaã sounds a bell underlining the slogan of the project "ring your bell" and also produces an image of connotation and apology to oral sex from the mask in AR.
Source: Make me up!

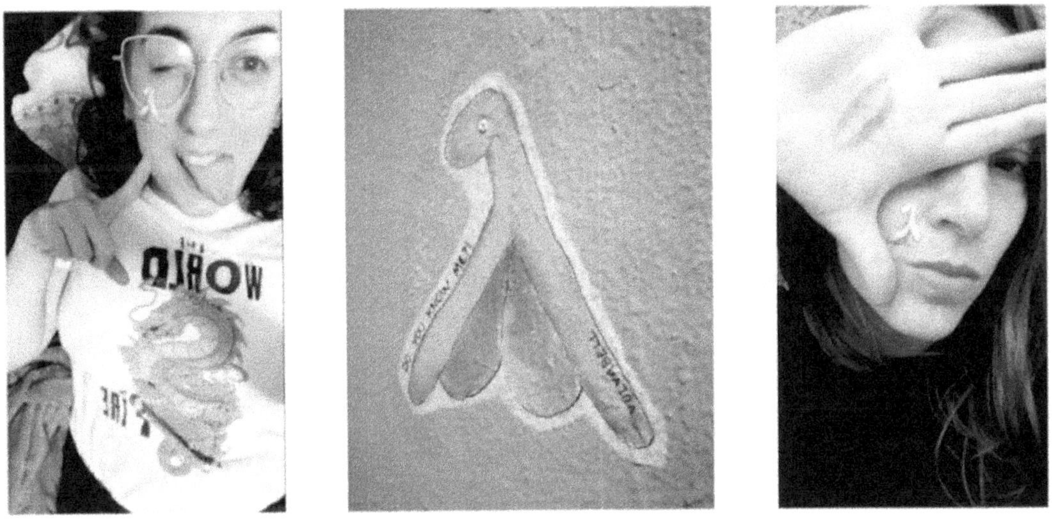

By glimpsing the re-signification of the artefact in new media, as in the case of video art and the exhibition scenario in *ARTECH 2021*, it was also given a new artistic function, and a discursive element was introduced to the project. In this sense, the audiovisual resource of the video performance was also illustrative of these effects, which revealed unexpected results and unexpected discourses in its use, enhanced by the aesthetics of each filter and the users' creativity. By idealizing the realization of this specific interactive installation in a public exhibition space, a double artistic composition was established to the conceptual circuit of *the body of the city-screen- body of the viewer-return to the body of the city*. Together with the art installation, live action of Brazilian visual artist Rafa Monteiro was also performed, offering the public the possibility of direct interaction with the artist. At the same time, she produced a *graffiti* version on canvas of the work *Toda forma de amor (2019) (All forms of love).*

Among the project's curatorial difficulties was the choice of works that could be transmedial without losing visual quality. Some experiences could not be constituted due to material factors that impacted the quality of the images of the works in public space in their digital transposition to augmented reality technology. Problems can be cited, such as the porosity of the original images applied on supports of various materials, such as old and deteriorated walls, low lighting for the production of high-quality photographs, and the wear and tear of the works due to the action of time and the meteorological impact in cities. Part of the images, as in the case of *Nossa Senhora Aparecida* and *Orgulho*, had to be used from its original digital format since the loss of quality of the street art image made the creation of the filter in *AR* unfeasible, and its activist aesthetic discourses were relevant for the creation of the first version of *Make me up!*. After the verification and approval of the functionality of the images, the subsequent challenge was to model the images and test their functionality in facial format and ensure cyber performativity that did not distort or interfere, *a priori*, in the original aesthetic discourse of the street art creations. These can be essential recommendations for future projects of this genre of filter creation via Spark AR Studio based on images and photographs taken in street artworks and for a second version of the artefact in question.

Figure 8. Collage of fruition of the work Toda forma de amor (2019) in live-action by Brazilian visual artist Rafa Monteiro during the installation of @makemeup.artproject to ARTECH 2021 Conference, at the Convention Centre in the touristic Portuguese city of Aveiro.
Source: Make me up!

FUTURE RESEARCH DIRECTIONS

The development of this academic research took into account the premises of Practice-Based Research (Candy, 2006) and Art-Based Research (Eisner, 2012), aiming to obtain new knowledge through artistic practice, as well as to produce not only original results but that could be demonstrated through the creation of artistic artefacts. The procedure following this research will be the conception of a new artistic artefact: a cyberperformance centered on the reinvention of the myth of Aretusa. The version of the myth of Aretusa that inspires the project and narrates an essential particularity of nature and the founding architecture of Ortigia, the island of Syracuse, in Sicily, Italy, is that in book V of the Metamorphosis by Ovid. The favorite nymph of the goddess Artemis, who, fleeing from Greece because of the obsession of Alpheus, one of the sons of Oceanus, reached the small island and became a fundamental element of this land: a homonymous water spring, from which Ortigia unfolds. The Aretusa Fountain has undergone incessant modifications over the centuries. However, it maintains part of the only wild population of Cyperus papirus in the western world and, especially in the European summer, is visited by tourists and travelers from all over the world. Decoding this myth, which crosses not only historical but also literary times, and which has in the mythical water spring a course that runs between fiction and reality, is one of the artistic intentions of the project. From experience with the development of the artefact *Make me up!* it is maintained in the creation of the new artefact not only the cyber-performative character but also the artivist aspect and that privileges the point of view of intersectional feminism.

CONCLUSION

In *Make me up!* digital AR technology was used to support aesthetic discourses that highlight artistic self- expression in cyberperformance and the network, having as substrate street artworks made under the perspective of intersectional feminist artivism. The use of the imperative verbal mode in its title was a call to the public to “reinvent” itself: an invitation to the digital performative gesture since it is understood that the relations of convergence between public space, street art and digital media-art have been changing substantially in the post-pandemic context. It can be concluded that the aesthetic discourses produced by the street artworks were maintained, enhanced and amplified with the transmediation of the creations in reality filters increases and from the interaction and cyberperformativity of users.

ACKNOWLEDGMENT

Thanks to the agents and institutions that support this research: FCT- Fundação para a Ciência e a Tecnologia de Portugal and CIAC-University of Algarve, in Portugal, through the PhD in Digital Media- Art-DMAD. [Research Grant UI/BD/150845/2021] and to the members of CyPET Arts - Performing Arts Study (CIAC- FCT), a project that aims at the theoretical and practical exploration of cyberperformance from creative, performative and communicational angles and intends to develop a new pedagogical model for its inclusion in the higher education curriculum.

REFERENCES

Azevedo, I. (2014). A arte urbana e seus possíveis entrelaçamentos: considerações sobre intervenções artísticas efêmeras. In *Revista DA Pesquisa*. Universidade do Estado de Santa Catarina. UDESC.

Bakhtin, M. (2003). *Estética da criação verbal* (3rd ed.). Martins Fontes.

Belleza, N. (1966) *Teatro grego e suas conseqüências: raízes do teatro*. Pongetti. Bento, B. (2017). *Transviad@s: gênero, sexualidade e direitos humanos*. EDUFBA.

Carvalho, I. (2016). Média-arte locativa e mapeamentos dinâmicos na compreensão de fluxos urbanos: o contributo do dispositivo móvel na atividade do caminhar como prática artística. Lisboa: UAb, UAlg.

Dawsey, J. (2008). *The uses of sidewalks: Women, art, and urban space, 1966–1980*. Stanford University.

Dixon, S. (2007). *Digital Performance - A history of New Media, Performance Art, and Installation*. The MIT Press.

Domingues, D., & Venturelli, S. (2008). Cibercomunicação cíbrida no continuum virtualidade aumentada e realidade aumentada: era uma vez… a realidade (Vol. 1). Revista ARS. Disponível em http// www.scielo.br/pdf/ars/v5n10/11.pdf /2011

Foster, H. (1996). The artist as ethnographer. In *The return of the real*. The MIT Press.

Franck, K. A., & Paxson, L. (1989) Women and Urban Public Space. In Public Places and Spaces. Human Behavior and Environment. Springer. doi:10.1007/978-1-4684-5601-1_6

Funari, P. (1999). Aspectos da cultura popular antiga: Apresentação, tradução e discussão de alguns graffitis pompeianos. *Revista Estudos de História*, *4*(1/2), 143–150.

Funari, P. (2003). *A vida quotidiana na Roma antiga*. Annablume.

Giannetti, C. (1998). Trespassar a Pele: o Teletrânsito. In ARS Telemática – Telecomunicação, Internet e Ciberespaço. Relógio D"#gua Editores.

Gitahy, C. (1999). *O que é graffiti*. Brasiliense.

Jamieson, H. V. (2008). *Adventures in cyberformance* [Doctoral dissertation]. Queensland University of Technology.

Janson, A. F., & Janson, H. W. (1996). *Iniciação à História da arte*. Martins Fontes.

Jr, B. N. (2017) De onde vem o poder das imagens que invadem nossas casas e corpos? In Ecologia da imagem e dos Media. Nova Vega.

Lauretis, T. D. (1994). A tecnologia do gênero. Tradução de Suzana Funck. In *Tendências e impasses: o feminismo como crítica da cultura*. Rocco.

Lehmann, H. T. (2013). Teatro Pós-dramático, doze anos depois / Postdramatic Theatre, 12 years later / Théâtre Postdramatique, douze ans plus tard. *Revista Brasileira de Estudos da Presença*, *3*(3), 859–878. doi:10.1590/2237-266039703

Lipovetsky, G., & Serroy, J. (2010). *O ecrã global: cultura mediática e cinema na era hipermoderna*. Edições 70.

Machado, A. (2007). *Arte e Mídia: aproximações e distinções. Galáxia*. Revista do Programa de Pós-Graduação em Comunicação e Semiótica.

Manovich, L. (2017). *Instagram & Contemporary Image*. Postdigital Aesthetics.

Manovich, L., & Tifentale, A. (2015). *Selfiecity: Exploring Photography and Self-Fashioning in Social Media*. Postdigital Aesthetics.

Marcos, A. (2016). *Média-Arte Digital: da centralidade do artefacto à transdisciplinaridade*. Congresso da Cibercultura. Braga: Universidade do Minho. https://repositorioaberto.uab.pt/bitstream/ 10400.2/6239/1/ CiberCult_AFMarcos-14-out2016.pdf

Modeo, S. (2018). *Dioniso in Sicilia. Mythos, Symposion, Hades, Theatron, Mysteria (!Mesogheia. Studi di storia e archeologia della Sicilia Antica)*. Edizioni Lussografica.

Nascimento, A., Leitão, A., Carvalho, A., & Gonçalves, T. (2019). *O caminhar é para todas?Uma abordagem de mulheres latino- americanas sobre derivas e flâneries na contemporaneidade*. ENANPUR.

Nesti, G. (2019). Mainstreaming gender equality in smart cities: Theoretical, Methodological and Empirical Challenges. *Information Polity Journal, 24*(3), 289–304. doi:10.3233/IP-190134

Okin, S. M. (2008). Gênero, o público e o privado. *Revista Estudos Feministas, 16*(2), 305–332. doi:10.1590/S0104-026X2008000200002

Ortega, V. (2015). El artivismo como acción estratégica de nuevas narrativas artístico- políticas. *Calle14, 10*(15), 100 – 111.

Osthoff, S. (2010). De musas a autoras: Mulheres, arte e tecnologia no Brasil. *ARS (São Paulo), 8*(15). Advance online publication. doi:10.1590/S1678- 53202010000100006

Parisi, V. (2015). The sex of Graffiti, Urban art, women and $gender perception#: testing biases inthe eye of the observer. *Street Art & Urban Creativity Scientific Journal, 1*(1), 53-62. https://scholar.google.com/citations?user=dJRo-ggAAAAJ&hl=en

Pereira, K. (2012). *Relações entre a cultura hacker e a intervenção urbana*. Dissertação de Mestrado pela Universidade Federal de Belo Horizonte: UFMG.

Portelinha, M. (2013). *Arte urbana: estratégias, contextos e técnicas*. IADE-U Creative University.

Rodrigues, A. (2018). *Performances urbanas: Formas artísticas e Intervenções Urbanas*. Rev.Cult. Ext. USP.

Ruiter, A. (2015). Imaging Egypt#s political transition in (post-)revolutionary street art: on the interrelations between social media and graffiti as media of communication. *Media, Culture & Society*. doi:10.1177/0163443714566901

Santos, I., & Jesus, P. (2006). Intervenções Urbanas: a problemática das artes e seus espaços na cidade de Salvador na contemporaneidade. In Encontros de Estudos Multidisciplinares em Cultura da Faculdade de Comunicação. UFBA.

Scolari, C. (2013). *Narrativas transmedia: cuando todos los medios cuentan*. Deusto.

Silva, E. (2015). *Intervenções Artísticas em espaços públicos e pegagocias da cidade: possibilidades de pesquisa. In 37a Reunião Nacional da ANPEd*. UFSC.

Stornaiuolo, A., & Thomas, E. E. (2018). Restorying as political action: Authoring resistance through youth media arts. *Learning, Media and Technology*, *43*(4), 345–358. doi:10.1080/17439884.2018.1498354

Suárez, M. L. (2018). *Azurduy por Anzoátegui: subjetividad femenina y espacio público en Huallparrimachi (1894)*. Pontificia Universidad Católica del Perú. doi:10.18800/lexis.201802.005

Tavares, M. (2015). Tracing Limits-Public and Private in the Cartography of Contemporary Cities: The Dialogue Boxes on Street Windows Project. In Resilient Territories Resilient Territories: Innovation and Creativity for New Modes of Regional. Cambridge.

Toft, T. (2016). *What Urban Media Art Can Do*. Avedition.

Valente, C. M. (2016). *A street art no feminino: o lugar da mulher na Arte Pública*. Dissertação de Mestrado, Faculdade de Letras, Universidade de Lisboa.

Venturelli, S. (2016). *Intervenção do artista no contexto urbano com projeção mapeada computacional. 25 Encontro ANPAP. Arte: seus espaços e/em nosso tempo*. ANPAP.

Wexel, J. (2012). *Medeia Vozes de Christa Wolf: a reinvenção polifônica do mito trágico*. Dissertação de Mestrado em Letras, Cultura e Regionalidade. Universidade de Caxias do Sul. UCS. https:// repositorio.ucs.br/handle/11338/772

Wexel, J. (2021). Theater, Audiovisual and Streaming: An Analysis of Theater-Making in Times of Pandemic Uncertainty in the Post-Dramatic Experience of the Play Waiting for Godette. *Rotura - Revista de Comunicação. Cultura e Artes.*, *1*, 39–46.

Wexel, J. (Ed.). (2021). *Video-Performance Make me up!* https://www.instagram.com/tv/CVAOiyP-KRqE/? utm_medium=copy_link

Zielinski, S. (2017). Pensar a arte após os media: a pesquisa como cultura da experimentação viva. In *Ecologia da imagem e dos Media*. Nova Vega.

ADDITIONAL READING

Jeannotte, M. S. (2021). When the gigs are gone: Valuing arts, culture and media in the COVID-19 pandemic. *Social Sciences & Humanities Open*, *3*(1), 2021. doi:10.1016/j.ssaho.2020.100097

Veiga, P. A. (2017). Generative Theatre of Totality. *Journal of Science and Technology of the Arts*, *9*(3), 33–43. doi:10.7559/citarj.v9i3.422

KEY TERMS AND DEFINITIONS

Augmented Reality: Augmented reality (AR) is a computer system technology that enables the integration of vision between physical and virtual environments. Augmented Reality creations offer interactive, sensory, and immersive experiences. An example of an AR product is QR Codes and Instagram filters.

Cyberperformance: It is about performance in the digital space, also known as digital theatre, online performance, telematic performance and networked performance. The term *cyberperformance* was first used by Helen Varley Jamieson in 2000 to identify a specific artistic genre in the performances and theatre where performers/actors and audiences meet synchronously in virtual space. While body expression is key, cyberperformance can include pre-recorded multimedia elements. Precursors of cyberperfomance can be found among experimental projects using strategies and technologies of participation and distribution, such as mail art, radio art, and satellite art, among others.

Digital Media Art: Art that uses digital media technology as a process (means) and/or as a product (result) where technology is a tool at the service of creative ingenuity (artistic, cultural, educational, playful, among others) or as an engine for innovation in terms of creating new forms and aesthetic discourses that explore the informative, sensorial, and interventional expressiveness of multimedia content.

Graffiti: The word *graffiti* has Italian origins and is the plural of graffito, which in its translation means writing done with charcoal. Also called *muralism*, it is one of the main categories of pictorial expression of street art, where artists use techniques such as spray art and stencil art to paint surfaces such as walls, façades, and other devices on public roads.

Instagrammism: Lev Manovich coined the word in the book !Instagram and Contemporary Image" (2016). The term deals with the contemporary phenomenon of aesthetic construction of images and their sharing, which emerged between 2010 and 2011, especially with the platforms *Instagram* and *Kinfolk*.

Intersectional Feminism: Theory-based perspective on understanding how oppressions in patriarchal society manifest themselves beyond gender inequalities since they also permeate social class and ethnicity, among others. Through the viewpoint of intersectional feminism, we analyse how these multiple aspects relate to each other and interfere in women's lives. The prism of intersectional feminism is also used as a theoretical-methodological tool in feminist movements as a resource for the analysis of socio-cultural structures such as the lack or presence of privileges, the dismantling of mechanisms based on structural machismo and the construction of egalitarian public policies of access to human rights.

Selfiecity: It is the name of the first systematic study of the selfie phenomenon worldwide, an initiative led by the digital culture theorist Lev Manovich and the American laboratory Software Studies Initiative since 2012. The project "Selfiecity" investigated the phenomenon in five global capitals (Bangkok, Hong Kong, Moscow, New York, and São Paulo) through the social network Instagram and updated questions about self- representation, "self-fashioning," and interpersonal communications by visual means.

Compilation of References

(2018). Introduction: Studying markets as spaces of contestation. InGonzález, S. (Ed.), *S. Gonzalez, Contested markets, contested cities. Gentrification and urban justice in retail spaces* (pp. 1–18). Routledge.

visit - castro of Monte Mozinho. (2022, January 20). *YouTube*. Retrieved August 18, 2022, from https://youtu.be/33jZ6tGZCyo

Aall, C., & Koens, K. (2019). The Discourse on Sustainable Urban Tourism: The Need for Discussing More Than Overtourism. *Sustainability*, *11*(15), 4228. doi:10.3390u11154228

Abbas, J. (2021). Exploring the impact of COVID-19 on tourism: transformational potential and implications for a sustainable recovery of the travel and leisure industry. *Current Research in Behavioral Sciences, 2.*

Abbas, J., Mubeen, R., Iorember, P., Raza, S., & Mamirkulova, G. (2021). Exploring the impact of COVID-19 on tourism: Transformational potential and implications for a sustainable recovery of the travel and leisure industry. *Current Research in Behavioral Sciences*, *2*. doi:10.1016/j.crbeha.2021.100033

Abbott, P., Barbosa, S. F. F. (2015). Usando tecnologia da informação e mobilização social para combater doenças. *Acta Paulista de Enfermagem*, *8*(1), iii-v.

Abreu, J. G. (2016). Tempos Românticos. Ou o Romantismo como Recorrência nas Artes. In *Atas do II Congresso O Porto Romântico*. CITAR.

Abreu, J. G. (2020). Letter on Statuary and Power. Statues as Political Statements. Cadernos de Arte Pública/Publica Art Journal, (2), 6-33.

Abreu, J. G. (2012). *A Escultura no Espaço Público do Porto. Classificação e Interpretação*. Universidade Católica Editora.

Acemoglu, D. (2022). *Working Paper 29247. Bureau of Economic Research*. https://www.nber.org/papers/w29247NATIONAL.

Acker, J. (2006). Inequality Regimes: Gender, Class, and Race in Organizations. *Gender & Society*, *20*(4), 441–464. doi:10.1177/0891243206289499

Acselrad, H. (2008). *Cartografias Sociais e Território*. IPPUR/UFRJ.

Adey, P. (2009). *Mobility*. Routledge. doi:10.4324/9780203875483

Adie, B. A., Falk, M., & Savioli, M. (2020). Overtourism as a perceived threat to cultural heritage in Europe. *Current Issues in Tourism*, *23*(14), 1737–1741. doi:10.1080/13683500.2019.1687661

Adu-Ampong, E. (2012). *Managing Tourism at Heritage Sites*. Lap Lambert Academic Publishing.

Agamben, G. (2007). *Qu'est-ce qu'un dispositif?* [What is a device?]. Rivages Poches, Petite bibliothèque.

Agamben, G. (2000). *Enfance et histoire* [Infancy and history]. Payot & Rivages. (Original work published 1978)

Aggarwal, A., & Gupta, M. (2000). *Sexual Harassment in the Workplace* (3rd ed.). Butterworths.

Akyol, P. K. (2020). Museum Activities During and After the Covid-19 Pandemic. *Milli Folklor, 127*, 72–86.

Albors-Garrigos, J., Peiro Signes, A., Val Segarra-Oña, M., & Garcia-Segovia, P. (2021). Breaking the glass ceiling in haute cuisine: the role of entrepreneurship on the career expectations of female chefs. *Tourism and Hospitality Management, 27*(3), 605–628. doi:10.20867/thm.27.3.8

Albuquerque, R. (2013). *Associativismo, capital social e mobilidade: contributos para o estudo da participação associativa de descendentes de imigrantes africanos lusófonos em Portugal* [Associativism, social capital and mobility: contributions to the study of the associative participation of descendants of Portuguese-speaking African immigrants in Portugal] (Vol. 41). Observatório da Imigração, ACIDI, IP.

Alis, D., Horts, C.-H. B., Chevalier, F., Fabi, B., & Peretti, J.-M. (2012). *Gestão de Recursos Humanos - Uma Abordagem Internacional*. Edições Piaget.

Alizada, N., Cole, R., Gastaldi, L., Grahn, S., Hellmeier, S., Kolvani, P., Lachapelle, J., Lührmann, A., Maerz, S. F., Pillai, S., & Lindberg, S. I. (2021). *Autocratization Turns Viral. Democracy Report 2021*. Varieties of Democracy Institute (V-Dem). https://www.v-dem.net/static/website/files/dr/dr_2021.pdf

Allnutt, C. (2021). Can Venice Reinvent Itself. *Financial Times*. https://www.ft.com/content/ff4c32eb-b34e-4638-a0c9-1c6656906b02

Almeida, I. D., & Abranja, N. A. (2009). Turismo e Sustentabilidade. *Revista Cogitur*, 15-31. https://revistas.ulusofona.pt/index.php/jts/article/view/506/409

Almeida, B. (2007). *Alberto Carneiro. Lição das Coisas*. Campo das Letras.

Ambrosini, F., & Cherici, S. (2022). Anthropogenic Venice: Surviving Life in a Lagoon. *Green European Journal*. https://www.greeneuropeanjournal.eu/anthropogenic-venice-surviving-life-in-a-lagoon/

Andersson, T. D., & Getz, D. (2008). Stakeholder management strategies of festivals. *Journal of Convention & Event Tourism, 9*(3), 199–220. doi:10.1080/15470140802323801

Andrade, M. (2018). *"Ficou tudo ao Deus dará.": S. Vicente de Paulo. Histórias e memórias de um bairro demolido na cidade do Porto*. Edições Afrontamento.

Andrade, P. (1981-1984). *Les groupes 'Excursionistes'-'De Déjeuners' Portugais et leur Art* [Portuguese Excursionist-Lunch Groups and their Art]. Paris: École des Hautes Études en Sciences Sociales.

Andrade, P. (1985). *La taverne, son boire et leurs savoirs: ruses et muses mystico-profanes* [The tavern, its drinking and their knowledge: mystical-profane tricks and muses]. Paris: École des Hautes Études en Sciences Sociales.

Andrade, P. (1986). A Arte Excursionista [The Excursionist Art]. *Colóquio-Artes, 68*, 5-11.

Andrade, P. (1991). A Taberna mediática, local reticular de negociações sociais e sociológicas [The Mediatic Tavern: reticular locality of social and sociological negotiations]. *Revista Crítica de Ciências Sociais, 33*, 265-286.

Andrade, P. (1993). Sociologia da viagem: deslocações diárias e anti-quotidiano nómada [Sociology of Travel: daily trips and nomad anti-everyday]. *Revista Crítica de Ciências Sociais, 37*, 51-77.

Andrade, P. (1997a). Navegações no cibertempo: viagens virtuais e vitualidades da ciberviagem [Navigations in cybertime: virtual journeys and cybertravel]. *Atalaia, 3*, 111-124.

Andrade, P. (1997b). Sociologia do Olhar: um novo olhar sobre a Sociologia? [Sociology of gaze: A new look at sociology?]. *Atalaia, 3*, 197-201.

Andrade, P. (2002). Some synthetic ideas to work with qualitative analysis software. *Atalaia/Intermundos, 10*(11), 153-165.

Andrade, P. (2006a). Net Art and Hybridosphere: the Hybrilog as Blogart [Arte da Net e Hibridosfera: o Hybrilog como Blogart]. In L. Dopico (Eds.), *Conferencia Internacional de Arte Electrónico e Dixital* (pp.137-143). Pontevedra: Artech.

Andrade, P. (2006b). Cybermuseus e cibermuseologia: o hybrilog como um Bloseum [Cybermuseums and cybermuseology: the hybrilog as a Bloseum]. In H. Gottesdiener & J. Vilatte (Eds.), *Culture and Communication* (pp. 553-557). Avignon: International Association of Empirical Aesthetics.

Andrade, P. (2006c). *A sub-version of writing and of citizenship: the blog's auto-writing and the hybrilog as trans-writing.* In VI Colóquio Internacional 'Discursos e práticas alquímicas', Guimarães.

Andrade, P. (2007a). O Alfabeto de Relações Universais (ARU) [The Alphabet of Universal relations (ARU)]. *Revista de Comunicação e Linguagens 'Mediação dos saberes', 38*, 143-55.

Andrade, P. (2007b). *Imagem e Web 2.0: algumas redes sócio-conceptuais mediadoras da Wikipédia* [Image and Web 2.0: some socio-conceptual networks mediating Wikipedia]. Paper presented at Conferência Internascional 'Imagem e Pensamento'. Centro Cultural de Belém, Lisboa.

Andrade, P. (2008a). A Sociedade da Investigação e do Jornalismo: boas práticas de cidadania participativa através da internet móvel e do social bookmarking [The Society of Research and journalism: good practices of participatory citizenship through the mobile internet and social bookmarking]. In *Anuário Internacional de Comunicação Lusófona* (pp. 307–312). Braga: SOPCOM.

Andrade, P. (2010a). Introdução. In P. Andrade (Ed.), *Museus, públicos e literacia científico-tecnológica: redes sociais de comunicação de significados no espaço interdimensional do museu* [Museums, publics and scientific-technological literacy: social networks for the communication of meanings within the interdimensional space of the museum] (pp.25-32). Lisboa: Ed Colibri. http://hdl.handle.net/1822/70456

Andrade, P. (2011a). A hibridação das literacias: métodos GeoNeoLógicos na Sociedade da Investigação [The hybridization of literacies: GeoNeoLogical Methods in the Research Society]. In S. Pereira (Ed.), *Atas do Congresso Nacional 'Literacia, Media e Cidadania'* (pp. 71-89). Braga: Centro de Estudos de Comunicação e Sociedade.

Andrade, P. (2011b). *Sociologia Semântico-Lógica da Web 2.0/3.0 na sociedade da investigação: significados e discursos quotidianos em blogs, wikis, mundos/museus virtuais e redes sociais semântico-lógicas* [Semantic-Logical Sociology of Web 2.0/3.0 in the Research Society: meanings and everyday discourses in blogs, wikis, virtual worlds/museums and semantic-logical social networks]. Lisboa: Edições Caleidoscópio.

Andrade, P. (2011c). *Novela GeoNeológica nº 1: um caso de Literatura Transmediática /1ª Novela da Web 3* [GeoNeoLogical Novel nº1: a case of Transmediatic Literature]. Lisboa: Edições Caleidoscópio.

Andrade, P. (2013a). Sociologie Sémantico-Logique des Ruines: pour une hermeneutique hybride de la ruine à la Web 2.0 et à la Web 3.0. *Sociétés, 115*, 105-119.

Andrade, P. (2013b). Sociological ontology of the digital public sphere: the case of Web 2.0/3.0. *Comunicação e Sociedade, 23,* 202 – 216. http://hdl.handle.net/1822/29484

Andrade, P. (2014). Post-Colonial Co-Ordinary Literature And Web 2.0/3.0: thinking back within transmediatic knowledge. In M. Pope (Ed.), *New Literary Hybrids in the Age of Multimedia Expression: crossing borders, crossing genres.* John Benjamin Press. http://hdl.handle.net/1822/39995

Andrade, P. (2015a). *Austerity History through Sociological Comics: A Guide to Social Media and Networks against Austeritocracy for use by all Generations*. Create Space.

Andrade, P. (2015b, June). *Discourses and local public opinion in the urban and virtual public sphere: The case of struggles on austerity at Lisbon, 2013*. Paper presented at II Simpósio Internacional EDiSo: Discursos e sociedades em movimento, 18-20 junho 2015, Coimbra.

Andrade, P. (2015c). 'Arts Numériques et Musées Virtuels', 'Jeux Sociologiques', 'Hybridologie', 'Ontologies Visuelles'. In M. Veyrat (Ed.), *110 Notions sur les Arts Numériques*. Paris: Les Éditions de l'Immatériel.

Andrade, P. (2016a). Introdução [Introduction]. In P. Andrade (Ed.), *Comunicação pública da arte* [Public Communnication of Art] (pp. 21-32). Lisboa: Edições Caleidoscópio. http://hdl.handle.net/1822/70454

Andrade, P. (2016b). Que teoria e metodologia para os Museum Studies? [What theory and metodology for Museum Studies?]. In P. Andrade (Ed.), *Comunicação pública da arte* (pp. 33-48). Lisboa: Edições Caleidoscópio. http://hdl.handle.net/1822/70463

Andrade, P. (2016c, June). *Os Sociological Comics: ou as lutas sociais contadas por histórias sociológicas* [Sociological Comics: or social struggles told via sociological stories]. Paper presented at VI Congresso da Associação Portuguesa de Antropologia, 2-4 Junho, Coimbra.

Andrade, P. (2016d). GeoNeoLogical Novel and other Hybrimedia Experiments: or how to use Hybrid Methods such as Hybrid Discourse Analysis (HDA) within a Knowledge Base. *MATLIT, 4*(2), 157-181.

Andrade, P. (2016e). Transcultural cinema debated in a knowledge network: postcolonial hybrid meanings within resistance cinema. *Comunicação e Sociedade, 29*, 395 – 411.

Andrade, P. (2018a). A urbe social e sociológica: ou a cidade entendida como contexto e enquanto co-texto quotidianos [The social and sociological city: or the city understood as everyday context and co-text]. In *Culturas Urbanas e Contemporaneidade* (pp. 103-110). Centro de Estudos Lusíadas.

Andrade, P. (2018b). Epistemologia e Metodologia do Turismo Cultural Urbano: o caso da Sociologia Artística das culturas móveis e da comunicação turística em redes sociais urbanas [Epistemology and methodology of urban cultural tourism: the case of the artistic sociology of mobile cultures and tourism communication in urban social networks]. *Revista de Comunicação e Sociedade, 33,* 243-269. http://hdl.handle.net/1822/55275

Andrade, P. (2018c). Innovative role of users within digital economy: the case of information / knowledge flows at social and semantic networks (web 2.0/3.0). In L. Carvalho & P. Isaías (Eds.), *Handbook of Research on Entrepreneurship and Marketing for Global Reach in the Digital Economy* (pp. 269-290). IGI Global. http://hdl.handle.net/1822/70088

Andrade, P. (2018d). Mobile art flows: tourism mobilities at the museum. In E. Araújo, R. Ribeiro, P. Andrade & R. Costa (Eds.), *Viver em|a mobilidade: rumo a novas culturas de tempo, espaço e distância. Livro de atas* (pp. 193-198). Braga: CECS. http://hdl.handle.net/1822/55107

Andrade, P. (2018e). Turismo cultural móvel. Espaços, tempos e logos da arte [Mobile Cultural Tourism. Art spaces, times and logos]. *Revista Lusófona de Estudos Culturais, 5*(2), 349 – 363. http://hdl.handle.net/1822/57459

Andrade, P. (2018g). Digital Literature and Big Knowledge: the case of Sociological Comics. *MATLIT, 6*(2), 138-147.

Andrade, P. (2018h). *Ativismo Sociológico através do Video Paper: artes e imagens públicas publicitadas por imagens sociológicas* [Sociological Activism through Video Paper: public arts and images disseminated via sociological images]. Paper presented at Congress 'Arte Urbana no Plural', Leiria.

Andrade, P. (2018i). Cultural e-Tourism depicted by Digital Discourse: The Case of Portuguese Networks of e-Heritage and Discursive Innovation. In *Innovative Perspectives on Tourism Discourse* (pp. 1-17). IGI Global. http://hdl.handle.net/1822/70087

Andrade, P. (2020a). Mobile culture for tourism communication. In *Encyclopedia of Organizational Knowledge, Administration, and Technology* (pp. 1638-1650). IGI Global. http://hdl.handle.net/1822/70086

Andrade, P. (2020b). Arte pública urbana e comunicação turística [Urban public art and tourism communication]. *Revista Lusófona De Estudos Culturais, 7*(1), 39-59. doi:10.21814/rlec.2119

Andrade, P. (2020c). Cultural resistance to Covid-19: an Encyclopedia of Public Art, for artists and tourism publics? *Cadernos de Arte Pública, 2*(2), 57-74. http://hdl.handle.net/1822/70084

Andrade, P. (2020d). Um autor de arte pública urbana: Luis Baldini [An author of urban public art: Luis Baldini]. *Revista Lusófona de Estudos Culturais, 7*(1), 259-279. doi:10.21814/rlec.2677

Andrade, P. (2020e). Pessoa effect within viral society: viral sociological poetry using apps and e-books anti-virus. In *Proceedings of 2nd International Conference on Transdisciplinary Studies in Arts, Technology and Society, ARTeFACTo2020* (pp. 96-103). Faro: Universidade do Algarve, Centro de Investigação e Comunicação. http://hdl.handle.net/1822/70089

Andrade, P. (2020f). Cultural tourism and mobile digital devices: an analysis of tourists in Lisbon through sociological comics and video paper. In C. Sarmento & M. Quincé (Eds.), *Intercultural Routes across Multicultural Spaces: From Theory to Dialogue* (pp. 23-37). Porto: Centro de Estudos Interculturais. http://hdl.handle.net/1822/70091

Andrade, P. (2020g). Rua Augusta, an intercultural space: tourist communication via locative media for territory sustainability. In *XII Congresso da Geografia Portuguesa: Geografias de Transição Para a Sustentabilidade, Livro de Atas* (pp. 326-331). Guimarães: Universidade do Minho. http://hdl.handle.net/1822/70090

Andrade, P. (2020h). Site *Viral & Intercultural Tourism City*. Available from: https://www.sites.google.com/view/viral-tourism-city

Andrade, P. (2021c) Viral Urban Tourism within Viral Society: Sociological Touring Guides using Apps and e-Books. In *International Conference Connections: Exploring Heritage, Architecture, Cities, Art, Media. Amps Proceedings Series 20.1* (pp. 125-134). University of Kent/AMPS. http://hdl.handle.net/1822/74116

Andrade, P. (2022). Cultural Mobile Tourism and Tourist Knowledge at Viral Society: Sociological Research Games Against Covid-19, Within A Big Knowledge Context. Amps Proceedings Series 25. Urban Assemblage: The City as Architecture, Media, AI and Big Data (pp. 99-114). University of Hertfordshire. Available at: http://architecturemps.com/wp-content/uploads/2022/02/Amps-Proceedings-Series-25.pdf

Andrade, P. (2022). Everyday Urban Life: Genealogy and Journal of Actors' Body in the Viral City. In Coelho, Z.; Pires, H., The City of the Senses: the Senses of the City (pp. 35-65). Braga: Uminho Editora. http://hdl.handle.net/1822/7038 https://ebooks.uminho.pt/index.php/uminho/catalog/view/51/127/1255-1

Andrade, P.(2022). Risk Tourism Within Viral Society: A Study Using Hybrid Discourse Analysis. [Lusophone Journal of Cultural Studies] Revista Lusófona de Estudos Culturais, 9,1, pp. 127-151. Available at: https://hdl.handle.net/1822/78577

Andrade, P. (2022). Dark Heritage Pedia: an Augmented Research Encyclopedia, for a Virtual Sociological Museum on the Insatiable/Unsafe/Insane Society. In Clara Sarmento (Ed.). Dark Heritage Tourism in the Iberian Peninsula: Memories of Tragedy and Death. Cambridge Scholars Publishing [in the press]."

Andrade, P. (2022b). Cultural Mobile Tourism and Tourist Knowledge at Viral Society: Sociological Research Games Against Covid-19, Within A Big Knowledge Context. In *Amps Proceedings Series 25. Urban Assemblage: The City as Architecture, Media, AI and Big Data* (pp. 99-114). University of Hertfordshire. Available at: http://architecturemps.com/wp-content/uploads/2022/02/Amps-Proceedings-Series-25.pdf

Andrade, P. (2022c). Everyday Urban Life: Genealogy and Journal of Actors' Body in the Viral City. In *The City of the Senses: the Senses of the City* (pp. 35-65). Braga: Uminho Editora. http://hdl.handle.net/1822/76038 https://ebooks.uminho.pt/index.php/uminho/catalog/view/51/127/1255-1

Andrade, P. (2022d). Risk Tourism Within Viral Society: A Study Using Hybrid Discourse Analysis [Lusophone Journal of Cultural Studies]. *Revista Lusófona de Estudos Culturais, 9*(1), 127-151. Available at: https://hdl.handle.net/1822/78577

Andrade, P. (2022e). Dark Heritage Pedia: an Augmented Research Encyclopedia, for a Virtual Sociological Museum on the Insatiable/Unsafe/Insane Society. In *Dark Heritage Tourism in the Iberian Peninsula: Memories of Tragedy and Death*. Cambridge Scholars Publishing.

Andrade, P., & Caeiro, M. (Eds.). (2020). Public Art for Tourist Communication. *Lusophone Journal of Cultural Studies, 7*(1). Available at: https://rlec.pt/index.php/rlec/issue/view/128/fullissue7.1EN

Andrade, P. (2012). *Público e privado na época contemporânea*. Centro de Estudos Comparatistas, Faculdade de Letras da Universidade de Lisboa.

Andrade, P. (2016). *Comunicação Pública da Arte: o caso dos museus de Arte locais e globais*. Caleidoscópio.

Andrade, P. (2022a). The Value of Products and Services in Tourism: the Case of Viral Society. In C. Sarmento (Ed.), *Concepts and Dialogues across Shifting Spaces in Intercultural Business* (pp. 83–106). Cambridge Scholar Publishing. Available at https://www.cambridgescholars.com/product/978-1-5275-7725-1

Andreotti, V. O., Stein, S., Pashby, K., & Nicolson, M. (2016). Social cartographies as performative devices in research on higher education. *Higher Education Research & Development, 35*(1), 84–99. doi:10.1080/07294360.2015.1125857

Aoqui, C., Añón, R. (2020). *Os primeiros 60 dias de COVID-19 no Brasil em 60 fatos, reflexões e tendências em filantropia, investimento social e o campo de impacto social* [The first 60 days of COVID-19 in Brazil in 60 Facts, reflections and trends in Philanthropy, Social investment and the field of socia impact]. ponteAponte.

Apadurai, A. (1996). *Modernity at large: Cultural dimensions of globalization*. University of Minnesota Press.

Appadurai, A. (2013). *Condition de l'Homme Global*. Payot.

Appadurai, A. (2013). *The future as cultural fact: Essays on the global condition*. Verso.

Arendt, H. (2001). A condição humana. Relógio d'Água. (Original publication 1958)

Arruda, E. P. (2020). Emergency Remote Education elements to Brazilian public education policies during Covid-19 [Educação remota emergencial: elementos para políticas públicas na educação brasileira em tempos de Covid-19]. Distance Learning Journal, 7(1), 257-275.

Art Fund. (2020). *COVID-19 impact. Museum sector research findings. Summary report*. https://www.artfund.org/assets/downloads/art-fund-covid19-research-report-final.pdf

Ascape, V. R. (n.d.). Retrieved August 18, 2022, from https://ascape.com/

Ashworth, G., Graham, B., & Tunbridge, J. (2007). *Pluralising pasts: Heritage, identity and space in multicultural societies*. Pluto Press.

Ashworth, G., & Page, S. J. (2011). Urban tourism research: Recent progress and current paradoxes. *Tourism Management, 32*(1), 1–15. doi:10.1016/j.tourman.2010.02.002

Askitas, N., Tatsiramos, B., & Verheyden, B. (2020). *Lockdown strategies, mobility patterns and COVID-19.* CESifo Working Paper (8338).

Assaré, P. D. (1978). *Sing there that I sing here* [Canta aí que eu canto aqui]. Voices.

Audretsch, D. B., & Belitski, M. (2022). A strategic alignment framework for the entrepreneurial university. *Industry and Innovation, 29*(2), 285–309. doi:10.1080/13662716.2021.1941799

Augé, M. (1997). L'impossible voyage, le tourisme et ses images [The impossible journey, tourism and its images]. Rivages Poches, Petite bibliothèque.

Avelino, F., Wittmayer, J. M., Pel, B., Weaver, P., Dumitru, A., Haxeltine, A., Kemp, R., Jørgensen, M. S., Bauler, T., Ruijsink, S., & O'Riordan, T. (2019). Transformative social innovation and (dis)empowerment. *Technological Forecasting and Social Change, 145*, 195–206. doi:10.1016/j.techfore.2017.05.002

Azevedo, I. (2014). A arte urbana e seus possíveis entrelaçamentos: considerações sobre intervenções artísticas efêmeras. In *Revista DA Pesquisa*. Universidade do Estado de Santa Catarina. UDESC.

Babii, A., & Nadem, S. (2021). *Tourism in a Post-Pandemic World.* https://www.imf.org/en/News/Articles/2021/02/24/na022521-how-to-save-travel-and-tourism-in-a-post-pandemic-world

Babou, I. (2009). Disposer de la nature, enjeux environnementaux en Patagonie argentine [Disposing of nature, environmental issues in Argentine Patagonia]. In Sociologie et environnement. L'harmattan.

Bai, X. (2007). Integrating global environmental concerns into urban management. The scale and readiness arguments. *Journal of Industrial Ecology, 11*(2), 15–29. doi:10.1162/jie.2007.1202

Bakhtin, M. (2003). *Estética da criação verbal* (3rd ed.). Martins Fontes.

Balaguer, J., & Cantavella-Jorda, M. (2002). Tourism as a long-run economic growth factor: The Spanish case. *Applied Economics, 34*(7), 877–884. doi:10.1080/00036840110058923

Balzac, H. D. (1965). Eugénie Grandet. Éditions GP – Presses de la Cité. (Original publication 1834)

Bánáthy, B. H. (1996). *Designing Social Systems in a Changing World.* Plenum. doi:10.1007/978-1-4757-9981-1

Bandarin, F., & Van Oers, R. (2012). *The historic urban landscape: managing heritage in an urban century.* John Wiley et Sons. doi:10.1002/9781119968115

Bandeira, M. L. B. (2020, Novembro 19). *Cordelista Day: who was Leandro Gomes de Barros?* [Dia do Cordelista: quem foi Leandro Gomes de Barros?]. Itaú Cultural. Retrieved October 15, 2021, from https://www.itaucultural.org.br/secoes/entrevista/cordelista-leandro-gomes-barros

Bandeira, M., Corais, F., & Pereira, F. (2021a). An urgent urban design challenge: time, sustainable mobility, and the humanization of collective space. In *Sustainable Urban Development. Topics, trends and solutions* (pp. 1–15). IOP Publishing Ltd. doi:10.1088/978-0-7503-3971-1ch3

Bandeira, M., Corais, F., & Pereira, F. (2021b). Coexistence of urban regeneration policies and sustainable mobility: reflections on, and practical cases of, urban acupunture. In *Sustainable Urban Development. Topics, trends and solutions* (pp. 1–11). IOP Publishing Ltd. doi:10.1088/978-0-7503-3971-1ch5

Banister, D., & Hickman, R. (2007). How to design a more sustainable and fairer built environment: transport and communications. *IEE Proceedings - Intelligent Transport Systems, 153*(4), 276–291. 10.1049/ip-its

Bardin, L. (1977). *Content Analysis* [Análise de conteúdo]. Lisbon: Editions 70.

Barreto, M. (2004). Relações entre Visitantes e Visitados: um retrospecto dos estudos socio antropológicos. *Turismo em Análise, 15*(2), 133-149. https://www.revistas.usp.br/rta/article/viewFile/62663/65458

Barthes, R. (2020). *Mitologias*. Edições 70. (Original publication 1957)

Barthes, R. (2014). *Mythologies* [Mythologies]. Seuil. (Original work published 1957)

Bartkiewicz, D. (2013). *APIs in Your Eyes with the Augmented Reality Wave of Bigger Data.* Available at http://drewbartkiewicz.sys-con.com/node/2569342

Bastidas, E. P., & Gonzales, C. A. (2009). Social Cartography as a Tool for Conflict Analysis and Resolution: The Experience of the Afro-Colombian Communities of Robles. *Peace and Conflict Studies*, *15*(2). Advance online publication. doi:10.46743/1082-7307/2009.1094

Bataille, G. (1970). *Œuvres Complètes* (Vol. 2). Gallimard.

Baudrillard, J. (1981). *Simulacres et simulation* [Simulacra and simulation]. Galilée.

Beck, D. (2020, August). Finding the Gaps about Uses of Immersive Learning Environments: A Survey of Surveys. *Journal of Universal Computer Science*, *26*(8), 1056-1057.

Belfiore, E. (2020). Whose cultural value? Representation, power and creative industries. *International Journal of Cultural Policy*, *26*(3), 383–397. doi:10.1080/10286632.2018.1495713

Belleza, N. (1966) *Teatro grego e suas conseqüências: raízes do teatro*. Pongetti. Bento, B. (2017). *Transviad@s: gênero, sexualidade e direitos humanos*. EDUFBA.

Benevolo, L. (1997). *História da Cidade*. Perspectiva.

Beni, M. C. (1990). Sistema de Turismo - SISTUR Estudo do Turismo face à Moderna Teoria de Sistemas. *Revista Turismo em Análise*. https://www.revistas.usp.br/rta/article/viewFile/63854/66610

Beni, M. C. (2020a). Turismo e Covid-19: algumas reflexões. Rosa dos Ventos Turismo e Hospitalidade, 12(3), 1-23. doi:10.18226/21789061.v12i3a02

Beni, M.C. (2020b). *Saturação e Rejeição ao Turismo nas Destinações Turísticas*. São Paulo: Revista Brasileira de Pesquisa em Turismo. doi:10.7784/rbtur.v14i2.1847

Benjamin, W. (1992). *Sobre arte, técnica, linguagem e política* [On art, technique, language, and politics]. Relógio d'Água.

Benjamin, W. (2008). *The Work of Art in the Age of Its Technological Reproducibility*. Belknap Press. doi:10.2307/j.ctv1nzfgns.6

Benjamin, W. (2016). *The Storyteller -Tales Out of Loneliness*. Ed.Verso.

Benjamin, W. (2017). *La obra de arte en la era de su reproducibilidad técnica*. La Moderna Ediciones.

Benjamin, W. (2019). *As passagens de Paris*. Assírio & Alvim.

Bereitschaft, B., & Scheller, D. (2020). How Might the COVID-19 Pandemic Affect 21st Century Urban Design, Planning, and Development? *Urban Science*, *4*(4), 56. doi:10.3390/urbansci4040056

Berger, J. (2018). *Modos de Ver*. Antígona.

Bergold, J., & Thomas, S. (2012). Participatory Research Methods: A Methodological Approach in Motion. *Historical Social Research (Köln)*, *37*(4), 191–222.

Berti, E. (2015). *Cultural Routes Management: From Theory to Practice*. Council of Europe Publishing.

Bertoni, A., Dubini, P., & Monti, A. (2021). Bringing Back in the Spatial Dimension in the Assessment of Cultural and Creative Industries and Its Relationship with a City's Sustainability: The Case of Milan. *Sustainability*, *13*(19), 10878. doi:10.3390u131910878

Bertrand Cabral, C., Pereira Roders, A., & Albuquerque, R. (2022). Formas de participação nos normativos patrimoniais. Ensaio de contextualização nos modelos da democracia participativa e deliberativa [Forms of participation in property regulations. Contextualization essay no participatory and deliberative democracy models]. Political Observer – Revista Portuguesa de Ciência Política, 16, 137-166. DOI: PCP2021.16 doi:10.33167/2184-2078.R

Bhabha, H. (1996). *Negotiating Rapture: The Power of Art to Transform Lives*. Museum of Contemporary Art. Braga Creative City of Media Arts. Accessed March 15, 2020: https://en.unesco.org/creative-cities/braga

Bhabha, H. (2004). *The location of culture*. Routledge.

Biagi, B., Ladu, M. G., Meleddu, M., & Royuela, V. (2020). Tourism and the city: The impact on residents' quality of life. *International Journal of Tourism Research*, *22*(2), 168–181. doi:10.1002/jtr.2326

Birringer, J. (2015). The New Digital Materialism. *PAJ: A Journal of Performance and Art*, *37*(3), 102-110. https://www.jstor.org/stable/10.2307/26935691

Black, R. E. (2012). *Porta Palazzo: The anthropology of an Italian market*. University of Pennsylvania Press. doi:10.9783/9780812205794

Blanchet, C. (2020). Crise de la Covid-19 dans le secteur du tourisme. Illustration de la couverture médiatique et académique. *I2D - Information, données & documents*, (3), 101-114.

Blascovich. (2002, November). Immersive Virtual Environment Technology as a Methodological Tool for Social Psychology. *Psychological Inquiry*, *13*(2), 110-111.

Bloch, E. (1978). Héritage de ce temps (J. Lacoste, Trans.). Payot.

Blöcher, K., & Alt, R. (2021). AI and robotics in the european restaurant sector: Assessing potentials for process innovation in a high-contact service industry. *Electronic Markets*, *31*(3), 529–551. doi:10.100712525-020-00443-2

Bogdan, R., & Biklen, S. (1994). *Qualitative Investigation in Education: an introduction to the theory and methods* [Investigação qualitativa em educação: uma introdução à teoria e aos métodos]. Porto Publisher.

Boletim ICOM-Portugal. (2020). Os museus de portas fechadas. *Os museus em tempo de Covid 19, 2*(14), 64-69.

Bonfanti, A., Vigolo, V., & Yfantidou, G. (2021). The impact of the Covid-19 pandemic on customer experience design: The hotel managers' perspective. *International Journal of Hospitality Management*, *94*, 94. doi:10.1016/j.ijhm.2021.102871 PMID:34866744

Borges, J. L. (1988). The unending gift. In Elogio da sombra. Obras Completas (1952- 1972). II. Teorema. (Original work published 1966)

Boto-García, D., & Mayor, M. (2022). Domestic tourism and the resilience of hotel demand. *Annals of Tourism Research*, *93*, 93. doi:10.1016/j.annals.2022.103352

Boudieu, P., & Darbel, A. (1992). *El amor al arte. Los museos europeos y su público.* Paidós.

Bougnoux, D. (2006). *La crise de la représentation* [The crisis of representation]. La Découverte.

Bourdieu, P. (2021). *O poder simbólico.* Lisboa: Edições 70.

Bourdieu, P. (1993). *La misère du monde.* Seuil.

Bourriaud, N. (2009). Estética relacional. Martins Fontes. (Original publication 1998)

Bovone, L. (2001). Os Novos Intermediários culturais – Considerações sobre a cultura pós-moderna. In Cidade, Cultura e Globalização. Oeiras: Celta Editora.

Boyer, M. (2002). El turismo en Europa, de la edad moderna ao siglo XX. *Historia Contemporánea, 25*, 13–34. https://eco.mdp.edu.ar/cendocu/repositorio/00924.pdf

Braga, J. (2021). A cidade como terreno comum: limiares urbanos e a invenção de novos horizontes do possível. In S. Stavrides (Ed.), Espaço comum. A cidade como obra colectiva (pp. 7–36). Orfeu Negro. (Original publication 2016)

Braun, V., & Clarke, V. (2006). Using thematic analysis in psychology. *Qualitative Research in Psychology, 3*(2), 77–101. doi:10.1191/1478088706qp063oa

BRINK XR - See the world through Virtual Travel Experiences. (n.d.). Retrieved August 18, 2022, from https://www.brinkxr.com/

Brito, B. R. (2002). *O Turista e o Viajante: Contributos para a conceptualização do Turismo alternativo e Responsável.* Lisboa: Associação Portuguesa de Sociologia. https://aps.pt/wp-content/uploads/2017/08/DPR462dea1a49422_1.pdf

Brochot, A. (2013). Maria Gravari-Barbas (dir.) 2005, Habiter le patrimoine: Enjeux, approches, vécu [Living in heritage: Issues, approaches, experiences]. *Strates, 14.* doi:10.4000/strates.6713

Brody, R. (2020, March 17). Lessons for the Movie Industry from the 1918 Influenza Pandemic. *New Yorker.* https://www.newyorker.com/culture/the-front-row/lessons-for-the-movie-industry-from-the-1918-influenza-pandemic

Brook, O., O'Brien, D., & Taylor, M. (2018). *There was no golden age: social mobility into cultural and creative occupations.* doi:10.31235/osf.io/7njy3

Brown, B., Nasiruddin, M., Dao, A., & Halabi, M. (2015). Responsible use of pop culture and communication in the face of Ebola virus. *PLoS Neglected Tropical Diseases, 9*(8), e0003890. doi:10.1371/journal.pntd.0003890 PMID:26248052

Brown, T. (2008). Design thinking. *Harvard Business Review, 86*(6), 84. PMID:18605031

Bruno, F. (2013). *Máquinas de ver, modos de ser: vigilância, tecnologia e subjetividade.* Sulina.

Buhalis, D., & Costa, C. (2006). Tourism Business Frontiers - consumers, products and industry. In D. Buhalis & C. Costa (Eds.), *Tourism Business frontiers consumers, products and industry.* Elsevier B.V.

Bulkeley, H., Broto, V. C., Hodson, M., & Marvin, S. (Eds.). (2010). *Cities and low carbon transitions* (Vol. 35). Routledge. doi:10.4324/9780203839249

Burke, P. (2017). *Testemunha Ocular. O Uso de Imagens como Evidência Histórica.* Editora Unesp.

Burnett, K. (2014). Commodifying poverty: Gentrification and consumption in Vancouver's Downtown Eastside. *Urban Geography, 35*(2), 157–176. doi:10.1080/02723638.2013.867669

Burnham, R., & Kai-Kee, E. (2011). *Teaching in the Art Museum.* J. Paul Getty Museum.

Burnshaw, S. (1995). *The Poem Itself: 150 of the Finest Modern Poets in the Original Languages*. University of Arkansas Press.

Butler, R. (2020). Tourism–resilient but vulnerable as "the times they are a changing" in the "new normality". *Worldwide Hospitality and Tourism Themes*, *12*(6), 663–670. doi:10.1108/WHATT-07-2020-0063

Cacciari, M. (2010). *A cidade*. Gustavo Gili.

Cadilhe, G. (2006). *A Lua pode esperar*. Oficina do Livro- Sociedade Editorial Lda.

Caeiro, M. (2014). *A Arte na Cidade. História Contemporânea*. Círculo de Leitores.

Calle-Vaquero, M. D. L., Garcia-Hernandez, M., & Mendoza de Miguel, S.Applications in Historic Centres. (2021). Urban planning regulations for tourism in the context of overtourism. Applications in historic centres. *Sustainability*, *13*(1), 70. doi:10.3390u13010070

Campbell, P., O'Brien, D., & Taylor, M. (2019). Cultural engagement and the economic performance of the cultural and creative industries: An occupational critique. *Sociology*, *53*(2), 347–367. doi:10.1177/0038038518772737

Camus, R. (2011). *Le Grand Remplacement*. Éditions David Reinharc.

Canclini, N. (2005). *Hybrid cultures: strategies for entering and leaving modernity*. University of Minnesota Press.

Cani, J. B., Sandrini, E. G. C., Soares, G. M., & Scalzer, K. (2020). Education and Covid-19: The art of reinventing the school by mediating learning "primarily" through DITC [Educação e covid-19: a arte de reinventar a escola mediando a aprendizagem "prioritariamente" pelas TDIC]. *IFES Science Journal*, *6*(1), 23–39.

Cao, L., Qu, Y., & Yang, Q. (2021). The formation process of tourist attachment to a destination. *Tourism Management Perspectives*. https://unctad.org/system/files/official-document/ditcinf2020d3_en.pdf

Capello, R., Cerisola, S., & Perucca, G. (2020). Cultural heritage, creativity, and local development: A scientific research program. In *Regeneration of the built environment from a circular economy perspective* (pp. 11–19). Springer. doi:10.1007/978-3-030-33256-3_2

Carayannis, E. G., & Campbell, D. F. (2011). Open Innovation Diplomacy and a 21st Century Fractal Research, Education and Innovation (FREIE) Ecosystem: Building on the Quadruple and Quintuple Helix Innovation Concepts and the "Mode 3" Knowledge Production System. *Journal of the Knowledge Economy*, *2*(3), 327–372. doi:10.100713132-011-0058-3

Carayannis, E. G., & Campbell, D. F. (Eds.). (2009). *Knowledge creation, diffusion, and use in innovation networks and knowledge clusters. A comparative systems approach across the United States, Europe, and Asia*. Praeger.

Carlomagno, M. C., & da Rocha, L. C. (2016). Como criar e classificar categorias para fazer análise de conteúdo: uma questão metodológica [How to create and classify categories to do content analysis: A Methodological Question]. *Revista Eletrônica de Ciência Política*, *7*(1).

Carneiro, A. (2007). *Das Notas para um Diário e outros Textos. Antologia*. Assírio & Alvim.

Carter, J. (2016, September 17) Holography: VR Without Glasses. *E-Magazine*. https://emag.directindustry.com/holography-vr-without-glasses/

Carus, P. (1892). What does anschauung mean? *The Monist, 2*(4), 527-532. Available at https://www.jstor.org/stable/27897002

Carvalho, A. & Matos, A. (2018). Museum Professionals in a Digital World: Insigths from a Case Study in Portugal. *Museum International*, *70*(277-278), 34-47. htpps:// doi:10.1111/muse.12191

Carvalho, I. (2016). Média-arte locativa e mapeamentos dinâmicos na compreensão de fluxos urbanos: o contributo do dispositivo móvel na atividade do caminhar como prática artística. Lisboa: UAb, UAlg.

Carvalho, L., Chamusca, P., Fernandes, J., & Pinto, J. (2019). Gentrification in Porto: floating city users and internationally-driven urban change. *Urban Geography*. https://www.tandfonline.com/doi/abs/10.1080/02723638.2019.1585139?journalCode=rurb20 doi:10.1080/02723638.2019.1585139

Carvalho, P., & Fernandes, J. L. (Eds.). (2012). *Património cultural e paisagístico. Políticas, intervenções e representações* [Cultural and landscape heritage. Policies, interventions and representations]. Imprensa da Universidade de Coimbra. doi:10.14195/978-989-26-0585-2

Casado-Aranda, L. (2021). Tourism research after the COVID-19 outbreak: Insights for more sustainable, local and smart cities. *Sustainable Cities and Society, 73*.

Castells, M. (1996). *The rise of the network society*. Blackwell Publishers.

Castells, M. (2002). *A Sociedade em Rede* (Vol. I). Fundação Calouste Gulbenkian.

Castells, M. (2004). *A Galáxia Internet. Reflexões sobre Internet, Negócios e Sociedade*. Fundação Calouste Gulbenkian.

Castells, M. (2007). A Era da Informação: Economia, Sociedade – *O Poder da Identidade*. Lisboa: Fundação Calouste Gulbenkian.

Castells, M. (2007). A Era da Informação: Economia, Sociedade. *Cultura, 2*.

Castro, R. (1909). *Obras Completas de Rosalía de Castro II. Cantares Gallegos*. Librería de los Sucesores de Hernando.

Castro-Nuño, M., Molina-Toucedo, J. A., & Pablo-Romero, M. P. (2013). Tourism and GDP: A meta-analysis of panel data studies. *Journal of Travel Research, 52*(6), 745–758. doi:10.1177/0047287513478500

CEDEFOP. (2020). *Skills developments and trends in the tourism sector*. Skills intelligence. https://www.cedefop.europa.eu/en/data-insights/skills-developments-and-trends-tourism-sector

Celan, P. (1996). O Meridiano [The Meridian]. In *Arte Poética. O Meridiano e outros Textos* (pp. 42–64). Colibri. (Original work published 1971)

Cerisola, S. (2019). A new perspective on the cultural heritage–development nexus: The role of creativity. *Journal of Cultural Economics, 43*(1), 21–56. doi:10.100710824-018-9328-2

Certeau, M. (1998). A invenção do quotidiano. 1. Arte de fazer. Petrópolis.

Chabert, G. (2012). Les espaces de l'écran. *MEI, Médiation et information, ecran et médias, 34*, 203-213. http://www.mei-info.com/wp-content/uploads/2015/03/MEI_34_14.pdf

Chafes, R. (2016). *O Meu Sangue é o Vosso Sangue*. Santa Casa da Misericórdia do Porto.

Chamola, V., Hassija, V., Gupta, V., & Guizani, M. (2020). A comprehensive review of the COVID-19 pandemic and the role of IoT, drones, AI, blockchain, and 5G in managing its impact. *IEEE Access: Practical Innovations, Open Solutions, 8*, 90225–90265. doi:10.1109/ACCESS.2020.2992341

Chamusca, P. (2021). Urban planning and policy in Portugal: An overview on the role of EU funds and guidelines. *Urban Research & Practice*, 1–22. Advance online publication. doi:10.1080/17535069.2021.1980607

Chamusca, P., Fernandes, J., Carvalho, L., & Mendes, T. (2019). The Role of Arbnb creating a "new"-old City Center; facts, problems and controversies in Porto. *BAGE, 83*(83). Advance online publication. doi:10.21138/bage.2820

Changsong, W., Kerry, L., & Marta, R. F. (2021). Film distribution by video streaming platforms across Southeast Asia during COVID-19. *Media Culture & Society*, *43*(8), 1542–1552. doi:10.1177/01634437211045350

Chatfield, T. (2013). *Como aproveitar ao máximo a era digital*. Alfragide: Lua de Papel.

Cheng, K. M. (2012). Tourism demand in Hong Kong: Income, prices, and visa restrictions. *Current Issues in Tourism*, *15*(3), 167–181. doi:10.1080/13683500.2011.569011

Chen, H. (2006). *Intelligence and security informatics for international security: information charing and data mining*. Springer. doi:10.1007/0-387-30332-4

Chen, H., Chiang, & Storey. (2012). Business intelligence and analytics: From big data to big impact. *Management Information Systems Quarterly*, *36*(4), 1165–1188. doi:10.2307/41703503

Chesbrough, H. (2003). *Open Innovation: The New Imperative for Creating and Profiting from Technology*. Harvard Business Press.

Choay, F. (2008). *Alegoria do património*. Coimbra: Edições 70.

Choay, F. (1992). *Alegoría del patrimonio*. Gustavo Gili.

Cho, B. (2000). Destination. In J. Jafari (Ed.), *Encyclopedia of Tourism* (pp. 144–145). Routledge.

Chu, F. L. (2014). Using a logistic growth regression model to forecast the demand for tourism in Las Vegas. *Tourism Management Perspectives*, *12*, 62–67. doi:10.1016/j.tmp.2014.08.003

Chung, N., Han, H., & Joun, Y. (2015). Tourists' intention to visit a destination: The role of augmented reality (AR) application for a heritage site. *Computers in Human Behavior*, *50*, 588–599. doi:10.1016/j.chb.2015.02.068

Cioffi, A., Lugi, C., & Cecannecchia, C. (2020). Apps for COVID-19 contact-tracing: Too many questions and few answers. *Ethics, Medicine, and Public Health*, *15*, 100575. doi:10.1016/j.jemep.2020.100575 PMID:32838002

Clair, J. (2011). *Malestar en los museos*. Trea.

CMB. (2015). *Plano Diretor Municipal de Braga*. Câmara Municipal de Braga.

CMB. (2019). *Síntese de resultados estatísticos do Turismo em Braga - 2019*. CMB - Divisão de Economia e Turismo.

CMB. (2020). *Síntese de Resultados Estatísticos do Turismo em Braga 2020*. CMB-Divisão de Economia e Turismo.

CMB. (2021). *Síntese de Resultados Estatísticos do Turismo em Braga - 2021*. CMB - Divisão de Economia e Turismo.

Cochrane, A. (2020). Urban policy. In A. Kobayashi (Ed.), *International Encyclopaedia of Human Geography* (pp. 93–96). Elsevier. doi:10.1016/B978-0-08-102295-5.10229-X

Cocola-Gant, A. (2016). Airbnb, buy-to-let investment and tourism-driven displacement: A case study in Lisbon. *Environment and Planning A: Economy and Space*. doi:10.1177/0308518X19869012

Cocola-Gant, A., Jover, J., Carvalho, L., & Chamusca, P. (2021) Corporate hosts: The rise of professional management in the short-term rental industry, *Tourism Management Perspectives*, *40*. https://www.sciencedirect.com/science/article/pii/S2211973621000921 doi:10.1016/j.tmp.2021.100879

Coelho, C. (2006). *Media Presence and Inner Presence: The Sense of Presence in Virtual Reality Technologies. In From Communication to Presence: Cognition, Emotions and Culture towards the Ultimate Communicative Experience. Festschrift in honor of Luigi Anolli*. IOS Press.

Colavitti, A. M., & Usai, A. (2016). How cultural and creative industries are redefining policies for the historic urban landscape. The experience of Cagliari City as ECoC 2019's finalist and Italian Capitals of Culture. *Cultural Management Education in Risk Societies-Towards a Paradigm and Policy Shift?!*, 66.

Collins-Kreiner, N., & Ram, Y. (2021). National tourism strategies during the Covid-19 pandemic. *Annals of Tourism Research*, *89*, 89. doi:10.1016/j.annals.2020.103076

Colomé, J. P. (2019). *Como você é espionado por seu celular Android sem saber*. https://brasil.elpais.com/brasil/2019/03/17/tecnologia/1552777491_649804.html

Cominelli, F., & Greffe, X. (2012). Intangible cultural heritage: Safeguarding for creativity. *City. Cultura e Scuola*, *3*(4), 245–250. doi:10.1016/j.ccs.2012.10.003

Condevaux, A., Djament-Tran, G., Gravari-Barbas, M. (2016). Avant et après le(s) tourismes(s). Trajectoires des lieux et rôles des acteurs du tourisme « hors des sentiers battus ». Une analyse bibliographique [Before and after tourism(s). Trajectories of the places and roles of the actors of "off the beaten track" tourism. A bibliographical analysis]. *Via*, *9*. doi:10.4000/viatourism.409

Consumer Electronic Show [CES]. (2019). *CES Fact Sheet*. Retrieved from https://cdn.ces.tech/ces/media/pdfs/ces-2019-fact-sheet.pdf

Corais, F., Bandeira, M., Silva, C., & Bragança, L. (2022). Between the Unstoppable and the Feasible : The Lucid Pragmatism of Transition Processes for Sustainable Urban Mobility : A Literature Review. *Future Transportation*, 86–114. doi:10.3390/futuretransp2010006

Corais, F. (2003). *Dinâmicas Territoriais na Aglomeração Urbana de Braga. Tese de Mestrado*. Faculdade de Arquitetura da Universidade do Porto.

Corais, F. (2021). *A Cidade a "Caminhar" para 2050. Braga como Laboratório para um Sistema Urbano Resiliente e Sustentável (Projeto Tese)*. Universidade do Minho.

Correia, F. C. (2009). *O mosteiro de Santo Tirso, de 978 a 1588: Estudo*. CMST.

Cosme, I., Santos, R., & O'Neill, D. W. (2017). Assessing the degrowth discourse: A review and analysis of academic degrowth policy proposals. *Journal of Cleaner Production*, *149*, 321–334. doi:10.1016/j.jclepro.2017.02.016

Council of Europe. (2005). *Council of Europe Framework Convention on the Value of Cultural Heritage for Society*. https://www.coe.int/en/web/conventions/full-list2?module=treaty-detail&treatynum=199

Coutinho, B., Baptista, M. M., Martins, M. L., & Rebelo, J. (2018). Portugal, land of tourism: Dissonances and touristic uses of the Estado Novo's heritage. *Lusophone Journal of Cultural Studies*, *5*(2), 213–251. doi:10.21814/rlec.347

Cranmer, E. E. (2020). Exploring the value of augmented reality for tourism. *Tourism Management Perspectives, 35*.

Cravidão, F. (2011). *Turismo, território e cultura: uma trilogia (sempre) em construção* (Vol. 29). Departamento de Geografia Humana da Universidade do Estado do Rio de Janeiro.

Crespi, F. (1997). *Manual de Sociologia da Cultura*. Editorial Estampa.

Crespo, J. (1992). *A História do Corpo*. Editorial Difel.

Crouch, D., Jackson, R., & Thompson, F. (2006). Introduction-The media and the tourist imagination. In M. Hall (Ed.), *Contemporary Geographies of Leisure, Tourism and Mobility* (pp. 1–13). Routledge.

Crouch, D., & Lübbren, N. (2003). *Visual culture and tourism*. Berg.

Crouch, G. I. (1994). Price elasticities in international tourism. *Hospitality Research Journal, 17*(3), 27–39. doi:10.1177/109634809401700304

Cummings, J., & Bailenson, J. (2015, May). How Immersive Is Enough? A Meta-Analysis of the Effect of Immersive Technology on User Presence. *Media Psychology, 19*(2), 272–309. doi:10.1080/15213269.2015.1015740

Cunha, L. (2008). Avaliação do Potencial Turístico. *COGITUR Journal of Tourism Studies*. http://jts.ulusofona.pt/turismo01/pdf01/artigo_liciniocunha.pdf

Cunha, L. (2012). Turismo de Portugal. Sucessos e Insucessos (1.ª ed.). Lisboa: Edições Universitárias Lusófonas.

Cunha, L. (1997). *Economia e Política do Turismo*. McGraw-Hill.

Cunha, L. (2017). *Turismo e Desenvolvimento – Realidades e Perspetivas*. Lidel – Edições Técnicas, Lda.

Cunha, L. (2020). *Desenvolvimento do Turismo em Portugal: Os Primórdios. Fluxos & Riscos n.º1*. Faculdade de Ciências Sociais e Humanas Universidade Lusófona de Humanidades e Tecnologias. https://core.ac.uk/download/pdf/48576342.pdf

Cuomo, M. T. (2021). Digital transformation and tourist experience co-design: Big social data for planning cultural tourism. *Technological Forecasting and Social Change, 162*.

Cyrulnik, N. (2017). Représenter le territoire, filmer la cité [Representing the territory, filming the city]. Habilitation à Diriger des Recherches, Alain Kiyindou garant, Université de Bordeaux - Montaigne.

Cyrulnik, N. (2020). Documentary narrative for new understanding of a stigmatized public space. *Revista Lusófona de Estudos Culturais / Lusophone Journal of Cultural Studies, 7*(1). https://rlec.pt/article/view/2121/2678

Cyrulnik, N. (2022). Les méthodes audio-visuelles qu'offre le cinéma documentaire [The audio-visual methods offered by documentary film]. In Méthodes visuelles en sciences humaines et sociales: de la captation à la diffusion des résultats: données, méthodes, résultats, valorisation. ISTE Open Science.

Cyrulnik, N., & Zénouda, H. (2014). La place des dispositifs socio-techniques d'information et de communication (Distic) dans les différentes situations de recherches [The place of socio-technical information and communication devices (Distic) in different research situations]. Cahiers de la Sfsic, 10, 133-224.

Cyrulnik, N. (2015). Le documentaire, un espace de liberté pour une nouvelle communauté [The documentary, a space of freedom for a new community]. In G. Gramaccia (Ed.), *Revue Française des Sciences de l'Information et de la Communication* (Vol. 7). doi:10.4000/rfsic.1744

Cyrulnik, N. (2018a). Le documentaire, vers une écologie filmique [The documentary, towards a filmic ecology]. In D. Urrutiaguer (Ed.), *Des expériences artistiques au prisme du développement durable, Revue Registres de l'Institut de Recherches en études théâtrales, Hors-série n°5. Presses Sorbonne Nouvelle.*

Cyrulnik, N. (2021). Questionner l'atmosphère au cinéma pour expérimenter l'interdisciplinarité, une méthode audiovisuelle [Questioning the atmosphere in cinema to experiment interdisciplinarity, an audiovisual method]. *Revue Française des Sciences de l'Information et de la Communication, 22*. Advance online publication. doi:10.4000/rfsic.10632

D'Alisa, G., Demaria, F., & Kallis, G. (2015). *Degrowth: A Vocabulary for a New Era* (G. D'Alisa, F. Demaria, & G. Kallis, Eds.). Routledge.

da Silva Lopes, H., Remoaldo, P. C., Ribeiro, V., & Martín-Vide, J. (2021). Effects of the COVID-19 Pandemic on Tourist Risk Perceptions—The Case Study of Porto. *Sustainability, 13*(11), 6399. doi:10.3390u13116399

Davidoff, P. (1965). Advocacy and Pluralism in Planning. *Journal of the American Institute of Planners, 31*(4), 331–338. doi:10.1080/01944366508978187

Davis, G. F., Morrill, C., Rao, H., & Soue, S. A. (2008). Introduction: Social movements in organizations and markets. *Administrative Science Quarterly*, *53*(3), 389–394.

Dawsey, J. (2008). *The uses of sidewalks: Women, art, and urban space, 1966–1980.* Stanford University.

de Kerckhove, (1998). *Inteligência conectiva: a emergência da cibersociedade* (A. Viseu, Trans.). Lisboa: Fundação para a Divulgação das Tecnologias de Informação.

De Luca, G., Shirvani Dastgerdi, A., Francini, C., & Liberatore, G. (2020). Sustainable cultural heritage planning and management of overtourism in art cities: Lessons from atlas world heritage. *Sustainability*, *12*(9), 3929. doi:10.3390u12093929

de Oliveira, E., Ens, R. T., Andrade, D. B. F., & de Mussis, C. R. (2003). Content Analysis and Research in Education [Análise de Conteúdo e Pesquisa na Área da Educação]. *Educational Dialogue Journal*, *4*(9), 1–17.

de Souza, L. R. D. S., & Passos, O. V. D. O. A. (2018). Cordel Literature: A pedagogical too [Literatura De Cordel: Um recurso pedagógico]. *FASETE Scientific Journal, 75.*

De Valck, M. (2008). Screening the Future of Film Festivals? A long tale of convergence and digitization. *Film International*, *6*(4), 15–23. doi:10.1386/fiin.6.4.15

de Walck, M. (2007). *Film festivals: history and theory of a European phenomenon that became a global network.* Amsterdam University Press. doi:10.1017/9789048506729

Debord, G. (1991). *A sociedade do espectáculo* [The society of the spectacle]. Mobilis in Mobile. (Original work published 1967)

Delaplace, M. (2021). Après la crise, un tourisme durable ? *L'économie politique*, *3*(91), 8-22.

Deleuze, G. (2002). Post-scriptum sobre as sociedades de controle. In Conversações (pp. 219-226). Editora 34.

Deleuze, G., & Guattari, F. (1972). *L'anti-oedipe* [Anti-oedipus]. Éditions de Minuit.

Deleuze, G., & Guattari, F. (1980). *Mille plateaux* [A thousand plateaus]. Editions de Minuit.

Deleuze, G., & Guattari, F. (1991). *Qu'est-ce que la philosophie?* [What is philosophy?]. Minuit.

Delgadillo, V. (2018). Markets of La Merced: new frontiers of gentrification in the historic centre of Mexico City. In S. Sarah González (Ed.), *Contested markets, contested cities: Gentrification and urban justice in retail spaces* (pp. 19–35). Routledge.

Delgado, G. N., & Carvalho, B. V. (2020) Breque dos Aplicativos: Direito de resistência na era digital [Applications Breque: Right of resistance in the digital age]. *Le Monde Diplomatique Brasil.*

Della Lucia, M., & Trunfio, M. (2018). The role of the private actor in cultural regeneration: Hybridizing cultural heritage with creativity in the city. *Cities (London, England)*, *82*, 35–44. doi:10.1016/j.cities.2018.05.003

Dewey, J. (2005). L'art comme expérience [Art as experience]. Gallimard. (Original publication 1934)

di Liberto, A. (2013). High skills, high growth: Is tourism an exception? *The Journal of International Trade & Economic Development*, *22*(5), 749–785. doi:10.1080/09638199.2011.603054

Dianoux, C., & Siadou-Martin, B. (2020). Eclairages sur la communication des entreprises lors de la crise de la COVID-19: que dire et comment le dire ? *Projectics, 3*(27), 9-20.

Dias, M. G. (2014). *Ao Volante pela Cidade. Dez Entrevistas de Arquitetura.* Relógio d'Água.

Didi-Huberman, G. (2017). *Supervivencia de las luciérnagas*. Abada editores.

Diez Tetamanti, I. J. M. (Ed.). (2012). Cartografía Social: investigaciones e intervención desde las ciencias sociales: métodos y experiencias de aplicación. Universitaria de la Patagonia.

Diez Tetamanti, I. J. M. (Ed.). (2014). Hacia una Geografía Comunitaria: Abordajes desde la cartografía social y los sistemas de información geográfica. EDUPA.

Disalvo, C., Lukens, J., Lodato, T., Jenkins, T., & Kim, T. (2014). Making public things: How HCI design can express matters of concern. In *Proceedings of Conference on Human Factors in Computing Systems* (pp. 2397-2406). Association for Computing Machinery. 10.1145/2556288.2557359

Dixon, S. (2007). *Digital Performance - A history of New Media, Performance Art, and Installation*. The MIT Press.

Dodds, R., & Butler, R. (Eds.). (2019). *Overtourism: Issues, realities and solutions* (Vol. 1). Walter de Gruyter GmbH & Co KG.

Domingues, D., & Venturelli, S. (2008). Cibercomunicação cíbrida no continuum virtualidade aumentada e realidade aumentada: era uma vez… a realidade (Vol. 1). Revista ARS. Disponível em http// www.scielo.br/pdf/ars/v5n10/11.pdf /2011

Douglas, M. (1991). *Pureza e Perigo. Ensaio sobre as noções de poluição e tabu*. Lisboa. Edições 70.

Doustaly, C., & Roy, V. (2022). A Comparative Analysis of the Economic Sustainability of Cultural Work in the UK since the COVID-19 Pandemic and Examination of Universal Basic Income as a Solution for Cultural Workers. *Journal of Risk and Financial Management, 15*(5), 196. doi:10.3390/ jrfm15050196

Du, J. (2017, October). Zero latency: Real-time synchronization of BIM data in virtual reality for collaborative decision-making. *Automation in Construction, 85*, 59-60. doi:10.1016/j.autcon.2017.10.009

Dufour-Coppolani, D. (2020). Une vision inclusive. *I2D - Information, données & documents*, (3), 7. https://www.cairn.info/revue-i2d-information-donnees-et- documents-2020-3-page-7.htm

Duro Juan, A., Perez-Laborda, A., Turrion-Prats, J., & Fernández-Fernández, M. (2021). Covid-19 and tourism vulnerability. *Tourism Management Perspectives, 38*. PMID:34873568

Duthion, B., & Mandou, C. (2016). *L'innovation dans le tourisme: Culture numérique et nouveaux modes de vie*. De Boeck.

Eck, E., Melik, R., & Schapendonk, J. (2020). Window on the Netherlands marketplaces as public spaces in times of the covid-19 coronavirus outbreak: First reflections. *Tijdschrift voor Economische en Sociale Geografie, 111*(3), 373–386. doi:10.1111/tesg.12431 PMID:32836485

Eidelman, J., & Céroux, B. (2009). La gratuité dans les musées et monuments en France: quelques Indicateurs de mobilisation des visiteurs. In *Culture études. Ministère de la Culture – DEPS*. CAIRN INFO.

Einarsen, S., Hoel, H., & Notelaers, G. (2009). Measuring exposure to bullying and harassment at work: Validity, factor structure and psychometric properties of the Negative Acts Questionnaire-Revised. *Work and Stress, 23*(1), 24–44. doi:10.1080/02678370902815673

Elias, F. (2013). *Big Data é oportunidade para o setor público*. http://www.guiadascidadesdigitais.com.br/site/pagina/big-data-oportunidade-para-o-setor-pblico

Elmorshidy, A. (2010, May). Holographic Projection Technology: The World is Changing. *Journal of Telecommunication., 2*(2), 201. https://arxiv.org/ftp/arxiv/papers/1006/1006.0846.pdf

Eltarabily, S., & Elghezanwy, D. (2020). Post-Pandemic Cities - The Impact of COVID-19 on Cities and Urban Design. *Architecture Research, 10*(3), 75–84. doi:10.5923/j.arch.20201003.02

ENACT. (2021). Available at: https://enactafrica.org/countries/human-trafficking

Equality and Humans Rights Commission. (2010). *The Equality Act: Guidance for small business: Understanding legal definitions and unlawful behaviour in the equality act.* Retrieved: 27.04.2021 from https://www.equalityhumanrights.com/sites/default/files/ea_legal_definitions_0.pdf

Ericksen, E. G. (1980). *The Territorial Experience: Human Ecology as Symbolic Interaction.* U. T. Press. doi:10.7560/780385

ETC (European Travel Commission). (2021). *European Tourism: Trends & Prospects (Q4/2020).* Brussels, Belgium: European Travel Commission. https://etc-corporate.org/uploads/2021/02/ETC-Quarterly-Report-Q4-2020_Public-1.pdf

Etzkowitz, H., & Leydesdorff, L. (2000). The dynamics of innovation: From National Systems and "Mode 2" to a Triple Helix of university-industry-government relations. *Research Policy, 29*(2), 109–123. doi:10.1016/S0048-7333(99)00055-4

European Economic Forecast. (2021). *Economic Forecast: From recovery to expansion, amid headwinds.* European Commission.

European Parliament. (2016). *Report on a coherent EU policy for cultural and creative industries (2016/2072(INI)).* https://www.europarl.europa.eu/doceo/document/A-8-2016-0357_EN.pdf

Eurostat. (2019). *Culture statistics — 2019 edition.* doi:10.2785/118217

Evans, N. (2001). Collaborative strategy: An analysis of the changing world of international airline alliances. *Tourism Management, 22*(3), 229–243. doi:10.1016/S0261-5177(01)00024-3

Everts, J., Jackson, P., & Juraschek, K. A. (2021). The socio-material practices of the transformation of urban food markets. *Area Royal Geographical Society, 53*(2), 389–397. doi:10.1111/area.12707

EY. (2015). *Cultural Times. The First Global Map of Cultural and Creative Industries.* Study on Behalf of the International Confederation of Societies of Authors and Composers (CISAC). https://unesdoc.unesco.org/ark:/48223/pf0000235710

Fabry, N., & Mugnier, P. (2020). Peut-on entrevoir les contours de la nouvelle industrie touristique ? *I2D - Information, données & documents*, (3), 85-89.

Fabry, N., & Zeghni, S. (2020). Au-delà de la Covid-19: la gouvernance des territoires touristiques. *I2D - Information, données & documents, 3*(3), 95-100.

Falanga, R. (2020). *Citizen participation during the COVID-19 pandemic: Insights from local practices in European cities.* Friedrich-Ebert-Stiftung. https://repositorio.ul.pt/handle/10451/45726?locale=en

Falk, J. H., & Dierking, L. D. (2018). *Learning from Museums.* Roman and Littlefield.

Fantaspoa – Festival Internacional de Cinema Fantástico de Porto Alegre. (2021). *Fantaspoa bate recorde de público e divulga seus premiados.* Fantaspoa. https://www.fantaspoa.com/2021/noticias/1/109/fantaspoa-bate-recorde-de-publico-e-divulga-seus-premiados

FCT – Fundação para a Ciência e Tecnologia. (2019). *Agenda Temática de Investigação e Inovação – Turismo, Lazer e Hospitalidade.* https://www.fct.pt/agendastematicas/docs/Agenda_Turismo_Final.pdf

Felice, M. (2016). *Entrevista concedida a Eduardo Weinhardt.* https://transobjeto.wordpress.com/2016/04/03/estrevista-massimo-di-felice/#more-1038

Felice, M. (2009). *Paisagens pós-urbanas. O fim da experiência urbana e as formas comunicativas do habitar.* Annablume.

Fernandes, J. R., & Chamusca, P. (2014). Urban policies, planning and retail resilience. *Cities (London, England), 36*, 170–177. doi:10.1016/j.cities.2012.11.006

Fernandes, J., Carvalho, L., Chamusca, P., Mendes, T., & Gago, A. (2018b). *O Porto e a Airbnb*. Book Cover.

Fernandes, J., Carvalho, L., Chamusca, P., Mendes, T., & Gago, A. (2019). *Lisboa e a Airbnb*. Book Cover.

Fernandes, J., Carvalho, L., Chamusca, P., & Pinto, J. (2018a). *Gentrification in Porto: problems and opportunities in the past and in the future of an internationally open city. Revista de Geografia e Ordenamento do Território (GOT), 15(Dezembro).* doi:10.17127/got/2018.15.008

Ferrão, J. (2014). *Políticas territoriais eficazes e participadas*. Lisboa: Instituto de Ciências Sociais, Universidade de Lisboa. https://www.animar-dl.pt/ponto-de-vista/704-politicas-territoriais-eficazes-e-participadas-joao-ferrao

Ferreira, I. (2016). *Criatividade nos Museus. Espaços Entre e Elementos de Mediação*. Edição Caleidoscópio.

Fields, D., & Rogers, D. (2019). Towards a critical housing studies research agenda on platform real estate. *Housing. Theory and Society*, 1–23. doi:10.1080/14036096.2019.1670724

Firmino, M. B. (2007). *Turismo – Organização e Gestão*. Escolar Editora.

Florida, R. (2019). *The Innovations of the Creative Class Affect a Rural Area's Fortunes*. https://www.bloomberg.com/news/articles/2019-08-15/rural-innovation-links-to-creative-class-presence

Florida, R. (2002). *The Rise of the Creative Class. And how it's transforming work, leisure and everyday life*. Basic Books.

Florida, R. (2008). *Who is your city? How the Creative Economy Is Making Where to Live the Most Important Decision of Your Life*. Basic Books, Random House.

Flusser, V. (2015). *Comunicologia. Reflexões sobre o futuro*. Martins Fontes.

Fojut, N. (2009). *The philosophical, political and pragmatic roots of the convention. In Heritage and beyond.* Council of Europe.

Fonseca, F. V. L., & Fonsêca, K. S. B. (2009). Contributions from Cordel Literature to Cartography teaching [Contribuições da Literatura de Cordel para o ensino de Cartografia]. *Geography, 17*(2), 123–132.

Fontcuberta, J. (2016). *La furia de las imágenes: Notas sobre la postfotografía*. Galaxia Gutenberg.

Foster, H. (1996). The artist as ethnographer. In *The return of the real*. The MIT Press.

Foucault, M. (1975/1984). *Surveiller et punir*. Gallimard.

Foucault, M. (1984). Of Other Spaces, Heterotopias. *Architecture, Mouvement, Continuité, 5*, 46–49.

France Europan Commission. (2020). *Strategic Foresigt Report*. https://ec.europa.eu

Franck, K. A., & Paxson, L. (1989) Women and Urban Public Space. In Public Places and Spaces. Human Behavior and Environment. Springer. doi:10.1007/978-1-4684-5601-1_6

Frantzeskaki, N., & Rok, A. (2018). Co-producing urban sustainability transitions knowledge with community, policy and science. *Environmental Innovation and Societal Transitions, 29*(June), 47–51. doi:10.1016/j.eist.2018.08.001

Friedberg, A. (2009). *The virtual window. From Alberti to Microsoft*. The MIT Press.

Friedman, G. S., & Whitman, J. Q. (2003). *The European Transinfromation of Harassment Law: Discrimination Versus Dignity* (Vol. 9). Columbia Journal of European Law.

Fuchs, M., Höpken, W., & Lexhagen, M. (2014). Big data analytics for knowledge generation in tourism destinations - A case from sweden. *Journal of Destination Marketing & Management*, *3*(4), 198–209. doi:10.1016/j.jdmm.2014.08.002

Funari, P. (1999). Aspectos da cultura popular antiga: Apresentação, tradução e discussão de alguns graffitis pompeianos. *Revista Estudos de História*, *4*(1/2), 143–150.

Funari, P. (2003). *A vida quotidiana na Roma antiga*. Annablume.

Fyall, A., Garrod, B., & Wang, Y. (2012). Destination collaboration: A critical review of theoretical approaches to a multi-dimensional phenomenon. *Journal of Destination Marketing & Management*, *1*(1–2), 10–26. doi:10.1016/j.jdmm.2012.10.002

Gadras, S., & Pailliart, I. (2013). Les territoires et les médias dans la conception de l'espace public [The territories and the media in the conception of public space]. In Médias et territoires, l'espace public entre communication et imaginaire territorial. Presses universitaires du Septentrion.

Gao, X., & Yu, J. (2020). Public governance mechanism in the prevention and control of the COVID-19: Information, decision-making and execution. *Journal of Chinese Governance*, *5*(2), 178–197. doi:10.1080/23812346.2020.1744922

Garcia Henche, B. (2018). Urban experiential tourism marketing: Use of social media as communication tools by the food markets of madrid. *Journal of Tourism Analysis*, *25*(1), 2–22. doi:10.1108/JTA-01-2018-0002

Garcia, A. G. (2016). *Gestão de Recursos Humanos na Indústria do Turismo em Portugal:Tendências Futuras do Capital Humano: um estudo exploratório em unidades hoteleiras no Distrito de Coimbra*. Tese de Mestrado, Instituto Politécnico do Porto.

García-Gómez, C., Demir, E., Díez-Esteban, J., & Bilan, Y. (2021). The impact of COVID-19 outbreak on hotels' value compared to previous diseases: the role of ALFO strategy. *Heliyon*, *7*(8). doi:10.1016/j.heliyon.2021.e07836

Garrigos, J. A. (2020). Gender discrimination in haute cuisine: A systematic literature and media analysis. *International Journal of Hospitality Management*, *89*, 102569. doi:10.1016/j.ijhm.2020.102569

Garrigos, J. A., Haddaji, M., Segovia, P. G., & Signes, A. P. (2020). Gender differences in the evolution of haute cuisine chef's career. *Journal of Culinary Science & Technology*, *18*(6), 439–468. doi:10.1080/15428052.2019.1640156

Garud, R., & Karnøe, P. (2003). Bricolage versus breakthrough: Distributed and embedded agency in technology entrepreneurship. *Research Policy*, *32*(2), 277–300. doi:10.1016/S0048-7333(02)00100-2

Gastenhaus, A. (1997). *Minds in Motion- Using Museum to Expand Creative Thinking*. Caddo Gap Press. https://www.cairn.info/revue-culture-etudes-2009-2-page-1.htm

Gauthier, G. (2010). Géographie sentimentale du documentaire [Sentimental geography of the documentary]. In Champs visuels. l'Harmattan.

Gauthier, G. (2004). *Un siècle de documentaristes français* [A century of French documentary filmmakers]. Armand Colin.

Gehl, J., & Svarre, B. (2013). Jan Gehl & Birgitte Svarre. In How to Study Public Life. doi:10.5822/978-1-61091-525-0

Gehl, J. (2010). *Cidade para pessoas*. Perspectiva.

Gehl, J. (2011). Life Between Buildings: Using Public Space. In *Landscape Journal*. IslandPress., doi:10.3368/lj.8.1.54

Genette, G. (1966). Frontières du récit [Boundaries of the narrative]. Communications, 8, 161.

Getz, D. (2007). *Event Studies: Theory, Research and Policy for Planned Events*. Butterworth-Heinemann.

Gherardini, F., Santachiara, M., & Leali, F. (2019). Enhancing heritage fruition through 3d virtual models and augmented reality: An application to romanArtefacts. *Virtual Archaeology Review*, *10*(21), 67–79. doi:10.4995/var.2019.11918

Giannetti, C. (1998). Trespassar a Pele: o Teletrânsito. In ARS Telemática – Telecomunicação, Internet e Ciberespaço. Relógio D"#gua Editores.

Giedion, S. (1951). *A decade of new architecture*. Girsberger.

Gielen, P. (2009). The Murmuring of the Artistic Multitude: Global Art, Memory and Post-Fordism. Antennae, 3.

Giffinger, R. (2007). *Smart Cities: Ranking of European medium-sized cities*. Centre of Regional Science.

Gil, J. (2020). *O tempo indomado*. Relógio d'Água.

Ginzarly, M., & Srour, F. J. (2021). Cultural heritage through the lens of COVID-19. *Poetics*, *101622*. Advance online publication. doi:10.1016/j.poetic.2021.101622 PMID:34725533

Giousmpasoglou, C., Marinakou, E., & Cooper, J. (2018). "Banter, bollockings and beatings": The occupational socialisation process in Michelin-starred kitchen brigades in Great Britain and Ireland. *International Journal of Contemporary Hospitality Management*, *30*(3), 1882–1902. doi:10.1108/IJCHM-01-2017-0030

Gitahy, C. (1999). *O que é graffiti*. Brasiliense.

Glass, R. (1964). *London: aspects of change*. MacGibbon & Kee.

Global Security Health Index. (2019). Available at: https://www.ghsindex.org/country/guinea-bissau/

Godovykh, M., Pizam, A., & Bahja, F. (2021). Antecedents and outcomes of health risk perceptions in tourism, following the COVID-19 pandemic. *Tourism Review*, *76*(4), 737–748. doi:10.1108/TR-06-2020-0257

Goeldner, C., & Ritchie, B. (2012). *Tourism. Principles, Practices, Philosophies*. John Wiley & Sons.

Goffman, E. (1959). *The Presentation of Self in Everyday Life*. Penguin Books Ltd.

Gomes, D., & Araújo, M. (2012). Oferta Turística Virtual: Un estudio del metaverso. *Estudios y Perspectivas en Turism. Volumen*, *21*, 876–903.

Gonia, A., & Jezierska-Thöle, A. (2022). Sustainable tourism in Cities—Nature reserves as a 'New' city space for nature-based tourism. *Sustainability (Switzerland)*, *14*(3), 1581. doi:10.3390u14031581

González Martínez, P. (2017). Built heritage conservation and contemporary urban development: The contribution of architectural practice to the challenges of modernisation. *Built Heritage*, *1*(1), 14–25. doi:10.1186/BF03545666

Gonzalez, S., & Dawson, G. (2015). *Traditional markets under threat: why it's happening and what traders and customers can do.* https://eprints.whiterose.ac.uk/102291/

González, J., Miguel, M., Rosso, I., Toledo López, A., & Toledo López, V. (2016). Mapeando el barrio construimos territorio. Experiencia de cartografía social en Villa Aguirre, Tandil. *Revista Másquedós*, *1*(1), 61–71.

González, S. (2011). Bilbao and barcelona 'in motion'. how urban regeneration 'models' travel and mutate in the global flows of policy tourism. *Urban Studies (Edinburgh, Scotland)*, *48*(7), 1397–1418. doi:10.1177/0042098010374510

González, S. (2020). Contested marketplaces: Retail spaces at the global urban margins. *Progress in Human Geography*, *44*(5), 877–897. doi:10.1177/0309132519859444

Gonzalez, S., & Waley, P. (2013). Traditional retail markets: the new gentrification frontier? Antipode: A Radical. *The Journal of Geography*, *45*(4), 965–983.

Goodger, D., & Ferran, K. (2020). *City tourism outlook and ranking: Coronavirus impacts and recovery*. Tourism Economics.

Goodwin, H. (2021). City destinations, overtourism and governance. *International Journal of Tourism in Cities*.

Google. (n.d.). *Cardboard - apps on Google Play*. Google. Retrieved August 18, 2022, from https://play.google.com/store/apps/details?id=com.google.samples.apps.cardboarddemo&hl=en_US+%E2%80%93+Google+Cardboard

Gössling, S., Scott, D., & Hall, C. M. (2021). Pandemics, tourism and global change: A rapid assessment of COVID-19. *Journal of Sustainable Tourism*, *29*(1), 1–20. doi:10.1080/09669582.2020.1758708

Gottlieb, E. E. (2009). "Somewhere better than this place/nowhere better than this place": The lifemap of Rolland G. Paulston (1929–2006). *Prospects*, *39*(1), 91–101. doi:10.100711125-009-9110-5

Gounder, A., & Cox, C. (2021). Exploring the role of tourism dependency on COVID-19 induced economic shock in the Small Island Developing States. *Current Issues in Tourism*, 1–18.

Gravari-Barbas, M., & Delaplace, M. (2015). Le tourisme urbain « hors des sentiers battus » [Urban tourism "off the beaten track"]. *Téoros*, *34*, 1-2. https://journals.openedition.org/teoros/2790

Gravari-Barbas, M. (2005). *Habiter le patrimoine: Enjeux, approches, vécu* [Living in heritage: Issues, approaches, experiences]. Presses Universitaires de Rennes. doi:10.4000/books.pur.2208

Gravari-Barbas, M. (2013). *Aménager la ville par la culture et le tourisme* [Developing the city through culture and tourism]. Le Moniteur.

Grazina, I., & Magalhaes, J. (2012). Moral harassment at work: The relationship with organizational culture. *Journal of Business and Retail Management Research, 7*(1).

Gretzel, U., Fuchs, M., Baggio, R., Hoepken, W., Law, R., Neidhardt, J., Pesonen, J., Zanker, M., & Xiang, Z. (2020). e-Tourism beyond COVID-19: A call for transformative research. *Information Technology & Tourism*, *22*(2), 187–203. doi:10.100740558-020-00181-3

Grotenhuis, F. J. B., & Fesel, M. (2020). *Perspectives on cultural and creative industries*. https://www.grotenhuisadviseert.nl/wordpress/wp-content/uploads/Frits-Grotenhuis-Perspectives-on-Cultural-and-Creative-Industries.pdf

Guasch, A. M. (2021). *Ensayos críticos sobre arte y pensamiento*. Akal.

Guerreiro, J. A. (2016). *A questão da autenticidade – do objetivismo à fenomenologia do turismo*. IX Congresso Português de Sociologia, Faro: Faculdade de Economia – Universidade do Algarve. https://www.researchgate.net/publication/305279908_A_questao_da_autenticidade_-_do_objetivismo_a_fenomenologia_do_turismo

Guibert, C., & Réau, B. (2021). Les travailleurs du tourisme dans la tourmente. *Economics and Politics*, 3(91), 36–46.

Guidi, G., Trocchianesi, R., Pils, G., Morlando, G., & Seassaro, A. (2010). A Virtual Museum for Design: New forms of interactive fruition. *16th International Conference on Virtual Systems and Multimedia*, 242-249. 10.1109/VSMM.2010.5665977

Guimarães Rosa, J. (2011). *Grande sertão – Veredas* [The devil to pay in the backlands]. Editora Nova Fronteira.

Guterres, A. (2020). *We are all in this together: Human Rights and Covid-19 response and recovery*. Available at: https://www.un.org/en/un-coronavirus-communications-team/we-are-all-together-human-rights-and-covid-19-response-and

Guttentag, D. A. (2010). Virtual reality: Applications and implications for tourism. *Tourism Management*, *31*(5), 637–651. doi:10.1016/j.tourman.2009.07.003

Habermas, J. (1978). *L'espace public, Archéologie de la publicité comme dimension constitutive de la société bourgeoise* [The public space, Archaeology of advertising as a constitutive dimension of bourgeois society]. Payot.

Haddaji, M., Albors-Garrigós, J., & García-Segovia, P. (2017). Women chefs' experience: Kitchen barriers and success factors. *International Journal of Gastronomy and Food Science*, *9*, 49–54. doi:10.1016/j.ijgfs.2017.06.004

Haddaji, M., Albors-Garrigós, J., & García-Sigovia, P. (2017). Women professional progress to chef's position: Results of an international survey. *Journal of Culinary Science & Technology*, 1–18. doi:10.1080/15428052.2017.1363109

Hall, C. M. (2008). Tourism Planning Policies, Processes and Relationships. In *Policies, Processes and Relationships* (2nd ed.). Pearson Prentice Hall.

Hall, C. M., Scott, D., & Gössling, S. (2020). Pandemics, transformations and tourism: Be careful what you wish for. *Tourism Geographies*, *22*(3), 577–598. doi:10.1080/14616688.2020.1759131

Hall, E. T. (1990). *The Hidden Dimension*. Anchor Books Editions.

Hamann, R., & April, K. (2013). On the role and capabilities of collaborative intermediary organisations in urban sustainability transitions. *Journal of Cleaner Production*, *50*(July), 12–21. doi:10.1016/j.jclepro.2012.11.017

Hampton, B. (2005). A History of the Movies. *Film History*, *17*(4), 466–485. doi:10.2979/FIL.2005.17.4.466

Han, B. (2012). *A sociedade da transparência*. Herder.

Han, B.-C. (2019). *Hiperculturalidade. Cultura e globalização*. Vozes.

Hanzlík, J., & Mazierska, E. (2021). Eastern European film festivals: streaming through the covid-19 pandemic. *Studies in Eastern European Cinema*. . doi:10.1080/2040350X.2021.1964218

Harbord, J. (2009). Film Festivals as Time-Event. In D. Iordanova & R. Rhyne (Eds.), Film Festival Yearbook 1: The Festival Circuit (pp. 40–46). Academic Press.

Harper, D. (2012). *Visual sociology*. Routledge.

Harris, D. A., & Giuffre, P. (2015). *Taking the heat: Women chefs and gender inequality in the professional kitchen*. Rutgers University. doi:10.36019/9780813571270

Harris, R., Williams, P., & Griffin, T. (2012). *Sustainable tourism*. Routledge. doi:10.4324/9780080498362

Harvey, D. (1973). *Social justice and the city*. Edward Arnold.

Harvey, D. (1989). *The Condition of Postmodernity*. Blackwell.

Harvey, D. (2009). *Condição Pós-Moderna*. Edições Loyola.

Harvey, D. (2019). *Ciudades rebeldes. Del derecho de la ciudad a la revolución urbana*. Akal.

Hasse, M. (1998). Oeiras e o Desenvolvimento de Novos Comportamentos de Lazer. A valorização de um novo mundo: o mar, a praia e as férias, Iº Ciclo de Estudos Oeirenses. *Oeiras – A Terra e os Homens*, 265-285.

Hasse, M. (1999). *O Divertimento do Corpo. Corpo, lazer e desporto na transição do séc. XIX para o séc. XX em Portugal*. Editora Temática.

Heidegger, M. (1986/2005). *Ser e Tempo. Parte I*. Editora Vozes.

Helbrecht, I. (2017). Gentrification and displacement. In Gentrification and Resistance: Researching Displacement Processes and Adaption Strategies. Springer Gachmedien Wiesbaden GmbH. doi:10.1007/978-3-658-20388-7

Henley, J. (2020, January 25). *Overtourism in Europe's historic cities sparks backlash.* https://www.theguardian.com/world/2020/jan/25/overtourism-in-europe-historic-cities-sparks-backlash

Hernández Cordero, A., & Andreeva Eneva, S. (2016). Mercados, museos o malls? La gentrificación de los mercados municipales en Barcelona y Madrid. EntreDiversidades. *Revista de Ciencias Sociales y Humanidades*, *6*, 143–173.

Hernández, A. (2019). *Turismo virtual y experiência turística: el caso de la zona de monumentos arqueológicos de Teotihuacán.* Tesina da Licenciatura em Turismo, Universidad Autónoma del Estado de México.

Higgins-Desbiolles, F. (2020). Socialising tourism for social and ecological justice after COVID-19. *Tourism Geographies*, *22*(3), 610–623. doi:10.1080/14616688.2020.1757748

Highmore, B. (2005). *Cityscapes. Cultural readings in the material and symbolic city.* Palgrave Macmillan.

Hillman, A. J., Withers, M. C., & Collins, B. J. (2009). Resource dependence theory: A review. *Journal of Management*, *35*(6), 1404–1427. doi:10.1177/0149206309343469

Hodges, C.B., Moore, S., Lockee, B.B., Trust, T., & Bond, M.A. (2020). *The difference between emergency remote teaching and online learning.* Academic Press.

Hofstede Insights. (2020). *Description and Country Comparison.* Retrieved from https://www.hofstede-insights.com/product/compare-countries/

Home. (n.d.). Retrieved August 18, 2022, from https://www.raiznewmedia.com/

Honey-Rosés, J., Anguelovski, I., Chireh, V. K., Daher, C., Konijnendijk van den Bosch, C., Litt, J. S., Mawani, V., McCall, M. K., Orellana, A., Oscilowicz, E., Sánchez, U., Senbel, M., Tan, X., Villagomez, E., Zapata, O., & Nieuwenhuijsen, M. J. (2020). The impact of COVID-19 on public space: An early review of the emerging questions – design, perceptions and inequities. *Cities & Health*, *00*(00), 1–17. doi:10.1080/23748834.2020.1780074

Honneth, A. (2010). La lutte pour la reconnaissance [The struggle for recognition]. In Passages. Cerf.

Howe, J. (2006). The rise of crowdsourcing. *Wired Magazine, 14*(6).

Hruby, H. (2019, July) Geovisualization with immersive virtual environments in theory and practice. *International Journal of Digital Earth, 12*(2), 123-136.

Hsu, S. Y., Woodside, A. G., & Marshall, R. (2013). Critical tests of multiple theories of cultures' consequences: Comparing the usefulness of models by Hofstede, Inglehart and Baker, Schwartz, Steenkamp, as well as GDP and distance for explaining overseas tourism behavior. *Journal of Travel Research*, *52*(6), 679–704. doi:10.1177/0047287512475218

Huang, S. S., Shao, Y., Zeng, Y., Liu, X., & Li, Z. (2021). Impacts of COVID-19 on Chinese nationals' tourism preferences. *Tourism Management Perspectives*, *40*, 100895. doi:10.1016/j.tmp.2021.100895 PMID:34642624

Husserl, E. (1950). Idées directrices pour une phénoménologie (P. Ricoeur, Trans.). Gallimard. (Original publication 1913)

Husserl, E. (2008). A Ideia da Fenomenologia. Edições 70, Lda.

Hutchinson, R. (Ed.). (2010). Encyclopaedia of Urban Studies. SAGE Publications, Inc. doi:10.4135/9781412971973

Huybrechts, E. (2018). The historic urban landscape and the Metropolis. *Built Heritage*, *2*(4), 20–30. doi:10.1186/BF03545681

Ianioglo, A., & Rissanen, M. (2020). Global trends and tourism development in peripheral areas. *Scandinavian Journal of Hospitality and Tourism*, *20*(5), 520–539. doi:10.1080/15022250.2020.1848620

IATA. (2022). https://www.iata.org/contentassets/8ba31bb109324ae3973bb8b6560829be/2021-09-30-02-fr.pdf

ICOM. (2020). *Cómo comunicarse a distancia con su público. Consejo Internacional de Museos*. https://icom.museum/es/COVID-19/recursos/como-comunicarse-a-distancia-con-su-publico/

ICOMOS. (2008). *Declaração de Québec sobre a preservação do "Spiritu loci"*. https://www.international.icomos.org/quebec2008/quebec_declaration/pdf/GA16_Quebec_Declaration_Final_PT.pdf

Ignarra, L. R. (2003). *Fundamentos do Turismo* (2nd ed.). Thomson Learning.

ILO. (2020). *COVID-19 and the media and culture sector*. ILO Sectoral Brief. https://www.ilo.org/sector/Resources/publications/WCMS_750548/lang--en/index.htm

INE - Instituto Nacional de Estatística. (2018). *Mobilidade e funcionalidade do território nas Áreas Metropolitanas do Porto e de Lisboa : 2017*. https://www.ine.pt/xportal/xmain?xpid=INE&xpgid=ine_publicacoes&PUBLICACOESpub_boui=349495406&PUBLICACOESmodo=2

INE. (2019). Visitantes dos Museus por região (NUTS II). Análise de resultados do Inquérito aos Museus e do Património Cultural Imóvel. *Estatísticas da Cultura 2019.*

INE. (2021). *Número de Dormidas nos estabelecimentos hoteleiros 2009-2020 em Portugal*. https://www.ine.pt/

INE. (2021a). *Territorial Statistics. Several indicators and years*. Available at https://www.ine.pt

INE. (2021b). *Tourism Satellite Account*. Available at https://www.ine.pt

INSEE. (n.d.). *Activité touristique* . https://www.insee.fr/fr/statistiques/4625624?sommaire=4625628

Irigoy, I. (2019). Rent gap reloaded: Airbnb and the shift from residential to touristic housing in the Palma Old Quarter in Mallorca, Spain. *Urban Studies (Edinburgh, Scotland)*, *56*(13), 2709–2726. doi:10.1177/0042098018803261

Italy's Bureau of Statistics. (2019). *Turismo in Italia nel 2018*. https://www.istat.it/it/archivio/232137

Ivanova, M., Ivanov, I. K., & Ivanov, S. (2021). Travel behaviour after the pandemic: The case of Bulgaria. *Anatolia*, *32*(1), 1–11. doi:10.1080/13032917.2020.1818267

J. W., MS, W. C., J. A., J. T., & MS, M. M. (2008a). Sexual harassment in the hospitality industry: Perceptions of restaurant employees. *Journal of Human Resources in Hospitality & Tourism.*

Jacobs, J. (1961). *The Death and Life of Great American Cities*. Random House.

Jacquot, S. (2018). Élargissement de la sphére patrimoniale. De la diversification des lieux de patrimoine à l`engagement populaire. *Espaces,* 340.

Jafari, J. (2005). El turismo como disciplina científica. Política y Sociedad, 42(1), 39–56. doi:10.5209/POSO.24139

Jamieson, H. V. (2008). *Adventures in cyberformance* [Doctoral dissertation]. Queensland University of Technology.

Jamil, J., & Puad, M. S. (2010). Challenges in managing infrastructure for urban tourism: A comparative study of three asian cities. *WIT Transactions on Ecology and the Environment*, *142*, 243–253. doi:10.2495/SW100231

Janson, A. F., & Janson, H. W. (1996). *Iniciação à História da arte*. Martins Fontes.

Jeannotte, M. S. (2021). When the gigs are gone: Valuing arts, culture and media in the COVID-19 pandemic. *Social Sciences & Humanities Open*, *3*(1), 100097. doi:10.1016/j.ssaho.2020.100097

Jelinčić, D. A. (2021). Indicators for Cultural and Creative Industries' Impact Assessment on Cultural Heritage and Tourism. *Sustainability*, *13*(14), 7732. doi:10.3390u13147732

Jenkins, H., Green, J., & Ford, S. (2014). Cultura da conexão. Criando valor e significado por meio da mídia programável. *Aleph (Jerusalem)*.

Jokilehto, J. (2014). *Evolution of the normative framework. In Reconnecting the City: The Historic Urban Landscape Approach and the Future of Urban Heritage*. Wiley.

Joo-Nagata, J., García-Bermejo, G., & Martínez-Abad, F. (2015). Patrimonio Virtual del Territorio: Diseño e implementación de Reursos Educativos en Realidad Aumentada y Navegación Peatonal Móvil. *VAEP-RITA*, *3*(1), 46–51.

Joron, P. (2006). La violence sacrificielle. *Les Cahiers de l'IRSA*, 245-261.

Joron, P. (2011). La sudation du quotidien ou les pores du réel médiatique. *Sociétés, 114*, 53-61.

Joron, P. (2015). Comment la sociologie est-elle possible? *Sociétés, 127*, 13-24.

Joron, P. (2012). La prime à l'émotionnel. In M. Maffesoli & B. Perrier (Eds.), *L'homme postmoderne* (pp. 23–32). Ed. François Bourin.

Joron, P. (2018). Georg Simmel et la sociologie du futile. Dans les anfractuosités du social et de l'intime. *Episteme*, *20*, 46–67.

Joron, P. (2019). Tourisme et terrorisme. Grand Tour et petites mains de la Terreur. *Sociétés*, *143*(143), 7–17. doi:10.3917oc.143.0007

Joron, P. (2020). Les réfléchissants. Gilets-Jaunes, ronds-points et nulle part ailleurs. In *Sociétés, 150* (pp. 67–81). De Boeck.

Joron, P. (2020). Sós todos juntos. Pele digital e fissuras digitais. In C. Freitas, J. Machado da Silva, & P. Joron (Eds.), *Laço social e tecnologia em tempos extremos Imaginário, Redes e Pandemia* (pp. 15–27). Sulina.

Joseph, J. (2010). *The Experience Effect: Engage Your Customers With a Consistent and Memorable Brand Experience*. Amacom.

Jr, B. N. (2017) De onde vem o poder das imagens que invadem nossas casas e corpos? In Ecologia da imagem e dos Media. Nova Vega.

Jünger, E. (1990). La mobilisation totale [Total Mobilization]. In *L'Etat Universel – suivi de La mobilisation totale*. Gallimard. (Original work published 1930)

Kallis, G., & March, H. (2015). Imaginaries of Hope: The Utopianism of Degrowth. *Annals of the Association of American Geographers*, *105*(2), 360–368. doi:10.1080/00045608.2014.973803

Kant, I. (2001). Crítica da Razão Pura (M. P. d. Santos & A. F. Morujão, Trans., 5th ed.). Lisboa: Fundação Calouste Gulbenkian.

Kaplan, D. (2021). "What Does It Matter Where My Body Happens to Be?" A Personal View of Online Art Museum Tours. *Journal of Museum Education*, *46*(4), 531–546. doi:10.1080/10598650.2021.1987768

Kaplan, S. L. (1984). *Provisioning Paris: merchants and millers in the grain and flour trade during the Eighteenth Century*. Cornell University Press. doi:10.7591/9781501731426

Kasni, N. W. (2021, November). The Multimodal Forms of Tourism Promotional Discourse in the Age of COVID-19. *International Journal of Linguistics, Literature and Culture*, *7*(6), 422–440.

Kemmis, S., & McTaggart, R. (2005). Participatory action research: communicative action and the public sphere. In N. K. Denzin & Y. S. Lincoln (Eds.), *The Sage Handbook of Qualitative Research* (pp. 559–603). Sage Publications.

Kemmis, S., McTaggart, R., & Nixon, R. (2014). *The Action Research Planner: Doing Critical Participatory Action Research*. Springer. doi:10.1007/978-981-4560-67-2

Kera, D., Rod, J., & Peterova, R. (2013). Post-apocalyptic citizenship and humanitarian hardware. In R. Hindmarsh (Ed.), *Nuclear disaster at Fukushima Daiichi: Social, political and environmental issues* (pp. 97–115). Routledge.

Kester, J. (2015). International Tourism Trends in EU28 Member States. *Current Situation and Forecasts for 2020-2025-2030*. https://ec.europa.eu/docsroom/documents/16845?locale=pt

Khalilur, R. M., & Gazi, M. (2021, September 1). Effect of Covid-19 pandemic on tourist travel risk and management perceptions. *PLoS One*, *16*(9), e0256486. Advance online publication. doi:10.1371/journal.pone.0256486

Khamung, R., & Hsu, P. S. (2021). Striving for a new normal after the Covid-19 pandemic: Taking small businesses at Bangsaen Beach as an example. *Journal of Tourism and Cultural Change*, 1–17. doi:10.1080/14766825.2021.1987447

Khlystova, O., Kalyuzhnova, Y., & Belitski, M. (2022). The impact of the COVID-19 pandemic on the creative industries: A literature review and future research agenda. *Journal of Business Research*, *139*, 1192–1210. doi:10.1016/j.jbusres.2021.09.062 PMID:34629569

Kindon, S., Pain, R., & Kesby, M. (2007). *Participatory Action Research Approaches and Methods: Connection People, Participation and Place*. Routledge. doi:10.4324/9780203933671

King, S. (1975). The Lawnmower Man. *Cavalier.*

Kist, C. (2020). Museums, challenging heritage, and social media during COVID-19. *Museum & Society*, *18*(3), 345–348. doi:10.29311/mas.v18i3.3539

Kock, F., Nørfelt, A., Josiassen, A., Assaf, A. G., & Tsionas, M. G. (2020). Understanding the COVID-19 tourist psyche: The Evolutionary Tourism Paradigm. *Annals of Tourism Research*, *85*, 85. doi:10.1016/j.annals.2020.103053 PMID:32921847

Koens, K., Postma, A., & Papp, B. (2018). Is Overtourism Overused? Understanding the Impact of Tourism in a City Context. *Sustainability*, *10*(12), 4384. doi:10.3390u10124384

Köhler, J., Geels, F. W., Kern, F., Markard, J., Onsongo, E., Wieczorek, A., & Wells, P. (2019). An agenda for sustainability transitions research: State of the art and future directions. *Environmental Innovation and Societal Transitions*, *31*(January), 1–32. doi:10.1016/j.eist.2019.01.004

Kohn, E. (2013). *How forests think. Toward an anthropology beyond the human*. California University Press.

Kokshagina, O. (2021). Open COVID-19: Organizing an extreme crowdsourcing campaign to tackle grand challenges. *R&D Management*.

Kolko, J. (2018). The divisiveness of design thinking. *Interaction*, *25*(3), 28–34. doi:10.1145/3194313

Komenda, E. (2020). COVID-19: Las Vegas Strip resorts are now closing, includes MGM, Wynn properties. *Reno Gazette Journal*. Retrieved from https://www.rgj.com/story/news/2020/03/15/las-vegas-strip-resorts-now-closing-first-up-wynn-and-encore/5056092002/

Koodsela, W., Dong, H., & Sukpatch, K. (2019). A holistic conceptual framework into practice-based on urban tourism toward sustainable development in Thailand. *Sustainability (Switzerland), 11*(24), 7152. doi:10.3390u11247152

Kotler, P., Kartajaya, H., & Setiawan, I. (2017). *Marketing 4.0: Mudança do tradicional para o digital.* Conjuntura Actual Editora – Edições Almedina.

Kovadloff, S. (2004). Sentido y riesgo de la vida cotidiana. Emecé Editores. (Original publication 1998)

Kraus, S., & Koch, N. (2021). Provisional COVID-19 infrastructure induces large, rapid increases in cycling. *Proceedings of the National Academy of Sciences of the United States of America, 118*(15), 1–6. doi:10.1073/pnas.2024399118 PMID:33782111

Kretz, A., & Sá, C. (2013). Third Stream, Fourth Mission: Perspectives on University Engagement with Economic Relevance. *Higher Education Policy, 26*(4), 497–506. doi:10.1057/hep.2013.32

Krippendorff, K. (2004). *Content analysis: An introduction to its methodology.* Sage publications.

Krueger, M. (1977). Responsive environments. In *Proceedings of the June 13-16, 1977, national computer conference. AFIPS '77.* Dallas, TX: ACM. 10.1145/1499402.1499476

Kucukkaya, A. G. (2004). Photogrammetry and remote sensing in archeology. *Journal of Quantitative Spectroscopy & Radiative Transfer, 88*(1-3), 83–88. doi:10.1016/j.jqsrt.2003.12.030

Kuklinski, H. P., & Cobo, C. (2020). *Expanding the university beyond emergency remote teaching: ideas for a post-pandemic hybrid model* [Expandir la universidad más allá de la enseñanza remota de emergencia: Ideas hacia un modelo híbrido post-pandemia] (1st ed., Vol. 1). Outliers School. https://outliersschool.net/wp-content/uploads/2020/05/Expandir_la_universidad.pdf

Kumar, N., Kumar, R. R., Patel, A., Hussain Shahzad, S. J., & Stauvermann, P. J. (2020). Modelling inbound international tourism demand in small Pacific Island countries. *Applied Economics, 52*(10), 1031–1047. doi:10.1080/00036846.2019.1646887

Kupi, M., & Szemerédi, E. (2021). Impact of the COVID-19 on the Destination Choices of Hungarian Tourists: A Comparative Analysis. *Sustainability, 13*(24), 13785. doi:10.3390u132413785

Labaki, A. (n.d.). *É tudo verdade.* Festival É Tudo Verdade. http://etudoverdade.com.br/br/pag/apresentacao

Labbé, M. (2019). *Reprendre place, contre l'architecture du mépris* [Reclaiming a place, against the architecture of contempt]. Payot.

Lamas, J. M. G. (1993). Os elementos morfológicos do espaço urbano - CAP2. In Morfologia Urbana e Desenho da Cidade (pp. 79–110). Academic Press.

LaMontagne, Smith, Louie, Quinlan, Shoveller, & Ostry. (2009). Unwanted Sexual Advances at Work: Variations by Employment Arrangement in a Sample of Working Australians. *Australian and New Zealand Journal of Public Health, 3*(2), 173–179. PMID:19413863

Landry, C. (2010). *Creativity, Culture & the City: A question of interconnection.* http://www.forum-avignon.org/sites/default/files/editeur/ECCE_report.pdf

Landry, C. (2016). Culture and the digital city: its impact and influence. *Culture urban future: Global report on culture for sustainable urban development*, 157-164.

Landry, C. (2006). *The Art of City Making.* Earthscan Publications.

Lanza, A., Temple, P., & Urga, G. (2003). The implications of tourism specialization in the long run: An econometric analysis for 13 OECD economies. *Tourism Management, 24*(3), 315–321. doi:10.1016/S0261-5177(02)00065-1

Las Vegas Convention and Visitors Bureau [LVCVA]. (2018). *Arriving visitors on direct international flights 2013-2017.* Retrieved from https://assets.simpleviewcms.com/simpleview/image/upload/v1/clients/lasvegas/2017_Top_Air_Markets_Intl_b930157b-ced0-4acd-9f31-ee90cb89e43a.pdf

Las Vegas Convention and Visitors Bureau [LVCVA]. (2019). The relative dependence on tourism of major U.S. economies. *Economic Impact Series Brief.* Retrieved from https://assets.simpleviewcms.com/simpleview/image/upload/v1/clients/lasvegas/EIS_Tourism_Dependence_Jan_2019_FINAL_e332cdc1-7e25-4179-b4e9-63e1890adefe.pdf

Latour, B. (2005). *Reassembling the social. An introduction to actor-network-theory.* Oxford University Press.

Latour, B. (2008). *What is the style of matters of concern? Two lectures in empirical philosophy.* Van Gorcum Amsterdam.

Latour, B., & Weibel, P. (2005). *Making things public: Atmospheres of democracy.* MIT Press.

Lauretis, T. D. (1994). A tecnologia do gênero. Tradução de Suzana Funck. In *Tendências e impasses: o feminismo como crítica da cultura.* Rocco.

Le Corf, J.-B. (2015). Le récit métropolitain, composante de la communication touristique. Les cas de Marseille et d'Istanbul face au label « capitale européenne de la culture [The metropolitan narrative, a component of tourism communication. The cases of Marseille and Istanbul in the face of the "European Capital of Culture" label]. In N. Cyrulnik & V. Meyer (Eds.), *Synergies Mondes Méditerranées* (Vol. 5, p. 148). Retrieved from https://gerflint.fr/Base/MondeMed5/le_corf.pdf

Leavy, P. (2017). *Research Design. Quantave, Qualitative, Mixed Methods, Arts-Based and COmmunity Based Participatory Research Approaches.* The Guilford Press.

Lebaron, F. (2020). Covid-19. L'anomie pandémique. *Savoir/Agir, 2*(52), 7-9.

Lecea, I. (2009). *Barcelona. Art Public.* Ajuntament de Barcelona i Àmbit Serveis Editorials.

Le, D., & Phi, G. (2021). Strategic responses of the hotel sector to COVID-19: Toward a refined pandemic crisis management framework. *International Journal of Hospitality Management, 94*, 94. doi:10.1016/j.ijhm.2020.102808 PMID:34785839

Lee, J., & Ingold, T. (2006). Fieldwork on foot: perceiving, routing, socializing. In Locating the field-space, place and context in Anthropology (pp. 67-86). Berg.

Lee, K. (2021). Critique of Design Thinking in Organizations: Strongholds and Shortcomings of the Making Paradigm. *She Ji: The Journal of Design, Economics, and Innovation, 7*(4), 497–515. doi:10.1016/j.sheji.2021.10.003

Lee, N., & Rodríguez-Pose, A. (2014). Creativity, cities, and innovation. *Environment & Planning A, 46*(5), 1139–1159. doi:10.1068/a46202

Lee-Peng, F., Mui-Yin, C., Kim-Leng, T., & Kit-Teng, P. (2021). The impact of COVID-19 on tourism industry in Malaysia. *Current Issues in Tourism, 24*(19), 2735–2739. doi:10.1080/13683500.2020.1777951

Lefebvre, H. (1991). The Production of Space (Nicholson-Smith, Trans.). Blackwell Publishing.

Lefebvre, H. (1986). La production de l'espace. *Anthropos. (Original publication 1974)*

Lefebvre, H. (2004). Rhythmanalysis: Space, time and everyday life. *Continuum.*

Lehmann, H. T. (2013). Teatro Pós-dramático, doze anos depois / Postdramatic Theatre, 12 years later / Théâtre Postdramatique, douze ans plus tard. *Revista Brasileira de Estudos da Presença, 3*(3), 859–878. doi:10.1590/2237-266039703

Leiper, N. (1990). Partial industrialization of tourism systems. *Annals of Tourism Research, 17*(4), 600–605. doi:10.1016/0160-7383(90)90030-U

Lemoine, B. (2020). Après la pandémie. *Villes en Parallèle, 49-50*(1), 446–452. doi:10.3406/vilpa.2020.1833

Lemos, A. (2008). Mobile Communication and New Sense of Places: A Critique of Spatialization in Cyberculture. *Galáxia (São Paulo), 16*, 91–108.

Lemos, A. (2021). *A tecnologia é um vírus*. Sulinas.

Lemos, A. (Ed.). (2004). *Cibercidade. As cidades na cibercultura*. Ed. E-papers.

Lenzi, C., & Perucca, G. (2020). Urbanization and subjective well-being. In *Regeneration of the Built Environment from a Circular Economy Perspective* (pp. 21–28). Springer. doi:10.1007/978-3-030-33256-3_3

Leonard, B. (Director). (1992). *The Lawnmower Man* [Movie]. Allied Vision.

Lerum, K. (2004). Power and Camaraderie in Service Work. *Gender & Society, 18*(6), 756–776. doi:10.1177/0891243204269398

Lessard, H. (2007). Backlash in the Academy: The Evolution of Campus Sexual Harassment Regimes. In D. E. Chunn, S. Boyd, & H. Lessard (Eds.), Reaction and Resistance: Feminism, Law, and Social Change (pp. 164–195). Academic Press.

Lévy, P. (2013). *O que é virtual? 2013. 1 vídeo (2,54min).* Publicado pelo canal Fronteiras do pensamento em 25 maio 2013. Disponível em: https://www.youtube.com/watch?v=sMyokl6YJ5U&feature=emb_logo

Li, C. (2021, December). Synthesizing Scene-Aware Virtual Reality Teleport Graphs. *ACM Trans. Graph., 40*(6).

Liebman, M., & Paulston, R. G. (1994). Social Cartography: A new methodology for comparative studies. *Compare: A Journal of Comparative Education, 24*(3), 233–245. doi:10.1080/0305792940240304

Liga Guineense dos Direitos Humanos (several years). (n.d.). *Relatórios e comunicados*. Bissau. Available at: http://www.lgdh.org/

Li, J. J., Bonn, M. A., & Ye, B. H. (2019). Hotel employee's artificial intelligence and robotics awareness and its impact on turnover intention: The moderating roles of perceived organizational support and competitive psychological climate. *Tourism Management, 73*, 172–181. doi:10.1016/j.tourman.2019.02.006

Li, J., Xu, L., Tang, L., Wang, S., & Li, L. (2018). Big data in tourism research: A literature review. *Tourism Management, 68*, 301–323. doi:10.1016/j.tourman.2018.03.009

Lima, J. M., de Sousa, J. M., & Germano, U. M. G. (2011). Cordel Literature as a means of science popularization: an intervention in Physics teaching [Os PCN e as potencialidades didático-pedagógicas do cordel].VIII ENPEC.

Lima, T. (2013). The national curriculum and the Cordel teaching and learning potential [Os PCN e as potencialidades didático-pedagógicas do cordel]. *Acta Scientiarum. Education, 35*(1), 133–139.

Lim, C., Zhu, L., & Koo, T. T. R. (2019). Urban redevelopment and tourism growth: Relationship between tourism infrastructure and international visitor flows. *International Journal of Tourism Research, 21*(2), 187–196. doi:10.1002/jtr.2253

Lim, J., & Won, D. (2020). preprint). How Las Vegas' tourism could survive an economic crisis? *Cities (London, England), 100*, 102643. Advance online publication. doi:10.1016/j.cities.2020.102643

Lioult, J.L. (2004). *A l'enseigne du réel, penser le documentaire* [In the sign of reality, thinking the documentary]. Publications de l'Université de Provence.

Lipovetsky, G., & Serroy, J. (2010). *O ecrã global: cultura mediática e cinema na era hipermoderna*. Edições 70.

Lipovetsky, G., & Serroy, J. (2007). *L'écran global*. Seuil.

Lipovetsy, G., & Juvin, H. (2011). *O Ocidentes Mundializado. Controvérsia sobre a Cultura Planetária*. Lisboa: Edições 70.

Lipovetsy, G., & Serroy, J. (2010). *A Cultura Mundo. Resposta a uma Sociedade Desorientada*. Lisboa: Edições 70.

Lits, M. (2015). Storytelling: réévaluation d'un succès editorial [Storytelling: reassessing an editorial success]. In Le storytelling, succès des histoires, histoire d'un succès. L'Harmattan.

Liu, X. (2013). A study on the competitiveness of urban tourism destinations taking loudi city in Hunan province for example. *Information Technology Journal*, *12*(14), 2905–2912. doi:10.3923/itj.2013.2905.2912

Li, X., Gong, J., Gao, B., & Yuan, P. (2021). Impacts of COVID-19 on tourists' destination preferences: Evidence from China. *Annals of Tourism Research*, *90*, 90. doi:10.1016/j.annals.2021.103258 PMID:34924648

Lobatón, S. B. (2009). Reflexiones sobre Sistemas de Información Geográfica Participativos (SIGP) y cartografía social. *Cuadernos de Geografía. Revista Colombiana de Geografía*, (18), 9–23. doi:10.15446/rcdg.n18.12798

Lofland, L. (1998). *The public realm. Exploring the city's quintessential social territory*. Aldine de Gruyter.

Lopa, J. (2020). Sexual harassment of hospitality interns. *Journal of Hospitality & Tourism Education*, *32*(2), 88–101. doi:10.1080/10963758.2020.1726767

Lopes, E. R. (2010). A constelação do Turismo na economia portuguesa-Portugal Desafios nos alvores do século XXI. Lisboa: Ed. O Sol é Essencial S.A.

López de Ávila, A. & García-Sánchez, S. (2013). Destinos turísicos inteligentes. *Harvard Deusto Business Review, 224,* 58-57.

López, L., Yagual, Patricia, C., & Aquino, J. (2016). El uso de la realidade virtual como herramienta tecnológica para fomentar el turismo em la península de Santa Elena. *3C Tecnología*, *5*(3), 53 – 67.

Loureiro, S. M. C. (2020). 20 years of research on virtual reality and augmented reality in tourism context: A text-mining approach. *Tourism Management, 77.*

Lourenço, H. M. (2021). *A Gestão de Recursos Humanos no Turismo: o caso da hotelaria na cidade de Lisboa*. Tese de doutoramento, Turismo, Universidade de Lisboa, Instituto de Geografia e Ordenamento do Território.

Lucas, L., & Moita, F. M. G. S. (2020). Emergency distance education: The impacts in the pedagogical practices during Covid-19 [Ensino Remoto Emergencial (ERE): impactos na prática pedagógica durante a Covid-19]. *Techonology Research in Education Journal*, *6*, e143320–e143320.

Lúcio, J. (2011). *A Mediação Sócio-Educativa na Construção da Cidade Educadora: O projeto "Massarelos, Freguesia Educadora"* (PhD essay). Universidade do Porto.

Lúcio, J., & Pereira, A. V. (2009). *Cartografia social do Porto*. Paper presented at the I Fórum de Investigação em Ciências da Educação, Lisboa.

Macamo, E. (Ed.). (2005). *Negotiating modernity: Africa's experience*. Zed Books Ltd.

MacCannell, D. (1976). *The Tourist: A New Theory of the Leisure Class*. Schocken Books.

MacCannell, D. (1992). *Empty Meeting Grounds: The Tourist Papers*. Routledge.

MacCannell, D. (2011). *The Ethics of Sightseeing*. University of California Press.

MacDonald, C. (2012). Understanding Participatory Action Research: A Qualitative Research Methodology Option. *Canadian Journal of Action Research, 13*(2), 34–50. doi:10.33524/cjar.v13i2.37

Machado da Silva, J. (2013). *A sociedade midíocre: passagem ao hiperspectacular.* Sulina.

Machado, A. (2007). *Arte e Mídia: aproximações e distinções. Galáxia.* Revista do Programa de Pós-Graduação em Comunicação e Semiótica.

Machado, L. J., de Moraes Almeida, I., & de Paula, L. B. (2021). Cordel Literature as a means of facilitating the teaching and learning process of chemistry [Literatura de cordel como recurso facilitador do processo ensino-aprendizagem em química]. *Brazilian Journal of Development, 7*(9), 86407–86424.

Maffesoli, M. (1987). *O conhecimento do quotidiano* (J. Lamy, Trans.). Vega.

Maffesoli, M. (1990). *Au creux des apparences. Pour une éthique de l'esthétique.* Plon.

Maffesoli, M. (2000). *L'instant éternel. Le retour du tragique dans les sociétés postmodernes* [The eternal instant. The return of the tragic in postmodern societies]. Denoël.

Maffesoli, M., & Martins, M. L. (2011). L'imaginaire des médias [The imaginary of the media]. *Sociétés, 111*(1), 5–9. doi:10.3917oc.111.0005

Magliacani, M., & Sorrentino, D. (2021). Reinterpreting museums' intended experience during the COVID-19 pandemic: Insights from Italian University Museums. *Museum Management and Curatorship.* Advance online publication. doi:10.1080/09647775.2021.1954984

Mahraz, M. I., Benabbou, L., & Berrado, A. (2019, July). A systematic literature review of digital transformation. *Proceedings of the International Conference on Industrial Engineering and Operations Management Toronto*, 917-931.

Main, A. (2021). Measuring Workplace Harassment Based on Gender Nonconformity. *International Journal of Business and Management Research, 9*(1), 11-19.

Manovich, L. (2017). *Instagram & Contemporary Image.* Postdigital Aesthetics.

Manovich, L., & Tifentale, A. (2015). *Selfiecity: Exploring Photography and Self-Fashioning in Social Media.* Postdigital Aesthetics.

Marcos, A. (2016). *Média-Arte Digital: da centralidade do artefacto à transdisciplinaridade.* Congresso da Cibercultura. Braga: Universidade do Minho. https://repositorioaberto.uab.pt/bitstream/ 10400.2/6239/1/CiberCult_AFMarcos-14-out2016.pdf

Marín, A. (2021). The immersive documentary: Understanding the phenomenon in nonfiction stories through a typology proposal. *El Profesional de la Información, 30*(4). https://doi.org/10.3145/epi.2021.jul.11

Marques Santos, A., Madrid, C., Haegeman, K., & Rainoldi, A. (2020). *Behavioural changes in tourism in times of Covid-19.* Publications Office of the European Union. doi:10.2760/00411

Marques, A. (2019). *A Importância da Qualificação dos Recursos Humanos no Turismo: O Caso dos Hostels da Cidade do Porto.* Tese de Mestrado, Instituto Politécnico do Porto.

Marr, B. (2020, December). The Future of Virtual Reality. *Forbes.* https://www.forbes.com/sites/bernardmarr/2020/12/18/the-future-of-virtual-reality-vr/?sh=dc008d127be8

Martínez Juárez, I. H., Puga Nathal, M. E., & Martínez-Miranda, J. (2019). Desarrollo de un guía virtual conversacional para museos en un entorno inmersivo de realidad virtual y aumentada. *Research in Computing Science, 148*(8), 387–395.

Martins, M. L. (2002). De animais da promessa a animais em sofrimento de finalidade [From animals of promise to animals in suffering for purpose]. *O Escritor, 18/20*, 351–354. Available at: http://hdl.handle.net/1822/1676

Martins, M. L. (2003). O quotidiano e os *media* [Everyday life and the media]. *Todas as Letras, 5*, 97–105. Available at: http://hdl.handle.net/1822/1064

Martins, M. L. (2007). Nota introdutória. A época e as suas ideias [Introductory note. The time and its ideas]. *Comunicação e Sociedade*, (12), 5–7. Available at: http://hdl.handle.net/1822/24115

Martins, M. L. (2009). Ce que peuvent les images. Trajet de l'un au multiple [What images can do. Journey from the one to the many]. *Les Cahiers Internationaux de l'Imaginaire, 1*, 158–162. Available at: http://hdl.handle.net/1822/24245

Martins, M. L. (2010a). Linguagem, verdade e conhecimento. As ciências da comunicação e o contemporâneo [Language, truth and knowledge. Communication Sciences and the contemporary]. In A. S. Silva, J. Cândido Martins, L. Magalhães, & M. Gonçalves (Eds.), *Comunicação, cognição e media* (vol. 1, pp. 76–86). Universidade Católica Portuguesa - Centro de Estudos Filosóficos e Humanísticos. Available at: http://hdl.handle.net/1822/24118

Martins, M. L. (2010b). A mobilização infinita numa sociedade de meios sem fins [The infinite mobilization in a society of means without ends]. In C. Álvares & M. Damásio (Eds.), *Teorias e práticas dos media. Situando o local no global.* Edições Lusófonas. Available at: http://hdl.handle.net/1822/24250

Martins, M. L. (2011a). *Crise no castelo da cultura. Das estrelas para os ecrãs* [Crisis in the castle of culture. From the stars to the screens]. Grácio Editor. Available at: http://hdl.handle.net/1822/29167

Martins, M. L. (2011b). Technologie et rêve d'humanité [Technology and the dream of humanity]. *Les Cahiers Européens de l'Imaginaire, 3*, 56–61. Available at: http://hdl.handle.net/1822/24245

Martins, M. L. (2019a). A comunicação e a informação na cultura [Communication and information in culture]. *Maremagnum*, (23), 33–40. Available at: http://hdl.handle.net/1822/61843

Martins, M. L. (2020). Tecnologia e literatura: As narrativas transmediáticas [Technology and literature: The transmedia storytelling]. *Letras de Hoje, 55*(1). doi:10.15448/1984-7726.2020.1.34786

Martins, M. L., Marques, M. A., & Ramos, R. (Eds.). (2020). Public discourses on migrants, refugees and minorities: the 21st century in debate. *Comunicação e Sociedade, 38*. https://revistacomsoc.pt/index.php/revistacomsoc/issue/view/154

Martins, M. L. (2012). Média digitais: Hibridez, interactividade, multimodalidade [Digital media: Hybridity, interactivity, multimodality]. *Revista de Comunicação e Linguagens, Lisboa, 43/44*, 49–60. doi:10.3917oc.111.0017

Martins, M. L. (2015). Cultural studies as the new humanities. *Lusophone Journal of Cultural Studies, 3*(1), 341–361. doi:10.21814/rlec.100

Martins, M. L. (2017). Prefácio: A circum-navegação de atmosferas e paisagens tecnológicas [Preface: The circumnavigation of technological atmospheres and landscapes]. In H. Pires, M. Curado, F. Ribeiro, & P. Andrade (Eds.), *Cibercultura: Circum-navegações em redes transculturais de conhecimento, arquivos e pensamento* (pp. 11–18). CECS. Available at http://hdl.handle.net/1822/49352

Martins, M. L. (2019b). The "refugee crisis" in Europe –The "refugee crisis" in Europe – between totality and the infinite. *Comunicação e Sociedade*, (Special Issue), 37–52. doi:10.17231/comsoc.0(2019).3059

Massey, D. (1995). The conceptualization of place. In M. Doreen & P. Jess (Eds.), *A place in the world? Place, culture and Globalization* (pp. 45–85). Oxford University Press.

Mateer, J. (2017, May). Directing for Cinematic Virtual Reality: How the traditional film director's craft applies to immersive environments and notions of presence. *Journal of Media Practice, 18*(1), 19–20. doi:10.1080/14682753.2017.1305838

Matthews, M. (2021). Expanding from the small screen – arts practice for affective digital presence. *International Journal of Art & Design Education, 40*(3), 492–507. doi:10.1111/jade.12359

Matulewitz, K. (2015). Law and the construction of institutionalized sexual harassment in restaurants. Canadian Journal of Law and Society, 30, 401-419.

Mauss, M. (1966). *The gift; forms and functions of exchange in archaic societies.* Cohen & West.

Mayrand, P. (2014). The New Museology Proclaimed. *Museum International, 66*(1-4), 115–118. doi:10.1111/muse.12048

Mbow. (2020). COVID-19 in Africa: Dampening the storm? *Science, 369*(6504). PMID:32764055

McCormick, K., Anderberg, S., Coenen, L., & Neij, L. (2013). Advancing Sustainable Urban Transformation. *Journal of Cleaner Production, 50*(July), 1–11. doi:10.1016/j.jclepro.2013.01.003

McCreery, M., Schrader, P. G., Krach, S. K., & Boone, R. (2013, March). A sense of self: The role of presence in virtual environments. *Journal of Computers in Human Behavior, 29*(4), 1635–1640. doi:10.1016/j.chb.2013.02.002

McGinley, C. A. (2007). Harassing Girls at the Hard Rock: Masculinities in Sexualized Environments. *University of Illinois Law Review, 27*(4), 1229–1277.

McIntyre, A. (2008). *Participatory Action Research.* Sage Publications. doi:10.4135/9781483385679

McKay, J., & Marshall, P. (2001). The dual imperatives of- action research. *Information Technology & People, 14*(1), 46–59. doi:10.1108/09593840110384771

McKercher, B., & Chon, K. (2004). The over-reaction to SARS and the collapse of Asian tourism. *Annals of Tourism Research, 31*(3), 716–719. doi:10.1016/j.annals.2003.11.002 PMID:32572281

McKercher, B., & du Cros, H. (2002). *Cultural Tourism – the partnership between Tourism and Cultural Heritage Management.* Routledge.

Mcluhan, M. (2007). *Os meios de comunicação como extensões do homem.* Cultrix.

McQuire, S. (2008). *The media city. Media, architecture and urban space.* Sage Publications. doi:10.4135/9781446269572

Medeiros, E. (2021). Urban participatory planning approaches in capital cities: The Lisbon case. *European Planning Studies*, 1–18.

Meiser, E. T., & Pantumsinchai, P. (2021). The normalization of violence in commercial kitchens through food media. *Journal of Interpersonal Violence.* PMID:33858241

Mello, D. M. C. (2020). O cinema durante a Covid e a memória ressignificada no experimento Odradek [Cindema during Covid and re-signified Memory in the Odradek experiment]. *Revista Z Cultural, 10.* http://revistazcultural.pacc.ufrj.br/wp-content/uploads/2020/10/O-cinema-durante-A-COVID-e-a-mem%C3%B3ria-ressignificada-no-Experimento-Odradek-%E2%80%93-Revista-Z-Cultural.pdf

Mendes, P. (2019). Os militares e a impunidade legítima. In C. Sangreman (Coord.), *Guiné-Bissau – Notas sobre o presente e o futuro* (pp. 129-151). Lisboa: CESA - ISEG/Universidade de Lisboa. Available at: https://www.researchgate.net/publication/337448432

Mendes, L. (2017). Gentrificação turística em Lisboa: neoliberalismo, financeirização e urbanismo austeritário em tempos de pós-crise capitalista [Tourist gentrification in Lisbon: neoliberalism, financialization and austeritarian urbanism in post-capitalist crisis times]. *Cadernos Metrópole*, *19*(39), 479–512. doi:10.1590/2236-9996.2017-3906

Mendes, R. M., & Miskulin, R. G. S. (2017). Content analysis as a methodology [A análise de conteúdo como uma metodologia]. *Research Notes*, *47*, 1044–1066.

Miège, B. (2010). L'espace public contemporain [The contemporary public space]. PUG.

Mihalic, T. (2020). Conceptualising overtourism: A sustainability approach. *Annals of Tourism Research*, *84*, 103025. doi:10.1016/j.annals.2020.103025

Milano, C. (2018). Overtourism, malestar social y turismofobia. Un debate controvertido. Pasos: Revista de Turismo y Patrimonio Cultural, 16(3), 551-564.

Ministério da Educação e da Ciência. (2015, May 18). Portaria nº 135/2015: Estatutos do Instituto Superior de Ciências Educativas do Douro. *Diário da República*. https://www.iscedouro.pt/Files/Pages/12/estatutos_iscedouro.pdf

Ministério do Planeamento. (2021). *PRR – Recuperar Portugal, Construindo o Futuro*. https://recuperarportugal.gov.pt/wp-content/uploads/2021/10/PRR.pdf

Ministry of Health of Ukraine. (2022). *Covid-19 pandemic in Ukraine: Current information about coronavirus and quarantine*. https://covid19.gov.ua/en

Mirbeau, O. (1907). *La 628-E8*. Bibliothèque Charpentier.

Mirzoeff, N. (2016). *Cómo ver el mundo: una nueva introducción a la cultura visual*. Paidós.

Moatti, S. (2021). Les vacances d'après. *Economics and Politics*, *2021*(3, n°91), 5–7.

Modeo, S. (2018). *Dioniso in Sicilia. Mythos, Symposion, Hades, Theatron, Mysteria (!Mesogheia. Studi di storia e archeologia della Sicilia Antica)*. Edizioni Lussografica.

Momigliamo, A. (2019). The Flooding of Venice: What Tourists Need to Know. *The New York Times*. https://www.nytimes.com/2019/11/20/travel/venice-flooding.html

Momigliano, A. (2020). Venice Tourism May Never Be the Same. It Could Be Better. *The New York Times*. https://www.nytimes.com/2020/07/02/travel/venice-coronavirus-tourism.html

Mondzain, M.-J. (2015). *Homo spectator*. Orfeu Negro.

Monnerot, J. (1946). *Les faits sociaux ne sont pas des choses*. Éditions de Minuit.

Montenegro, A., & Ushenko, A. (2021, October) Installation Pictorial Representation of the Wedding Chamber. In *Artech 2021 ACM Conference proceedings* (Vol. 1, pp. 726-729). ACM.

Montenegro, M. D. S. M. (2014). *Manoel Monteiros and the multifaceted Cordel texts* [Manoel Monteiro e as várias faces do texto de cordel] [Thesis]. Federal University of Paraíba, João Pessoa.

Montezuma, J., & McGarrigle, J. (2019). What motivates international homebuyers? Investor to lifestyle 'migrants' in a tourist city. *Tourism Geographies*, *21*(2), 214–234. doi:10.1080/14616688.2018.1470196

Montgomery, C. (2015). Happy city. Transforming our lives through urban design. Penguin Books. (Original publication 2013)

Morand, J.-C., & Mollard, B. (2008). *Tourisme 2.0*. M21 Editions.

Morar, C., Tiba, A., Basarin, B., Vujičić, M., Valjarević, A., Niemets, L., Gessert, A., Jovanovic, T., Drugas, M., Grama, V., Stupariu, M., Stoica, A., & Lukić, T. (2021). Predictors of Changes in Travel Behavior during the COVID-19 Pandemic: The Role of Tourists' Personalities. *International Journal of Environmental Research and Public Health, 18*(21), 11169. doi:10.3390/ijerph182111169 PMID:34769688

Moreira, A., & Carneiro, A. (2015). *Museu Internacional de Escultura Contemporânea de Santo Tirso 1990-2015*. CMST.

Moreira, J. A., & Schlemmer, E. (2020). For a new concept and paradigm of online digital education [Por um novo conceito e paradigma de educação digital onlife]. *UFG Journal, 20*(26).

Moreno-Melgarejo, A., & Sariego-López, I. (2017). Relaciones entre Turismo y Arqueología: El Turismo Arqueológico, una tipología turística propia. *PASOS. Revista de Turismo y Patrimonio Cultural, 15*(1), 163–180.

Morère, N. (2007). Análisis del turismo arqueológico en España: Un estado de la cuestión. *Estudios Turísticos, 171*, 115–139.

Morin, E. (2014). Pensée complexe et pensée globale. In Conférence "Tout se Transforme… (pp. 26–31). Fondation Calouste Gulbenkian. Délégation en France. doi:10.3917h.342.0026

Morin, E. (1972). *Les stars*. Seuil.

Morin, E. (2005). *Introduction à la pensée complexe* (Vol. 534). Edition Points.

Morozov, E (2019). *A cidade inteligente. Tecnologias urbanas e democracia*. Ubu.

MPT. (2019). *Plano de Mobilidade Urbana Sustentável de Braga*. Author.

Mumford, L. (2004). A cidade na história. Suas origens, transformações e perspectivas. Martins Fontes. (Original publication 1982)

Municipality of Penafiel. (2019, May 8). *5º Simpósio Internacional de Arqueologia Virtual*. https://www.cm-penafiel.pt/5-o-simposio-internacional-de-arqueologia-virtual/

Naemura, T. (2002). Real-Time Video Based Modeling and Rendering of 3D Scenes. *IEEE Computer Graphics and Applications Journal*, 70-71.

Nascimento, A., Leitão, A., Carvalho, A., & Gonçalves, T. (2019). *O caminhar é para todas?Uma abordagem de mulheres latino- americanas sobre derivas e flâneries na contemporaneidade*. ENANPUR.

Nel.lo, O. (2018). *A cidade em movimento. Crise social e responsabilidade cidadã*. Livraria Tigre de Papel.

NEMO. (2020). *Survey on the Impacto of the COVID-19 Situation on Museums in Europe: Final Findings and Recommendations*. https://bit.ly/2X5NOI3

Nery, M. B., Souza, A. A. L., & Adorno, S. (2019). *Os padrões urbano-demográficos da capital paulista*. doi:10.1590/s0103-4014.2019.3397.002

Nesti, G. (2019). Mainstreaming gender equality in smart cities: Theoretical, Methodological and Empirical Challenges. *Information Polity Journal, 24*(3), 289–304. doi:10.3233/IP-190134

Netto, A., & Nechar, M. C. (2014). Epistemologia do turismo: escolas teóricas e proposta crítica. *Revista Brasileira de Pesquisa em Turismo, 8*(1), 120-144. http://www.spell.org.br/documentos/download/30514

Netto, P. A., & Gaetra, C. (2010). *Turismo da Experiência*. SENAC.

Nielsen, B., Glase, L., & Einarsen, S. (2017). Exposure to workplace harassment and the Five Factor Model of personality: A meta-analysis. *Personality and Individual Differences, 104.*

Nikjoo, A., & Hernández-Lara, A. B. (2021, April). The flag-bearers of change in a patriarchal Muslim society: Narratives of Iranian solo female travelers on Instagram. *Tourism Management Perspectives*, *38*, 100817. doi:10.1016/j.tmp.2021.100817

Niney, F. (2002). La poétique documentaire comme forme de connaissance [Documentary poetics as a form of knowledge]. *Etats généraux du film documentaire.* http://www.lussasdoc.com/etatsgeneraux/2002/sem_poetique.php4

Niney, F. (2000). *L'épreuve du réel à l'écran. Essai sur le principe de réalité documentaire* [The test of reality on the screen. Essay on the principle of documentary reality]. De Boeck Université.

Niney, F. (2009). *Le documentaire et ses faux-semblants.* Klincksieck.

Noble, K. (2021). Challenges and Opportunities: Creative Approaches to Museum and Gallery Learning during the Pandemic. *International Journal of Art & Design Education*, *40*(4), 676–689. doi:10.1111/jade.12380

Nóbrega, R., Jacob, J., Coelho, A., Weber, J., Ribeiro, J., & Ferreira, S. (2017). *Mobile location-based augmented reality applications for urban tourism storytelling.* Paper presented at the EPCGI 2017 - 24th Encontro Portugues De Computacao Grafica e Interacao. 10.1109/EPCGI.2017.8124314

Noehrer, L., Gilmore, A., Jay, C., & Yehudi, Y. (2021). The impact of COVID-19 on digital data practices in museums and art galleries in the UK and the US. *Humanit Soc. Science Communication*, *8*(1), 236. doi:10.105741599-021-00921-8

Nora, P. (1984) Les Lieux de Mémoire. Gallimard.

O'Brien, P. W. (2010). The careers of chef: a study of swiss expatriates working in Thailand. Bulletin of Faculty of Contemporary International Studies, 6.

Odin, R. (1995). *Le Film de famille. Usage privé, Usage public* [The Family Film. Private use, Public use]. Méridiens Klincksieck.

OECD. (2018). *OECD Tourism Trends and Policies 2018.* OECD Publishing. doi:10.1787/tour-2018-

OECD. (2020). *Culture shock: COVID-19 and the cultural and creative sectors.* OECD. https://www.oecd.org/coronavirus/policy-responses/culture-shock-COVID-19-and-the-cultural-and-creative-sectors-08da9e0e/

OECD. (2020). *OECD Tourism Trends and Policies 2020.* OECD Publishing. doi:10.1787/6b47b985-

Offer, A. (1997). Between the gift and the market:the economy of regard. *The Economic History Review*, *L*, 3, 450–476.

Oh, H., Fiore, A. M., & Jeoung, M. (2007). Measuring Experience Economy Concepts: Tourism Applications. *Journal of Travel Research, 46*, 119-132. doi:10.1177/0047287507304039

Okin, S. M. (2008). Gênero, o público e o privado. *Revista Estudos Feministas*, *16*(2), 305–332. doi:10.1590/S0104-026X2008000200002

Oldenburg, R., & Brissett, D. (1982). The third place. *Qualitative Sociology*, *5*(Winter), 265–284. doi:10.1007/BF00986754

Oliveira, A.L. (2011). *Esculturas de Arlindo. Quando a escultura passou a ser apenas escultura.* Tese de Mestrado, FBAUL.

Oliveira, C. S. de. (2018). Mobilidades em português : paradigma, cultura e potencialidades. *Centro de Estudos Comunicação e Sociedade*, 10–21.

Oliveira, V. (2014). Diferentes abordagens no estudo da forma urbana. *Revista de Morfologia Urbana*, 3–4. http://pnum.fe.up.pt/files/7614/0188/9302/RMU 2.1 Editorial.pdf

Oliveira, V. (2018). Article. *Revista de Morfologia Urbana*, *2018*(6), 1–75.

OMT - Organização Mundial do Turismo. (1999). *O Código Mundial de Ética do Turismo*. Santiago do Chile: OMT. https://ethics.unwto.org/sites/all/files/docpdf/portugal.pdf

Ooi, C., & Strandgaard Pedersen, J. (2010). City branding and film festivals: Re-evaluating stakeholder's relations. *Place Branding and Public Diplomacy*, *6*(4), 316–332. doi:10.1057/pb.2010.30

OPSTP (One Planet Sustainable Tourism Programme). (2021). *Glasgow Declaration: A commitment to a decade of climate action*. https://www.oneplanetnetwork.org/programmes/sustainable-tourism/glasgow-declaration

Orîndaru, A., Popescu, M.-F., Alexoaei, A. P., Căescu, S.-C., Florescu, M. S., & Orzan, A. O. (2021). Tourism in a Post-COVID-19 Era: Sustainable Strategies for Industry's Recovery. *Sustainability*, *2021*(13), 6781. doi:10.3390u13126781

Ortega, V. (2015). El artivismo como acción estratégica de nuevas narrativas artístico- políticas. *Calle14*, *10*(15), 100–111.

Ortigão, R. (1944). *Banhos das Caldas e Águas Minerais*. Clássica Editora.

Ost, C., & Saleh, R. (2021). Cultural and creative sectors at a crossroad: From a mainstream process towards an active engagement. *Built Heritage*, *5*(1), 1–16. doi:10.118643238-021-00032-y

Osthoff, S. (2010). De musas a autoras: Mulheres, arte e tecnologia no Brasil. *ARS (São Paulo)*, *8*(15). Advance online publication. doi:10.1590/S1678- 53202010000100006

Ou, J., & Wong, I. A. (2021). Strategic crisis response through changing message frames: A case of airline corporations. *Current Issues in Tourism*, *24*(20), 2890–2904. doi:10.1080/13683500.2020.1849051

OWHC. (2022a). *Introduction and mission*. Organization of World Heritage Cities. https://www.ovpm.org/all-about-owhc/introduction-and-mission/

OWHC. (2022b). *COVID-19 Pandemic Initiatives*. Organization of World Heritage Cities. https://www.ovpm.org/COVID-19/initiatives/

Pacheco, E. (2001). *Alteração das Acessibilidades e Dinâmicas Territoriais na Região Norte: expectativas, intervenções e resultantes*. Tese de mestrado. FLUP.

Pagès, D. (2013). Passages de l'image: d'Extramuros aux cahiers de la Métropole, l'imporbable figure de l'habitant métropolitain? [Qualitative analysis in the humanities and social sciences humanities and social sciences public]. In Médias et territoires, l'espace public entre communication et imaginaire territorial. Presses universitaires du Septentrion.

Paillé, P., & Mucchielli, A. (2005). *L'analyse qualitative en sciences humaines et sociales public* [Qualitative analysis in the humanities and social sciences humanities and social sciences public]. A. Colin.

Paquette, G. (2010). Visual knowledge modelling for semantic web technologies: models and ontologies. Hershey. doi:10.4018/978-1-61520-839-5

Paquot, T. (2009). L'espace public [The public space]. La découverte.

Paquot, T. (2021). Demeures terrestres, enquête vagabonde sur l'habiter [Demeures terrestres, vagabond investigation on living]. Terre urbaine, l'esprit des villes collection.

Paquot, T. (2014). *Le voyage contre le tourisme* [Travel versus tourism]. Eerotopia.

Parisi, V. (2015). The sex of Graffiti, Urban art, women and $gender perception#: testing biases inthe eye of the observer. *Street Art & Urban Creativity Scientific Journal, 1*(1), 53-62. https://scholar.google.com/citations?user=dJRoggAAAAJ&hl=en

Patah, M.O.R.A., Abdullah, R., Naba, M.M., Zahari, M.S.M., & Radzi, S.M. (2010). Workplace Bullying Experiences, Emotional Dissonance and Subsequent Intention To Pursue A Career In The Hospitality Industry. *Journal of Global Business and Economics, 1*(1).

Patrão Neves, M. (2021). Os desafios da COVID-19 aos Direitos Humanos. *Cadernos Ibero-Americanos de Direito Sanitário, 10*(3), 66-82 Available at: https://www.cadernos.prodisa.fiocruz.br/index.php/cadernos/article/view/825/836

Paulston, R. G. (1996). *Social Cartography: Mapping Ways of Seeing Social and Educational Change*. Garland Publishing.

Pearce, P. L. (2005). The role of relationships in the tourist experience. In F. T. William (Ed.), *Global Tourism* (3rd ed., pp. 103–122). Butterworth-Heinemann. doi:10.1016/B978-0-7506-7789-9.50012-1

Peeters, P., Gössling, S., Klijs, J., Milano, C., Novelli, M., Dijkmans, C., Eijgelaar, E., Hartman, S., Heslinga, J., Isaac, R., Mitas, O., Moretti, S., Nawijn, J., Papp, B., & Postma, A. (2018). Research for TRAN Committee - Overtourism: impact and possible policy responses. European Parliament, Policy Department for Structural and Cohesion Policies, Brussels.

Peixoto, P. (2000). *O património mundial como fundamento de uma comunidade humana e como recurso das indústrias culturais urbanas, Coimbra.* https://www.ces.uc.pt/publicacoes/oficina/ficheiros/155.pdf

Pereira Roders, A. (2019). The Historic Urban Landscape approach in action: Eight years later. In *Reshaping Urban Conservation* (pp. 21–54). Springer. doi:10.1007/978-981-10-8887-2_2

Pereira, I. (2015). As 10 tendências do futuro do turismo. Publituris. https://www.publituris.pt/2015/12/30/as-10-tendencias-do-futuro-do-turismo/

Pereira, K. (2012). *Relações entre a cultura hacker e a intervenção urbana*. Dissertação de Mestrado pela Universidade Federal de Belo Horizonte: UFMG.

Pereira, C. S. (2017). Mercados públicos municipales: Espacios de resistencia al neoliberalismo urbano. *Ciudades, 114*, 39–46.

Pereira, R. H. M., Braga, C. K., Servo, L. M., Serra, B., Amaral, P., & Gouveia, N. (2020). *Urban mobility and access to the healthcare system by patients with suspected and severe cases of COVID-19 in the 20 largest cities of Brazil*. IPEA.

Pérez, X. P. (2009). *Turismo Cultural - Uma visão antropológica.* Santa Cruz de Tenerife: PASOS. https://repositorio.utad.pt/bitstream/10348/.../1/livro%20tc%20xerardo.pdf

Pérez-García, A., & García Abad, L. (2018). Turismofobia: Presencia, impacto y percepción del concepto a través de los medios de comunicación impresos [Turismofobia: Presense, impact and perception of concept through the printed media]. *adComunica – Scientific Journal of Strategies, Trends and Innovation in Communication, 16*, 201–219.

Perkis, A., & Timmerer, C. (2020, May). QUALINET White Paper on Definitions of Immersive Media Experience (IMEx). *European Network on Quality of Experience in Multimedia Systems and Services, 14th QUALINET Meeting*. https://arxiv.org/abs/2007.07032

Perkumienė, D., & Pranskūnienė, R. (2019). Overtourism: Between the right to travel and residents' rights. *Sustainability, 11*(7), 2138. doi:10.3390u11072138

Perniola, M. (2004). *O sex appeal do inorgânico* [Sex appeal of the inorganic]. Ariadne. (Original work published 1994).

Perniola, M. (1993). *Do sentir* [From feeling]. Presença. (Original work published 1991)

Pessoa, F. (1982). *The Book of Disquiet*. Penguin Classics.

Petersen, I.-H., & Kruss, G. (2021). Universities as change agents in resource-poor local settings: An empirically grounded typology of engagement models. *Technological Forecasting and Social Change*, •••, 167.

Pfeffer, J., & Salancik, G. (1978). *The external control of organizations: a resource dependence perspective*. Harper & Row Publishing.

Pimentel, H. (2016). O papel da Gestão de Pessoas no Turismo. R-LEGO - Revista Lusófona de Economia e Gestão das Organizações, 4, 81-111.

Pimentel, H., & Duarte, I. (2016). Gestão de Recursos Humanos no Turismo. *II International Meeting of ISSOW*, 416-429.

Pine, B. J., & Gilmore, J. H. (2011). The experience economy (updated edition). Harvard Business Review Press.

Pine, B. J., & Gilmore, H. J. (1999). *The Experience Economy: Work is Theatre & Every Business a Stage*. Harvard Business School Press.

Pine, B. J., & Gilmore, J. H. (1998). Welcome to the experience economy. *Harvard Business Review, 76*(4), 97–105. PMID:10181589

Piñeiro-Naval, V., & Serra, P. (2018). *Cultura, património e turismo na sociedade digital: Uma perspetiva ibérica* [Culture, heritage and tourism in the digital society: An Iberian perspective] (Vol. I). UBI – Labcom.

Pinto-Coelho, Z., & Pinto, R. (Eds.). (2018). Discourses and images of cultural tourism. *Lusophone Journal of Cultural Studies, 5*(2). https://rlec.pt/index.php/rlec/issue/view/99

Pirolli, B. (2018). *Informing tourists in the digital age*. Routledge.

Plan de modernización y competitivida de sector turístico. (2021). *Gobierno de España*. https://www.lamoncloa.gob.es/temas/fondos-recuperacion/Documents/16062021-Componente14.pdf

Plan de Recuperación. Transformación y Resiliencia. (2021). *Gobierno de España*. https://planderecuperacion.gob.es/

Plan Nacional e Integral de uturismo 2012-2015. (n.d.). *Secretaría de Esado de Turismo, Instituto de Turismo de España*. https://turismo.gob.es/es-es/servicios/Documents/Plan-Nacional-Integral-Turismo-2012-2015.pdf

Plan Turístico de Extremadura 2017-2020. (n.d.). *Dirección General de Turismo, Consejería de Economía e Infraestrcutura. Junta de Extremadura*. http://www.juntaex.es/filescms/con02/uploaded_files/planes/Plan_Turistico_Extremadura.pdf

Planetoscope. (2017a). https://www.planetoscope.com/tourisme/415-nombre-de-touristes-en-france.html

Planetoscope. (2017b). https://www.planetoscope.com/Avion/1837-passagers-transportes-par-avion-dans-lemonde.;html

Platão. (1986). *O banquete* [Symposium]. Europa-América.

Polycarpou, C. (2018). *The ViMM Definition of a Virtual Museum*. https://www.vi-mm.eu/2018/01/10/thevimm-definition-of-a-virtual-museum/

Portelinha, M. (2013). *Arte urbana: estratégias, contextos e técnicas*. IADE-U Creative University.

Porter, M. (1990). The Competitive Advantage of Nations. *Harvard Business Review*. https://hbr.org/1990/03/the-competitive-advantage-of-nations

Porter, M. E. (1998). Clusters and the new economics of competition. Watertown-Massachusetts. *Harvard Business Review, 76*, 77–90. PMID:10187248

Postma, A., & Schmuecker, D. (2017). Understanding and overcoming negative impacts of tourism in city destina-tions: Conceptual model and strategic framework. *Journal of Tourism Futures*, *3*(2), 144–156. doi:10.1108/JTF-04-2017-0022

Poulot, D. (2011). *Museo y museología*. Abada Editores.

Prata, T. (2020, May 6). Caos silencioso: acostumada a sediar grandes festivais, Tiradentes sofre com prejuízos e a pandemia [Silent Chaos: Sued to hosting major festivals, Tirsadentes suffers from losses and the Pandemic]. *Hoje em Dia*. https://www.hojeemdia.com.br/entretenimento/caos-silencioso-acostumada-a-sediar-grandes-festivais-tiradentes-sofre-com-prejuizos-e-a-pandemia-1.785625

Prats-Joaniquet, C. (2009). Museos y territorio: Del recurso cultural al producto turístico. *HER&MUS*, *2*, 80–89.

Quadrilátero. (2013). *Programa estratégico de cooperação*. https://www.cm-braga.pt/docs/CMB/Documentos/PEC-Quadrilatero.pdf

Que Ver e Fazer, O. (n.d.). Retrieved August 18, 2022, from https://www.rotadoromanico.com/pt/experiencias/o-que-ver-e-fazer/

Quétel, C., Bordin, G., Abreu, A., Lemi, I., & Sangreman, C. (2021). On the nature and determinants of poor households' resilience in fragility contexts. *Journal of Human Development and Capabilities*, 1–18.

Quivy, R., & Campenhoudt, L. (2008). *Manual de Investigação em Ciências Sociais*. Gradiva Lda.

Ramalho, O. (1966). *As Praias de Portugal*. Livraria Clássica Editora.

Ramos, D., & Costa, C. (2017). Turismo: Tendências de evolução. *PRACS: Revista Eletrônica de Humanidades do Curso de Ciências Sociais da UNIFAP*, *10*(1), 21–33. doi:10.18468/pracs.2017v10n1.p21-33

Ramos, S. P. (2021). Sobre os desavessos da pandemia e o medo da anormalidade: diálogos com Boaventura de Sousa Santos. In *Sociedade em pandemia(s)* (p. 10). EDUFAL.

Ram, Y. (2018). Hostility or hospitality? A review on violence, bullying and sexual harassment in the tourism and hospitality industry. *Current Issues in Tourism*, *21*(7), 760–774. doi:10.1080/13683500.2015.1064364

Rancière, J. (2010). *Le partage du sensible, esthétique et politique* [The sharing of the sensitive, aesthetic and political]. La Fabrique éditions. (Original publication 2000)

Rastegar, R., Higgins-Desbiolles, F., & Ruhanen, L. (2021). COVID-19 and a justice framework to guide tourism recovery. *Annals of Tourism Research*, *91*, 2021. doi:10.1016/j.annals.2021.103161 PMID:33583982

Rather, R. (2021). Demystifying the effects of perceived risk and fear on customer engagement, co-creation and revisit intention during COVID-19: A protection motivation theory approach. *Journal of Destination Marketing & Management*, *20*. doi:10.1016/j.jdmm.2021.100564

Ravenscraft, E. (2021, November 25). What is the Metaverse Exactly? *Wired*. https://www.wired.com/story/what-is-the-metaverse/

Raz, N. (2016). Inaugural Hainan Airlines flight from Beijing to Las Vegas lands at McCarran. *Las Vegas Review Journal*. Retrieved from https://www.reviewjournal.com/business/tourism/inaugural-hainan-airlines-flight-from-beijing-to-las-vegas-lands-at-mccarran/

Real Decreto-Ley 463/2020, de 14 de marzo, por el que se declara el estado de alarma para la gestión de la situación de crisis sanitaria ocasionada por la COVID-19. *BOE*, 67/2020. Ministerio de la Presidencia, Relaciones con las cortes y Memoria Democrática. BOE-A-2020-3692

Recht, R. (2014). *Pensar el patrimonio. Escenificación y ordenación del arte*. Abada Editores.

Reisinger, Y. (2009). *International tourism, cultures, and behaviour*. Butterworth-Heinemann.

Rengifo-Gallego, J. I., & Sánchez-Martín, J.-M. (2021). Conceptos básicos del turismo. Aproximación a la evolución reciente de las políticas turísticas en Extremadura. In Recursos turísticos, territorio y sociedad en Extremadura: catalogación, nuevos usos y perspectivas. Universidad de Extremadura, Servicio de Publicaciones.

Report, E. S. (2020). *Global Imbalances and the COVID-19 Crisis*. International Monetary Fund.

Reyero, C. (1999). *La Escultura Conmemorativa En España: La* Edad De Oro *Del Monumento Publico, 1820-1914*. Cátedra Caduernos Arte.

Ribas, M. C. C., & da Silva Malafaia, R. (2021). Cordel Literature and Edication: an interartistic mosaic [Literatura de Cordel e Educação: um mosaico interartístico]. *PÓS: Arts Postgraduate Program Journal from EBA/UFMG, 11*(21), 61-89.

Ribeiro, P. O. (2014). *Cartografia social: (Re)Conhecer perceções; Mapear territórios*. Paper presented at the 1st Research Methodology School – What needs to be done? Coimbra University.

Ribeiro, T. (1863). Sociedade Typographica Franco-Portuguesa. Academic Press.

Richards, G. (2011). *Creativity and Tourism: The State of the Art*. http://www.forum-avignon.org/sites/default/files/editeur/ECCE_report.pdf

Richards, G. (2018). Cultural tourism: A review of recent research and trends. *Journal of Hospitality and Tourism Management, 36*, 12–21. doi:10.1016/j.jhtm.2018.03.005

Richards, G. (2020). Designing creative places: The role of creative tourism. *Annals of Tourism Research, 85*, 102922. doi:10.1016/j.annals.2020.102922

Richards, G., & Raymond, C. (2000). Creative Tourism. *ATLAS News, 23*, 16–20.

Richter, L. K. (2003). International tourism and its global public health consequences. *Journal of Travel Research, 41*(4), 340–347. doi:10.1177/0047287503041004002

Rickly, J., Hannam, K., & Mostafanezhad, M. (Eds.). (2016). *Tourism and Leisure Mobilities: Politics, Work, and Play*. Routledge. doi:10.4324/9781315686660

Ricoeur, P. (2004). Parcours de la reconnaissance [Recognition course]. Gallimard.

Ricœur, P. (1983). *Temps et récit* [Time and narrative]. Le Seuil.

Riegl, A. (1999). *El culto moderno a los monumentos: caracteres y origen*. Visor.

Rincón García, W. (2013). Escultura del siglo XIX en Zaragoza. De la imagen devocional al monumento conmemorativo. In M. C. Lacarra Ducay (Ed.), *Arte del siglo XIX* (pp. 13–79). Instituto Fernando el Católico.

Rivlin, P., & González, S. (2018). Public markets: Spaces for sociability under threat? The case of Leeds' Kirkgate Market. In S. González (Ed.), *Contested markets contested cities. gentrification and urban justice in retail spaces* (pp. 133–149). Routledge.

Robinson, W. I. (2002). Remapping Development in Light of Globalisation: From a Territorial to a Social Cartography. *Third World Quarterly, 23*(6), 1047–1071. doi:10.1080/0143659022000036658

Robles, J. (2008). Comercio urbano em espacios metropolitanos. Mercados, vidas y barrios. *Distribución y Consumo*, (Sep-Oct), 19–31.

Rocha, P. M. A. (2015a). *O Constructo da informação no espaço.* 9º Congresso da Associação Portuguesa de Ciências da Comunicação (SOPCOM) – Comunicação e transformações Sociais. Coimbra.

Rocha, P. M. A. (2015b). *Augmented Reality - The Augmented Construct of Communication.* International Conference on Communication and Light, 2-4 November 2015, Universidade do Minho, Braga, Portugal.

Rodrigues, J. C. B. (2012). *A Implantação do Automóvel em Portugal (1895-1910).* Lisboa: Dissertação para obtenção do Grau de Doutor em História, Filosofia e Património da Ciência e da Tecnologia. Faculdade de Ciências e Tecnologia. Universidade Nova de Lisboa. https://run.unl.pt/bitstream/10362/9676/1/Rodrigues_2012.pdf

Rodrigues, A. (2018). *Performances urbanas: Formas artísticas e Intervenções Urbanas.* Rev.Cult. Ext. USP.

Rodwell, D. (2018). The historic urban landscape and the geography of urban heritage. *The Historic Environment: Policy & Practice, 9*(3-4), 180–206. doi:10.1080/17567505.2018.1517140

Rogers, A. P. (2017). Historic urban landscape approach and living heritage. *Conserving Living Urban Heritage: Theoretical Considerations of Continuity and Change.*

Romão, F. (2021). O Turismo Receptivo no "Novo Normal" – a Perspectiva da HighSun DMC. In *Turismo Mundial, Crise Sanitária e Futuro: novos tempos, novos ritmos - World Tourism, Health Crisis and Future: new times, new rhythms* (pp. 59–66). Instituto Politécnico de Tomar. http://www.cda.ipt.pt/download/Livros/novos-tempos-novos-ritmos-ebook_compressed.pdf

Romero-Dexeus, C., & Paton, J. (2020). *Innovación turística y especialización inteligente en España. Palancas imprescindibles para la recuperación.* Segittur.

Romney, J. (2020, Sep 12). Venice film festival 2020 roundup – against all the odds, a triumph. *The Guardian.* https://www.theguardian.com/film/2020/sep/12/venice-film-festival-2020-review-human-voice-almodovar-tilda-swinton-vanessa-kirby-cate-blanchett-nomadland-new-order

Rončák, M., Scholz, P., & Linderová, I. (2021). Safety Concerns and Travel Behavior of Generation Z: Case Study from the Czech Republic. *Sustainability, 13*(23), 13439. doi:10.3390u132313439

Roque, M. I. (2018). *O dilema do museu: apenas real ou também virtual? In Património Cultural e transformação Digital. Povos e Culturas 22. CEPCEP.* Universidade Católica Editora.

Rosa, H. (2014). *Aliénation et accélération. Vers une théorie critique de la modernité tardive.* La Découverte.

Rosetti, I., Bertrand Cabral, C., Pereira Roders, A., Jacobs, M., & Albuquerque, R. (2022). Heritage and Sustainability: Regulating Participation. *Sustainability, 14*(3), 1674. doi:10.3390u14031674

Rosler, M. (2017). *Clase cultural: arte y gentrificación.* Caja negra.

Rotmans, J., van Asselt, M., & Vellinga, P. (2000). An integrated planning tool for sustainable cities. *Environmental Impact Assessment Review, 20*(3), 265–276. doi:10.1016/S0195-9255(00)00039-1

Roxborough, S. (2020, Jul 28). Venice Boss Alberto Barbera on Hosting the First Post-Lockdown Festival. *The Hollywood Reporter.* https://www.hollywoodreporter.com/movies/movie-news/venice-film-festival-boss-alberto-barbera-hosting-first-post-lockdown-fest-1304607/

Rubio-Tamayo, J. (2017, September). Immersive Environments and Virtual Reality: Systematic Review and Advances in Communication, Interaction and Simulation. *Multimodal Technologies and Interact, 1*(21), 9-20.

Rui Barbosa House Foundation. (n.d.). *Cordel: popular Literature in verse* [Cordel: Literatura popular em verso]. Retrieved October 15, 2021, from http://antigo.casaruibarbosa.gov.br/cordel/apresentacao.html#

Ruiter, A. (2015). Imaging Egypt#s political transition in (post-)revolutionary street art: on the interrelations between social media and graffiti as media of communication. *Media, Culture & Society*. doi:10.1177/0163443714566901

Ruiz-Torres, P. (2021). Covid-19: Teletrabajo en tiempos de pandemia. *Historia y comunicación social, 26*, 11-18. doi:10.5209/hics.74237

Sacco, P. (2011). *Culture 3.0: A new perspective for the UE 2014-2020 structural funds programming*. European Expert Network on Culture.

Sacco, P. L., Ferilli, G., & Tavano Blessi, G. (2018). From Culture 1.0 to Culture 3.0: Three Socio-Technical Regimes of Social and Economic Value Creation through Culture, and Their Impact on European Cohesion Policies. *Sustainability, 10*(11), 3923. doi:10.3390u10113923

Sacramento, J. P. (2021, Jan 14). *24ª Mostra Tiradentes abre calendário audiovisual brasileiro e homenageia a cineasta Paula Gaitán. Mais vertentes* [Mostra Tiradentes opens the Brazilian Audiovisual Calendar and pays tribute to Filmmaker Paula Gaiatn. More Strands]. https://www.maisvertentes.com.br/noticia/1985/24-mostra-tiradentes-abre-calendario-audiovisual-brasileiro-e-homenageia-a-cineasta-paula-gaitan

Saelens, B. E., Sallis, J. F., Black, J. B., & Chen, D. (2003). Neighborhood-Based Differences in Physical Activity: An Environment Scale Evaluation. *American Journal of Public Health, 93*(9), 1552–1558. doi:10.2105/AJPH.93.9.1552 PMID:12948979

Said, E. (2008). *On late style: Music and literature against the grain*. Vintage.

Salas-Olmedo, M. H., Moya-Gómez, B., García-Palomares, J. C., & Gutiérrez, J. (2018). Tourists' digital footprint in cities: Comparing big data sources. *Tourism Management, 66*, 13–25. doi:10.1016/j.tourman.2017.11.001

Salazar, A. (2013). *The European Crisis*. Orfeu.

Salmon, C. (2007). *Storytelling, la machine à fabriquer des histoires et à formater des esprits* [Storytelling, the machine for making stories and formatting minds]. Poche.

Salti, R. (2020). Do Not Go Gentle into That Good Night: Film Festivals, Pandemic, Aftermath. *Film Quarterly, 74*(1), 88–96. doi:10.1525/fq.2020.74.1.88

Samara, D., Magnisalis, I., & Peristeras, V. (2020). Artificial intelligence and big data in tourism: A systematic literature review. *Journal of Hospitality and Tourism Technology, 11*(2), 343–367. doi:10.1108/JHTT-12-2018-0118

Samaroudi, M., Echavarria, R., & Perry, L. (2020). Heritage in Lockdown: Digital Provision of Memory Institutions in the UK and US of America During the COVID-19 Pandemic. *Museum Management and Curatorship, 35*(4), 337–361. doi:10.1080/09647775.2020.1810483

Sánchez-Barrioluengo, M., & Benneworth, P. (2019). Is the entrepreneurial university also regionally engaged? Analysing the influence of university's structural configuration on third mission performance. *Technological Forecasting and Social Change, 141*, 206–218. doi:10.1016/j.techfore.2018.10.017

Sánchez-Oro, M. (2021). Aproximación sociológica al turismo de Extremadura. In Recursos turísticos, territorio y sociedad en Extremadura: catalogación, nuevos usos y perspectivas. Universidad de Extremadura, Servicio de Publicaciones.

Sánchez-Oro, M., García-García, Y., & Ramos-Vecino, N. (2021). *Informe sobre turismo Urbano*. Observatorio de Turismo, Extremadura & Universidad de Extremadura.

Sandell, R., & Janes, R. R. (2019). *Museum Activism*. Routledge.

Sangreman, C., Barreto, N., Barros, I., & Sanhá, C. (2020). *Observando Direitos na Guiné – Bissau, educação, saúde, habitação, água, energia, justiça, igualdade de género.* Projeto Observatório de Direitos, ACEP com LGDH e CEsA, Lisboa e Bissau. Available at: https://www.repository.utl.pt/handle/10400.5/20866

Sansot, P. (1993). *Les gens de peu.* PUF.

Santaella, L. (2014). *Dos não-lugares à cidade senciente.* https://periodicos.sbu.unicamp.br/ojs/index.php/rua/article/view/8638917

Santaella, L. (2007). *Linguagens líquidas na era da mobilidade.* Paulus.

Santaella, L. (2013). A cidade e o corpo como interfaces. In *Comunicação ubíqua. Repercussões na cultura e na educação* (pp. 55–70). Paulus.

Santaella, L. (2016). *Cidades inteligentes. Por que, para quem?* Educ.

Santos, I., & Jesus, P. (2006). Intervenções Urbanas: a problemática das artes e seus espaços na cidade de Salvador na contemporaneidade. In Encontros de Estudos Multidisciplinares em Cultura da Faculdade de Comunicação. UFBA.

Santos, P. P. P. D. O., Lopes, B. B. G., & de Souza, N. (2021). Covid-19 and cordel literature: health education through folkcommunication [Covid-19 e literatura de cordel: educação em saúde pela via da folkcomunicação]. *International Journal of Folkcommunication, 19*(43), 247-265.

Santos, T. H., & Ramalho, G. (2021). Escuta em tempos de pandemia: Participação em museus a partir da experiência do Museu da Imigração do Estado de São Paulo. *Simbiótica. Revista Eletrônica, 8*(2), 92–114. doi:10.47456imbitica.v8i2.36380

Saramago, J. (1985). *Viagem a Portugal.* Editorial Caminho.

Sartre, J. P. (1996). *L'existentialisme est un humanisme.* Gallimard. (Original work published 1946)

Scaltsa, M., Nitsiou, P., & Georgaki, P. (2016). Educating a new generation of tour guides for the forthcoming era, at the Aristotle University of Thessaloniki. *Tourismos, 11*(2), 240–261. doi:10.26215/tourismos.v11i2.475

Scheiner, T. (2020, January/July). Museologia, hipercultura, hipertextualidade: Reflexões sobre o Museu do Século 21. *Museologia & Interdisciplinaridade., 9*(17), 46–63. doi:10.26512/museologia.v9i17.31592

Schneider, F., Kallis, G., & Martinez-Alier, J. (2010). Crisis or opportunity? Economic degrowth for social equity and ecological sustainability. Introduction to this special issue. *Journal of Cleaner Production, 18*(6), 511–518. doi:10.1016/j.jclepro.2010.01.014

Schrock, A. R. (2012). *Mobile Interface Theory: Location-Aware Mobile Devices in Urban Environments.* https://ijoc.org/index.php/ijoc/article/viewFile/1727/783

Schuurman, D., De Marez, L., & Ballon, P. (2016). The impact of living lab methodology on open innovation contributions and outcomes. *Technology Innovation Management Review, 6*(1), 7–16. doi:10.22215/timreview/956

Scolari, C. (2013). *Narrativas transmedia: cuando todos los medios cuentan.* Deusto.

Scott, J. C. (2013). A dominação e a arte da resistência. Discursos ocultos. Letra Livre. (Original publication 1992)

Secchi, B. (2019). *A cidade do rico e a cidade dos pobres.* Âyiné.

Selçuk Kurtuluş, S., Kılıç, B., & Kurnaz, A. (2018). Evaluation of women chefs in professional kitchens. *Journal of Tourism and Gastronomy Studies, 6*(3), 119–132. doi:10.21325/jotags.2018.275

Sennet, R. (2018). Construir e habitar: Ética para uma cidade aberta [Build and Inhabit: Ethics for an open city]. *Record.*

Sennett, R. (2017). The fall of public man. W.W. Norton & Company. (Original publication 1974)

Sennett, R. (2018). *Building and dwelling. The ethics for the city.* Penguin Books.

Serraballe, F. (2019). Augmented reality in the tourism industry: A multi-stakeholder analysis of museums. *Tourism Management Perspectives, 32*(October). https://doi.org/10.1016/j.tmp.2019.07.002

Serravalle, F. (2019). Augmented reality in the tourism industry: A multi-stakeholder analysis of museums. *Tourism Management Perspectives, 32.*

Sesic, M. D. & Stefanovic, M. (2021). The Ethics and Aesthetics of Museum Practices in Serbia During the Pandemic and The Digital Turn. *IPSI BGD Transactions on Internet Research, 17*(2), 33-44.

Sharma, A., Shin, H., Santa-María, M. J., & Nicola, J. L. (2020). Hotels' COVID-19 innovation and performance. *Annals of Tourism Research, 88*, 2021.

Sharma, G. D., Thomas, A., & Paul, J. (2021). Reviving tourism industry post-COVID-19: A resilience-based framework. *Tourism Management Perspectives, 37*, 100786. doi:10.1016/j.tmp.2020.100786 PMID:33391988

Shaviro, S. (2003). *Connected. Or what it means to live in the network society.* University of Minnesota Press.

Shaw, J. (1989). *The Legible City.* Interactive Installation, Interface and Computer Monitor Projection.

Shaw, J. (2018, October). Platform real estate: Theory and practice of new urban real estate markets. *Urban Geography, 17.* Advance online publication. doi:10.1080/02723638.2018.1524653

Sheller, M., & Urry, J. (2006). *Mobile Technologies in the City.* London: Routledge.

Sheller, M., & Urry, J. (2004). *Tourism mobilities: Places to play, places in play.* Routledge. doi:10.4324/9780203340332

Sheller, N., & Isaacson, M. (2013). *Tourist mobility and advanced tracking technologies.* Routledge.

Shen, S. (2022). Exploring the factors influencing the adoption and usage of Augmented Reality and Virtual Reality applications in tourism education within the context of COVID-19 pandemic. *Journal of Hospitality, Leisure, Sport & Tourism Education.*

Shepard, M. (2011). *Sentient city. Ubiquitous computing, architecture, and the future of urban space.* The MIT Press.

Shin, D. (2019). How does immersion work in augmented reality games? A user-centric view of immersion and engagement. *Information Communication and Society, 22*(9), 1212–1229. https://doi.org/10.1080/1369118X.2017.1411519

Sigala, M. (2020). Tourism and COVID-19: Impacts and implications for advancing and resetting industry and research. *Journal of Business Research, 2020*(117), 312–321. doi:10.1016/j.jbusres.2020.06.015 PMID:32546875

Silva, M. (2018). *Bienal de Cerveira (1978-2007). Memória e singularidade.* PhD thesis, FBAUP

Silva, A. (2000). *Ensaios sobre Cultura e Literatura Portuguesa e Brasileira I.* Âncora Editora.

Silva, E. (2015). *Intervenções Artísticas em espaços públicos e pegagocias da cidade: possibilidades de pesquisa. In 37a Reunião Nacional da ANPEd.* UFSC.

Silvano, F. (2007). Antropologia do espaço. Uma introdução. Celta Editora. (Original publication 2001)

Silva, S. (2013). *Turismo interno. Uma Visão Integrada.* Lidel.

Simas, T., Le Cocq de Oliveira, S., & Cano-Hila, A. (2021). Tourismophobia or touristification? An analysis of the impacts of tourism in Poblenou, Barcelona. *Ambiente Construído, 21*(3), 117–131. doi:10.15901678-8621202100030054I

Simmel, G. (2019). Digression sur l'étranger. L'École de Chicago, naissance de l'écologie urbaine, 53-77. (Original publication 1908)

Simmel, G. (1988). Pont et porte [Bridge and door]. In *La Tragédie de la Culture et autres essais* (pp. 161–168). Rivages.

Simmel, G. (1990). Digressions sur l'étranger. In Y. Grafmeyer & I. Joseph (Eds.), *L'école de Chicago* (pp. 53–59). Aubier. (Original work published 1908)

Simmel, G. (1992). *Le conflit*. Circé.

Simmel, G. (1997). A metrópole e a vida do espírito. In C. Fortuna (Ed.), *Cidade, cultura e globalização. Ensaios de sociologia* (pp. 31–43). Celta. (Original work published 1903)

Simmel, G. (2013). As grandes cidades e a vida do espírito. In A. Botelho (Ed.), *Essencial Sociologia* (pp. 311–329). Companhia das Letras.

Simmons, A. (2009). *The Story Factor: Inspiration, Influence, and Persuasion through the Art of Storytelling*. Basic Books.

Singh, T. (2018). Is over-tourism the downside of mass tourism? *Tourism Recreation Research, 43*(4), 415–416. doi:10.1080/02508281.2018.1513890

Sitte, C. (1996). *L'Art de bâtir les villes: l'urbanisme selon ses fondements artistiques*. Seuil.

Skare, M., & Riberio, S. D. (2021). Impact of COVID-19 on the travel and tourism industry. *Technological Forecasting and Social Change*, 163.

Sloterdijk, P. (2000). La mobilisation infinite [The infinite mobilization]. Christian Bourgois Ed.

Smith, B., & Ripp, M. (2020). *Guidelines for Sustainable Cultural Tourism*. Heritage Europe, OWHC. https://www.ovpm.org/wp-content/uploads/2020/07/tourismguidelines2020.pdf

Söderström, O., Paasche, T., & Klauser, F. (2014). Smart cities as corporate storytelling. *City, 18*(3), 307–320. doi:10.1080/13604813.2014.906716

Soeiro, T. (2008). A Arqueologia Histórica no Município de Penafiel. *Oppidum Atas do I Encontro de Arqueologia das Terras de Sousa*, 195-209.

Sol, H., de Brito, M.P., Coelho, J.P., Pratt, C., & Lopes, E.R. (2017). Rooting a new event in its place: the case of Festa Templária, Tomar, Portugal. *International Journal of Event and Festival Management, 8*(3), 324-345. doi:10.1108/IJEFM-10-2016-0068

Soliku, O., Kyiire, B., Mahama, A., & Kubio, C. (2021). Tourism amid COVID-19 pandemic: impacts and implications for building resilience in the eco-tourism sector in Ghana's Savannah region. *Heliyon, 7*(9). doi:10.1016/j.heliyon.2021.e07892

Sorel, G. (1990). Réflexions sur la violence. Seuil. (Original publication 1908)

Soulez, G. (2011). *Quand le film nous parle, Rhétorique, Cinéma, Télévision* [When film speaks to us, Rhetoric, Cinema, Television]. PUF, Coll. Lignes d'art.

Sousa, J. P. (2021). *Desafios da Gestão de Recursos Humanos na Geração Millennial - Dificuldades no recrutamento e retenção de talento - Aplicação ao setor do turismo e hotelaria*. Dissertação Mestrado em Gestão de Recursos Humanos, U. Lusófona.

Sousa, V. (2020). Tempo, espaço e Covid-19: unidade, fragmentação, dissincronia, velocidade, Einstein, Kafka, nova-normalidade e incerteza no futuro. In *C. Sales, E. Araújo & R. Costa (Orgs.), Tempo e sociedade em suspenso* (pp. 203–216). CIES/ISCTE.

Souza, E., Silva, A., & Frith, J. (2012). *Mobile interfaces in public spaces: Locational Privacy, Control, and Urban Sociability*. Routledge.

Spyer, J. (2007). *Conectado. O que a internet faz com você e o que você pode fazer com ela*. Zahar.

Steffen, W., Richardson, K., Rockström, J., Cornell, S. E., Fetzer, I., Bennett, E. M., Biggs, R., Carpenter, S. R., de Vries, W., de Wit, C. A., Folke, C., Gerten, D., Heinke, J., Mace, G. M., Persson, L. M., Ramanathan, V., Reyers, B., & Sörlin, S. (2015). Planetary boundaries: Guiding human development on a changing planet. *Science*, *347*(6223), 1259855. Advance online publication. doi:10.1126cience.1259855 PMID:25592418

Stenou, K. (2002). *UNESCO Universal Declaration on Cultural Diversity: a vision, a conceptual platform, a pool of ideas for implementation, a new paradigm*. UNESDOC Document code: CLT/2002/WS/11.

Stephenson, N. (1992). *Snow Crash*. Bantam Books.

Steyerl, H. (2012). *Os Condenados da Tela*. Caja Negra editora.

Stornaiuolo, A., & Thomas, E. E. (2018). Restorying as political action: Authoring resistance through youth media arts. *Learning, Media and Technology*, *43*(4), 345–358. doi:10.1080/17439884.2018.1498354

Strachan, H. (2021). *Why are women still the target? How harassment in kitchens has led women to not assume or explore roles in the kitchen and even stop work completely*. Bachelor dissertation.

Strandgaard Pedersen, J., & Mazza, C. (2011). International Film Festivals: For the Benefit of Whom. *Culture Unbound*, *3*(2), 139–165. doi:10.3384/cu.2000.1525.113139

Suárez, M. L. (2018). *Azurduy por Anzoátegui: subjetividad femenina y espacio público en Huallparrimachi (1894)*. Pontificia Universidad Católica del Perú. doi:10.18800/lexis.201802.005

Suh, S., & Prophet, J. (2018, April). The state of immersive technology research: A literature analysis. *Computers in Human Behavior Journal*, 77-90.

Sumartojo, S., & Pink, S. (2019). *Atmospheres and the experiential world, theory and methods - ambiances, atmospheres and sensory experiences of spaces*. Routledge.

Sundbo, J., & Darmer, P. (2008). *Creating Experiences in The Experience Economy*. Edward Elgar. doi:10.4337/9781848444003

Susca, V. (2011). *Joie tragique. Les formes élémentaires de la vie numériques*. CNRS Éditions.

Svendson, L. (2006). Filosofía del tedio. Tusquets Editores. (Original publication 1999)

Svensson, P., Klofsten, M., & Etzkowitz, H. (2012). An entrepreneurial university strategy for renewing a declining industrial city: The Norrköping way. *European Planning Studies*, *20*(4), 505–525. doi:10.1080/09654313.2012.665616

Tacussel, P. (1984). *L'attraction sociale*. Méridiens Klincsieck.

Tanga, M. (2021). Let's Imagine a New Museum Staff Structure. *Journal of Conservation & Museum Studies*, *19*(1), 7. doi:10.5334/jcms.197

Tarachucky, L. (2021). *Framework conceitual para apoiar a formulação de estratégias de design de interação urbana envolvendo o uso de recursos de mídia locative digital* [Doctoral dissertation]. Federal University of Santa Catarina. Repositório UFSC. https://repositorio.ufsc.br/handle/123456789/226801

Tarachucky, L., Sabatini-Marques, J., Yigitcanlar, T., Baldessar, M. J., & Pancholi, S. (2021). Mapping hybrid cities through location-based technologies: A systematic review of literature. *Cities (London, England), 116*, 103296. doi:10.1016/j.cities.2021.103296

Tavares, M. (2015). Tracing Limits-Public and Private in the Cartography of Contemporary Cities: The Dialogue Boxes on Street Windows Project. In Resilient Territories Resilient Territories: Innovation and Creativity for New Modes of Regional. Cambridge.

Taylor, K. (2020). The ideology of the urban cultural landscape construct. In *The Routledge Handbook on Historic Urban Landscapes in the Asia-Pacific* (pp. 48–66). Routledge.

Terzić, L. (2018). How tourism destination competitiveness contributes to GDP growth –The case of selected European economies. *Acta Oeconomica Pragensia, 2018*(3), 24–34. doi:10.18267/j.aop.606

Theobald, W. F. (2012). *Global tourism*. Routledge. doi:10.4324/9780080478043

Thibault, L. (2019, November). Is it Real? Measuring the Effect of Resolution, Latency, Frame rate and Jitter on the Presence of Virtual Entities. *ISS'19,* 10–13.

Thirumaran, K., Mohammadi, Z., Pourabedin, Z., Azzali, S., & Sim, K. (2021). COVID-19 in Singapore and New Zealand: Newspaper Zahra portrayal, crisis management. *Tourism Management Perspectives, 38*, 38. doi:10.1016/j.tmp.2021.100812

Thomazeau, F. (2009). Les cités mal aimées se rebiffent [The unloved cities rebel]. *Le Monde*. https://www.lemonde.fr/europe/article/2009/12/31/les-cites-mal-aimees-se-rebiffent_1286303_3214.html

Tierney, T. F. (2013). *The public space of social media: connected cultures of the network society*. Routledge. doi:10.4324/9780203093863

Tilden, F. (2008). *Interpreting Our Heritage* (3rd ed.). The University of North Carolina Press. https://is.muni.cz/el/1421/podzim2017/MUI_338/Interpreting_Our_Heritage__Chapel_Hill_Books_.pdf.

Tilden, F. (1957). *Interpreting Our Heritage*. University of North Carolina Press.

Timur, S., & Getz, D. (2009). Sustainable tourism development: How do destination stakeholders perceive sustainable urban tourism? *Sustainable Development, 17*(4), 220–232. doi:10.1002d.384

Toft, T. (2016). *What Urban Media Art Can Do*. Avedition.

Tönnies, F. (1988). *Community and society*. Routledge. (Original work published 1887)

Tourism, T. (2017). *Berlin declaration on "transforming tourism"*. http://www.transforming-tourism.org/fileadmin/baukaesten/sdg/downloads/Berlin_Declaration.pdf

Townley, B., Beech, N., & McKinlay, A. (2009). Managing in the creative industries: Managing the motley crew. *Human Relations, 62*(7), 939–962. doi:10.1177/0018726709335542

Townsend, A. M. (2013). *Smart cities. Big data, civic hackers, and the quest for a new utopia*. W.W. Norton & Company.

Trencher, G., Yarime, M., & Kharrazi, A. (2013). Co-creating sustainability: Cross-sector university collaborations for driving sustainable urban transformations. *Journal of Cleaner Production, 50*(July), 40–55. doi:10.1016/j.jclepro.2012.11.047

Trescack, T. (2019, November 17). The main problem with virtual reality? It's almost as humdrum as real life. *The Conversation*. https://theconversation.com/the-main-problem-with-virtual-reality-its-almost-as-humdrum-as-real-life-126761

Tribe, J. (2011). *The Economics of Recreation, Leisure and Tourism*. Elsevier. doi:10.4324/9780080890517

Truong, N. (2020). *Jürgen Habermas: «Dans cette crise, il nous faut agir dans le savoir explicite de notre non-savoir»*. Le Monde Online.

Tsionas, M. G. (2021). COVID-19 and gradual adjustment in the tourism, hospitality, and related industries. *Tourism Economics*, *27*(8), 1828–1832. doi:10.1177/1354816620933039

TUB. (2021). *Relatório de Contas 2020-TUB*. Transportes Urbanos de Braga. https://tub.pt/templates/frontoffice/enterprise/pdf/rl2020.pdf

Tulumello, S. (2016). Reconsidering neoliberal urban planning in times of crisis: Urban regeneration policy in a "dense" space in Lisbon. *Urban Geography*, *37*(1), 117–140. doi:10.1080/02723638.2015.1056605

Turismo 2020, Plan del Turismo Español Horizonte 2020. (2017). *Ministerio de Industria, turismo y Comercio*. https://www.tourspain.es/es-es/Conozcanos/Documents/HistoricoPoliticaTuristica/PlanTurismoEspanolHorizonte2020.pdf

Turismo de Portugal. (2021). *Reativar Turismo | Construir Futuro - Plano de ação - Plano de ação apresentado pelo Ministro de Estado, da Economia e da Transição Digital*. https://www.turismodeportugal.pt/SiteCollectionDocuments/estrategia/plano-acao-reativar-o-turismo-construir-o-futuro.pdf

Turner, C., Hutabarat, W., Oyekan, J., & Tiwari, A. (2016, December). Discrete Event Simulation and Virtual Reality Use in Industry: New Opportunities and Future Trends. *IEEE Transactions on Human-Machine Systems*, *46*(6), 888–889. doi:10.1109/THMS.2016.2596099

Tussyadiah, P., Wang, D., & Jung, T. M. C. (2018) Virtual reality, presence, and attitude change: Empirical evidence from tourism. *Tourism Management, 66,* 140-154. Disponível em: https://www.sciencedirect.com/science/article/abs/pii/S0261517717302662?via%3Dihub

Tussyadiah, I. (2020). A review of research into automation in tourism: Launching the annals of tourism research curated collection on artificial intelligence and robotics in tourism. *Annals of Tourism Research*, *81*, 102883. Advance online publication. doi:10.1016/j.annals.2020.102883

Tussyadiah, I. P., Wang, D., Jung, T. H., & tom Dieck, M. C. (2018). Virtual reality, presence, and attitude change: Empirical evidence from tourism. *Tourism Management*, *66*, 140–154. doi:10.1016/j.tourman.2017.12.003

UN – United Nations. (2020). *The Sustainable Development Agenda*. https://www.un.org/sustainabledevelopment/development-agenda/

UNESCO & The World Bank. (2021). *Cities, Culture, Creativity: Leveraging Culture & Creativity for Sustainable Urban Development & Inclusive Growth*. https://www.worldbank.org/en/topic/urbandevelopment/publication/cities-culture-creativity

UNESCO & UNDP. (2013). *Creative Economy Report 2013. Special Issue*. http://www.unesco.org/culture/pdf/creative-economy-report-2013.pdf

UNESCO. (1946). *UNESCO Constitution*. http://portal.unesco.org/en/ev.php-URL_ID=15244&URL_DO=DO_TOPIC&URL_SECTION=201.html

UNESCO. (1976). *Recommendation concerning the Safeguarding and Contemporary Role of Historic Areas*. UNESCO. http://portal.unesco.org/en/ev.php-URL_ID=13133&URL_DO=DO_TOPIC&URL_SECTION=201.html

UNESCO. (1980). *Recommendation concerning the Status of the Artist.* https://en.unesco.org/creativity/governance/status-artist

UNESCO. (2005). *Convention on the Protection and Promotion of the Diversity of Cultural Expressions.* https://en.unesco.org/creativity/convention

UNESCO. (2010). *The Executive Board of UNESCO: 2010 edition.* UNESCO. UNEDOC Document code: SCX-2010/BROCH/CONSEIL EXECUTIF.

UNESCO. (2011). *Recommendation on the Historic Urban Landscape* (36 C/ Resolution 41). UNESCO World Heritage Centre, UNESCO. https://whc.unesco.org/en/hul

UNESCO. (2011). Recommendation on the Historic Urban Landscape. *Records of the General Conference, 36th session, Paris, 25 October - 10 November 2011, v. 1: Resolutions. Resolution 41,* 50-55. UNESCO. https://unesdoc.unesco.org/ark:/48223/pf0000215084

UNESCO. (2013). *Culture and UNDAF: A UNESCO Handbook.* UNESDOC Document code: CLT.2013/WS/9.

UNESCO. (2015). *Global Report 2015 - Re|Shaping Cultural Policies.* https://unesdoc.unesco.org/ark:/48223/pf0000242866

UNESCO. (2016). *Culture: urban future; global report on culture for sustainable urban development.* UNEDOC Document code: CLT-2016/WS/18.

UNESCO. (2018a). *Global Report 2018 - Re|Shaping Cultural Policies - Advancing creativity for development.* https://unesdoc.unesco.org/ark:/48223/pf0000260592

UNESCO. (2018b). *Investing in Creativity.* UNESCO. https://en.unesco.org/creativity/publications/investing-creativity

UNESCO. (2019a). *Culture & working conditions for artists.* https://en.unesco.org/creativity/publications/culture-working-conditions-artists

UNESCO. (2019b). *Culture | 2030 Indicators.* https://unesdoc.unesco.org/ark:/48223/pf0000371562

UNESCO. (2020). *Incidences socio-économiques et culturelles du Covid-19 en Afrique: quelles réponses pour l'UNESCO?* Available at: https://unesdoc.unesco.org/ark:/48223/pf0000373904

UNESCO. (2020). *Museums around the world in the face of COVID-19.* https://unesdoc.unesco.org/ark:/48223/pf0000373530

UNESCO. (2020a). *UNESCO for Sustainable Cities.* https://en.unesco.org/unesco-for-sustainable-cities

UNESCO. (2020b). *COVID-19* Response. https://en.unesco.org/creativity/COVID-19

UNESCO. (2020c). *Explore initiatives & stories from UNESCO networks.* https://en.unesco.org/covid19/initiatives

UNESCO. (2020d). *Freedom & Creativity: Defending art, defending diversity.* https://unesdoc.unesco.org/ark:/48223/pf0000373357

UNESCO. (2020e). *Culture in Crisis: Policy guide for a resilient creative sector.* https://unesdoc.unesco.org/ark:/48223/pf0000374631

UNESCO. (2020f). *Creative Cities Network.* https://en.unesco.org/creative-cities/

UNESCO. (2020g). *Resolution of the General Assembly: 8.GA 9.* https://ich.unesco.org/en/Resolutions/8.GA/9

UNESCO. (2020h). *UNESCO Creative Cities' Response to COVID-19*. https://unesdoc.unesco.org/ark:/48223/pf0000374264?posInSet=7&queryId=1f26fc80-270a-4c54-bb11-17e83ea8888b

UNESCO. (2020i). *Mission Statement*. Creative Cities Network. https://en.unesco.org/creative-cities/sites/default/files/Mission_Statement_UNESCO_Creative_Cities_Network_1.pdf

UNESCO. (2021a). *INF.5A.2 Impact of the COVID-19 pandemic on World Heritage and responses by the Secretariat*. Document WHC/21/44.COM/INF.5A.2. https://whc.unesco.org/document/187761

UNESCO. (2021b). *UNESCO report: museums around the world in the face of COVID-19*. https://whc.unesco.org/en/cities

UNESCO. (2022). *Re|shaping policies for creativity- addressing culture as a global public good*. https://unesdoc.unesco.org/ark:/48223/pf0000380474.locale=en

UN-Habitat (2021). *Cities and pandemics: Towards a more just, green and healthy future*. Author.

UNHABITAT. (2020). *World Cities Report 2020*. https://unhabitat.org/sites/default/files/2020/10/wcr_2020_report.pdf

United Kingdom Legislation. (2010). *Equality Act 2010*. Retrieved 05.04.2021 from https://www.legislation.gov.uk/ukpga/2010/15/section/26

United Nations Department of Economic and Social Affairs. (2018). *World urbanization prospects: The 2018 revision*. Author.

United Nations World Tourism Organization [UNWTO]. (2020). *World tourism barometer, No. 18*. Retrieved from https://unwto.org/world-tourism-barometer-n18-january-2020

United Nations. (2018). *2018 revision of world urbanization prospects*. https://www.un.org/development/desa/en/news/population/2018-revision-of-world-urbanization-prospects.html

United Nations. (2019). World Urbanization Prospects. The 2018 Revision (ST/ESA/SER.A/420). Department of Economic and Social Affairs. Population Division.

United Nations. (2020). *Policy Brief: COVID-19 and transforming tourism*. https://www.un.org/sites/un2.un.org/files/sg_policy_brief_covid-19_tourism_august_2020.pdf

United States Department of State. (2021). *Country Reports on Human Rights Practices for 2020 Bureau of Democracy, Human Rights and Labor - Guinea-Bissau 2020 Human Rights Report*. Available at: https://www.state.gov/wp-content/uploads/2021/03/GUINEA-BISSAU-2020-HUMAN-RIGHTS-REPORT.pdf

United States Department of State. (2021). *Trafficking in persons report*. Available at: https://www.state.gov/reports/2021-trafficking-in-persons-report/

UNSDG. (2020). *Policy Brief: COVID-19 and Transforming Tourism*. https://unsdg.un.org/resources/policy-brief-COVID-19-and-transforming-tourism

UNTWO. (2021). *The Economic Contribution of Tourism and the Impact of COVID-19*. UNTWO. doi:10.18111/9789284423200

UNWTO & IPSOS. (2019). *Global survey on the perception of residents towards city tourism: Impact and measures*. Madrid: UNWTO. doi:10.18111/9789284421107

UNWTO & OAS. (2018). *Tourism and the Sustainable Development Goals – Good Practices in the Americas*. Madrid: UNWTO. doi:10.18111/9789284419685

UNWTO & UNDP. (2017). *Tourism and the Sustainable Development Goals – Journey to 2030, Highlights.* Madrid: UNWTO. doi:10.18111/9789284419340

UNWTO INSTO. (2022). *INSTO Observatories.* https://insto.unwto.org/observatories/

UNWTO. (2018). *'Overtourism'? Understanding and Managing Urban Tourism Growth beyond Perceptions, Executive Summary.* World Tourism Organization Centre of Expertise Leisure, Tourism & Hospitality; NHTV Breda University of Applied Sciences; and NHL Stenden University of Applied Sciences, UNWTO. . doi:10.18111/9789284420070

UNWTO. (2019). *The Future of Work and Skills Development in Tourism – Policy Paper.* Madrid: UNWTO. doi:10.18111/9789284421213

UNWTO. (2020). *International Tourism Highlights – 2020 Edition.* https://www.e-unwto.org/doi/epdf/10.18111/9789284422456

UNWTO. (2020, April 1). *Message from Madrid: Tourism and Covid-19.* https://www.unwto.org/news/madrid-tourism-covid-19

UNWTO. (2021). *International Tourism Highlights 2020.* https://www.e-unwto.org/doi/epdf/10.18111/9789284422456

UNWTO. (2021). *UNWTO Inclusive Recovery Guide. Sociocultural Impacts of COVID-19. Issue 2: Cultural Tourism.* UNWTO. https://www.e-unwto.org/doi/book/10.18111/9789284422579

UNWTO. (2021a). *International Tourism Highlights* (2020 Edition). UNWTO., doi:10.18111/9789284422456

UNWTO. (2022). *Tourism grows 4% in 2021 but remains far below pre-pandemic levels.* UNWTO Barometer. https://www.unwto.org/taxonomy/term/347

Urry, J. (1995). *Consuming places.* Routledge.

Valente, C. M. (2016). *A street art no feminino: o lugar da mulher na Arte Pública.* Dissertação de Mestrado, Faculdade de Letras, Universidade de Lisboa.

Van roof, M. (2016). Late-Breaking Work: Games & Playful Interaction. *CHI 2016*, 1990-1991.

Vardoulakis, S., Salmond, J., Krafft, T., & Morawska, L. (2020). Urban environmental health interventions towards the Sustainable Development Goals. *The Science of the Total Environment*, *748*, 141530. doi:10.1016/j.scitotenv.2020.141530 PMID:32827895

Veblen, T. (1994). The Theory of the Leisure Class: An Economic Study of Institutions. Penguin Books. (Original publication 1899)

Vedana, V. (2018). Escutar no som: Gravação e edição de etnografias sonoras a partir de um paradigma ecológico. *Ilha*, *20*(1), 117–144. doi:10.5007/2175-8034.2018v20n1p117

Veldpaus, L., Pereira Roders, A. R., & Colenbrander, B. J. F. (2013). Urban Heritage: Putting the Past into the Future. *The Historic Environment*, *4*(1), 3–18. doi:10.1179/1756750513Z.00000000022

Velotta, R. (2012) International tourists in Las Vegas outspend domestic travelers, analyst says. *Vegas Inc.* Retrieved from https://vegasinc.lasvegassun.com/business/real-estate/2012/jul/10/international-tourists-las-vegas-outspend-domestic/

Venturelli, S. (2016). *Intervenção do artista no contexto urbano com projeção mapeada computacional. 25 Encontro ANPAP. Arte: seus espaços e/em nosso tempo.* ANPAP.

Viallon, P. (2013). La communication touristique, une triple invention. *Mondes du Tourisme*, *7*.

Viallon, P., & Gardère, E. (2020). *Médias dits sociaux, médias dissociants*. De Boeck.

Vidal, L. (2020). *Les Hommes Lents. Résister à la modernité XV – XX Siècle*. Flammarion.

Villaça, F. (2012). *Reflexões sobre as cidades brasileiras* [Reflections on Brazilian cities]. Nobel.

Violier, P. (2021). La pandémie du COVID, une parenthèse dans la mondialisation touristique. *Economics and Politics*, 3(91), 23–35.

Virilio, P. (2000). *A velocidade de libertação*. Relógio D'Água.

Visitar Penafiel. (n.d.). Retrieved August 18, 2022, from https://www.cm-penafiel.pt/visitar-penafiel/

Wadley, D. (2020). *The City of Grace An Urban Manifesto*. Springer Nature Singapore Pte Ltd. doi:10.1007/978-981-15-1112-7

Wassler, P., & Talarico, C. (2021). Sociocultural impacts of COVID-19: A social representations perspective. *Tourism Management Perspectives*, 38.

Watercuter, A. (2022, Jan 22). Film Festivals Are Evolving for the Better. *Wired Magazine*. https://www.wired.com/story/film-festivals-hybrid/

Watson, S. (2009). The magic of the marketplace: Sociality in a neglected public space. *Urban Studies (Edinburgh, Scotland)*, *46*(8), 1577–1591. doi:10.1177/0042098009105506

Watson, S., & Studdert, D. (2006). *Markets as sites for social interaction. Spaces of diversity*. Policy Press.

Weber, M. (2022). *Economia e sociedade. Fundamentos da sociologia compreensiva*. Edições 70. (Original publication 1920)

Wehbe, R. (2016, November) Above Water: Extending the Pay Space for Health. *ISS Conference*, 499. 10.1145/2992154.2996882

Wen, J., Kozak, M., Yang, S., & Liu, F. (2020). COVID-19: Potential effects on Chinese citizens' lifestyle and travel. *Tourism Review*, *76*(1), 74–87. https://doi.org/10.1108/TR-03-2020-0110

Weston, R., Guia, J., Mihalič, T., Prats, L., Blasco, D., Ferrer-Roca, N., Lawler, M., & Jarratt, D. (2019). *Research for TRAN Committee – European tourism: recent developments and future challenges*. European Parliament, Policy Department for Structural and Cohesion Policies, Brussels. https://research4committees.blog/2019/10/29/european-tourism-recent-developments-and-future-challenges

Weston, R., Guia, J., Mihalič, T., Prats, L., Blasco, D., Ferrer-Roca, N., Lawler, M., & Jarratt, D. (2019). *Research for TRAN Committee – European tourism: recent developments and future challenges*. European Parliament, Policy Department for Structural and Cohesion Policies, Brussels. https://research4committees.blog/2019/10/29/european-tourism-recent-developments-and-future-challenges/

Wexel, J. (2012). *Medeia Vozes de Christa Wolf: a reinvenção polifônica do mito trágico*. Dissertação de Mestrado em Letras, Cultura e Regionalidade. Universidade de Caxias do Sul. UCS. https:// repositorio.ucs.br/handle/11338/772

Wexel, J. (Ed.). (2021). *Video-Performance Make me up!* https://www.instagram.com/tv/CVAOiyPKRqE/? utm_medium=copy_link

Wexel, J. (2021). Theater, Audiovisual and Streaming: An Analysis of Theater-Making in Times of Pandemic Uncertainty in the Post-Dramatic Experience of the Play Waiting for Godette. *Rotura - Revista de Comunicação. Cultura e Artes.*, *1*, 39–46.

White, C., & Hardemo, L. (2010). Sexual harassment in kitchens. *Journal of Foodservice Business Research.*

WHO. (2021, June 29). *Listings of WHO's response to COVID-19.* https://www.who.int/news/item/29-06-2020-covid-timeline

Williams, A. M., Chen, J. L., Li, G., & Baláž, V. (2022). Risk, uncertainty and ambiguity amid Covid-19: A multi- national analysis of international travel intentions. *Annals of Tourism Research, 92*, 92. doi:10.1016/j.annals.2021.103346 PMID:35013630

Winkin, Y. (2001). *Anthropologie de la communication – de la théorie au terrain* [Anthropology of communication - from theory to field]. Le Seuil.

Witt, S. F., Newbould, G. D., & Watkins, A. J. (1992). Forecasting domestic tourism demand: Application to Las Vegas arrivals data. *Journal of Travel Research, 31*(1), 36–41. doi:10.1177/004728759203100108

Wolf, D. (2021, June). *Augmenting Teleportation in Virtual Reality with Discreet Rotation Angles.* arXiv:2016.04257v1 [cs. HC].

World Bank. (2021). *Poverty and shared prosperity. Reversals of fortune.* Available at: https://www.worldbank.org/en/publication/poverty-and-shared-prosperity

World Bank. (2022). *World Development Report 2022: Finance for an Equitable Recovery.* Washington, DC: World Bank. doi:10.1596/978-1-4648-1730-4

World Economic Forum. (2021) *Which 5 Countries Have The Best National Brand?* https://www.weforum.org/videos/which-5-countries-have-the-best-national-brand

World Health Organization (WHO)‡ (2020). *Modes of transmission of virus causing COVID-19: Implications for IPC precaution recommendations: Scientific brief.* World Health Organization.

World Tourism Barometer. (2020). *UNWTO.* doi:10.18111/wtobarometereng

World Tourism Organization. (2022) *'Overtourism'? Understanding and Managing Urban Tourism Growth beyond Perceptions.* WTO. https://www.e-unwto.org/doi/pdf/10.18111/9789284420070

Wut, T. M., Xu, J. B., & Wong, S. M. (2021). Crisis management research (1985–2020) in the hospitality and tourism industry: A review and research agenda. *Tourism Management, 85*, 104307. https://doi.org/10.1016/j.tourman.2021.104307

Wynn, J. (2010). City Tour Guides: Urban Alchemists at Work. *City & Community, 9*(2), 145–163. doi:10.1111/j.1540-6040.2010.01322.x

Xu, Z., Zhang, H., Hu, C., Liu, Y., Xuan, J., & Mei, L. (2017). Crowdsourcing-based timeline description of urban emergency events using social media. *International Journal of Ad Hoc and Ubiquitous Computing, 25*(1-2), 41–51. doi:10.1504/IJAHUC.2017.083481

Yang, Y., Zhang, C. X., & Rickly, J. M. (2021). A review of early COVID-19 research in tourism: Launching the Annals of Tourism Research's Curated Collection on coronavirus and tourism. *Annals of Tourism Research, 91*, 103313. https://doi.org/10.1016/j.annals.2021.103313

Yang, Y., Zhang, C. X., & Rickly, J. M. (2021). A review of early COVID-19 research in tourism: Launching the Annals of Tourism Research's Curated Collection on coronavirus and tourism. *Annals of Tourism Research, 91*, 91. doi:10.1016/j.annals.2021.103313 PMID:34611371

Yoo, M., Singh, A. K., & Ghaharian, K. (2017). Foreign exchange rate effect on international gaming demand: An examination from an upscale Las Vegas casino. *UNLV Gaming Research & Review Journal, 21*(1), 1–23.

YouTube. (n.d.). *Home - youtube VR*. YouTube. Retrieved August 18, 2022, from https://vr.youtube.com/

Yubero, C., Condeço-Melhorado, A. M., García-Hernández, M., & Fontes, A. C. (2021). Comparing spatial and content analysis of residents and tourists using Geotagged Social Media Data. The Historic Neighbourhood of Alfama (Lisbon), a case study. *Investigaciones Turisticas,* (22), 95-120. doi:10.14198/INTURI2021.22.5

Yung, R., & Khoo-Lattimore, C. (2019). New realities: A systematic literature review on virtual reality and augmented reality in tourism research. *Current Issues in Tourism*, *22*(17), 2056–2081. https://doi.org/10.1080/13683500.2017.1417359

Zelizer, V. (1994). *The social meaning of money*. Basick Books.

Zelizer, V. (2005). *The purchase of intimacy*. Princeton University Press.

Zhang, C. X., Wang, L., & Rickly, J. M. (2021). Non-interaction and identity change in Covid-19 tourism. *Annals of Tourism Research*, *89*, 89. doi:10.1016/j.annals.2021.103211

Zhang, T., Bufquin, D., & Lu, C. (2019). A qualitative investigation of microentrepreneurship in the sharing economy. *International Journal of Hospitality Management*, *79*, 148–157. doi:10.1016/j.ijhm.2019.01.010

Zhan, L., Zeng, X., Morrison, A. M., Liang, H., & Coca-Stefaniak, J. A. (2020). A risk perception scale for travel to a crisis epicentre: Visiting Wuhan after COVID-19. *Current Issues in Tourism*, *25*(1), 150–167. https://doi.org/10.1080/13683500.2020.1857712

Zhaohui, L. (2014, October 29). Embedding perspective cue in holographic projection display by virtual variable-focal-length lenses. *Proceedings Optoelectronic Imaging and Multimedia Technology III. SPIE/COS Photonics Asia*.

Zheng, D., Luo, Q., & Ritchie, B. W. (2021). Afraid to travel after COVID-19? Self-protection, coping and resilience against pandemic 'travel fear'. *Tourism Management*, *83*, 104261. https://doi.org/10.1016/j.tourman.2020.104261

Zielinski, S. (2017). Pensar a arte após os media: a pesquisa como cultura da experimentação viva. In *Ecologia da imagem e dos Media*. Nova Vega.

Zimmerman, T. G., Lanier, J., Blanchard, C., Bryson, S., & Harvill, Y. (1987). A hand gesture interface device. *SIGCHI Bull.*, *17*(SI), 189–192. doi:10.1145/30851.275628

Zook, M., Graham, M., Shelton, T., & Gorman, S. (2010). Volunteered geographic information and crowdsourcing disaster relief: A case study of the Haitian earthquake. *World Medical & Health Policy*, *2*(2), 7–33. doi:10.2202/1948-4682.1069

Zuboff, S. (2018). Big other: Capitalismo de vigilância e perspectivas para uma civilização de informação. In F. Bruno, B. Cardoso, M. Kanashiro, L. Guilhon, & L. Melgaço (Eds.), *Tecnopolíticas da vigilância: Perspectivas da margem* (pp. 17–68). Boitempo.

Zuk, M., Bierbaum, A. H., Chapple, K., Gorska, K., & Loukaitou-Sideris, A. (2018). Gentrification, Displacement, and the Role of Public Investment. *Journal of Planning Literature*, *33*(1), 31–44. doi:10.1177/0885412217716439

Zuniga-Teran, A. A., Orr, B. J., Gimblett, R. H., Chalfoun, N. V., Going, S. B., Guertin, D. P., & Marsh, S. E. (2016). Designing healthy communities: A walkability analysis of LEED-ND. *Frontiers of Architectural Research*, *5*(4), 433–452. doi:10.1016/j.foar.2016.09.004

About the Contributors

Pedro Andrade is a Sociologist and Researcher at University of Minho. Phd in Sociology of Culture at FCSH, New Univ. of Lisbon (2002). Previously, he taught at University of Coimbra - Faculty of Economy and at University of Lisbon - Faculty of Fine Arts, in the areas of Sociology of Culture, Communication, Sociological Methods and Digital Humanities. Major areas of research: urban cultures, art communication, art and science museums, literacies, digital social networks (Web 2.0/Web 3.0), methodologies and hypermedia. Coordinator of several research projects and teams e.g. Scientific-Technological Literacy and Public Opinion: the case of science museums; Public Communication of Arts: the case of local/ global art museums, both funded by Foundation for Science and Technology, Lisbon. Participation within international university webs e.g. Virginia Comm. Univ., USA; member of project Manifesto Art and Social Inclusion in Urban Communities, coordinated by Mel Jordan at Univ. Loughborough, and including King's College, Chelsea School of Arts, Universities of Plymouth, Leuven, Utrech, Gronigen, Minho. ORCID ID: https://orcid.org/0000-0003-4550-0562

Moisés de Lemos Martins is a Full Professor at the Department of Communication Sciences of the University of Minho. He is the former director of Communication and Society Research Center, which he founded in 2001. He is also the founder and the former editor of the journals Comunicação e Sociedade, Lusophone Journal of Cultural Studies, and Vista. He holds a PhD from the University of Strasbourg in Social Sciences (Sociology). He has published in the area of Sociology of Culture, Social Semiotics, Sociology of Communication, Visual Semiotics, Intercultural Communication, Scientific Policies, and Lusophone Studies. He directed for ten years the Institute of Social Sciences, University of Minho. He was president of Sopcom – Portuguese Association of Communication Sciences, from 2005 to 2015; Lusocom – Federation of Lusophone Associations of Communication Sciences, from 2011 to 2015; and Confibercom – Ibero-American Confederation of Scientific and Academic Communication Associations, from 2012 to 2015. Orcid: https://orcid.org/ 0000-0003-3072-2904.

* * *

José Guilherme Abreu, PhD Contemporary Art History (FCSH-UNL). Invited auxiliary lecturer in Escola das Artes - Universidade Católica Portuguesa, senior researcher of CITAR (https://citar.artes.porto.ucp.pt/en/). Since 2018, he is Director of Public Art Information, Research, and Intervention Network. Since 2017, he is Secretary of Association Raymond Abellio de Recherches et Études, in Paris. He is author of books and articles in national and international scientific journals. He has been directing MA and Phd thesis. His main research domains are public art and modern and contemporary art history. In 2009, his PhD thesis got Ignasi de Lecea Public Art Award, by Barcelona's University Research Centre POLIS.

Rosana Albuquerque is a professor in the Department of Social Sciences and Management at Open University Lisbon. PhD in Sociology, specializing in Sociology of Migration and Master in Intercultural Relations. As a researcher at the Center for the Study of Migration and Intercultural Relations, has devoted special attention to issues related to the descendants of immigrants, to associativism, to civic participation and to citizenship. Other areas of interest: intersectional theory to analyze discrimination and inequalities and the interconnection between sexism and racism; social policies in their relationship with the promotion of equality, equity and citizenship; cooperation and education for development.

Maria Baldessar is a PhD in Communication Sciences from the University of São Paulo (2006), Master in Political Sociology from the Federal University of Santa Catarina (1999), coordinator of the Research Group Geographies of Communication and MidiaCon - Media and Convergence, certified by CNPq. She is an Associate Professor at the Federal University of Santa Catarina (UFSC). She is a permanent lecturer in the Postgraduate Programs in Knowledge Engineering and Management and in Translation Studies. She teaches courses at the undergraduate degree in Journalism related to online journalism, media economics and textual production. She has experience in the area of Communication, with an emphasis on Specialized Journalism (Community, Rural, Business, Scientific, Institutional), working mainly on the following topics: journalism, internet, hypermedia, online journalism, media economy, media history, digital convergence, usability and journalism education. She is part of the Interactive Digital Television Center.

Miguel Bandeira was Councillor (2013-21) of urban regeneration, traffic and mobility, heritage, university connection, urbanism, and urban planning, in City Council of Braga, Portugal. Associate Professor with Habilitation, in the Institute of Social Sciences (ICS), University of Minho-Pt. President of ICS (2010-2013) and Vice-President (2003-2010). He graduated in Geography at the University of Porto (FLUP-1982), held a master's degree at the University of Coimbra (FLUC-1993) and completed a PhD at the Institute of Social Sciences, University of Minho (2002). It was Secretary General of Foundation, Studies Centre Euro-Regional's Galicia (Sp) - North Portugal (CEER) (Pt). Effective researcher of the Center for Studies in Communication and Society / University of Minho (CECS). He is interested in particular in the fields of Urban Studies and Urban Planning, Cultural Studies/History, Cultural and Natural Heritage. He teaches in the Undergraduate Architecture, Planning, and other Social Sciences in University of Minho. Member of the Board of the Portuguese Association of Municipalities with Historical Center, and the Portuguese Association of Municipal Housing (2013/17). He is voluntary and an active member of various communities associations. President of the Municipal Transport Authority. President of the Administration of the - Urban Transport of Braga, (TUB) E. M. Member of the Administration of the public Library, Lúcio Craveiro da Silva; Curator of the Bracara Augusta Foundation.

Cecília Baptista is a professor at the Departmental Engineering Unit of the Polytechnic Institute of Tomar. Director of the Degree in Chemical and Biochemical Engineering and Degree in Chemical Technology. Researcher at the Techn&Art Research Center - Center for Technology, Restoration and Enhancement of the Arts. PhD in Chemistry and Master in Science and Technology of Paper and Forest Products from the University of Beira Interior. Degree in Pharmaceutical Sciences from the Faculty of Pharmacy of the University of Lisbon.

Ângela Berlinde is an artist, curator and researcher at CICANT, Center for Research in Applied Communication, Culture and New Technologies, Lisboa and Member of the research group NANO Organisms of New Arts at the School of Fine Arts at UFRJ, Brazil. Research about indigenous and digital electronic arts. She has a PhD in Visual Communication from the University of Minho, with a FCT scholarship dedicated to indigenous studies; Master in Photography, from the Utrecht School of Arts-Netherlands and a Law Degree from the University of Minho. She has a Postdoctoral research in Visual Arts, in the field of Interdisciplinary Poetics at the School of Fine Arts of UFRJ, Brazil. She is guest Adjunct teacher at Polytechnic Institute of Porto in Portugal where she coordinated the Pos graduation in Photobooks and the International Summer Schools dedicated to Media Arts. She is also guest teacher at Universidade Lusofóna de Lisboa at ECATI. She has worked as a curator of national and international exhibitions, published intensely and had lectured on contemporary arts throughout Europe, Latin American Countries and Asia. ORCID: https://orcid.org/0000-0003-4550-0562.

Mihaela Bogdan is a cinematography, photography, and media undergraduate student at the Faculty of Theatre and Film in Cluj-Napoca, Romania. In addition, as part of the exchange program Erasmus+, student within Production of Interactive Content and Multimedia at Instituto Superior de Ciências Educativas do Douro, Penafiel, Portugal.

Clara Bertrand Cabral is an anthropologist, Master in Anthropological Sciences and currently PhD candidate in Social Sustainability and Development at Open University Portugal. Carried out research in the areas of ethnography, ethnotechnology, museology and heritage, as well as on participatory processes in the UNESCO Convention for the Safeguarding of Intangible Cultural Heritage and the World Heritage Convention. Programme Specialist for Culture at the Portuguese National Commission for UNESCO since 2005. Professor of Intangible Cultural Heritage Methodologies at Lusophone University in Lisbon since 2011.

Pedro Chamusca is Geographer, University Research Associate at the Communication and Society Research Centre (CECS) of the University of Minho and the President of the Portuguese Association of Geographers. He holds a PhD in Geography at the Faculty of Arts and Humanities of the University of Porto and is professionally qualified in Geographic Information Systems (GIS). He is the author of more than 80 scientific publications; developing research on topics related to urban geography, territorial cohesion, GIS, governance, planning, tourism and spatial planning. He has several pedagogical collaborations in higher education, namely in teaching classes in the courses of geography of the University of Porto, University of Coimbra and Federal University of Pelotas as well as at Porto Business School. He is in charge of national/international initiatives such as the Geography Olympics (since 2018) or the 1st edition of the Night of Geography (2018), and coordinates/participates in various projects of knowledge application, in areas associated with strategic planning, spatial planning, development and commercial urbanism.

Cinzia Colapinto, Ph.D. in Business and Corporate Finance (University of Milan, Italy), Associate professor at the Department of Management, Ca' Foscari University of Venice (Italy), and Senior Researcher in Strategy and Entrepreneurship at IPAG Business school (France). Her research interests focus on Entrepreneurship, Innovation, Strategy. In particular, she is interested in the role played by digital technologies on business model transformations and on their impact on the achievement of sustainable development goals by SMEs. Her main publications are in Management International Review,

Management Decision, European Journal of Operational Research, Energy Policy, Annals of Operations Research. She is the author of several monographs, including Adaptive Decision Making and Intellectual Styles (Springer), and is part of international research networks on strategic innovation and entrepreneurship. She is part of the Urban GoodCamp Project, a Horizon Plus initiative that aims to empower Higher Education Institutions (HEIs) and their urban stakeholders to work together on pressing urban challenges, by developing co-design approaches.

Fernando R. Contreras is Professor of Cyberculture in the Degree of Journalism at the University of Seville and Director of the Research Group on Technology, Art and Communication of the Andalusian Research Plan, Development and Innovation (PAIDI HUM 868). He has a PhD in Philosophy, a PhD in Information Sciences and a Bachelor in Fine Arts (Design and Print). He has studied industrial design and audiovisual design. His research work is focused on Visual and Cultural Studies, Art and Technology (Media Art) and Digital Humanism.

Filipa Corais is currently developing a PhD at the School of Architecture of University of Minho. She has a degree in Architecture from the University of Coimbra (12.1998) and a master's degree in "Planning and Design of the Urban Environment", from the FAUP/FEUP of the University of Porto (7.2004). She has been the Head of the Mobility Division at the Braga City Council since 6.2016. Up to that date and since 2000 she was a Senior Technician in the Planning Division of the Municipality of Braga. Among the responsibilities performed in Braga City Council stand out: participation in the development of the Braga PDM review (2015); the coordination of the project "Braga Urban Innovation Laboratory Demonstrator (BUILD)", that frames the SchoolBus project and the Braga Urban Innovation Laboratories, that integrates a multidisciplinary team of scholarship holders and whose main partners are the University of Minho and the International Iberian Nanotechnology Laboratory (INL); coordination, by the municipality: of the development of the Sustainable Urban Mobility Plan (PMUS); of the Municipal Transport Authority of Braga; of the partnerships and networks of cities for Mobility (e.g. Quadrilátero, URBACT, CIVITAS, Eurocities, among others) and of the several national and European applications for the promotion of mobility. She is a visiting professor at the ESA-IPVC, since 2008 and she was an invited teacher in undergraduate and master's classes at the University of Minho, at the School of Engineering and Architecture (2015; 2018; 2019; 2020).

Antónia Correia is a professor at the Universidade do Algarve, affiliated at CEFAGE, Research center award by national science foundation. She is also the President of KIPT Colab, the first collaborative laboratory in tourism. She holds a Bachelor's degree from University of Algarve (Portugal), a Master from ISEG, University of Lisbon (Portugal), a Ph.D. from University of Algarve in tourism demand (Portugal) and an aggregation from University of Algarve; all relating to tourism and hospitality management. She has published a wide range of articles in top-tier journals such as (i.e. Annals, TM, JTR, IJHM, JTTM, among others), conference papers in more than 40 countries and chapter books released by Elsevier, Routledge, CABI, Emerald, Cambridge Scholars Publishing. She acts as an editorial board member of JTR, JBR, Anatolia, TA, IJCTHR. In addition, she served as an ad-hoc reviewer for many other tourism, hospitality and leisure journals. She is also involved in several tourism-based national and international research projects, particularly with her partners based in Europe. Developing such partnerships has resulted in publishing quality journal papers, conference presentations and book chapters. She was also invited as a guest lecturer in STHM, Hong Kong Polytechnic, at Stonehill College, Easton, MA, USA, and at Cartagena University, at international level, at national level she collaborates with ISEG,

Aveiro University, IPL and ESHTE. She has also organized conferences or acting as part of the scientific committee worldwide, in ATMC and CPTHL. Her work has received over one hundred citations in ISI-referenced journals and over one thousand citations counted by Google Scholar. Recently, she was ranked among the top 100 tourism researchers in terms of the number of publications from 1975-2014 (McKercher, 2015) nominated as an outstanding reviewer by Emerald (2014), among other awards and recognitions. Her research interests entail tourism consumer behavior, marketing, and image. Tourism economics and models are also one of her research passions.

José da Cunha Barros is an anthropologist working as a correspondent researcher at the Laboratory of Tourism (L-tour) of the Polytechnic Institute of Tomar (IPT). He holds a PhD in Social Sciences (specialising in cultural anthropology) from the Technical University of Lisbon (UTL). He taught at ISCSP, institution of the UTL having integrated some of its management bodies and participated as a researcher in one of its research centres (CAPP). He has published two books, on thermalism and thermal tourism (2002), and on tourism and leisure (2004). He has participated, with communications, in national and international congresses and seminars. He has published articles in scientific journals, mainly with peer review, indexed in global bibliographic data bases. The anthropology of tourism, anthropology of history, leisure and the implications of new technologies on culture are some of his research areas. He is registered in ORCID as a researcher: https://orcid.org/0000-0003-0570-3170.

Natacha Cyrulnik is a documentary filmmaker, professor at the University of Aix-Marseille, and responsible for the master's degree "production and direction" in the SATIS (Science, Art and Technic of Image and Sound) department and researcher at the UMR 7061 PRISM (Perception, Representations, Image, Sound, Music). Her research focuses on the representation of territories through documentary film, film pedagogy and audio-visual methods.

Cécile Dolbeau-Bandin is a teacher in information and communication sciences. She teaches at the University of Caen Normandy. Her research topics are: digital technologies, social robotics, e-health. University of Caen (Normandy, France) et chaire Unesco (Univesity of Strasbourg).

Jane de Almeida is an Associate Professor at the Graduate Studies in Education, Art and History of Culture (MfA and Ph.D.) at Mackenzie University and an interdisciplinary researcher and independent curator. Currently she is a researcher at the Getty Foundation and at the Los Angeles Filmforum under a PST 2021-2024 grant. She has published several articles and book chapters such as "On Elephants, Telescopes, Microscopes, Cartography, Aliens, and Computers: Notes on Scalability in Alexander Kluge's Works" In: Stichwort: Kooperation. Keiner ist alleine schlau genug (Alexander Kluge-Jahrbuch, Verlag, 2017), "The dot and the line" for A companion to Contemporary Drawing (Wiley, 2020) and organized the books CineGrid: cinematic futures (Cinusp, 2017) and Harun Farocki: Programming the Visible (bilingual, Cinusp, 2018) with articles by Thomas Elsaesser, Erika Balsom, Alfredo Suppia, Harun Farocki, Patricia Moran among others. She curated the first 4K films exhibition in Brazil at the FILE (Electronic Language International Festival) and organized the first transmission from Latin America to the US (UCSD) and Japan (Keio University) of a 4k film over a 10 Gbit/s photonic network at FILE 2009 Transcontinental. Currently she coordinates the Advanced Visualization Working Group (Content) for the Brazilian National Education and Research Network (RNP). Jane is the director of the short films 2014k (4k 3D, 2010, 10 mins) screened at Casa Brasil during the FIFA Soccer World Cup in Johannesburg and San Diego), StereoEssays: Five or Six Essays in Search of a Narrative (4k 3D,

2011, 16 min), which has been exhibited in venues such as the Maison de L'Amérique Latine in Paris, CineGrid in San Diego, the Sesc Film Festival of Arts in São Paulo and the Tecnofagias Art Exhibition (in loop for 2 months) at the Tomie Ohtake Institute for Contemporary Art. She has also directed the short films Openmouthed (4k, 2014, 3 mins), The Diving (4k, 2014, 2 mins) and Pixel Race (4k, 2015, 9 mins), which were screened at CineGrid in San Diego and Tokyo. She is also organizer of the Science Film Festival (FeFiCi).

José Dias Lopes has a PhD in Management, specialty Operations Management. He graduated in Physical and Materials Engineering and started his activity in the industry as project engineer and then as industrial engineer. He has been a professor in higher education for about 30 years and is currently researcher-coordinator at KIPT, a Collaborative Laboratory for tourism professions.

Marta Margarida Santos Dionísio is currently an Associate Professor at the Polytechnic Institute of Tomar. She has 22 years of experience as a lecturer at a higher education level. She obtained her Ph.D. in American Studies with the thesis entitled Oprah Winfrey – the Art of Communicating from the University of Lisbon, Faculty of Letters, in 2015, in the scientific field of Studies of Literature and Culture; her Master in American Studies entitled Lucille Ball - an American Woman from the University of Lisbon, Faculty of Letters, in 2005; and her BA in English and German Teaching from the University of Aveiro, in 1997. She is also member of the Board of Directors of the Techn&Art Research Centre - Technology, Restoration and Arts Enhancement Centre being an Integrated Researcher since 2018. She is a researcher in two FCT Projects (FesTab and Insignia) and has published some articles in specialized magazines.

Sérgio Eliseu holds a PhD in Art and Design from the Faculty of Fine Arts of the University of Porto; Master in Contemporary Artistic Creation by the Department of Communication and Art at the University of Aveiro; Degree in Art History from the Faculty of Arts of the University of Coimbra; He is a Guest Professor at the Department of Communication and Art at the University of Aveiro, as well as an Adjunct Professor at ISCE Douro: Penafiel, Porto. Senior Designer at AbsoluteBPC. Coordinator NIAM - Research Group in Arts and Multimedia at Isce Douro in Penafiel. Integrated member of ID+, Institute for Research in Design Media and Culture.

Raquel Faria has a PhD degree in High Studies in Contemporary History - Contemporary Times, university of Coimbra (2014). Master's degree in Political Science from the University of Aveiro (2010). Graduated in Public Administration - political science branch (2008). As the issue, participates in the projects "Memory of Africa and the East", CESA/ISEG (since 2009); "That other world that is the world: the world of media and the world of development", CEsA, among others (2014 - 2016). From 2012 to 2014, he was a researcher in the project "The cluster as a theoretical and practical instrument of international cooperation for Portuguese: the case of Mozambique, Timor-Leste, São Tomé and Príncipe and Angola", CEsA. In 2016, she worked as a lecturer and tutor in the e-learning course "Introduction to International Cooperation for Development" at Camões - Institute of Cooperation and Language (ICL) and Association for Professional Training and Research of the University of Aveiro (UNAVE), Portugal). The most recent publications list as follows: "The Yaoundé and Lomé Conventions as the first Cooperation Agreements between the European and African continents" (2015). Work paper no. 133. Lisbon: CEsA; Sangreman, C. & Faria, R., 2015. "The values of Portuguese International Cooperation for Development". Journal of Social Sciences for political implications, vol. 3, no. 2, pp. 47-62; and, Faria, R., 2016. "Portuguese Cooperation and its performance in African health: the financing of the last

10 years". In Évora, I. & Frias, S. (coord.), In Progress: 2nd social seminar and development in Africa (pp. 47-66). Lisbon: CEsA; and, Oliveira, A. & Faria, R., 2016. Development in the Media: perceptions and visions of journalists and development professionals. Lisbon: ACEP.

Luis Manuel Mota do Santos Figueira has a PhD in Art History from the University of Coimbra, MA in Art History from the University of Lisbon and a degree in Art History from the University of Lisbon. He conducted post-doctoral internship in tourism, between 2017 and 2010 at the University of Aveiro. Coordinating Professor at the Polytechnic Institute of Tomar, the Departmental Area of Social Sciences, teaches and investigates the Tourism course Culture and Management in: Research Methodology in Tourism and Culture; Tourism Systems; Tourism Heritage and Local Development; Applied Tourism Project. He is member of the research unit on Governance, Competitiveness and Public Policy (GOVCOPP)-University of Aveiro. It also teaches in the development course of Cultural Products tourism: Cultural Heritage; Applied Research Methods; Stage –Thesis. Director of L-tour.ipt –Tourism Laboratory–Polytechnic Institute of Tomar, since 2014. Director of the Agricultural Museum of Riachos -Torres Novas, since 1994. Lines of research / action: tourism, museology, museography for tourism; cultural tourism and territorial development; ethnoarchaeology and history; design tourist routes and history of art.

Vladi Finotto, PhD in Management, Associate professor in Entrepreneurship and Strategy at the Department of Management, Ca' Foscari University of Venice. He has been working on themes such as business model reconfiguration and strategic innovation induced by changes in consumer preferences and behaviors since 2008. On the topic he published in international management, innovation and strategy journal and a chapter on the strategic implication of consumer community dynamics in an edited Oxford Handbook on creative industries. His research has focused on a variety of industries with particular attention to mature, medium-tech ones with an increasing focus on the food industry. He coordinated the local unit in a EU/FP6 project, took part in a number of national high impact projects and is part of international research networks on innovation and strategy. Currently he teaches strategy and innovation for the agri-food industries in Ca' Foscari's Master's programme in Food and Wine Culture. At Ca' Foscari he is the delegate for technology transfer, entrepreneurship and industrial liaisons and coordinates a variety of industry-university initiatives.

Regina Frank, born in 1965, has exhibited her installations under the title The Artist is Present internationally from 1989-2010 in windows, museums, and public spaces. In 2010 she changed her Motto to The HeArt is Present. The performer's visual language, spoken mainly through text and textiles, reveals a meditative process of exploring inner and outer networks, including political, cultural, and spiritual issues. Her works encompass a field of tension between virtual and real, analog and digital. She has been investigating the phenomena of Social Software and Software installation, Augmented Reality, NFT, and creative data processing. For the past 22 years, she is focusing on environmental issues with public installations working with schools, fairs, science museums, GOs, and NGOs. While in the nineties Regina Frank mainly dealt with social problems, using the textile industry as an example, and created interactive platforms in which visitors could share poems and texts, from 2000 her work concentrated more and more on environmental issues. Thus, from 2002, she already worked with solar energy in her installations and researched and published extensively on climate change. Regina Frank has lectured extensively, including at the School of the Museum of Fine Arts in Boston, New York University, the

Metropolitan Museum, School of the Art Institute in Chicago, M.I.T. Boston, UPV Universitat Politècnica de València, Chienku Technology University (CTU), Changhua, Taiwan, in Taiwan and taught several masterclasses at Bauhaus Dessau. Her works have been exhibited at the New Museum of Contemporary Art and the Bronx Museum in New York, at Serpentine Gallery in London, Reina Sofia in Madrid, MOCA in Los Angeles, Spiral Wacoal Art Center in Tokyo and at the 1996 Olympic Games in Atlanta and 1998 ARCO in Madrid in the project section. She has participated (with performances) in the 1999 Venice Biennial, the EXPO 2000 in Hannover, 2001 she showed at EXIT in Helsinki, the Intermediale in Mainz, the Performance Studies Conference 7, House of Press in Berlin, and Muffathalle in Munich. The San Diego Museum of Art commissioned and acquired Whiteness in Decay for their permanent collection in 2003. Frank was Artist in Residency at SAIR in Sapporo, Japan in 2004, with exhibitions at Contemporary Art Institute and Sapporo Museum of Modern Art. 2005 she was Artist in Residency in Taiwan and created a comprehensive art project for the Umweltbundesamt UBA (Federal Environmental Agency, FEA) UBA, Berlin/Dessau, and exhibited at UNESCO Headquarters and Fondapol in Paris. At Sally and Don Lucas Artist Program in Silicon Valley, she developed her Dream Weaving Software project in 2007 with support from the Gulbenkian Foundation culminating in a catalog, a CD, and tapestries. 2008 she participated in the biannual Waldkunst.com working with Solar Energy and LED Technology. Since 2011 she is performing and installing versions of her comprehensive environmental research iLAND. It began at the Venice Biennale, traveled in 2012 to Rapid Pulse Festival in Chicago, in 2013 to Haus der Künstler in Munich, "Infr'action", (during the Venice Biennial,) to the Helsinki Gallery Augusta, Suomenlinna, Finland, Musée Paul Valerie, Séte, France, and to Live 4 Festival at YouYou Contemporary Art Centre, Guangzhou, China. Her solo exhibition The HeArt is Present in 2015, featured 7 different solo projects from 7 phases of creation and started in the Museum Schloss Messkirch, where Regina was born 7x7 years ago. In 2018 her project iLAND was exhibited at MAAT (Museum of Art Architecture and Technology) in Lisbon, the Kunsthalle Hannover, TheNewArtFest18, Websummit, and MUNHAC (Museum for Natural History and Science) Lisbon, BioArt in Seoul, Korea, and in 2020 at the Pavilhão do Conhecimento (Science Museum) in Lisbon. She created PTS Pandemic and Mother Mandala at the International Forest Art Path and the LEW1 in 2020. Later that year, PTS Pandemic was exhibited at MAIA Cultural Center and CCB (Cultural Center Belem). Her solo exhibition Silenced Science (2021) was at the Natural History Museum and Silenced Sides at the CPS (Centro Portugues de Seriegraphia) and Galeria Prates. New Art Fest 21 showed PTS Pandemic at MUNHAC in Lisbon. Regina Frank's monograph: The Artist Is Present, was published by the Senate for Cultural Affairs in Berlin, Germany in 1999, her second book: The HeArt is Present, in November 2015, features performance installations, third book: The Art is Present drawings and collages. Frank's work featured in many publications: including the cover of Sculpture Magazine, Screen Multimedia, Being-on-Line (Lusitania), Parade Magazine, Harpers, The New York Times, Rethinking Marxism, the Village Voice, Neo, Time Magazine, Vogue, Cosmopolitan, Frankfurter Rundschau, Frankfurter Allgemeine, Japan Times, Contemporary Art, Art Forum, Art Papers, Art Examiner, Art Press, Arte y Parte, Public Art, and Ecology, On the Issues Magazine, AvantArt, FAZ Frankfurter Allgemeine Zeitung, Yale University Radio, FRESH ART INTERNATIONAL, Avant Art Magazine, and the Observador. She received grants, prizes, residencies, and commissions, among them from the Senate for Cultural Affairs, the IFA (Institut für Auslandsbeziehungen), the Goethe Institute in Japan, Canada, USA, Israel; the German Embassy in the USA, China, Canada, and Japan; Stiftung Kulturfond (Foundation Culture Fund); IBM Interactive Media Program, Pixelpark, Kade Collaborative Works Program, Arts International, Montalvo Artist in Residency Program, USA, Kio-A-Thau Artist in Residency Program, Taiwan, SAIR Sapporo Artist in Residency Program, Japan.

Manuela Hasse is an Anthropologist and Researcher at Lisbon University, FMH. PHD Completion in 1994, on Body Anthropology and History, in the area of Human Motricity at Lisbon University FMH. She has been teaching at Lisbon University, FMH on Body Anthropology and History, Leisure Sociology and Leisure History at Lisbon University, Coimbra University and Rio Maior Santarém High School Politechnic Institut. She taught at Lusíada University Luanda-Angola. She earned the award in 1999 by the Portugal Libraries Institut. In 2010 she received a Research Grant from Montpellier University. Her major areas of research are: body and health anthropology and history, medicine anthropology and history, body visual culture, leisure visual history. Currently she's working on Angola body anthropology and history related with colonizing process and health. She's working as well on professional and scientific biographies. She is the author of one book (sold out) and co-author of several books and scientific papers published on several scientific proceedings both on European and Brazilian scientific societies. She oversaw the scientific leadership coordination and co-orientation of several Brazilian PHD works as well as Masters works related with body anthropology and history. Published in international and national journals with peer review.

Maria Graciete P. Reis Henriques Honrado (Graciete Honrado) is an Adjunct Professor at the Departmental Unit of Social Sciences, at the Escola Superior de Gestão, Instituto Politécnico de Tomar, since 2000 and is a Corresponding Researcher of L-tour.ipt. He holds a bachelor's degree in Business Management, a degree in Business Management, a Master's in Organizational Behavior and a PhD in Management, in the scientific area of Human Resource Management.

Cicero Inacio da Silva is an interdisciplinary researcher and professor in the field of digital media technologies applied to science, culture, arts, health and media. Cicero was a Visiting Professor at the Arthur C. Clarke Center for Human Imagination and at the Speculative Design major at University of California. He is an associate member of the Cultural Analytics Lab at the City University of New York (CUNY) led by digital media researcher Lev Manovich and served as a member on the Interdisciplinary Assessment Committee of the Canada Foundation for Innovation (CFI). George Landow, in his latest book Hypertext 3.0 stated that Cicero's work Plato Online is one of the "...most exciting projects and published examples of hypermedia". Currently he is an Associate professor and researcher at the Educational Design course at the Federal University of São Paulo (Unifesp). Since 2006, he coordinates Brazil's Software Studies initiative and the Walkingtools Lab, both affiliated with the University of California, San Diego (UCSD) and CUNY. He was Visiting Scholar at the University of California (between 2006-2010 – supported with a grant from CAPES agency – Coordination for the Improvement of Higher Education Personnel), where he developed his postdoctoral research about Ted Nelson, inventor of the Hypertext and Hypermedia concepts and sponsored by Noah Wardrip-Fruin (digital media, software studies and games author and researcher) and at Brown University (2005 – grant from CAPES), where he did part of his PhD research affiliated with George Landow, Noah Wardrip-Fruin and Roberto Simanowski. He was a visiting professor at the Center for Research in Computing and the Arts (CRCA) – University of California, San Diego (UCSD) between 2008 and 2010. In 2006 Cicero developed a research at the MediaLab Prado about netart and digital media. In 2001 he was a researcher at the University of Porto, where he developed research about Hypermedia language. In 2011 he was a visiting researcher at the Centro Multimedia at CENART in Mexico City. Cicero is one of the organizers of CineGrid Brasil. Cicero da Silva holds a B.Sc. degree in Psychology and Master and Ph.D. degrees in Semiotics and communication. Cicero is founder of CineGrid Brasil and FeFiCi.

Philippe Joron is Full Professor and "Professeur des Universités" [Professor of Universities] in Sociology at Paul-Valéry University - Montpellier III. He is Director of LEIRIS - Laboratory for Interdisciplinary Studies on the Real and the Social Imaginary. He is former Director of Foreign and Regional Languages and Cultures Faculty in Paul-Valéry University - Montpellier III and also Vice-President of this University. He did his PhD in Social Sciences at Paris-Descartes University (Sorbonne). He has authored many books and papers on matters around social theory, imaginary and social representations, aesthetics and advertising communication, digital identities. His main publication, translated into several languages, is La vie improductive. Georges Bataille et l'hétérologie sociologique [Unproductive life. Georges Bataille and sociological heterology], 2009. Other signifiant books: La fête à pleins bords. Bayonne: fêtes de rien, soif d'absolu [Full-throttle party. Bayonne: parties of nothing, thirst for the absolute], 2012; Violences et communication. Approches franco-brésiliennes des altérités communicationnelles [Violences and communication. Franco-Brazilian approaches to communicational alterities], 2006.

Viola Krebs is a Member of the University of Strasbourg Chair of UNESCO, "Pratiques journalistiques et médiatiques. Entre mondialisation et diversité culturelle". She has extensive international teaching experience as a professor and lecturer in International Business, Communications and International Relations, having taught for University of Geneva and University of Business and International Studies, among others. She further has experience with academic coordination, accreditations and publications. Prior to a full-time academic career, she worked both in the private and public sectors, including the World Economic Forum (WEF), Merrill Lynch and UBS. Founder of ICVolunteers, an international non-profit organization focusing on communications, where she led international projects in Europe, Africa, Asia and the Americas. Has formally been involved in different UN summits and processes (World Summit on the Information Society, Strategy Council of the United Nations Global Alliance for ICT and Development, etc.). Graduate of University of Strasbourg and University of Geneva.

Ângela Lopes is currently a researcher at KIPT CoLAB – Knowledge to Innovate Professions in Tourism, the first collaborative laboratory in tourism. She has been working in the area of consultancy and research in tourism and marketing since 2010 in several regional, national and international projects, mainly in strategic and marketing plans for tourism development. More recently, she held a research grant at CEFAGE, University of the Algarve. The main working areas are strategic planning, tourism marketing, destination management and governance, culture and events and tourist behavior.

Leandro Lucas is approved in the Doctorate in Teaching, by the Northeast Teaching Network (RENOEN-2021), Master in Science Teaching and Mathematics Education (2018), linked to the Scientific Culture, Technology, Information and Communication Research Line, Specialist in Education Fundamentals: Practices Interdisciplinary Pedagogics (2014), and Graduated in the Full Degree in Mathematics (2009), training carried out at the State University of Paraíba (UEPB). He has experience in the area of Mathematics and Mathematics Education, he is currently a professor of Basic Education at the State Education Network of Paraíba and a member of the TDAC (Digital Technology and Knowledge Acquisition) research group at the State University of Paraíba (UEPB).

Alba Marín received her MS in Communication and Information Science from the University of Sevilla in 2014 and her PhD in Communication Science in 2019. In 2018 she was hired as a Researcher and professor by the Communication and Hypermedia Department at The University of Savoie Mont Blanc. She is currently a Postdoc Researcher at the University of Extremadura and the University of

Sevilla. She writes and presents widely on visual studies and visual methods for social sciences research, digital and audiovisual journalism, and interactive and immersive nonfiction storytelling.

Davi Marques Camargo de Mello is a filmmaker, professor, and researcher in the field of Cinema and Audiovisual. Currently he is a Ph.D. candidate at the Graduate Studies in Education, Art and History of Culture at Mackenzie Presbyterian University advised by Prof. Jane de Almeida. Davi holds a Master degree in Communication from Universidade Anhembi Morumbi (UAM, PROSUP-CAPES scholarship) and a bachelor in Cinema and Audiovisual by the same institution (UAM-2014). His research involves cinematographic ethics and aesthetics in post-war Polish cinema, as well as the diverse scene of experimental films and audiovisual installations. As a director and screenwriter, he has award-winning short films that total more than 100 participations in national and international festivals, including selections at the Rio Festival, São Paulo International Short Film Festival, Fest New Directors | New Films (Portugal), Bogoshorts (Colombia), Fantaspoa, Festival Internazionale del Cinema di Salerno (Italy), Izmir International Short Film Festival (Turkey), among others. In 2020, he created the virtual installation 'Experimento Odradek-Domicílio Incerto' alongside filmmaker Deborah Perrotta, a co-production between Brazil and Italy. Some films from the installation were awarded and selected for important events, such as the In Out Festival, an audiovisual exhibition that takes place annually in the city of Gdansk (Poland), at the Laznia Contemporary Art Center. Since 2020, he has been part of the faculty of Universidade São Judas Tadeu (São Paulo) in the courses of Cinema and Audiovisual, Radio and TV and Audiovisual Production.

Filomena Moita has a PhD in Education in the area of concentration in Education, Communication and Culture from the Federal University of Paraíba, CAPES scholarship holder did a sandwich doctorate at the University of Lisbon. Professor Dr. Associate D with aggregation from the State University of Paraíba. Professor at the Department of Education at UEPB-Campus I. She is part of the permanent staff of the Postgraduate Master's in Science and Mathematics Education and Professional Master's Programs and of the RENOEN-Doctoral Program of Rede Nordeste. With publications in specialized journals, works in annals of national and international events, book chapters and published books. Master's and doctoral advisor with emphasis on technology and knowledge acquisition, technology and science and mathematics teaching, ICT and teacher training, digital literacy, science didactics and methodology. TDAC (Digital Technology and Knowledge Acquisition) research group coordinator.

Andres Montenegro is an animator, and Associate Professor of 3D computer animation, 3D modeling, and advanced rendering in the Department of Art and Design at Purdue University. Fort Wayne. His work develops immersive environments using real time 3D animations while integrating Augmented Reality and Virtual Reality interactive animations. He has an extensive experience with software and hardware oriented toward the generation of different styles of rendered images using painting as the main source of research. His personal work uses metaphors and symbolic narrative to assemble a conceptual language depicting cross-cultural references. His work of art begins in the 80´s regarding painting as the main vehicle of production, and experimentation being the framework for mixing language, aesthetics, and conceptual venues based on Art History as well. His digital work begins in 1990 with early experiences using 3D animation software. From 1983 – 2009 he presented one man shows and collective exhibitions in Chile, South America, Europe and the United States. He received his BFA in Art and Education from University of Chile in 1986, his MA from University of Playa Ancha Chile in 1996, and his MFA in Digital Arts from University of Oregon in 2006. While he was studying there, he was awarded the Clarice Krieg Scholarship, and University of Oregon Scholarship in 2004, 2005, and 2006.

Luiz Felipe Napole de Haro Alvares is an architect and urban planner, with a master's degree in Intelligence Technologies and Digital Design from PUCSP and a doctoral candidate from the same post-graduate program at PUCSP. Member of the Comunidata research group, he has also worked in architecture and urbanism since 1998.

Ana Cristina B. Nata dos Santos (Ana Nata) is graduated, master and PhD in Pure Mathematics from University of Coimbra since 1997, 2001 and 2011, respectively. Currently is Adjunct professor at Department of Mathematics and Physics of the Polytechnic Institute of Tomar Portugal, member of the Centre of Matematics of the University of Coimbra, Portugal, and a corresponding researcher of L-tour.ipt. She is also member of the Laboratory of Pedagogical Innovation and Distance Education of the Polytechnic Institute of Tomar, and member or the Diversity and Inclusion Team of the Polytechnic Institute of Tomar working in a strategy program on this topics in the workplace.

Kim Nehls serves as Assistant Dean of External Relations for Lee Business School at UNLV, where she is also a Lecturer of International Business and Marketing. Kim has a PhD in higher education from UNLV, an MBA from Fuqua School of Business at Duke University, and a bachelor's degree from University of Illinois. She is also a PhD student in business administration at University of Denver's Daniels College of Business. In general, Kim's research interests focus on marginalized populations in business, education, and philanthropy.

Pedro Soares Neves, PhD, 1976, Lisbon. Multidisciplinary and post graduate academic training in design and architecture (Barcelona and Rome), and Lisbon Faculty of Fine Arts research Center CIEBA scholar. Specialized in participatory methodologies and informal visual signs in public space (eg, Graffiti and Street Art). Lecturer and urban designer / consultant at metropolitan and city scale (eg, project CRONO), co-responsible for adapting to Lisbon's exhibition "The Street Belongs to all of us" (IVM, Françoise Asher) and winner of the first prize for the "No Rules Great Spot" Oporto. Co-creator and professor of "Non commissioned Public Art" workshop and scholar in Science and Technology Management in Faculty of Fine Arts University of Lisbon. Founder of the AP2 (Portuguese chapter of IAP2, International Association for Public Participation), APAURB (Portuguese Urban Art association), and of the Lisbon Street Art & Urban Creativity Conference and ongoing associated International Research Topic (urbancreativity.org). Editor of the Street Art & Urban Creativity Scientific Journal among other book publications on the topic.

Marta Noronha, with a diversified professional experience, in several areas related to communication, professional training and scientific research, is currently dedicated mainly to teaching. In 2018, she completed her PhD in Communication Sciences at the University of Minho, with the thesis entitled "The Propagation of Stories Through Multiple Media: For a typology of transmediation, adaptation and other derivative forms". Since 2018, she has been a professor at ISCE Douro, in Penafiel. Currently, her preferred areas of interest are transmedia communication, narratives, fiction in particular, pop culture and new audio-visual and multimedia dynamics and formats.

Daniela Pavan, Research Fellow at the Department of Management, Ca' Foscari University of Venice (Italy), Consultant in the field of Marketing & Communication, Design Thinking, and Internationalisation, Co-Founder of the storytelling platform and production boutique Creative Pois-On LLC, New York (USA). Her research interests and work focus on Entrepreneurship, Innovation, Urban Studies, Digital storytelling and Creativity as a means to develop innovative and unconventional business ideas. She was part of the team that was awarded with the Communicator Awards of Distinction in 2015 for the category "Mobile" for the creation, design, implementation of the mobile experience for a Satellite TV Service in the United Kingdom, and the Retail System Awards-Mobile Retailer of the Year in 2013 for the design, implementation and communication approach for a fashion mobile experience in the United States. US Forbes contributor until 2020, she is the author of several articles that explore how creativity and business can work together. She is part of the Urban GoodCamp Project, an Horizon Plus initiative that aims to empower Higher Education Institutions (HEIs) and their urban stakeholders to work together on pressing urban challenges, by developing co-design approaches.

Jérémy Picot is an economist who works on several fields of research: education, health, politics, social choice, information and communication, etc. He is a specialist of socio-economic evaluation of public policies. He also has co-authored some publications on fake news during the Covid-19 pandemic.

Jorge Ricardo Pinto is a coordinator professor at ISCET - Instituto Superior de Ciências Empresariais e do Turismo, and a researcher at CEGOT (Centro de Estudos de Geografia e Ordenamento do Território). His main research areas are Urban (Historical) Geography, Urban Tourism and History of Tourism.

Zara Pinto-Coelho is an Associate Professor in the Department of Communication Sciences at the University of Minho. She holds a Ph.D. in Communication Sciences, Semiotics, from the University of Minho in 2003. She teaches in the areas of Discourse, Image, and Sociology of Communication Studies. Her current research interests intersect issues associated with gender inequalities and political participation through discourse and image. ORCID: https://orcid.org/0000-0001-6651-3720.

Helena Pires is an Associate Professor in the Department of Communication Sciences, at the Institute of Social Sciences, University of Minho. She got a PhD in Communication Sciences, in the area of Communication Semiotics, from the University of Minho, in 2007. She teaches in the areas of Advertising, Arts, Semiotics and Communication. She has published and developed research work on semiotics of the (urban) landscape, urban culture as well as in the area of communication and visual culture. ORCID: https://orcid.org/ 0000-0002-5533-4687

Fabio Pranovi, full professor in Ecology at the Environmental Sciences, Informatics and Statistic Department of the Ca' Foscari University of Venice; main research interests are in the field of the ecosystem approach implementation within the context of renewable resources management, possible effects of climate changes in the structure and functioning of marine ecosystems and ecosystem services assessment, both in marine and terrestrial environments. Teaching activities in the 'Environmental Sciences' and 'Environmental Humanities' courses. Authors of about 130 papers in scientific journals.

José Alberto Rio Fernandes is a Full Professor President of Geography Department at Porto University President of General Assembly of the Portuguese Geographers Association More than 20 books and 100 scientific texts. Teacher and researcher in urban geography, public policies and planning.

Pedro M. Azevedo Rocha graduated in Lic. Electrical Engineering - Electronics and Computers (2000, ISEP, Porto), and MSc. Bioengineering (2002, UPC, Barcelona). He was a research fellow at INETI (2003/10, Lisbon), research assistant at the Sleep Lab. (Hospital de Santa Maria (2003/5, Lisbon)), and taught Electronics and Digital Systems at ESTG (2009/10, Leiria). Professionally was systems administrator (2015/17, Fernando Pessoa Univ., Porto), now is a senior developer at a digital healthcare startup (2017, Virtualcare, Porto). Furthermore, he is a student member to CECS (2016, Minho Univ.), a research collaborator (2022, CTEC / UFP, Porto), and member of STELLAR Science Advisory Board (2022, Porto). He attended in full in a Management PhD (2006/11, Univ. Évora), but not concluded it. He initiated a Communication Sciences PhD (2016, Minho Univ.): now on a leave period. He graduated from EDIT (Porto) in "Front-end & Responsive Web Development" (2017), and in "Digital Art Direction & Creativity" (2018/19). Early he participated in the exhibition "À Luz de Einstein" (2005/6, Calouste Gulbenkian), and latter graduated as therapist and Reiki Master (2016/21, EPRTC). Done some projects in social cartography, e.g. for the "Reclaim the Future" project in Campanhã (2017, Visões Úteis). Also published articles on information systems, social semiotics, and semiotics of space (eg Agrootics). Has interest in mitogeometry, and ethnography and archaeology cartography.

Ana Roders is a professor of Heritage and Values at the Department of Architectural Engineering and Technology at TU Delft. Graduated as Architect and obtained a PhD in Building Technology at Eindhoven University of Technology. Worked as Visiting Professor at VU Amsterdam, KU Leuven and ULiege (Belgium), Durham University (UK), and Universidade Lúrio (Mozambique). Founding co-editor of the Journal of Cultural Heritage Management and Sustainable Development (Emerald, UK). Since 2008, cooperates closely with UNESCO World Heritage Centre. Led the creation of global online platforms such as Protected Urban Planet and the Global Observatory on the Historic Urban Landscape.

Filipe Romão is a Board Member at HighScore - Professional Qualification & Board Member at HighSun DMC & Tour Operator Portugal Corresponding Researcher of L-tour.ipt.

Carlos Sangreman is a PhD in Interdisciplinary African Studies in Social Sciences, with the theme "Structural adjustment policies and the well-being of urban families in the city of Bissau, 1986 - 1994 - 1998 - 2001". Teacher at University of Lisbon and Aveiro, Portugal; international consultant, from 1985 to the present, with missions in all African speaking countries and Eats-Timor for International Organizations such as the World Bank, UNDP, IOM and European Union, governments of Portugal, Sweden, Cape Verde and Guinea-Bissau, foundations such as Calouste Gulbenkian and Portugal-Africa. Between 1986 and 1987, statistical adviser to the Minister of the Guinea-Bissau Plan; between 1998 and 2003 advisor to the Minister of Solidarity and Social Security for the organization of the Department of Development Cooperation; and, between 2008 and 2012, advisor to the Secretary of State for Foreign Affairs and Cooperation to boost the Civil Society Forum for Development Cooperation. Researcher author of articles, books and book chapters on Cooperation, Human Rights, Migration and Public Politicians, responsible for research projects and cooperation with public and private funding, national and international, expert in Cooperation in the EEC of the European Union.

Lucia Santaella is a researcher 1A at CNPq, with a PhD in Literary Theory (PUC-SP, 1973) and in Communication Sciences (ECA/USP, 1993). She is full professor at the PhD in Communication and Semiotics and in Technologies of Intelligence and Digital Design (PUC-SP) and Director of the latter. Santaella has been an invited professor at several universities in Europe and Latin America. She published 53 books and organized 29, in addition to publishing almost 500 articles in Brazil and abroad. She received the Jabuti awards (2002, 2009, 2011, 2014), the Sergio Motta award (2005) and the Luiz Beltrão award (2010).

Ana Cláudia Leal Marques Pires da Silva Mendes Pinto is a PhD in Economics from the Universidad de Extremadura. She is an Associate Professor at the Polytechnic Institute of Tomar, director of the departmental unit of Social Sciences at the Polytechnic Institute of Tomar (IPT), Member of the Academic Council of the IPT, researcher at the Techn&Art research center, Researcher at GEDITEC, coordinator of the project "The Trays Festival, Cultural Heritage and the Community (FesTab)" funded by the FCT, researcher in the project "Military Tourism Route: model for the enhancement of the national historical-military heritage (Insignia)" financed by the FCT. She has published some articles in specialized magazines. She recently won an award for best communication at the International Tourism Congress. She works in the area of Social Sciences specifically Economics.

Rafael Simão has a Master's degree in architecture and Urbanism, currently a PhD. candidate at the Knowledge Engineering and Management Program at the Federal University of Santa Catarina (UFSC). Studies the impacts of new media on urban management, how local governments use social media and mobile devices as tools for crowdsourcing and e-participation, and the use of social media by politicians and public institutions. Currently part of the editorial team of the International Journal of Knowledge Engineering and Management, and working with the Humane, Smart and Sustainable Cities Lab (LabCHIS) on a series of workshops based on these concepts in four cities in the state of Mato Grosso, Brazil. ORCID: https://orcid.org/0000-0001-5566-2456.

João Tomaz Simões is a Tourism and Leisure specialist and was born in the Templar city of Tomar, where he began his studies in Communication. He later attended a degree in Management of Tourism Operations and began his career as a tourism consultant and advisor. It was at ISLA Santarém in 2008 that he finished his degree in Tourism (recognized by Turismo de Portugal, I.P.). Before starting his teaching career in Social and Human Sciences (Portuguese Language, History and Geography of Portugal, Administration and Tourism), he performed as a Director of F&B, while finishing in 2012 the Master in Development of Cultural Tourism Products, in the School of Management in Tomar of the Polytechnic Institute of Tomar. He is currently completing his PhD in Tourism at the Institute of Geography and Spatial Planning (IGOT) of the University of Lisbon, in partnership with the Higher School of Hotel Management and Tourism of Estoril (ESHTE). At the professional level, he divides his teaching activity in tourism between the courses of Nuno de Santa Maria School Grouping in Tomar, the Polytechnic Institute of Tomar and the Techn&Art Investigation Center.

Jorge Simões is Associated Teacher at the Polytechnic Institute of Tomar (IPT), Tomar, Portugal. He taught at the Polytechnic Institute of Tomar (IPT), on introduction to management and entrepreneurship. His academic background includes a PhD in Management from the University of Beira Interior (UBI) and a master's degree in management, Accounting and Administration from the University of Minho

(UM). Currently, he is the Director of Master Program in Management at IPT. He is a research fellow in Techn&Art (IPT). Author of several scientific papers published in international and national journals with peer review, indexed in global bibliographic databases (Web of Science, Scopus). His research interests include entrepreneurship, business creation and Knowledge Management. ORCID: https://orcid.org/0000-0002-7967-5216.

Alfredo Luiz Paes de Oliveira Suppia holds a degree in Communication (journalism) from the Pontifical Catholic University of Campinas (PUC-Campinas), with a specialization in Portuguese Language and Brazilian Literature from Centro Universitário Padre Anchieta (UNIANCHIETA, 2000), specialization in Scientific Journalism from the Laboratory of Studies Advanced in Journalism, State University of Campinas (LABJOR-UNICAMP, 2006), a master's (2002) and a Ph.D. (2007) in Multimedia from the State University of Campinas (Arts Institute, UNICAMP). He is also a "Livre-docente" from the Dept. of Art History, Federal University of São Paulo (UNIFESP), Campus Guarulhos (obtained in 2020). Member of the Brazilian Society for Cinema and Audiovisual Studies (SOCINE) since 2001. Member of the Science Fiction Research Association (SFRA) since 2007. A film critic, he has professional experience in pre-production, production and post-production in cinema and audiovisual. Author of the books A Metrópole Replicante: Construindo um Dialogo entre Metropolis and Blade Runner (Juiz de Fora: Ed. UFJF, 2011) and Atmosfera Rarefeita: A Ficção Científica no Cinema Brasileiro (São Paulo: Devir, 2013). Alfredo organized the volumes Cinema(s) Independentes: Cartografias para um Fenômeno Audiovisual Global (Juiz de Fora: Ed. UFJF, 2013). Representative of Brazil on the board of the Science Fiction Research Association (SFRA), he is a member of the advisory board of the World Science Fiction Studies collection by Peter Lang publisher and a researcher at Science Fiction Working Group. SF, Fantasy & Utopian Studies at the University of Florida, USA (https://sciencefiction.group.ufl.edu/). He has experience in the areas of Arts and Communication, with an emphasis on Cinema and Audiovisual, working mainly on the following subjects: History and Theory of Cinema, World Cinema, Film Criticism and Film Analysis, Brazilian and Latin American Cinema, Film and Audiovisual, Screenwriting, Film and Audiovisual Genres, Independent Film, Expanded Film, Comparative Literature, Intersemiotic Translation/Adaptation, Digital Cinema, 3D Cinema, GPS Cinema (GPS Films), Locative Media, Cultural Journalism and Science Journalism. He organizes the Science Film Festival (FeFiCi).

Hannah Strachan graduated from Varna University of Management in 2021 with a bachelor's degree in Culinary Arts.

Laryssa Tarachucky is an Architect and Urban Planner interested in participatory design processes. She holds a PhD on Knowledge Management and Engineering (PPGEGC/UFSC) and she is currently a postdoctoral fellow at the University of São Paulo (USP) Advanced Studies Institute (IEA). Laryssa is a research member of MídiaCon, LabCHIS and Children's Books Going Mobile (UFSC), and she coordinates A Cidade e a Névoa, an independent research group. She is the founder and coordinator of LabIC Novale, a civic laboratory in Jaraguá do Sul aimed at supporting local citizens and grassroots communities cocreate solutions for issues of shared concern. Her overarching research career is to investigate means of using digital technologies to meaningfully engage citizens in the long term with issues of common concern as well as exploring how to support citizens to develop the skills that they need to appropriate technology for their own purposes. ORCID: https://orcid.org/0000-0003-2073-4151.

Mirian Nogueira Tavares is an associate professor at the University of Algarve, Portugal. She studied Communication Sciences, Semiotics, and Cultural Studies (earning a Ph.D. in Communication and Contemporary Cultures from the Federal University of Bahia), and has researched and produced theoretical work in film and art aesthetics. As a professor at the University of Algarve, she participated in the creation of the graduation project in Visual Arts, the master's degree and Ph.D. in Communication, Culture, and Arts, and the Ph.D. in Media-Digital Art. Currently, she is the coordinator of CIAC (Centre for Research in Arts and Communication) and coordinator of the Ph.D. in Media-Digital Art. She has curated several exhibitions since 2005, such as O Século XX passou por aqui (within the scope of Faro Capital Nacional da Cultura, 2005), Dialogue Boxes on Street Windows (co-curated with Xana, within the scope of the Allgarve project, 2009), Politically Excited (2012), and Arte: Um Assunto de Mulheres (Trem Gallery, Faro, 2015). As co-curator at the Trem Gallery, since 2009, she has written all the texts in the exhibition hand-outs and some for artists' catalogs, such as Pedro Cabrita Reis, Pedro Cabral Santo, or Manuel João Vieira.

Bubacar Turé is a Guiné-Bissau jurist, Lusophone University at Lisbon. During his professional career, he developed multiple activities, maintaining the central scope of defending Human Rights in Africa, in connection with international institutions, NGOs, associations and communities, and by participating in different civil society movements. For example, he worked at UNDP as gender equality consultant, and also as Legal Advisor to the Youth Association for Social Reintegration, from 2007 to 2020. Moreover, he was nominated for several prestigious positions, such as Vice President of the Guinean League of Human Rights, from 2014 till now. He has co-authored various books and papers on topics and questions around civil rights, peace, gender equality, social reintegration and global development, the more recent one about COVID and Human Rights in the city of Bissau.

Philippe Viallon is a professor, researcher, and teacher in information and communication sciences at the University of Strasbourg (France). He is also responsible for the Unesco Chair in Journalism and Media Practices, which brings together some thirty researchers. Author or co-author of 13 books, more than 50 book chapters or articles in scientific journals and nearly 100 conferences in France and abroad, he works on understanding digital culture and social media in particular. He claims a global approach to these phenomena which, like all media, can only be beneficial for individuals if they are educated in their use.

Juliana Wexel is a multimedia artist, journalist, researcher (FCT), and doctoral student in Digital Media Art at the Arts and Communication Research Center at the University of Algarve, Faro, Portugal (CIAC-DMAD). She is Master of Letters (Literature) and Culture, was a lecturer at the Department of Languages, Literature and Modern Cultures at the University of Bologna, Italy, and Assistant Professor of Journalism at Universidade Veiga de Almeida, in Rio de Janeiro, Brazil. Juliana is the author of the audiovisual series on street art in Rio de Janeiro and São Paulo produced by Canal Futura and connected to the UN Why Poverty?. In 2020 she has created ivagination, a digital installation with an autobiographical narrative in a site-specific model and vulva art genre, in Lisbon. She intersects scientific investigation based on artistic practice, narratives, digital technologies, feminism, and urban space, as well as associating research with the Writing Urban Places.

Silvena Yordanova is an associate professor in social management. Having more than 10 years experience as lecturer in Management, Strategic management, Organizational behavior. Consultant and lecturer in Electronic marketing. Scientific interests- electronic commerce, virtual communities in tourism. Monograph published "Virtual communities in tourism", 2016. Graduated Univeristy of economics-Varna, Bulgaria. Ph.D. in Social Management. Lectured Electronic commerce in Italy, Norway, Finland. Lecturing in English, Bulgarian. Speaks Bulgarian, English, Russian.

Index

F

G

H

I

L

M

O

P

Q

R

S

T

V

W

www.ingramcontent.com/pod-product-compliance
Lightning Source LLC
LaVergne TN
LVHW080357011026
849161LV00009B/2327